ADHESIVE CHEMISTRY

DEVELOPMENTS AND TRENDS

POLYMER SCIENCE AND TECHNOLOGY

Recent volumes in the series:

Volume 16 POLYMERIC SEPARATION MEDIA
Edited by Anthony R. Cooper

Volume 17 POLYMER APPLICATIONS OF RENEWABLE-RESOURCE MATERIALS
Edited by Charles E. Carraher, Jr., and L. H. Sperling

Volume 18 REACTION INJECTION MOLDING AND FAST POLYMERIZATION REACTIONS
Edited by Jiri E. Kresta

Volume 19 COORDINATION POLYMERIZATION
Edited by Charles C. Price and Edwin J. Vandenberg

Volume 20 POLYMER ALLOYS III: Blends, Blocks, Grafts, and Interpenetrating Networks
Edited by Daniel Klempner and Kurt C. Frisch

Volume 21 MODIFICATION OF POLYMERS
Edited by Charles E. Carraher, Jr., and James A. Moore

Volume 22 STRUCTURE PROPERTY RELATIONSHIPS OF POLYMERIC SOLIDS
Edited by Anne Hiltner

Volume 23 POLYMERS IN MEDICINE: Biomedical and Pharmacological Applications
Edited by Emo Chiellini and Paolo Giusti

Volume 24 CROWN ETHERS AND PHASE TRANSFER CATALYSIS IN POLYMER SCIENCE
Edited by Lon J. Mathias and Charles E. Carraher, Jr.

Volume 25 NEW MONOMERS AND POLYMERS
Edited by Bill M. Culbertson and Charles U. Pittman, Jr.

Volume 26 POLYMER ADDITIVES
Edited by Jiri E. Kresta

Volume 27 MOLECULAR CHARACTERIZATION OF COMPOSITE INTERFACES
Edited by Hatsuo Ishida and Ganesh Kumar

Volume 28 POLYMERIC LIQUID CRYSTALS
Edited by Alexandre Blumstein

Volume 29 ADHESIVE CHEMISTRY: Developments and Trends
Edited by Lieng-Huang Lee

ADHESIVE CHEMISTRY

DEVELOPMENTS AND TRENDS

EDITED BY

LIENG-HUANG LEE

Webster Research Center
Xerox Corporation
Webster, New York

PLENUM PRESS • NEW YORK AND LONDON

Library of Congress Cataloging in Publication Data

Symposium on Recent Developments in Adhesive Chemistry (1983: Seattle, Wash.)
Adhesive chemistry, developments and trends.

(Polymer science and technology; v. 29)
"Proceedings of a Symposium on Recent Developments in Adhesive Chemistry, held March 21–23, 1983, in Seattle, Washington"—T.p. verso.
Includes bibliographies and indexes.
1. Adhesives—Congresses. I. Lee, Lieng-Huang, 1924– . II. Title. V. Series.
TP967.S96 1983 668′.3 84-24977
ISBN 0-306-41888-6

Proceedings of a symposium on Recent Developments in Adhesive Chemistry, sponsored by the Division of Organic Coatings and Plastics Chemistry of the American Chemical Society, held March 21–23, 1983, in Seattle, Washington

A Division of Plenum Publishing Corporation
233 Spring Street, New York, N.Y. 10013

Printed in the United States of America

PREFACE

Since the first symposium on Recent Advances in Adhesion, held September, 1971 in Washington, D.C., this Division of the American Chemical Society has continuously sponsored several symposia on adhesion and adhesives. The chemists have gradually realized the importance of adhesion in various fields of science and technology. During these years, the science of adhesion has steadily grown along with progress in surface science and fracture mechanics. Moreover, new adhesives have been invented and applied in actual structures, for example, structural and aerospace adhesives.

In response to socio-economic demands, new forms of adhesives have been introduced to combat the problems of pollution and to promote energy-conservation. The developments of hot-melt adhesives, waterborne adhesives, and radiation-curable adhesives are vivid examples of successes in solving some of the problems. As chemists, our natural desire is to understand how those new adhesives and new forms of adhesives are made. Thus, we are interested in learning about the chemistry of adhesives so that we may create new generations of materials to satisfy future needs. It was based on this common interest that we set forth to organize this Symposium on Recent Developments in Adhesive Chemistry. It was held from March 21 through 23, 1983 in the Westin Hotel, Seattle, Washington. The Symposium was very well attended. As a matter of fact, for the first two sessions, we had to move from the smaller Mt. St. Helen Room to the larger Grand Ballroom, not because of superstition, but because of sheer necessity to accommodate a larger than expected attendance.

As a record of this successful Conference, all papers (after revision) presented to the Symposium and the discussions during the meeting are now published in these Proceedings. In addition, the contributions shortly after the Conference in projecting future trends of adhesive research are also included. As a result, the new title for the volume becomes "Adhesive Chemistry - Developments and Trends." The six parts following the original order of the Conference are:

I. Chemistry of Adhesion - The Importance of Interface

II. Physics of Adhesion - Characterization of Surface and Bulk of Adhesive Systems

III. Radiation-curable Adhesives

IV. High Temperature Adhesives

V. Anaerobic and Structural Adhesives

VI. Trends in Adhesive Research

As Chairman of the Symposium, I would like to take this opportunity to thank our contributors; their names and biographies are at the end of this volume. And, I thank Session Chairmen: Dr. L. H. Sharpe, Dr. V. D. McGinniss, Dr. K. C. Stueben, and Professor J. P. Wightman. I sincerely appreciate the skill and patience of Ms. Jeraldine T. Allen and Ms. Nancy MacDonald in typing the entire Proceedings in a camera-ready format.

Acknowledgement is also made to the donors of the Petroleum Research Fund, administered by the American Chemical Society, for their partial support of this Symposium.

Lieng-Huang Lee
June, 1984

CONTENTS

PART ONE: Chemistry of Adhesion - The Importance of Interface

PART TWO: Physics of Adhesion - Characterization of Surface and Bulk of Adhesive Systems

PART ONE:

CHEMISTRY OF ADHESION–THE IMPORTANCE OF INTERFACE

Introductory Remarks

Lieng-Huang Lee

Webster Research Center

Xerox Corporation

Webster, New York 14580

Adhesion is essential in the structure of matter, in the living body and in mechanical and technological devices which enrich our lives. In a living body, many tissues are bound together through cell adhesion. Both normal and abnormal living processes are controlled by the biological adhesion. In medicine, surgical adhesives have been used to mend broken parts or to attach artificial materials to living organs.

Adhesives and sealants have become inseparable parts in mechanical and technological devices. Automobiles, airplanes and other transportation media cannot be built without structural adhesives, high-temperature adhesives and various sealants. Many mechanical devices are held together with bolts, nuts, and adhesives. More adhesives are going to replace metallic fasteners.

Besides bulk adhesives, many new specialty adhesives have been developed to meet various socio-economical needs, e.g., time-saving, energy-saving, non-polluting, etc. As a result, we have seen several important transitions for the developments in adhesives chemistry:

1. From natural products to synthetic resins.

2. From re-formulation of old lines of material to the design of new products.

3. From organic-solvent-base to water-base.

4. From low-solid to high-solid, e.g., hot-melt adhesives.

5. From slow-cure to rapid-cure for thermosetting systems.

6. From thermal-curing to radiation-curing.

As a whole, adhesive technologies have gradually transformed from art to science. Industry-wise, automation and robotics have started to replace old-fashioned operations. Both productivity and profits have been improved. As a result, as recession hit the bottom in late 1982, while many industries suffered from financial losses, the specialty adhesive industries made some impressive gains.

In this Session, we shall discuss the chemistry aspects of recent developments of adhesive industry. I shall give a brief account of the latest developments. Dr. Maugis originally planned to be here to discuss the fundamentals of adherence phenomena, unfortunately he is unable to come. Dr. Michael Lee of GM Research Laboratories will present an analytical method for determining the surface energies of solid polymers. Two papers on surface treatment to enhance adhesion will be given by Dr. P. Dreyfuss of the University of Akron and Dr. E. J. Berger of Owens-Corning Fiberglas Corporation.

Recent Developments in Adhesive and Sealant Chemistry

Lieng-Huang Lee

Webster Research Center
Xerox Corporation
Webster, New York 14580

ABSTRACT

In this paper, we briefly review recent developments in adhesive and sealant chemistry. The subjects discussed are the: chemistry of adhesion, physics of adhesion, radiation-curable adhesives, high-temperature adhesives, anaerobic and structural adhesives, hot-melt adhesives, film adhesives, waterborne adhesives, aerospace structural adhesives, conventional sealants, advanced aerospace sealants, and adhesives and sealants for solar collectors.

This review emphasizes new materials for adhesives and sealants. Thus, we discuss both new adhesives and new forms of adhesives. As a result of socio-economical demands, energy-reduction and pollution-control are two main reasons for the development of radiation-curable, hot-melt, film, and waterborne adhesives. We shall examine some of these new forms of adhesives in detail. The applications of new adhesive materials to aerospace and solar-energy industries will also be briefly mentioned.

I. INTRODUCTION

The use of synthetic adhesives in the past twenty-five years (1) has grown, particularly the use of eight classes of polymers: polyvinyl acetate, polyolefins, styrenic block copolymers, acrylics, cyanoacrylates, anaerobics, polyurethanes, and epoxy resins. Some of these polymers are still in high demand as specialty adhesives (2). During the last several years, however, other polymers have been added to this list, e.g., polyamides, polyimides, and polyesters. Today, synthetic adhesives account for 75% of the adhesives produced and 85% of the sales, while the market share of natural products has steadily declined.

The growth in the use of structural adhesives (3) will remain at 3-4% a year throughout the 1980's. The use of new "high-technology" structural adhesives, especially for the aerospace industry, may increase at a higher rate in light of the new developments discussed in this paper. For example, recent applications of robotics could enhance the large-volume use of adhesives for primary-bonded structures in aircraft. Thus, along with other recent developments, the share of the fastening and joining materials market attributed to adhesives could increase from 22% in the 1970's to 35% in the 1980's.

In this paper, we shall describe the development of adhesives during the last several years and the trends for further advances. In general, traditionally used, solvent-borne systems (4) will be gradually replaced by waterborne systems, hot-melts, nonvolatile solid (or liquid) systems, two-part adhesives, radiation-curable adhesives, and powder and reactive liquid systems (Table 1).

In the following sections, we shall discuss: adhesion chemistry, adhesion physics, radiation-curable adhesives, high-temperature adhesives, anaerobic and structural adhesives, hot-melt adhesives, film adhesives, waterborne adhesives, aerospace structural adhesives, conventional sealants, advanced aerospace sealants, and adhesives and sealants for solar collectors.

II. CHEMISTRY OF ADHESION - THE IMPORTANCE OF INTERFACE

Since my last review of adhesion chemistry (5) several

Table 1. Growth Market Technology Profile*

Systems	Number of Growth Opportunities
Water-borne	56
Hot-melts	36
Non-volatile solid or liquid	14
Two-part	10
Radiation-cured	5
Reactive-liquid	2
Powder	2
Solvent-borne	0
Total	125

*Data from the paper by H. Ellerhorst, Jr., Adhesives Age 25, 42 (1982).

books have been published on this subject (6-9) and Kinloch has written two reviews that cover both the surface chemistry (10) and the mechanics of adhesive joints (11). Despite the progress made in fracture mechanics, basic surface chemical interactions still govern adhesion at the monolayer level. Without these, the joint has no foundation. For example, it has been established (12) that the threshold failure energy θ_o approaches the thermodynamic work of adhesion $\overline{W}_A$. Recently, surface energetics have been related to fracture surface energetics of polymers (13,14). A new interpretation of the surface energies of polymers will be described by Lee (15).

The importance of surface modification to adhesion has been recognized by many authors. A microwave plasma treatment (16) has been used to modify the acid-base balance

*Registered trademark of DuPont Company.

of the polymer-filler surface. Reactive groups such as isocyanate, acyl chloride, and azo (17) have been introduced onto polymer surfaces to modify surface properties and adhesion. The surface of Kevlar® has been modified (18) to provide reactive amine sites for enhancing its adhesion to an epoxy resin and the water sensitivity of many polymer surfaces can be reduced by plasma polymerization (19). These are only a few examples of how surface modification can influence adhesion. The characterization of these modified surfaces requires the use of new techniques (20) such as ESCA, to be discussed in the following section.

III. PHYSICS OF ADHESION - THE CHARACTERIZATION OF SURFACE AND BULK

Adhesion may not be regarded just as a phenomenon related to the contact of two solid bodies. Recently, physicists have applied the concept of adhesion to understand the structure of basic particles at the subatomic level (21). New terms, such as "glue balls" or "gluons," have appeared in particle and nuclear physics literature. The phenomena of adherence have been studied by Maugis and Barquin (22) and Maugis (23).

New instrumental surface techniques (24) have been introduced to study adhesion during the last ten years. Among them, ESCA (or electron spectroscopy for chemical analysis) techniques (25) have given us insights about the structure of the polymeric interface within the first 50A. New applications of ESCA are discussed by Briggs (26). A combination of ESCA and AES (27) has been used to investigate the interfacial bonding between aluminum and chromium(III) fumarato-coordination compounds.

Not only new techniques deserve attention, but also any new developments in old techniques, such as scanning electron microscopy (28). Scanning electron microscopy, for example, can now enhance the examination of the adhesive interface in greater detail. Two other old techniques have also found new applications in adhesive chemistry. One is dynamic mechanical analysis (29,30), which has been accepted for the study of pressure-sensitive adhesives and the curing mechanism of epoxy resins (31,32). The other is the use of a fluorescence probe to examine the curing mechanism (33).

An exciting contribution by Dickinson (34) is the application of fracto-emission in studying the phenomena of adhesive failure. The emission of electrons (EE) and positive ions (PIE), as well as free radicals, has been vividly shown during the fracture processes. In addition, the energetics distribution of the emission can be accurately measured.

IV. RADIATION-CURABLE ADHESIVES

The contributions of physics to adhesion are further manifested by the extension of UV- and electron-beam curing of adhesives to low temperatures. Radiation-curable adhesives were originally intended to eliminate solvents in the systems. However, the use of certain toxic acrylate monomers presented some problems. So far, radiation-curable systems have become technologically challenging, but economically less attractive than most thermally cured systems. Since one of the earliest papers on photo-adhesives by Nakano (35) many patents have been issued on this subject. Two reviews have been published by Stueben (36).

Free-radical, radiation-curable systems generally consist of monomers, oligomers, photoactivators, other resins, and fillers or tackifiers. An ultraviolet source or electron-beam generator is used to cure the systems. As an illustration Stueben (37) describes a typical UV-cured PSA system containing acrylates and polyvinyl ether. McGinniss (38) discusses formulation design related to UV-curable systems. The cationic photoinitiators are shown in Fig. 1.

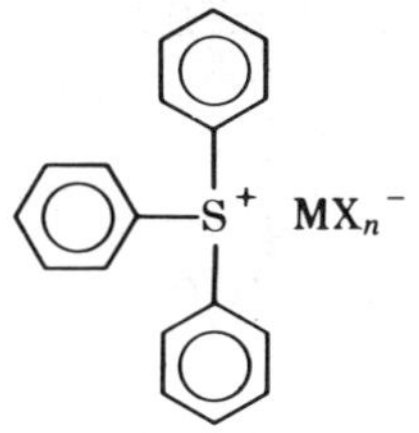

where MX_n^- is BF_4^-, PF_6^-, AsF_6^-, SbF_6^-, ClO_4^-, etc.

Fig. 1. Photoinitiators for Cationic Polymerization

(Ref: J. V. Crivello and H. W. Lam, J. Polym. Sci., Polym. Chem. 18, 2677 (1980).)

Ledwith (40) has recently discussed his new findings related to the polymerization mechanisms involving cation radicals (Fig. 2). By the same type of polymerization, the addition of acrylate to epoxide (41) created a product superior to that derived from epoxide or acrylate alone. The radicals generated during the cationic polymerization are presumably used for grafting acrylate onto the epoxide (42).

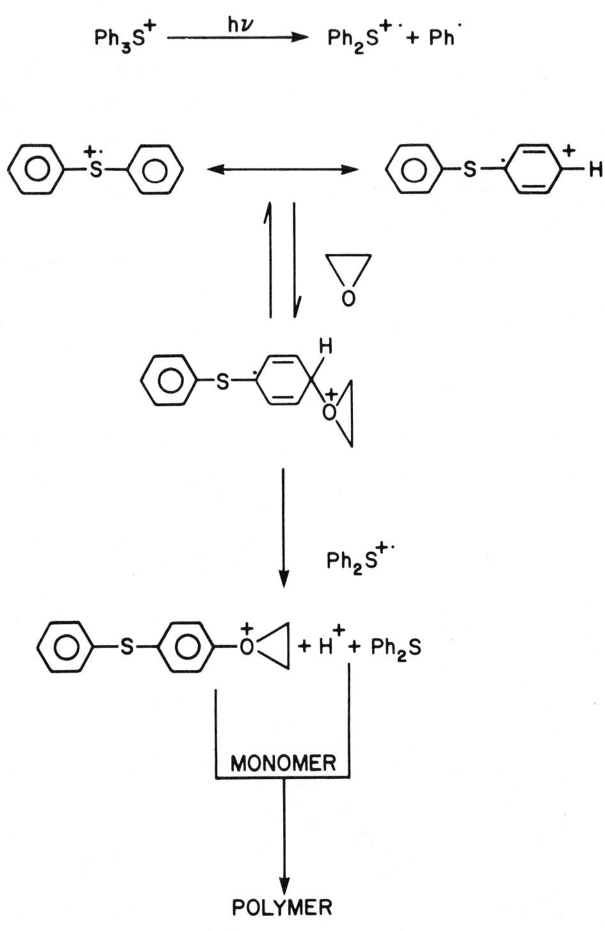

MECHANISM OF CATIONIC PHOTOPOLYMENZATION

Fig. 2. Cation-radical Mechanism for Photopolymerization

(Ref: A. Ledwith, a paper presented to the Northeast Regional ACS Meeting, Rochester, N.Y., Oct. 1981.)

In fact, the importance of the cationic photopolymerization lies beyond epoxides (39). Many oxygen-containing compounds, e.g., vinyl ethers, tetrahydrofuran, oxetane, lactones, trioxane, and some unsaturated compounds (Fig. 3) can be polymerized by the same mechanism to form adhesives or coating materials. Crivello (43) reviews the scope of cationic photopolymerization, providing us with a perspective on this promising process.

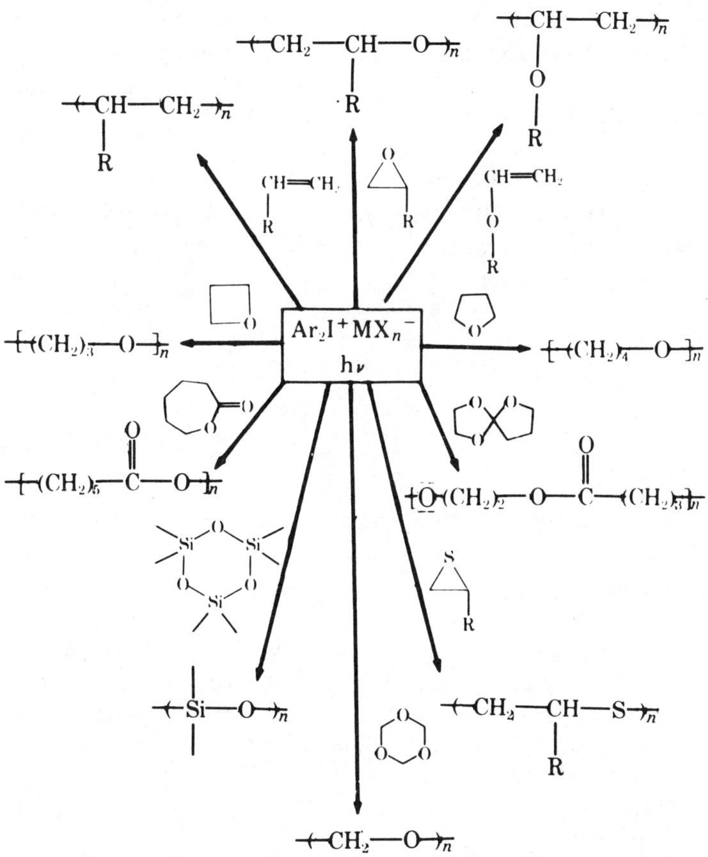

Fig. 3. Photoinitiated Cationic Polymerization using Diaryliodonium Salt

(Ref: J. V. Crivello, CHEMTECH, 10, 624 (1980).)

In recent years, not only has UV-curing been accepted by the adhesive industry, but electron-beam (E-B) curing is also being applied to continuous operation (44). E-B curing uses low energy electrons (<300 kV) generated from a filament, as shown in Fig. 4. The electrons are then accelerated in a field of 100-300 kV before hitting a target on a moving web, which is only 1 to 2 inches from the window. Since electrons move at a very high velocity, very little energy is lost when they move through the window.

The depth of electron penetration depends on the density of the material. Thus, the penetration ability of the processor is quoted as thickness of unit density product or weight per unit area, e.g., gram per square meter or pound per ream (1 lb/ream=1.6 g/m^2). The relationship between the electron dose (% front surface) versus penetration is shown in Fig. 5. The only way to increase the penetration depth is to increase the accelerating voltage.

For EB curing, the important processing parameters are the beam current I_e, the voltage V_e, and the product speed S. E-B processor performance is characterized by the yield factor (k), which relates I_e, V_e, and S through the dosage, D:

$$D = \frac{k \cdot I_e}{S} \qquad (1)$$

The units of the parameters are mrads for D, mA for I_e, and ft/min for S. At a given preset dosage, line speed is the primary process parameter in a slave mode of operation.

The advantage of E-B is its high-speed curing of adhesives on paper, film, foil, fabric, and other flexible substrates. For E-B-cured, hot-melt adhesives, the modulus, plasticity, creep, tack, etc., can be controlled *in situ*. Though the capital investment is high, a cost advantage can be realized for a large-volume, high-speed curing operation.

V. HIGH-TEMPERATURE ADHESIVES

The aerospace industry has developed strong demands for high-temperature adhesives. The adhesives used on missiles

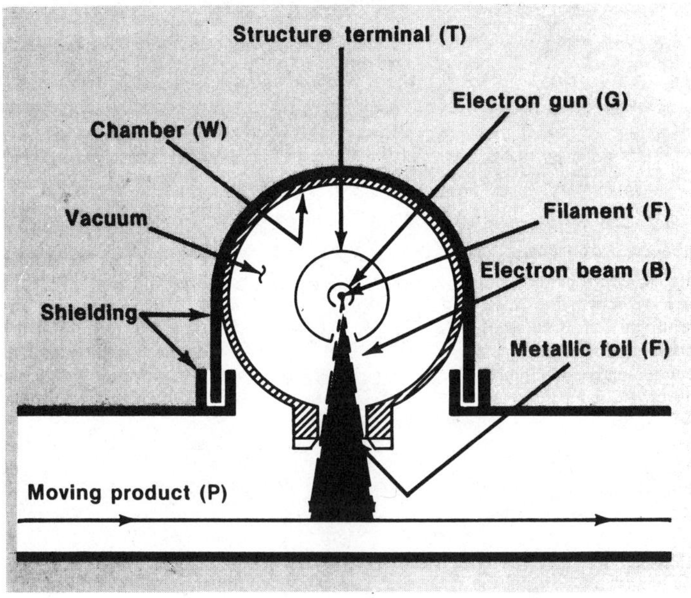

Fig. 4. Schematic of Electron Processor

(Ref: R. Kardashian and S. V. Nablo, Adhesives Age, 25, No. 12, 25 (1982).)

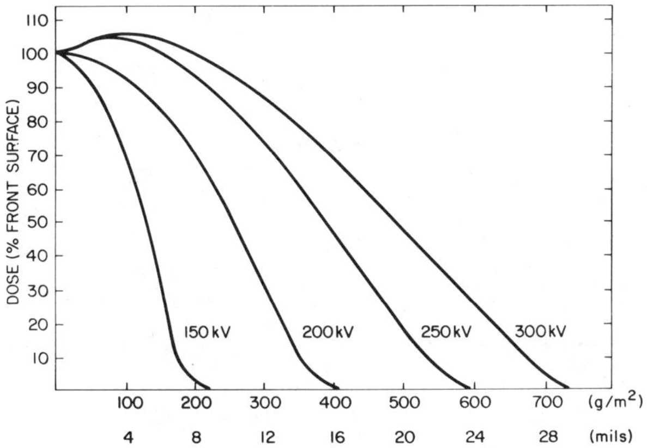

Fig. 5. Processor Penetration Performance in Unit Density Product

(Ref: R. Kardashian and S. V. Nablo, Adhesives Age, 25, No. 12, 25 (1982).)

(45,46) require mechanical performance for less than one minute but at temperatures greater than 538°C (Fig. 6). Thus, organic polymers with exotic structures have been designed to meet such high-temperature requirements (47).

For adhesive applications, thermal stability alone is insufficient. Processability of high-temperature polymers is of great importance. These polymers should be processed under moderate conditions with no volatile evolution. The adhesives should be compatible with adherends and should have good, durable properties. For broader uses, the manufacturing cost should be low because cost, so far, has been the major obstacle to developing high-temperature adhesives.

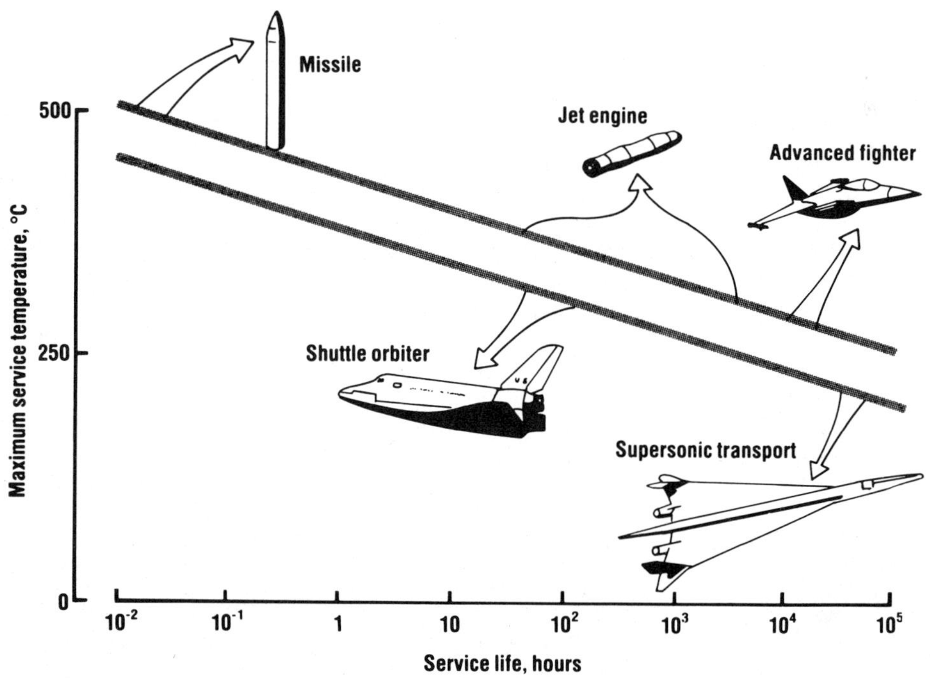

Fig. 6. Service Lives of High Temperature Adhesives versus Maximum Service Temperatures.

(Ref: A. K. St. Clair, NASA Technical Memorandum 81884, Oct. 1980.)

During the last several years, a thermoplastic condensation polyimide (PI) was developed at NASA Langley Research Center (49). From the reaction between BTDA (3,3', 4,4'-benzophenonetetracarboxylic acid dianhydride) and DABP (3,3'-diamino benzophenone), the polyimide LARC-TPI (Fig. 7) was prepared by Bell et al. (49). Later, St. Clair (50) used LARC-TPI as an adhesive for titanium. The adhesion of titanium and PI has been studied by Wightman et al. (51) with XPS (ESCA) and FT-IR techniques.

Besides the condensation PI, there have been two new types of addition PI's: ethynl (acetylenic) and norbornene (nadic) end-capped PI. The acetylene-terminated PI (ATI) was first reported by Bilow and co-workers (52) and the nadic-terminated PI (NTI) (Fig. 8) was derived from the work of Lubowitz (33). An NTI prepared at NASA Langley Center has been designated as LARC-13 (34). When LARC-13 was compared with ATI as the adhesive for titanium, the lap shear strength of LARC-13 after 1000 hours at 232°C was lower than that of ATI.

Many polyimides are brittle. To toughen a brittle PI, special elastomers (55) have been added. Silastic rubber, vinyl fluorosiloxane, and Sylgard, divinyl-dimethylsiloxane, can be directly blended with PI, e.g., LARC-13. Two amino-functional elastomers have been used successfully: (1) ATBN, aniline-terminated butadiene-nitrile rubber, and (2) ATS, aniline-terminated dimethylsiloxane.

Polyphenylquinoxalines (PPQ's) were first prepared in 1967 by Hergenrother (56). In 1975, Koval et al. (57,58) introduced the acetylene-terminated PPQ (ATPQ) (Fig. 10), which is the addition type of PPQ. Resembling ATPI, ATPQ also has shortcomings. The ethynl group hinders the flow of the polymer chain, and the cured ATPQ is less thermal-oxidatively stable than the comparable PPQ. The detailed status of PPQ, as well as other high-temperature adhesives, is discussed by Hergenrother (45).

The properties of several typical high-temperature adhesives are listed in Table 2 (59). Note that one listed is FM-34 condensation polyimide from the Bloomingdale Division of American Cyanamid Company. The contributions of the above high-temperature adhesives to the aerospace structural adhesives will be discussed later.

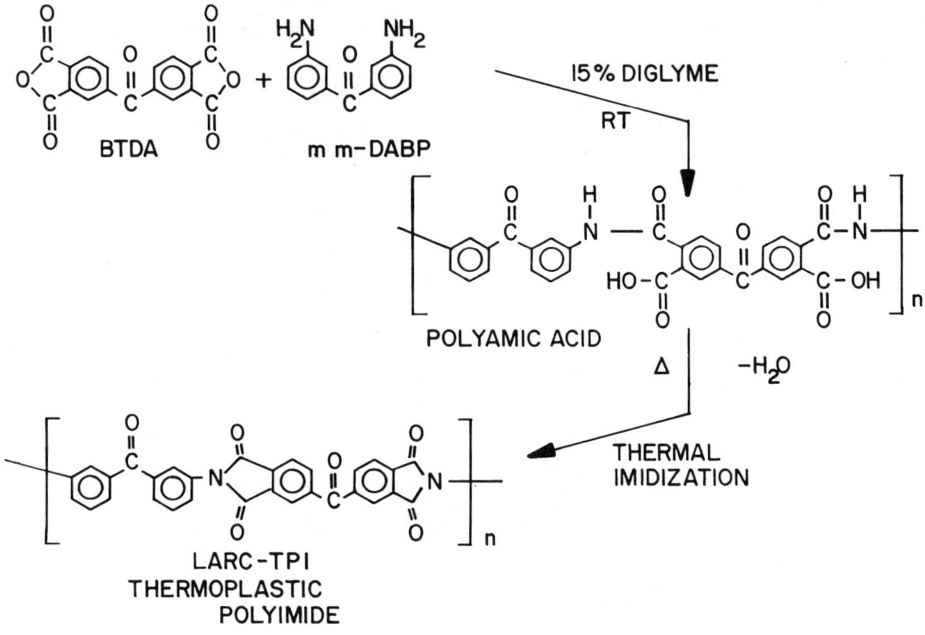

Fig. 7. Preparation of LARC-TPI Adhesive

(Ref: V. L. Bell, B. L. Stump, and H. Gazer, J. Polym. Sci., Polym. Chem. Ed. 14, 2245 (1976).)

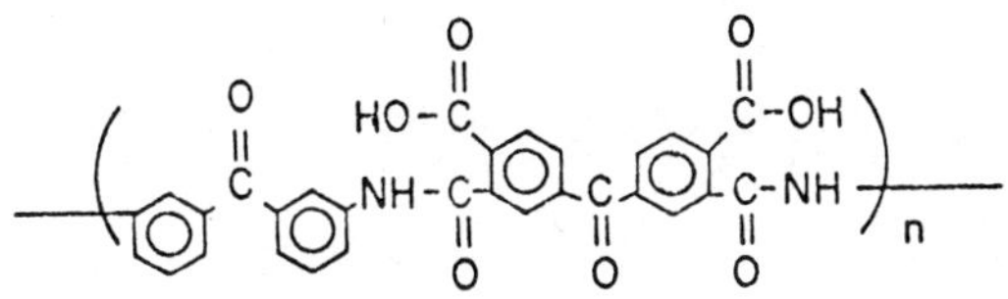

LARC-2 POLYAMIC ACID ADHESIVE

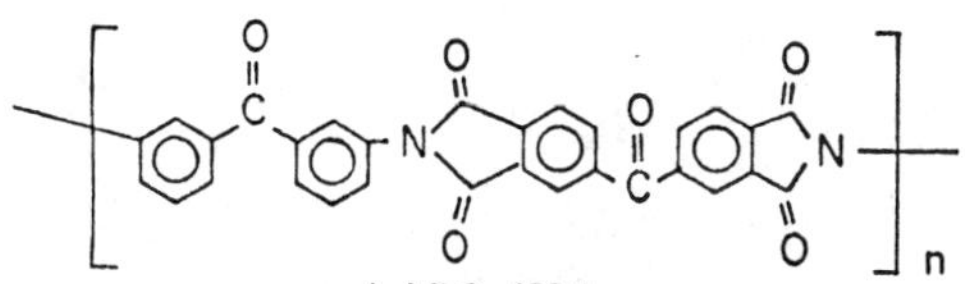

LARC-TPI
THERMOPLASTIC
POLYIMIDE

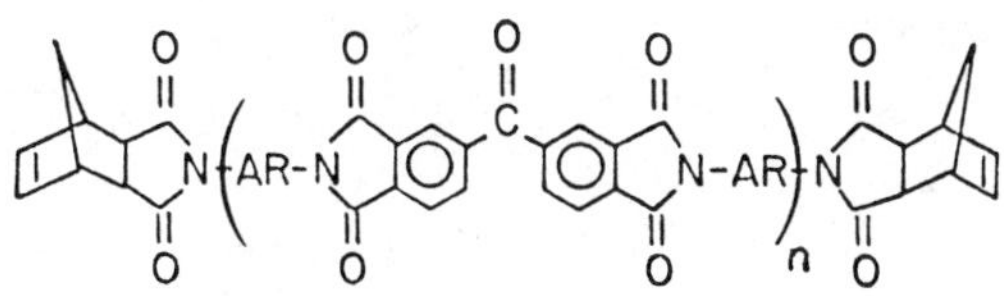

NADIC-TERMINATED IMIDE (NTI)

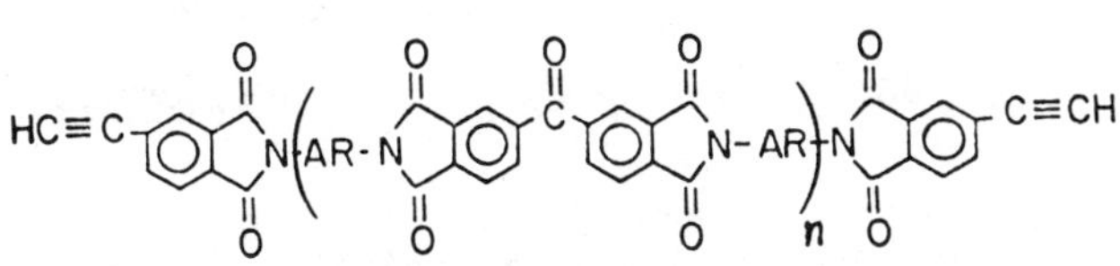

ACETYLENE-TERMINATED IMIDE (ATI)

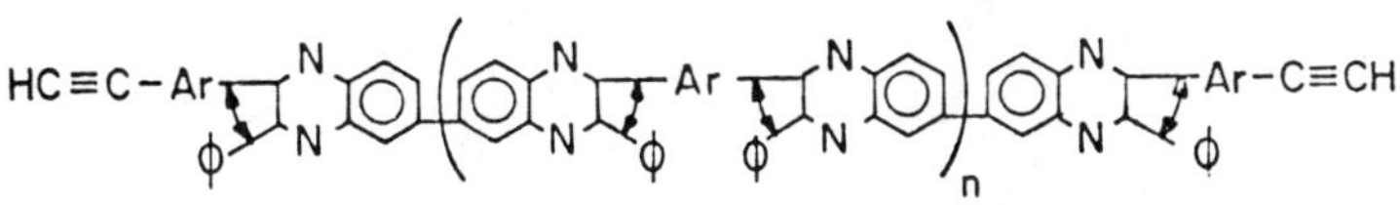

Acetylene-terminated PPQ (ATPQ)

$Ar = \phi-O-\phi$

Fig. 8. Structures of High Temperature Adhesives

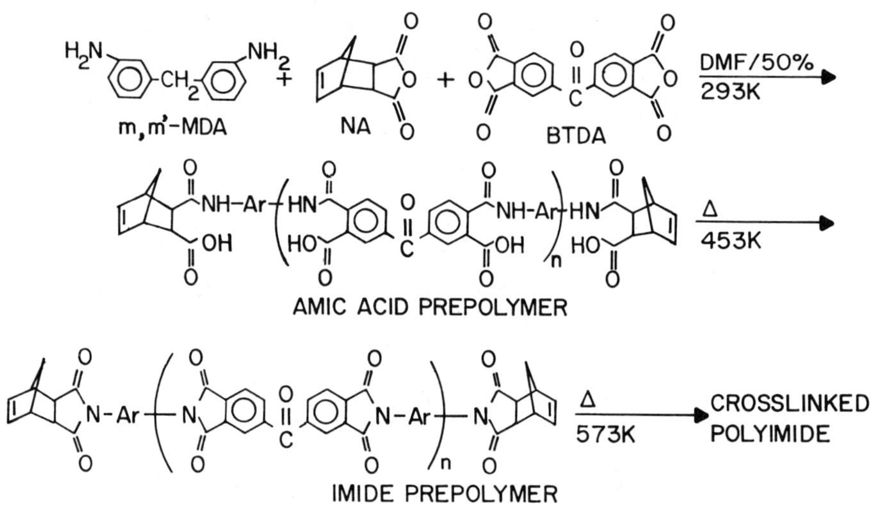

Fig. 9. LARC-13 Adhesive Chemistry

(Ref: A. K. St.Clair and T. L. St.Clair, Polym.and Eng. Sci., 22, No. 1, 9, (1982).)

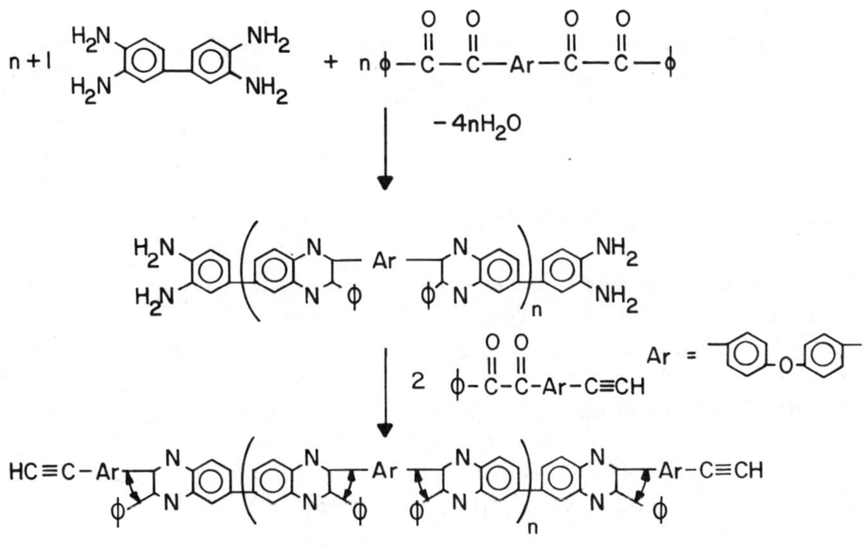

Fig. 10. Preparation of ATPQ Oligomer

(Ref: P. F. Kovar, G. F. L. Ehlers, and F. E. Arnold, Polym. Chem. Preprints, 16, (2) 246 (1975)).

Table 2. High Temperature Adhesive Tape Properties

Adhesive	FM-34	NR-056X	LARC-13	PPQ
Filler	Yes	Yes	Yes	No
Tack	Yes	No	No	No
Drape	Yes	No	No	No
Flow	Good	Good	Good	Fair
Shelflife	Poor	Poor	Good	Good
Volatiles, %	~14	~11	<1	<1

(Ref: P. M. Hergenrother, "High-temperature Adhesives," lecture notes for the course on "Fundamentals of Adhesion," State University at New Paltz, New York, Nov. 1983.)

VI. ANAEROBIC AND STRUCTURAL ADHESIVES

Acrylic polymers are one of the important structural adhesives (Fig. 11). These polymers can be classified as anaerobic structural acrylics (60), second generation acrylics (60,61), and aerobic acrylics (61,62). Most anaerobic adhesives are based on dimethacrylic esters, e.g., TEGDMA (triethylene glycol dimethacrylate) (Fig. 12). Anaerobic adhesives (63) containing maleimides have been reported to have improved thermal resistance.

Recently, the curing characteristics of anaerobic sealants and adhesives (Fig. 13) were reported by Lees (60) and Stamper (64). The present systems can be cured rapidly in most metallic substrates, yet possess shelflives in excess of one year. The factors affecting curing are the: cure system, surface preparation, substrate, temperature, and bondline thickness.

Cyanoacrylates (65) are no longer the wonder-glues for small applications. In general, newer, modified cyano-acrylates can be used on porous surfaces, oil-contaminated metals, and even wood. Some recently introduced grades can have a service temperature as high as 150°C, and are less sensitive to humidity. Furthermore, impact resistance has been improved to qualify the adhesives for applications such as mounting door trims on autos.

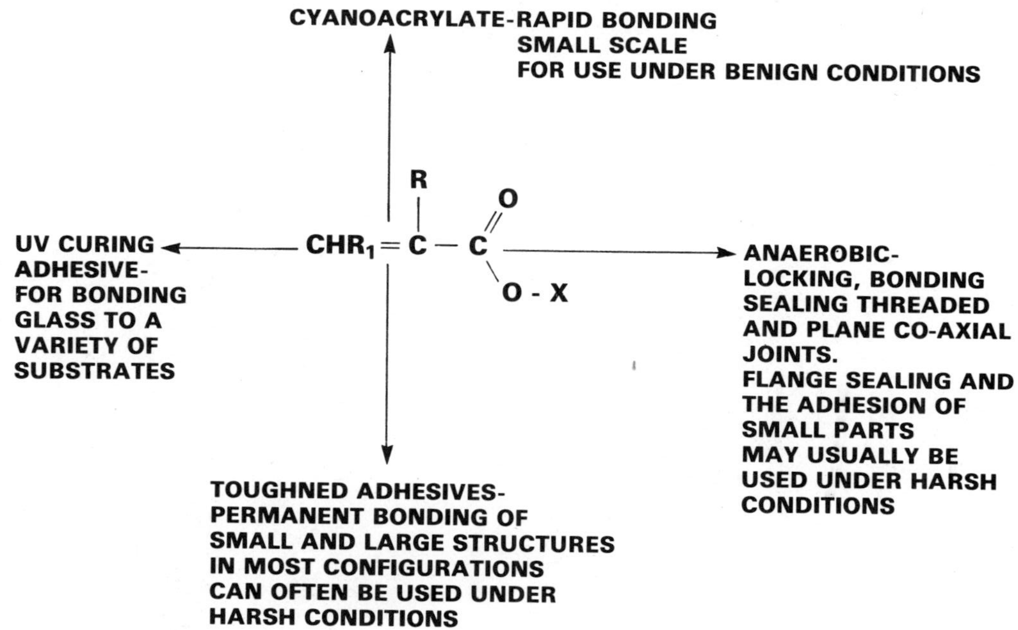

Fig. 11. New Developments in Acrylic Adhesives

(Ref: W. A. Lees, Brit. Polym. J., 11, 64 (1979).)

TEGDMA AND 'HOMOLOGUES' ARE REMARKABLE MOLECULES

$$CH_2{=}CH{-}\overset{O}{\overset{\|}{C}}{-}O{-}\left[CH_2CH_2{-}O\right]_3{-}\overset{O}{\overset{\|}{C}}{-}CH{=}CH_2$$

OR FOR THE 'HOMOLOGUES' —

Ac — X — R

WHERE X IS ANY ESTER-BASED GROUP(S) AND R IS USUALLY, H, OH OR A FURTHER ACRYLIC ESTER.

Fig. 12. Anaerobic Adhesives

(Ref: W. A. Lees, Brit. Polym. J., 11, 64 (1979).)

The second generation acrylics have also been called "reactive acrylics", "toughened acrylics," or "modified acrylics." These adhesives can be bonded rapidly at room temperature and are somewhat tolerant to an oil-contaminated surface. The bonds are flexible and give high peel and impact strengths. Generally, these adhesives have good low temperature properties, good heat resistance, and excellent solvent resistance. For formulation, they have a long open-time and provide flexibility for specific end-use requirements. The acrylics approach the tensile-lap shear-strength level of the epoxies - 4000 psi (28 MPa) 2024-T3 aluminum-to-aluminum for acrylics versus 4,740 to 6,000 psi (34-41 MPa) for a typical epoxy.

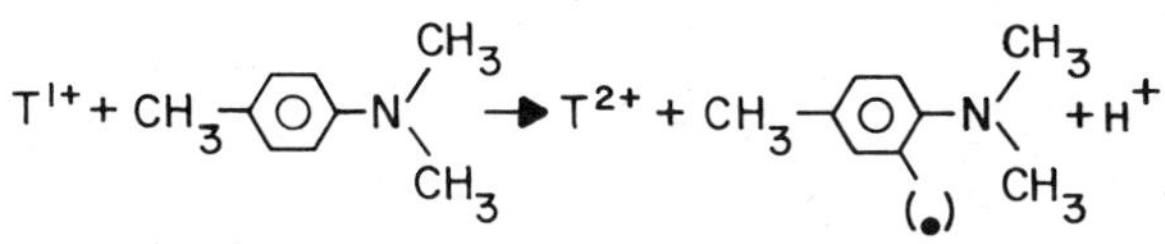

HENCE:

$$RO_2R + T^{2+} \rightarrow T^{3+} + RO^{\bullet} + RO^{-}$$

$$T^{3+} + AH \rightarrow T^{2+} + A^{\bullet} + H^{+}$$

ADD $RO_2R + T^{2+} + T^{3+} + AH \rightarrow T^{3+} + RO^{\bullet} + RO^{-} + T^{2+} + A^{\bullet} + H^{+}$

SUBTRACT $T^{2+} + T^{3+}$

$$R_2OR + AH \xrightarrow[T^{2+} \rightleftharpoons T^{3+}]{\text{CATALYTIC}} RO^{\bullet} + ROH + A^{-}$$

WHERE: T IS A TRANSITION METAL
AH IS THE AMINE
R IS ANY SUITABLE GROUP
A IS A STABLE STRUCTURE

Fig. 13. Curing Mechanism of Anaerobic Adhesives

(Ref: W. A. Lees, Brit. Polymer J., 11, 64 (1979).)

Improved versions of the reactive acrylics have been developed and are called "aerobic" acrylics (61,62). These aerobic acrylics have diminished sensitivity to air inhibition and can be cured between two surfaces, regardless of the presence or absence of air. The adhesives are low in odor, nonflammable, and low in toxicity. The fast curing rate, e.g., 10 to 60 seconds, and the solvent-diluted activator minimize the movements in the substrates prior to fixture. The compositions, as claimed in patents (66,67), contain an elastomer that is colloidally dispersed in a monomer or in a monomer/oligomer/polymer solution. On the general subject of the rubber-reinforcement of acrylics, Drake et al. have written a recent review (68). However, they emphasize nitrile rubbers.

Several recently developed epoxides deserve mention. Dow has introduced a series of new experimental resins based on triglycidylether of tris(hydroxyphenyl)methane isomers and higher oligomers (73):

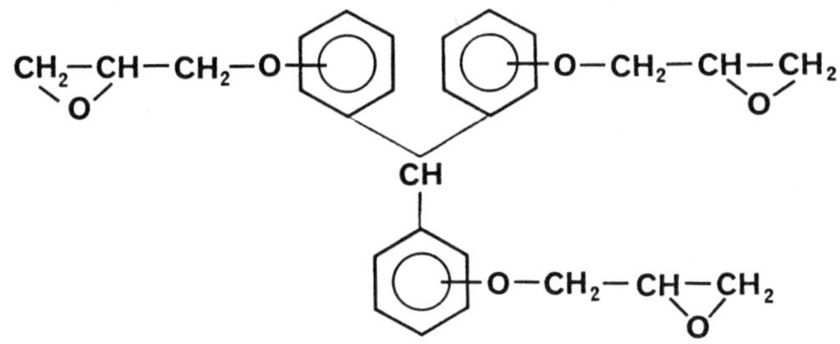

The cured epoxies have high heat distortion temperatures and improved thermal stability compared to similar epoxy resins. As for adhesives, Dow XD-7342.00L semisolid resin gives toughness, high wet strength, and long-term high-temperature oxidative resistance.

The literature describes a room-temperature curing epoxide made by converting a part of the epoxy function into episulfide by the addition of NMBTT (N-methyl-benzothiazole-2-thione) (74).

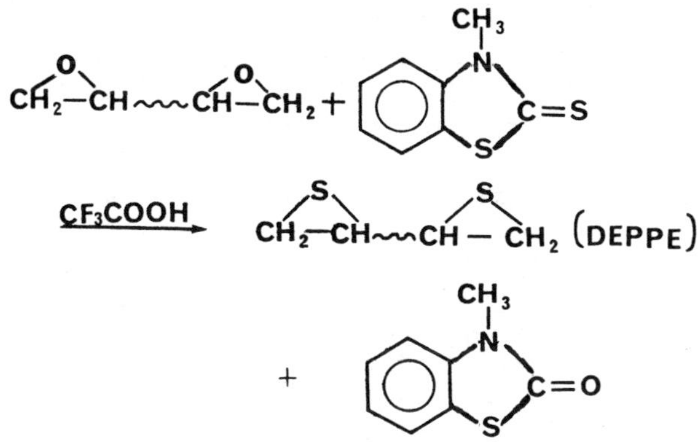

The resultant product (74) can be cured into a "bond cement" in 5-20 minutes at room temperature.

Another interesting low-temperature cure system (75) contains an epoxy resin - triglycidyl p-minophenol (TGPAP)

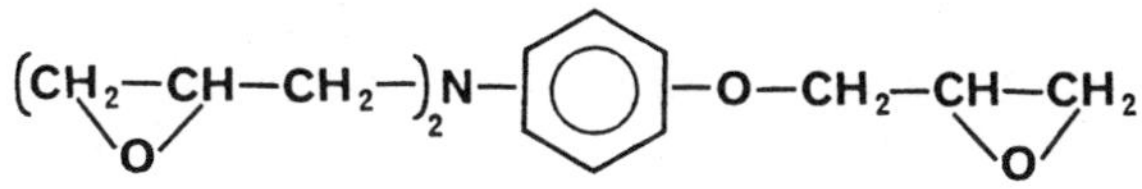

and BTDA as the curing agent.

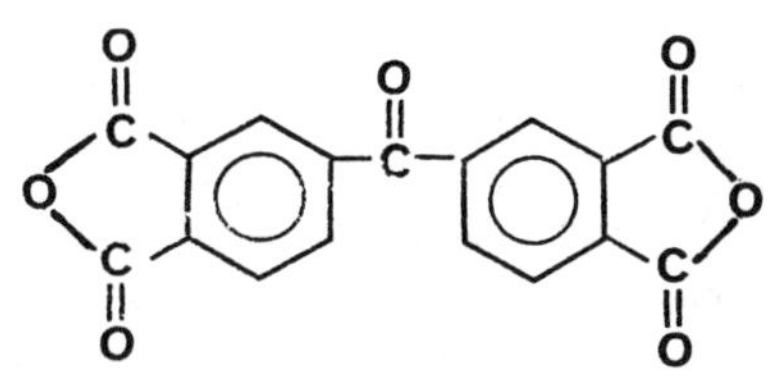

No other catalyst is needed. Solubilization of BTDA in the epoxide may be aided by slightly warming the mixture or by the addition of a mono anhydride, such as maleic anhydride (MA). The cured resin has high-temperature properties.

Among new epoxides are two families containing fluorine and phosphorous, respectively. The fluoro-epoxides prepared by Griffith (76,77) at the Naval Research Laboratory have the following general structure:

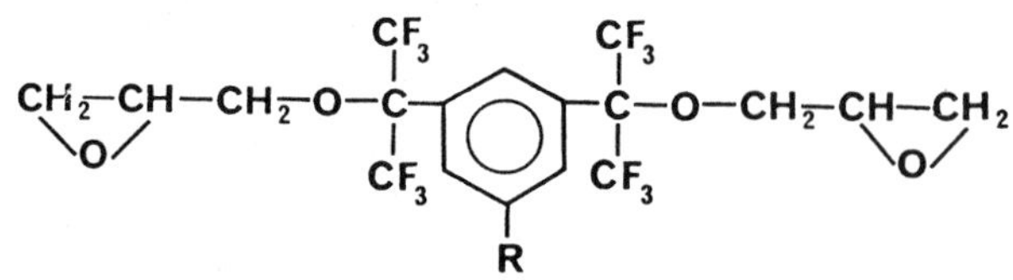

These epoxies are extremely low in water absorption and show good resistance to biological fouling. Thus, they are suitable for marine coating and underwater composites (78), in addition to being water-resistant adhesives.

Besides fluoroepoxides, fluoroanhydrides (76,79) have been synthesized. The structure of the substituted phtalic anhydride is as follows:

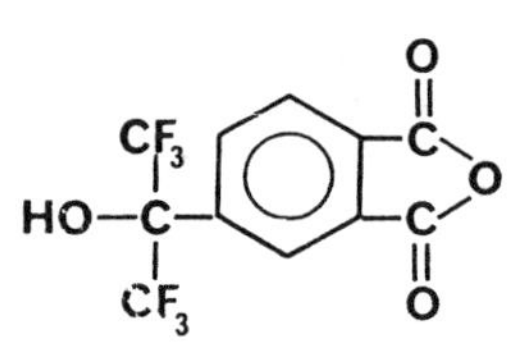

Other fluoroanhydrides can be derived from trimellitic and pyromellitic anhydrides.

The phosphorous-containing epoxides (80) are good, flame-resistant materials, especially when used for structural adhesives. A typical phosphorylated epoxide is based on bis-(3-glycidyloxyphenyl)methyl phosphine oxide (BGDDPMPO):

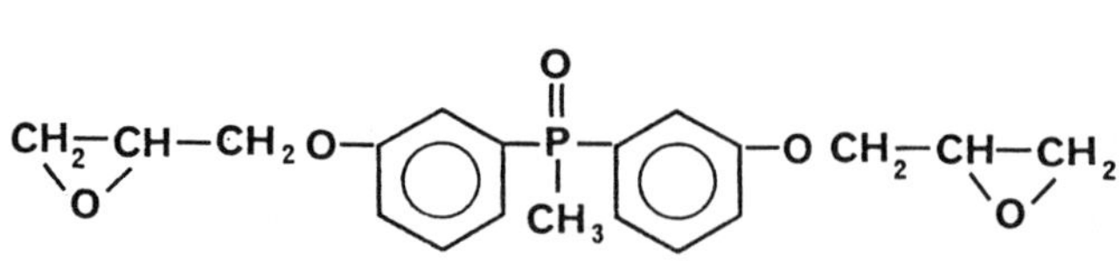

The phosphorylated epoxide is low in smoke evolution, high in char yield, and high in oxygen index.

The structural adhesives discussed in this section are acrylics, anaerobics, cyanoacrylates, and epoxies. Among them, epoxies still are the most important. With new innovations, epoxies will be more readily cured at room temperature to compete with cyanoacrylates and anaerobics. In the following sections, we shall discuss several new forms of adhesives, e.g., hot-melt adhesives, film adhesives and waterborne adhesives. Several structural adhesives will be discussed again in relation to new forms of adhesives.

VII. HOT-MELT ADHESIVES

Hot-melt adhesives are currently one of the fastest growing segments of the industry (Fig. 14). No solvent discharge is the best answer to the pollution problem and, in comparison with the waterborne systems, hot-melts require less energy. From a capital investment standpoint, the installation of a hot-melt coater line does not require a large oven to remove water or solvent from the system. As a result, hot-melt adhesives have been widely used, e.g., the PSA's (81). Some of the hot-melts can be cured by UV radiation or electron beams.

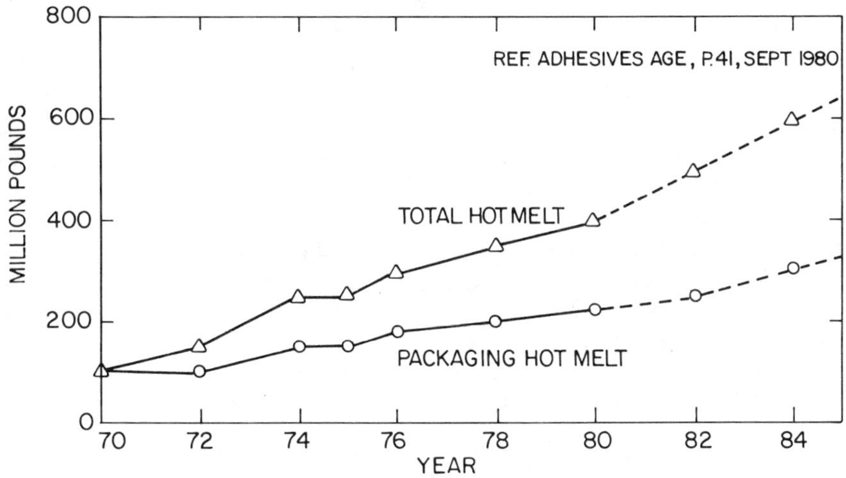

Fig. 14. Growth of Hot-Melt Adhesives (1970-1984)

The adhesives industry recognizes several advantages (81) in using hot-melt PSA's (HMPSA's):

. Increased production rate via elimination of the vehicle removal step

. Lower process energy demand

. Potential for heavy deposition and easy pattern coating

. Ready compliance with environmental protection regulations

However, the early systems had three major problems:

. Excessive thermoplasticity

. UV, heat, oxidation degradation susceptibility

. Limited plasticizer resistance

Significant improvements have eliminated the first two problems. Research is still being undertaken to solve the third problem.

Most hot-melts are based on polyolefins. The best known hot-melt is ethylene-vinyl acetate copolymer (EVA), e.g., duPont Elvax. In 1983, duPont added a new line of resins (82) designated as Elvax II ethylene copolymers. This family of copolymers contains a substituted alkyl group and an acid functionality. These resins are thermally stable and compatible with essentially all materials currently used in hot-melts. The acid functionality improves adhesion of the resin and enhances mechanical properties, e.g., toughness. Furthermore, the resins are cross-linkable through the acid groups.

Acrylic PSA's offer UV stability, color stability, clarity, and good adhesion. The recent development of hot-melt acrylics for PSA's deserves attention. However, the problems with using acrylic HMPSA's are many (83). For example, good cohesive strength is difficult to achieve in acrylic HMPSA's, especially at elevated temperatures, in a product that can be applied satisfactorily in melt form.

Some problems of acrylic HMPSA's are circumvented by

formulation. For example, the addition of vinyl toluene/alpha-methyl styrene tackifier resin, Piccotex LC (Hercules), can enhance peel adhesion, and reduce melt viscosity with no loss in UV resistance. New lines of acrylic HMPSA's have been reported (83), but further development is needed to warrant acrylic HMPSA as a competitive adhesive.

One of the new developments in hot-melts is the application of foam technology. Several years ago, the Nordson Laboratory (84,85) discovered that the foamed hot-melts are easier to spread than the non-foamed hot-melts. Later, foamability (86) studies were carried out for the following hot-melt adhesives: polyethylene, polypropylene, polyamide, ethylene-vinyl acetate copolymer, and polyester. Air, CO_2, and nitrogen were used as the blowing gases; various types of surfactants were used for dispersion.

A two-step gear pump (Fig. 15) was used to introduce the gas into the hot-melt adhesive. Foamability was later found to vary with the polymers used. Polyethylene is excellent for foaming. Polypropylene, polyester, and 12-polyamide (Nylon 12) are also very good. Styrene-butadiene and ethylene-vinyl acetate are good and dimer acid polyamide is fair.

Hot-melt foams should have many useful applications in: appliances, automobiles, bag manufacture, book manufacture, cans and closures, case forming and sealing, manifold business forms, disposables, roll wrapping, and metal building fabrication. The greatest potential use for hot-melt foams is in new applications where hot-melt adhesives have not been used before.

A new technological development for hot melts is their application by robotics (87). Robotics have been used to apply hot melts to structural components, e.g., fiberglas-reinforced components. In one case, structural adhesives are applied at 480°F (249°C) with a specially designed hot-melt dispenser and gun marketed by Monsanto Co. Robotics have also been used to apply hot-melts to make PSA's. Robotics are useful in applying hot-melts to gaskets and seals. In general, robotics provide higher productivity, improved quality, and reduced cost.

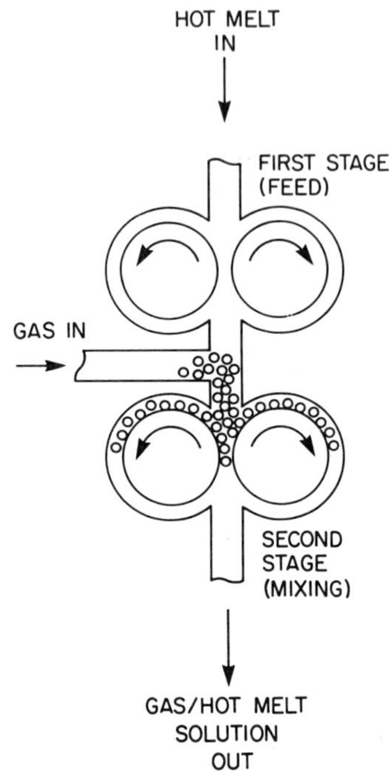

Fig. 15. Hot-Melt Foaming Process

(Ref: F. T. Hughes, Adhesives Age, 25, No. 9, 25 (1982).)

VIII. FILM ADHESIVES

In addition to hot-melt adhesives, film or sheet adhesives (88,89) have been used for the same purpose in eliminating solvent pollution. The adhesive-bearing films, such as PSA, can be cut to size and applied between mating surfaces. Curing, if necessary, can be carried out in the usual manner. Thus today, coated film adhesives can be used for large structural parts to replace solvent-borne adhesives.

Hot-melts, radiation-curable adhesives, and epoxies can all be applied to a substrate in film or sheet form. The substrate can be metal, plastic, or paper. In addition, the substrate can be woven, nonwoven, or knot. The porous substrates can be used to manufacture filter aids or medical adhesives, e.g., to attach prostheses to the skin. An interesting form of substrate is a foam that joins surfaces of irregular gaps. The foam can also be formed instantaneously during curing to fill gaps.

A recent development is the use of more than one adhesive on the same substrate. For example, a cross-linkable acrylic adhesive can be applied on a close-cell polyurethane foam, and the sheet adhesive can be used to mount autobody side-moldings. A two-sided film or sheet can be made to contain a thermoplastic adhesive on one side to bond to the auto-carpet fibers and another adhesive on the opposite side to bond to the polypropylene body and console panels.

The use of film adhesive in the aerospace industry will be discussed in a forthcoming section.

IX. WATERBORNE ADHESIVES

As discussed in the Introduction, in addition to hot-melt adhesives, one of the most important forms of adhesives for the future is the waterborne system. In general, an advantage of the waterborne system is that it is free from the solvent-pollution problem. For this reason, we shall describe several new waterborne systems in detail: acrylics, modified ethylene-vinyl acetate, polyurethanes, phenolics, epoxies, isocyanates, and tackifiers. Applications of these systems range from PSA's contact adhesives, and laminating adhesives to wood adhesives.

Acrylics

For PSA's, acrylic latexes (83) appear to be the most promising compounds because of their UV and thermal stability, and high resistance to the plasticizer migration from vinyl substrates. Acrylics also offer advantages in lower formulating and manufacturing costs.

A new acrylic family (83) was developed specifically for low surface energy surfaces, e.g., polypropylene. A new form of cure, a "low energy cure" or LEC polymer, can take place at low temperatures. The LEC catalysts are generally alkaline, e.g., sodium sesquicarbonate. The LEC polymers provide rapid water release and unique adhesion to glass surface. These polymers can also be used for lamination.

For contact adhesives, acrylic latexes compare favorably to Neoprene A latex as well as to the solvent-based Neoprene. As expected, the latexes dry slower than the solvent-based Neoprene; however, the drying time can be shortened through formulation. In terms of lap shear adhesion, acrylic latexes are better than Neoprene A latex. In addition, acrylic latexes offer comparable performance, freeze-thaw stability, nonflammability, ease of handling, and ease of cleanup.

For lamination, waterborne adhesives have actually been used in manufacturing flexible packaging and industrial laminates (91). Other latexes used besides acrylic are vinyl acetate, ethylene-vinyl acetate, vinylidene chloride, etc. The flexible packaging systems using these adhesives are: PVDC-Cello/LDPE, OPP/OPP, polyester/LDPE, and metallized polyester/LDPE. Similarly, industrial laminates using these adhesives are PE/textile/PE, OPP/printer paper, and polyester/printed Tyvek (PE).

Polyvinyl acetate

A conventional emulsion polymer based on vinyl acetate has been modified by carboxylation (92). The carboxylated ethylene-vinyl acetate adhesive increases the adhesion to metal and polymer surfaces and the resistance to oil, grease, and water. The films are also acid, alkali, and UV resistant. Furthermore, the polymers are cross-linkable through the carboxylic groups. These adhesives can react with aminoplasts, phenolics, and epoxy resins for increased water and creep resistance.

Polyurethanes

One of the most interesting waterborne systems is polyurethane (93). Polyurethane dispersions can be made by the following processes (94): acetone, prepolymer mixing, melt-dispersion (95), self-dispersion of solids, ketimine process (96), and ketazine process (97). The acetone process involves the formation of an ionomer first in organic solvent and then in a dispersion in the acetone-H_2O mixture (8:2). In the prepolymer-mixing process, a prepolymer containing an ionic group terminated with isocyanate groups can form a spontaneous dispersion upon the addition of water. The polyurethane ionomer dispersion is illustrated in Fig. 16.

In the melt-dispersion process (95), an NCO-terminated prepolymer is prepared next from a polyester or polyether diol, a diisocyanate, and a glycol containing a potential ionic group. The NCO/OH is between 1.2 and 1.8. Then, the prepolymer is terminated with urea to form biurets, which are further alkylated to give an ionomer as shown in Fig. 17. The bis-biuret can be considered to be somewhat hydrophilic to be methylolated with formaldehyde. Upon dilution with water, a stable dispersion is formed.

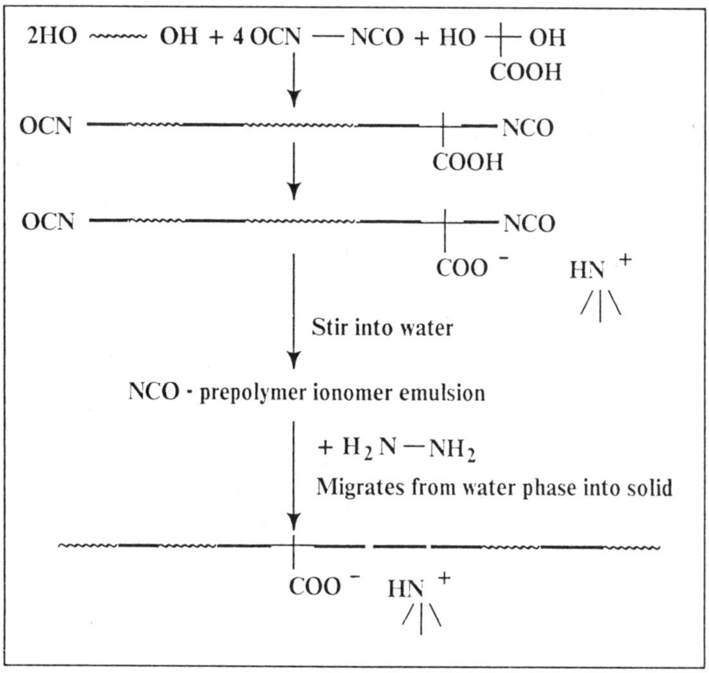

Fig. 16. Prepolymer-mixing Process for Polyurethane ionomer Dispersion

(Ref: D. Dietrich, Prog. Org. Coatings, 9, 281 (1981).)

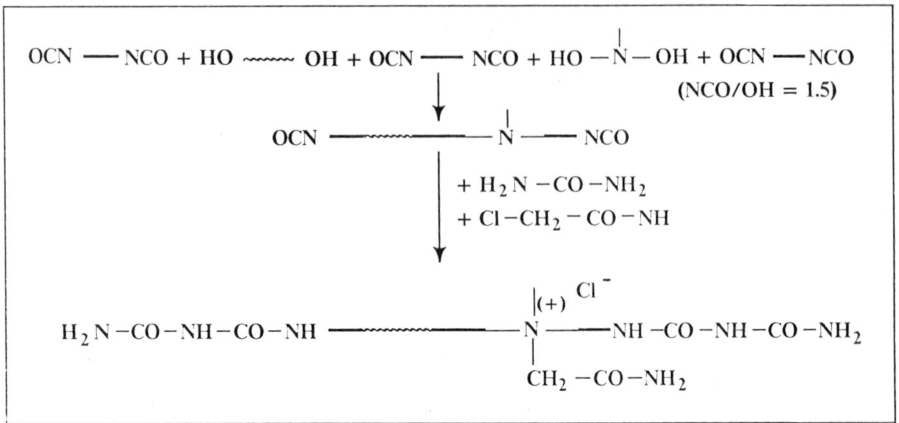

Fig. 17. Formation of Self-Dispersible Oligomer

(Ref: D. Dietrich, Prog. Org. Coatings, 9, 281 (1981).)

The ketimine process combines the quality and versatility of the acetone process with the simplicity and economics of the prepolymer mixing process. Ketimines are formed from a ketone and diamines (96).

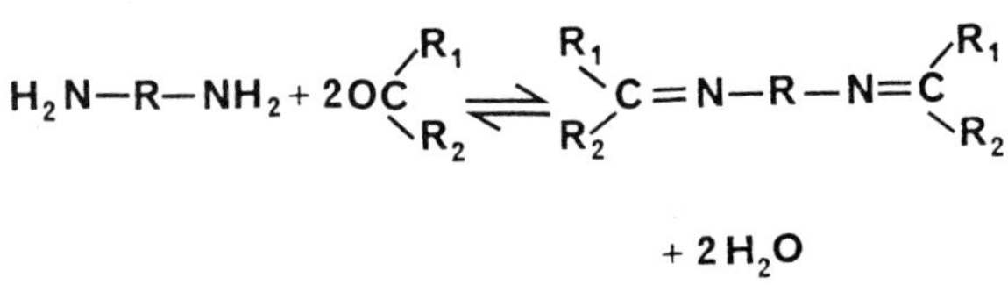

The ketimines can then be mixed with NCO prepolymers without inducing a reaction. However, upon the addition of water, a dispersion results and, about the same time, the ketimine is hydrolyzed to free the amine. The amine then undergoes chain extension to produce a material that can also be obtained by the acetone process.

In the ketazine process (97), the NCO prepolymer is mixed with a ketazine, aldazine, or hydrazone before dispersing. Upon the addition of water, hydrazine is liberated to promote chain extension:

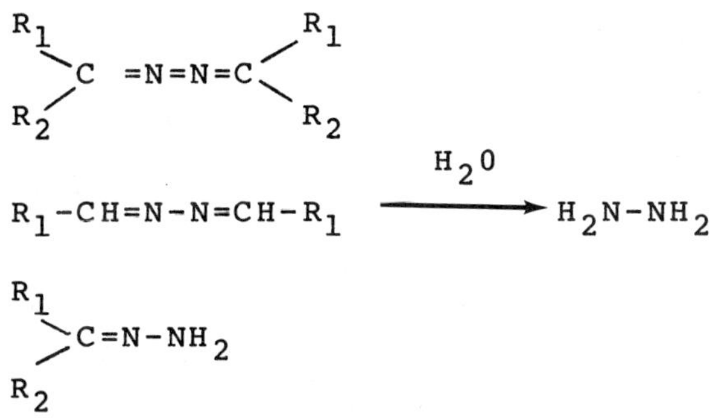

<u>Phenolics</u>

In the waterborne phenolic adhesives, a new dispersion system (98) has been reported. This system uses the solubility of some thermoplastics in phenol, and the subsequent polymerization takes place upon the addition of formaldehyde, an emulsifier, and water. The dispersion is then formed *in situ*. The versatility of this process is indicated by the use of many polymers with a wide range of properties, e.g., polyvinyl formal, polyvinyl butyral, acrylonitrile-butadiene copolymer, bisphenol-A resins, polyvinyl acetate, polycarbonate, polymethyl methacrylate, and nylons.

The resultant polyblends are generally two-phased and contain discrete particles dispersed in the phenolic matrix. With the reinforcements, the waterborne phenolic can compete with the solvent-borne counterparts. Thus, some of them can be used as metal-bonding adhesives.

Epoxy Resins

Waterborne epoxy resin (99) has been used for lamination. The epoxy copolymers PD-6049 and PD-4070 are supplied by Celanese at 69% solids in glycol ether co-solvents, which can readily be dispersed with water. When the dispersion is formulated with an aminoplast catalyst, curing takes place in several seconds at 450°F (232°C). The dispersion has excellent adhesion to paper, PET film, and aluminum foil.

Isocyanates

As wood adhesives, isocyanates have found applications as binders for composition board (100). The advantages of isocyanates are many: high adhesive and cohesive strengths, flexibility in formulation, versatility of various cure temperatures and curing rates, excellent structural properties, ability to bond with material having moisture content, and lack of formaldehyde emission. The most important advantage is their ability to form waterborne adhesives.

Emulsion-polymer isocyanates (EPI's) were introduced by Ashland Chemical Co. as better performing wood adhesives (101). EPI's are equivalent to phenol-resorcinol and cross-linkable polyvinyl acetate adhesives. EPI's are more advantageous than others because they have shorter press times, no chalking temperature, a low-temperature curing, no formaldehyde emission, better resistance to high temperatures and accelerated aging, environmental attractiveness, an ability to determine performance by adjusting hardener level, and more tolerance to variations in process conditions.

Waterborne Tackifiers

For PSA's with the resin-rubber systems, several types of latexes are being used in place of rubber for the waterborne adhesives. Styrene-butadiene (102) latexes and natural rubber latexes (103) have been mentioned in the literature. In addition, waterborne tackifiers (104) have been developed for PSA's. A family of emulsifier-free, high solids, completely freeze-thaw stable tackifying resins was developed by Rohm and Haas. The new products are viscous resins based on the replenishable tall-oil rosin. The resins can spontaneously disperse in water without the addition of a surfactant or a solvent. Other waterborne tackifying resins developed by Hercules have been discussed by Jones (105).

Phenolic tackifiers (106) in the form of dispersions can be used to formulate neoprene or acrylic latex into waterborne contact adhesives. For example, a solid ratio of latex and phenolic dispersion (Union Carbide-BKUA-2370 or BKUA-2260) of 3:1 can form a stable adhesive at pH 6.0-6.5.

The increasing amount of research in waterborne systems should some day overcome technological difficulties. However, the change-over from some solvent-borne to waterborne systems has presented difficulties (107), for example, poor coating quality, excessive foam, insufficient water resistance, and even some corrosion and clean-up problems. Despite the problems, some waterborne systems have already been adopted for the manufacture of PSA's, contact adhesives, laminating adhesives, and wood adhesives. However, few applications of the waterborne systems as structural adhesives, especially as aerospace structural adhesives have been found.

X. AEROSPACE STRUCTURAL ADHESIVES

A special study on aerospace structural adhesives was conducted in 1976 by the National Materials Advisory Board of the National Research Council. The final report was published as one of the series in Treatise on Adhesion and Adhesives (108). That report discussed primarily phenolic and epoxy-based materials, e.g., vinyl-phenolics, nitrile-phenolics, epoxy-phenolics, nylon-epoxies, and elastomer-epoxies. Polyimide and polybenzamidazole adhesives were barely mentioned. Most of the high-temperature adhesives discussed in one of the preceding sections were either not invented or not developed at that time.

Today with the advance of structural and high temperature adhesives, aircraft, spacecraft, and missles (109) are built with secondary-bonded and some primary-bonded structures. The PABST (Primary Adhesive Bonding Structures Technology) program (110) was carried out at the Air Force Materials Laboratory between 1976 and 1981. That project led to the construction of a totally adhesive-bonded 42-foot-long circumferential fuselage section of a

McDonnell Douglas YC-15 military aircraft. Epoxy adhesives replaced panel bolts and rivets; the bonded panels were cured in an autoclave. After 120,000 hours of testing, the flaws did not grow into catastrophic proportions as in the case of riveted structures.

Bonded fuselages were not the only primary structures studies; bonded fuel-carrying wings were investigated in the same laboratory. One was the epoxy-bonded aluminum laminate for the lower skin of a seven-foot wing section of F-16 fighter and another was the AWT (Advanced Wing Technology) for the 10-foot-long section of a bomber wing that used epoxy-bonded aluminum laminate in the lower skin. The latter was tested with simulated fuel inside the "wet" wing. The test showed that rivetless wing bottoms enhanced the fuel containment of the wings and minimized fuel leak, just as riveted structures did. The adhesive bond did not fail even after other failures occurred at the end of the aircraft two-lifetime testing. Furthermore, the adhesive-bonded structures were at least 10% lighter than the riveted structures.

At the end of the PABST program, there were several questions unanswered about: (1) the ease and cost of maintaining the adhesive-bonded structures, (2) the durability in actual service environments where temperature cycles and moisture attack could reduce the life of the bonded joints, and (3) the feasibility and cost of curing an immense primary-bonded structure in an autoclave.

To solve the third problem, other approaches have been considered. One of them is the application of weld-bonding (111). The metal parts, with a paste adhesive sandwiched in between, are spot-welded by two electrodes that exert pressure on both sides. Then a current is passed through the electrode to cause the neat resin to melt and weld the parts at the electrode contact point. The curing process can then be continued in a regular oven. This process has actually been used by the Air Force to build an epoxy-bonded 100-square-foot forward fuselage section for the C-130H transport aircraft. After eight-years monitoring over 6,000 flight hours, the structure performed very well.

For film adhesives, ultrasonic welding (112) at a frequency of approximately 15,000 kHz can be used instead of weld-bonding. The subsequent curing can also be completed in a regular oven. This technique has been applied to small access panels for the Air Force A-10 attack aircraft at Fairchild Republic Co. Plans are underway to treat large fuselages and fuselage doors by the same weld-bonding method.

In the future, some of these new techniques will be used to cure high-temperature adhesives for building aircraft without rivets or bolts. In addition, with the advance of many low-temperature curing methods, the processing may be made simpler for metal-to-metal, metal-to-composite, or composite-to-composite structures.

One of the major problems with structural adhesives is durability. Many publications have been written on this subject. Recently, Kinloch (113) published a very timely book about this topic. A symposium proceedings volume on Adhesive Joints (114) is due to appear sometime this year. Thus, it is not our intention to devote any length of this review to this topic; however, there are several papers that need to be mentioned related to high-temperature adhesives. Hendricks and Hill (115) reported the evaluation of high temperature structural adhesives for extended service. Helminiak and Jones (116) studied the influence of the molecular weight of polyphenylquinoxaline thermoset on fracture behavior. Burks and St. Clair (117) mentioned the effect of molecular weight on the melt viscosity and fracture energy of BDSDA/APB, polyphenylene ether sulfideimide. Some of these polymers could be the aerospace structural adhesives of the future.

In addition to adhesives, the aerospace industry also needs conventional sealants and special sealants to withstand the temperature extremes of the aerospace environments. Both these subjects will be discussed in the following sections.

XI. CONVENTIONAL SEALANTS

High quality, conventional sealants (118) will dominate the industries in the 80's. Energy-efficient, water-proof, and durable sealants will be in demand by the construction and automotive industries. Commonly used sealants are polysulfides, polyurethanes, silicones, oil-based caulks, butyls, and solvent or latex acrylics. However, judged by their performances and durability (Table 3), the first three will be the most important sealants for this decade.

Table 3. Comparison of High Performance Sealants

	2-Part Polysulfide	1-Part Polysulfide	2-Part Urethane	1-Part Urethane	2-Part Modified Urethane	1-Part Silicone
Cure rate	M – hr	S – several hr or days	M – hr	S – hr	S – days	Fa – min or hr
Service temp range, $^{\circ}F$	$-20,-40^{\circ}$ $+180^{\circ}$	$-20,-40^{\circ}$ $+180^{\circ}$	$-20,-50^{\circ}$ $+180^{\circ}$	$-20,-50^{\circ}$ $+180^{\circ}$	-65° $+180^{\circ}$	-65° $250-425^{\circ}$
Movement capability based on joint dimensions	$+12.5$ ± 25	$+12.5$ ± 25	$+25$ ± 30	$+25$ ± 35	$+25$ ± 35	$+25$ $+40$ ± 50 $\pm 100/-50$
Abrasion	G	G	E (very tough)	E (very tough)	E (very tough)	F
Weatherability for sealing & caulking	G	G	E	E	E	E
Other unique features	Some are acceptable for underwater use on concrete.		Now being tested for underwater use on concrete – each will probably require special primer.			
	Compatible to all or most substrates.		Compatible to all or most substrates.		Most are compatible to all or most substrates. Some, the acid cure, cannot be used on concrete, marble, and certain metals.	

Note: E = Excellent; G = Good; F = Fair; M = Moderate; S = Slow; Fa = Fast
(Ref: J.M. Klosowskii, Adhesives Age 24, No. 11, 32 (1981).)

For durability, the polysulfides are designed to last for 10 to 20 years in outdoor environments. The polyurethanes should last 20 years, and silicones should have a life longer than 20 years. One of the problems with polysulfides is that they have the largest compression set of the three, causing joint failure, especially in winter months. With the trend toward narrower and prettier joints, the polysulfides (65 to 75 percent recovery) are being outperformed by the polyurethanes (85-95 percent recovery) and the silicones (90-100 percent recovery).

Besides durability, premium sealants are judged by special properties as shown in Table 4. The ability to take on greater elongation and compression is measured by movement capability in terms of joint width. The stability to UV exposure is important for those glazing and insulation compounds used in modern high-rise structures. Thermal stability is in demand for solar collectors, or for other structural materials. On the basis of these evaluations, we can foresee future trends of sealants as shown in Table 4. Silicones appear to out-perform others. In the meantime, technical advances will provide low-modulus polysulfides, and better movement ability for both polysulfides and polyurethanes. Their cure time will be decreased and the UV stability will be improved to match or compete with silicones. All three will be developed for better adhesion under the un-primed conditions.

For highway sealants (119), the desired properties are: adhesion (without primer), weatherability (adequate retention of properties for more than 10 years), resilience, temperature stability, hardness range for hot or cold (firm enough to resist incompressibles, puncture and tear resistant), good uncured consistency, solvent resistance, repairability, and abrasion resistance. Silicones, especially those of low-modulus, offer nearly all the properties desired as a good highway sealant.

Now many composite materials are used in aircrafts (Fig. 18). Graphite/epoxy and graphite/Kevlar are commonly used. For these materials, polysulfides (120) have been used as sealants. To provide airframe integrity, everything has to be sealed, preferably with flexible sealants that allow joint movement, fuel and weather resistance, and corrosion prevention. Today airframes are guaranteed to have a service life of 20 years without any corrosion damage. For this reason, the inhibited polysulfides are used to prevent corrosion between aluminum and graphite/epoxy composite parts.

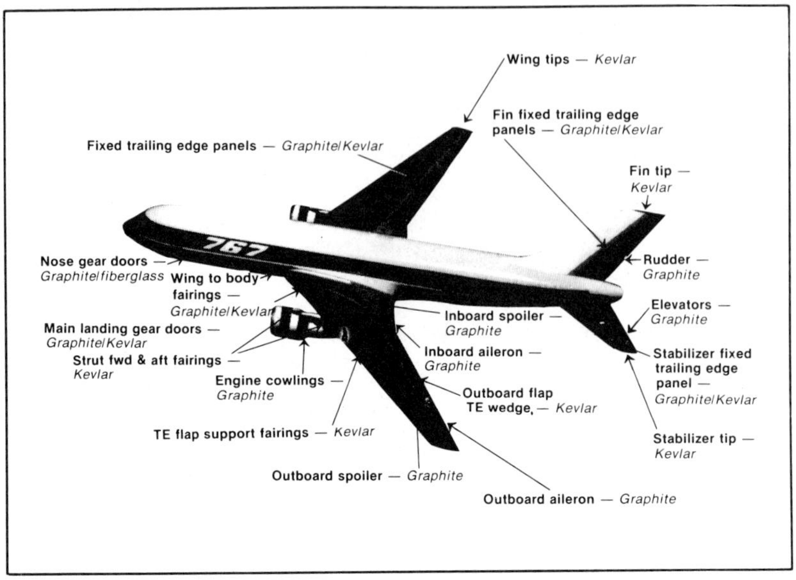

Fig. 18. Composite Applications for the Boeing 767.

(Ref: B. Silverman and A. Norrbom, Adhesives Age, 26, No. 7, 28 (1983).

Most conventional, polysulfide sealants are based on disulfides. Recently, a new family of sulfur-containing sealants (121,122) has been developed based on thioethers (Fig. 19). The thioether structure is thermally more stable than the polysulfide, thus the upper operating temperature of the former is 60°-70°F (26.6°-31°C) higher than the latter. Moreover, the thioethers show better hydrolytic stability and chemical resistance than the disulfides. A product based on thioether, Permapol-3, is now available from Products Research & Chemical Corp. Permapol-3 can also react with urethanes or epoxies to form blended adhesives and sealants.

Though thioether sealants are somewhat more thermally stable than polysulfides, they are still not qualified as the advanced sealants for aerospace applications in severe environments. In the following section, we shall discuss some of the advanced sealants being considered for future applications.

Table 4. 10-Year Forecast for Sealants

	Gain	Loss	Maintain	Overall
Silicones	Expansion joints and perimeter sealing market. Insulated glass market – secondary seal. Structural glazing – no competition. Solar glazing and sealing markets. Do-it-yourself market. Highway sealant market. Industrial adhesive/sealing markets.	–	Form-in-place gasket market share. Aerospace and specialty sealing market share.	Some fast and some steady gains in volume and market share.
Polyurethanes	Expansion joint and general sealing markets. Structural adhesive seal – non-glazing.	–	Pedestrian traffic sealing market.	Gain volume and market share, mostly from polysulfides.
Polysulfides	Volume gains in insulated glass.	Market share for insulated glass. Glazing market. General sealing and expansion joint market.	Specialty market - industrial, fuel tank, for example.	Slow decline in most areas.
Butyls	Insulated glass primary and secondary seal. Do-it-yourself market from oil-based caulks.	Mobile home sealing. Do-it-yourself market to silicones and urethanes.	–	–
Acrylics Solvent	–	Will show decline in all areas.	–	–
Latex	Do-it-yourself market gain from oil based loss to silicones.		–	Steady market – no growth.

*All gains and losses refer to gains in both volume and market shares except where specifically noted.
(Ref: Adhesives Age, November 1981.)

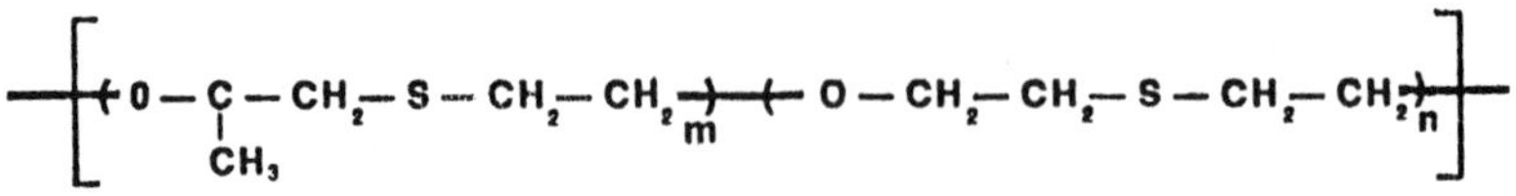

PERMAPOL P-3 POLYMER

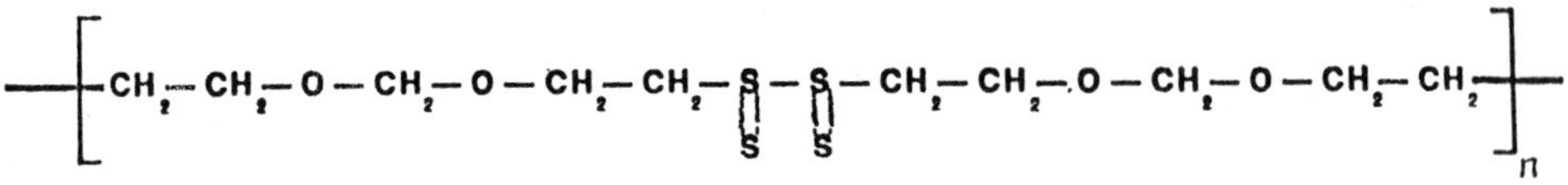

POLYSULFIDE POLYMER

Fig. 19. Structures of Polythioether and Polysulfides

(Ref: J.W. Huff and H. Singh, technical publication by Products Research & Chemical Corp.)

XII. ADVANCED SEALANTS FOR AEROSPACE APPLICATIONS

For aerospace applications, several advanced sealants have been mentioned: cyanosilicones (123), fluoroalkylarylenesiloxanylene (FASIL) (124), phosphonitrilic fluoroelastomers (PNF®) (125,126), flexible polyimide (127), tetrafluoroethylene oxide phenylquinoxaline elastomer (FEX) (128), perfluoroalkyl ethers (129), etc. These advanced sealants have been used for fuel tank sealing, channel sealing, filleting, aerodynamic smoothing, and other assemblies. Basically, they should be high-temperature serviceable, fault-tolerant, and adhere well to metals and/or composites.

Cyanosiloxane Sealants

For space shuttles, a special type of sealant, cyanosiloxane (123) was found to be unique and operational. Cyanosiloxanes can be synthesized by the catalytic addition of an unsaturated nitrile to a methyl hydrogen polysiloxane (Fig. 20). At the end of the reaction, at least one unreacted -Si-H (hydride) bond remains per molecule. This bond is subsequently hydrolyzed to -SiOH, which then undergoes a condensation to give a high molecular weight, but not cross-linked, polymer (Fig. 21). The polymer is a clear, dark brown, thick liquid with a viscosity of 5,000 poise. At room temperature, a liquid-gel transformation can take place but the gel can reversibly transform into the liquid under shear. The polymer is resistant to hot hydrocarbon fuels, but soluble in polar solvents.

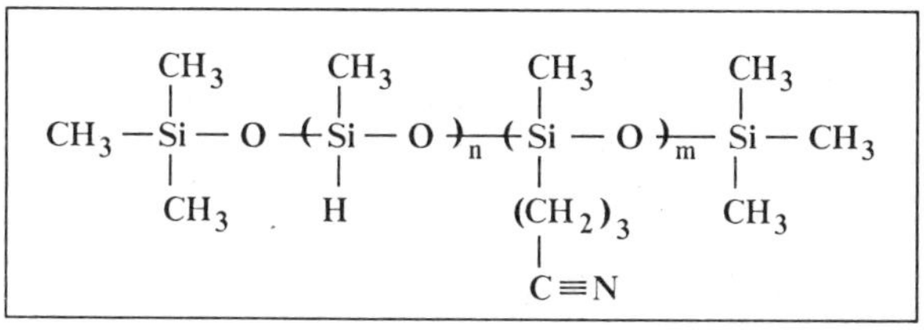

Fig. 20. Structure of Polycyanosiloxane
(Ref: H. Singh, Adhesives Age, 21, No. 4, 24 (1978).)

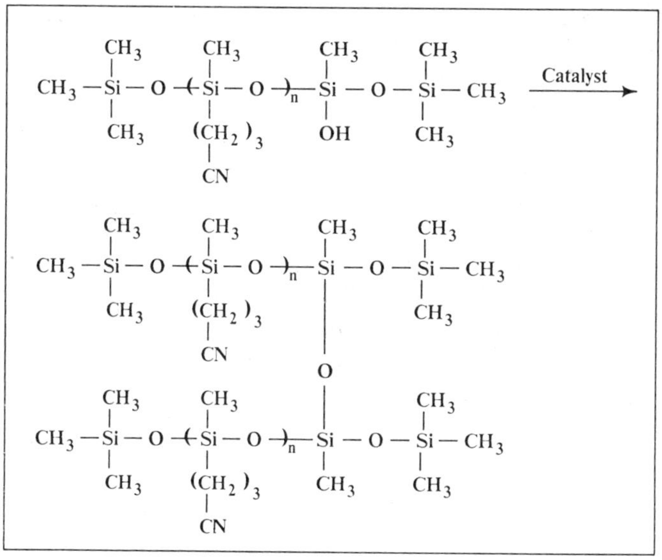

Fig. 21. Reaction to Increase Molecular Weight of the Polycyanosiloxane Polymer
(Ref: H. Singh, Adhesives Age, 21, No. 4, 24 (1978).)

The glass temperature of this polymer is -60°F (-51°C). When exposed to 350°F (177°C) for 48 hours, no curing takes place. The sealants from this polymer are also not cured even in the presence of oxygen at 450°F (232°C). They appear to be somewhat rubbery and remain flexible to -60°F (-51°C).

Two products, PR-711 and PR-719, are manufactured by Products Research Chemical Corp. PR-711 was used for environmental sealing of structural faying surfaces of the first two flights of the space shuttle orbital airframe. PR-719 is designed to seal integral fuel tanks of aircraft. The sealant has to be a non-curing type so that the channel can be resealed without disassembling the structure.

Fluoroalkylarylenesiloxanylene (FASIL) Sealant

FASIL was developed at the Air Force Materials Laboratory. This elastomer has a broad temperature servicing range: -54°C $\longrightarrow$ 260°C (-66°F $\rightarrow$ 500°F). It has excellent adhesion to Ti and Al; it is JP-4 fuel resistant and hydrolytically stable.

Two classes of FASIL elastomers have been chosen for preliminary evaluation: the methyl- and 3,3,3-trifluoropropyl-substituted poly(m-xylylene siloxanylenes) (129) and the corresponding poly(m-phenylene siloxanes) (124). The general formula of FASIL is as follows:

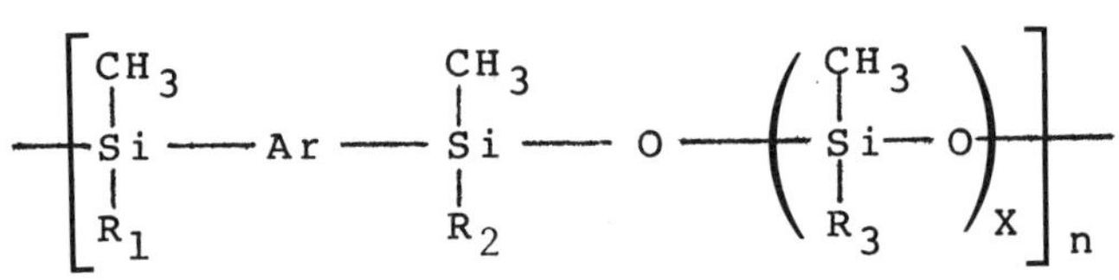

where R_1, R_2, and R_3 are either methyl or 3,3,3-trifluoropropyl and X=0, 1 or 2; Ar is either m-phenylene or m-xylyene.

Phosphonitrilic Fluoroelastomer (PNF®)

Another fluorine-containing polymer, phosphonitrilic fluoroelastomer (PNF®) (125,126) is prepared from a soluble precursor, chlorophosphazene and a mixture of fluorinated alkoxides containing 65% trifluoroethoxide and 35% telomer fluoroalkoxides. The structure of PNF is:

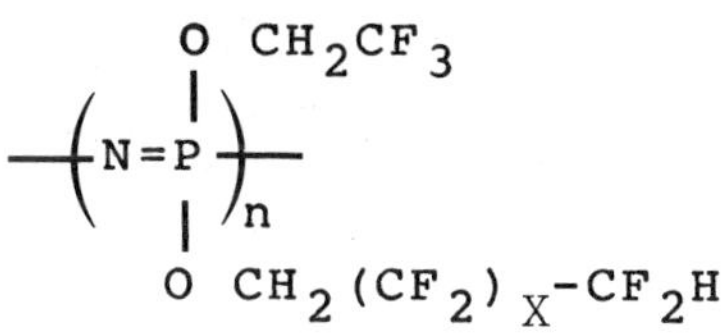

where X can be 1, or 3, or 5......

This polymer contains approximately 55% fluorine and a small number of unsaturated sites for cross-linking. The fluorine content imparts outstanding resistance to fuels, including JP-4 fuel, oils, most hydraulic fluids and chemicals. The glass temperature is $-68^{\circ}C$ ($-91^{\circ}F$) and thus it allows the service temperatures to be as low as $-65^{\circ}C$ ($-85^{\circ}F$). The upper service temperature is $175^{\circ}C$ ($347^{\circ}F$).

Since the structure contains very few weak links, it is oxidative and ozone resistant. The unique mechanical properties are low compression set, high modulus, excellent flex fatigue, and good abrasion resistance. Thus, it has been evaluated as an advanced sealant for aerospace applications.

Flexible Polyimide Fuel Tank Sealants

The requirements for the fuel tank sealant (127) were:

- Compliant elongation to break (>50 percent)
- High-temperature stability [$450^{\circ}F$ ($232^{\circ}C$) to $600^{\circ}F$ ($316^{\circ}C$)]
- Low-temperature flexibility [$-50^{\circ}F$ ($-46^{\circ}C$) to $-60^{\circ}F$ ($-51^{\circ}C$)]
- Titanium compatibility and adhesion
- Aliphatic/aromatic (JP) fuel stability
- Low-cure temperature $<400^{\circ}F$ ($204^{\circ}C$)
- Useable at high resin solid content in non-toxic solvents (>80% w/w)

To meet the above requirements, a synthetic approach based on the Diels-Alder addition was used. The key monomers were chosen from the three monomers shown in Fig. 22.

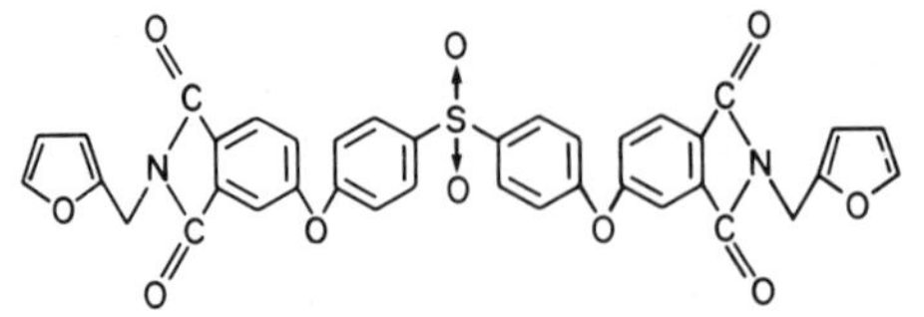

Bis (furfuryl) imide of bis [4--(3,4--dicarboxyphenoxy) phenyl] sulfone dianhydride (BFBSI)

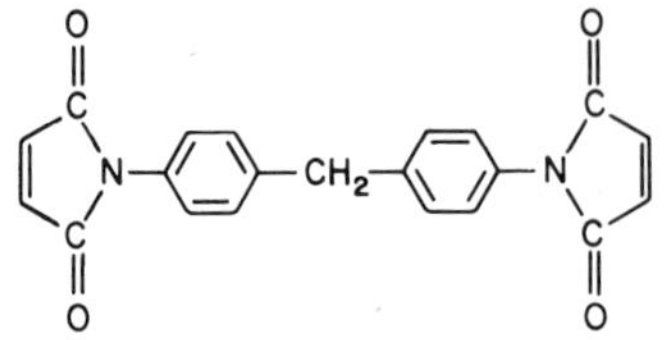

Bis (4--maleimidophenyl) methane (BMPM)

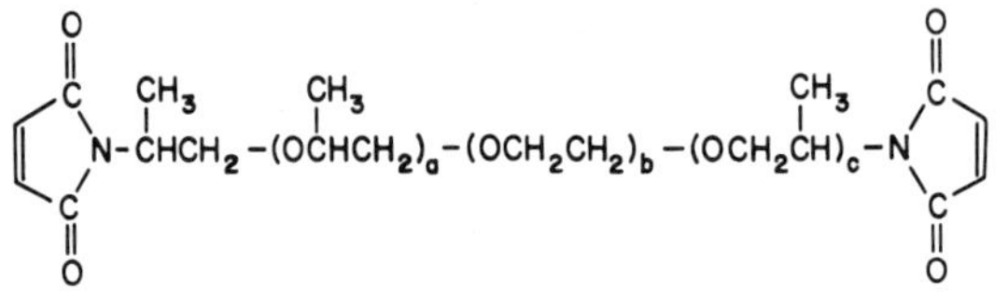

Bis (maleimide) of Jeffamine ED* (BMPM)

*Repeat units a, b, an c are integers yielding molecular weights of approximately 600, 900, and 2000

Fig. 22. Monomer for Flexible Polyimides
(Ref: R.J. Jones and H.N. Cassey, Adhesives Age, 22, No. 11, 33 (1979).)

The reaction product between BFBSI and BMJEF gave a flexible polyimide sealant with the following properties:

- High-temperature fuel stability (500°F) (260°C)
- Low-temperature flexibility (-50°F)(-46°C)
- Excellent JP fuel resistance
- High adhesion to AL and Ti substrates
- No stress-corrosion imparted to Ti surface
- Facile cure and processability
- Available in 100 percent solid formulary
- Excellent tensile properties

Adhesives and sealants used for the aerospace structures must endure severe environments. Another family of adhesive materials has also been developed for solar collectors, presumably based on the same ability to endure severe environments. Though there are different kinds of severe environments, conventional polymers generally do not survive in the solar collector environment. We shall discuss those adhesive materials used for solar applications in the following section.

XIII. SOLAR COLLECTOR ADHESIVES AND SEALANTS

Adhesives for Solar Collectors

Adhesives (128) play an important role in the construction of a solar parabolic trough reflector. The adhesives should be compatible with silvered glass mirrors and should be able to carry structural loads. Moreover, the adhesives should be durable for at least twenty years without losing their properties. Adhesives screened for solar collectors were epoxies, urethanes, and acrylics. Three types of stress conditions were identified as the most likely causes of failure in the trough modules:

1. Thermal cycling
2. High temperature, low humidity: 160°F (71°C), low R.H.: (e.g., R.H. 8%)
3. High temperature, high humidity: 160°F (71°C), high R.H.: (e.g., R.H. 80%)

When a flat solar collector is used, a polymer film based on fluoropolymers, e.g., Tedlar ® polyvinylidene fluoride, has been used to make a laminate with an adhesive. The solar collector, so formed, is light-weight and low-cost.

Sealants for Solar Collectors

The sealant materials used for the solar collectors must be durable under harsh environments (132). They must be stable to air at high temperatures, which can exceed 200°C (392°F) for a short period or several weeks inside the cell. The temperature can vary between 125°C (257°F) and 200°C (392°F). In addition, high humidity, ozone, and ultraviolet radiation could shorten the service life.

Fluoroelastomers, silicone rubbers, butyl rubbers, ethylene-propylene terpolymer (Nordel), ethylene-acrylic copolymer (Vamac), and polyacrylate (Hycar 4054) were evaluated for the PS (preformed seal or gasket) compounds. After 32 days at 225°C (437°F), only the fluoroelastomer, Viton, appeared not to suffer any loss of tensile properties.

® Tedlar is a registered trademark of duPont Company.

For the intermediate caulking compounds (SC), acrylics still displayed a significant elongation after 32 days at 125°C (257°F). However, the accelerated hydrolytic stability of the acrylic is poor in comparison with that of the silicone (e.g., RTV 103).

The outgassing of sealants in the collector system is a serious problem (128). For this reason, the polymer was tested at 150°C for several days. The condensate was collected and identified. The results indicated that only fluorocarbon did not outgas under the test conditions. In the case of silicones, during the outgassing, cyclic siloxanes were released through thermal degradation.

There may be some disadvantages to the fluorocarbons. The first is relatively high cost and the second is a then high compressive set at temperatures of 10°C and lower, which could present problems in colder climates (134). Another study (135) of the edge seal indicated that butyl rubbers, if stabilized against UV radiation, were better than silicones and might be substituted for the high-cost fluoropolymers for that particular application.

XIV. CONCLUSIONS

In this brief review, we have attempted to demonstrate the importance of chemistry to the developments of new (or new forms of) adhesive materials. Much success has been achieved in the syntheses of polymers for aerospace adhesives and sealants. New or modified structural adhesives have been developed with improved mechanical properties and durability. However, there is still room for innovation for developing new adhesive materials for use under severe environments.

To counter energy-saving problems, low-temperature curing by radiation has gained acceptance. For a long-term solution of pollution problems, hot-melt adhesives and the waterborne systems are being increasingly used as a substitute for the solvent-borne adhesives. In the future, the waterborne systems could dominate the entire adhesive industry.

REFERENCES

1. A. Brief and I. Skeist, "The Adhesives Industry - 25 Years of Growth," Adhesives Age, 26, No. 6, 12 (1983).

2. P. L. Layman, Chem. and Eng. News, 8, Nov. 1, 1982.

3. J. C. Bittence, "Adhesives Update," Mat. Eng. 34, Aug. 1983.

4. H. Ellerhorst, Jr., "Industrial Adhesive Markets - A Profile for the 1980's," Adhesives Age, 25, No. 1, 42 (1982).

5. L. H. Lee, "Significant Advances and Developments in Adhesion and Adhesives," in Adhesion Science and Technology, L. H. Lee, Editor, Vol. A, 1, Plenum, New York (1975).

6. W. C. Wake, Adhesion and the Formulation of Adhesives, Applied Science Publishers, London (1976).

7. W. C. Wake, Developments in Adhesives, Applied Science Publishers, London (1977).

8. L. H. Lee, Adhesion and Adsorption of Polymers, Plenum, New York (1979).

9. "Adhesion and Adhesives: Science, Technology and Applications," International Conference Proceedings, Sept. 1980,published by Plastics and Rubber Institute, London (1981).

10. A. J. Kinloch, "The Science of Adhesion - Part 1. Surface and Interfacial Aspects," J. Mater. Sci., 15, 2141 (1980).

11. A. J. Kinlock, "The Science of Adhesion - Part 2. Mechanics and Mechanisms of Failure," J. Mater. Sci., 17, 617 (1982).

12. E. H. Andrews and A. J. Kinloch, "Mechanics of Adhesive Failure. I," Proc. Roy. Soc. (London), A, 332, 385 (1973).

13. L. H. Lee, "Surface Free Energies and Fracture Surface Energies of Glassy Polymers," in Physical Aspects of Polymer Surfaces, K. L. Mittal, Editor, Vol. 1, p. 523, Plenum, New York (1983).

14. L. H. Lee, "Adhesive Chemistry - Developments and Trends," L. H. Lee, Editor, Plenum, New York (1984).

15. M. C. H. Lee, "An Analytical Method for Determining the Surface Energy of Solid Polymers," in Ref. 14.

16. H. P. Schreiber, C. Richard, and M. R. Wertheimer, "Acid-Base Considerations of Surface Interactions in Polymer Systems; Control by Microwave Plasma Treatment," in Physicochemical Aspects of Polymer Surfaces, Vol. 1, 739, Plenum, New York (1981).

17. R. K. Samal, H. Iwata, and Y. Ikada, "Introduction of Reactive Groups onto Polymer Surfaces," in p. 801 of Ref. 16.

18. T. S. Keller, A. S. Hoffman, B. D. Ratner, and B. J. McElroy, "Chemical Modification of Kevlar [R] Surfaces for Improved Adhesion to Epoxy Resin Matrices: I. Surface Characterization," p. 861 in Ref. 16.

19. H. Yasuda, "Adhesion of Plasma Polymerized Films (A Model Study on Water Sensitivity of Adhesion)," in Adhesion Aspects of Polymeric Coatings, K. L. Mittal, Editor, p. 193, Plenum, New York (1981).

20. A. Dilks and A. VanLaeken, "An ESCA Investigation of the Plasma Oxidation of Poly(p-Xylylene) and its Chlorinated Derivatives," p. 749 in Ref. 16.

21. K. Ishikawa, "Glueballs," Scientific American, 247, No. 5, 142, Nov. 1982.

22. D. Maugis and M. Barquins, "Fracture Mechanics and Adherence of Viscoelastic Solids," in Adhesion and Adsorption of Polymers, L. H. Lee, Editor, Vol. 1, 203, Plenum, New York (1979).

23. D. Maugis, "Overview of Adherence Phenomena," in Ref. 14.

24. L. H. Lee, Editor, Characterization of Metal and Polymer Surfaces, Academic Press, New York (1977).

25. D. T. Clark, "The Application of ESCA to Studies of Structure and Bonding of Polymers," in Advances in Polymer Friction and Wear, L. H. Lee, Editor, Vol. 1, 241, Plenum, New York (1974).

26. D. Briggs, "New Applications of ESCA," in Ref. 14.

27. YANG Qi-xiang and ZHOU Qin-li, "The ESCA and AES Studies of the Interfacial Chemical Bonding between Aluminum and Chromium(III) Fumarato-Coordination Compound," in Ref. 14.

28. R. Buchanan, "Some Recent Developments in Scanning Electron Microscopy," in Ref. 14.

29. B. C. Copley, "Dynamic Mechanical Properties of Silicone Pressure Sensitive Adhesives," in Ref. 14.

30. J. B. Class and S. G. Chu, "Viscoelastic Properties of Rubber-Resin Pressure Sensitive Adhesive Formulations," in Ref. 14.

31. Y. Hazony, S. J. Stadnicki, J. K. Gillham, "Low Frequency Thermomechanical Spectrometry of Polymeric Materials: Computerized Torsional Braid Experiments. I. Overview," Polymer Preprints, 15 No. 2, 549 (1974).

32. W. Alex Lee and M. J. Oliver, "Study of Cure of Epoxy Resins by Torsional Braid Analysis," Brit. Polym. J. 15, 40 (1983).

33. R. L. Levy and D. P. Ames, "Monitoring Epoxy Cure Kinetics with a Viscosity-Dependent Fluorescence Probe," in Ref. 14.

34. J. T. Dickinson, "Fracto-Emission Accompanying Adhesive Failure," in Ref. 14.

35. Y. Nakano, "Photosensitive Adhesives and Sealants," Adhesives Age, 16, No. 12, 28 (1973).

36. K. C. Stueben, "Radiation Curing of Pressure-Sensitive Adhesives: A Literature Review," Adhesives Age, 20, No. 5, 16 (1977).

37. K. C. Stueben, "Ultraviolet Cured Pressure Sensitive Adhesives," in Ref. 14.

38. V. D. McGinness, "Polymer and Formulation Design Characteristics for Developing Adhesive Bonding Radiation-Curable Systems," in Ref. 14.

39. J. V. Crivello and J. H. W. Lam, "Photoinitiated Cationic Polymerization by Dialkyl-4-Hydroxyphenyl Sulfonium Salts," J. Polym. Sci., Polym. Chem. Ed., 18, 2677, 1980 and 2697, 1980 and J. V. Crivello, "Cationic Polymerization Initiated by Light," CHEMTECH, 10, 624 (1980).

40. A. Ledwith, a paper presented to the 11th Northeast Regional ACS Meeting, Oct. 19, 1981; also A. Ledwith, S. Al-Kass, D. C. Sherrington, and P. Bonner, "Ion Pair Dissociation Equilibria for Iodenium and Sulphonium Salts Useful in Photoinitiated Cationic Polymerization," Polymer, 22, 143 (1981).

41. J. H. Tsao and A. D. Ketley, U.S. 4,156,035 (to W. R. Grace Co.), 1979.

42. W. C. Perkins, "New Developments in Photo-induced Cationic Polymerizations," J. Rad. Curing, 8, No. 1, 16, Jan. 1981.

43. J. V. Crivello, "Recent Progress in Photoinitiated Cationic Polymerization," in Ref. 14.

44. R. Kardashian and S. V. Nablo, "Electron Beam Curing Equipment," Adhesives Age, 25, No. 12, 25 (1982).

45. P. M. Hergenrother, "Status of High-temperature Adhesives," in Ref. 14.

46. A. K. St. Clair, "Polyimides: Thermally Stable Aerospace Polymers," NASA Technical Memorandum 81884, October 1980.

47. P. E. Cassidy, Thermally Stable Polymers, Syntheses and Properties, Marcel Dekker, New York (1980).

48. V. L. Bell, B. L. Stump, and H. Gager, J. Polym. Sci., Polym. Chem. Ed. 14, 2275 (1976).

49. V. L. Bell, U.S. Pat. 4,094,862 (1978) (to NASA).

50. A. K. St.Clair, W. S. Slemp, and T. L. St.Clair, Sci. Adv. Mat'l. Proc. Eng. Series, 23, 113 (1978).

51. J. P. Wightman and S. Dias, "The Application of Thermodynamics and Spectroscopic Techniques to Adhesion in the Polyimide/Ti 6-4 and Polyphenylquinoxaline/Ti 6-4 Systems," in Ref. 14.

52. N. Bilow, A. L. Landis, and L. J. Miller, U.S. Pat. 3,845,018 (1974) (to Hughes Aircraft Co.).

53. H. R. Lubowitz, U.S. Pat. 3,528,950 (1970) (to TRW, Inc.).

54. A. K. St.Clair and T.L. St.Clair, "Addition Polyimide Adhesives Containing Various End Groups," Polym. Eng. and Sci., 22, No. 1, 9 (1982).

55. A. K. St.Clair and T. L. St.Clair, "Elastomer-Modified Polyimide Adhesives," in Ref. 14.

56. P. M. Hergenrother and H. H. Levine, "Polyquinoxalines. I. Synthesis and Preliminary Mechanical Properties as Laminating and Adhesive Resins," J. Appl. Polym. Sci., 14, 1037 (1970).

57. P. F. Kovar, G. F. L. Ehlers, and F. E. Arnold, "Addition Cured Acetylene Terminated Polyphenylquinoxaline Oligomers," Polym. Chem. Preprints, 16, 21, 246 (1975).

58. R. F. Kovar and F. E. Arnold, U.S. Pat. 3,966,729 (1976) (to U.S. Air Force).

59. P. M. Hergenrother, "High-Temperature Adhesives," a lecture for the course "Fundamentals of Adhesion" held at the State University of New York at New Paltz, New York (Nov. 1983).

60. W. A. Lees, "The Science of Acrylic Adhesives," Brit. Polym. J., 11, 64 (1979).

61. A. G. Bachmann, "'Aerobic' Acrylic Adhesives," Adhesives Age, 25, No. 8, 19 (1982).

62. A. G. Bachmann, "'Aerobic' Acrylic Adhesives - New Technology in Acrylic Adhesives," in Ref. 14.

63. B. M. Malofsky and L. J. Baccei, "Anaerobic Adhesives Containing Maleimides Having Improved Thermal Resistance," in Ref. 14.

64. D. J. Stamper, "Curing Characteristics of Anaerobic Sealants and Adhesives," Brit. Polym. J., 15, 34, Mar. 1983.

65. H. Lee, Editor, Cyanoacrylate Resins - The Instant Adhesives, T.C. Publication, Div. of Technology Conferences, El Segundo, CA (1981).

66. U.S. Pat. 3,616,049 (1971) (to Loctite Corp.).

67. U.S. Pat. 3,890,407 (1975) (to duPont Company).

68. R. S. Drake and A. R. Siebert, "Reactive Butadiene/Acrylonitrile Liquid and Solid Elastomers: Formulating Acrylic, Anaerobic, and Radiation-curable Adhesives," in Ref. 14.

69. R. E. Brown and R. E. McCrea, "Competition, Chances for Growth of Epoxy Adhesive Markets," Adhesives Age, 25, No. 2, 21 (1982).

70. E. W. Garnish, "Epoxide Resins as Adhesives: Past and Present," Brit. Polym. J., 11, 72, June 1979.

71. A. Hardy and R. T. Agger, "Polyurethane Adhesives," Brit. Polym. J., 11, 54, June 1979.

72. R. S. Drake and A. R. Siebert, "Reactive Butadiene/Acrylonitrile Liquid and Solid Elastomers: Formulating Epoxy Structural Adhesives," in Ref. 14.

73. K. L. Hawthorne, F. C. Henson, and R. Pinzelli, "High Performance Tris(hydroxylphenyl)methane Based Epoxy Resins," Org. Coatings and Appl. Polym. Sci. Proc., 46, 493 (1982).

74. W. Ku and J. P. Bell, "Fast Curing Epoxy-Episulfide Resin for use at Room Temperature," Org. Coatings and Appl. Polym. Sci. Proc., 46, 498 (1982).

75. J. A. Graham and J. E. O'Conner, "Epoxy with Low-Temperature Cure and High-Temperature Properties Developed," Adhesives Age, 21, No. 7, 20 (1978).

76. J. R. Griffith, "Epoxy Resins Containing Fluorine," CHEMTECH, 12, 290 (1982).

77. J. R. Griffith and J. G. O'Rear, "Silicone Amine Cured Fluoroepoxy Resins," in Resins for Aerospace, C. A. May, Editor, American Chemical Society, Washington, D.C. (1980).

78. C. M. Thompson and R. Y. Ting, "A Study of Some Epoxy Polymers for Underwater Acoustic Use," Org. Coatings and Appl. Polym. Sci. Proc. 46, 661 (1982).

79. J. R. Griffith, J. G. O'Rear, and J. P. Reardon, "A Fluoro-anhydride Curing Agent for Heavily Fluorinated Epoxy Resins," in Adhesion Science and Technology, L. H. Lee, Editor, 9A, 429 Plenum, New York (1975).

80. D. A. Kourtides, J. A. Parker, T. W. Giants, N. Bilow, and M-T. Hsu, "Thermophysical and Flammability Characterization of Phosphorylated Epoxy Adhesives," Adhesives for Industries, 92, Technology Conferences, El Segundo, California, June 25, 1980.

81. J. Fries, "Hot Melt PSAs are Finding a More Demanding Niche," Adhesives Age, 26, No. 10, 39 (1983).

82. E. F. Eastman, "Ethylene Copolymers - Improved Performance for an Old Molecule," Adhesives Age, 26, No. 10, 32 (1983).

83. W. J. Sparks, "Advances in Hot Melt and Waterborne Acrylic PSAs," Adhesives Age, 25, No. 3, 38 (1982).

84. D. Crosby, "Foamed Hot Melt Adhesives and Applicating Equipment," in Adhesives for Industry, 131, Technology Conferences, El Segundo, California, June 1980.

85. F. T. Hughs, "Foamed Hot Melt Adhesives," Adhesives Age, 25, No. 9, 25 (1982).

86. W. H. Cobbs, Jr., "Foamability of Hot Melt Adhesives," Adhesives Age, 22, No. 12, 31 (1979).

87. B. Stoops and P. Ferrier, "Merging Two Technologies - Robotics and Hot Melt Adhesives," Adhesives Age, 26, No. 4, 22 (1983).

88. E. Rowand and T. D. MacGrandee, "The Age of Film Adhesives," Adhesives Age, 14, No. 9, 34 (1971).

89. J. C. Bittence, "Adhesives Update," Materials Engineering, 34, Aug. 1983.

90. D. R. Gehman, F. T. Sanderson, S. A. Ellis, and J. J. Miller, "New Acrylic Polymers for Aqueous Contact Adhesives," Adhesives Age, 20, No. 12, 23 (1977).

91. J. A. Fries, "The Growing Uses of Waterborne Adhesives," Paper, Film & Foil Converter, 49, Dec. 1981.

92. F. M. Rosenblum, "Carboxylation Increased Versatility of Polyvinyl Acetate-Based Adhesives," Adhesives Age, 22, No. 4, 19 (1979).

93. D. Dietrich and J. N. Rieck, "Aqueous Polyurethane Systems and Their Possible Uses," Adhesives Age, 21, No. 2, 24 (1978).

94. D. Dietrich, "Aqueous Emulsions, Dispersions and Solutions of Polyurethanes; Synthesis and Properties," Prog. Org. Coatings, 9, 281 (1981).

95. G. B. Guise and M. B. Jackson, U.S. Pat. 3,898,197 (1972).

96. J. Pedain, K. Nachtkamp, K. Noll, and J. Grammel, DOS 2,725,589 (1978) (assigned to Bayer, A.G.).

97. K. Nachtkamp, J. Pedain, and J. Grammel, DOS 2,811,148 (1978) (assigned to Bayer, A.G.)

98. C. G. Demmer, E. W. Garnish, and D. J. R. Massy, "Thermoplast Phenolic Resin Aqueous Dispersion," Brit. Polym. J., 15, 76, March 1983.

99. F. A. Hudock and R. B. Graver, "Applications for Aqueous Dispersions of Epoxy Resins," Org. Coatings and Appl. Polym. Sci. Proc., 46, 665 (1982).

100. J. B. Wilson, "Isocyanate Adhesives as Binders for Composition Board," Adhesives Age, 24, No. 5, 41 (1981).

101. H. Pagel and E. R. Luckman, "Emulsion Polymer Isocyanate: A Durable, Water-based Wood Adhesives," Adhesives Age, 24, No. 10, 34 (1981).

102. A. D. Hickman, "The Design of S/B Latex Pressure Sensitive Adhesives," a paper presented to the TAPPI, 1983 Paper Synthetics Conference, Sept. 1983.

103. R. C. Oldack and R. E. Bloss, "Compounding Natural Latex in Water-based PSAs," Adhesives Age, 22, No. 4, 38 (1979).

104. J. M. Evans and K. E. Krajca, "Waterborne Adhesive Tackifiers," Adhesives Age, 25, No. 3, 25 (1982).

105. M. J. Jones, "Elastomers and Resin Modifiers for Water-based Adhesives," in Ref. 14.

106. J. S. Fry and R. J. Stregowski, "Environmentally Compatible Coatings and Adhesives Based on Novel Phenolic Resins," Polymer Preprints, 24, No. 2, 203 (1983).

107. R. A. Bafford, "Converting to Water-based Adhesives: One Manufacturer's Experience," Adhesives Age, 22, No. 12, 21 (1979).

108. R. L. Patrick, Editor, "Structural Adhesives - with Emphasis on Aerospace Applications," Vol. 4, Marcel Dekker, New York (1976).

109. G. Graff, "The Glued-together Airplane," High Technology, 67, Sept. 1983.

110. E. W. Thrall, Jr., "PABST Program Test Results," Adhesives Age, 22, No. 10, 22 (1979).

111. R. W. Vaughn, C. H. Sheppard, and R. Baucom, "Polyimide Adhesives for Weld-Bonding Titanium," Adhesives Age, 20, No. 7, 19 (1977).

112. R. D. Tuttle, "Ultrasonic Plastics Assembly," in Adhesives for Industry, 269, Technology Conferences, El Segundo, California, June 1980.

113. A. J. Kinloch, Durability of Structural Adhesives, Applied Polymer Science, London (1984).

114. K. L. Mittal, Editor, Adhesive Joints: Formation, Characteristics, and Testing, Plenum, New York (1984).

115. C. L. Hendricks and S. G. Hill, "Evaluation of High Temperature Structural Adhesives for Extended Service," in Ref. 14.

116. T. Helminiak and W. B. Jones, "Influence of Molecular Weight on Polyphenylquinoxaline Thermoset on Fracture Behavior," in Ref. 14.

117. H. D. Burks and T. L. St.Clair, "The Effect of Molecular Weight on the Melt Viscosity and Fracture Energy of BDSDA/APB," J. Appl. Polym. Sci., 29, 1027 (1984).

118. J. M. Klosowski, "'Premium' Sealants of 1980s," Adhesives Age, 24, No. 11, 32 (1981).

119. J. B. Thorton, "Georgia Highway Joint Sealing," Adhesives Age, 26, No. 9, 24 (1983).

120. B. Silverman and A. Norrbom, "Use of Polysulfide Sealants in Aircraft Composite Structures," Adhesives Age, 26, No. 7, 28 (1983).

121. H. Singh, J. W. Huff, M. E. Williams, and V. Nuys, U.S. Pat. 4,366,307 (1982) (assigned to Products Research & Chemical Corp.).

122. J. W. Huff and H. Singh, "New High Temperature and Fuel Resistant Sealants Based on Permapol ® P-3 Polythioether Polymers," a technical publication by Products Research & Chemical Corp., Glendale, California.

123. H. Singh, "Cyanosiloxane Sealants Used for Space Shuttle Orbiter," Adhesives Age, 21, No. 4, 24 (1978).

124. H. Rosenberg and E-W Choe, "Synthesis and Properties of Fluoroalkylarylene Siloxanylene (FASIL) High Temperature Polymer," Org. Coatings and Plastics Chem., 40, 793 (1979).

125. PNF ® Elastomer by Firestone, Firestone Rubber and Tire Co., Oct. 1979.

126. D. F. Lohr and J. A. Beckman, "PNF ® Phosphonitrilic Fluoroelastomer: Properties and Applications," a publication by Firestone.

127. R. J. Jones and H. N. Cassey, "Flexible Polyimide Fuel Tank Sealants," Adhesives Age, 22, No. 11, 33 (1979).

128. G. F. Schmitt, "Spacecraft, Aircraft, and Missiles: Pushing the Limits of Adhesives, Sealants and Coatings," a paper presented at Case Center for Adhesives, Sealants, Coatings Inaugural Symposium, Jan. 11, 1983.

129. H. Rosenberg and E-W. Choe, "Methyl- and 3,3,3-Trifluoropropyl-substituted m-Xylylenesiloxanylene Polymers," Org. and Plastics Preprints, 37, No. 1, 166 (1977).

130. N. H. Clark, D. Clements, and V. Grasso, "Adhesives Used in Reflector Module of Troughs-Effect of Environmental Stress," Polymer Preprints, 23, No. 1, 233 (1982).

131. W. G. Wilhelm, "The Use of Polymer Film and Laminate Technology for Low Cost Solar Energy Collectors," Polymer Preprints, 23, No. 1, 244 (1982).

132. M. A. Mendelsohn, R. M. Luck, F. A. Yeoman, and F. W. Navish, Jr., "Sealants for Solar Collectors," Org. Coatings and Plastics Chem., 42, 232 (1980).

133. R. M. Luck and M. A. Mendelsohn, "The Degradation and Outgassing of Polymeric Sealants and Plastics and Their Effects on Solar Collector Efficiency," Polymer Preprints, 23, No. 1, 235 (1982).

134. M. A. Mendelsohn, F. W. Navish, Jr., R. M. Luck, and F. A. Yeoman, "Effects of Thermal, Oxidative and Hydrolytic Degradation on Physical Properties of Solar Collector Sealants," Polymer Preprints, 23, No. 1, 211 (1982).

135. K. B. Wishman, "Protective Coatings and Sealants for Solar Applications," Polymer Preprints, 23, No. 1, 246 (1982).

Overview of Adherence Phenomena

Daniel Maugis

Equipe de Recherche de Mécanique des Surfaces
du CNRS, LCPC, 58, Bd Lefebvre
75732 Paris Cedex 15, France

ABSTRACT

The aim of this paper is to clear up the various concepts involved in the adherence of solids, and to discuss the manner how surface effects, geometry, elasticity and rheological effects are entangled. Adhesive contact of elastic solids is firmly established on the basis of thermodynamics, theory of elasticity or fracture mechanics. Extension to viscoelastic solids is straightforward using fracture mechanics, and experimental evidence is given. Tack force, environmental effects and stick-slip are discussed. Adherence of metals is reviewed, and particular attention is paid to cold welding, microcontacts, and high temperature effects. The mode of action of adhesives is briefly examined.

ADHESION AND ADHERENCE

A number of well known attractive forces (ionic, covalent, metallic, hydrogen, van der Waals) ensures the cohesion of solids. The work needed to break these bonds along an imaginary plane of unit area and to reversibly separate the two parts of the solid is two times the surface energy γ of the solid. Contribution of long range van der Waals forces is generally negligible with respect to other short range forces. Metallic bonding for example is formed at a distance of about 2Å and contributes to about 90 per cent of the surface energy of metals (1). Surface energy is thus characteristic of the nature of bonds ensuring the cohesion of the solid through this imaginary plane. So metals and covalent solids have high surface energy (from 1,000 to 3,000 $mJ.m^{-2}$), ionic crystals (100 to 500 $mJ.m^{-2}$) and molecular solids ($\gamma < 100$ $mJ.m^{-2}$) have lower surface energies.

The work 2γ is the area under the force-displacement curve, $\sigma - \Delta a$, of interatomic bonds. Assuming a linear force law, one has $2\gamma = 1/2\,\sigma\,(\Delta a)_{max} = \sigma^2 a/2E$;

hence the Orowan formula $\sigma \simeq \sqrt{\gamma E/a}$ for the theoretical stress to break the bonds (theoretical strength), where E is the Young's modulus and a the equilibrium bond length. But this stress is higher than actual values by some two orders of magnitude, and the exact knowledge of the force law does not change the order of magnitude. The discrepancy is due to the fact that a solid never ruptures as a rigid body; elastic energy is progressively stored around a pre-existing flaw, and then a crack propagates, breaking the bonds one by one as a zip fastener does. Griffith's analysis shows that in a purely elastic solid the mean stress to transform this flaw into a propagating crack is

$$\sigma = \sqrt{\gamma E/c} \qquad (1)$$

where c is the size of the flaw. Pre-existing flaws of the order of $10^4 a - 10^5 a$ explain the observed discrepancies.

The same story can be told for adhesion of solids. <u>Adhesion</u> between two solids may be due to ionic, covalent, metallic, hydrogen or van der Waals forces. To cut these bonds and to reversibly separate the two solids 1 and 2 in contact on unit area, energies γ_1 and γ_2 are needed to create the unit areas 1 and 2, whereas the excess energy γ_{12} (interfacial energy) is recovered. The quantity $w = \gamma_1 + \gamma_2 - \gamma_{12}$ is the <u>Dupré energy of adhesion</u>, or the <u>thermodynamic work of adhesion</u>. For a single crystal one has $\gamma_{12} = 0$; between two grains of a polycrystal γ_{12} is the grain boundary energy ($\gamma_{GB} \simeq \gamma/3$) or the twin boundary energy ($\gamma_{TB} \simeq \gamma/50$). The "adhesivity" of surface can be increased or decreased by appropriate chemical treatment, that amounts to promote or hinder the bonding between the surface. Anyway, all these chemical aspects can be blocked in the term w.

But the knowledge of w, or even the knowledge of the force law, is insufficient to predict the force to separate the two solids after a contact has been made (<u>adherence force</u>). In adherence,a number of additional factors are involved. As above, the separation never occurs as a whole, but by progression of a crack (so that the adherence force may be independent of the area of contact as in peeling or in lap joints with large overlaps). During this propagation interfacial bonds are broken, elastic energy is released and irreversible work is dissipated at the crack tip. Generally, the adherence force between two solids will depend on the nature of the interfacial bonds, the geometry of the system, the rheological properties of the materials, the stiffness of the measuring apparatus and the method used (fixed grips, fixed load, or fixed velocity conditions). Thus the calculation of interfacial bonds by quantum mechanics or other methods cannot give an adherence force, but only a theoretical strength of adhesion or the Dupré energy of adhesion w that will be considered hereafter as a material constant for the contacting solids, as Young's modulus and Poisson's ratio are.

ADHERENCE OF ELASTIC SOLIDS

Adhesive contact of two elastic solids can be studied from two points of view: thermodynamics and theory of elasticity. They give the same result, but they enlighten the problem differently.

Let us consider two elastic solids in contact over an area A. A force P (either compressive or tensile) can be applied to the two elastic solids, either by a dead load as in Fig. 1a or by a spring of stiffness k_m as in Fig. 1b. The area of contact is allowed to vary at fixed load P or at fixed displacement δ (infinite stiffness) or more generally at fixed Δ. The edge of the contact area can be considered as a fracture tip in mode I that recedes or advances accordingly as the area of contact decreases or increases.

A thermodynamic analysis (2) shows that the Helmholtz and Gibbs free energies of the system are

$$dF = -SdT + Pd\delta + (G-w)dA \tag{2}$$

$$d\mathcal{G} = -SdT - \delta dP + (G-w)dA \tag{3}$$

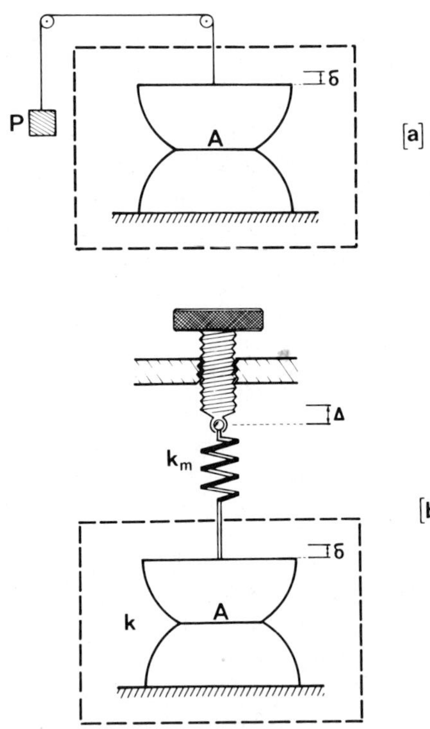

Fig. 1. Equilibrium contact of two elastic solids: a) dead load, b) testing machine with finite stiffness k_m.

where G is the strain energy release rate defined by

$$G = (\partial U_E/\partial_A)_\delta = (\partial U_E/\partial_A)_P + (\partial U_P/\partial_A)_P \quad (4)$$

where U_E and U_P are the elastic and potential energies of the system. Equilibrium at fixed load condition (dP = 0) or fixed grips condition ($d\delta = 0$) respectively corresponds to maximum of $\mathcal{G}$ or F. In either case equilibrium is given by the relation

$$G = w \quad (5)$$

(Griffith's criterion for equilibrium), that links two of the three variables δ, P, A so that the equilibrium curves $\delta(A)$, A(P), $P(\delta)$ are function of w.

If $G \neq w$ the area of contact will spontaneously change so as to decrease the thermodynamic potential. If $G < w$, A increases and the crack recedes; if $G > w$, A decreases, and the crack extends; GdA is the mechanical energy released when the crack extends by dA. The breaking of interfacial bonds requires the energy wdA and the excess (G-w)dA is changed in kinetic energy if there is no dissipative factor.

The equilibrium defined by Eq. 5 can be stable, unstable or neutral. A thermodynamic system under a given constraint (e.g. $P = C^{te}$, $\delta = C^{te}$, ...) is stable if the corresponding thermodynamic potential is minimum. The more general condition of stability is

$$(\partial G/\partial A)_\Delta > 0; \; d\Delta = 0$$

so that the stability can depend on the stiffness k_m of the measuring apparatus. In this case it can be shown (2) that the stability range monotonically increases from the fixed load case ($k_m = 0$) to the fixed grips case ($k_m = \infty$).

Figure 2 shows the variation of Gibbs free energy $G = U_T = U_E + U_P - wA$ with the area of contact A and the load P for the case of two contacting spheres. The equilibrium curve G = w is the locus of extrema. For $P > P_c$ equilibrium is stable, and the crack only advances if the load P is varied: one is dealing with controlled rupture of an adhesive joint. At $P = P_c$ the equilibrium becomes unstable and the crack accelerated under P_c, until complete separation occurs.

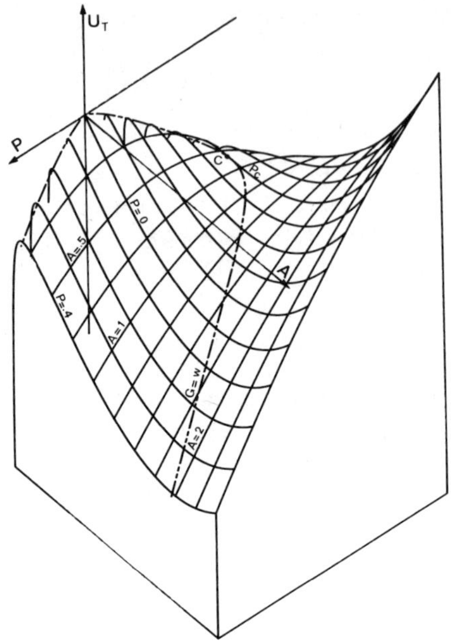

Fig. 2. Total energy stored in the adhesive contact of two spheres as a function of the load and the area of contact (in normalized coordinates). Equilibrium points corresponding to the extrema of U_T are in dashed line. Equilibrium at fixed load becomes unstable at C, and $P_c/3\pi wR = -.5$ is the adherence force.

Thus the load corresponding to:

$$G = w$$

$$(\partial G/\partial A)_P \lesseqgtr 0$$

will be called the elastic adherence force for experiments at fixed load. At fixed grip, or with an apparatus of stiffness k_m the thermodynamic potential would be different and the adherence force would be more negative.

Evaluation of elastic energy in systems with surface energy needs special care, for besides external forces, there are forces due to molecular attraction that cause elastic deformation and lead to elastic energy storage.

dU_E is no longer equal to $Pd\delta$ but is

$$dU_E = Pd\delta + GdA \tag{6}$$

Johnson et al. (3) have shown how the elastic energy can be computed in the case of a spherical punch: by adding a rigid body displacement to the classical elastic solution. In fact, this method is quite general for punches of any shape and can be proved by the general theory of elasticity (4): stresses and discontinuities of displacement are those of fracture mechanics:

$$\sigma = K_I/\sqrt{2\pi\rho} \tag{7}$$

$$[Uz] = (4(1-\nu^2)/E)\ K_I\sqrt{\rho/2\pi} \tag{8}$$

where ρ is the distance to the crack tip, and K_I the stress intensity factor related to G by:

$$G = 1/2\{(1-\nu_1{}^2)/E_1 + (1-\nu_2{}^2)/E_2\}\,K_I{}^2 \quad \text{for plane strain} \tag{9a}$$

$$G = 1/2\ (1/E_1 + 1/E_2)\ K_I{}^2 \quad \text{for plane stress} \tag{9b}$$

For three-dimensional cracks (e.g. circular cracks), plane strain formula must be applied.

Thus, the elastic adherence force P_c can be computed for a number of geometries either with the G method by an energy balance or with the K_I method by stress analysis at the crack tip. When using the energy method, one has to verify that the solution correctly gives stress singularity at the crack tip according to Eq. 7, together with the correct crack profile. With beam or plate theories, special precautions must be taken to have the proper result: rotation of built-in sections, elastic foundations.

Adherence of spheres and flat punches was studied by Johnson et al. (3), Kendall (5), and Maugis and Barquins (2,6). For an axisymmetric rigid punch contacting an elastic half-space, under a load P, over an area of radius a, it can be shown (7,4) that

$$G = \{(1-\nu^2)/E\}\ ((P_1-P)^2/8\pi a^3) \tag{10}$$

where P_1, the apparent load, is $P_1 = 0$ for a flat punch, $P = 4Ea^3/3(1-\nu^2)R$ for spheres, $P = \pi Ea^2 tg\,\beta/2(1-\nu^2)$ for cone (8). Note that the adherence forces are not necessarily proportional to w; in the present case they are respectively:

$P_c = -\left[\{(8\pi E/(1-\nu^2)\}a^3 w)\right]^{1/2}$	for flat punch at fixed grips or fixed load
$P_c = -(3/2)\pi wR$	for sphere at fixed load
$P_c = -(5/6)\pi wR$	for sphere at fixed grips
$P_c = -54(1-\nu^2)w^2/\pi E tg^3\beta$	for cone at fixed load
$P_c = -6(1-\nu^2)w^2/\pi E tg^3\beta$	for cone at fixed grips.

i.e. proportional to $\sqrt{w}$, w and w^2. But the mean stress at separation is still given by Eq. 1 where c is the radius a of the external crack, and γ is replaced by w.

Double cantilever beam (DCB) is another geometry where G and K_I have been independently evaluated (9-13)

$$G = K_I{}^2/E \simeq (12\,P^2L^2/Eb^2h^3)(1 + 2h/3L)^2 \qquad (11)$$

where b, h, L are the width, height and length of the cantilever beam. It must be pointed out that at fixed grips $(\partial G/\partial A)_\delta > 0$ at equilibrium, so that no elastic adherence force can be defined.

For peeling at an angle θ, an energy balance easily gives (14,15)

$$G = (P/\bar{b})(1-\cos\theta) + (P/\bar{b})^2\ 1/2(Eh) \qquad (12)$$

leading to an indifferent equilibrium under the elastic adherence force P_c, but the equivalence with fracture mechanics has not yet been proved.

The blister test was analyzed by the finite element method (16). To evaluate G, Andrews and Stevenson (17) have tentatively added the elastic energy computed by plate theory (far field) to that computed for an internal crack in an infinite medium (near field) for the same radius and pressure. But the closed form solution (analog to the JKR solution (3) for spheres, and the Kanninen solution (12) for DCB) is not yet known.

From the point of view of concepts, the problem of the adherence of elastic solids is completely solved: only three ingredients are needed: geometry of the system, elastic properties, and energy for adhesion. Eventually a fourth ingredient must be added: the rigidity of the measuring apparatus that can change the stability of equilibrium.

<u>Influence of Pre-stresses or Residual Stresses</u>

One advantage of the energy balance method is that influence of pre-stresses or residual stresses can easily be studied. These stresses increase the stored elastic energy, hence the available energy for crack propagation: they reduce the quasistatic force of adherence. This point has been theoretically and experimentally shown by Kendall for peeling (18): when an elastic strip (thickness h, width b) is stretched to $\varepsilon = dL/L$ before placed on a rigid plane, the elastic energy stored is $U_E = 1/2(\varepsilon^2 EhA)$ and thus the quantity $1/2(\varepsilon^2 Eh)$ released when the crack advance by dA must be addded to G. For $\pi/2$ peeling, for example, one has

$$G = F/b + (1/2)\varepsilon^2 Eh \tag{13}$$

so that the elastic adherence force (for G = w) is reduced by $1/2\,(\varepsilon^2 Eh)$ and disappears (spontaneous peeling) when the elastic energy stored in the strip is equal to the total energy of adhesion wA at the interface. If such a deformation appears by thermal dilatation, the elastic strip must be as thin as possible to avoid spontaneous peeling.

A special case of residual stress effect is the adherence of rough surfaces: higher asperities are compressed, more elastic energy than on smooth surface is stored and the adherence decreases and may disappear (19).

ADHERENCE OF VISCOELASTIC SOLIDS

In fact purely elastic solids do not exist. Solids like elastomers are viscoelastic; they lose energy when subjected to a cycle of deformation, and particularly at a crack tip where stresses and strain rates are high. $G - w$ is the crack extension force applied to the crack tip; under this force the crack takes a limiting speed v, instead of continuously accelerate as for elastic solids, and one can write (6)

$$G - w = w\,\phi(a_T v) \qquad (14)$$

The following points must be emphasized:

1. The losses in the hinterland are negligible compared to those at the crack tip, so that gross displacements are elastic and G can be computed, as above, by the theory of elasticity (with relaxed Young's modulus).

2. The right — hand member is a viscous drag proportional to the thermodynamic work of adhesion: that means that losses only arise if the interface itself is capable of withstanding stresses. Reducing the work of adhesion increases the crack speed. Surface properties are thus completely de-coupled from elastic properties, geometry and loading conditions included in G, and from viscoelastic properties described by $\phi(a_T v)$.

3. The dimensionless factor $\phi(a_T v)$ is thus a material property for a given mode of crack propagation (e.g. mode I). Results at various temperatures can be shifted to a reference temperature by using the WLF shift factor a_T. At the present time the function ϕ must be experimentally determined, but it is hoped that in the future it will be computed from the knowledge of $E'(\Omega)$ and $E''(\Omega)$, the frequency-dependent real and imaginary parts of the loss modulus.

4. Experiments at equilibrium ($v = 0$, i.e. $\phi = 0$) give the thermodynamic work of adhesion w.

Once the function $\phi(a_T v)$ is known (e.g. by peeling

experiments), the Eq. 14 allows to predict any feature, such as kinetics of detachment at fixed load, fixed grips or fixed crosshead velocity, for any geometry. Note that Eq. 14 is quite different from an expression such as $P = P_o f(a_T v)$ where P_o would be a quasistatic force of adherence; for spheres, for example, $P_o = (3/2)\pi wR$ is a constant, whereas G depends both on the load and the radius of contact. It is also different from a Griffith's formula $\sigma \sim (Ew/a)^{1/2}$ where w would be a rate-dependent quantity.

Let us make some examples with adherence of glass ball on polyurethane. The Dupré energy of adhesion ($w = 65\ mJ.m^{-2}$) was determined by measuring the equilibrium radii of contact a under various loads P (with a dwell time of 10 min.), and using Eq. 10 with $P_1 = a^3K/R$ (JKR theory). Figure 3 from Ref. 6 shows the variation of radius of contact

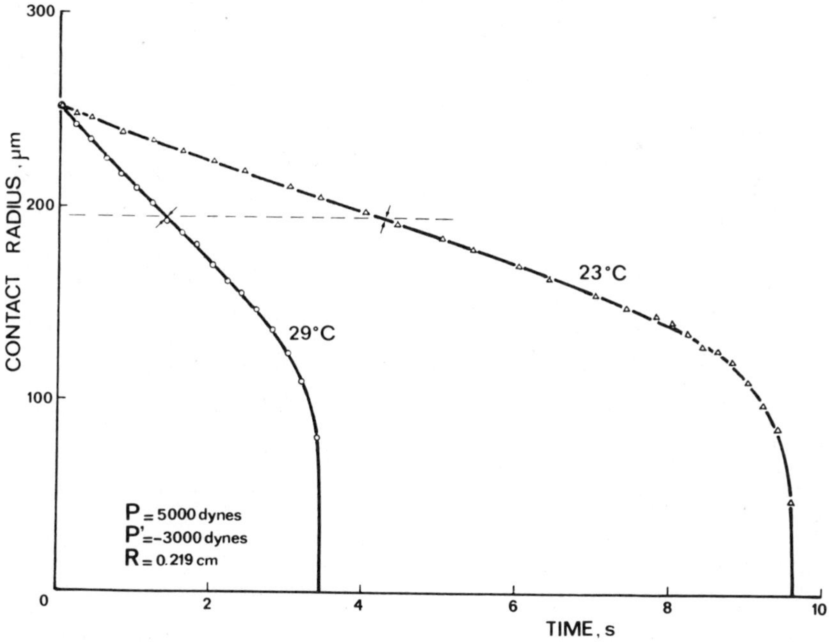

Fig. 3. Glass ball on polyurethane. Unloading curves from P = 5,000 to P' = -3,000 dynes at two different temperatures (from ref. 6).

during unloading at two different temperatures: at every data point G and v = da/dt can be determined, and thus the function $\phi(v)$, or $\phi(a_T v)$ using the WLF shift factor with $T_g = -50°C$ for the glass transition temperature. The same master curve is obtained for other geometries such as flat punch, peeling, flat-ended sphere, or other experimental condition and is shown in Fig. 4 from ref. 6. It varies as $(a_T v)^{0.6}$ for a large range of crack speeds, a result often found (18,20,21) for peeling of rubber-like materials. Figure 5 compares theory and experiment for unloading of a smooth sphere or a flat-ended sphere (22). Figure 6 shows unloading of a glass ball for the same imposed displacement Δ and various stiffnesses (23); here also continuous lines are theoretical curves.

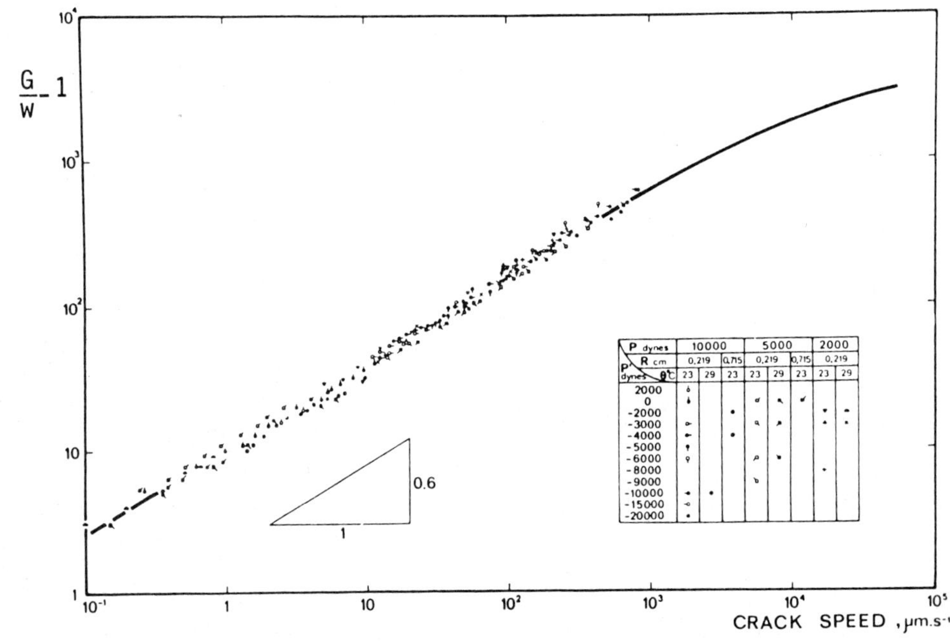

Fig. 4. Master curve for the kinetics of adherence of glass on polyurethane. The results at various temperatures are shifted to T = 23°C by the WLF transformation. This curve gives the dissipative function $\phi(a_T v)$ of the polyurethane and is valid for any geometry (from ref. 6).

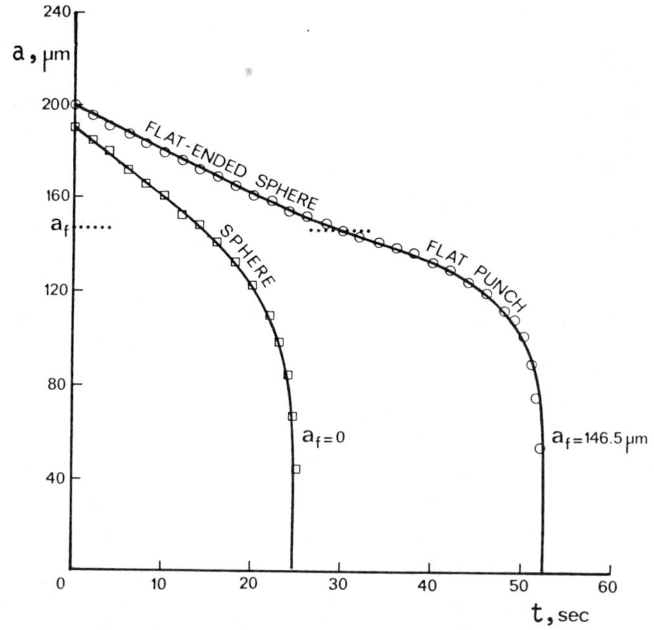

Fig. 5. Kinetics of unloading of a smooth sphere and a flat-ended sphere (radius of the sphere R = 2.19 mm, radius of the flat a_f = 146.5 µm) on polyurethane. Unloading from P = 20 mN to −30 mN. Such a small flat increases the time to rupture by a factor two (from ref. 22).

Adsorption generally decreases the energy of adhesion, hence the viscoelastic losses, according to Eq. 14. This point was verified for water adsorption by measuring the rolling resistance T of a glass cylinder of length L rolling on the sample (G ≠ T/L) as a function of speed for various humidity contents (2,21,24). Figure 7 shows the vertical translation of the G(v) curves with water adsorption (24).

The influence of pre-stress on peeling was verified by Kendall (18) for rubber on glass. Eq. 14 becomes:

$$F/b = w - 1/2\ \varepsilon^2 Eh + w\alpha v^{0.6}$$

and Fig. 8 clearly shows the vertical translation with pre-stress. Note that the rubber can have a peel resistance under kinetic conditions, whereas the elastic adherence force (quasistatic conditions) has disappeared.

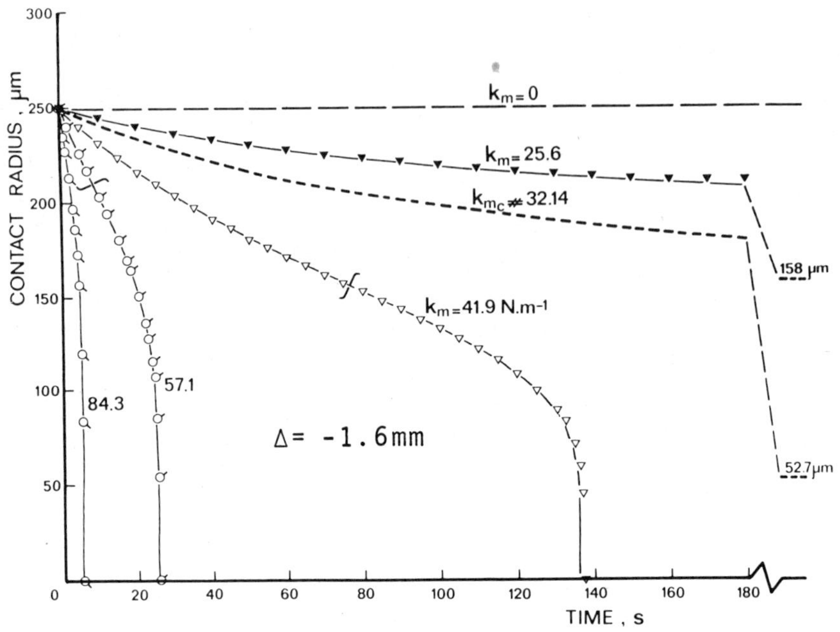

Fig. 6. Kinetics of unloading of a glass ball on polyurethane for the same imposed displacement and various stiffnesses (from ref. 23).

Tackiness

The adherence of solids is more often measured with a tensile test machine than at constant load, but the kinetics of separation is less easy to interpret due to the competition between increasing δ at constant a, and decreasing a at constant δ for increasing G:

$$dG/dt = (\partial G/\partial \delta)_a \, d\delta/dt + (\partial G/\partial a)_\delta \, da/dt \qquad (15)$$

The recorded force first increases, then decreases; the maximum value, called the tack force is a measure of the adherence in this experimental condition; the area under the curve, called the tack energy, is equal to the work GdA of the singular stress at the crack tip. Tackiness refers to the ability of an elastomer to adhere instantaneously to a solid surface, or to itself, after a brief time of contact under low pressure. Probe tack testing procedure can be analyzed by Eq. 14 and tack curves obtained by computer integration closely coincide with experimental ones (25).

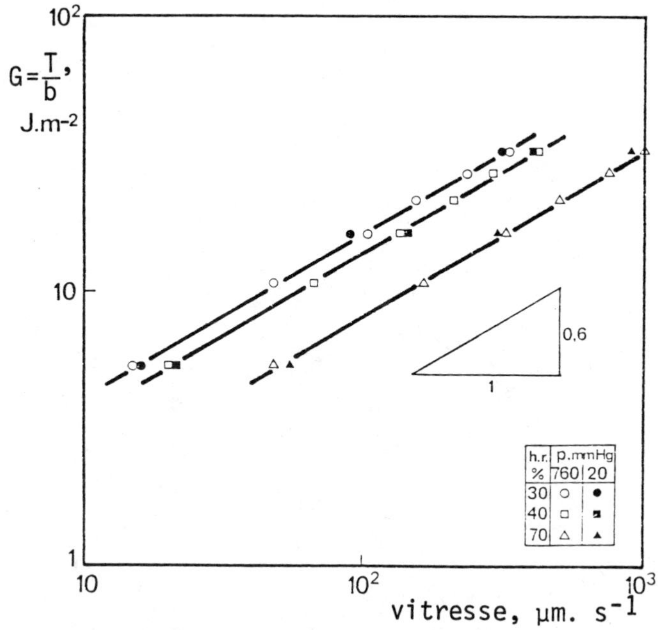

Fig. 7. Rolling resistance of a glass cylinder on polyurethane for various room humidity content (from ref. 24).

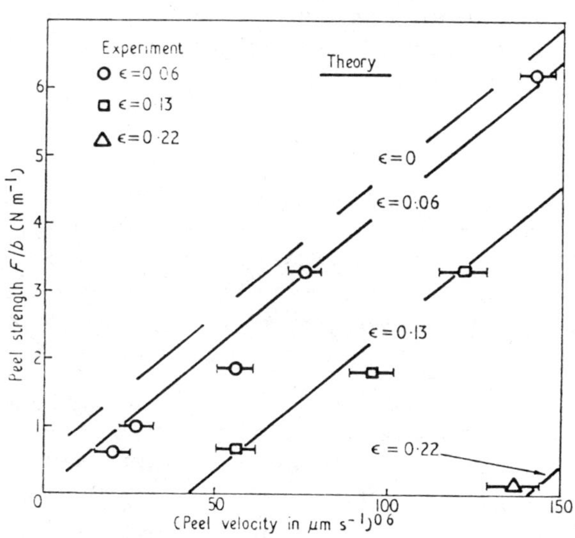

Fig. 8. Peel strength of rubber-glass bonds; variation with shrinkage strain (from ref. 18).

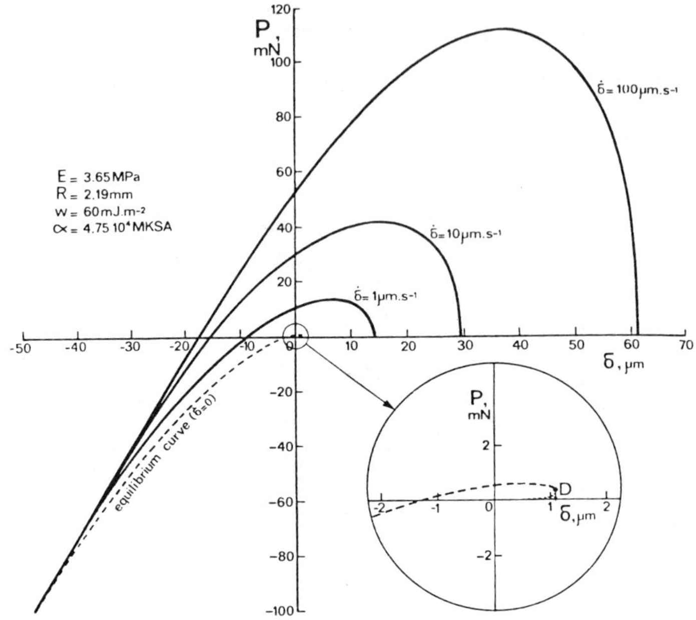

Fig. 9. Glass ball on polyurethane; influence of withdrawal speed on the tack force (from ref. 25).

Figure 9 is for spherical probe and shows that even at very low crosshead velocity the viscoelastic effects considerably increase the adherence force compared to the elastic (or quasistatic) adherence forces at fixed load (point C) or fixed grips (point D).

Dwell Time

Dwell times generally increase the adherence force, and Fig. 10 shows the increase of w with contact time between glass and polyurethane (26). Between two polymers (autohesion) this effect is ascribed to interdiffusion of chains by reptation; theory and experiments (27-29) lead to a G_c (hence a Dupré energy of adhesion, see below) increasing as $t^{1/2}$. However this mechanism is most unlikely for contacts between glass and elastomer, and when computed from time to rupture under constant load (delayed rupture), w seems to increase as $t^{0.1}$ (26). In fact, it is G - w that decreases with time, and a more likely explanation is a decrease of G instead, by viscoelastic relaxation of stresses in compressed asperities (26).

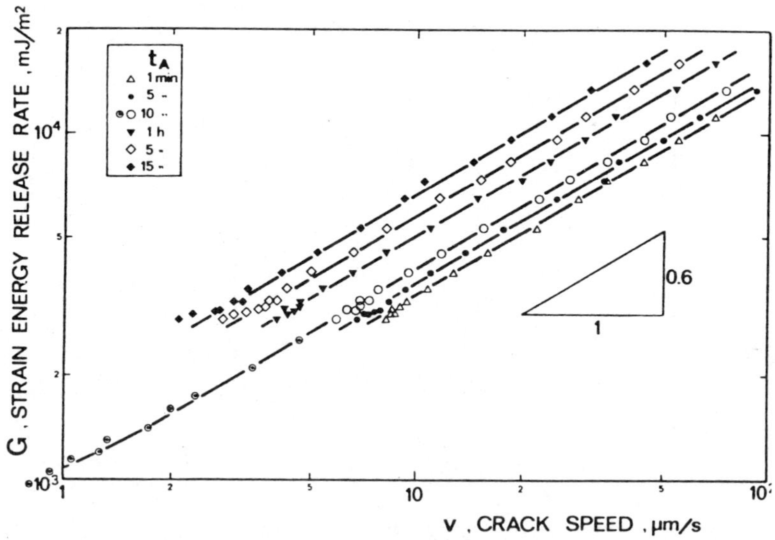

Fig. 10. Glass ball on polyurethane; relation between strain-energy release-rate and crack speed for various dwell times (from ref. 26).

Crack Closure

When G approaches w, the crack speed decreases and is zero for G = w. For G < w, the crack speed reverses and the area of contact increases. This point was studied by loading spheres on polyurethane (Fig. 11). As G cannot be negative, crack closure generally begins with propagation at G = 0; the contact is thus Hertzian with a tangential connection; there is no stress singularity at the crack tip (K_I = 0), so viscoelastic losses are negligible and crack propagation is very fast. Once the Hertzian area of contact is reached, stress singularity and discontinuity of displacement reappears, and the mode I crack very slowly recedes to its equilibrium position with a driving force w - G varying from w to 0. So, if the equilibrium radii of contact seem to be more quickly reached by loading than by unloading (30), it is because most of the area of contact is instantaneously reached, and because after a time, longer dwell time seems to have no effect on the area of contact.

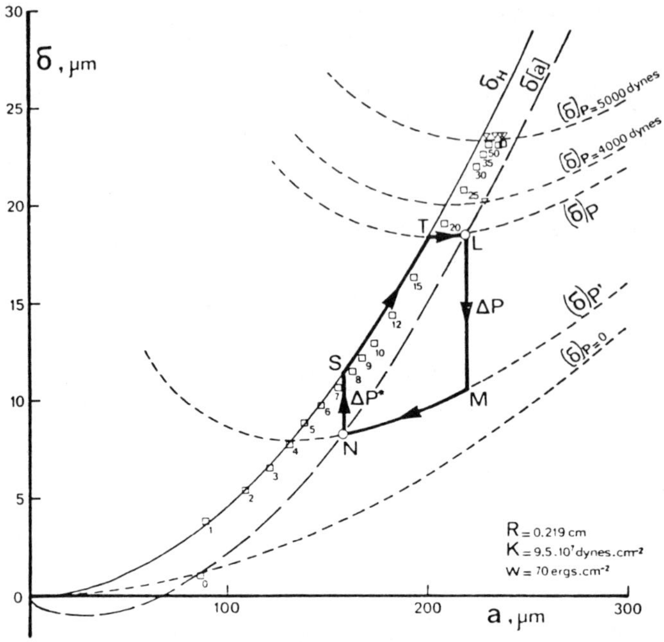

Fig. 11. Relation between elastic displacement δ and radius of contact a of two spheres in an unloading-loading experiment. After an unloading from P to P´ (path LMN), a loading from P´ to P leads to an instantaneous displacement first at constant a (branch NS) and then along the Hertz's curve (branch ST). This path (NST) is followed by an evolution along the curve (δ)P towards the equilibrium point L. Experimental results for loading step from P´ = 0 and P´ = 4000 dynes to P = 5000 dynes are given (from ref. 2).

A Model for Stick-slip and Embrittling Effects

It is well known that in a viscoelastic body, energy losses decrease at high frequencies, so that the behavior becomes elastic again. It can be anticipated that the function $\phi(a_T v)$ first increases, then decreases when the crack speed increases. However crack speed is limited to the Rayleigh waves speed, and another loss mechanism (such as acoustic emission) must occur at these high speeds. Thus, the $\phi(a_T v)$ curve would appear S shaped, with a maximum and a minimum. In this model (31) the branch between the maximum and the minimum has a negative slope ($d\phi/dv < 0$) and

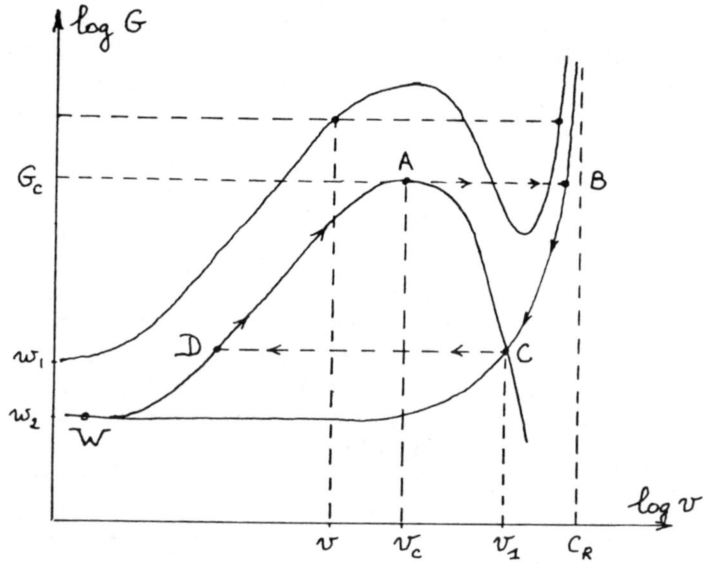

Fig. 12. Implication of an unstable branch (negative slope) in the G - v curve: stick-slip and adsorption induced embrittlement (from ref. 31).

corresponds to an unstable process. Let us examine the implications of this unstable branch for adhesive joints (Fig. 12). If a constant G is applied, the crack takes a constant speed v(G) as long as $G < G_c$. When $G = G_c$ (point A) the speed jumps from the value v_c to $v \gg v_c$ (some orders of magnitude higher) on the second stable branch: the joint appears brittle. When a mean speed v with $v_c < v < v_1$, is imposed (with a tensile machine) the hysteresis cycle ABCDA would be observed, leading to stick-slip motion. (The suddenly starting and stopping crack giving rise to acoustic emission.) For peeling, calculation leads to "wavelength" increasing with the peeled length, as observed by Aubrey (32) (this stable or unstable crack propagation defined by $d\phi/dv \gtrless 0$ must not be confused with stable or unstable equilibrium defined by $(\partial G/\partial A) \gtrless 0$). Increasing now the temperature, the stick-slip disappears.

This model leads also to interesting comparison with embrittlement by surface active liquids (Rehbinder effect) or by intergranular segregation. As viscoelastic losses are proportional to w, a reduction of the Dupré energy of adhesion by adsorption or segregation lowers the whole G(v) curve, and a loading giving a subcritical crack growth for high w values, can give a catastrophic crack speed at lower w values (Fig. 12).

ADHERENCE OF METALS

Clean metals have high surface energies, and when in contact very strong metallic bonds are immediately formed. The force per unit area needed to separate them is the ultimate tensile strength of the softer of the materials if the metallic contact are sufficiently large to avoid deep notch effects. The rupture is ductile and the rheological properties of metals are thus involved. If however impurites are adsorbed on the surfaces, metallic bonding is prevented, adherence abruptly decreases, and the rupture is adhesive instead of cohesive. So, contact of clean metal surfaces must be avoided in friction, but is necessary for cold welding.

Cold Welding

It is the merit of Milner and his colleagues (from 1959 to 1969) to have explained the mechanisms of cold welding (33,34). It is the mechanical disruption of oxide films and the extrusion of virgin materials through cracks that explains cold welding by rolling (35) butt pressure or twist compression. Adherence is stronger as the proportion of clean virgin metal in contact is high. A large area of virgin material is instantaneously contaminated by impurities trapped at the interface and has a low adherence. Unless working in vacuum (36), large surface deformations (varying from 10% to 90% with metals) are needed to disperse the contaminants. This threshold deformation is lowered by scratch-brushing (producing a heavily worked layer) or reducing the oxide film thickness (37). An important point is that oxide films break up coherently (double blocks of oxide) exposing face to face underlying metal, at high friction coefficients and thin oxide layers (37,38). An elegant demonstration was given by using plasticine models with brittle coatings (39).

Adherence of Metals at High Temperature (40)

Experimental data show that above a threshold

temperature, the adherence of metals increases with temperature and contact time. This threshold temperature decreases with load but a representative value is about 0.3 T_m, where T_m is the absolute melting point. Therefore adherence experiments at room temperature are high temperature experiments for indium, tin and lead.

When two metals come into contact under a given load, asperities are pressed against one another and undergo plastic flow and creep; the area of contact increases with temperature and contact time, as in a hot hardness mutual indentation. Then, when the stress has dropped sufficiently the area of contact predominantly increases by surface or boundary diffusion as in sintering experiments. As these various mechanisms are thermally activated, the increase of area of contact with temperature can be written:

$$A = A_o e^{-Q_A/RT}$$

where the activation energy Q_A can be deduced from the activation energies of the above mechanisms.

In high temperature contact strong metallic bonds are easily developed and the separation generally occurs by creep rupture of the material (cohesive rupture), so that the adherence force, as in cold welding, can be written as

$$F = kA\sigma_u$$

where k is constraint factor decreasing linearly from 2 to 1 as the load-bearing area is increased. The proportionality between U.T.S. and hardness ($\sigma_u \simeq H/3$) holds at high temperature, so that σ_u has the same activation energy Q_H for hot hardness. In creep rupture, as in hot hardness, there is equivalence between time and temperature and master curves can be obtained, using the Sherby-Dorn parameter $\Theta = t\exp(-Q/RT)$ where Q is the activation energy for self-diffusion. Cohesive rupture is thus time dependent, and the adherence force decreases as the time to rupture increases.

High temperature adherence is thus characterized by the competition between increasing the area of contact by creep or sintering with time and temperature, and decreasing the mechanical properties. In fact the problem is rendered complex by phenomena such as oxidation, oxide dissolution, segregation of impurities, interdiffusion, or phase transformation. Mutual solubility is not a condition of high adherence, and insoluble pairs such as Ag-Fe, Ag-Ni, Pb-Au can adhere better than soluble ones such as Ag-Au. Interdiffusion only changes the mechanical properties of the created junction; if the junction is hardened the adherence increases, but the formation of brittle intermetallic compounds or the development of porosity by Kirkendall effect reduce it.

Adherence of Microcontacts

Due to roughness effects, adherence of metals at moderate temperature and pressure is difficult to analyze. When roughnesses undergo plastic deformation, the true area of contact is proportional to the applied load P, and the adherence force F is often proportional to the load (hence the definition of an adhesion coefficient α = F/P), and independent of the apparent area of contact. These two "Laws of adhesion" (41) are similar to Amonton's laws of friction. As shown by Gilbreath (42) the adhesion coefficient is very sensitive to adsorption. More precise experiments by Buckley (43,44) on single crystals in ultrahigh vacuum have shown that the adherence force does not increase linearly with the load, and that the position of the knees depends on the adsorption as if the effectively applied load depended on adsorption.

To avoid complications due to roughness a number of adhesion experiments have been performed on single asperity contact, but the situation is still complex. The contact of the tip can be elastic, elastoplastic or full plastic as in hardness experiments; the separation can occur at the interface ("brittle" or adhesive rupture) or in the softer material ("ductile" or cohesive rupture). In the latter case the adherence force is the force to rupture a deeply notched bar (45)

$$F_{pl} = \pi a^2 H \tag{16}$$

where a is the radius of contact, and H the hardness related to the elastic limit Y by H = 3 Y.

On an elastoplastic material the various adherence forces can be computed. If the contact is elastic, the "brittle" adherence force is given by the JKR theory (3) $F_{el} = 3/2(\pi wR)$and cohesive rupture is very unlikely. If the contact is fully plastic, the elastic adherence force is given by the same equation, but taking into account the elastic relaxation (45,46).

$$F_{el} = (9\pi w^2K^2/4\pi H^3)^{1/2} \sqrt{P} \quad (17)$$

where K is an apparent Young modulus. This force can be observed if it is lower than the ductile force, i.e. (45) if

$$P > 9w^2K^2/4\pi H^3 \quad (18)$$

A point of importance is the adhesion-induced plastic deformation observed in sintering and in microcontacts. In absence of surface forces, plastic deformation is initiated under a sphere when the mean pressure $P/\pi a^2$ reaches a value of 1.1Y. Due to surface forces, the area of contact increases and the stress distribution under the sphere is modified. The stress tensor for contact of adhesive spheres and the conditions of incipient plasticity have been derived by Barquins and Maugis (4).

GLUING TWO SOLIDS TOGETHER

Adherence between two dirty and not molecularly smooth solids is generally negligible; but inserting a viscous fluid between them leads to high adherence as in the case of gage blocks (47).

The adherence force between two disks of radius R separated by a liquid of thickness h is a function of the time t for separation:

$$F = 2\pi R^2 \gamma \cos\theta/h + (3\pi\eta R^4/4h^2) 1/t \quad (19)$$

where γ and η are the surface energy and the viscosity of the liquid and θ the wetting angle. The first term is the quasistatic adherence force (Laplace law); the second is the rheological term derived from the Stefan law of viscous flow; it explains the mechanics of "pressure sensitive adhesives" as sticking wetted gummed labels or stamps on a wettable surface. However the surface energy term is of importance from the fact that if the liquid does not wet the

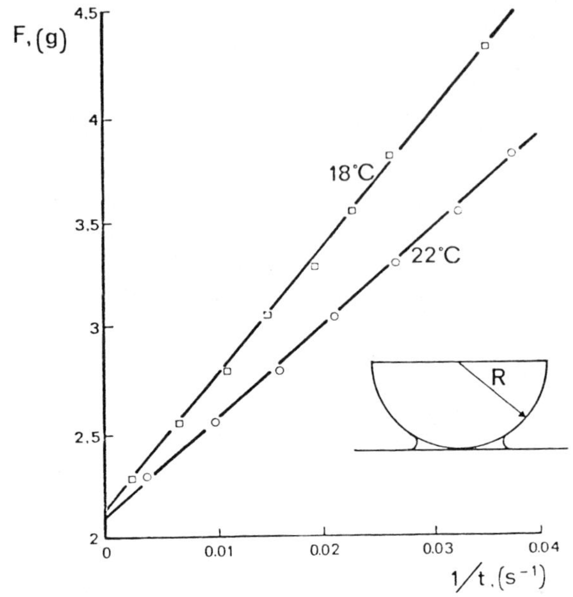

Fig. 13. Adherence of a glass ball (R = 5.042 cm) on a glass plate through castor oil. Influence of rupture time (from ref. 48).

surface, the liquid pushes the solids away from one another, just as soon as the pressure is removed.

In the case of a liquid film between a sphere of radius R and a plane, the quasistatic term is (48,49)

$$F_o = 2\pi\gamma R\,(\cos\theta_1 + \cos\theta_2) \qquad (20)$$

and experiment verifies it is proportional to the radius R and independent of the film thickness (down to 0.1 μm). For castor oil, Fig. 13 shows the linear variation of the adherence force with 1/t; extrapolation to 1/t = 0 exactly gives the value 4 $\pi\gamma$ R corresponding to this oil (γ = 33 $mJ.m^{-2}$). The temperature dependence of adherence reflects the temperature dependence of the viscosity; the higher the viscosity, the higher the adherence.

It is easily understood that a good adhesive must wet and clean the adherend, have a high Dupré energy of adhesion, a high viscosity, and no residual stress. To wet the adherend the adhesive is often applied as a liquid and then solidified by cooling, evaporation, oxidation, polymerization, etc. Wetting can be improved by increasing the roughness of the adherend, for the contact angles θ' and θ of rough or smooth surfaces are different according to the Wenzel law:

$$\cos\theta' = r\cos\theta \tag{21}$$

where r is the roughness. (In brazing metals with low melting alloys, this wetting and cleaning action is obtained by adding flux agents to the alloy.) The liquid must have low viscosity and high surface energy in order to increase capillary flow into pores and crevices, and to avoid the formation of gas pockets. That high surface energy must be compatible with low contact angle, i.e. the Dupré energy of adhesion must be high, according to the Young-Dupre equation:

$$w = \gamma(1 + \cos\theta)$$

for $\theta<\pi$ (the condition $w > 2\gamma$ is the spreading condition). To maximize the Dupré enegy of adhesion is the chemical aspect of the science of adhesives: choice of surface treatments of the adherend, or reactive terminal groups to have chemical bonds. When "solidifed", the behavior of the adhesive layer is in fact that of a viscoelastic solid. Residual stresses usually developed on solidification can be avoided by adjusting the thermal expansion coefficient to that of the adherend.

When the adhesive layer is very thin, the elastic energy that can be stored in the adhesive is negligible, and the quasistatic adherence force can be computed by fracture mechanics as above (i.e. Eq. 10 for a butt joint, Eq. 11 for DCB). But viscoelastic losses at the crack tip involve only the adhesive, and the Dupré energy of adhesion is two times the surface energy if the crack propagates in the adhesive. A very elegant model was studied by Burns and Lawn (50): they inserted a wetting liquid between two glass strips and ruptured the film by driving in a wedge as in a double cantilever experiment at fixed grips. The stable equilibrium was correctly given by $G = 2\gamma$, where G is the classical strain energy release rate for DCB computed from the glass arms only and γ the surface energy of the liquid. They noted the influence of viscosity: for water, periods of several minutes were required to reach equilibrium, and for heavy oils, viscosity prevented equilibrium conditions from being reached within a period of days. Another example is the determination of surface energy of naphthalene crystals by Skvortsov et al. (51): the thin crystal was attached to a glass block on one side and to a metal plate on the other side and then cleaved by bending the metal plate; surface energy was derived using the only mechanical properties of the bended plate (in fact their experiment was analogous to a tack force measurement; and as dissipative effects were not taken into account, the thermodynamic value was not obtained).

Wang et al. (52) have given a careful analysis of the crack tip stress field in DCB adhesive fracture specimens by a finite element method. The main results are the following: 1) the singular field is localized in the adhesive, so that there is no K_I in the adherend. 2) G is determined primarily by the adherend (Young's modulus E_2) and is relatively insensitive to the presence of the adhesive. 3) The K_I in the adhesive is given by $K_I^2 = GE_1$, where E_1 is the Young's modulus of the adhesive. 4) The effect of adhesive thickness on K_I is negligible. Comparison with Burns and Lawn experiments is striking: the stress intensity factor can be very low in the adhesive, but the balance of energy is still given by Eq. 14. If G_c corresponds to the peak losses as suggested above, it can be understood that a glued joint has sometimes higher strength than the adherend.

CONCLUSION

In adherence of solids, surface effects, rheological effects and fracture mechanics are mixed in an intricate manner, and no experiment can be performed involving only the chemical aspects. The adherence of elastic solids is now clear; that of viscoelastics is in progress; for other materials, the constitutive equation must be known and taken into account.

It must be pointed out that the thermodynamic work of adhesion can only be reached in stable equilibrium conditions such as sphere on half-space, or DCB at fixed grips, and that the least crack motion gives an apparently increased work of adhesion, impossible to reconcile with thermodynamic evaluations.

REFERENCES

1. J.E. Inglesfield, "Adhesion Between Al Slabs and Mechanical Properties," J. Phys. F. Metal Phys., 6, 687 (1976).

2. D. Maugis and M. Barquins, "Fracture Mechanics and Adherence of Viscoelastic Solids," in Adhesion and Adsorption of Polymers L.H. Lee, ed., Part A, Plenum, N.Y. (1980), p. 203.

3. K.L. Johnson, K. Kendall, and A.D. Roberts, "Surface Energy and the Contact of Elastic Solids," Proc. Roy. Soc. A 324,301 (1971).

4. M. Barquins and D. Maugis, "Adhesive Contact of Axisymmetric Punches on an Elastic Half-Space: The Modified Hertz-Huber's Stress Tensor For Contacting Spheres," J. Mech. Theor. Appl., 1, 331 (1982).

5. K. Kendall, "The Adhesion and Surface Energy of Elastic Solids," J. Phys. D. Appl. Phys., 4, 1186 (1971).

6. D. Maugis and M. Barquins, "Fracture Mechanics and Adherence of Viscoelastic Bodies," J. Phys. D. Appl. Phys., 11, 1989 (1978).

7. H.M. Pollock, D. Maugis, and M. Barquins, "The Force of Adhesion Between Solid Surfaces in Contact," Appl. Phys. Lett., 33, 798 (1978).

8. D. Maugis and M. Barquins, "Adhesive Contact of a Conical Punch on an Elastic Half-Space," J. Physique, lettres, 42, 295 (1981).

9. J.E. Strawley and B. Gross, "Stress Intensity Factors for Crack-Line-Loaded Edge-Crack Specimens," Mat. Res. and Std., 7, 155 (1967).

10. S.M. Wiederhorn, A.M. Shorb, and R.L. Moses, "Critical Analysis of the Theory of Double Cantilever Method of Measuring Fracture Surface Energies," J. Appl. Phys., 39, 1569 (1968).

11. P.P. Gillis, "Discussion on the Theory of Double Cantilever Cleavage," J. Appl. Phys., 39, 2149 (1968).

12. M.F. Kanninen, "An Augmented Double Cantilever Beam Model for Studying Crack Propagation and Arrest," Int. J. Fracture, 9, 83 (1973).

13. D.J. Chang, R. Muki, and R.A. Westman, "Double Cantilever Beam Models in Adhesive Mechanics," Int. J. Solid Structures, 12, 13 (1976).

14. R.S. Rivlin, "The Effective Work of Adhesion," Paint Technol., 9, 215 (1944).

15. K. Kendall, "Thin Film Peeling--The Elastic Term," J. Phys. D Appl. Phys., 8, 1449 (1975).

16. S.H. Bennet, K.L. Devries, and M.L. Williams, "Adhesive Fracture Mechanics," Int. J. Fracture, 10, 33 (1974).

17. E.H. Andrews and A. Stevenson, "Fracture Energy of Epoxy Resin Under Plane Strain Conditions," J. Mat. Sci., 13, 1680 (1978).

18. K. Kendall, "Shrinkage and the Peel Strength of Adhesive Joints," J. Phys. D. Appl. Phys., 6, 1782 (1973).

19. K.N.G. Fuller and D. Tabor, "The Effect of Surface Roughness on the Adhesion of Elastic Solids," Proc. Roy. Soc., A 345, 327 (1975).

20. A.N. Gent and R.P. Petrich, "Adhesion of Viscoelastic Materials to Rigid Substrates," Proc. Roy. Soc., A 310, 433 (1969).

21. A.D. Roberts, "Looking at Rubber Adhesion," Rubber Chem. Technol., 52, 23 (1979).

22. D. Maugis and M. Barquins, "Adhesive Contact of Sectionally Smooth-Ended Punches on Elastic Half-Spaces. Theory and Experiment," J. Phys. D: Appl. Phys. 16, 1843 (1983).

23. M. Barquins, "Influence of the Stiffness of Testing Machine on The Adherence of Elastomers," J. Appl. Polymer Sci., 28, 2647 (1983).

24. M. Barquins, Thèse Paris (1980).

25. M. Barquins and D. Maugis, "Tackiness of Elastomers," J. Adhesion, 13, 53 (1981).

26. M. Barquins, "Influence of Dwell Time on the Adherence of Elastomers," J. Adhesion, 14, 63 (1982).

27. S.S. Voyutskii, "Autoadhesion and Adhesion of High Polymers," Interscience, N.Y. (1963).

28. P.G. de Gennes, "Sur la Soudure Des Polymères Amorphes," CR Ac. Sc., B 291, 219 (1980).

29. K. Jud, H.H. Kausch, and J.G. Williams, "Fracture Mechanics Studies of Crack Healing and Welding of Polymers," J. Mat. Sci., 16, 204 (1981).

30. A.D. Roberts and A.G. Thomas, "The Adhesion and Friction of Smooth Rubber Surfaces," Wear, 33, 45 (1975).

31. D. Maugis, "Adherence of Solids," in Microscopic Aspects of Adhesion and Lubrication, J.M. Georges, ed., Elsevier, Amsterdam (1982).

32. D.W. Aubrey, "Viscoelastic Basis of Peel Adhesion," in Adhesion 3, K.W. Allen, ed., Appl. Sc. Publ., London (1979).

33. D.R. Milner and G.W. Rowe, "Fundamentals of Solid-Phase Welding," Metallurgical Rev., 7, 433 (1962).

34. R.F. Tylecote, "The Solid Phase Welding of Metals," E. Arnold, London (1968).

35. L.R. Vaidyanath, M.G. Nicholas, and D.R. Milner, "Pressure Welding by Rolling," Brit. Weld. J., 6, 13 (1959).

36. W.C. Sherwood and D.R. Milner, "The Effect of Vacuum Machining on The Cold Welding of Some Metals," J. Inst. Metals, 97, 1 (1969).

37. N.A. Cantalejos and G. Cuminsky, "Morphology of the Interface of Roll-Bonded Aluminum," J. Inst. Metals, 100, 20 (1972).

38. J.A. Cave and J.D. Williams, "The Mechanism of Cold Pressure Welding by Rolling," J. Inst. Metals, 101, 203 (1973).

39. J.R. Osias and J.H. Tripp, "Mechanical Disruption of Surface Films on Metals," Wear, 9, 388 (1966).

40. D. Maugis, "Creep, Hot Hardness and Sintering in the Adhesion of Metals at High Temperature," Wear, 62, 349 (1980).

41. O.L. Anderson, "The Role of Surface Shear Strains in the Adhesion of Metals," Wear, 3, 253 (1960).

42. W.P. Gilbreath, "Definition and Evaluation of Parameters Which Influence the Adhesion of Metals," in *Adhesion or Cold Welding of Materials in Space Environments*, ASTM, STP 431 (1967).

43. D.H. Buckley, "Effect of Sulfur, Oxygen, and Hydrogen Sulfide Surface Films on the Adhesion of Clean Iron," Int. J. Nondestructive Testing, 2, 171 (1970).

44. D.H. Buckley, "Adhesion of Metals to a Clean Iron Surface Studied with LEED and Auger Emission Spectroscopy," Wear, 20, 89 (1972).

45. K.L. Johnson, "Adhesion at the Contact of Solids," in *Theoretical and Applied Mechanics*, W.T. Koiter, ed., North Holland, Amsterdam (1977).

46. D. Maugis, G. Desalos-Andarelli, A. Heurtel, and R. Courtel, "Adhesion and Friction on Al Thin Foils Related to Observed Dislocation Density," ASLE Trans., 21, 1 (1978).

47. H.M. Budgett, "The Adherence of Flat Surfaces," Proc. Roy. Soc., A 86, 25 (1911).

48. J.S. Mc Farlane and D. Tabor, "Adhesion of Solids and the Effect of Surface Films," Proc. Roy. Soc., A 202, 224 (1950).

49. W.J. O'Brien and J.J. Hermann, "The Strength of Liquid Bridges Between Dissimilar Materials," J. Adhesion, 5, 91 (1973).

50. S.J. Burns and B.R. Lawn, "A Simulated Crack Experiment Illustrating the Energy Balance Criterions," Int. J. Fracture Mech., 4, 339 (1968).

51. A.G. Skvortsov, E.A. Sinevitch, N.V. Pertsov, E.D. Shchukin, and P.A. Rebinder, "Determination of the Surface Energy of Naphthalene Crystals by the Cleavage Method," Sov. Phys. Doklady, 15, 669 (1971).

52. S.S. Wang, J.F. Mandell, and F.J. McGarry, "An Analysis of the Crack Tip Stress Field in DCB Adhesive Fracture Specimens," Int. J. Fracture, 14, 39 (1978).

An Analytical Method for Determining the Surface Energy of Solid Polymers

Michael C. H. Lee

Polymers Department
General Motors Research Laboratories
Warren, MI 48090-9055

ABSTRACT

A phenomenological theory for adhesion has been derived. This theory provides us with a new analytical method for determining the many important adhesion variables, such as the solid/vapor surface energy, the solid/liquid surface energy, the initial spreading coefficient and the equilibrium work of adhesion. The theory also provides significant correlations with many existing theories. Among them are the Zisman equation, the Wenzel equation, the Good-Girifalco equation, and the Good-Girifalco-Fowkes equation. Excellent experimental substantiation was obtained using data obtained at our laboratory and data from the literature for different solid polymers. Based on the foregoing theory, two new thermodynamic criteria for the spreading and for the cohesive failure of an adhesive system have also been established. Furthermore, the effects of temperature and roughness on the adhesion properties can now be determined.

INTRODUCTION

The knowledge of adhesion science is an important part for many other sciences and technologies. Among them are biology, medicine, engineering technologies, chemical processes, material science and electronic applications. In order to effectively apply the knowledge of adhesion science to the foregoing fields, a theory that combines the many important variables such as roughness and temperature into a unifying form is needed. In this paper, a thermodynamic theory for adhesion has been established with experimental substantiation. This theory also provides us with new methods and new criteria for determining and for evaluating the many important adhesion properties, such as solid/vapor surface energy (γ_{SV}), liquid/solid surface energy (γ_{LS}), equilibrium work of adhesion (W_A) and initial spreading coefficient (λ_{LS}), directly from equilibrium contact angle measurements.

Thermodynamic Equation of State for Adhesion

Consider putting a drop of liquid on a solid substrate as shown in Fig. 1. The input of this system is represented by adhesion tension defined as ($\gamma_{SV}-\gamma_{LS}$). At an equilibrium condition, the output of the system can be measured using the cosine of contact angle ($\cos\theta$). Using Young's equation, (1) adhesion tension equals the product of $\cos\theta$ and liquid/vapor surface energy (γ_{LV}). Since $\cos\theta$ is measurable, γ_{LV} and $\cos\theta$ can then be considered phenomenologically as the input and output of an equilibrium contact angle measurement. The value of $\cos\theta$ depends on the temperature of the experiment and the roughness of the substrate. A plot of the output ($\cos\theta$) versus the input (γ_{LV}) is shown in Fig. 2. This is known as the Zisman plot (2). The results obtained by Zisman (2) clearly show nonlinear relationships between $\cos\theta$ and γ_{LV}. This nonlinear behavior leads to the first hypothesis of the present theory.

(1) The thermodynamic input of the system, F, is defined in terms of the normalized adhesion tension as

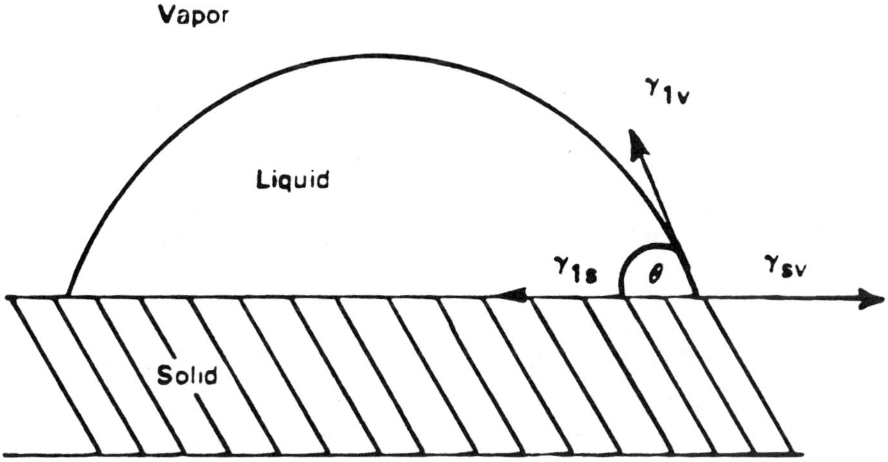

Fig. 1. Contact Angle Between a Drop of Liquid and a Flat Solid Surface

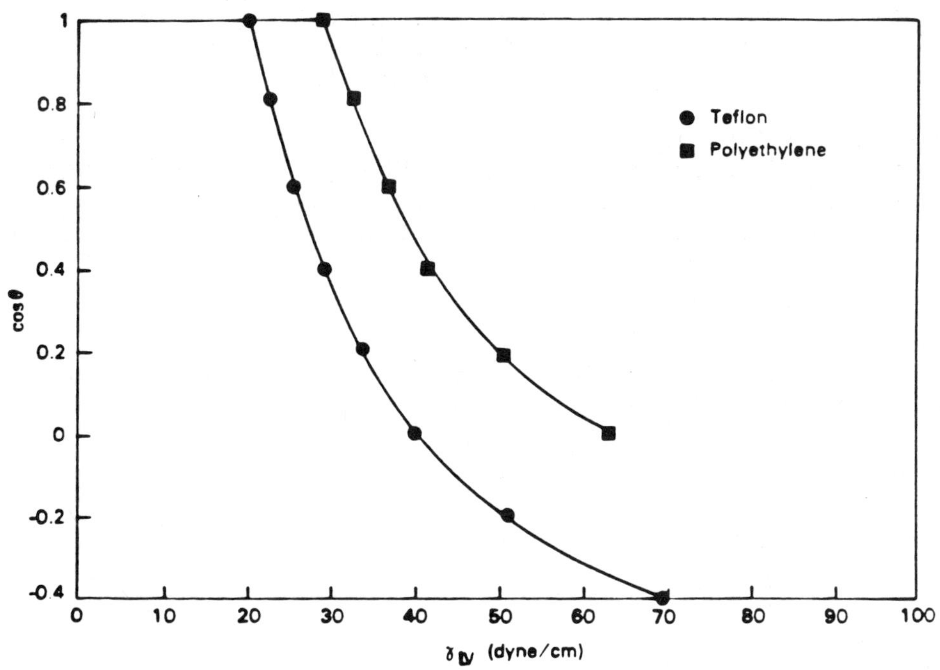

Fig. 2. Classical Zisman Plot of Cosθ versus γ_{LV} for Polyethylene and Teflon Wetted by Water/Glycerol Series (2) at 298K

$$F = K_B\ \ell n\left(\frac{\gamma_{SV}-\gamma_{LS}}{\gamma_S^*}\right) = K_B\ \ell n\left(\frac{\gamma_{LV}\cos\theta}{\gamma_S^*}\right) \tag{1}$$

where γ_S^* is the surface energy of solid in vacuum and K_B is the Boltzmann constant.

(2) The thermodynamic output of the system, R, is defined in terms of the cosine of contact angle as

$$R = K_B\ \ell n(\cos\theta) \tag{2}$$

(3) The structure roughness of the system S is defined as

$$S = K_B\ \ell n(\delta) \tag{3}$$

where δ is the surface roughness state, which can be shown to be proportional to Wenzel number (15). This point will be discussed later in the text.

(4) The variable, which accounts for the temperature effect of the system is defined as Q_c/T, where Q_c is a steady state activation enthalpy for adhesion and its value depends on the chemical characteristics of the liquid and the substrate and T is temperature.

The second hypothesis of this theory is that the output (RT) is a state function of F, S, and Q_c/T. This defines the characteristics of the equation of state. Forming the Paffian differential of R, we obtain

$$dR = \hat{\alpha}dF + \hat{\beta}d\left(\frac{Q_c}{T}\right) + \hat{\gamma}dS \tag{4}$$

For a partially wetting system, $\alpha \neq 0$. Eq. (4) can be inverted with respect to F as

$$dF = H_1 dS + H_2 dR + H_3 d\left(\frac{Q_c}{T}\right) \tag{5}$$

where $H_1 = -\hat{\gamma}/\hat{\alpha}$; $H_2 = 1/\hat{\alpha}$ and $H_3 = -\hat{\beta}/\hat{\alpha}$.

Let $Y_1 = S$, $Y_2 = R$, and $Y_3 = Q_c/T$, then we obtain

$$H_i = (\partial F/\partial Y_i)_{Y_{j\neq i}} = \hat{H}_i(Y_1, Y_2, Y_3);\ i=1,2,3 \tag{6}$$

Forming the Paffian differential of H_i, we obtain

$$dH_i = \Sigma \Delta_{ij} dY_j \qquad (7)$$

Since H_i is a state variable, this means $dH_i = 0$ is integrable. In other words, the Paffian differential equations, $\Sigma\Delta_{ij}dY_j = 0$, should be integrable. The necessary and sufficient condition (3) that the foregoing Paffian differential equations should be integrable is that

$$\underset{\sim}{Z} \bullet \underset{\sim}{\nabla} \times \underset{\sim}{Z} = 0 \qquad (8)$$

where $\underset{\sim}{Z} = \left[\Delta_{i1}, \Delta_{i2}, \Delta_{i3}\right]$ and $\underset{\sim}{\nabla} = \left[\partial/\partial Y_1, \partial/\partial Y_2, \partial/\partial Y_3\right]$. The simplest functional form for Δ_{ij} which satisfies the foregoing condition is

$$\Delta_{ij} = K_j \exp\left[\Sigma K_m Y_m + C_i\right] ; m=1,2,3 \qquad (9)$$

Combining Eq. (9) with Eq. (7) and integrating, we obtain the <u>equation of state of adhesion</u> shown below

$$\cos\theta = 1 + b\, \ell n\left(\gamma_C/\gamma_{LV}\right) \qquad (10)$$

where $b = \delta_{AP}^{-1} \exp\left(\frac{-Q_c}{K_B T}\right)$ = extent of interaction

and δ_{AP}^{-1} = apparent surface roughness (11)

$$\gamma_C = \gamma_S^* \exp(-C_4) \bullet \delta_{AP}^{-1} \bullet \left(\frac{1}{b}\right) \exp\left(1 - \frac{1}{b}\right) \qquad (12)$$

The detailed mathematical derivations of Eq. (10) are shown in the appendix. Eq. (10) implies that wetting a solid substrate with liquids will yield a linear relationship between the $\cos\theta$ and $\ell n\gamma_{LV}$. The intercept of this linear relationship at $\cos\theta=1$ defines the critical surface energy of wetting (γ_C). The slope of this linear relationship defines the extent of interaction (b).

We will now use Eq. (10) to establish the mathematical relationship to determine the liquid/solid surface energy (γ_{LS}) and the solid/vapor surface energy (γ_{SV}) from the contact angle measurements. The relationship between the equilibrium contact angle and the three energy terms (γ_{LS}, γ_{SV}, and γ_{LV}) is described by Young's equation (1).

$$\gamma_{LV}\cos\theta = \gamma_{SV} - \gamma_{LS} \tag{13}$$

Combining Eq. (10) and Eq. (13) yields the expression for γ_{LS} as

$$\gamma_{LS} = \gamma_{SV} - \left[\gamma_C \exp\left(\frac{1-\cos\theta}{b}\right)\right]\cos\theta \tag{14}$$

Equation (14) implies that at $\cos\theta = 1$, the following is true

$$\gamma_{LS} = \gamma_{SV} - \gamma_C \tag{15}$$

Taking the first order and the second order derivatives of γ_{LS} with respect to $\cos\theta$, we get

$$\left.\frac{d\gamma_{LS}}{d\cos\theta}\right|_{\gamma_{SV}} = -\left[\gamma_C \exp\left(\frac{1-\cos\theta}{b}\right)\right]\left[1 - \frac{\cos\theta}{b}\right] \tag{16}$$

and

$$\left.\frac{d^2\gamma_{LS}}{d\cos\theta^2}\right|_{\gamma_{SV}} = \gamma_C\left(\frac{2}{b} - \frac{\cos\theta}{b^2}\right)\exp\left(\frac{1-\cos\theta}{b}\right) \tag{17}$$

Setting $\left(\frac{d\gamma_{LS}}{d\cos\theta}\right)$ equal to zero, we obtain

$$\cos\theta = b \tag{18}$$

and

$$\frac{d^2\gamma_{LS}}{d\cos\theta^2} = \gamma_C\left(\frac{1}{b}\right)\exp\left(\frac{1}{b} - 1\right) > 0$$

In other words, γ_{LS} will have a minimum value, γ_{LS}^{min}, when $\cos\theta$ equals b. γ_{LS}^{min} has the following expression

$$\gamma_{LS}^{min} = \gamma_{SV} - \left[\gamma_C \exp\left(\frac{1}{b} - 1\right)\right]b \tag{19}$$

Good and Elbling (17) suggested that the minimum value of liquid/solid interface energy will not in general be zero regardless of the existence of zero contact angle. The above conclusion was obtained based on the assumption that the Good bonding coefficient, Φ_{12}, in Good's equation (17) is not a function of the surface energies of solid and liquid. It has also been shown by Rhee (14) that γ_{LS}^{min} equal to zero is a proper assumption for both polymer and ceramic systems. Assuming that γ_{LS}^{min} equals zero, then Eq. (19) yields

$$\gamma_{SV} = \gamma_C \bullet b \bullet \exp(\frac{1}{b} - 1) \tag{20}$$

Eq. (20) can be used to calculate the solid/vapor surface energy (γ_{SV}) from contact angle measurements. Substituting Eq. (20) into Eq. (10), we obtain

$$\cos\theta = b[1 - \ell nb + \ell n(\frac{\gamma_{SV}}{\gamma_{LV}})] \tag{21}$$

Combining Eq. (21) with Eq. (14), we obtain the relationship for γ_{LS} in terms of γ_{SV} and γ_{LV} as

$$\gamma_{LS} = \gamma_{SV} + \gamma_{LV} - \gamma_{LV}\left[1 + b\exp(1 - \frac{1}{b}) + b\bullet\exp(1 - \frac{1}{b})\bullet\ell n(\frac{\gamma_{SV}}{b\gamma_{LV}})\right] \tag{22}$$

Rearranging Eq. (20), we obtain

$$\gamma_C = \Phi^2\gamma_{SV} \tag{23}$$

where

$$\Phi^2 = \frac{1}{b}\exp(1 - \frac{1}{b}) \tag{23a}$$

Eq. (23) has been suggested by previous workers (10,11). The parameter Φ is the Good's bonding coefficient (17). Comparing Eq. (23) and Eq. (12), we obtain

$$\gamma_{SV} = \exp(-C_4) \bullet \delta_{AP}^{-1} \bullet \gamma_S^* \tag{24}$$

Eq. (24) implies that the solid/vapor surface energy, γ_{SV}, determined from contact angle measurements is proportional to the roughness state of the substrate and the surface energy of solid in vacuum.

A summary of derived relationships for the present theory is shown in Table 1. When the extent of interaction (b) equals one, the Good's bonding coefficient (Φ) also equals one. Under the conditions that $(\gamma_{SV}/\gamma_{LV})^{1/2} \simeq 1$, which means the power-series expansion $\ell n(\gamma_{SV}/\gamma_{LV})^{1/2} \simeq$

Table 1.

Summary of Derived Relationships from the Present Theory

- **Equation of State**

$$\cos\theta = 1 + b\,\ell n(\gamma_C/\gamma_{LV})$$

- **Extent of Interaction**

$$b = \delta_{AP}^{-1}\exp[-(Q_C)/(K_BT)]$$

- **Critical Surface Energy of Wetting**

$$\gamma_C = [(1/b)\exp(1-1/b)]\gamma_{SV}$$

$$\gamma_C = \Phi^2\,\gamma_{SV} \qquad \text{Good's Equation (1964)}$$

- **Good's Bonding Coefficient**

$$\Phi = [(1/b)\exp(1-1/b)]^{1/2}$$

- **Polarity**

$$[(1-P_S)(1-P_L)]^{1/2} + (P_SP_L)^{1/2} = [(1/b)\exp(1-1/b)]^{1/2}$$

Table 1. (Continued)

- **Solid/Vapor Surface Energy**

$$\gamma_{SV} = \gamma_C[b\ \exp(1/b-1)] = \gamma_S^* \cdot \exp(-C_4) \cdot \delta_{AP}^{-1}$$

- **Liquid/Solid Surface Energy**

$$\gamma_{LS} = \gamma_{SV} + \gamma_{LV} - \gamma_{LV}\Big[1 + b\ \exp(1-1/b) + b\ \exp(1-1/b) \cdot \ln\Big(\frac{\gamma_{SV}}{b\ \gamma_{LV}}\Big)\Big]$$

$$\gamma_{LS}^{\theta=0} = \gamma_{SV}-\gamma_{LV}$$

- **Equilibrium Work of Adhesion**

$$W_A = \gamma_{LV}[2 + b\ \ln(\gamma_C/\gamma_{LV})]$$

$$\text{or } W_A = \gamma_{LV}\Big[b\ \ln\Big(\frac{\gamma_{SV}}{\gamma_{LV}}\Big) + b + 1 - b\ \ln b\Big]$$

- **Harkin's Spreading Coefficient**

$$\lambda_{LS} = \gamma_{LV}\ \Big\{b\ [\exp(1-1/b)]\Big[1 + \ln\Big(\frac{\gamma_{SV}}{b\gamma_{LV}}\Big)\Big] - 1\Big\}$$

- **Activation Enthalpy Difference**

$$\Delta Q = Q_{C2} - Q_{C1} = K_B T\ \ln(b_1/b_2)$$

$(\gamma_{SV}/\gamma_{LV})^{1/2}-1$ is valid, and b=1, Eq. (21) can be reduced to the familiar Good-Girifalco-Fowkes (GGF) equation (4,5) shown below

$$\cos\theta = -1+2\,(\gamma_{SV}\gamma_{LV})^{1/2}/\gamma_{LV} \tag{25}$$

Similarly, γ_{LS} shown in Eq. (22) can be reduced to the GGF equation (4,5) as

$$\gamma_{LS} = \gamma_{SV} + \gamma_{LV} - 2(\gamma_{LV}\gamma_{SV})^{1/2} \tag{26}$$

Analyses of Equilibrium Contact Angle Measurements

As has been discussed, the new equation of state implies that a plot of $\cos\theta$ versus $\ell n\gamma_{LV}$ will exhibit a linear relationship as shown in Fig. 3. The intercept at $\cos\theta=1$ defines the critical surface energy (γ_C), and the value of the slope of the line determines the extent of interaction (b). Eq. (10) has been verified using data from our laboratories and data in the literature (2,6-8). The first example shown in Fig. 4 is a plot of $\cos\theta$ versus $\ell n\gamma_{LV}$ of Zisman's data (2), which has been shown in Fig. 2. The second example, shown in Fig. 5, is a plot of $\cos\theta$ versus $\ell n\gamma_{LV}$ of Wu's data (6) measured for Teflon wetted by different liquids. Both examples clearly show linear or piecewise linear relationships between $\cos\theta$ and $\ell n\gamma_{LV}$. It is important to note that Eq. (10) cannot predict a piecewise linear behavior in the plots between $\cos\theta$ and $\ell n\gamma_{LV}$. The piecewise linear behavior may be attributed to different mechanisms occurring in the adhesion system. To explain the piecewise linear behavior, molecular theories are required. Nevertheless, the linear regime of the plots between $\cos\theta$ and $\ell n\,\gamma_{LV}$ close to $\cos\theta=1$ can still be utilized for determining the adhesion properties of the solid.

Table 2 shows calculated adhesion variables bases on this new theory using data obtained by Dann (7) for many polymer substrates wetted by formamide/2-ethoxyethanol liquids. The adhesion variables obtained are γ_{LV}, γ_{LS} and Φ. Table 3 shows the effects of wetting liquid on adhesion variables γ_C, γ_{SV} and Φ of a polystyrene substrate. In

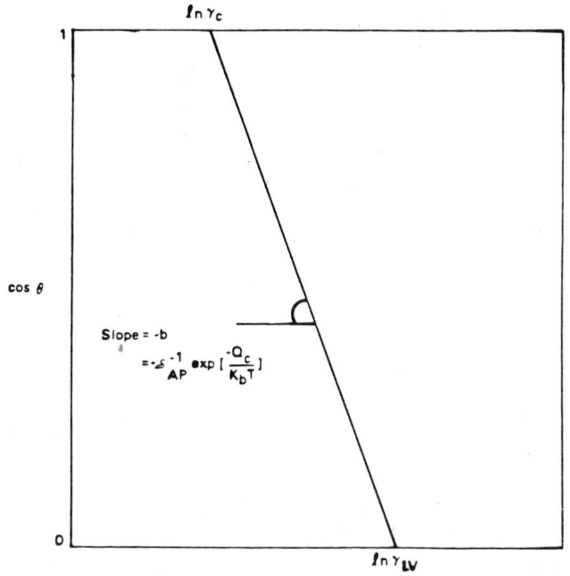

Fig. 3. Schematic Plot of $\ell n\, \gamma_{LV}$ versus $\cos\theta$ at Fixed Temperature and Fixed Roughness (δ_{AP}^{-1}).

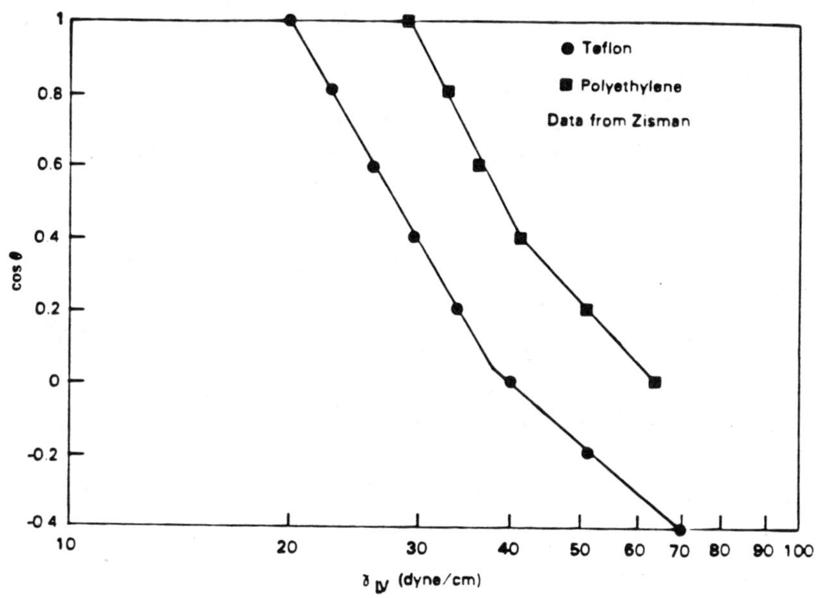

Fig. 4. Plot of versus $\ell n\, \gamma_{LV}$ for the Data (2) Shown in Fig. 2.

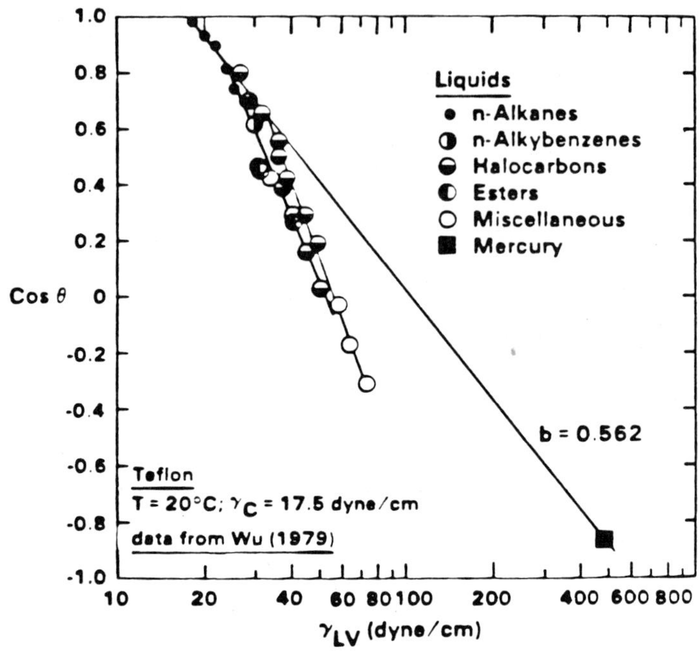

Fig. 5. Plot of cos θ versus $\ln\gamma_{LV}$ for the Data Measured by Wu (5).

Table 3, $\bar{\gamma}_C$ is the critical surface energy of PS measured using the classical Zisman plot. The difference between $\bar{\gamma}_C$ and γ_C is less than 12%.

The use of both γ_C and b to calculate other important adhesion properties is fundamentally significant. Thermodynamically, γ_C and b is a set of Legendre transformation (18) of the equation of state Eq. (10). Therefore, using the value of γ_C and b is equivalent to using the complete contact angle measurements.

Effects of Temperature and Roughness on Adhesion Properties

The effects of temperature and roughness on adhesion are described in Eq. (11). Schematically, Eq. (11) implies that a plot of $\ln b$ vs. 1/T for a fixed roughness state of the substrate will result in a linear relationship. The

Table 2

Calculated Adhesion Properties using the Present Theory. Data Obtained from Dann (7) for Formamide/ 2-Ethoxyethanol Liquids

Polymers*	γ_C	b	γ_{SV}	$\gamma_{LS}^{\theta=0}$	Φ
PE	28.20	1.36	29.45	1.25	0.98
PS	31.50	1.67	35.26	3.76	0.95
PMMA	33.00	0.68	36.08	3.08	0.96
PVC	29.15	0.48	41.42	12.27	0.84
PCDMT	32.25	0.80	33.12	0.87	0.99
N11	31.00	0.95	31.04	0.04	1.00
PTFE	25.20	2.20	32.33	7.13	0.88
Paraffin	25.45	1.77	29.13	3.68	0.94
PET	42.10	2.03	51.50	9.40	0.90

*PE: Polyethelyne
PS: Polystyrene
PMMA: Poly(methyl methacrylate)
PVC: Poly(vinyl chloride)
PCDMT: Poly(1,4-cyclohexanedimethylene terephthalate)
N11: Nylon 11
PTFE: Teflon
PET: Poly(ethylene terephthalate)

Table 3

Comparisons of Adhesion Properties of PS Determined using Different Liquids at T = 298°K

Liquids	γ_C	$\bar{\gamma}_C$	γ_{SV}	Φ	Reference
Formamide/ 2-ethoxyethanol	31.50	30.0	35.26	0.95	Dann (7)
Polyglycol	35.50	31.5	35.82	1.00	Dann (7)
Mixed Glycerol	32.35	30.5	32.97	0.99	Dann (7)
Ethanol/Water	27.00	27.0	29.18	0.96	Dann (7)

intercept at 1/T = 0 defines the apparent roughness state ($\ell n\ \delta_{AP}^{-1}$) and the slope equals ($-Q_C/K_B$). It can be shown mathematically that the apparent roughness state (δ_{AP}^{-1}) shown in the equation of state is proportional to Wenzel's number, (15) which is defined as the total surface area divided by the apparent planar area. An example is shown in Table 4 using the data measured by Neumann, Renzow and Haage (8) for n-alkanes on Teflon at various temperatures. Adhesion variables such as γ_{SV}, Φ, P_S, γ_{SV}^P and $\gamma_{LS}^{\theta=0}$ can be determined for different temperatures using the new theory. Based on the data in Table 4, a plot of ℓnb versus 1/T is shown in Fig. 6. The results indicate a linear relationship. The apparent roughness δ_{AP}^{-1} is 123 and the activation enthalpy for adhesion (Q_C) is 3.22 kcal/mole.

Criteria for Spreading and Cohesive Failure of an Adhesion System

The Harkins initial spreading coefficient (9) is defined as

$$\lambda_{LS} = \gamma_{SV} - \gamma_{LV} - \gamma_{LS} > 0; \text{ for spreading} \tag{27}$$

The criterion for cohesive failure is defined as

$$W_A/\gamma_{LV} > 2 \tag{28}$$

Combining the foregoing definitions with the new theory, we obtain the following new criteria for cohesive failure and spreading of an adhesion system.

$$(1)\quad \frac{(\gamma_{SV}/\gamma_{LV})}{\exp[\frac{1}{b} - 1 + \ell nb]} > 1\ ;\quad \text{for cohesive failure} \tag{29}$$

$$(2)\quad \frac{(\gamma_{SV}/\gamma_{LV})}{b\ \exp[\frac{1}{b}\ \exp(\frac{1}{b} - 1) - 1]} > 1\ ;\quad \text{for spreading} \tag{30}$$

In Eqs. (29) and (30), b is the extent of interaction and can be determined from the slope of the line in the plot of $\cos\theta$ against $\ell n\gamma_{LV}$. The above new criteria are shown in Fig.

Table 4

Calculated Adhesion Properties Using Equations Shown in Table 1. Data Obtained from Neumann, Haage and Renzoe (8) for n-Alkanes on Teflon at Various Temperatures

Temperature	γ_C	b	γ_{SV}	$\gamma_{LS}^{\theta=0}$	Φ	P_S	γ_{SV}^P
303°K	15.92	0.61	18.45	2.53	0.93	0.14	2.52
313°K	15.95	0.67	17.55	1.60	0.95	0.09	1.60
323°K	16.71	0.81	17.12	0.41	0.99	0.02	0.41
333°K	16.80	0.98	16.80	0.00	1.00	0.00	0.00
343°K	16.43	1.07	16.46	0.00	1.00	0.00	0.00

7 by plotting γ_{SV}/γ_{LV} versus b. Several important findings can be summarized as follows: (1) This theory predicts a non-spreading/adhesive failure region for $\gamma_{SV}/\gamma_{LV} \leq 0.88$ instead of $\gamma_{SV}/\gamma_{LV} \leq 1$ predicted from the GGF theory (5). (2) For $\gamma_{SV}/\gamma_{LV} > 0.88$, there exist four regions of different characteristics. They are nonspreading/adhesive failure zone, non-spreading/cohesive failure zone, spreading/ cohesive failure zone and spreading/adhesive failure zone. (3) Increasing temperature, decreasing activation enthalpy for adhesion and increasing roughness of the substrate can change the characteristics of an adhesion system. (4) By changing the value of γ_{SV}/γ_{LV} through processing conditions, the adhesion characteristics can also be altered.

Determination of Adhesion Properties for Polymer-Polymer and Polymer-Filler Systems

Some applications require adhesion between two polymers. Typical examples are systems of polymer blends and structural adhesives. Based on Eqs. (10) and (23), we obtain the equation of state for polymer i in contact with polymer j as

$$\cos\theta_{ij} = 1 + b_{ij}\,\ln\left(\frac{\Phi_{ij}^2\gamma_j}{\gamma_i}\right) \tag{31}$$

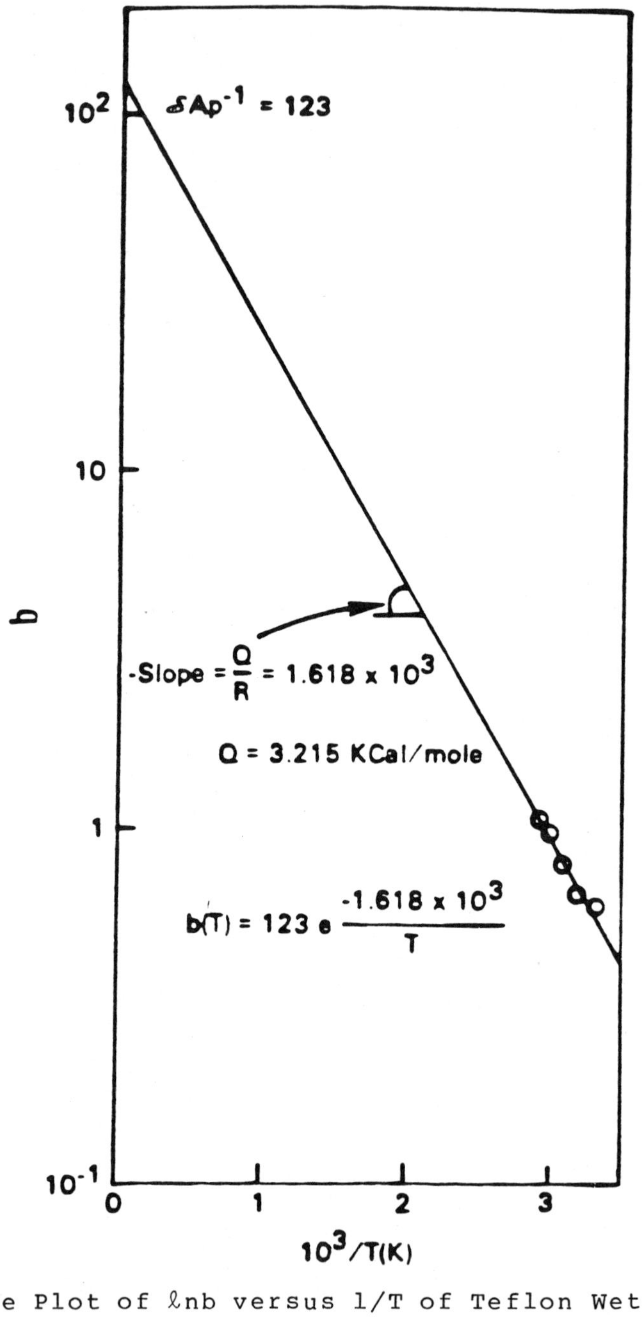

Fig. 6. The Plot of $\ell n b$ versus 1/T of Teflon Wetted by n-Alkanes

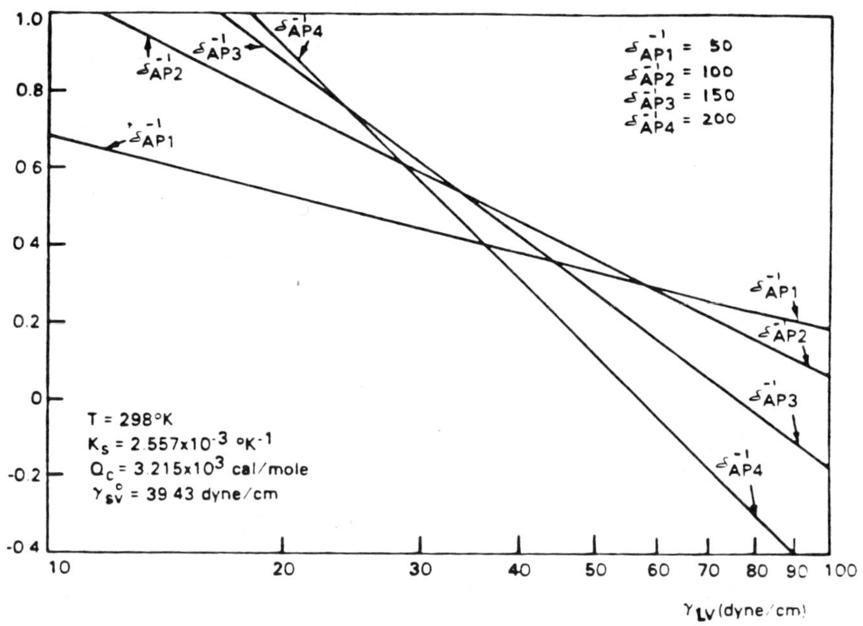

Fig. 7. Prediction of the Different Adhesion Characteristics in the Plot of $(\gamma_{SV}/\gamma_{LV})$ versus the extent of interaction (b).

and the equation of state for polymer j in contact with polymer i as

$$\cos\theta_{ji} = 1 + b_{ji}\,\ln\left(\frac{\Phi_{ji}^2\gamma_i}{\gamma_j}\right) \qquad (32)$$

Since $b_{ij} = b_{ji} \equiv b$ and $\Phi_{ij} = \Phi_{ji} \equiv \Phi$, we obtain the following relationship

$$\cos\theta_{ij} - \cos\theta_{ji} = 2b\,\ln\left(\frac{\gamma_j}{\gamma_i}\right) \qquad (33)$$

Eq. (33) can be used to calculate the value of the extent of interaction (b) for systems in which contact angle measurements using solvents are too complex. The experimental results of several polymer pairs by Wu (12) were used to calculate the values of the extent of interaction (b), the bonding efficiency factor (Φ), and the critical surface energy of wetting (γ_c). The results are shown in Table 5. It is important to note that the values of γ_c for PE, PMMA and PS are in agreement with the values measured by Dann (7) (see Table 2). Using the adhesion properties shown in Table 5 and the relationships listed in Table 1, we can determine the work of adhesion and spreading characteristics of polymer-polymer systems.

The equation of state, Eq. (10), can also be employed for determining the surface energy of fillers. Examples for the fillers investigated are an N330 carbon black and an attrited N330 carbon black. The attrition process of the carbon black used consists of two parts. We first heat the carbon black in a vacuum oven at 150°C for 24 hours. We then ball-milled the carbon black at 125°-225°C for 48 hours. The amount of carbon black put into the ball mill container was about one third of the volume of the container.

Contact angle measurements were conducted at room temperature using a filler column method in which the steady flow of wetting liquids induced by capillary pressures was measured for a vertical filler bed (16). The wetting liquids used in this study were decane, benzyl alcohol, aniline, formamide and water. Using Washburn's equation (13), the analyses of the flow measurements lead to the data for the advancing contact angle of different wetting liquids on the carbon blacks. The plots of cos Θ vs. γ_{LV} for N330 and the attrited N330 carbon blacks are shown in Fig. 8. Again, linear relationships are obtained. The critical surface energy of wetting (γ_c) and the extent of interaction (b) were then determined for each carbon black. The solid/vapor surface energy of each black (γ_{SV}) and other related adhesion variables were then calculated using the equations shown in Table 1. Some important properties of the carbon blacks are summarized in Table 6.

We will now use the adhesion properties for polychloroprene (12) shown in Table 5, namely, γ_{SV} = 33.2 dyne/cm and b = 0.72, to determine the possible characteristics of polymer-filler interactions of carbon black filled polychloroprene. The geometrical mean for calculating the extent of interaction is used in this case. The adhesion properties of N330 and attrited N330 filled

Table 5

Adhesion Properties of Polymer-Polymer Systems Determined Using Contact Angle Measurements Obtained by Wu (12)

Polymer* Pair 1/2	γ_1/γ_2	$\cos\theta_{12}-\cos\theta_{21}$	$2\ln(\frac{\gamma_2}{\gamma_1})$	b	Φ	γ_{c1}	γ_{c2}
CR/B-PE	33.2/27.3	-0.28	-0.39	0.72	0.97	31.20	25.65
PMMA/L-PE	32.0/28.8	-0.18	-0.21	0.83	0.99	31.43	28.28
PnBMA/L-PE	24.1/28.8	0.35	0.36	0.97	1.00	24.09	28.78
PS/L-PE	32.1/28.8	-0.21	-0.22	0.95	1.00	32.06	28.77
PVA/L-PE	28.6/28.8	0.01	0.01	0.99	1.00	28.60	28.80
CR/PnBMA	33.2/24.1	-0.32	-0.64	0.50	0.85	24.25	17.60
CR/PDMS	33.2/14.1	-0.76	-1.71	0.44	0.80	21.29	9.04
CR/PS	33.2/32.1	-0.05	-0.07	0.73	0.97	31.37	30.33
PMMA/PnBMA	32.0/24.1	-0.28	-0.57	0.50	0.85	23.31	17.56
PMMA/PS	32.0/32.1	-0.05	0.01	0.88	1.00	31.74	31.84
PVA/PnBMA	28.6/24.1	-0.26	-0.34	0.75	0.98	27.33	23.03

* CR: Polychloroprene
B-PE: Branched polyethylene
PMMA: Poly(methyl methacrylate)
L-PE: Linear polyethylene
PnBMA: Poly(n-butyl methacrylate)
PS: Polystyrene
PVA: Poly(vinyl acetate)
PDMS: Polydimethylsiloxane

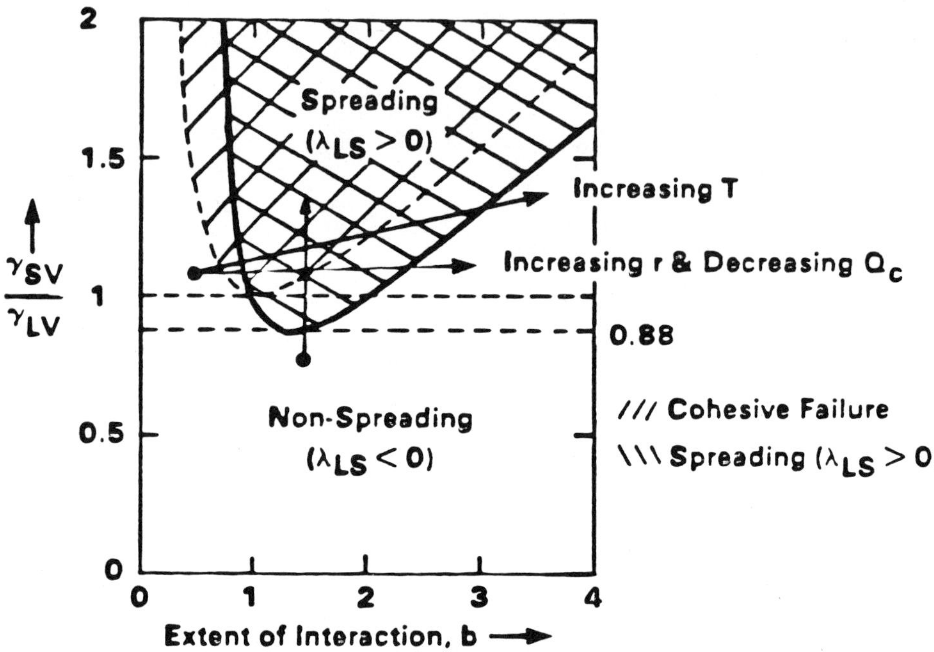

Fig. 8. Cos θ versus γ_{LV} for the Attrited and Non-attrited Carbon Blacks

Table 6

Changes in the Properties of Carbon Black After Attrition Determined at T = 20°C

Properties	Before Attrition	After Attrition
Particle size (Å)	240	150
pH of slurry	8.1	3.4
Critical surface energy of wetting (γ_c)($\frac{dyne}{cm}$)	40.0	32.6
Extent of interaction (b)	1.509	1.368
Solid/vapor surface energy (γ_{SV})($\frac{dyne}{cm}$)	43.1	34.1

polychloroprenes are summarized in Table 7. The adhesion characteristics for both systems are plotted in (γ_{SV}/γ_{LV}) versus b and are shown in Fig. 9. The results show that the N330 filled polychloroprene is inside the spreading/cohesive failure zone, while the attrited N330 filled polychloroprene is on the boundary of the spreading/cohesive failure zone. Based on the foregoing findings, the results suggest that the polymer-filler interaction for the N330 filled polychloroprene is stronger than that for the attrited N330 filled polychloroprene. Consequently, the spreading characteristics for polychloroprene on both N330 and attrited N330 carbon blacks are also quite different. The values of Harkin's spreading coefficient indicate that spreading of the polymer chain on the surface of N330 carbon black is more complete than that in the case of attrited N330 carbon black.

The foregoing method can also be employed to determine the adhesion properties of other filled polymer systems. The procedures involved in this analytical method are summarized as follows: (1) Use proper experiments to obtain the contact angle measurements for both the filler and the polymer, (2) Plot the contact angle measurements as $\cos\theta$ versus $\ln\gamma_{LV}$, and determine the critical surface energy of wetting (γ_c) at $\cos\theta = 1$ and the extent of interactions (b) from the magnitude of the slope of the line, (3) Use equations shown in Table 1 to obtain the adhesion properties, such as γ_{SV}, γ_{LS} and Φ, for both the polymer and the filler, (4) Use these adhesion properties to determine the adhesion characteristics of a given polymer-filler system by plotting γ_{SV}/γ_{LV}, which is equivalent to $\gamma_{filler}/\gamma_{polymer}$, versus the extent of interaction (b), which is the geometric mean value of b_{filler} and $b_{polymer}$ (see Figs. 7 and 9).

Some Remarks on the Analytical Method

In this work we have established a new analytical method for determining the many important adhesion properties of solid polymer systems. This method contains many useful relationships shown in Table 1. Examples have also been given to illustrate the usage of this method in different experiments. To further verify the validity of this method, different adhesion systems and different adhesion

Table 7

Adhesion Properties of N330 and Attrited N330 Filled Polychloroprene (CR)

Properties	CR-N330	CR-A330*
b	1.04	0.99
γ_{SV}/γ_{LV}	1.30	1.03
$\gamma_{LS}(\frac{dyne}{cm})$	0.272	0.274
$W_A(\frac{dyne}{cm})$	76.01	67.03
$\lambda_{LS}(\frac{dyne}{cm})$	9.63	0.63
W_A/γ_{LV}	2.29	2.01

* CR-A330 represents the attrited N330 filled polychloroprene.

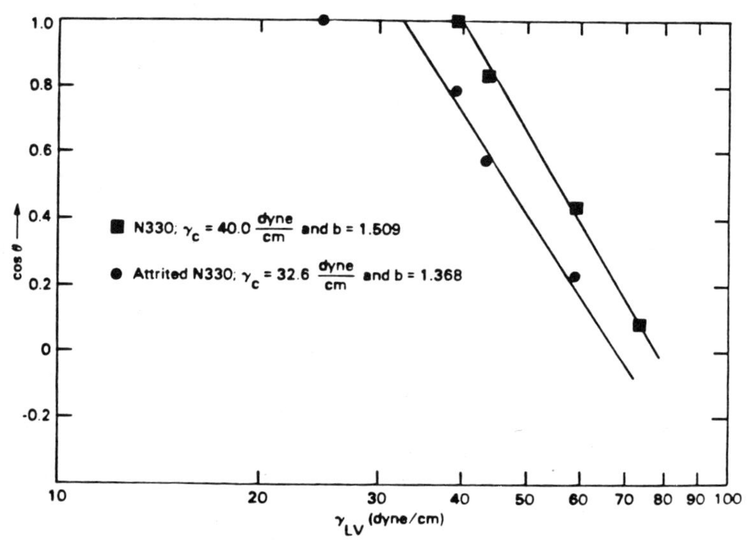

Fig. 9. The plots of $\frac{\gamma_{SV}}{\gamma_{LV}}$ versus the extent of interaction (b) for both N330 filled and attrited N330 filled polychloroprenes at 20°C.

conditions should be considered for designing future experiments. Many new phenomena observed in this work also need further investigations. For example, the contact angle data plotted based on the equation of state clearly show the knee points as shown in Figs. 4 and 5. A molecular explanation of this phenomenon should be investigated. Dynamic (advancing and receding) contact angle measurements for filler systems will also be an important area for further investigation.

CONCLUSIONS

A new analytical method for determining the surface energy of solid polymers has been developed based on a thermodynamic theory for adhesion. This theory, which has been substantiated experimentally, contains an equation of state describing the relationship between the cosine of contact angle ($\cos\theta$) and the liquid/vapor surface energy (γ_{LV}). Using this theory, new criteria for spreading and cohesive failure of an adhesion system are also established in terms of the (γ_{SV}/γ_{LV}) and the extent of interaction (b). These criteria can now be used to determine the adhesion characteristics of filled polymers and polymer blends.

Based on the results shown in this paper, we now can determine the many different adhesion properties of solids from contact angle measurements.

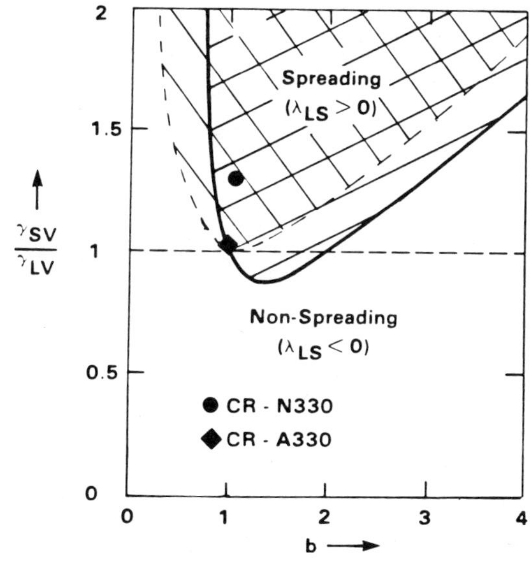

Figure 10

ACKNOWLEDGMENT

The author wishes to acknowledge Drs. Phillip Weiss, G. Dale Cheever and Thomas J. Dearlove for their helpful comments. The author is also indebted to Prof. Robert J. Good and Dr. Lieng-Huang Lee for their valuable comments.

REFERENCES

1. T. Young, Phil. Trans. Roy.Soc. (London), 95, 65 (1805).
2. W.A. Zisman, Adhesion and Cohesion, P. Weiss, Ed., Elsevier, Amsterdam, p. 176 (1962).
3. I.N. Sneddon, Elements of Partial Differential Equations, McGraw-Hill, New York, p. 28 (1957).
4. L.A. Girifalco and R.J. Good, J. Phys. Chem., 61, 904 (1957).
5. F.M. Fowkes, J. Phys. Chem., 66, 382 (1962).
6. S. Wu, J. Adhesion, 71, 605 (1979).
7. J.R. Dann, J. Colloid & Interface Sci., 32, 2, 302 (1970)
8. A.W. Neumann, G. Haage and D. Renzow, J. Colloid & Interface Sci., 35, 379 (1971).
9. W.D. Harkins, The Physical Chemistry of Surface Films, Reinhold, New York (1952).
10. D.H. Kaelble, Physical Chemistry of Adhesion, Wiley-Interscience, New York, NY, pp. 137-189 (1971).
11. R.J. Good, Adv. Chem, Ser., 43, 74 (1964).
12. S. Wu, J. Adhesion, 5, 39 (1973).
13. E.W. Washburn, Phys. Rev., 17, 273 (1921).
14. S.K. Rhee, J. Material Sci. and Eng., 11, 311 (1973).
15. R.N. Wenzel, Ind. Eng. Chem., 28, 988 (1936).
16. V.T. Crowl and W.D.S. Wooldridge, in Wetting, S.C.I. Monograph No. 25, Gordon and Breach, Science Publishers, Inc., New York, p. 200 (1967).
17. R.J. Good and E. Elbling, Ind. Eng. Chem., 62, 54 (1970).
18. H.B. Callen, "Thermodynamics", John Wiley, New York, NY, p. 90, (1960).

APPENDIX: Derivation of the Equation of State for Adhesion

As indicated in the text, the simplest functional form for Δ_{ij}, which satisfies the condition shown in Eq. (8) is

$$\Delta_{ij} = K_j \exp[\sum K_m Y_m + C_i']; \; m = 1,2,3 \tag{9}$$

where $C_i' = \ell n[CK_i]$

Combining Eq. (7) with Eq. (9) and integrating the resulting form, we obtain

$$H_i = \exp[\sum_{m=1}^{3} K_m Y_m + \ell n(CK_i)] + C_i \tag{A-1}$$

The explicit form for the criterion of exact differential, shown in Eq. (8) is

$$[\partial H_i/\partial Y_m] = [\partial H_m/\partial Y_i] \tag{A-2}$$

Differentiating Eq. (A-1) with the criterion shown in Eq. (A-2), we arrive at

$$dH_i = K_i[dF - \sum_{m=2}^{3} C_m dY_m] \tag{A-3}$$

or

$$H_i = K_i \; [F - \sum_{m=2}^{3} C_m Y_m + \bar{C}_4] \tag{A-4}$$

Since H_1 is the change of F with respect to the change of S at fixed R and Q_c/T, this means H_i must be a non-zero function. Consequently, the constant C_1 in Eq. (A-1) is zero. Using this condition and rearranging Eq. (A-1) leads to the following:

$$H_i = (K_i/K_1)H_1 + C_1 \tag{A-5}$$

Combining Eq. (A-5) with Eq. (A-4), we obtain

$$H_1 = K_1[F - \sum_{m=2}^{3} C_m Y_m] + \bar{C}_4 - (K_1/K_i)C_i \quad \text{(A-6)}$$

Equation (A-6) implies that

$$K_1 = KC_i \quad ; \quad i = 2,3 \quad \text{(A-7)}$$

Combining Eq. (A-4) with Eqs. (A-1) and (A-7), we finally arrive at

$$K_1 = K \quad \text{(A-8)}$$

and

$$K(F - C_2R - C_3Q_c/T + \bar{C}_4) = 1 + KC\exp[K(S + C_2R + C_3Q_c/T)] \quad \text{(A-9)}$$

or in the explicit form

$$K[\ell n(\gamma_{LV}\cos\theta/\gamma_S^*) - C_2\ell n(\cos\theta) - C_3Q_c/T + \bar{C}_4]$$
$$= 1 - (S_{AP})^K(\cos\theta)^{KC_2}\exp[KC_3Q_c/(K_BT)] \quad \text{(A-10)}$$

where $$S_{AP} = [(-KC)^{1/K} S] \quad \text{(A-11)}$$

Assume that $C_2 = C_3 = K = 1$ as a special case for Eq. (A-10), then Eq. (A-10) can be reduced to the equation of state for adhesion shown in Eq. (10) in the text.

Effect of Amine Surface Treatment on the Adhesion of a Polyurethane to the Surface

Fasi Liang and Patricia Dreyfuss

Institute of Polymer Science
The University of Akron
Akron, Ohio 44325

ABSTRACT

Amines, in general, and 3-aminopropyltriethoxysilane, in particular, were shown to improve the adhesion of a polyurethane to glass. At equivalent concentrations of amine in the solutions used to treat the glass the order of increase in the work of adhesion was 3-aminopropyltriethoxysilane >> 1,4-diaminobutane $\simeq$ p-phenylenediamine > piperazine >> aniline. The polyurethane was prepared from ARCO's hydroxy-terminated polybutadiene, R-45HT, toluene diisocyanate, trimethylolpropane and N,N-bis(2-hydroxypropyl)aniline. The ratio ([-NCO]/[total OH]) was kept constant and equal to 1.0. The swelling ratio, after curing, also remained constant. As the -NCO content compared to polymer-OH increased, the adhesion of the polyurethane to glass increased initially, passed through a maximum for prepolymers with 6% excess -NCO, and then decreased again. Elongation at break behaved similarly but ultimate tensile strength reached a maximum and then stayed constant. The significance of these results is discussed. Adhesion of the polyurethane to chrome-plated steel did not show similar effects.

INTRODUCTION

This paper is a continuation of our investigation of the role amines play in adhesion. The overall investigation originated in repeated observations by ourselves and others that the presence of amine groups at an interface between a substrate and an adherend has a positive effect on the adhesion between the substrate and the adherend (1-4). In filled systems, mechanical properties, compared to filled systems where the amines are absent, are enhanced also. Previous work by Eckstein and Dreyfuss (4-8) on peroxide-cured bulk polybutadiene systems revealed both that the improvement in properties can be explained by the formation of chemical bonds through the amine group between the surface and the polymer and that the degree of improvement is related to the structure of the amine used.

In some applications of interest to us, notably solid fuel rocket propellants, polybutadiene is used as the soft segment of a polyurethane rather than as bulk polymer. Hence, our studies of the role of amines in adhesion have been extended to include systems with polyurethanes derived from functionalized liquid polybutadiene.

In polyurethane synthesis, amines are used both as chain extenders and as curing agents. The chemical reactions that occur and the products that form are well known (9-11). The questions that this study addressed were two. How does the structure of the amine used affect the adhesive properties of the resulting polyurethane and how can the observed effects be explained? This paper describes the synthesis and characterization of a new elastomeric polyurethane that shows remarkably high adhesion to glass and other substrates. A positive effect of amine surface treatment of the substrates on the adhesion of the polyurethane to the substrates is also reported. The results are interpreted in terms of the structure of the amines and of the polyurethane formed.

EXPERIMENTAL

Materials

The materials used were: 1,4-diaminobutane, anhydrous piperazine *p*-phenylenediamine, and 2-ethyl-2-(hydroxymethyl)-1,3-propanediol from Aldrich Chemical Company,

Inc.; aniline ("Baker Analyzed" Reagent) from J.T. Baker Chemical Co.; 3-aminopropyltriethoxysilane from Petrarch Systems Inc.; TDI (80/20 mixture of 2,4 and 2,6 isomers of toluene diisocyanates) from BASF Wyandotte Corporation; Isonol-100 (N,N-bis-(2-hydroxypropyl)aniline) from Upjohn Polymer Chemicals; and hydroxy-terminated polybutadiene resin (R-45HT) from ARCO Chemical Co. ARCO lists the properties given in Table 1 for the latter resin.

The amines were the same as those used in earlier studies (4-8). Their properties are given in Table 2.

Table 1. Properties of ARCO's R-45HT

Nonvolatile material	wt%	99.9
Viscosity	poise, 30°C	50
Hydroxyl value	meq/gm	.083
Molecular weight	No. Average	2800
Moisture	wt%	0.05
Iodine number		398
Trans-1,4 60%,	Cis-1,4 20%,	Vinyl-1,2 20%

Table 2. List of Amines Used in Experiments

Amine	Structure	b.p.(°C)	m.p.(°C)	Symbol
1,4-Diaminobutane	$NH_2(CH_2)_4NH_2$	159	27	AB
Piperazine	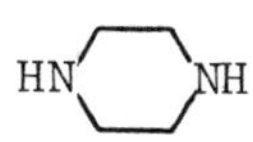	145	109	P
Aniline	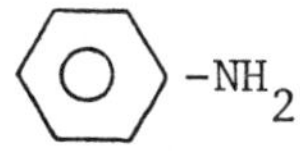	184	-6.3	A
p-Phenylenediamine	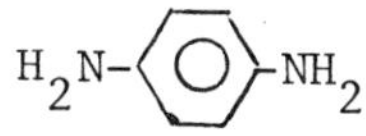	267	140	PDA
3-Aminopropyl-triethoxysilane	$NH_2(CH_2)_3Si(OEt)_3$	122/30mm		AS

Preparation of Polyurethane

The polyurethane was prepared from ARCO's hydroxy-terminated polybutadiene (R-45HT), toluene diisocyanate (TDI), trimethylolpropane and N,N-bis(2-hydroxypropyl)-aniline using a slightly modified two-stage procedure from that described by ARCO (12,13). The ratio ([-NCO]/ total OH]) was kept constant and equal to 1.0 but the ratio [-NCO] to hydroxyl number of R-45HT was varied in some experiments. For adhesion studies, a 6% excess [-NCO] to hydroxyl number of R-45HT was used. The ratio of the substituted aniline to the triol was about 10 to 1 in most experiments. In one series, the latter ratio was varied.

The following equations were used to calculate the required amounts of diisocyanate and hydroxyl compounds.

a) Amount of the TDI for selected % free isocyanate in first stage:

$$\frac{\text{Wt of TDI}}{\text{100g polybd}} = \frac{\text{100(Equiv. wt of TDI)}}{\text{(Equiv. wt of polybd)}} \times \frac{\text{100(Equiv. wt of NCO)} + \text{(Equiv. wt of polybd) (\% Free NCO)}}{\text{100(Equiv. wt of NCO)} - \text{(Equiv. wt of TDI) (\% Free NCO)}}$$

Where Equivalent weight of polybd = 100/hydroxyl value
Equivalent weight of NCO = 42
Equivalent weight of TDI = 87

b) Amount of 2-ethyl-1,3-hexanediol (glycol) needed to react with prepolymer containing selected % free isocyanate from first stage:

$$\frac{\text{Wt of Glycol}}{\text{100g prepolymer}} = \frac{\text{(\% Free NCO in prepolymer)} \times \text{(Equiv. wt of glycol)}}{42}$$

$$\text{Where Equivalent weight of glycol} = \frac{1000 \times 56.1}{\text{hydroxyl number}}$$

$$= \frac{\text{Molecular weight of glycol}}{2}$$

Surface Treatment of Substrates

Precleaned Opticlear soda lime microscope slides (75x25x1.06 mm) from KIMBLE were heated at 140°C for 1 hr., and kept in a desiccator over $CaCl_2$ until treated with the desired reagent. Amines used for surface treatment included p-phenylenediamine (PDA), 1,4-diaminobutane (AB), piperazine (P), aniline (A), and 3-aminopropyltriethoxysilane (AS). Glass slides were treated with amine at room temperature usually by immersing the slides for 5 min. in an amine-absolute ethanol solution. After removing the glass slides from the amine solution, the slides were dried in air and then heated for 30 min. at 60°C under vacuum before applying the elastomer layer. Pure ethanol treatment was carried out in the same manner. 3-Aminopropyltriethoxysilane was similarly applied except that a 50/50 mixture with water by volume was used instead of pure ethanol and after air drying the slides were heated at 100°C for 1 hr.

Small plates cut from Apollo mirror-finish chrome ferro-type plates were similarly treated with amines. Before treatment the plates were cleaned with acetone and dried for 30 min. at 140°C.

Application of Elastomer Layer and Curing

For adhesion studies, appropriately treated substrates were placed in a Teflon-coated compression mold and the prepolymer containing the glycol was poured on. The top plate of the mold was covered with a sheet of washed, dried and pressed cotton cloth. The sample was cured at 100°C for 4 hrs. in a heated press. Rubber sheets were cured in a vertical mold in a vacuum oven at 100°C for 5 hrs.

Measurement of Work of Adhesion (W_a)

180° peeling tests were carried out on strips of cloth-backed elastomer layer after trimming them to a uniform width of 2 cm on the substrate. The cloth-backed elastomer layer was peeled off the substrate at a constant rate of 0.5 cm/min. The work of adhesion per unit area of interface was calculated from the time average of the peel force P per unit width of the detaching layer: $W_a = 2P$.

Tensile Tests

Tensile tests were carried out at room temperature and a crosshead speed of 50 cm/min using dumbbell specimens. Breaking elongation was calculated from the ratio $(L_b-L_o)/L_o$, where L_o and L_b were the initial length and the length at break, respectively.

Swelling Ratio Measurements

Samples were cut from molded rod. The weight of a rod about 2.5 cm long and 0.6 cm in diameter was measured before, W_o, and after, W_f, immersion in dioxane solvent at room temperature for 70 hrs. The swelling ratio, calculated from the relationship $(W_f-W_o)/W_o$ remained constant throughout all experiments.

Differential Scanning Calorimetry

Differential scanning calorimetry was carried out with a du Pont 990 Thermal Analyzer and its accessories, using a normal scanning speed of 10^oC/min and a sample weight of approximately 14 mg.

X-Ray Diffraction

Wide-angle X-ray diffraction patterns were taken with a Phillips Universal Flat-Plate Camera type PW 1030. The X-ray facility consisted of a Phillips ultrastable generator model PW 1310/PW 1320.

RESULTS AND DISCUSSION

Effect of Free -NCO Content of Prepolymer on Properties of Polyurethane

According to ARCO, polybutadiene resin, R-45HT, has the following chemical structure (12-13):

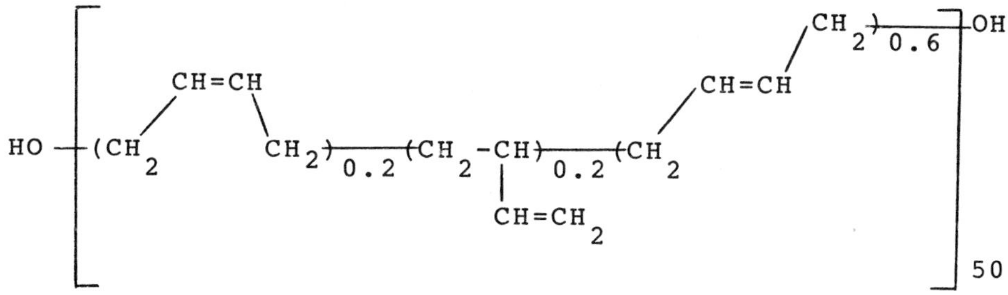

The hydroxyl functionality is slightly greater than 2.0. When excess isocyanate (TDI) is used in the preparation of a prepolymer in a two-stage process, the hydroxyls are converted to urethane groups, R'N(H)C(=O)OR', which can then react with the excess isocyanate to form allophanates, $R'N(CO_2R)C(=O)$-NRH.

Since 2-9% excess isocyanate was used to prepare the polymers in this study, the polymers undoubtedly contained some allophanate groups. The effect of excess -NCO content compared to hydroxyl number of the polybutadiene on tensile properties of the polyurethane are shown in Figure 1. As the -NCO content increased, the ultimate tensile strength, σ , passed through a maximum at about 6% excess -NCO and then stayed constant. As shown in Table 3, Shore A Hardness increased continuously with % free-NCO content.

Table 3. Effect of Free-NCO Content of Prepolymer on Swelling Ratio and Shore A Hardness of Polyurethane[a]

Free-NCO Content of Prepolymer (%)	2	3	4	6	7	9
Swelling Ratio	--	4.02	4.05	4.08	4.13	3.94
Shore A Hardness	33	36	43	60	64	76

[a]The ratio of glycol to triol was about 10 to 1 in these experiments.

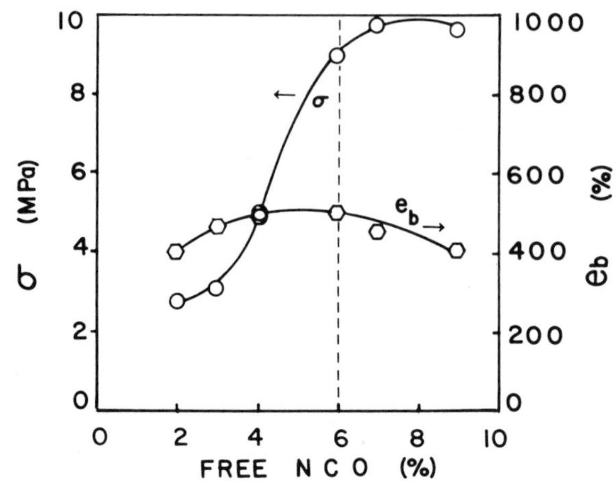

Fig. 1. Effect of free -NCO content of prepolymer on tensile strength (σ) and breaking elongation (e_b).

Polymerizations were completed by chain extension and cross-linking using N,N-bis-2-hydroxyaniline and trimethylolpropane, usually in a 10:1 ratio. For these polymers, as shown in Table 3, the swelling ratio remained constant. (The (total OH)/NCO ratio was constant and equal to 1.0.) As can be seen in Fig. 1, elongation at break, e_b, passed through a maximum at about 6% excess -NCO and then decreased again.

Figure 2 shows the results of differential scanning calorimetry studies of these polymers. The glass transition temperature of the polybutadiene resin, R-45HT, is about -75°C. The glass transition temperature increases as the percentage of excess TDI used to make the polyurethane is increased. The behavior of T_g is similar to that of Shore A Hardness described above.

Wide-angle X-ray photographs of a polyurethane whose prepolymer was prepared using 9% excess isocyanate groups did not show any evidence of crystallinity.

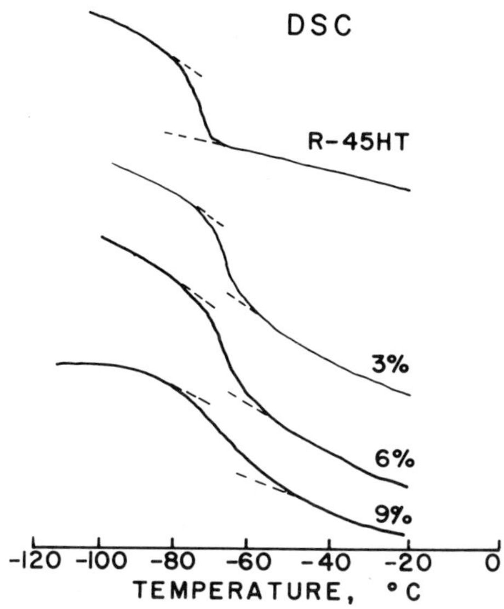

Fig. 2. Effect of concentration of excess TDI on glass transition temperature of resulting polyurethanes.

These effects occur because increasing the amount of TDI and chain extender with respect to the polybutadiene increases the relative concentration of urethane, allophanate, and other aromatic groups. This results in increased possibility for intermolecular hydrogen bonding, increased M_c, increased rigidity and increased elastomer strength.

Effect of Free -NCO Content of Prepolymer on Adhesion

The effect of free -NCO content of prepolymer on W_a to a glass substrate is shown in Fig. 3. The results indicate that W_a increases initially as the -NCO content increases, reaches a maximum at $\sim$ 6% free -NCO content and then decreases with further increases in -NCO content. As has already been maintained above, increasing the free -NCO content of the prepolymer leads to an increase in the concentration of urethane groups in the elastomer backbone and results in increased intermolecular attractive forces in the elastomer. Reegen and Ilkka (14) have made similar

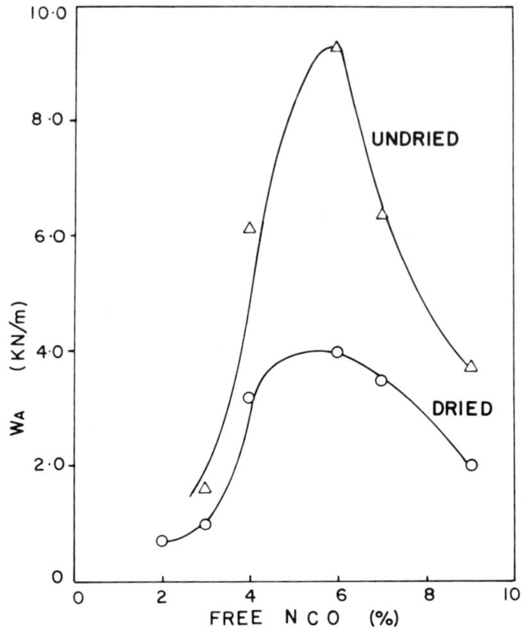

Fig. 3. Effect of free -NCO content of prepolymer on work of adhesion (W_a) of polyurethane to glass. △ undried, ○ dried.

observations. They attributed the higher peel strength observed with increasing concentration of urethane groups to two factors: first, an increase in the true adhesion of the polyurethane and second, a more rigid and stiffer elastomer, which requires a greater force to bend it to an angle of 180°. In our experiments when the free -NCO content was over 6%, the rigidity of the elastomer increased so much that e_b and W_a decreased.

Effect of Water and Ethanol on Adhesion

A glass surface which has been exposed to the atmosphere is covered with a water film (15,16). When our polyurethane was applied to such a glass surface as shown in Fig. 3 a much higher work of adhesion was obtained in comparison with a preheated glass surface. This can be explained as follows. The -NCO group can react with the water on the glass surface to form a primary amine, which on further reaction leads to urea linkages instead of the urethanes formed from alcohols. Urea groups are more polar than urethane groups and therefore promote higher adhesion than urethane groups (14).

The formation of urea groups may not be the only explanation for the greatly enhanced adhesion of undried surfaces. The values of W_a observed for this polyurethane are substantially higher than those previously reported in the literature for other polyurethanes (14,17). We attribute the increased adhesion to the presence of tertiary amine groups from the N,N-bis(2-hydroxypropyl)-aniline in the backbone of our polyurethane. These amine groups potentially can react with the acidic silanol groups present in the glass surface or perhaps even with surface moisture to form ionic bonds:

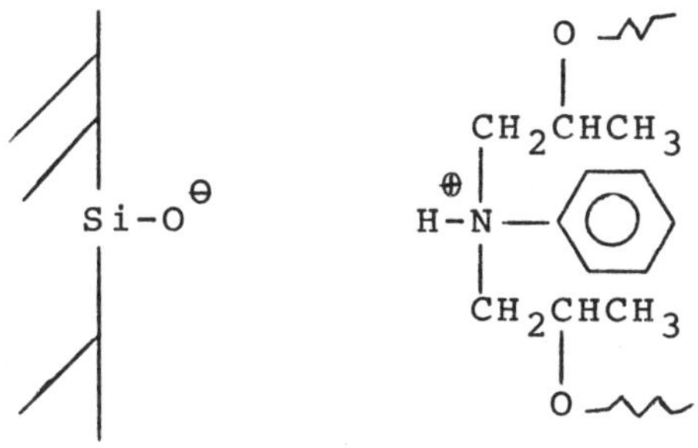

The presence of ionic groups in polyurethanes is known to drastically change the mechanical properties of polyurethanes (18), but the remarkable effects on adhesion have not been reported previously. The presence of such ionic bonds is consistent with the fact that adhesion in these systems is markedly reduced by soaking a few days in water, which would solvate the ionic clusters (18) and reduce their effectiveness in promoting adhesion. It is noteworthy that strong adhesive bonds reform on drying (19).

Experiments intended to illustrate the effect of the concentration of tertiary amine groups on adhesion led to the remarkable changes in W_a shown in Fig. 4. In this series of experiments the ratio of diol(tertiary amine) and triol was varied while a 6% excess isocyanate was maintained. Polymers with the greater percentage of diol were stronger and their swelling ratio was higher. Such systems have both an increased number of tertiary amine groups and are more chain extended than those prepared with the triol alone. Ahagon and Gent (20), among others, have shown that W_a

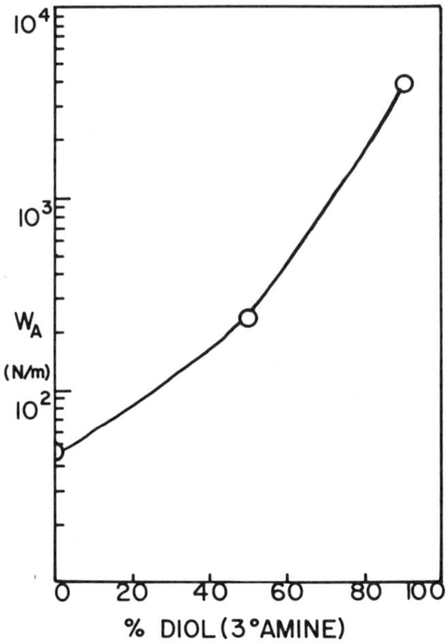

Fig. 4. Effect of concentration of diol (tertiary amine groups) on adhesion.

decreases as the cross-link density of the polymer increases and the molecular weight between cross-links decreases. Thus some of the increase in W_a with increased percent of diol is undoubtedly due to the decrease in cross-link density and polymer stiffness that accompanies the increase in diol. But the magnitude of the increase is greater than expected from this change alone. As stated above, we attribute the high strength of adhesion to the presence of ionic bonds. This interpretation is consistent with the observations of Kozakiewicz and Penczek (21), who reported enhanced adhesion in systems shown to contain ionic bonds.

Preheated glass slides which had been immersed in absolute ethanol also led to a somewhat higher work of adhesion than untreated, preheated slides, possibly due to an increased number of urethane groups at the interface.

Effect of Amines on Adhesion

The values of the work of adhesion of our polyurethane to glass slides pretreated with different amine solutions are shown in Figs. 5 and 6. The work of adhesion compared to untreated slides was increased in all cases except that of aniline. At equivalent concentration of amines, the order of enhancement in the work of adhesion was AS>>PDA∿AB>P>>A. Cohesive failure of the adhesive bonds was observed for slides pretreated with AS. The results can be explained in terms of the different structures of the amines and the reactions that can occur with the amines. We conclude that just like in the work of Eckstein and Dreyfuss referred to above (4-8), chemical bonds form from the surface through the amine to the substrate: The degree of enhancement of adhesion is related to the number of bonds that can form between the surface and the adhesive. Aniline does not lead to reinforcement because it is monofunctional and the ring does not become part of the backbone.

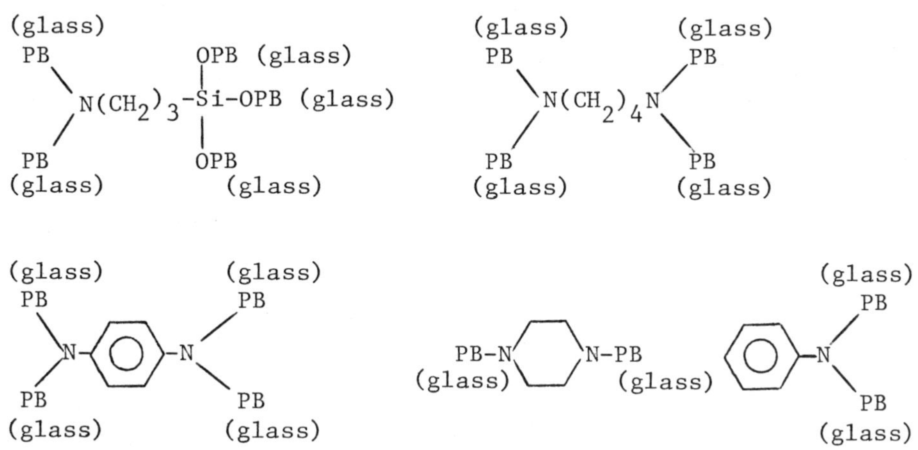

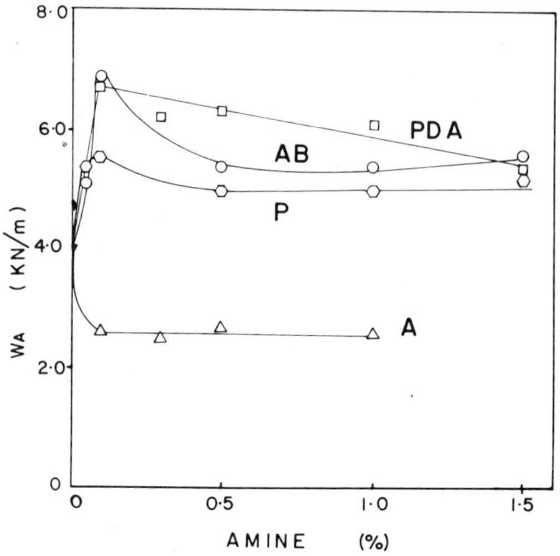

Fig. 5. Effect of concentration of amine on work of adhesion (W_a) of polyurethane to glass. ○ AB, □ PDA, ⬡ P, △ aniline, ● ethanol, ▼ dried and untreated.

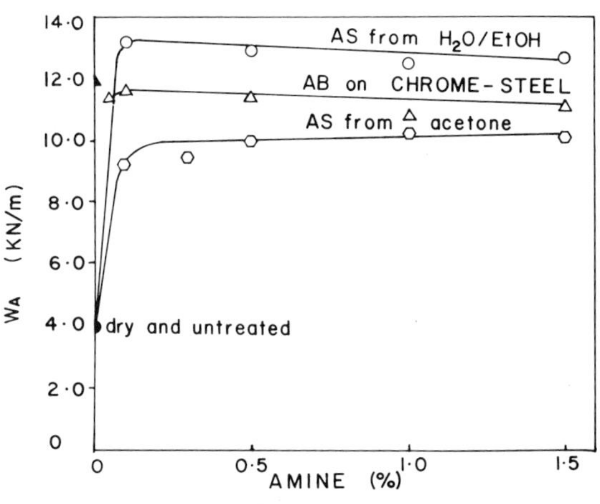

Fig. 6. Effect of concentration of AS on work of adhesion (W_a) of polyurethane to substrates. ○ glass treated with AS from 50/50 by volume water/ethanol, △ chrome steel treated with AB, ⬡ glass treated with AS from acetone, ▲ chrome steel treated with absolute ethanol, ● dried and untreated glass.

The structures shown are intended to indicate that the amino groups can, in all cases, be bonded _either_ to the glass _or_ to the polybutadiene. The structures are not intended to suggest that bonding of the amine to the glass occurs only through the polybutadiene.

In the present case the chemical reactions occurring between the polybutadiene and the amines or glass are quite different from those described by Eckstein and Dreyfuss (4-8), because in the present case no peroxide is present. All reactions occur among the functional endgroups. Bonding of the polybutadiene to the amines takes place as a result of the reaction of the isocyanate endgroups of the prepolymer with the amine to form a urea:

$$OCN\text{-}PB\text{-}NCO + NH_2RNH_2 \rightarrow OCN\text{-}PB\text{-}NHCO\text{-}NHRNH\text{-}CONH\text{-}PB\text{-}NHCO\text{-}NHRNH_2$$

Bonding of the polybutadiene to the glass may occur by reaction of the prepolymer with silanol groups in the glass surface:

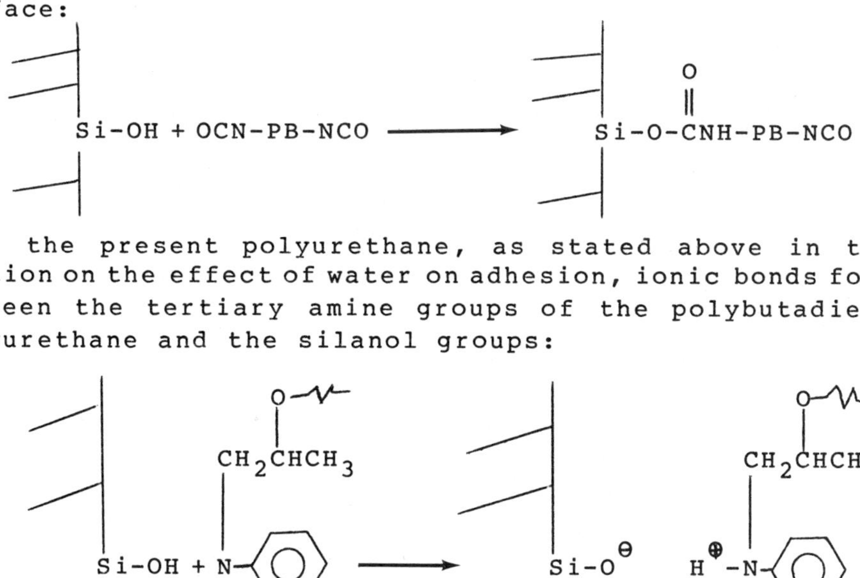

With the present polyurethane, as stated above in the section on the effect of water on adhesion, ionic bonds form between the tertiary amine groups of the polybutadiene polyurethane and the silanol groups:

The reactions that occur between the glass and the amines are probably the same as those already described by Eckstein and Dreyfuss (4-8). All of the amines can react with the silanol groups on the glass surface to form ionic bonds similar to those shown above for the tertiary amine. Reactions of 1,4-diaminobutane and 3-aminopropyltriethoxysilane are probably facilitated by prior formation of carbamates. In addition, covalent bonds of a presently undefined structure probably form between the glass and the amine.

Adhesion of Polyurethane to Chrome Steel

Metals have a higher surface energy than glass. After treatment with ethanol, W_a of our polyurethane to preheated and clean chrome steel surfaces was much higher than to preheated glass surfaces. Treating the surface of the steel with amines before applying the elastomer did not enhance adhesion. Dried and untreated surfaces had W_a of about 3 kN/m. The results are included in Fig. 6. Some preliminary experiments have been carried out on chrome steel surfaces treated with AS. W_a is not significantly increased but the bonds that form are extremely stable in water. They can be soaked for a month with no apparent change in W_a (19).

The chemical reactions occurring are probably similar to those described above for glass (22). Supporting evidence that amino compounds are chemisorbed by metal oxide surfaces has recently been obtained by inelastic tunnelling spectroscopy (23). Further investigations are being carried out to find explanations for the differences observed in the effects of amine surface treatment of glass and of metal on the adhesion of this polyurethane to the substrate.

ACKNOWLEDGEMENTS

This work forms part of a program of research on the adhesion of elastomers supported by a research grant from the Office of Naval Research. One of us (F.L.) is grateful to the Education Ministry of the People's Republic of China for opportunity to carry out this work.

REFERENCES

1. P. Dreyfuss, A.N. Gent, and J.R. Williams, J. Polym. Sci.: Polym. Phys. Ed., 18, 2135 (1980).

2. M.W. Ranney and C.A. Pagano, Rubber Chem. Technol., 44, 1080 (1971).

3. M.W. Ranney, S.E. Berger, and J.G. Marsden, Chapter 5 in Composite Materials, 6, E.P. Plueddemann, Ed., Academic Press, N.Y., N.Y., 1974, p. 131.

4. E.P. Plueddemann, Adhesives Age, 18, 38 (1975).

5. Y. Eckstein and P. Dreyfuss, J. Adhesion, 15, 193 (1983).

6. Y. Eckstein and P. Dreyfuss, J. Adhesion, 15, 179 (1983).

7. Y. Eckstein and P. Dreyfuss, J. Adhesion, 15, 163 (1983).

8. Y. Eckstein and P. Dreyfuss, J. Adhesion, 13, 303 (1982).

9. P. Wright and A.P.C. Cumming, Solid Urethane Elastomers, MacClaren and Sons, London (1969).

10. K.C. Frisch and S.L. Reegen, Eds., Advances in Urethane Sciences and Technology, Technomic Publishing Co., Stamford, Conn. (1971).

11. R.J. Athey, Contribution No. 138, Division of Rubber Chemistry, American Chemical Society, Cincinnati, Ohio (1958).

12. "Poly bd [R] liquid resins in urethane elastomers," Product Bulletin BD.3, ARCO Chemcial Co., October, 1974.

13. "Poly bd [R] resins," General Bulletin, ARCO Chemcial Co., January, 1978.

14. S.L. Reegen and G.A. Ilkka, in Adhesion and Cohesion, P. Weiss, Ed., Elsevier Pub. Co., Amsterdam, N.Y., 1962, p. 159.

15. E. Warburg and T. Ihmori, Ann. d. Phys., 27, 481 (1886).

16. T.H. Elmer, "Glass Surfaces" in Silylated Surfaces, D.E. Leyden and W. Collins, Eds., Gordon and Breach Science Pub., 1980.

17. A.N. Kuksin, Yu. S. Lipatov, L.M. Sergeeva, and T.D. Kadurina, Chapter 26 in Soviet Urethane Technology, Soviet Progress in Polyurethanes Series, 1, A.M. Schiller, Ed., Technomic Pub. Co., Westport, Conn., 1973, p. 171.

18. R. Rembaum, in Advances in Urethane Science and Technology, 2, K.C. Frisch and S.L. Reegen, Eds., Technomic Pub. Co., Stanford, Conn., 1971, p. 109.

19. F. Liang and P. Dreyfuss, to be published.

20. A. Ahagon and A.N. Gent, J. Polym. Sci.: Polym. Phys. Ed., 13, 1285 (1975).

21. J. Kozakiewicz and P. Penczek, Angew. Makromol. Chem., 78, 89 (1979).

22. F.J. Boerio, ACS Organic Coatings and Plastics Chemistry Preprints, 44, 625 (1981).

23. T. Furakawa, N.K. Eib, K.L. Mittal, and H.R. Anderson, Jr., SIA, Surf. Interface Anal., 4(6), 240 (1982).

Effect of Silane on Glass/Resin Adhesion Strength

Yona Eckstein and Elisabeth J. Berger

Owens-Corning Fiberglas Corporation
Technical Center
Granville, OH 43023

ABSTRACT

Glass fiber/resin adhesion, in terms of interfacial shear strength, was determined in a system consisting of silane coated single filaments embedded in a polymeric resin. Wettability of the fibers in the resin was measured with a Wilhelmy wetting balance. The interfacial shear strength and the resin wetting were found to be dependent on the concentration of the silanes in the coating solution. The dependence on silane concentration is similar in character for both properties, though different for different silane compounds. Studies by GPC, DMA and FTIR of the coatings suggested deposition of a polymeric, low molecular weight silane product by dilute solutions, and a high molecular weight, highly cross-linked polymer by concentrated solutions. It is concluded that the different properties of silane coatings are responsible for the observed differences in resin wetting, which in turn affect the interfacial bond strength and its degradation in the presence of water. Also, at a given concentration of a coupling agent, better adhesion and water resistance might be achieved by increasing the number of glass and/or resin reactive groups on the silane compound.

INTRODUCTION

The glass fiber/matrix interface has long been considered the most sensitive region in fiber reinforced plastic (FRP) products. The presence of moisture is known to degrade the composite strength, presumably by attacking the glass/resin interfacial bond. To enhance and preserve the bond strength, particularly against moisture attack, most composites with an organic matrix employ a "coupling agent." The most commonly used among these agents are the silane compounds, having the general formula $R\text{-}Si\text{-}(OR')_3$, where the -OR' are alkoxy functions and R is referred to as the organofunctional group. According to the prevalent theories (1), the silane coupling agents bond to the resin by reaction of the organofunctional group with the resin molecules, and bond to the glass surface through interaction of the alkoxy groups with the surface silanols.

In FRP products, the silanes are usually applied to the fibers from an aqueous solution. Earlier studies carried out in our laboratory (2) indicated a strong dependence of composite mechanical properties on the concentration of silane in the coating solution, as well as on the silane organofunctionality. Our investigations focus on the effect of silane solution concentration and functionality on the nature and properties of silane coatings deposited from such solutions. Fourier transform infrared spectroscopy (FTIR), gel permeation chromatography (GPC), and dynamic mechanical analysis (DMA) were used for evaluation of the molecular structure of silane coupling agents on the surface of glass fibers. Measurements of wettability in epoxy resin, methylene iodide, and water on the silane-treated fibers provided insight into the surface chemistry of silane-treated glass fibers, as well as into the mechanisms of glass/resin interactions. Based on these insights, we examined the effect of these phenomena on the strength of the glass/resin interface, as measured by a single-filament adhesion test.

EXPERIMENTAL

Materials

The silanes used in this study and their suppliers are listed in Table 1. Single E glass filaments (formed in house) were coated directly under the one-hole bushing.

Table 1. Silanes Used in this Work

Silane	Chemical Structure	Source	Trade Name
methyltriethoxysilane	$CH_3Si(OEt)_3$	Union Carbide	A163
methacryloxyproplyl-triethoxysilane	$Me_2C{=}CMe\overset{O}{\overset{\|}{C}}O(CH_2)_3Si(OEt)_3$	Union Carbide	A174
γ-aminopropyltriethoxysilane	$NH_2(CH_2)_3Si(OEt)_3$	Union Carbide	A1100
trimethoxysilylproply-diethylenetriamine	$NH_2(CH_2)_2NH(CH_2)_2NH(CH)_3Si(OMe)_3$	Petrarch	T2910
bis[3-(triethoxysilyl)propyl]amine	$NH[(CH_2)_3Si(OEt)_3]_2$	Petrarch	B2492
N-2-aminoethyl-3-amino-propyltrimethoxysilane	$NH_2(CH_2)_2NH(CH_2)_3Si(OMe)_3$	Petrarch	A0700
γ-aminopropyltriethoxysilane	$NH_2(CH_2)_3Si(OEt)_3$	Petrarch	A0750
aminopropyldimethylethoxy-silane	$NH_2(CH_2)_3Me_2Si(OEt)$	Petrarch	A0735
octyltriethoxysilane	$CH_3(CH_2)_7Si(OEt)_3$	Petrarch	09835

Model Studies of Siloxane Films

In model studies of silane coatings, the as-received silanes were dissolved in aqueous solution and acidified with acetic acid to pH=3.5. Equal volumes of silane solution were then poured into teflon-coated aluminum dishes. The silane precipitated upon heating and was dried at 110°C for two hours. A Fourier transform infrared spectrometer (Nicolet 7000) was used for measurements of the absorption spectra of the polymerized silanes. The dry samples were incorporated into KBr pellets using infrared quality KBr powder (Harshaw). Relative molecular weight distributions of the silane polymer species dissolved in tetrahydrofuran were determined with gel permeation chromatography.

Properties of Siloxane Coatings on Glass Surface

Dynamic mechanical properties of silane coatings were measured with a DuPont DMA-981 instrument. Specimens made of heat cleaned woven glass cloth tape from Columbia Fiberglass Mutual Industries, 0.5" wide and 0.01" thick were immersed for 5 seconds in an aqueous solution of a silane compound, removed and cured for one hour at 110°C and then mounted in the DMA instrument for characterization of the coating.

Wetting of Glass Fibers and Glass/Resin Adhesion

Previous work done in this laboratory (3) measured the thermodynamic work of adhesion, W_a, of epoxy resin (Dow Chemical DER-331), of water, and of methylene iodide on single filaments of E glass coated with dilute (0.2 to 2.5 percent by weight) aqueous solutions of silanes. W_a was calculated from contact angle (Θ) measurements using the formula

$$W_a = \gamma_{LV}(1+\cos\Theta) \tag{1}$$

where γ_{LV} is the surface tension of the wetting liquid.

The glass/resin interfacial shear strength was measured in specimens consisting of a single filament embedded in a resin (3) both before and after immersion in boiling water for six hours (denoted dry strength and wet strength respectively).

RESULTS AND DISCUSSION

Molecular Composition and Structure of Silane Films

In an attempt to infer the nature of silane coatings on the glass surface, our study characterized polymerized silane films, deposited from silane solutions of different concentrations. Since methyltrimethoxysilane (MS) has the simplest chemical composition among the silane compounds used as coupling agents, polymerized silane films of the MS were chosen as a model system. The nature of other silane coatings on the glass surface is inferred on the basis of this model system.

An example of the mid-infrared adsorption spectra of the silane deposited from 0.2 and 2.5 wt.% solutions of MS compared to the spectrum of the undiluted monomeric silane is shown in Fig. 1. The greatest differences between the spectra are in the ranges of $-OCH_3$ (2840 cm^{-1}), Si-O-Si (1030 cm^{-1}), and Si-OH (592 cm^{-1}) absorptions. These

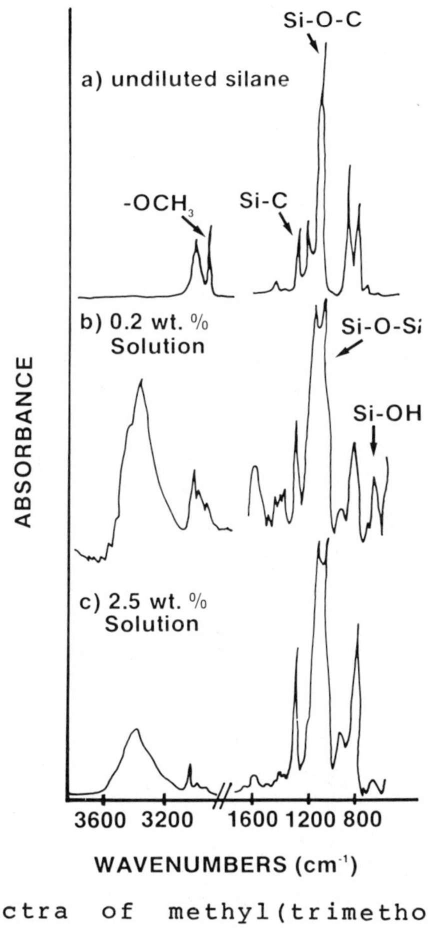

Fig. 1. IR Spectra of methyl(trimethoxy) silane (a) undiluted and deposited from aqueous solution (b) 0.2 wt%, (c) 2.5 wt%.

differences can be explained by the sequence of chemical reactions taking place in a water/silane system upon drying, as shown in Eq. (2).

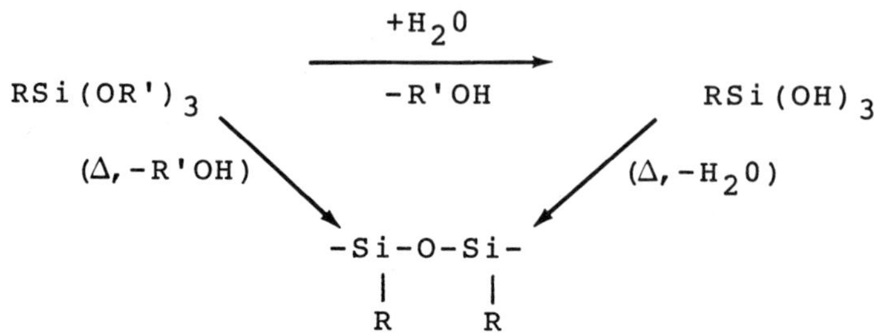

(2)

In the initial stage, the monomeric silane is hydrolyzed, partially or totally, by water. During drying, it undergoes condensation or oligomerization. In the case of MS, the hydrolysis reaction is expected to reduce the intensity of the $-OCH_3$ absorption, characteristic of the monomeric, nonhydrolyzed species, giving rise at the same time to a new absorption from the Si-OH group, because of the formation of silanols. Condensation of the silane upon drying should result in a decreased concentration of $-OCH_3$ and/or Si-OH groups, and the appearance of a new band because of the Si-O-Si absorption of the oligomerized silane. In contrast, the intensity of the Si-C absorption is independent of the changes in composition of the silane molecule as a result of the reactions described by Eq. 2. Thus, the Si-C bond can be used as a reference in studies of the extent of reaction occurring in the water/silane system.

The ratios of intensities, I(Si-O-Si)/I(Si-C) and I($-OCH_3$)/I(Si-C), reflect the relative degree of polymerization and hydrolysis, respectively. Both polymerization and hydrolysis reduce the concentration of $-OCH_3$ groups in the dry product. The relative molecular weight of the polymerized silane is directly proportional to the ratio I(Si-O-Si)/I(Si-OH). The I(Si-OH) represents the concentration of -Si-OH end groups in the dried deposit.

In Fig. 2, the intensity ratios of $-OCH_3$/Si-C, Si-O-Si/Si-C and Si-O-Si/Si-OH for the polymerized MS are plotted as a function of concentration of the silane in the solution from which it was deposited. The plot of I($-OCH_3$)/I(Si-C) reaches a maximum, while that of I(Si-O-Si/I(Si-C) shows a minimum for the deposit obtained from 1.0 wt% solution. Above the 1.0 wt%, the behavior exhibited by the two plots follows that expected from Eq. 2, but the data below this concentration are puzzling. Interestingly, studies by Ishida, et al. (4) of hydrolysis of vinyltrimethoxysilane (VS) as a function of the silane concentration showed a similar break near 1.0 wt% of VS.

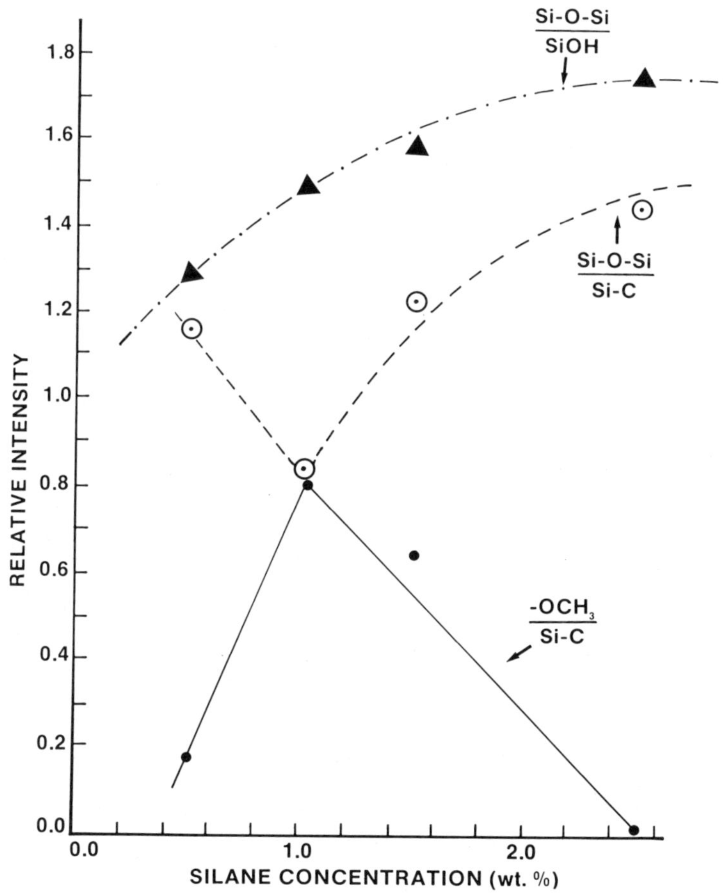

Fig. 2. Ratio of IR intensities for polymerized methyl(trimethoxy)silane as a function of concentration.

(Since the molecular weights of MS and VS are similar, the 1.0 wt% concentration corresponds to similar molar concentrations of both silanes.) The break was ascribed to the onset of the formation of monomer aggregates from isolated, hydrolyzed monomers via hydrogen bonding. Above this concentration, the rate of oligomer formation should be faster. Thus, the differences in behavior below 1.0 wt% observed in our dry specimens may be ascribed to the different relative rates of hydrolysis and polymerization in the water/silane system.

Apparently, below 1.0 wt%, not all hydrolyzed species undergo condensation. This is reflected by the low I(-OCH_3)/I(Si-C) ratio and the presence of residual -SiOH end groups in the cured products obtained from the dilute solutions. Allen, et al. (5) proposed that the residual uncondensed groups are more likely to be held within ring or spiral structures. Infrared spectra of MS films showing the presence of methoxy groups in all cured samples, regardless of concentration, indicate incomplete hydrolysis of silanes in their solutions. A similar observation was reported by Allen, et al. (6) for glycidoxypropyltrimethoxysilane (γ-MPS). This suggests the existence of equilibrium concentrations of nonhydrolyzed and hydrolyzed silane species.

The monotonic increase in the ratio I(Si-O-Si)/I(Si-OH) with silane concentration (Fig. 2), implies formation of a lower molecular weight product from dilute solutions and a higher molecular weight product from more concentrated solutions. This conclusion is supported by GPC of the polymerized material dissolved in tetrahydrofuran (Fig. 3). High retention times (low molecular weights) are characteristic of films obtained from dilute solutions (0.2 - 0.5 wt%), while the polymer obtained from the 2.5 wt% solution was eluted in shorter times, and had a distinct bimodal distribution of molecular weight. Such bimodality is characteristic of two different groupings of polymeric species in the specimen.

DMA damping curves of glass cloth treated either with 2.5 wt% solutions of MS or γ-MPS (Fig. 4) displayed similar bimodality to that shown by GPC curves, though temperatures of phase transitions shown by the two silanes were different (-93°C and -63°C in MS and -120°C and -80°C in γ-MPS).

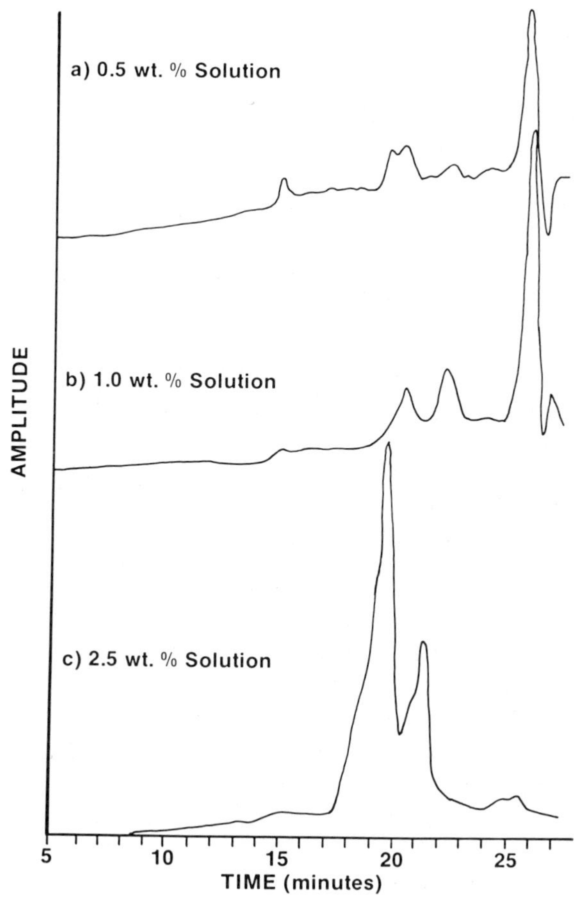

Fig. 3. Gel permeation chromatographs of polymerized methyl(trimethoxy)silane dissolved in THF.

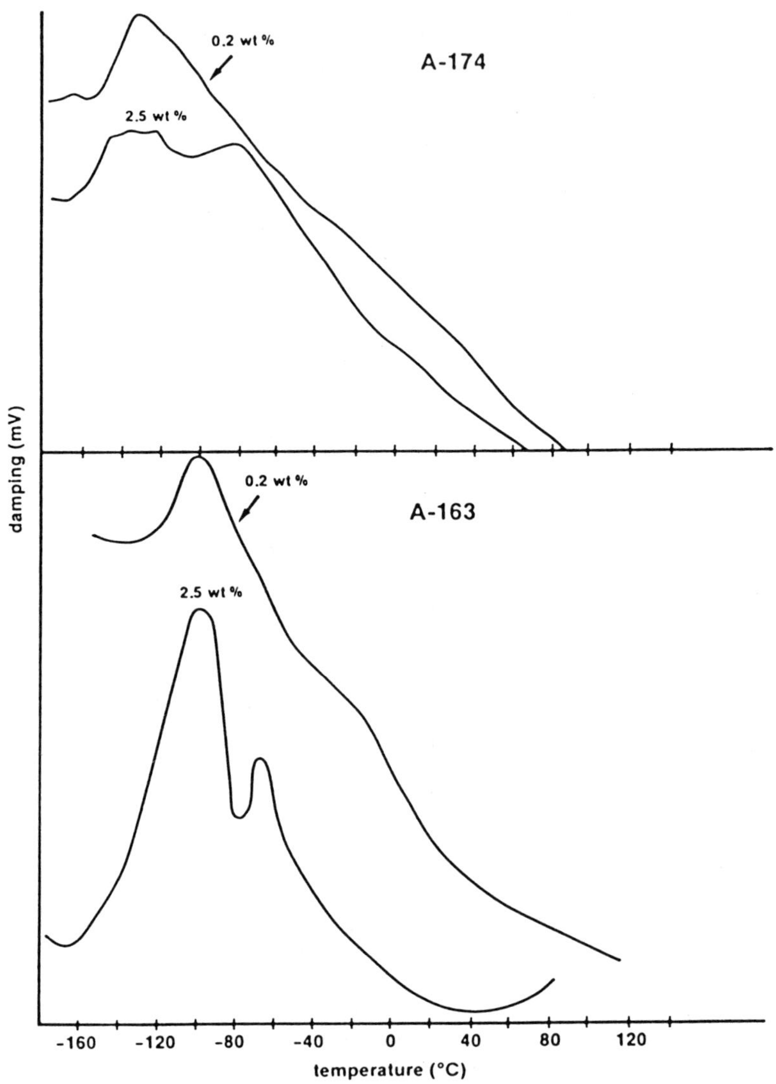

Fig. 4. DMA damping curves of glass cloth treated with solution of MS or γ-MPS.

Damping curves of glass cloth treated with a dilute solution of 0.2 wt% exhibited only one low temperature phase transition (-95°C in MS and -120°C in γ-MPS). Generally, the temperature of a phase transition in polymers, such ass the glass transition (T_g), is an indicator of the relative mobility of the polymer segments, with the lower T_g values corresponding to the higher mobility. Cross-linking hinders the segmental motion and, therefore, raises the T_g of a polymer. Thus, the higher temperature damping peaks, shown by the glass treated with a concentrated solution of either MS or γ-MPS, suggests the presence of a more cross-linked silane coating compared to that obtained from dilute solution. Flexible -Si-O-Si- chains with free RSiO- groups might be responsible for the lower temperature transitions, while the higher temperature transitions arise from hindered cross-linked siloxane chains. The fact that both transitions appear in coatings deposited from concentrated solutions suggests that there is still a substantial proportion of RSiO- units which are neither cross-linked nor participating in a ring structure.

It seems likely that the free siloxane polymers deposited on teflon surfaces are different in structure from the relatively thin films deposited on the glass surface when using siloxanes as coupling agents. However, it is likely that they have many features in common, including the effect of silane solution concentration on the relative molecular weight of the cured siloxane coating, as evidenced by the DMA studies.

We conclude, therefore, that the concentration of silanes in aqueous solution has a significant effect on the molecular composition and structure of deposited films. Higher concentration of silane in the solution results in higher molecular weight and a higher degree of cross-linking of the polymerized silane coatings.

Surface Properties of Silane Coatings

Measurements of surface energy and wettability of a surface by different liquids may provide an additional insight into surface properties of silanized glass surfaces, the structure of the coatings, and the mechanism of interfacial interactions. Theoretically, solid surface tension energy is composed of two additive terms:

$$\gamma_s = \gamma_s^d + \gamma_s^p \quad (3)$$

where the superscripts d and p denote dispersive and polar components, respectively, of the surface tension γ_s. In our previous work (3), we calculated these components for glass surfaces treated with silane solutions. The results for γ-MPS and γ-aminopropyltriethoxysilane (γ-APS) are summarized in Table 2.

The surface tension of the γ-MPS-treated glass fibers is independent of the concentration of the silane in the treating solution, although it is higher than that of glass treated with water only. The surface tension of glass coated with γ-APS increases from as low as 29.5 to 44.1 ergs/cm^2 with increasing silane concentration. Examination of Table 2 reveals enhanced polar contributions to the solid surface tension with increasing γ-APS concentration.

The observed differences in surface tension of γ-MPS- and γ-APS-treated glass fibers are probably caused by the

Table 2. Dispersion (γ_s^d) and Polar (γ_s^p) Contributions to Surface Tension (γ_s) of Silanized Glass

Treatment	In solution, wt%	γ_s	γ_s^d	γ_s^p
Water (pH=3.5)	--	35.4	12.7	22.6
γ-MPS*	0.2	41.1	13.0	28.1
	1.0	42.1	8.5	33.6
	2.5	40.9	8.7	32.2
γ-APS**	0.2	29.5	13.5	16.0
	0.5	34.6	10.8	23.8
	1.0	36.2	13.0	23.2
	2.5	44.1	9.6	34.5

* γ-MPS denotes γ-methacryloxypropyltriemethoxy silane.
** γ-APS denotes γ-aminopropyltriethoxy silane.
Surface free energy is in units of ergs/cm^2, errors are $\pm$ 1 ergs/cm^2.

different organofunctionality of the silanes, which in turn may affect the molecular structure of the coatings. Ishida and Koenig (7) concluded that for γ-MPS-coated fibers a layer of hydrocarbon chains exist because of polymerization involving C=C bonds. The existence of such external layers may account for the constant value of the surface tension, even though the polar component of the surface tension is higher than that expected from hydrocarbon.

Measurements made on γ-MPS coated flat plates of E glass (as opposed to fibers) showed water contact angles increasing from 64° for 0.2 wt% to 78° for 2.5 wt% of silane. Water contact angles on the same coatings aged for two days in air at room temperature declined to a constant value of 30° to 34° for the whole range of concentrations studies. The changes in properties of the silanized surfaces were irreversible; heating for one hour at 120°C in vacuum increased the contact angle only slightly. These results imply interaction of the γ-MPS coating either with atmospheric water or carbon dioxide to form an external layer of stable adduct compound. It is possible that at the temperature of the resin cure (100-120°C) the adduct decomposes, leaving polymerized, organic layer. Existence of this type of adduct may account for the high polarity of the γ-MPS treated fibers.

The trend in surface free energy displayed by γ-APS-coated fibers can be understood in view of the tendency of the aminosilane to form intra- and inter-molecular hydrogen bonded structures (8). A molecular orientation of the silane, with amino group facing the surface hydroxyls was reported by Boerio, et al. (9) and Jen (10). Orientation of the nonpolar part of the molecule towards the vapor phase is known to minimize the surface energy (11). In thicker coatings of γ-APS, a structural gradient was noted by Boerio, et al. (12). Thus, higher surface energy might be caused by increased presence of high energy amino groups at the surface.

Fig. 5 shows the wettability in water and epoxy resin of the glass fibers in terms of work of adhesion, W_a, as a function of concentration of γ-MPS and γ-APS silanes in the coating solutions. The wettability in water of γ-MPS-treated fibers is independent of the concentration of the silane in the coating solution. This is expected from the presence of an adduct compound in the external layer. In epoxy, mutual solubility and interdiffusion of the epoxy resin with the silane is possible (5) due to the similarity of their solubility parameters, resulting in better molecular contact between the interacting surfaces. Such interdiffusion is more feasible in the presence of short polymeric chains and is restricted by cross-linking and higher molecular weight polymers. Thus, the coating produced by 0.2% γ-MPS will interdiffuse more easily with the epoxy resin, giving a higher value of W_a than the higher concentrations of γ-MPS.

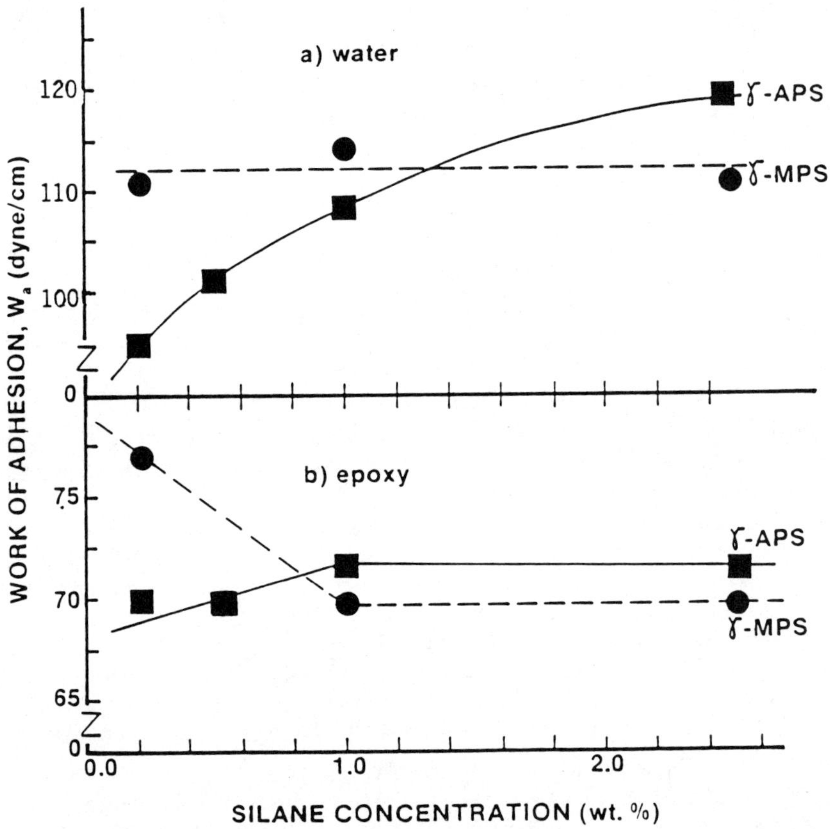

Fig. 5. Wettability of glass fibers in water and epoxy as a function of silane concentration in solution.

The wettability in water and epoxy of γ-APS-coated fibers increases with increasing concentration of the silane in solution. Enhanced polarity of the γ-APS coating may be responsible for the effect.

From the above, it can be stated that the concentration of silanes in an aqueous solution affects the surface free energy and wettability of the silane coated fibers. The observed trends, however, depend on the organo-functionality of the silane compound as it affects the molecular composition and structure of the coatings and the nature of the interfacial interactions.

Silane Coated Fibers/Epoxy Resin Interfacial Adhesion Strength

In order to determine the effect of silane concentration on interfacial adhesion, the extent of glass fiber/epoxy adhesion, as expressed by the magnitude of the interfacial shear stress, was plotted against the concentration of γ-MPS and γ-APS in the coating solution (Fig. 6). The strength of interfacial adhesion is a function of the silane concentration. γ-MPS-coated fibers give the best interfacial bond strength at the lowest concentration tested (0.2 wt%), while γ-APS-coated glass reaches a maximum at 1.0 wt%. Noteworthy is the fact that the trend displayed by γ-MPS-coated samples correlates with the wettability of the fibers by Dow DER 331 epoxy resin (Fig. 5). Similarly, a correlation exists for γ-APS-coated fibers up to 1.0 wt% solution of the silane. Above this concentration, the interfacial adhesion declines, while the wetting in epoxy stabilizes.

These correlations should be considered in terms of both the resin/silane interdiffusion or interpenetration as mentioned above, and chemical interactions. In both γ-MPS-coated fibers and in γ-APS samples, the adhesion is achieved through interdiffusion and amine-epoxy or silanol-epoxy polar interactions. For γ-APS chemical interaction may lead to the formation of interfacial covalent bonds as shown in the following reactions:

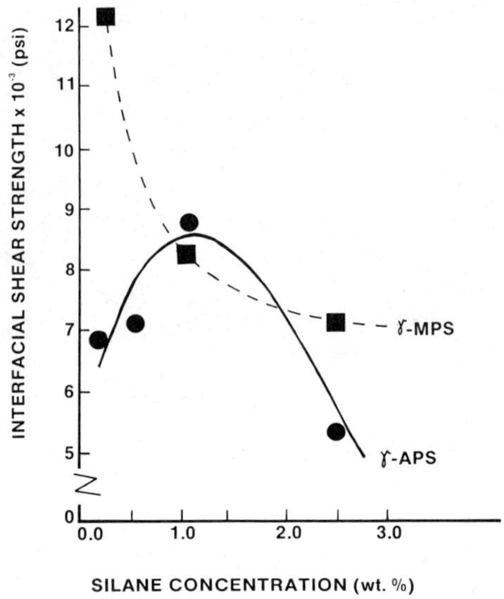

Fig. 6. Interfacial shear strength as a function of silane concentration in solution.

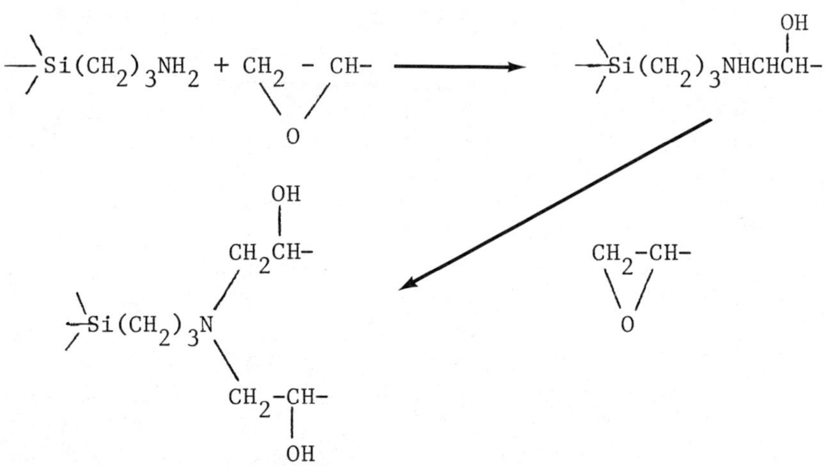

(4)

and/or

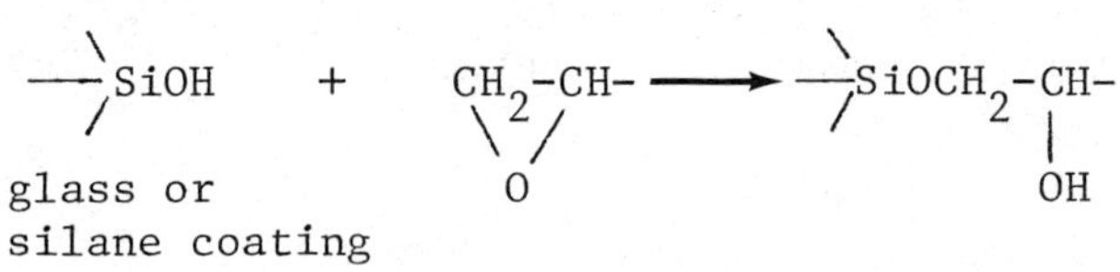

(5)

γ-MPS may bond via reaction 5, but not by reaction 4.

The better interfacial adhesion with the 0.2 wt% γ-MPS coated fibers, compared to 0.2 wt% γ-APS coated fibers results from the higher polarity and interdiffusion of the γ-MPS samples. Above 0.2 wt%, interdiffusion declines because of the increased molecular weight and cross-linking density of the silane coating. Thus, in γ-MPS samples above 0.2 wt%, reduced interdiffusion combined with decreased concentration of -SiOH groups on the surface accounts for the deterioration of the interfacial adhesion. In contrast, in γ-APS samples, the effect of reduced interdiffusion is masked by the positive effect of the higher concentration of amino groups at the glass surface. The increased opportunity for bonding, as per reaction 4, results in a net improvement of adhesion above 0.2 wt% γ-APS. Only above 1 wt% γ-APS does the effect of reduced interdiffusion become pronounced.

As interpenetration will not occur with an uncoated, smooth glass surface, the high interfacial shear strength obtained in samples of glass fibers in the absence of the silane coating can be understood only if assuming formation of interfacial covalent bonds, according to reaction 5.

Earlier, Lowenstein (14) also suggested that the reactivity of the ethylene oxide groups with water is responsible for the excellent adhesion of epoxides to glass surfaces.

Measurements of the glass/epoxy interfacial shear strength carried out after six hours of immersion of the specimens in boiling water, referred to in the following as wet strength, demonstrated degradation of the interfacial bonding for the two silanes. The wet strength of γ-MPS-coated fibers declined with decreasing silane concentration, while the wet strength of γ-APS-treated fibers showed an opposite effect (Fig. 7). The opposite trends displayed by the two silanes are in agreement with Ishida and Koenig's observation (7) that amine groups are subjected to gradual hydrolysis, while methacryloxy groups can homopolymerize through the C=C bond, giving hydrolytically stable layers.

It was found that the wet strength data did not directly correlate with the wettability of the fibers in epoxy or water. However, we found that for γ-APS-coated fibers, a correlation exists between the wet strength and the ratio of

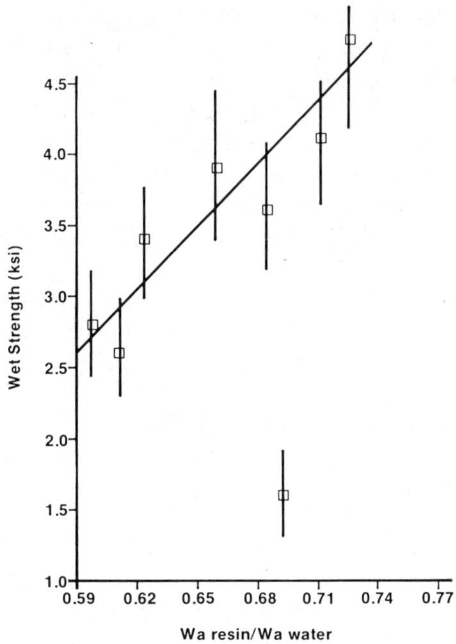

Fig. 7. Single filament interfacial shear strength after water boil vs. concentration of silanes in solution.

wetting in epoxy to wetting in water, W_a(resin)/W_a(water) (Fig. 8). Since polar molecules, such as water, can compete for the glass surface with the epoxy resin molecules, this ratio of wettings reflects the relative resistance of the interfacial adhesion to degradation by water. The trend in the relationship between the W_a(resin)/W_a(water) ratio and the concentration of γ-APS is similar to that between wet strength of composites and the concentration of γ-APS in coating solutions (compare Fig. 7 with Fig. 8). No such correlation was found to hold for γ-MPS coatings with increasing thickness of the silane deposit. Thus, the observed enhancement of the wet strength with concentration of γ-MPS solution can be ascribed to the reduced solubility of this silane coating in water.

A plot of all the wet strength data versus W_a(resin)/W_a(water) ratio (Fig. 9), disregarding organo-functionality of the silane compound, shows a general increase in wet strength as the ratio increases. Such a correlation between wet strength and W_a(resin)/W_a(water) ratio points to the silane/epoxy interface as the most susceptible for water degradation.

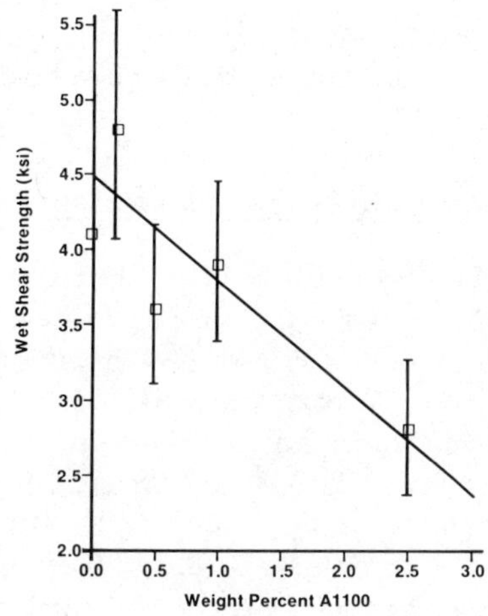

Fig. 8. Ratio of resin wetting to water wetting vs. silane solution concentration.

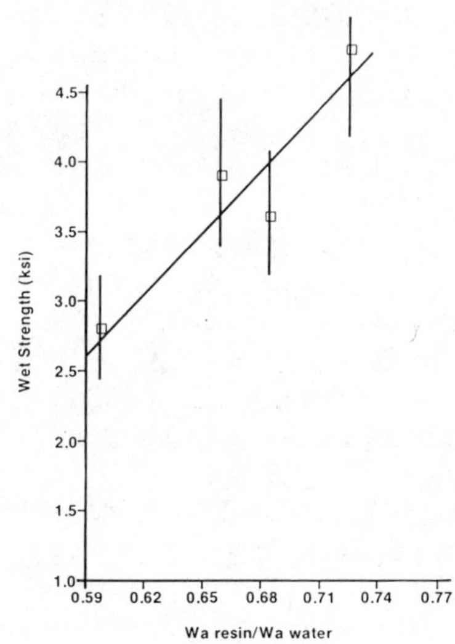

Fig. 9. Single filament interfacial shear strength after water boil vs. the ratio of work of adhesion of epoxy resin on glass filaments to the work of adhesion of water on the filaments.

It should be noted that the 0.2 wt% γ-MPS point does not fit the trend of the rest of the data. We are confident that this is a real phenomenon. It is quite possible that for a low concentration of the silane (which produces low molecular weight deposit), silane-epoxy interdiffusion increases the W_a(resin)/W_a(water) ratio. Apparently, in the absence of chemical bonding, interdiffusion alone cannot improve resistance of the adhesive joint to degradation by water.

Role of Silane/Epoxy Chemical Bonding in Adhesion

According to Plueddemann (15), increased cross-linking and increased rigidity of the interphase are essential for good adhesion in a glass/resin system. Eckstein and Dreyfuss (16) reported improvement in glass/polybutadiene rubber adhesion with increased resin functionality of an amino compound. Chiang and Koenig (17) postulated better silane/epoxy adhesion with multi-amino silane coupling agents. Emadipour, et al. (18) demonstrated improved adhesion when using diamino- over monoamino-silane.

In this work, we wish to elucidate the role of the glass/silane and silane/resin covalent interfacial bonds in adhesion and to evaluate their contribution to the resistance of the interface to degradation by water. Therefore, a series of aminosilanes with increasing resin-reactive and/or glass-reactive functionalities were applied from water solutions to glass fibers.

To facilitate comparison, the molar concentration of all the solutions was constant, 0.045 M. In the case of γ-APS, this concentration corresponds to a solution of 1.0 wt%. The pH of the solutions was 3.5, acidified with acetic acid. The compounds and the measured dry and wet interfacial shear strengths are summarized in Table 3.

Generally, better dry and wet adhesion are achieved by increasing the functionality of both the resin-reactive groups (in this case, the amino groups) and/or the glass-reactive alkoxy and/or silanol groups. However, the degree of degradation by water of the glass/resin interfacial bond is independent of the number of resin-functional groups in the silane compound, while it is strongly affected by the concentration of glass/silane interfacial bonds (Fig. 10). Thus hydrolytic stability of adhesive bonds can be improved

Table 3. Effect of glass and resin functionality of amino silanes on interfacial shear strength.

Silane coating	# glass reactive	# resin reactive	Interfacial Shear, ksi dry	wet	% wet retention
Water, no silane	--	--	13.4	3.6	27
$NH_2(CH_2)_3Si(OEt)_3$*	3	1	9.3	8.4	90
$NH_2(CH_2)_2NH$-$(CH_2)_3Si(OMe)_3$**	3	2	12.7	9.8	77
$NH_2(CH_2)_2NH(CH_2)_2$-$NH(CH_2)_3Si(OMe)_3$**	3	3	13.6	12.8	94
$NH_2(CH_2)_3(CH_3)_2SiOEt$*	1	1	9.7	3.8	45
$NH[(CH_2)_3Si(OEt)_3]_2$*	6	1	12.8	14.6	114
$(NH)_2[(CH_2)_3Si(OEt)_3]_2$*	6	2	15.5	16.4	105
$PhNH(CH_2)_2Si(OMe)_3$** ***	3	1	11.4	12.5	110

* Et = ethyl
** Me = methyl
*** Ph = phenyl

by increasing the number of glass-reactive groups in the compound. Alternatively, the degree of degradation by water can be reduced by decreasing the ratio of resin-reactive (amino) groups to glass-reactive (alkoxy or silanol) groups in the silane compound (Fig. 11). As shown in Eq. 4, both hydrogens of a primary amino group may participate in reaction with epoxy. In Fig. 11, we plotted, therefore, the ratio of the number of hydrogens of amino group (N)H to alkoxy SiOX groups in the silane compound as a function of strength retention in wet specimens. A decreased (N)H/SiOX ratio implies a lower magnitude of polar silane-epoxy interactions or a higher hydrophobicity of the silane coating. As demonstrated by n-phenyl-aminopropyltrimethoxysilane treatment, increased hydrophobicity of the silane coating may diminish degradation by water.

We conclude, therefore, that while the existence of silane/epoxy chemical bonding plays an important role in the formation of strong interfacial adhesion, the contribution of glass/silane bonding to the adhesion cannot be overlooked. The degree of degradation of the glass/resin bond by water seems to be dependent on the magnitude of polar interaction or hydrophobicity of the silane coating.

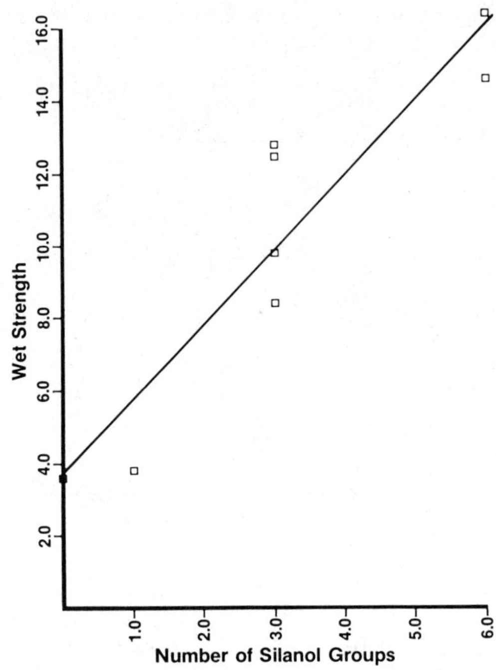

Fig. 10. Effect of the number of silanol groups of amino silanes on the interfacial adhesion after five hour water boil.

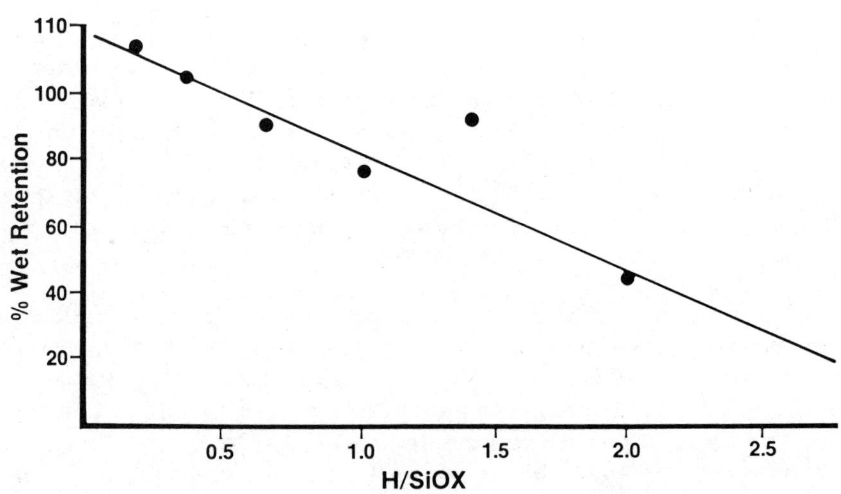

Fig. 11. Ratio of number of hydrogens bound to nitrogens to number of alkoxy groups.

CONCLUSIONS

1. The concentration of a silane solution is a significant factor in determining the molecular structure of a film deposited from that solution. Hydrolysis and polymerization take place in solutions at all concentrations; however, more dilute solutions form lower molecular weight products than do solutions at higher concentrations. For the more concentrated solutions, both an uncross-linked species and a relatively highly cross-linked species are present.

2. In glass fiber/epoxy composites, silane concentration of the fiber coating solution affects interfacial adhesion and thermodynamic wetting by liquid epoxy resin. The optimum concentration with respect to maximum shear strength varies with the silane organofunctionality. The affinity of the resin for the fiber may lead to diffusion of the resin into the silane coating, creating an interlocking network. This would explain the high interfacial adhesion of low concentrations of γ-MPS.

3. As the work of adhesion of glass single filaments in epoxy resin increases, so does the interfacial glass/resin adhesion.

4. The ratio between the resin wetting and the water wetting of a fiber can be related to the wet strength, indicating a competition between resin and water for the glass surface.

5. Stronger interfacial adhesion is achieved by increasing the amine or the alkoxy functionality in a series of aminosilane compounds. While the amine functionality does not strongly affect the degree of degradation of the interfacial adhesion by water, the alkoxy functionality does, with more alkoxy groups, giving better hydrolytic stability.

6. Enhanced interdiffusion and chemical bonding seem to be the most important mechanisms in formation of strong interfacial adhesion. On the other hand, reduced silane-resin polar interaction and increased hydrophobicity of the silane coatings are of major consequence in producing glass/resin interfaces resistant to degradation by water.

ACKNOWLEDGMENTS

The authors wish to thank M.G. Wisnosky and C. Gramelt for their valuable assistance.

REFERENCES

1. E.P. Plueddemann, "Silane Coupling Agents," Plenum Press, 1982.
2. R. Wong, J. Adhesion, 4, 171 (1972).
3. E.J. Berger and Y. Eckstein, "Epoxy Resin Wetting of E Glass Single Filaments as it Relates to Shear Strength," ACS National Meeting, September 1982.
4. H. Ishida and J.L. Koenig, Appl. Spectros., 32 (5), 462 (1978).
5. K.H. Allen, A.K. Hansrani and W.C. Wake, J. Adhesion, 12, 199 (1981).
6. K.H. Allen, A.K. Hansrani and W.C. Wake, Adhesion and Adhesives: Science, Technology and Applications, Intern. Conf., Gray College, Durham, Plastics and Rubber Inst., London.
7. H. Ishida and J.L. Koenig, J. Polym. Sci., Polym. Phys. Ed., 18, 1931 (1980).
8. H. Ishida, C. Chiang and J.L. Koenig, Polymer, 23, 251 (1982).
9. F.J. Boerio and J.E. Greivenkamp, Proc. 32nd Ann. Tech. Conf. Reinforced Plastics/Composite Div., SPI, Sec 4A (1977).
10. J. Jen, Ceramic Engineering & Science Proceedings, Vol 9 and 10 (1982).
11. D.H. Kaelble in: Physical Chemistry of Adhesion, Wiley Interscience, New York, p. 124 (1971).
12. F.J. Boerio, S.Y. Cheng, L. Armogan, J.W. Williams and C. Gosselin, Proc. 35th Ann. Techn. Conf. Reinforced Plastics/Composites Div., SPI, Sec. 23C (1980).

13. D.H. Kaelble, Physical Chemistry of Adhesion, Wiley-Interscience, New York, Chapter 5 (1971).

14. K.L. Loewenstein, in: Composite Materials, L. Holliday, ed., Elsevier Publ. Co., p. 150 (1966).

15. E.P. Plueddemann, in Composite Materials, Vol. 6, Ch. 6, Academic Press, New York, p. 173 (1974).

16. Y. Eckstein and P. Dreyfuss, J. Adhesion, 15, 193 (1983).

17. C. Chiang and J.L. Koenig, J. Polym. Sci.: Polym. Phys. Ed., 20, 2135 (1982).

18. H. Emadipour, P. Chiang and J.L. Koenig, Res. Mechanics, 5, 165 (1982).

Discussion

On the Paper by L. H. Lee

M. Mathur (IBM Corp., Tucson): Water-based coating systems may cause environmental problem due to atmospheric water assisted decomposition especially of isocyanate systems, discussed in your talk.

Are you concerned about it? Please comment.

L. H. Lee (Xerox Corp.): The water-base adhesive systems are rather new. There have been problems associated with the change-over from the solvent-base to the water-base systems. The pollution problem of effluents should be addressed properly. However, in recent years, this problem has been carefully considered by most industries. I do not think that the isocyanate waste can ever be discharged without any proper treatments.

R. C. Ihlint (Lockheed Missles Co.): In connection with fracture mechanics, and crack radius, is the radius related to shear strength, block tensile strength or peel strength?

L. H. Lee: The crack radius is related to fracture energy G_c. Thus, it is not directly related to any of the above.

R. C. Ihlint (Lockheed Missles Co.): Do you foresee waterborne adhesives used in structural adhesives?

L. H. Lee: Yes. Whether the industries want them or not, OSHA will force us to go that direction. The waterborne systems currently developed are epoxies, phenolics, polyurethanes, etc. These are already structural adhesives. Besides cyanoacrylates and anaerobics, most structural adhesives available today can be readily transformed into waterborne systems.

J. A. Simms (DuPont): What are the technical problems that need to be solved to allow one large growth of waterborne adhesives to occur?

L. H. Lee: We do not foresee any major technical problems before the large growth of waterborne adhesives to occur. There are more waterborne adhesives on the market today than four or five years ago.

For a changeover from a solvent-base to a waterborne system, there may be some technical or operational problems, e.g., foaming, leveling, cleaning, disposing of unusable adhesives or sludges in the storage tank. However, the changeover is generally welcomed by workers in the plant. Furthermore, the energy cost for drying is generally less than those for solvent systems. Once the adhesive industries realize that there are indeed many advantages of the waterborne systems, the growth will be phenomenal.

An article entitled "Converting to Water-based Adhesives: One Manufacturer's Experience" (Adhesives Age 22, No. 12, 21 (1979)), describes clearly some of the problems mentioned in the above paragraph.

On the Paper by M. C. H. Lee

J. M. Fowkes (Lehigh University, Bethlehem, PA): The contact angle θ is a measure of the interaction of three interfacial tensions (γ_{LV}, γ_{SV}, γ_{SL}); $\gamma_{LV}\cos\theta$ then measures the difference between γ_{SV} and γ_{SL}. There is not enough information to determine either γ_{SV} or γ_{SL}. The γ_{SV} values in the literature, supposedly determined by contact angles, depend on a further assumption (which is not always obvious) that γ_{SL} is zero when $\theta = 0$. This assumption sometimes seems quite reasonable, but cannot be proven. It is clearly incorrect for contact angles on mercury (where γ_{HgL} and γ_{HgV} are measureable), and for contact angles on mineral sulfides or on talc. I recommend that you make your assumptions quite clear.

M. C. H. Lee (GM Research Laboratories, Warren, MI): One of the purposes of this theory is designed to obtain relationships for determining γ_{SV} and γ_{LS} from the contact angle measurement, namely, γ_{LV} and cosθ. In order to accomplish this, firstly, I derived an equation of state for adhesion and the hypotheses for this equation of state are clearly stated in the text. Using this equation of state in conjunction with Young's equation, I then obtained mathematical forms for determining γ_{LS} and γ_{SV}. In this stage of derivation, I used the assumption that the minimum value for γ_{LS} is zero, which was also suggested by Rhee (14).

N. Levy (Bell Laboratories): How can a contact angle measurement with relatively poor precision be transformed into a precise prediction of bonding efficiency?

M. C. H. Lee: Calculation shows only trend.

On the Paper by F. Liang and P. Dreyfuss

W. V. Cohen (Heminway and Bartlett): Would amine containing silanes increase adhesion of urethanes to other surfaces than glass or metals? How about increased adhesion improvement on synthetic polymers, specifically on polyamides, e.g., Nylon?

P. Dreyfuss (University of Akron): Adhesion of urethanes on Nylon would be expected to be high in any case. However, amino containing silanes might further improve this adhesion. We have no data with synthetic fibers.

H. Doniuk (Israel Ministry of Defense): Have you tried to prove the formation of chemical bond by spectroscopic methods such as infrared, etc.?

P. Dreyfuss: In some earlier work with Eckstein, we have done extensive work on the reactions of amines with triethanol silanol and/or fumed silica. Evidence from infrared, nmr, mass spectrometry, elemental analysis and preparation of model compounds all consistently indicate that chemical bonds do form. Amines are known to react in a variety of ways in the course of the synthesis of polyurethanes, as I'm sure you know. Thus we have no doubt that chemical bonds can and do form. References to our earlier work can be found both in the preprint and in our paper on the present work.

On the Paper by Y. Eckstein and E. J. Berger

D. D. Wheeler (Allied Corp.): Since the silane concentration in the treating solution often effects the amount picked up by the fiber, was the amount of silane applied to the fiber determined at the various concentrations?

Y. Eckstein (Owens-Corning Fiberglas): The amount on the fiber was not determined.

PART TWO:

PHYSICS OF ADHESION–CHARACTERIZATION OF SURFACE AND BULK OF ADHESIVE SYSTEMS

Introductory Remarks

Louis H. Sharpe

Bell Laboratories

Murray Hill, New Jersey 07974

The papers in this session are quite varied as to subject matter. The first two apply to techniques by means of which it is possible to learn something about the chemistry of what I like to call the interphase region of an adhering system, and something about failure processes. Another makes use of an interesting effect of viscosity of a medium on the fluorescence of a component of that medium to learn something about the kinetics of cure of an adhesive system.

The fourth and fifth papers have to do with properties of pressure-sensitive adhesives. In particular, the matter of how the materials composing pressure-sensitive adhesives (rubbers and resins) interact and phase separate to produce the phenomenon of tack or pressure-sensitivity is addressed. Both studies use dynamic mechanical measurements to uncover phasing - one in a silicone and the other in natural and styrene-butadiene rubber systems tackified with various resins.

The final paper, a contribution from The Institute of Chemistry in Beijing, covers the development and some of the properties of a CTBN-toughened epoxy adhesive cured with an imidazole.

In the paper by Briggs, derivatization techniques and the use of ESCA lead to detailed knowledge about the chemistry of the surface of polyethylene which has been "corona" treated. Correlations are made between the chemical species ("functional groups") on the surface of treated polyethylene and the strength of joints made with that polyethylene. Detailed information about the surface composition of treated polyethylene is, of course, interesting. However, there is some concern about whether it is proper to interpret such a simple cause and effect relationship in such a broad manner. It is questionable whether the modification of the polymer by corona discharge is confined strictly to the surface. It is far more likely that the polymer is modified to a depth of many tens or hundreds of angstroms below the surface. Such a modified region (an "interphase") will influence the failure process in a joint. Because of this, and because an adhesive joint is a <u>system</u> of materials with coupled responses, we should be very careful about broad-based extrapolation of simple cause and effect relationships which derive from application of a single analytical technique.

The paper by Dickinson, on fracto-emission, describes and illustrates a phenomenon which involves the emission of electrons, positive and negative ions, neutral species and photons during, and subsequent to, fracture of certain materials. The utility of the technique is based on the fact that the type(s) and flux of particles emitted is dependent upon the materials(s) being fractured and the details of fracture. The author says that the method is useful for distinguishing between cohesive and "adhesive" failure. One would expect that this could be so. However, the evidence that the author presents is not convincing, inasmuch as the detection of "adhesive" failure was based on SEM photographs and, as the author states, his criterion for "adhesive" failure was that the surface examined was, "...free of macroscopic pieces of polymer..." However, despite the fact that there are certain experimental difficulties with the technique - particularly that fracture and measurement must be carried out in high vacuum (10^{-6} to 10^{-9} torr) - the technique merits further development. Is energy resolution, i.e., fracto-emission spectroscopy, being contemplated?

The third paper, by Levy and Ames, presents an interesting new way to monitor cure kinetics. They measure the viscosity of an epoxy system as it cures by use of a fluorescent probe the intensity of which increases as the viscosity of the medium increases. Since the viscosity of the system is a function of molecular weight, the increase in intensity of the fluorescence can be used to get at the kinetics of the curing reaction. In addition, they found that one of the epoxy resins which they used in the study had self-probe capability, i.e., _it_ fluoresced. Thus, by using two probes, it may be possible not only to study the micro-rheological environment of the reacting epoxy molecule but, also, the rotational restrictions which occur as the epoxy molecule is reacted into the polymer network. In addition, the method shows promise for the study of the rheology of polymerizing systems over wider ranges of viscosity than is now possible without fear of damage to measuring equipment.

The paper by Copley presents a study of two types of silicone pressure-sensitive adhesives, of known composition, by dynamic mechanical and thermal analysis. The materials were prepared from a siloxane resin and two silicone gums: poly(dimethylsiloxane) and poly(dimethyl-co-phenylsiloxane).

The dynamic mechanical analysis shows that both pressure-sensitive materials are heterogeneous blends, i.e., they consist of a gum-rich phase which coexists with a resin-rich phase of resin and gum. In this respect, they are similar to other pressure-sensitive adhesives containing other resins and rubbers.

In a complementary paper by Class and Chu, model resins - polystyrene and poly(vinylcyclohexane) - in combination with natural and styrene-butadiene rubbers, were used to study effects of resin structure, molecular weight and concentration on viscoelastic properties of pressure-sensitive adhesives resulting from these combinations.

They reach the general conclusion that the compatibility of the various resins with the various rubbers impacts on the viscoelasticity of the blends and therefore on their pressure-sensitive performance. To the extent that resin structure and molecular weight determine compatibility, they are the variables of importance. They also reach the conclusion that the tan delta peak temperature for compatible rubber-resin blends can be estimated by the inverse temperature law of Fox.

The final paper in the session, by Yu Yunchao, et al., has to do with CTBN-toughened epoxy adhesives which use 2-ethyl, 4-methyl imidazole (EMI) as the curing catalyst. The advantage of this catalyst is that it permits the curing of the material at relatively low temperature, e.g., 80-120C, to give respectably tough adhesives with high lap shear strength. It should be mentioned, however, that these materials are not shelf-stable at room temperature - a disadvantage - and that there are available relatively shelf-stable, one-part materials, cured with dicyandiamide and an accelerator, which cure at the top of the above-mentioned range, 125°C. Another point worth mentioning is that non-adducted systems, that is those in which the epoxy resin and the CTBN are not pre-reacted (as in the present case), are not stable at room temperature because the adducting reaction (esterification) will occur at room temperature.

Editor's Note: The final paper by Yu Yunchao is included in Part V. The contribution by T. C. Ward, A. W. Brinkley, M. Sheridan and R. Konig is included instead.

New Applications of ESCA

D. Briggs

ICI Petrochemicals and Plastics Division
Wilton, Middlesbrough, UK

ABSTRACT

The importance of strong specific interactions in polymer adhesion (eg., covalent bonding, acid-base interaction) is well-known. However, these interactions are difficult to study directly, particularly for multifunctional surfaces. Derivatisation techniques have been developed which allow X-ray photoelectron spectroscopy (ESCA) to quantify functional group populations and to assess the success of specific modification techniques. Derivatised surfaces can then be used to study the role of specific interactions.

I. INTRODUCTION

A principal goal of adhesion science is to understand the interfacial interactions which give rise to high levels of practical adhesion. Empirically it is observed that for a large number of substrates some kind of surface pretreatment is required before adequate adhesion can be achieved and this raises the related questions: what does the pretreatment do to the surface, i.e., what is the modification mechanism? and what is the mechanism by which adhesion is improved? Experiments aimed at addressing these questions have, over the last twenty years or so, given rise to a great deal of controversy. The reasons for this are not hard to find. Many factors affect adhesion, e.g., surface chemistry and morphology, the thermodynamics and kinetics of wetting, the presence of low molecular weight 'contamination' (often referred to as a 'weak boundary layer'). For a complete understanding of any adhesion situation all these factors need to be characterised and most experiments tend to concentrate on one factor at a time.

In a discussion of the controversy surrounding adhesion to the hydrocarbon polymers - polyethylene and polypropylene, Brewis and Briggs (1) pointed out the improvement in understanding which had come from the application of ESCA to studies of polyolefin surface modification. A systemic re-examination of those processes, both of commercial and academic interest, which improve adhesion to polyolefin surfaces has been carried out by Briggs and co-workers (2-10). In these studies ESCA was combined with traditional techniques such as contact-angle, reflection-infrared spectroscopy and scanning electron microscopy together with practical joint strength measurements of an appropriate kind. The processes studied were chromic acid treatment (2,3), flame treatment (4), electrical ('corona') discharge treatment (5-8), melt extrusion (9-10), and ozone treatment (10). The results, summarised in Reference 1, demonstrated the crucial importance of surface chemical analysis by a technique sensitive to the composition of the uppermost layers (ESCA typically samples 20-100 Å into a material compared with several microns for reflection IR) but essentially uninfluenced by microroughness (a drawback of contact angle measurements).

During the last few years there has been an increasing awareness of the importance of specific interactions in adhesion, largely brought about by the work of Fowkes and co-workers (11). Their demonstration of the important role of acid-base interactions (which includes H-bonding) covers a wide range of adhesive situations. The ESCA work mentioned above had led us to conclude that a detailed analysis of oxygen containing functional groups introduced into polyolefin surfaces during pretreatments was necessary to investigate the importance of specific interactions, especially H-bonding, in adhesion, but that this was impossible without development of the ESCA technique. How this was achieved for discharge-treated polyethylene, and why, is described below.

II. ELECTRICAL DISCHARGE TREATMENT

Electrical ('corona') discharge treatment of polyethylene and polypropylene film has been used for many years in order to render these surfaces printable and suitable for lamination or coating. In this process the polymer film is passed over a grounded metal roller covered with a "dielectric" (insulating) material. Separated from the film by $\simeq$2 mm is an electrode bar to which a high voltage is applied (typically 15 kV at 20 K Hz). Air in the film-electrode gap is ionised, the corona discharge thus formed is stable and this "treats" the film surface.

Discharge treatment of low-density polyethylene (LDPE) has been studied by many workers, particularly with reference to the autoadhesion enhancement effect. LDPE autoadheres when two surfaces are contacted under pressure at temperatures above $\simeq 90^{o}C$. However, after fairly low levels of discharge treatment, treated surfaces will autoadhere at significantly lower temperatures ($\simeq 70^{o}C$). Two theories had been proposed to account for this effect. The first, due to Canadian workers (12), suggested that electret formation was involved - the resulting increased adhesion being electrostatic in nature. The evidence for this was that discharge treatment in both "active" (air, O_2) and "inert" (N_2, Ar, He) gases gave the effect, its magnitude being related to the power dissipated in the discharge irrespective of which gas. Moreover the maximum effect was achieved in oxidising atmospheres before ATR-IR spectroscopy could detect any surface oxidation. The second, diametrically opposed theory due to Owens (13), suggested that hydrogen bonding between polar groups formed by the discharge treatment was responsible. The effects of reacting treated surfaces with specific chemical agents and noting subsequent adhesive characteristics suggested a specific interaction between carbonyl (keto, aldehyde) and tautomeric enol functions across the interface. However, only LDPE treated in air was studied.

In our work (5) discharge treatment was carried out in a model apparatus with static film samples and at low frequency (50 Hz). As Fig. 1 demonstrates, ESCA showed (5) that treatment in air, N_2 and Ar (at atmospheric pressure) leads to surface chemical changes including oxidation in all cases. Clearly treatment in Ar requires longer time but power dissipation measurements confirmed the earlier result (12) that the power required to achieve a given level of autohesion (or peel strength) was independent of the gas. It is also clear from Fig. 1 that this corresponds roughly to the same degree of surface oxidation. On the basis of these results the electret theory can be discounted but the Owen's theory is given a firmer foundation. Fig. 2 shows typical spectra for LDPE discharge treated in air (5). A simple deconvolution of the high binding energy shoulder on the C1s peak gives three peaks which can be ascribed to $-CH_2O-$ (e.g., alcohol, ether, ester, hydroperoxide) at $\simeq$286.5 eV, $>C=O$ (e.g., aldehyde, ketone) at $\simeq$288.0 eV and

$$-\overset{\displaystyle O}{\overset{\|}{C}}-O-$$

(carboxylic acid or ester) at $\simeq$289.5 eV. The O1s peak is even less informative. Most oxygen functional groups give O1s BE's of $\simeq$532 eV, the exception is the ester oxygen in carboxyl groups at $\simeq$533.5 eV. The shift in the O1s peak, shown in Fig. 2, with increasing oxidation reflects the increasing relative concentration of carboxyl groups. Clearly this information is not specific enough to probe the detailed structure of the oxidised layer or to allow correlations to be made with the autohesion results. To

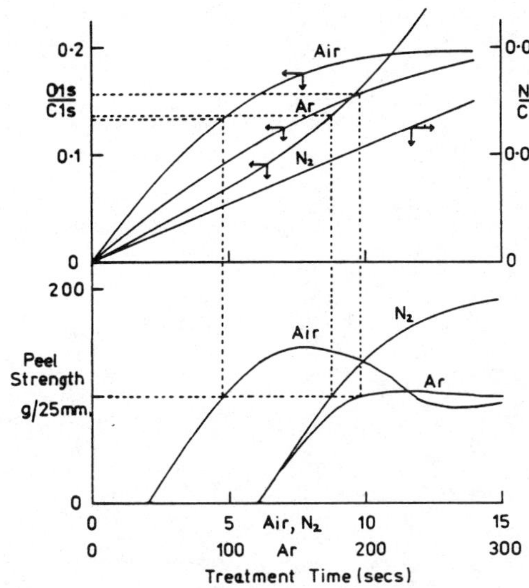

Fig. 1. Variation of autohesion of LDPE (peel strength for seals made at 75°C for 2 secs at 15 psi) and surface chemistry with discharge treatment time in air, nitrogen (50 Hz, 13.7 kV peak) and argon (50 Hz, 2.2 kV). The O1s : C1s intensity ratio is a qualitative measure of surface oxidation. Note the similar values of this parameter for surfaces giving 100g peel strength - dotted lines. The N1s : C1s ratio was measured for films treated in N_2 only (Ref. 1).

overcome this problem a series of derivitisation reactions have been devised to label specific functional groups.

III. DERIVATISATION

The criteria for successful derivatisation are rather stringent and can be listed as follows:

1) the reagent must undergo a specific reaction with a particular functional group and introduce into the surface a heteroatom "label", preferably an element with a high cross section for X-ray photoemission;

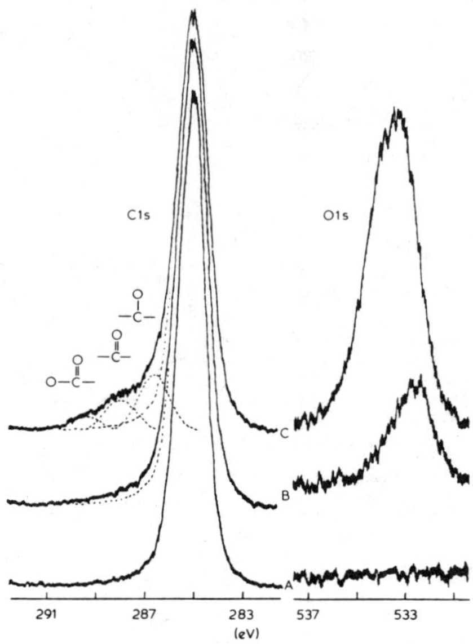

Fig. 2. C1s and O1s peaks for LDPE treated at 50 Hz, 13.7 kV in air for (A) 0 sec, (B) 8 sec and (C) 30 sec respectively (Ref. 5).

2) the derivatising reaction should proceed rapidly under mild conditions;

3) solvents, if necessary, must be benign.

Conditions 2) and 3) are the most difficult to meet. Reactions which proceed rapidly at room temperature in the solution phase are often sterically hindered in the polymer surface layers. Solvents which permeate into the polymer are likely to aid reaction but may, at the same time, give rise to surface reorganisation, e.g., functional group migration into the bulk.

These effects have recently been studied by Everhart and Reilley (14). Solvents may also extract lower molecular weight material produced during the surface modification process.

The solvent-type reaction conditions described above are to some extent a compromise between achieving complete derivatisation of the modified surface (which angular variation ESCA experiments show to have a thickness comparable with the sampling depth of the technique) and avoiding solvent attack on the very material under study. They are the result of many experiments, which also showed that reproducibility was very dependent on the use of fresh solutions of the derivatising agent. We believe that derivatisation has reached a high degree of completion (compared with the "infinity" result under the conditions used).

Fig. 3 summarises the reactions taking place with the reagents used in this study. The derivatisation of aldehyde- or keto-carbonyl groups with pentafluorophenylhydrazine (PFPH) or hydrogen atoms in methylene groups adjacent to these carbonyls (α - H) with Br_2 and of carboxylic acid groups with NaOH have been discussed previously (6). Bromine also reacts with unsaturation in the polymer and in quantification the "blank" value for Br uptake was subtracted from that of the modified polymer (assuming discharge treatment does not affect double bonds). Full experimental details for the other reactions can be found in Ref. 7.

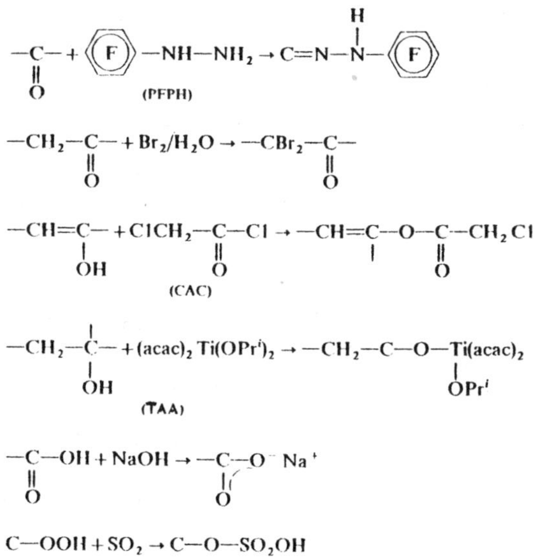

Abbreviations:
PFPH = pentafluorophenylhydrazine
CAC = chloroacetylchloride
TAA = di-isopropoxytitanium bisacetylacetonate.

Fig. 3. Derivatisation reactions employed.

Acid chlorides react potentially with alcoholic -OH and also with the enol tautomer of the above-mentioned carbonyl groups (the equilibrium would normally be heavily in favour of the carbonyl). Under the conditions used with chloroacetyl chloride (CAC) it seems that the -OH reaction is predominantly with the enolic -OH (see below). The search for a mild derivatising reagent for alcoholic -OH eventually led to the reagent di-isopropoxide titanium bis-acetylacetonate (TAA) which is sold as an adhesion promoting, OH cross-linking agent. The evidence from carrying out multiple reactions, discussed below, is that this reagent does not derivatise enolic -OH. Prior reactions of a surface with NaOH to remove carboxylic -OH did not markedly affect the TAA reaction, so the reagent does not apparently react with carboxylic acid groups either.

The specific reaction of SO_2 with hydroperoxides has previously been described (15) and this is used, with IR, as a diagnostic tool for bulk R-OOH assay. Being a gas-phase reaction it is admirably suited to the ESCA derivatisation technique.

The level of discharge treatment used in this study is quite high, but not by any standards excessive. (Experimental details can be found in Ref. 7.) It corresponds to the level of maximum autohesion under the heat-sealing conditions employed in this study (Fig. 3). It also corresponds to a level at which the film used would pass a searching printability test. The surface energy measured using the ASTM wipe-test was 55 dyne cm^{-1}.

Both chain scission and cross-linking take place. The key intermediate is the hydroperoxide group, whose stability and decomposition has been the subject of much research. Hydroperoxides in polyethylene can have long lifetimes so if this mechanism is correct these groups should be detectable. The SO_2 reaction is positive identification and, we believe, the first direct evidence for this mechanism. Of the groups likely to be produced by hydroperoxide decomposition, derivatisation techniques have therefore identified -C=O, C-OH and COOH.

The consensus view in the literature for the likely mechanism of oxidation during discharge treatment of LDPE is the following:

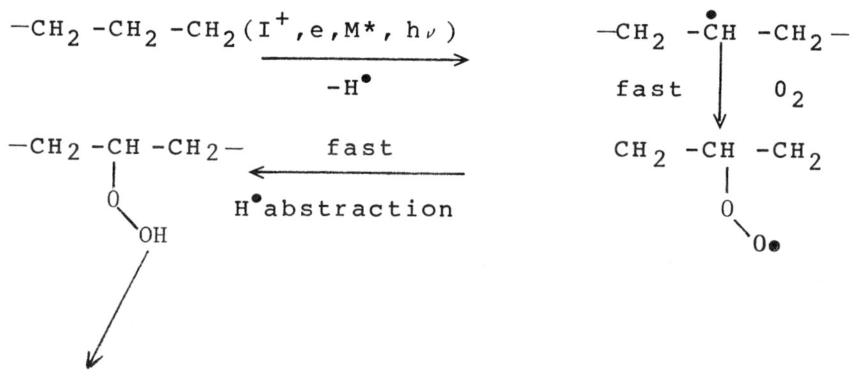

Products (-C=O, C-OH, C-O-R, -COOH, -COOR, etc)

The ESCA data can be quantified, as previously described (7) to give the data in Table 1. The value for the population of $CH_2C=O$ groups assumes that on average two α-H atoms will be replaced during bromination. Since this group can tautomerise to give one enol -OH the population of $-CH_2C=O$ assessed by Br_2 uptake and enolic -OH assessed by the CAC reaction should be comparable, as is observed. The raw C1s spectra tend to show broadly similar intensities for the C-OH (etc), >C=O and -COOH regions which is also borne out by these results. The total assay of >C=O, C-OH and COOH groups would give an atomic O:C ratio of 5.7×10^{-2} where "C" is the carbon atoms in the original surface. This compares with the actual value of 8.7×10^{-2} from the discharge treated surface. Considering that ether and ester groups are also likely to be present, in numbers comparable to the groups which have been derivatised, this assay is seen to be entirely reasonable. The apparent internal consistency of these results is additional evidence for the essential reliability of the derivatisation procedures used.

Table 1. Quantification of Functional Groups

REACTION	ESCA intensity ratio (core level/Cls)[a]		Atomic ratio (element/carbon)	Number of functional groups per original surface $-CH_2-$	
PFPH	(Fls)	0.205	5.5×10^{-2}	>C=O	1.1×10^{-2}
Br_2/H_2O	(Br3d)	3.6×10^{-2}	10.6×10^{-3}	$CH_2C=O$	5.3×10^{-3}
CAC	(Cl2p)	1.3×10^{-2}	6.0×10^{-3}	C-OH	6.0×10^{-3}
TAA	($Ti2p3_{/2}$)	6.2×10^{-2}	1.5×10^{-2}	C-OH	1.5×10^{-2}
NaOH	(Nals)	8.8×10^{-2}	1.1×10^{-2}	-COOH	1.1×10^{-2}
SO_2	(S2p)	7.6×10^{-3}	4.7×10^{-3}	C-OOH	4.7×10^{-3}
None	(Ols)	0.209	8.7×10^{-2}		

[a]Estimated error ± 5% for a given sample, ± 15% for the complete experiment.

II. SPECIFIC SITE BLOCKING AND ADHESION MEASUREMENTS

The derivatisation technique as developed for the extension of ESCA analytical capacity has the added bonus that the same surfaces can be used to study specific interactions in adhesion, in the spirit of Owens' approach (13), but with a built-in monitor of surface composition. This overcomes the possibility that artefacts, such as contamination, loss of surface material or solvent-induced re-orientation effects could be responsible for the observed adhesion changes - artefacts that Owens' experiments could not rule out.

We previously used these techniques to show (6) that hydrogen bonding between carbonyl and enol functions on opposite discharge treated LDPE surfaces was responsible for autohesion, as postulated by Owens (13). The results also showed that the required keto-enol tautomerisation occurred. In this work we demonstrated (6) that the blocking of enol functions prevented adhesion between discharge treated LDPE and a commercially obtainable printing ink. We now extend this investigation of the role of specific interactions using the autohesion between discharge treated LDPE surfaces.

It should be emphasised that under the conditions of heat sealing used (at 85°C) untreated LDPE surfaces do not autohere, in other words, an interdiffusion mechanism is unlikely.

Table 2 shows the effect on adhesion of reacting the surfaces with PFPH and TAA. As reported earlier PFPH prevents adhesion by eliminating enolisable carbonyl groups. On the other hand, reaction with TAA actually increases adhesion above that of the discharge treatment alone. This would be consistent with the opening of another specific interaction "channel", namely cross-linking of OH groups via the Ti complex. Since the above ESCA analysis suggests that enolic and alcoholic OH groups can be separately derivatised, then reaction with PFPH and TAA should give independent control over these two sites for specific interaction. This is clearly seen in the case of sequential reaction with the two reagents. Despite the increased error involved in carrying out two solvent based derivatisation reactions the ESCA data are reasonably self-consistent also.

Table 2. ESCA and Autohesion Measurements for Derivatised Surfaces.

TREATMENT	Cls ($3x10^4$)[a]	Ols (10^4)	Fls (10^4)	Ti2p3/2 ($3x10^3$)	Peel strength[b] g/25 mm
DT	20.4	12.8	-	-	250
DT-PFPH	20.0	7.5	12.3	-	0
DT-PFPH-TAA	18.7	11.2	9.7	15.2	241
DT-TAA	19.8	12.9	-	15.3	390
DT-TAA-PFPH	18.8	15.0	9.1	11.6	180

a Counts sec^{-1} full scale of range on which peak was measured

b Estimated error ± 5% for discharge treated samples, ± 10-15% for derivatized samples

Table 3 gives similar results for the reactions with CAC and TAA. This time CAC reacts directly with the enolic -OH. Also shown are the water contact angles. Clearly there is no correlation between contact angle and adhesion. Broadly speaking the contact angle hysteresis is constant (the error here is $\pm 4^{o}$) indicating that no major changes in surface roughness or heterogeneity are introduced by the chemical treatments.

Although the adhesion values are from single experiments many series of experiments confirmed the observed trends. Both the discharge treatment and the heat sealing stages of the experiment are subject to significant variability and best results were always obtained for a complete set of data obtained on any single day. It should be noted from Tables 2 and 3 how consistent the ESCA and peel strength data are when these are comparable. The TAA reaction is the most novel extension of this investigation and Fig. 4 shows that the adhesion improvement is achieved across a wide range of discharge treatment level. Incidentally, we believe the smoothing out of the dip in the original autohesion curve is due to the solvent removal of the low-molecular-weight oxidised material during the derivatisation reaction.

Table 3. ESCA, Contact Angle and Autohesion Measurements for Derivatised Surfaces.

TREATMENT	Cls $(3x10^4)$	Ols (10^4)	Cl2p$3/_2$ (10^3)	Ti2p$3/_2$ $(3x10^3)$	θa (deg)[a]	θr	Peel strength g/25 mm
None					104	96	0
DT	20.4	12.8	-	-	68	63	220
DT-TAA	19.4	12.7	-	11.6	86	73	393
DT-CAC	20.0	11.4	7.9	-	90	87	12
DT-CAC-TAA	18.0	10.6	8.2	11.8	79	73	283

a Advancing and receding water contact angles

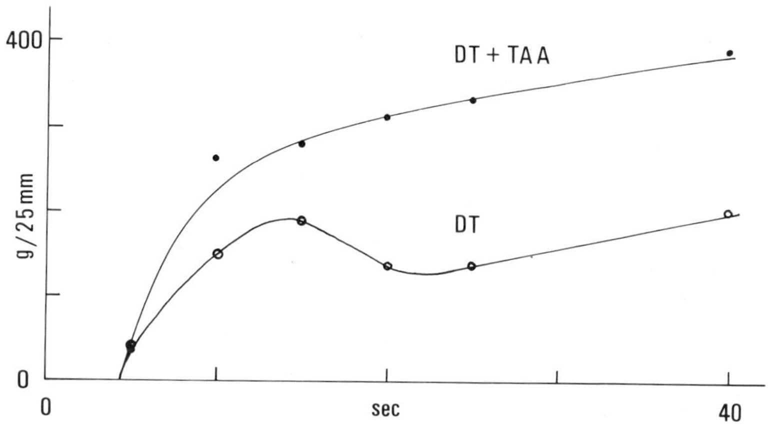

Fig. 4. Variation of autohesion of LDPE (peel strengths for seals made at 85°C for 2 secs at 15 psi) after discharge treatment in dry air at 12.6 mW cm^{-2} (DT) and after subsequent treatment with diisopropoxytitanium bisacetylacetonate (DT + TAA), as a function of discharge treatment time.

Similar experiments with SO_2 reacted surfaces showed that this had no effect on autohesion. Carley and Kitze (16) have recently speculated on the role of peroxidic groups in discharge treated surfaces. Their analysis of these groups, by reaction with diphenylpicrylhydrazyl (DPHH) and subsequent colorimetry, excluded the possibility of hydroperoxide reactions since these groups could not be detected by ATR. The results presented here must cast some doubt on the validity of their method.

In our previous paper we concluded (6) that NaOH reaction with discharge treated surfaces did not affect adhesion; thus carboxylic acid groups are not implicated in adhesion mechanisms. Although Owens observed (13) a similar result we now find that the situation is much more complex. Depending on the conditions of the NaOH reaction. (Concentration, exposure time and washing procedure), almost any value of adhesion from zero to an undiminished value could be obtained even though ESCA showed derivatisation had taken place. We are therefore inclined to agree with Everhart and Reilley (17) that this reaction is unreliable and a better derivatising agent needs to be found for examining the role of carboxylic acid groups.

The quantitative ESCA data shows that only a fraction of the polar groups which lead to the increase in surface energy (or wettability) of LDPE need be involved in giving adhesion values as high as 400g/25 mm. In the case of hydrogen bonding via enolic -OH only 3×10^{12} groups cm^{-2} are involved (assuming 5×10^{14} carbon atom cm^{-2} in the LDPE surface). In the case of cross-linking via alcoholic -OH twice as many groups are involved.

It is interesting to note that whilst the enhanced autohesive bonds resulting from discharge treatment are immediately disrupted by the presence of water (5,13) the use of TAA as a cross-linking agent gives some hydrolytic stability, as might be anticipated. In the presence of water peel strengths of the order of 60-100 g/25 mm are retained throughout the peel test.

V. INTRODUCTION OF SINGLE FUNCTIONAL GROUPS INTO POLYETHYLENE SURFACES

The experiments described above illustrate ways in which the role of individual functional groups in a multifunctional surface can be assessed in a particular adhesion system. A rather different approach, which in principle could yield information of more general utility, would be to produce surfaces containing only one functionality whose adhesive properties could be assessed in a wide range of situations. Polyethylene offers an ideal starting point for such experiments since it is essentially non-adhesive. Modification experiments aimed at introducing single functionalities have been attempted previously but have suffered from problems associated with the inability of the techniques used (e.g., ATR, contact angle) to fully characterise the modified surfaces. We felt that the use of ESCA and associated derivatisation techniques might allow this approach to be re-investigated fruitfully.

The following results are preliminary and it is not appropriate here to give experimental details. It should, however, be emphasised that a great deal of effort was required to eliminate artefacts, such as unwanted side reactions and surface contamination, and to delineate procedures for ensuring reproducibility. These details will be published in due course (18).

The original strategy aimed to introduce isolated Br atoms into LDPE by reaction with Br_2 under UV irradiation. These reactive sites could then be converted into a range of functional groups using specific synthetic procedures. Table 4 gives results for LDPE and HDPE reacted with Br_2/UV, then with tributyl-tin hydride (TBTH) to debrominate back to the original polyolefin and finally rebrominated as in the first step. Attempted rebromination after step-two but in the dark gave rise to minimal Br incorporation (by ESCA) and certainly no more than for the original polymer showing that the TBTH reaction did not produce unsaturation. We believe these results are the first to demonstrate reversible chemical modification of a polymer surface leading to equally reversible changes in adhesive performance. It should be emphasised that rigorous procedures adopted to prevent any oxygen incorporation maintained O:C ratios below 0.7 atomic % in all surfaces, only slightly higher than for the untreated surfaces.

Attempts at replacing Br by nucleophilic substitution-type reactions usually resulted in Br elimination if a reaction actually occurred. This is presumably due to steric effects at the polymer surface. A clean reaction for near-total dehydrobromination leading to the production of unsaturation was found to be the reaction with n-butylamine in tetrahydrofuran (THF). Theoretically rebromination (in the dark) of this unsaturated surface should double the surface Br concentration compared with the initial bromination reaction, and this situation is more-or-less achieved in the case of HDPE, as shown in Table 5. It is interesting to note the significant increase in adhesion of the unsaturated surface compared with the original polyolefin and the fact that similar surface behaviour of LDPE and HDPE, as monitored by contact angle, is not necessarily reflected in similar surface composition measured by ESCA. These results will be the subject of further study. The unsaturation centres introduced by

Table 4.

SUBSTRATE			Joint Strength MNm^{-2} $\pm$ st dev[a]	Failure[b]	θ_e deg[c]	Br:C at %[d]
LDPE	- untreated	(1)	1,4 $\pm$ 10%	I	96	0
(1)	- brominated	(2)[e]	12,7 $\pm$ 13%	I/M	76	11
(2)	- TBTH reduced	(3)	1,8 $\pm$ 7%	I	94	0
(3)	- rebrominated	(4)[f]	10,5 $\pm$ 11%	I	78	8,3
HDPE	- untreated	(5)	0,70 $\pm$ 9%	I	96	0
(5)	- brominated	(6)[e]	11,5 $\pm$ 14%	M	74	5,9
(6)	- TBTH reduced	(7)	0,76 $\pm$ 13%	I	94	0
(7)	- rebrominated	(8)[f]	10.0 $\pm$ 9%	M	72	3,8

a Composite butt joints with an epoxy adhesive, mean of six determinations

b I = apparent interfacial, M = material

c Equilibrium water contact angle

d Derived for ESCA measurements

e 2 hours irradiation in atmosphere of saturated Br_2/N_2 for LDPE, 3.5 hours irradiation in same system for HDPE

f Using same procedure as in (e)

these reactions are ideally suited to further specific conversion to oxygen containing functional groups, e.g., -OH, -COOH, -CHO, -OOH and >C=O. Success with these conversions will result in the monofunctional polyethylene surfaces we desire. Initial experiments with these conversion processes look promising in several cases, but further developments in derivatisation techniques may be required to prove conclusively that the surfaces are truly monofunctional.

Table 5.

SUBSTRATE		Joint Strength MNm^{-2} $\pm$ st dev[a]	Failure[b]	θ_e deg[c]	Br:C at %[d]
LDPE - untreated	(1)	1.4 $\pm$ 10%	I	96	0
(1) - brominated	(2)[e]	6.7 $\pm$ 7%	I	73	4.5
(2) - dehydrobrominated	(3)	3.7 $\pm$ 11%	I	65	0.14
(3) - rebrominated in dark	(4)	6.3 $\pm$ 13%	I	58	3.5
HDPE - untreated	(5)	0.7 $\pm$ 9%	I	96	0
(5) - brominated	(6)[e]	11.2 $\pm$ 12%	I	75	5.3
(6) - dehydrobrominated	(7)	2.9 $\pm$ 13%	I	67	0.5
(7) - rebrominated in dark	(8)	15.9 $\pm$ 6%	M	55	8.3

a - d see Table 4

e 1 hour irradiation for LDPE, 3.5 hours irradiation for HDPE (see Table 4)

ACKNOWLEDGMENT

The work on electrical discharge treatment and derivatisation was carried out in collaboration with Dr. A.R. Blythe and Mr. C.R. Kendall (ICI). The work on monofunctionalisation of polyethylene is the postgraduate thesis work of Miss A. Chew, part of a collaborative project with Dr. D.M. Brewis and Dr. R.H. Dahm of Leicester Polytechnic.

REFERENCES

1. D.M. Brewis and D. Briggs, Polymer, 22, 7 (1981).

2. D. Briggs, D.M. Brewis and M.B. Konieczko, J. Mater. Sci., 11, 1270 (1976).

3. D. Briggs, V.J.I. Zichy, D.M. Brewis, J. Comyn, R.H.

Dahm, M.A. Grees, and M.B. Konieczko, Surf. Interface Anal., 2, 107 (1980).

4. D. Briggs, D.M. Brewis, and M.B. Konieczko, J. Mater. Sci., 14, 1344 (1979).

5. A.R. Blythe, D. Briggs, C.R. Kendall, D.G. Rance, and V.J.I. Zichy, Polymer, 19, 1273 (1979).

6. D. Briggs and C.R. Kendall, Polymer, 20, 1053 (1979).

7. D. Briggs and C.R. Kendall, Int. J. Adhesion and Adhesives, 2, 13 (1982).

8. D. Briggs, C.R. Kendall, A.R. Blythe, and A.B. Wootton, Polymer 24, 47 (1983).

9. D. Briggs, D.M. Brewis, and M.B. Konieczko, J. Mater. Sci., 12, 429 (1977).

10. D. Briggs, D.M. Brewis, and M.B. Konieczko, Euro. Polym. J., 14, 1, (1978).

11. F.M. Fowkes, in Adhesion and Adsorption of Polymers Polymer Science and Technology, Vol. 12A, L.H. Lee (Ed), Plenum, 1980, p 43.

12. M. Stradal and D.A.J. Goring, Can. J. Chem. Eng., 53, 427 (1975).

13. D.K. Owens, J. Appl. Polym. Sci., 19, 265 (1975).

14. D.S. Everhart and C.N. Reilley, Surf. Interface Anal., 3, 126 (1981).

15. J. Mitchell Jr., and L.R. Perkins, Appl. Polym. Symposia, 4, 167 (1967).

16. J.F. Carley and P.T. Kitze, Polym. Eng. Sci., 20, 230 (1980).

17. D.S. Everhart and C.N. Reilley, Anal. Chem., 53, 665 (1981).

18. A. Chew, R.H. Dahm, D.M. Brewis, and D. Briggs, in preparation.

Fracto-Emission Accompanying Adhesive Failure

J. T. Dickinson

Department of Physics
Washington State University
Pullman, WA 99164-2814

ABSTRACT

For a wide range of materials the emission of electrons (EE), positive ions (PIE), neutral species (NE), and photons (phE) has been observed accompanying fracture. We refer collectively to these emissions as fracto-emission. In this paper we review our work on fracto-emission where the fracture event involves interfacial or adhesive failure. The interfaces to be discussed include the following: brittle materials/epoxy, glass/elastomers, and brittle materials/pressure sensitive adhesives. Results are presented for both large (1 cm^2) planar surfaces as well as a few micron fibers (e.g., E-glass, S-glass, Kevlar™, and Graphite) and small (10-500 μm) particles in polymer matrices. When adhesive failure occurs it appears to lead to highly reactive surface species, perhaps free radicals, and charge separation. Subsequent chemical reactions in the presence of surface charge produces intense, long lasting (several minutes) emission of charged particles and excited neutral species. The energies of the charged particle emission can be several hundred electron volts. In this paper we describe measurements of the time dependence, energy distributions, crack velocity dependence, and spatial distributions of the FE. We also discuss correlations between the various FE components and present results on determining the mass of the PIE.

I. INTRODUCTION

Crack propagation through an insulating material or at an interface produces regions of high electronic and chemical activity at the freshly created surfaces. This activity causes the emission of particles, i.e., electrons, ions, and neutral species, as well as photons, from the surfaces both during and after crack propagation. This emission is called fracto-emission (FE). Electron emission from deformed materials for particular variations of mechanical stimulation has also been called tribo-stimulated exo-emission and mechano-emission. Photon emission during deformation and/or crack propagation is often referred to as triboluminescence.

In this paper we would like to review our work on FE at Washington State University concerning systems involving adhesive or interfacial failure. Work of other researchers involving FE accompanying adhesive failure can be found in References 1-6.

The primary goals of our research have been to characterize fracto-emission from various materials, to further our understanding of the FE mechanisms, and to examine the dependence of FE on the fracture event and material properties. These studies (References 7-20) have included fracture of oxide coatings on Al (7-12), measurements of neutral molecule emission accompanying fracture (9), the examination of FE accompanying adhesive failure (13,15), measurements of the dependence of electron emission (EE) on crack velocity in filled elastomers (14), measurements of the mass of the positive ion emission (PIE) accompanying fracture (16-18), an examination of the time correlation between the EE and PIE being emitted from the same sample (17), and preliminary studies of imaging EE and PIE coming from fracture surfaces.

FE studies are still quite new and considerable work is necessary to further characterize FE and to understand emission mechanisms. Nevertheless, we can speculate to some extent concerning potential applications to studies of fracture in general, and adhesive failure in particular. The most obvious application is the sensitive detection of crack formation and growth where broken bonds are in communication with the surrounding vacuum (so particles can be detected). The formation of surface cracks, crazing, micro-cracking, and finally a running catastrophic crack are examples of detectable events using FE. The motion of a crack as a function of time can be measured over a relatively wide range of time scales. In many cases, particle counting techniques used by nuclear physicists can be employed. We have examined, for example, the fracture of filled elastomers and epoxies with 1 μs time resolution. Thus, FE may serve as a convenient way to measure the instantaneous crack velocity. Imaging of charged particles using image intensifier technology or sensitive photon detection promises the use of FE to determine the location of damage as well as temporal information during straining of a sample.

Examination of the emission kinetics immediately after the separation of the two fracture surfaces may serve as a way to measure the surface temperature at the crack tip. Models we have constructed for the post-fracture or after-emission all require a temperature rise with fracture that decays exponentially with a decay time of a few seconds.

Furthermore, FE has considerable potential as a probe of the locus of fracture in a multi-component system such as a composite. The intensity, time dependence, and species emitted appear to be sensitive to where fracture has occurred. For example, FE may be useful for determining when delamination has occurred and to what degree. In conjunction with acoustic emission, the failure mechanisms may be better understood (perhaps at the molecular level) through studies of FE.

Finally, FE may relate in important ways to fracture mechanics parameters and chemical bonding phenomena. If reliable connections can be made to such parameters, FE could be used to measure them on an instantaneous and microscopic scale.

Let us briefly examine the major characteristics of FE that we have observed:

1. FE is a wide-ranging phenomenon. We have observed charged particle emission from inorganic crystalline materials, ceramics, glasses, glassy polymers, filled and unfilled elastomers, fiber-reinforced composites, single crystals of molecular solids, and the peeling of a number of adhesives.

2. The occurrence of all charged particle and photon emission that we have observed tends to correlate with some form of crack growth and/or bond breaking. This includes small signals due to crack formation processes or larger signals accompanying dynamic crack propagation. Emission is generally most intense during crack growth. In polyethylene and high-impact polystyrene we have evidence that FE precedes fracture and may be due to crazing. In elastomers, we have observed substantial neutral emission prior to crack growth, perhaps due to evolution of absorbed gases.

3. FE intensity has been observed to correlate with fracture energy (11,12).

4. In filled elastomers, the EE intensity grows exponentially with crack velocity, suggesting a strong dependence of the production of reactive species (which are responsible for EE) on the crack velocity (14).

5. The masses of the PIE which we have obtained to date indicate that the particles are atomic and molecular fragments of the material being fractured rather than impurities. For example, from anodized aluminum coatings, either O^+ or OH^+ is observed to be emitted during fracture (16,17).

6. Interfacial failure between polymers and glass, graphite, Kevlar, and metals, and peeling of pressure sensitive adhesives produces very intense, long-lasting, energetic EE and PIE (13,15). This is thought to be because of the production of a high concentration of surface free radicals and a high degree of surface charge due to charge separation. The EE and PIE distributions which we have observed for these systems are broad, slowly decreasing functions peaking near 0 eV but extending to >1 keV. This is in agreement with measurements by Deryagin, et al. (1), showing EE energies on the order of 10^3 to 10^4 eV for polymers peeled from glass plates. However, they attribute the EE they observe to a field emission mechanism which leads to several contradictions with our observations.

7. A wide variety of materials triboluminesce; i.e., yield photon emission (phE) during fracture. Chandra and Zink (21) classify materials according to the emission

spectra: type a) resembles discharge spectra of surrounding gases (due to the breakdown of gases in the crack), type b) resembles photoluminescence spectra, and type c) resembles a superposition of types a) and b). Since most of our work is done in vacuum, we are most interested in materials of types b) and c). Our primary interest in phE is its relationship to other forms of FE. We have found that in the cases of mica and adhesive failure, phE is very intense during fracture, but decays much more rapidly than EE or PIE. Krutyakova and Smirnov (22) examined the phE from NaCl and LiF at atmospheric pressure and found it to be of type b), suggesting that excitations in the ions and atoms of the crystal itself are involved in at least some of the triboluminescence observed.

8. The EE and PIE time distributions tend to be identical and frequently of the same order of magnitude. For a number of systems, the EE and PIE show a substantial component in coincidence (17) (within 0.5 μs - due to the finite time-of-flight of the ion), suggesting that the EE and PIE share the same de-excitation process.

9. EE and PIE are very sensitive to temperature. Both are activated processes. The effects of temperature on FE should tell us a great deal about the mechanisms involved.

10. For filled polybutadiene, we have shown the existence of an excited neutral species, also emitted in coincidence with the electrons (17).

11. The neutral emission (NE) species are sensitive to the chemical composition of the fracture material. For example, from anodized aluminum coatings, O_2 is the predominant species emitted (19).

12. The molecular crystals sucrose, PETN, and HMX have been shown to emit EE and PIE. This implies that fracture of such crystals involves bond-breaking on a significant scale.

13. Simple imaging experiments with PIE show that the particles are being emitted from the fracture surface. In the case of peeling pressure sensitive adhesives from solid substrates, we find that the adhesive face of the tape is the major emitter of both EE and PIE.

14. The general conceptual model we have constructed for explaining these features of EE, PIE, and perhaps phE consists of the following steps:

A. Fracture produces highly reactive, displaced species X and Y (e.g., atoms, radicals, and ions; bulk defects like F centers, V_k centers, etc., and electrons in high-lying traps).

B. Diffusion-limited mass transport brings reactive species together where an energetic reaction occurs, creating an excited intermediate:

$$(X+Y) \longrightarrow (X+Y)^*.$$

C. This intermediate decays producing emitted particles

$$(X+Y)^* \rightarrow \begin{cases} h\nu & \text{chemiluminescence (23,24),} \\ e^- & \text{chemi-emission (23,24),} \\ +\text{Ion} & \text{chemi-ionization (25), stimulated desorption (26).} \end{cases}$$

Such de-excitations often require the participation of available electrons; e.g., trapped electrons near the conduction band.

D. Kinetics dictated by rate-limiting step; usually B. To a first approximation, the emission rate would be given by the Waite Equation (27):

$$i = -\frac{dc}{dt} = -(4\pi r_o D)\left[1 + \frac{r_o}{(\pi D t)^{1/2}}\right] C^2,$$

where

- i = emission rate,
- C = concentration of A and B (assumed equal),
- D = $D_A + D_B$ (diffusion constants),
- r_o = capture radius (if A and B are within a distance r_o, they react virtually instantaneously).

E. The surfaces from which the emission occurs are frequently charged due to charge separation. Most often, this occurs in charge patches. The energy of repulsion that results from being released near these charge patches produces the observed kinetic energies. Crudely, the observed energy distributions can be translated into surface charge density distributions.

This model has been tested primarily on polymer systems (a more detailed treatment will be presented in a publication in preparation) but appears to explain to a great extent the observed features of FE, in particular, the time dependence of EE and PIE. In terms of applications involving adhesive failure, such a model would be useful to relate the measured FE characteristics to fracture phenomena of interest; i.e., the fragment species, density of trapped electrons, initial concentrations (before decay) and their rate of production all should be closely related to processes occurring at the crack tip. Although the FE model and characteristics outlined above have been observed in a limited number of situations, there appears to be evidence for considerable generality.

A number of the above features that relate to adhesive and interfacial failure will be discussed in greater detail in later sections.

In Table 1, we summarize the FE characteristics we study, possible mechanisms, and parameters that are known or suspected to influence FE.

Most of the characteristics of FE mentioned in Table 1 are self-explanatory except perhaps for "time distributions relative to fracture". Consider Figure 1 which represents an idealized log FE intensity vs. time to illustrate the features of interest. Let t_o = the time stress is first applied to the sample, t_1 = the time for the onset of crack growth, and t_2 = the instant the two fracture surfaces separate (i.e., the completion of crack propagation).

We focus on the following three time regions:

τ_1: $t_o < t < t_1$, the time interval during which no crack growth is occurring. The emission we observe during

Table 1. FE Characteristics, Mechanisms, and Parameters

<u>Characteristics of FE Investigated</u>

Types of particles emitted: electrons, ± ions, ground state neutrals, excited neutrals, photons

Species of ± ions, neutrals

Energies of electrons, ions, neutrals, photons

Time distributions relative to fracture

<u>Possible Mechanisms Contributing to FE</u>

Production of highly reactive species; e.g., free radicals displaced atoms and molecules, defects

Production of trapped electrons

Recombination of reactive species; annihilation of defects

Electronic transitions during bond breaking and during recombination/defect annihilation

Charge separation yielding less tightly bound electrons and intense E-fields; accelerates ejected charge

Increased temperatures at the crack-tip due to crack propagation - enhances thermally stimulated processes

<u>Parameters Influencing FE</u>

Material (e.g., composition, structure)

Sample history (e.g., exposure to radiation, temperature cycles, strain-cycles)

Temperature

External electric field

Gaseous environment

Nature of fracture (e.g., fracture mode, crack velocity, locus of fracture in multi-phase systems)

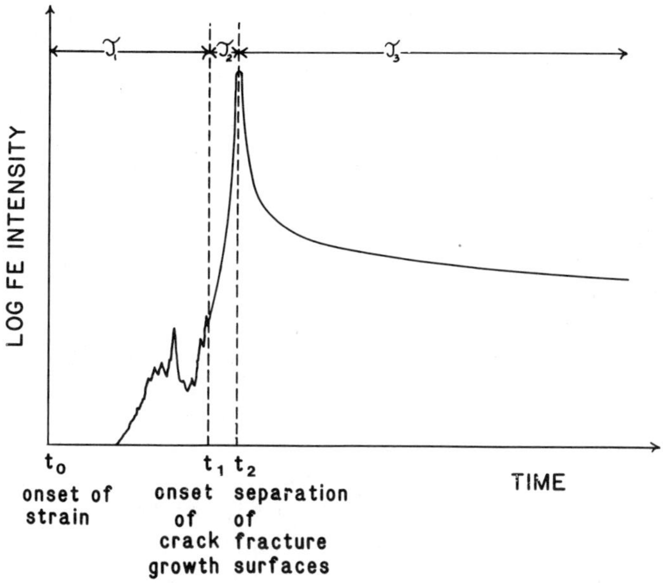

Fig. 1. Idealized emission curve for FE accompanying fracture of a material. The time intervals shown are before, during, and after crack growth.

the time τ_1 for a number of systems has been attributed to pre-fracture phenomena such as micro-cracking, crack formation, crazing, and delamination (e.g., in composites). The potential of using FE in this time region as a nondestructive testing tool should obviously be examined.

τ_2: $t_1 < t < t_2$, the time interval during crack growth. We generally observe a rapidly rising intensity peaking at the instant of separation. A number of interesting projects concerning dynamic crack growth, e.g., crack velocity studies, are underway, one of which we will discuss here.

τ_3: $t > t_2$, the time interval immediately following crack growth. Generally we see an immediate decay of emission after the instant of separation t_2. This decay is non-exponential and can vary in duration from fractions of a microsecond to thousands of seconds.

Concerning mechanisms, there is no unique mechanism for all the FE components (e.g., electrons, ions, neutrals, photons). In fact, the fundamental processes to which one could attribute the emission phenomena are in general poorly understood. In the case of triboluminescence accompanying adhesive failure, Deryagin and others (1) have attributed the photoemission to a gas discharge arising in the gap between the separated surfaces. Neutral emission accompanying fracture of polymers has been attributed to two fundamental sources by Grayson and Wolf: a) the release of entrapped (indigenous) volatiles, and b) fracture-induced decomposition, presumably a type of thermal degradation occurring on or near the fracture surface.

The electron emission (EE) mechanisms have been more evasive, in that for many materials, those involving adhesives in particular, the emission is so long-lasting. Thus, the EE involves fracture-produced intermediate states that have considerable lifetimes at or near room temperature (at which our experiments are usually performed).

Two types of intermediates are likely candidates:

1. Highly reactive species/defects which upon reaction, recombination, or annihilation give off localized energy to produce ejected electrons. In polymers, for example, bond breaking is known to produce free radicals which frequently must diffuse and react via second-order reactions. Chemical reactions at surfaces are known to be accompanied by electron emission (called chemi-emission)(23,24) via nonradiative electronic transitions.

2. Electrons trapped in shallow surface states produced and filled during fracture. These electrons are thermally stimulated into the conduction band of the material. If there is sufficient thermal energy to overcome the electron affinity of the material, then electrons may be

emitted from the surface. Electron-hole recombination could also lead to electron emission via multi-electron nonradiative transitions, the rate limiting step being the mobility of the charge centers. This process would be similar to that described under item 14 above.

Both these mechanisms generally involve the consequences of broken bonds. The electron emission thus may provide a measure of the degree of bond breaking that has occurred in producing the fracture surface.

The positive ion emission (PIE) we see has been shown to follow closely the EE in terms of time correlations on time scales ranging from microseconds to minutes. In fact, we found a large component of EE and PIE being emitted in coincidence (17). This suggests that, again, one of two mechanisms could be responsible for PIE:

1. The chemi-emission mechanism for EE is accompanied by the release of an ion; e.g., chemi-ionization.

2. An electron emitted by either EE mechanism mentioned above can strike a surface and stimulate the emission of an ion. In surface physics, electron-stimulated desorption (ESD) is a widely studied phenomenon that has recently been extended to organic adsorbates (28). The energetics of this mechanism would demand that electrons strike the surface with a minimum energy of several eV, but this is conceivable considering the charging of the fracture surfaces that frequently occurs and the EE energy distributions observed. If this latter mechanism is correct, the flux of EE measured÷the flux that strikes the surfaces would have to be 10^{-6} - 10^{-10} due to the small cross sections for ESD.

Possible Parameters Influencing FE - The parameters we have investigated on a limited number of materials include material, strain rate, exposure to radiation and temperature cycles, temperature (at or immediately following fracture), gaseous environment, crack velocity, and locus of fracture. In all cases investigated, these parameters had a noticeable influence on the FE. Some of these studies will be mentioned below. As an overview of the phenomena, we show in Table 2 a survey of

Table 2 Survey of Materials Investigated for EE and PIE

ELECTRONS

Materials	Approx. Decay Times of Fracto-Emission	Electrons Detected/ cm^2 of Crack Wall
INORGANIC		
Sapphire	<1 s, minutes	10^3
Alumina	<1 s, minutes	10^4
Al_2O_3 Anodized Layer	.1 - 20 μsec	10^5
BN	<1 s, minutes	10^6
Quartz	<1 s, minutes	10^6
LiF	<1 s	10^4
Mica (Muscovite)	<1 s, minutes	10^6
NaCl	<1 s	10^5
MgO	<1 s	10^5
Fused Silica	Several ms	10^3
Soda Lime Glass	Several ms	10^3
PZT	<1 s	10^6
Graphite	<1 s	10^2
ORGANIC CRYSTALS		
HMX	<1 s	10^4
PETN	<10 ms	10^4
Sucrose	Minutes	10^6
FIBERS		
Kevlar	45 μs	10^8
Graphite	10 μs	10^8
E-Glass	10 μs	10^8
S-Glass	10 μs	10^8
SiC	<10 μs	10^8
Al_2O_3	<1 ms	10^8
PLASTICS		
Epoxy (DER 332)	25 μs	10^3
Polymide	<1 s	10^5
PMMA	<50 μs	10^2
Lucite	<2 ms	10^2
PET Fibers	<1 s	10^5
Polystyrene	500 μs, 12.3 μs	10^3
Polyethylene	<1 s	10^4
PVF_2	<1 ms	10^3
ELASTOMERS		
Neoprene	<1 s	10^2
Viton	<1 s	10^3
Buna N	<1 s	10^2
Natural Rubber	<1 s	10^3
Natural Rubber (abraded)	Minutes	10^7
Silicone Rubber	<1 s, minutes	10^5
Solithane	<.2 s	10^4
Vinyl Rubber-filled	<1 s, minutes	10^4
Polybutadiene	0.04 s, minutes	10^3
Polybutadiene-filled	<1 s, minutes	10^7
Nylon-66	<1 s	10^4
Isoprene	<1 s	10^6
Amber Rubber	<1 s	10^5
BAMO	<1 s	10^4

Table 2 (continued)

POSITIVE IONS

Material	Approx. Decay Times of Fracto-Emission	Ions Detected/ cm^2 of Crack Wall
FIBER-EPOXY COMPOSITE		
Graphite	<1 ms	10^5
Al_2O_3	<1 s	10^7
Kevlar	<1 s	10^6
E-Glass	<1 s	10^6
INORGANIC		
Mica (Muscovite)	1 s, minutes	10^6
FIBERS		
Kevlar 49	45 μs	10^8
Carbon	10 μs	10^8
E-Glass	10 μs	10^7
S-Glass	11 μs	10^8
PLASTICS		
Epoxy (DER 332)	25 μs	10^3
Polyurethane	<1 s	10^6
Lucite	<2 msec	10^2
Polystyrene	35 μs	10^4
Nylon 66	<1 s	10^4
ELASTOMERS		
Buna N	<1 s, minutes	10^3
Natural Rubber	<1 s	10^4
Natural Rubber (abraded)	Minutes	10^7
Silicone Rubber	<1 s, minutes	10^3
Solithane	<.1 s	10^6
Vinyl Rubber-filled	<1 s, minutes	10^5
Polybutadiene	<.04 s, minutes	10^5
Polybutadiene-filled	<.2 s, minutes	10^6
SBR-filled	<1 s	10^6

the materials we have examined for EE and PIE. It should be noted that we have not tested comprehensively for PIE. The decay times are frequently unmeasured for relatively fast decay, and we simply note upper limits.

II. EXPERIMENTAL TECHNIQUES

In this article, we will briefly discuss the general experimental techniques we have used in our FE studies. Because we frequently are dealing with the detection of particles such as electrons, ions, and neutral molecules, it is necessary to work in a vacuum. We have two vacuum systems that use liquid nitrogen trapped diffusion pumps to attain pressures of $1 - 10 \times 10^{-7}$ torr within a few minutes. We have one ultrahigh vacuum system capable of reaching pressures below 10^{-9} torr with a light bake-out. The latter system is used in neutral emission studies to minimize background "noise" in the mass spectra and to investigate the influence of the gaseous environment on the charged particle/photon emissions. The work presented in this paper has shown no evidence of a dependence on background gases in these pressure ranges. However, a word of caution is necessary here. When working with metals, we (7-12,29) and others (23,24) have found that when clean surfaces are exposed to reactive gases such as O_2, H_2O, and halogens, electron and photon emission occurs. This phenomenon is known as chemi-emission and chemiluminescence and has been attributed to excited intermediates occurring during the reaction of the gas with the clean substrate.

For FE studies, our vacuum systems are equipped with devices to stress samples in various ways including tension, flex, and compression, while measuring stress and/or strain. Most of our experiments are performed in tension. Fig. 2 shows a typical experimental arrangement.

The detectors used for charged particles are channeltron electron multipliers (CEM) which produce fast (10 ns) pulses with approximately 90% absolute detection efficiency for electrons and nearly 100% efficiency for positive ions. The gains of the CEM's used were typically $10^6 - 10^8$ electrons/incident particle. The detectors were positioned 1 - 4 cm away from the sample with a bias voltage on the front cone of the CEM to attract the charged particles

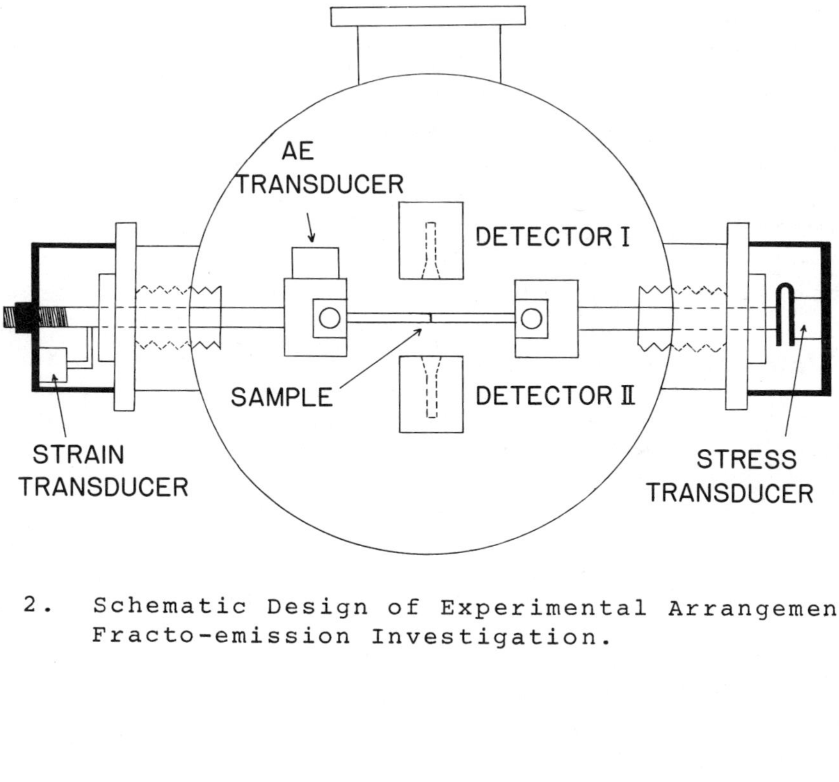

Fig. 2. Schematic Design of Experimental Arrangement of Fracto-emission Investigation.

of interest. Background noise counts ranged from 1 to 10 counts/second. Standard nuclear physics data acquisition techniques were employed to count and store pulses, normally as functions of time. The time scales of interest are submicrosecond to several second intervals, which we can easily cover with commercial electronics.

Photons are detected using a Bendix BX 754A Photon Counter tube with an S-20 photosensitive surface and a background count rate of 10 - 20 counts per second. For experiments done in air, the emission of photons was detected using an RCA 6217 (cooled with dry ice and acetone to reduce the noise level) with an analog output amplified with an electrometer whose output was fed to a strip chart recorder.

Energy distributions of charged particles can be examined by placing fine mesh screens in the region between the fracture zone and the CEM detector and performing retarding potential analysis. The derivative of the current (counts) vs. retarding potential is the energy distribution n(E)dE, where dE is a constant energy interval used in the measurement, typically 2-5 volts. As an example, Fig. 3a shows a retarding potential curve, and Fig. 3b is a differential curve for the cleavage of mica. This particular energy analysis shows typical results; namely, a peak near 0 eV and a long tail extending to several hundred eV. In addition to the above capabilities, we also have a UHV system for the measurement of the neutral molecule emission accompanying fracture. It employs a quadrupole mass spectrometer to measure selected mass peaks and a Bayart-Alpert gauge to measure total neutral emission.

For studies of composites and ceramics, we frequently mount an acoustic emission transducer on the sample for measurements of acoustic emission simultaneous with measurements of various FE components.

The materials to be discussed in this article include the following:

Filaments of glass, graphite and Kevlar TM-49, typically 10 micrometers in diameter, mounted so that single filaments could be broken sequentially. Epoxy strands of these materials consisting of approximately 250 filaments in DOW DER 332 epoxy (Bisphenol-A Type). Also

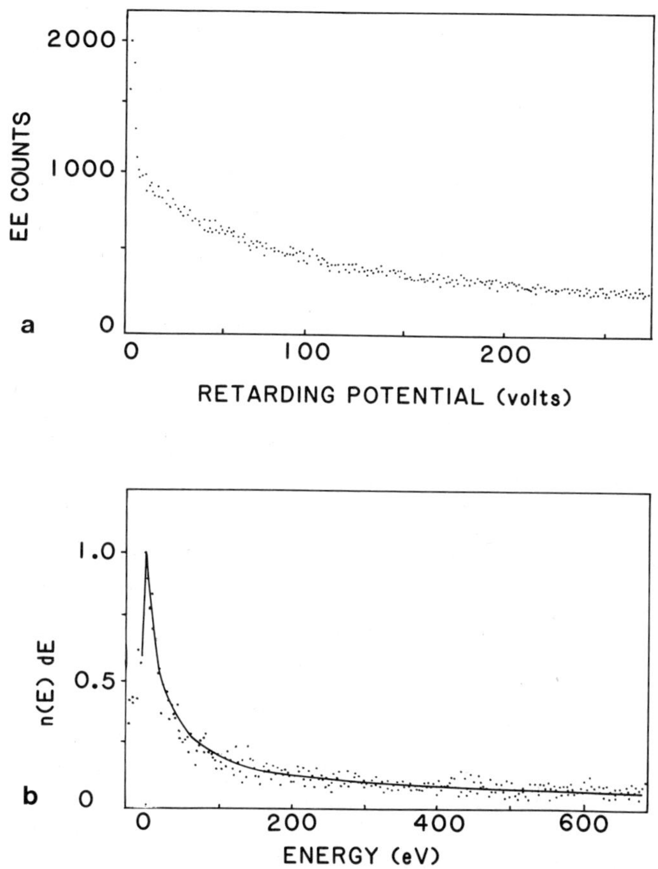

Fig. 3. a) The Variation of EE with Retarding Potential and b) the EE Energy Distribution for MICA Obtained by Differentiating a Curve Similar to a. Different Ranges of Energy are Shown.

samples of uniaxial, 16-ply, 0° graphite-epoxy samples were tested.

Particulate-filled epoxy - Al_2O_3 particles, typically 10 m in diameter in EPON 828 (Z-Hardener).

Filled elastomers - polybutadiene containing small untreated glass beads 30 - 95 μm in diameter, 0 - 34% by volume, cross-linked with dicumyl peroxide. We also examine the peeling detachment of polybutadiene bonded to a macroscopic flat glass surface.

Commercial 50 Durometer Red Silicon Rubber (SI) filled 50% by volume with silica particles of irregular shape, typically 10 μm in diameter, and very fine Fe_2O_3 powder.

3M Magic Tape, peeled from a number of substrates. Pretreatment of the surface consisted simply of washing in acetone.

3M Filament Tape - delamination or separation of the polyester backing from the glass filaments which adhere by means of a natural rubber-based adhesive.

III. COMPOSITES

An early discovery we made involving adhesive failure and its effect on charged particle emission concerned the fracture of composites (a more detailed discussion of this work can be found in References 13, 15, and 20). Starting with the constituents of a composite, namely, the filaments and neat resin, we found that the electron emission following their fracture appeared as shown in Fig. 4. The most important point to observe is that the emission is very rapid, decaying with time constants on the order of 10 - 100 microseconds. Early identical emission curves are observed for the PIE from the same materials.

When we fracture the corresponding filament-epoxy strands made from the same filaments and epoxies, we find emission curves that are entirely different. Fig. 5 shows, on a log scale, EE and PIE curves for Kevlar-epoxy taken simultaneously with two detectors. On the time scale shown, the duration of fracture is on the order of 1 - 2 channels. The curves are seen to be of very high peak intensity which

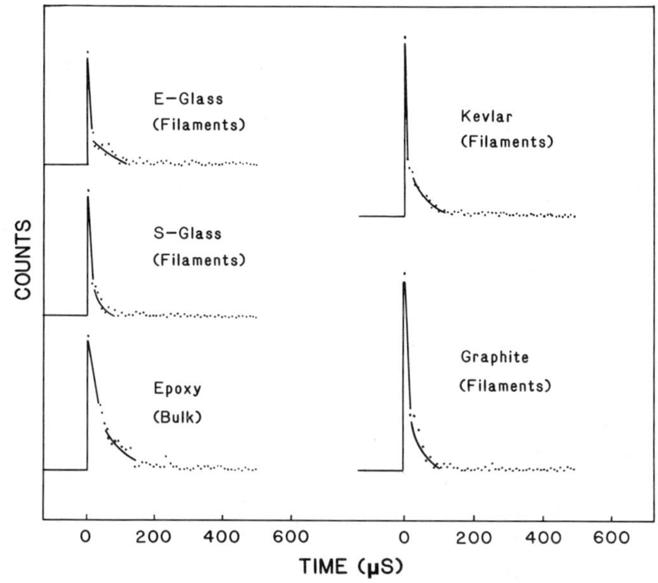

Fig. 4. The time distribution of EE due to the fracture of graphite, E-glass, Kevlar filaments and bulk epoxy (DOW DER 332/T304). Note the fast time scale.

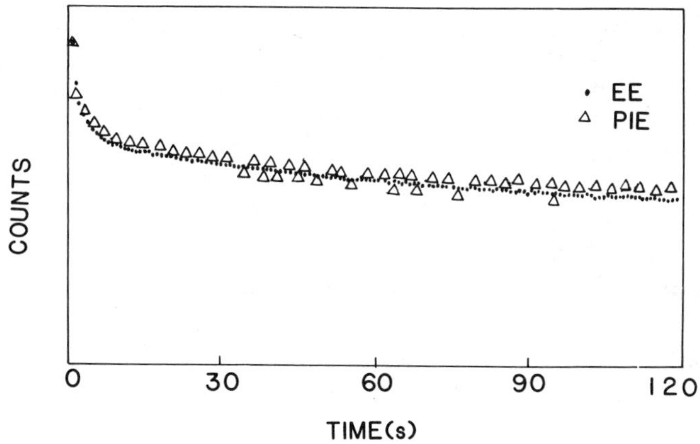

Fig. 5. EE and PIE from the fracture of Kevlar/epoxy strands.

decays in a complicated fashion and lasts for hundreds of seconds. We have watched the decay on some samples for as long as two hours. In the data of Fig. 5, we have normalized the PIE to the EE at a single point. Generally with our detector geometry, we find EE to be greater than PIE by a factor of approximately two. Note that the curves fall on top of one another. This suggests that the two types of emission share a common rate-limiting step.

SEM photographs of the strands show considerable delamination and free fiber surface due to separation of the epoxy from the fibers. We thus suspect that the interfacial failure accompanying the composite fracture was indeed responsible for the intense, long-lasting EE and PIE observed. In support of this, in Figures 6a and 6b, we compare the EE for two types of epoxy strands made from 20 μm diameter E-glass filaments and 10 μm S-glass filaments embedded in DOW DER 332 epoxy. Note the different time scales used for the two different materials. Examination under the SEM shows that there is considerably more delamination and separation of the filaments in the case of E-glass than for S-glass, which apparently results in considerably higher count rates and longer-lasting emission. By far the predominant emission is coming from the surfaces created by the separation of the filaments from the matrix.

To further investigate the FE from composites, we placed 16-ply 0° graphite-epoxy specimens in flex and measured EE, acoustic emission (AE), stress, and strain as the sample deformed. The samples were tested with a span to depth ratio of 30:1 and a strain rate of 0.064 mm/s. The AE count rate was determined by the number of bursts and the number of "rings" each burst created above a discriminator level. Fig. 7 shows the resulting data as a function of strain (time) where the EE and AE count rates are plotted on a log scale. One notes the gradual rise of AE prior to failure, presumably due to interlaminar shear and delamination internal to the sample, and possibly some internal fiber failure. The large AE peaks accompanying failure of successive plys are clearly visible. The accompanying EE shows a slow buildup (starting much later than the AE) prior to first ply failure, which indicates possible microfacture and crazing occurring on the outer surface in tension, and clear correlations of the large

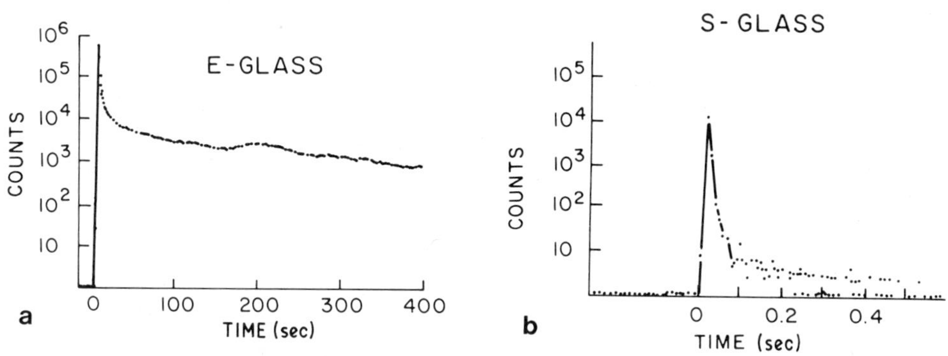

Fig. 6. Electron emission during and after fracture of a) E-glass/epoxy and b) S-glass/epoxy strands. Note the different time scales (15).

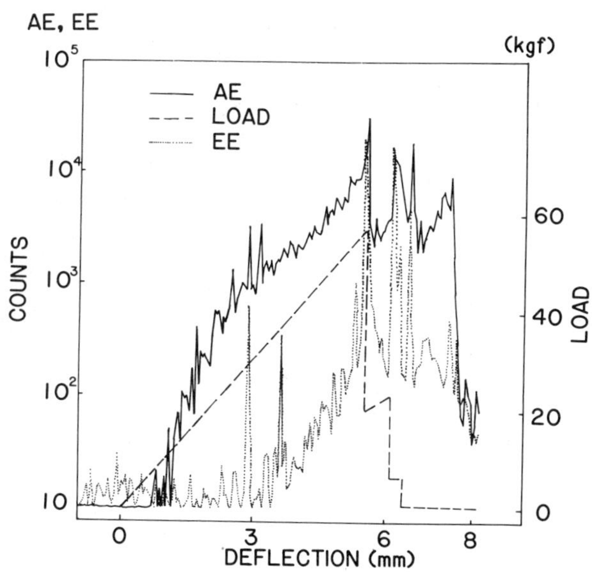

Fig. 7. EE, AE, and load accompanying the flexural straining of 16 layer unidirectional graphite-epoxy composite (20).

bursts of EE with ply failure. EE thus can detect evidence of pre-failure phenomena and indicate the onset of these processes in a very sensitive way.

Because the EE and PIE from systems involving interfacial failure frequently were intense and long-lasting, we were able to take measurements of energy distributions, n(E), using retarding potential analysis. The curves in Fig. 8 represent n(E)dE for both EE and PIE, where dE is 2 eV, plotted on a log scale and normalized to unity at the peak. Both curves are very similar (within our experimental error, they are identical), showing a peak near 0 eV and a significant number of higher energy particles in the tail. It is well known that charge separation is a common occurrence with adhesive failure and can leave the surfaces in a highly charged state. The probable cause of the high energy particles is the release of the charges in physical proximity of charge patches of the same sign, yielding an acceleration of the particles to the observed energies.

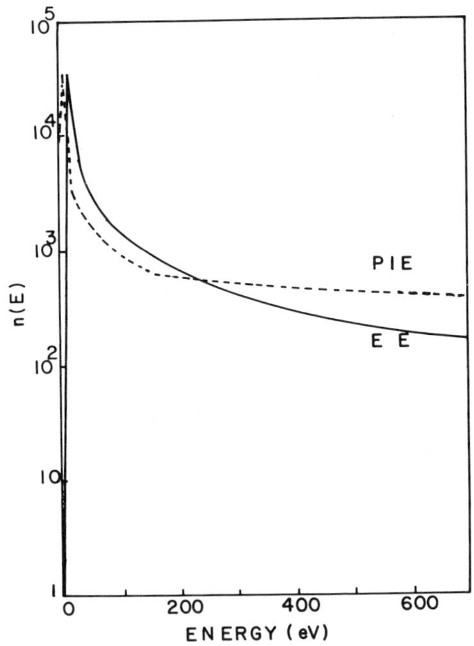

Fig. 8. Energy distribution on a log scale for EE and PIE from the fracture of Kevlar/epoxy strands (15).

Preliminary experiments involving fracture of the filaments and neat resin alone do not seem to yield particles having such high energies. Thus, we appear to have a distinct indicator of interfacial failure in a composite system:

a) intense, long-lasting EE and PIE,

b) the presence of high energy EE and PIE.

Retarding grids could easily reject the low energy particles and thus obtain a signal which is entirely due to interfacial failure. Proper steps to quantify these measurements could allow a precise determination of the degree of delamination/interfacial failure that has occurred during a fracture event.

Another composite structure we have investigated recently is a particulate-filled epoxy. The epoxy is EPON 828 (Z-hardener) filled with irregularly shaped alumina particles with an average diameter of approximately 10 μm. This material is quite strong and brittle so we fractured most of the samples in a three-point flexure mode. The cross section of the sample was 2 mm x 6 mm. A typical EE curve plotted on a log scale is shown in Fig. 9, where t = 0 corresponds to the instant of failure. The material for this emission curve is filled at an Al_2O_3/epoxy ratio α of 3 to 1 by weight.

The emission intensity is strongly influenced by the concentration of filler particles. Just taking the counts in the first channel (0.8 seconds per channel) as a measure of the initial EE count rate or peak height vs. (α), one sees this dependence in Fig. 10. The total emission (measured over several hundred seconds) follows basically the same curve. Compared to the unfilled material (α = 0) the EE intensity rises rapidly as α increases, and peaks near α = 1. This is followed by a slower decline.

These results are preliminary, and we are not entirely sure why the EE intensity depends on α in this fashion. Optical inspection of the fracture surface indicates that alumina particles are indeed being exposed, although SEM micrographs are far less convincing, so we are still seeking measurements of the degree of interfacial failure that is occurring. Secondly, as α increases, the mechanical

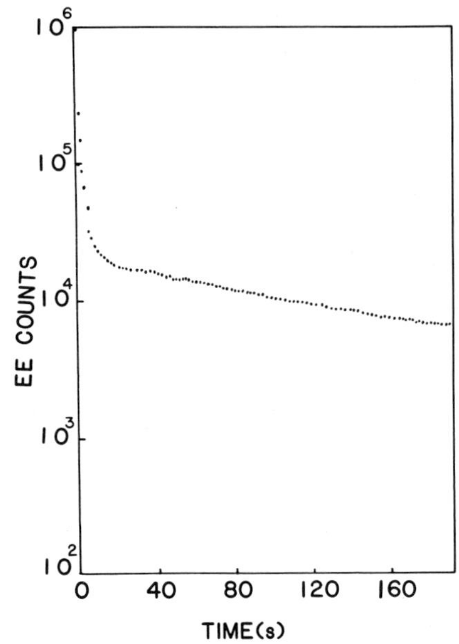

Fig. 9. Typical EE curve plotted on a log scale from the fracture of an alumina particle filled epoxy.

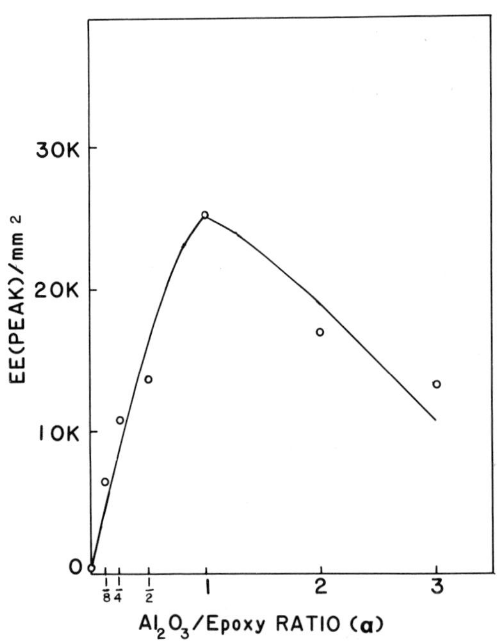

Fig. 10. Peak EE as a function of the Al_2O_3/epoxy ratio, α.

parameters such as fracture energy, surface energy, and the degree of interfacial failure are bound to change, which in turn may effect the emission. We are obviously interested in correlating these mechanical properties with the resulting EE.

Photon emission (phE) accompanying fracture appears to be of a different character. We have performed in air a number of experiments on the phE from epoxy strands of filaments with a strand cross section of 0.5 mm^2. Fig. 11 shows the visible light emission accompanying the straining and failure of epoxy strands of Kevlar, E-glass, and graphite. Several show phE prior to failure, possibly due to crack formation on a surface visible to the photomultiplier, or to chemiluminescence as observed by George and Pinkerton (30), and Fanter and Levy (31). The decay that we observe for the phE following fracture is within the time constant of the electrometer used to measure the photon detector current. Although the cause of the major burst of phE during fracture is unknown, we suspect, as with many cases of triboluminescence, that breakdown is occurring in the crack tip due to the high potentials produced by charge separation. This will be particularly intense at instances when delamination and adhesive failure are occurring. Further experiments need to be carried out to confirm this.

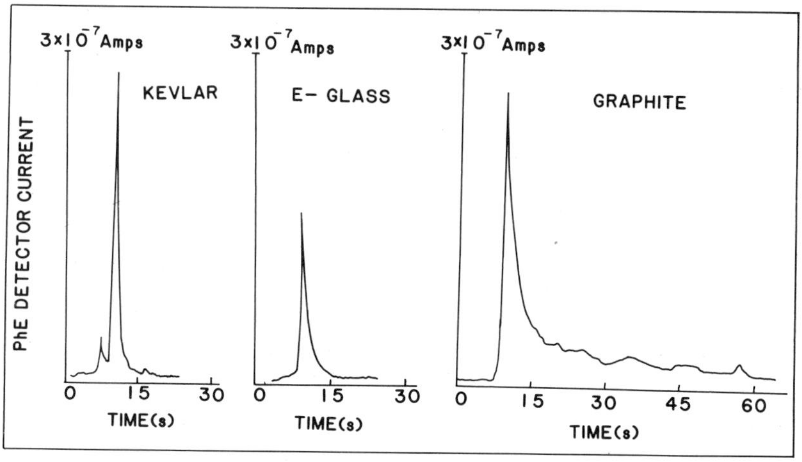

Fig. 11. Photon-emission accompanying the fracture of Kevlar, E-glass, and graphite epoxy strands.

PhE was also measured during T peel tests of two-ply Kevlar-epoxy panels. The entrance to the photomultiplier was approximately 2 cm from the "crack," and directed toward it. PhE was observed only during separation of the plys and decayed immediately upon release of the stress. For a constant area of new fracture surface (5 cm^2), the intensity of phE per unit area of fracture surface was found to depend strongly on the crack velocity, defined as the linear rate of the creation of new surface (cm/s). Fig. 12 shows the phE for a typical delamination. Fig. 13 shows this dependence where the ordinate represents the area under the emission curves for various velocities. The light intensity tends to increase for more rapid separation of the two surfaces, with a saturation occurring at a velocity of 10^{-1} cm/s.

We intend to perform similar experiments in vacuum and compare the phE with other types of FE to advance our understanding of the processes that are occurring at the crack tip during interfacial failure.

IV. FILLED ELASTOMERS

Another important adhesive bond, technologically, is the bond between hard filler particles and an elastomer matrix. When a material such as polybutadiene (BR) filled with untreated glass beads is fractured, one finds that the fracture surface appears as shown in Fig. 14. Clearly, as the resolution shown, the fracture involves a high degree of interfacial failure. The EE and PIE for filled (34% by volume) and unfilled BR are shown in Fig. 15. The peak intensity created during fracture and the post fracture or after emission are considerably more intense when the glass beads are present. The detachment of the beads from the polymer tends to happen very quickly and this interfacial failure is responsible for the enhanced emission. Fig. 16 shows the dependence of the total emission (counts accumulated in 200 seconds) on the quantity of glass beads in BR. Although we have drawn straight lines through the data, one sees a curvature suggestive of a peak, as was seen in the alumina-filled epoxy (Fig. 10). In fact, similar data taken with larger glass beads (250-500μm in diameter), as seen in Fig. 17, definitely show peaks in both EE and PIE intensities. The larger beads greatly reduce the strength of the material at a lower concentration. As a result, a crack begins moving through the material at lower

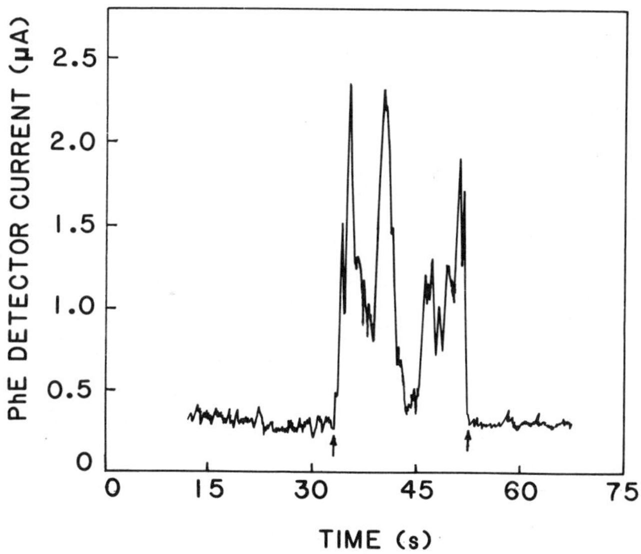

Fig. 12. Photon-emission from the delamination of a Kevlar/epoxy composite.

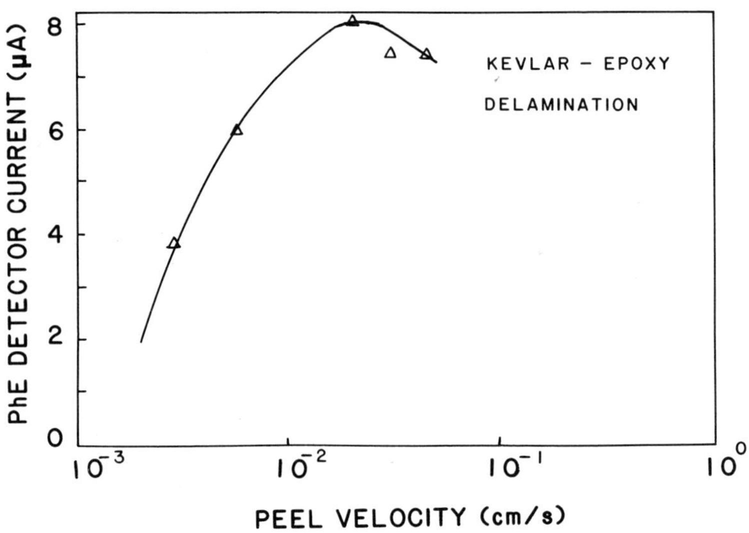

Fig. 13. Photon-emission from Kevlar/epoxy delamination as a function of various peel velocities.

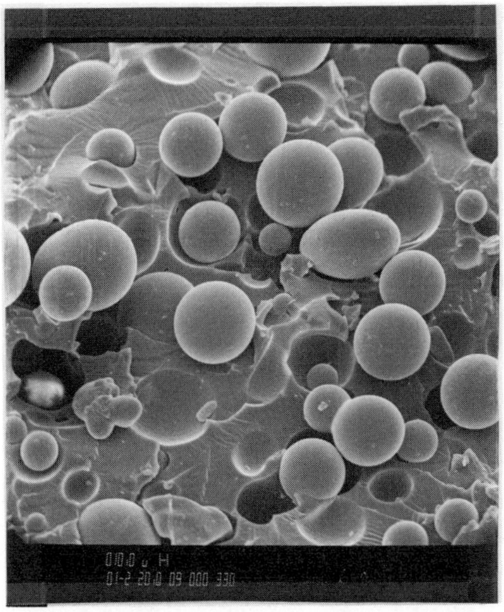

Fig. 14. SEM photograph of the fracture surface of polybutadiene filled with glass beads (30-95 μm in diameter) showing the high degree of interfacial failure.

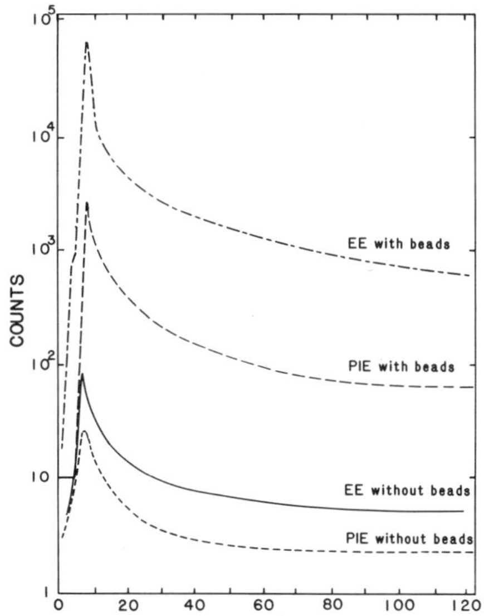

Fig. 15. EE and PIE from the fracture of polybutadiene with and without glass beads (15).

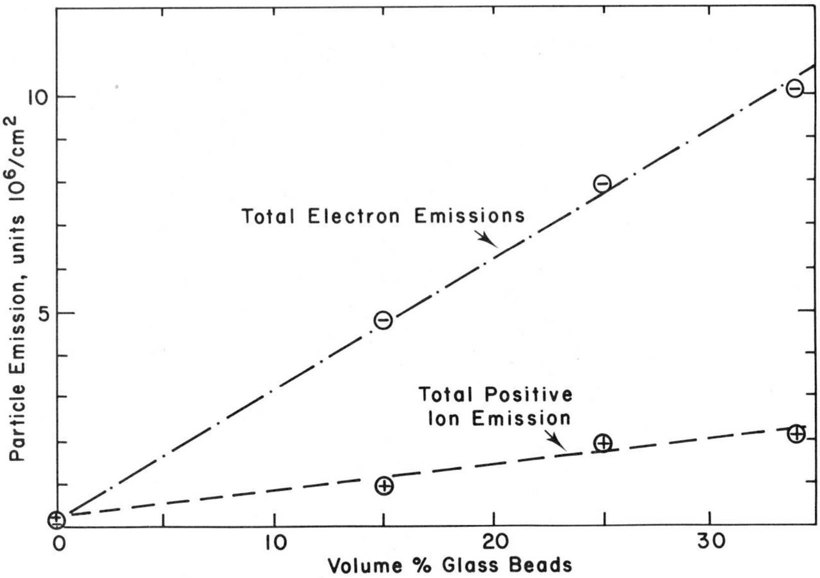

Fig. 16. The total emission (counts accumulated over 200 s) as a function of the percent of glass beads (30-95 μm in diameter) in polybutadiene (15).

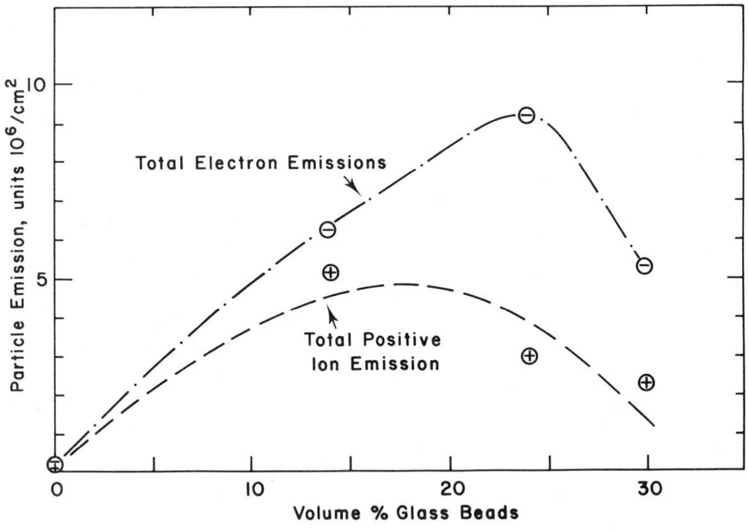

Fig. 17. The total emission (counts accumulated over 200 s) as a function of the percent of glass beads (250-500 μm in diameter) in polybutadiene.

strain and at lower crack velocity. As we shall show in the following paragraphs, this rate of separation of the interfaces can greatly influence the emission intensity.

We examined the dependence of EE intensity on the instantaneous crack velocity, V_c, of two filled elastomers: BR and silicone rubber (SI) by measuring the crack-tip position and EE intensity simultaneously. The former was done by using a video recorder with the camera imaging the region where the crack was propagating. Measurements of the crack position vs. time from each frame (1/60 second) allowed calculation of V_c to values up to 20 cm/s. At a constant strain rate, the crack would accelerate until separation. Figs. 18 - 19 show the resulting EE counts per channel (1 ms) vs. crack velocity for both BR and SI. Fig. 18 represents the initial data for low velocities (≤ 7 cm/s) where nonlinear rise is observed. Fig. 19 shows both low and high velocities. Comparing the two curves, we see the remarkably strong dependence of EE intensity on crack velocity, namely, nearly an exponential dependence at higher velocities.

Qualitatively, we have observed in a number of systems involving adhesive failure a strong dependence of FE intensity on the rate the surfaces separate, i.e., crack velocity. This suggests the interface is left in a higher state of non-equilibrium, perhaps because of a higher degree of bond breaking and a higher concentration of free radicals and/or trapped electrons near the conduction band. This higher concentration may be a consequence of the fact that more rapid loading does not allow local molecular stresses in the region of the interface to be relieved via viscoelastic relaxation mechanisms, leading to a higher degree of bond scissions. These results suggest FE may be a sensitive probe of the microscopic events occurring during fracture.

As a verification that the enhanced, long-lasting EE and PIE in the BR filled with glass beads was indeed due to interfacial failure, we obtained samples of the same polymer cross-linked with 0.075 percent dicumyl peroxide in contact with untreated soda lime glass plates (intended to simulate the surface of the soda lime glass beads). The elastomer layer, which was only weakly adhering to the glass surface, could be peeled from the glass surface and the FE

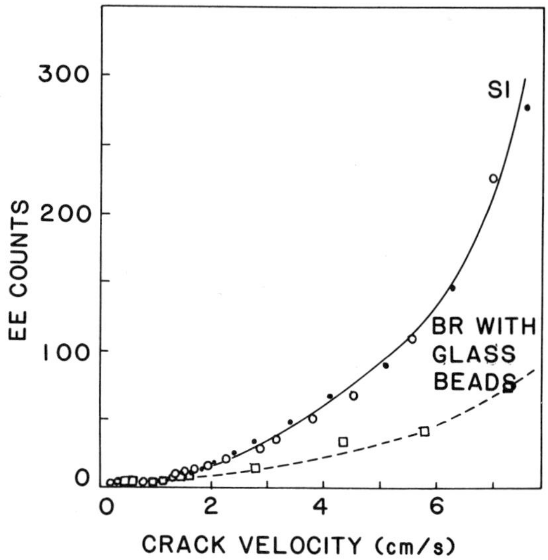

Fig. 18. EE vs. crack velocity on a linear scale for the first part of the velocity scale. The data shown for SI are from two samples (14).

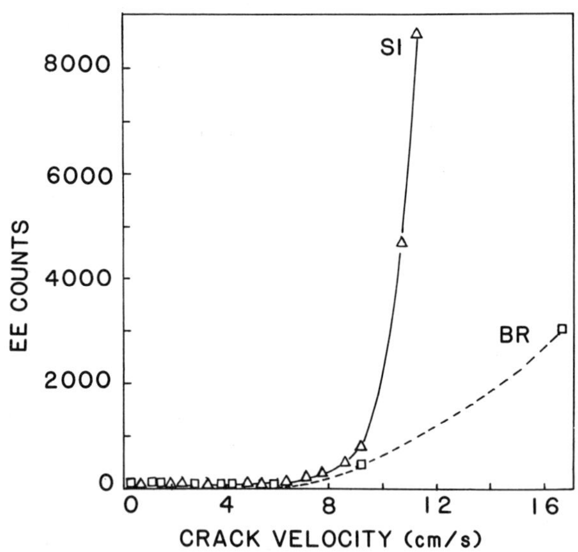

Fig. 19. EE vs. crack velocity on a linear scale for the complete range of low and high velocities (14).

monitored with a particle detector. Typical PIE results are shown in Fig. 20, where the time τ_2 is the interval over which the peeling occurred; approximately 2 cm^2 of area was exposed during peeling. The emission, again, is intense and long lasting. Qualitatively, the duration of the decay curve and the total counts obtained in a 400 second time interval increased with increasing peel velocity. Results for EE are essentially identical. We thus conclude that the interfacial failure in the BR filled with glass beads is indeed responsible for the intense, long-lasting emission observed and that the rate of surface separation is an important parameter effecting FE.

Similar to a number of materials examined, in the BR filled with glass beads, we found identical intensity vs. time curves for EE and PIE when the two were normalized at a single point. This result led us to ask if the two types of particles, electrons and ions were possibly in coincidence on a faster time scale. Use of two detectors, as shown in

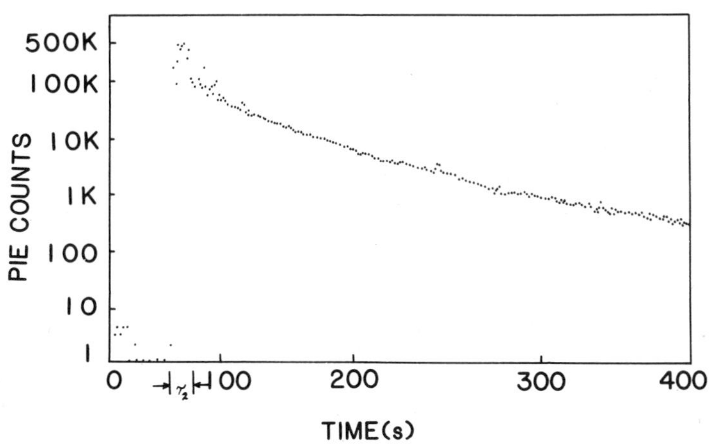

Fig. 20. PIE from the peeling of polybutadiene from a soda lime glass plate surface.

Fig. 21a, and associated pulse circuitry (e.g., coincidence circuits) showed that indeed on time scales of sub-microseconds there was a high degree of real coincidence between EE and PIE. A time correlation could also be found by use of time-interval circuits to measure the time between the detection of an electron and the next positive ion. The resulting time-delay spectrum is shown in Fig. 22 for two different voltages applied to the front cone of the PIE detector. The first peak (0-0.5μs) is due to positive ions that are in near coincidence with an electron. The width and shift of this peak is due to the finite time-of-flight (TOF) of the ion relative to the negligible TOF of the electron. The peak at 1.5μs which does not shift with PIE accelerating potential has been assigned to a neutral molecule in an excited state (e.g., a metastable state), perhaps a re-neutralized ion. Thus it appears that a good fraction of the EE and PIE are sharing a common mechanistic step.

The coincidence between the EE and PIE can be exploited to measure the mass-to-charge ratio of the PIE using a TOF method. Fig. 21b is a schematic diagram of the TOF arrangement with a drift tube of 25 cm in length. By using the electron pulse as a start, the arrival time of the positive ion can be measured easily. A TOF spectrum of the PIE from BR filled with glass beads for a drift tube potential of -2kV is seen in Fig. 23. Four major peaks are observed where positions in time can be shifted with changes in the voltage, V, applied to the drift tube. Analysis of these peak positions for the four peaks led to the M/q values given in Table 3. Also shown are structures of fragments from BR which could give such M (assuming q = e). It appears that PIE may include fragments produced from the polymer during fracture.

Fig. 24 is a similar TOF curve from fracture of a Kevlar-epoxy strand (as discussed in Section III) again for a 25 cm drift tube length and a tube voltage of -2kV. The peaks over the time interval 1 to 5 μs are due to heavier ions; we are still in the process of trying to identify them. The large feature in the first channel (0-0.25μs) is also a positive ion (it can be shifted slightly with different voltage on the tube). For reasonable initial kinetic energies (less than 4 keV), the only M capable of reaching the detector that fast is mass 1 or 2, i.e., H or H_2. We conclude, therefore, that either H^+ or $H_2{}^+$ is a predominant component of the PIE

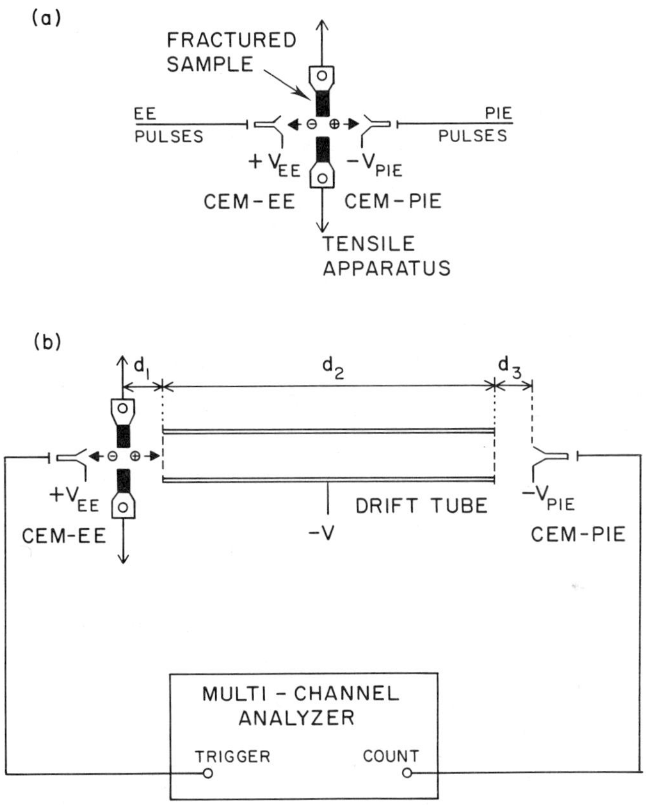

Fig. 21. Schematic illustration of the experimental apparatus for a) time correlations and b) time-of-flight (TOF) measurements. The distances are $d_1 = d_3 = 1$ cm, $d_2 = 25$ cm.

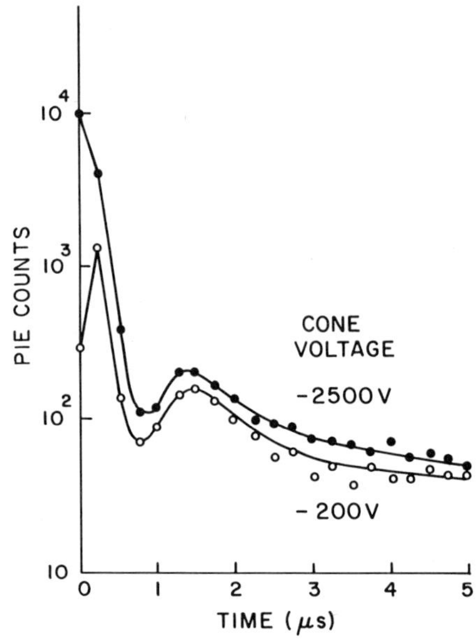

Fig. 22. Time delay spectrum of the particles detected at the CEM-PIE relative to the electrons detected at the CEM-EE. The two curves are for two voltages applied to the front cone of the CEM-PIE. The peak at 1.5 μs did not shift with this voltage, implying an excited neutral molecule.

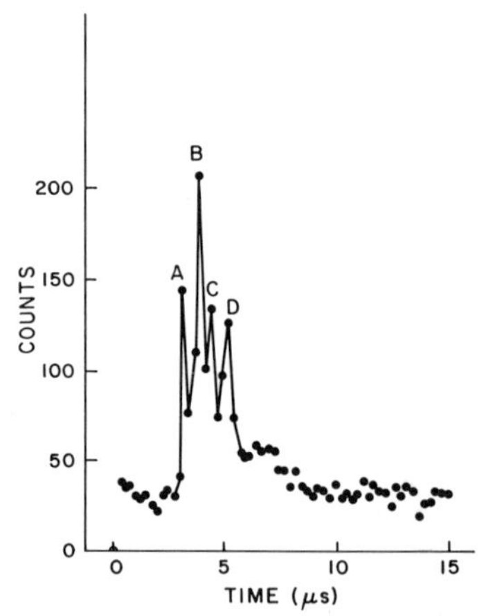

Fig. 23. The typical TOF distribution for PIE from polybutadiene filled with glass beads. The drift tube was -2 kV. Four major peaks labeled A,B,C,D are observed.

Table 3. Ions from Polybutadiene

Peak	$\frac{M}{q}$ $\frac{amu}{e}$		Possible Fragments
A	85 ± 6	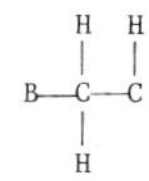	(81)
B	123 ± 6	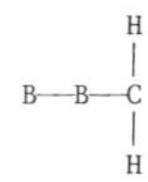	(122)
C	170 ± 6	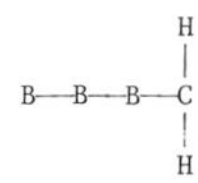	(176)
D	230 ± 6	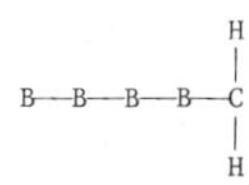	(230)

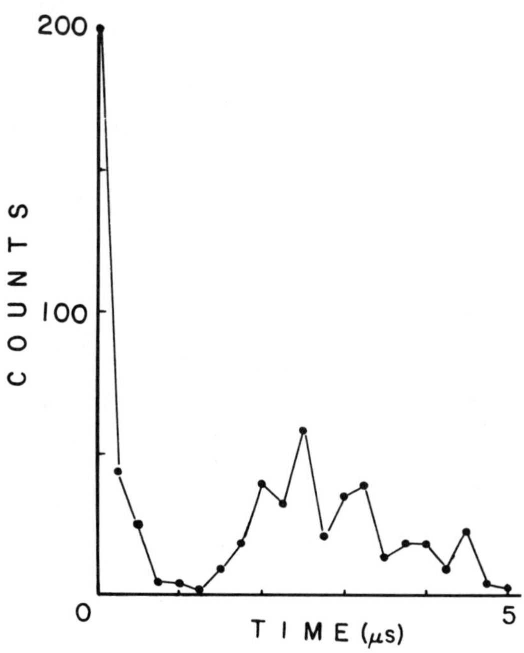

Fig. 24. The TOF for PIE from the fracture of Kevlar/epoxy strands. The major peak near 0μs is attributed to H^+ or H_2^+.

from this particular material. Of course, both the epoxy resin and the Kevlar filaments contain abundant quantities of hydrogen.

V. PRESSURE SENSITIVE ADHESIVES

During the peeling of the tapes from surfaces in air, it is well known that visible light can be seen. Deryagin (1) has attributed this "luminous radiation" to electrical breakdown in the gap between adhesive and adherend due to the high degree of charge separation that occurs during peeling. The photon emission (phE) measured in air (see Section III) during the peeling of 3M Filament Tape from its own polyester backing material at an average peeling velocity, V_p, (linear displacement of crack tip per unit time) of 0.25 cm/s is shown in Fig. 25a. The velocity is not uniform and presumably this changing velocity is the cause of the somewhat erratic signal. A much smoother and more intense phE curve was obtained during the delamination in air of the fibrous filaments from the polyester backing of 3M Filament Tape as seen in Fig. 25b. If the Filament Tape was folded in such a way to place the adhesive sides together, the separation of these surfaces (adhesive-adhesive) in a T-peel test yielded negligible phE unless V_p was increased to 1 cm/sec which yielded intense emission (Fig. 25c). The failure appeared to be entirely cohesive in this case. Why there is such a high velocity threshold for phE in this type of failure is not understood. The other modes of failure described above also showed peel velocity dependence, generally increasing with V_p, and saturating near $V_p = 10^{-2}$ to 10^{-1} cm/s, suggestive of a dependence of phE on peel strength. In all cases, as soon as the peeling was stopped, the phE disappeared; no measurable decay curve was observed.

EE and PIE from pulling adhesive tape from solid surfaces, measured in vacuum is significantly different from the phE in air. We surveyed the charged particle emission from the peeling of 3M Magic Transparent Tape from a number of surfaces. The surface preparation consisted only of cleaning in acetone - no attempt was made to remove natural oxide layers or other contaminants that would not be removed with acetone. The surface area created by peeling was 50 mm^2 at a peeling velocity of 0.3 cm/s. A typical EE

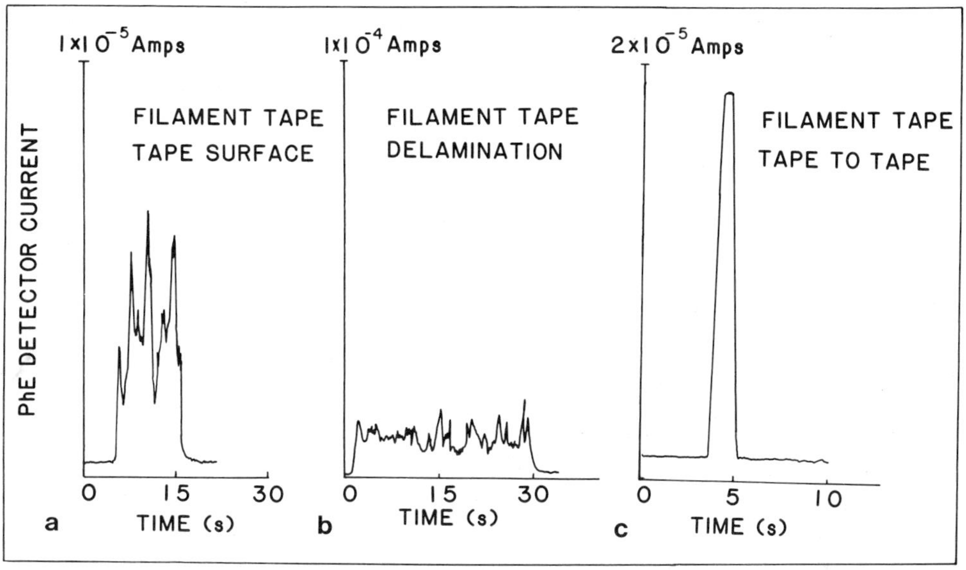

Fig. 25. Photon-emission from 3M Filament Tape during adhesive failure involving a) peeling the tape from its backing material, b) delamination of the filaments from the backing material, and c) peeling the adhesive from the adhesive at a high peel velocity.

curve for Magic Tape peeled from a Cu surface is shown in Fig. 26. Table 4 presents a summary of the peak count rate (during peeling) and total counts accumulated in 400 seconds for EE and PIE from a variety of substrates, ranked in order of peak count rate. The amount of EE is quite sensitive to the nature of the substrate, perhaps due to variations in the transfer of charge from the substrate to the adhesive from contact. Work on clean, well characterized substrates might clarify such effects. The energies of the charged FE are quite high due to the high degree of surface charging that occurs during separation.

Energy distribution curves for EE and PIE for the delamination of the filaments from the polyester backing of 3M Filament Tape are shown in Fig. 27 for the energy range 0 to 700 eV. Again, high energy particles are seen to occur for another case of interfacial failure.

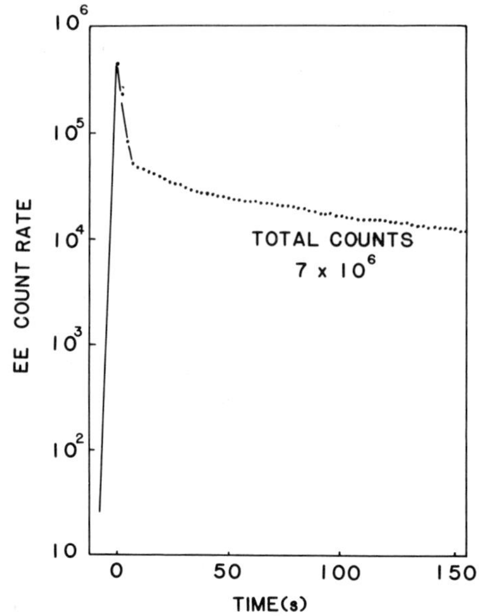

Fig. 26. Typical EE curve from the peeling of 3M Magic Tape off the Cu surface.

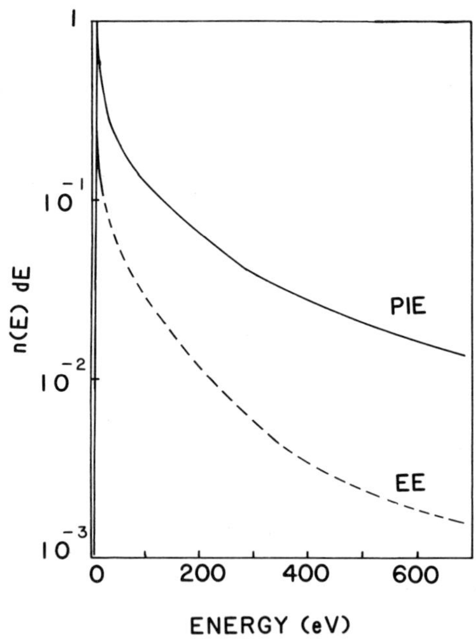

Fig. 27. EE and PIE energy distribution from the delamination of 3M Filament Tape from its polyester backing.

Table 4 FE from Peeling 3M Magic Tape

ELECTRONS

Adherend	Peak Count Rate (counts per second)	Total Counts (400 s)
Teflon	20	600
Magic Tape (adhesive)	250	1,000
Polybutadiene	300	14,000
LiF	3,600	38,000
Si Rubber	5,000	240,000
Mylar	8,000	240,000
$CaCO_3$	9,000	125,000
Stainless Steel	12,000	224,000
Glass	16,000	70,000
DER 332 Epoxy	18,000	1,000,000
PMMA	20,000	300,000
Magic Tape (back)	40,000	1,400,000
LiF (irradiated)	50,000	1,300,000
Polystyrene	54,000	5,000,000
Copper	300,000	6,000,000
Brass	350,000	5,000,000
μ-metal	400,000	25,000,000

POSITIVE IONS

Adherend	Peak Count Rate (counts per second)	Total Counts (400 s)
Teflon	10	100
Magic Tape (adhesive)	30	200
DER 332 Epoxy	50	700
Mylar	150	2,500
Magic Tape (back)	5,000	46,000
Brass	9,500	33,000
μ-metal	11,000	128,000
$CaCO_3$	12,400	144,000
LiF	13,400	227,000
LiF (γ-irradiated)	20,500	2,000,000

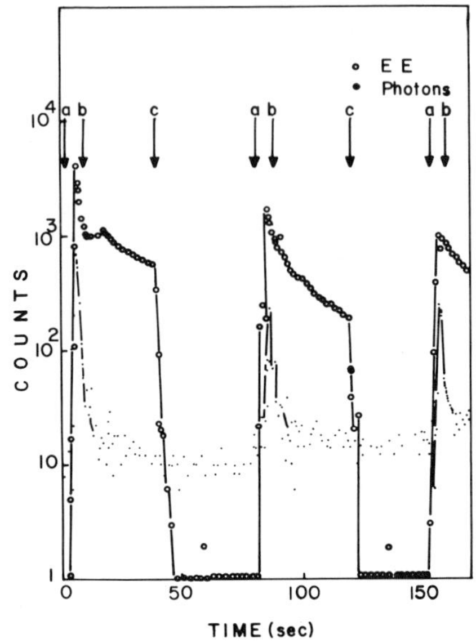

Fig. 28. EE and photon-emission during and after delamination of the filaments from the polyester backing of 3M Filament Tape (15).

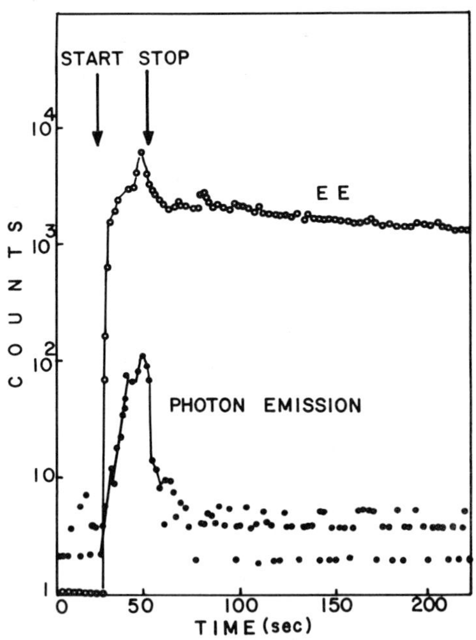

Fig. 29. EE and photon-emission from the peeling of 3M Magic Transparent Tape from a PMMA surface (15).

In vacuum we can measure simultaneously the EE and phE coming from similar systems. Fig. 28 shows EE and phE during and after delamination of the fiber filaments from the polyester backing of 3M Filament Tape. Between a and b, we have delaminated the surfaces and then re-attached the surfaces at c (which causes the EE to disappear), then repeated the process, showing that EE and phE are produced by the second and third detachment. The EE is very high during peeling and decays very slowly with relatively high emission lasting many minutes. The phE, similar to what was observed in air, shows intense emission only during the peeling. In spite of the vacuum environment, we know that gases are given off from the materials during peeling and so the local pressure inside the gap between the separating surfaces may be high enough to allow breakdown to occur and thus produce phE. The makeup of these gases should be quite different from air (predominantly N_2), so that differences in emission spectra should exist. It would be an interesting measurement to pursue. In comparing the EE and phE from the peeling of Magic Tape, we conclude that although the mechanical detachment is the initiator of both processes, the actual mechanisms are quite different.

Similar results are seen from the peeling of 3M Magic Transparent Tape from PMMA as seen in Fig. 29. The arrows indicate when peeling was started and stopped.

One fundamental question which arises in these peeling experiments is: from which surface are the particles being emitted? Two experiments we have performed helped to answer this question. The first employed a charged particle imaging device called a micro-channel plate or Channel Electron Multiplier Array (CEMA), which produces on a phosphor screen an intensity pattern of the charged particles incident upon the plate. The patterns observed from peeling 3M Magic Tape from a solid surface depended on the orientation of the peel test and substrate relative to the CEMA.

To capture the time dependence of these patterns, we recorded the images on the phosphor screen on video tape. Photographs of the images were then made from the TV monitor.

Two orientations of the substrate relative to the CEMA were used, as shown in Fig. 30: a) the exposed substrate parallel to and facing the CEMA, and b) the substrate normal to the CEMA. Due to the types of electric potentials necessary to achieve good imaging, we chose to investigate the PIE images. The initial patterns during the peeling were totally indistinguishable due to saturation from the intense emission; the image was everywhere white. However, a few seconds after the peel, the patterns would emerge. Fig. 30 also shows typical images for a) parallel and b) normal orientations of a Cu substrate. The predominant ion emission is coming from the edges of the tape on the adhesive side. The center of the tape is at most emitting weakly. The substrates in these cases are dark. We would suggest that stress concentration at the edges during the peel is responsible for the inhomogeneous emission seen coming from the adhesive surface.

A second non-imaging experiment, performed with a variety of substrates, consisted of carefully monitoring the FE from peeling 3M Magic Tape in two situations: a) where the freshly created adhesive surface is visible to a particle detector, and b) where, after peeling, the tape is pulled into a metal box through a small slit and wound on a spool in such a way that the adhesive surface is on the inside. Fig. 31 shows the resulting PIE curves for these two cases for a stainless steel substrate. When we are left with just the metal substrate, we see that after the emission during peeling (which is unavoidable) the removal of the tape leaves us with negligible PIE. Similar experiments have been carried out for both EE and PIE on stainless steel, copper, PMMA, and glass with identical results. In all cases, the EE and PIE came from the adhesive side of the tape and the substrate did not appear to be emitting.

A preliminary experiment done with 3M Magic Tape peeled from a metal surface shows that the EE following the peeling is a thermally stimulated process. Following peel (which starts at A) in Fig. 32, the tape surface was heated indirectly by radiation from a nearby metal surface. Because of the difficulty of measuring the temperature of the adhesive surface, we did not measure the surface temperature vs. time. Nevertheless, at point B (Fig. 32) where the temperature was increased, the EE rose rapidly to a peak and then fell gradually to very low emission in response to this heating. A rise in temperature enhanced the rate of emission and depleted the "source" or concentration of active species. A major goal for future

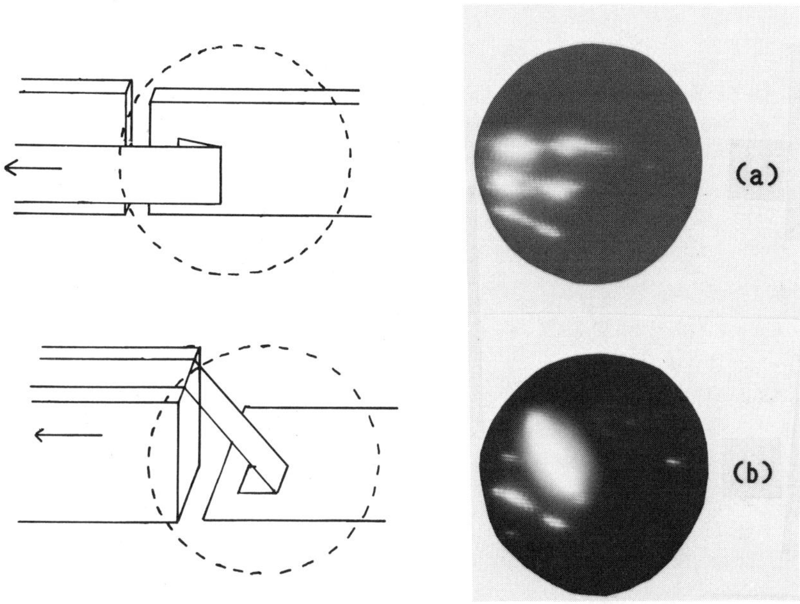

Fig. 30. Images of PIE from the peeling of 3M Magic Tape off a metal surface. Substrate a) parallel, and b) normal to the CEMA.

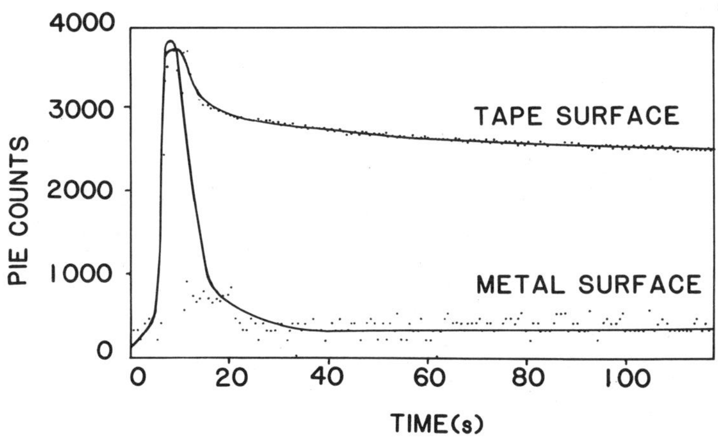

Fig. 31. PIE from the adhesive and metal surface during and after peeling.

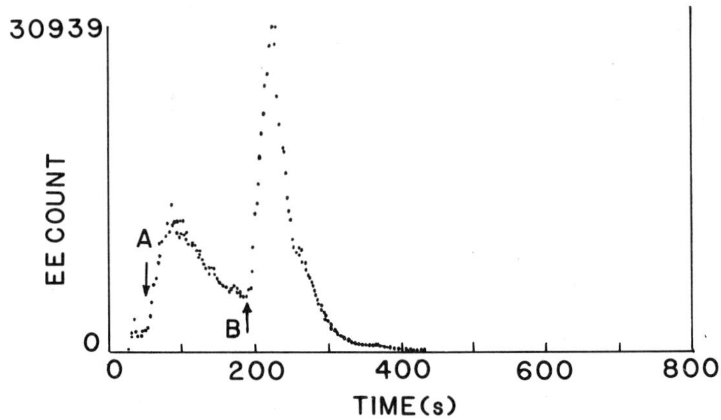

Fig. 32. Evidence for the rise in EE with increasing temperature.

work is careful temperature vs. time measurements so that mechanistic models can be rigorously tested.

V. ALUMINUM-EPOXY INTERFACES

Our initial work in fracto-emission examined a number of details of the EE and PIE accompanying fracture in oxide coatings which served as surface preparations for the adhesive bonding of aluminum. This work can be found described in References 7-12 and will not be described here. However, we would like to present recent results on the EE accompanying T-peel tests of adhesively bonded Al (Al606T-T6/M11113 adhesive) where 1 cm^2 of fresh surface was created. In Fig. 33, curve (a) represents the EE observed when the crack moved entirely along the interfaces leaving behind a relatively clean "metal" surface. Curve (b), on the other hand involved large patches of cohesive failure with some interfacial failure. The differences in the peak emission during peeling and the decay curves reflect the enhanced emission one would expect from interfacial failure compared to "mixed" failure.

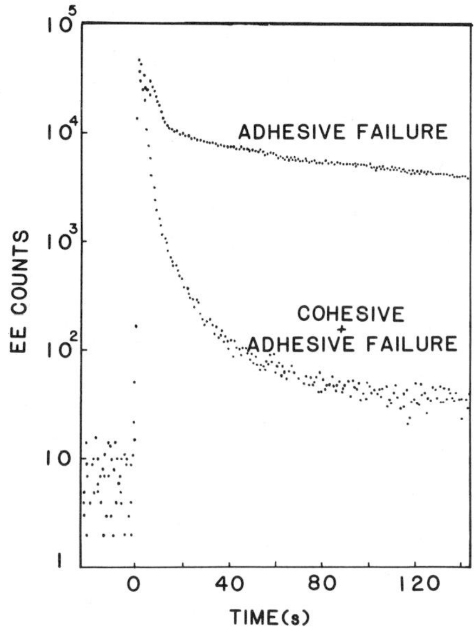

Fig. 33. EE from T-peel testing of Aluminum/epoxy interface for adhesive and adhesive plus cohesive failure.

VI. DISCUSSION AND CONCLUSIONS

I have tried to show a wide variety of FE results on a number of systems involving adhesive failure at interfaces and indicate some of the parameters that are influencing this emission. The need for careful studies of the physics and chemistry of these phenomena is obvious. The usefulness of FE as a tool for NDT or for investigation of failure mechanisms requires a broad based attack combining fracture mechanics, materials science, and fundamental fracto-emission studies on materials of mutual interest. Since the field is relatively unexplored, I will conclude by speculating on some potential areas of usefulness of FE, many of which depend critically on further understanding of FE itself.

POTENTIAL APPLICATIONS OF FE

1. Probe of crack growth on an extremely wide range of time scales. This need not be catastrophic fracture and might involve crazing, microcracking, linking of microcracks, and other pre-failure events.

2. A probe of the departure of the fracture surface from equilibrium. As the charge and defects/free radicals decay away, FE may well be a measure of the initial concentrations and the rate of surface reactions such as defect recombination.

3. The energies of the FE components may serve as a measure of the density of the charge distributions created on the fracture surface.

4. FE may serve as a way to measure the surface temperature at the crack tip by careful modeling of the emission curves at short times after fracture. Our modeling to date has required an elevated temperature of fracture that decays quickly away.

5. FE may serve as a means of measuring instantaneous crack velocity.

6. FE may serve as a probe of the locus of fracture in composite materials and in illuminating failure mechanisms. The roles of contact charging and chemical bonding in adhesion may be determined by FE studies.

7. FE may serve as an NDT tool, perhaps in conjunction with acoustic emission. FE would be particularly useful when sensitivity to events near the surface is desired.

8. FE may be related in important ways to fracture mechanics parameters such as surface energy, fracture strength, or fracture toughness. If reliable connections could be made to such parameters, FE might be used to measure them.

ACKNOWLEDGMENTS

I wish to first acknowledge the important contributions to this work by my Washington State University colleagues, Les Jensen, Ahmad Jahan-Latibari, and Ed Donaldson. I also wish to thank those people who have contributed specimens used in these studies, particularly R. L. Moore, Lawrence Livermore Laboratory, for samples of filaments and fiber/epoxy strands, and A. N. Gent, Institute of Polymer Science, University of Akron, for all the polybutabiene samples. We are also appreciative of interest and advice from O. Ishai, A. Gray, and L. C. Clements of NASA-Ames

Research Center, and W. D. Williams, Sandia National Laboratories.

This work was supported by the Office of Naval Research contract N00014-80-C-0213, Sandia National Laboratories, National Science Foundation, NASA-Ames Research Center, and a grant from the M. J. Murdock Charitable Trust.

REFERENCES

1. B. V. Deryagin, N. A. Krotova, and V. P. Smilga, Adhesion of Solids (English Translation), Consultants Bureau, New York (1978).

2. B. V. Deryagin, V. I. Anisimova, V. A. Klyuev, and N. A. Krotova, Doklady Akademii Nauk SSSR, 222, 644 (1975).

3. Yu A. Khrustalev, L. I. Prokhorova, N. K. Baramboim, Yu F. Kurdnbov, and N. A. Krotova, Kolloidnyi Zhurnal, 38, 814 (1976).

4. V. A. Klyuev, V. I. Anisimova, Yu P. Toporov, N. A. Krotova, and B. V. Deryagin, Kolloidnyi Zhurnal, 40, 244 (1978).

5. M. I. Molotskii, Sov. Phys. Dokl. SSSR, 243, 1438 (1978).

6. B. V. Deryagin, V. A. Klyuev, V. I. Anisimova, Yu P. Toporov, and N. A. Krotova, J. Colloid and Interface Sci., 77, 472 (1980).

7. J. T. Dickinson, P. F. Braunlich, L. Larson, and A. Marceau, Appl. Surf. Sci., 1, 515 (1978).

8. D. L. Doering, T. Oda, J. T. Dickinson, and P. F. Braunlich, Appl. Surf. Sci., 3, 196 (1979).

9. L. A. Larson, J. T. Dickinson, P. F. Braunlich, and D. B. Snyder, J. Vac. Sci. Technol., 16, 590 (1979).

10. J. T. Dickinson, D. B. Snyder, and E. E. Donaldson, J. Vac. Sci. Technol., 17, 429 (1980).

11. J. T. Dickinson, D. B. Snyder, and E. E. Donaldson, Thin Solid Films , 72, 225 (1980).

12. J. T. Dickinson, E. E. Donaldson, and D. B. Snyder, J. Vac. Sci. Technol., 18, 238 (1981).

13. J. T. Dickinson, E. E. Donaldson, and M. K. Park, J. Mater. Sci., 16, 2897 (1981).

14. J. T. Dickinson and L. C. Jensen, J. Polymer Sci. Polymer Phys. Ed., 20, 1925 (1982).

15. J. T. Dickinson, M. K. Park, E. E. Donaldson, and L. C. Jensen, J. Vac. Sci. Technol., 20, 436 (1982).

16. J. T. Dickinson, L. C. Jensen, and M. K. Park, J. Mater. Sci., 17, 3173 (1982).

17. J. T. Dickinson, L. C. Jensen, and M. K. Park, Appl. Phys. Lett., 41, 443 (1982).

18. J. T. Dickinson, L. C. Jensen, and M. K. Park, Appl. Phys. Lett., 41, 827 (1982).

19. M. H. Miles and J. T. Dickinson, Appl. Phys. Lett., 41, 924 (1982).

20. J. T. Dickinson, A. Jahan-Latibari, and L. C. Jensen, to appear in Proceedings of the "Symposium on Composites, Interfaces," American Chemical Society Meeting, Seattle, 1983.

21. B. P. Chandra and J. I. Zink, Phys. Rev., B 21, 816 (1980).

22. V. P. Kratyakova and V. N. Smirnov, Zhurnal Prikladnoi Spektroskopii, 30, 846, 777 (1979).

23. J. Harris, B. Kasemo, and E. Tornqvist, Chem. Phys. Letters, 52, 538 (1977).

24. B. Kasemo, E. Tornqvist, J. K. Norskov, and B. I. Lundqvist, Surf. Sci., 89, 554 (1979).

25. M. S. B. Munson and F. H. Field, J. Am. Chem. Soc., 88, 2621 (1966).

26. M. L. Knotek and P. J. Feibelman, Phys. Rev. Letters, 40, 964 (1978).

27. T. R. Waite, Phys. Rev., 107, 463 (1967).

28. J. A. Kelber and M. L. Knotek, Surface Science, 121, L499 (1982).

29. L. Larson, T. Oda, P. Braunlich, and J. T. Dickinson, Solid State Communications, 32, 347 (1979).

30. G. A. George and D. M. Pinkerton, "Proceedings of A Critical Review of Characterization of Composites", June 8-10, 1981, Massachusetts Institute of Technology (Office of Naval Research, 666 Summer St., Boston, MA (1981).

31. D. L. Fanter and R. L. Levy, in ACS Symposium Series No. 95, "Durability of Macromolecular Materials", R. K. Eby, Ed., p. 211, American Chemical Society, Washington, DC (1979).

Monitoring Epoxy Cure Kinetics with a Viscosity-Dependent Fluorescent Probe

R. L. Levy and D. P. Ames

McDonnell Douglas Research Laboratories
McDonnell Douglas Corporation
St. Louis, MO 63166

ABSTRACT

Properties of carbon-epoxy composites are strongly dependent on the epoxy cure-cycle. Availability of techniques for *in situ* monitoring of cure kinetics would enhance control of optimized production of composite structures. Experiments are described which exploit the microviscosity dependence of the fluorescence yield of p-(N,N-dialkylamino)benzylidene malononitrile (R. Loutfy, Macromolecules 14, 270 (1981)) for monitoring epoxy cure kinetics. Epoxy resins based on either diglycidylether of bisphenol-A (DGEBA) or tetraglycidyl-diaminodiphenyl methane (TGDDM) containing 0.5 wt % of the malononitrile as a solute probe show an increase of the fluorescence intensity (I_f) at the probe's emission band maxima (495 nm) as a function of the degree of cure. Fluorescence-derived profiles for cure kinetics (I_f as a function of cure time) closely follow the kinetic profiles derived by differential scanning calorimetry (DSC). The fluorescence yield of the TGDDM monomer was found to be viscosity-dependent and therefore, possesses a unique "self-probing" capability which permits monitoring of its viscosity during cure.

*This work was performed under the McDonnell Douglass Independent Research and Development Program.

INTRODUCTION

Properties of carbon-epoxy composites are dependent on the chemical and rheological events occurring during the epoxy cure cycle (1,2). The ability to conduct continuous *in situ* measurements of the epoxy cure states, cure kinetics, and viscosity of the resin matrix would enhance control of composite structures. Rheological, thermo-analytical, and spectroscopic techniques that measure these variables exist and are frequently used in laboratory studies of epoxy cure behavior. Physical restrictions, however, prevent the application of such techniques for *in situ* measurements during autoclave curing of composite structures in the factory setting.

Dielectric spectroscopy techniques (3) represent an exception to this limitation. In fact, Senturia et al. (4) recently described a microdielectrometric method for *in situ* measurement of epoxy cure. Using a novel micro-dielectric probe, they studied isothermal cures of digylcidyl ether bisphenol-A (DGEBA) epoxy in which the low-frequency dielectric relaxation time, τ, is correlated with the bulk viscosity (4). In this paper we report preliminary results on the utilization of a viscosity-dependent fluorescence probe as an alternative (and possibly complementary) new approach with potential for *in situ* monitoring of epoxy cure kinetics. We also report the observation of a unique fluorescence "self-probe" capability of tetraglycidyl-diaminodiphenyl methane (TGDDM) epoxy which has not been reported previously.

The fluorescence quantum yields of certain compounds exhibit a strong dependence on the viscosity of the medium. As early as 1913, Stark (5) noticed that some dyes which do not fluoresce in ordinary solvents will, however, fluoresce strongly in highly viscous media such as glycerol at low temperatures. A number of studies on the viscosity-dependence of the fluorescence quantum yields of various compounds have appeared (6-12). Particularly noteworthy among these are the work of Oster and Nishijima (6), Forster and Hoffman (7), and Sharafy and Muszkat (8). However, Loutfy (13,14) was the first to exploit this effect for monitoring polymerization reactions. He applied the viscosity-dependent fluorescence probe approach for study of polymerization of vinyl monomers and discussed the potential of such probes for the study of polymeric systems

(13,14). In other studies of a broader scope, Loutfy and Law (15,16) reported on the spectroscopy of fluorescence probes in polymeric matrices and on their intramolecular charge-transfer (ICT) interactions. Law (17) also proposed application of the viscosity-dependent fluorescence dyes as sensitive probes for measuring polarity, micro-viscosity, and structural changes in a micro- environment. More recently, Loutfy and Teegarden (18) have exploited the unique properties of julolidinemalononitrile (a "molecular rotor" probe with viscosity-dependent fluorescence) for determination of polymer chain tacticity.

EXPERIMENTAL

The experimental sequence for fluorescence probe monitoring of epoxy cure kinetics or characterization of cure states is schematically illustrated in Fig. 1. In a typical epoxy cure experiment, the fluorescence emission spectrum of a probe-containing specimen is recorded at room temperature after each curing interval at the selected cure temperature. A parallel series of spectra is recorded of an epoxy specimen which does not contain the fluorescence probe. This reference series of fluorescence emission spectra is recorded to ascertain that the observed increase of fluorescence results solely from the increase of the fluorescence quantum yield of the probe. In some series of experiments the "degree of cure" (DOC) reached after each curing time interval was determined by differential scanning calorimetry (DSC) as schematically shown in Fig. 1.

Fluorescence-monitored epoxy curing sequences were performed with two different resin systems: DGEBA cured at room temperature with 11 wt % diethylene triamine (DETA) and TGDDM cured at 150°C with 24 wt % diaminodiphenyl sulfone (DDS). The fluorescecnce probe used "polyester yellow" dye (13)

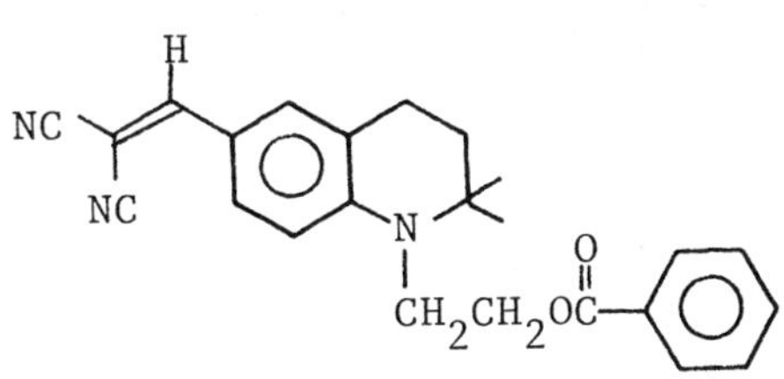

is one of a series of p-(N,N-dialkylamino)benzylidene malononitriles with viscosity-dependent fluorescence

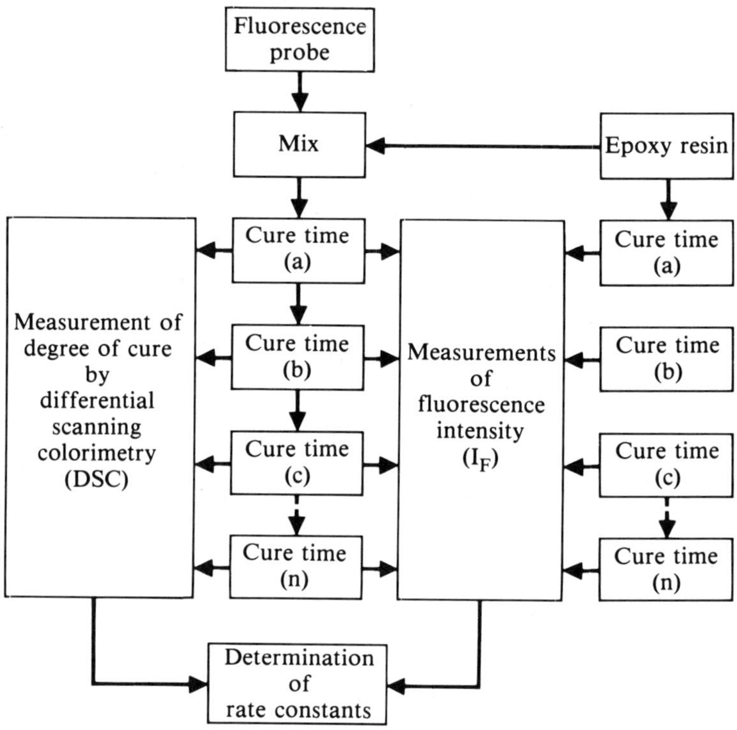

Fig. 1. Experimental sequence for fluorescence-probe monitoring of epoxy cure kinetics and comparison of fluorescence and DSC data.

yields (13-15). The probe (0.5 wt %) was dissolved in the epoxy resin at 80°C prior to adding the curing agent. TGDDM (95-97% pure) was obtained by a preparative-LC separation procedure described by Hagnauer (19) using the Ciba-Geigy MY720 epoxy resin as the starting material.

The epoxy specimens were cured in disposable fluorescence/transmission cells (Fig. 2) made from two microscope-slide glass plates cut to fit the width and height dimensions of the standard cell holder (12 x 45 mm) with a light path of 1.5 mm determined by the thickness of a silicone rubber spacer/sealant strip.

Fluorescence measurements were performed using a spectrophotometer (Cary 17D) equipped with fluorescence accessory (Cary model 1712) operating in the front surface illumination mode with a filter monochromator providing the fluorescence excitation (λ_{ex} = 450 nm). The spectrophotometer is interfaced with microprocessor (Digital Equipment Corp. MINC 11-23) laboratory microcomputer which controls the instrument and processes the data. Both instrument control and data processing functions are based on software developed by MDRL.

The fluorescence emission spectra of the epoxy specimens in the 450-600 nm range were recorded at room temperature after curing for different intervals at the desired temperature. Fluorescence intensities I_F at the band maxima (λ_{max} = 495-505 nm) were determined for each cure state and plotted as function of cure time (Fig. 3).

The degree of cure (DOC) of TGDDM/DDS epoxy reached as a result of each interval of cure was determined by differential scanning calorimetry (DSC). Ten TGDDM/DDS specimens (10-20 mg) were cured in DSC sample cups together with the TGDDM/DDS specimen on which fluorescence measurements were made. Following each cure interval, one of the specimens was subjected to DSC analysis. The degree of cure values for each cure state were determined from

$$DOC = \left\{(A_o - A_t)/A_o\right\} x\ 100,$$

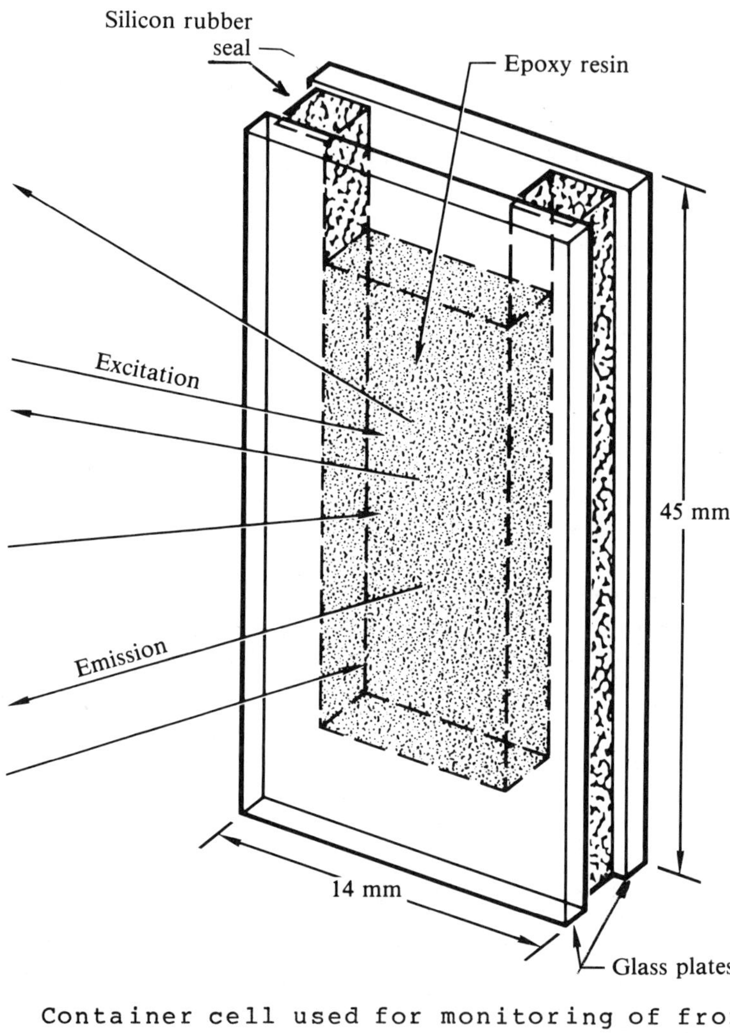

Fig. 2. Container cell used for monitoring of front-face fluorescence intensities at different epoxy cure states.

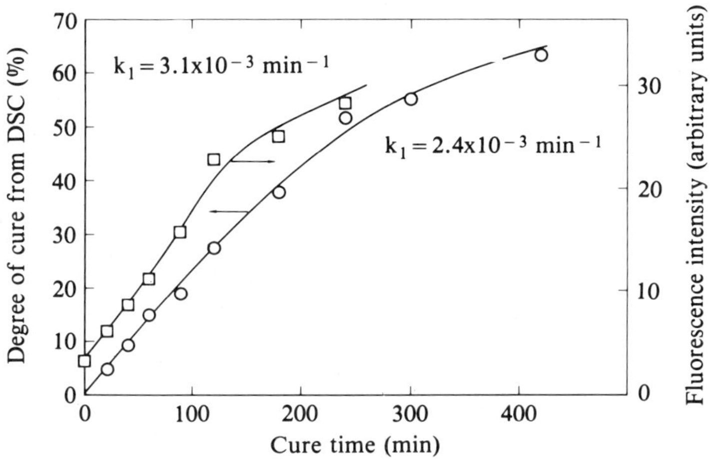

Fig. 3. Degree of cure and fluorescence intensity as a function of TGDDM/DDS epoxy cure time at 150°C.

where A_o and A_t are the normalized areas under the DSC exotherms for samples before cure (A_o) and samples after cure time t (A_t).

RESULTS AND DISCUSSION

Fluorescence-probe-containing DGEBA/DETA epoxy cured at room temperature shows an approximate 16-fold increase in the initial fluorescence intensity (I_F) during cure, resulting in an I_F vs cure-time kinetic profile similar to the profile shown in Fig. 3 for TGDDM/DDS epoxy. The fluorescence intensity of DGEBA/DETA that did not contain the fluorescence probe shows no change of I_F as function of cure time. The observed increase in I_F as function of cure time in the probe-containing specimen can, therefore, be attributed solely to the increase in the fluorescence quantum yield of the probe.

Specimens of probe-containing TGDDM/DDS resin cured at 150°C produced I_F vs cure time profiles similar to the profile shown in Fig. 3. Surprisingly, the parallel series of spectra of the TGDDM/DDS epoxy that did not contain the fluorescence probe (Fig. 1) produced the I_F vs cure-time profile shown in Fig. 3, which is almost identical to the profile obtained from the probe-containing specimens.

A series of experiments was performed to determine the origin of the cure-induced increase of the probeless TGDDM/DDS fluorescence. A summary of the experiments performed and the observed results is given in Table 1. The results of these experiments conclusively indicate that the observed increase in the TGDDM/DDS fluorescence originates in the TGDDM resin. Furthermore, the increase in the fluorescence yield of uncured TGDDM resin observed when the temperature is lowered, thus leading to an increase of its viscosity, strongly suggests that the fluorescence yield of TGDDM is viscosity-dependent. This conclusion is further strengthened by published data (6) which indicates that the fluorescence yield of tetramethyl diaminodiphenyl ketone (TMDDK) a structural analog to TGDDM, is viscosity-dependent.

Table 1. Summary of Experiments Designed to Determine the Origin of the TGDDM/DDS Cure-Induced Increase of Fluorescence Yields

Description of Experiment*	Observed Result	
	I_F increases	I_F unchanged
DGEBA cured with 11% DETA		X
DGEBA cured with 10% DDS		X
TGDDM cured with 24% DDS	X	
TGDDM cured with 10% DETA	X	
TGDDM cured with 10% DETA	X	
TGDDM cured with 1% DETA	X	
TGDDM cooled to -50°C	X	

*conducted without a fluorescence probe

The structural similarity of TGDDM and TMDDK and the similarity in their potential modes of internal or torsional rotations which provide pathways for nonradiative deactivation of the excited state are illustrated in Fig. 4. These torsional relaxation modes are analogous to the relaxation modes illustrated by Loutfy (13) for the malononitrile probes and are in agreement with Tredwell and Osborne's (10) discussion on the correlation between the rate of internal conversion and the mobility of the $-NR_1R_2$ group about the C—N bond which adds support to the conclusion of this study.

Since the monomer exhibits viscosity-dependent fluorescence we term this unique capability as self-probing. In the case of solute probes, the viscosity-dependence of the fluorescence yield is attributed to the restriction of rotational, nonradiative decay of the excited state by the medium, thus leading to an increase of the fluorescence yield in media of higher viscosities (14-17).

In the case of the self-probe behavior of TGDDM reported here, the restriction of internal rotation is due to incorporation of the monomer in the polymer network in addition to the direct effect of the medium's microviscosity. Thus, it is possible that a significant difference in the overall behavior of the self-probe and solute-probes will emerge.

A comparison of the kinetic profiles of the TGDDM/DDS degree of cure by DSC and the self-probe fluorescence intensity as function of cure-time is shown in Fig. 3. Also given are the rate constants calculated from the corresponding data by the Guggenheim method (20,21) with an appropriate interpolation to meet the fixed time-interval Δt between data points required by the method. This comparison shows that the fluorescence-derived profile for cure kinetics closely follows the profile derived by DSC.

CONCLUSIONS

The findings described here corroborate the use of viscosity-dependent fluorescence probes for monitoring polymerization reactions. Fluorescence-derived rate constants for epoxy curing produce values comparable to those derived by DSC. The self-probe fluorescence of TGDDM epoxy observed in this study provides unique opportunities for _in situ_ monitoring of epoxy cure and detection of changes in epoxy networks.

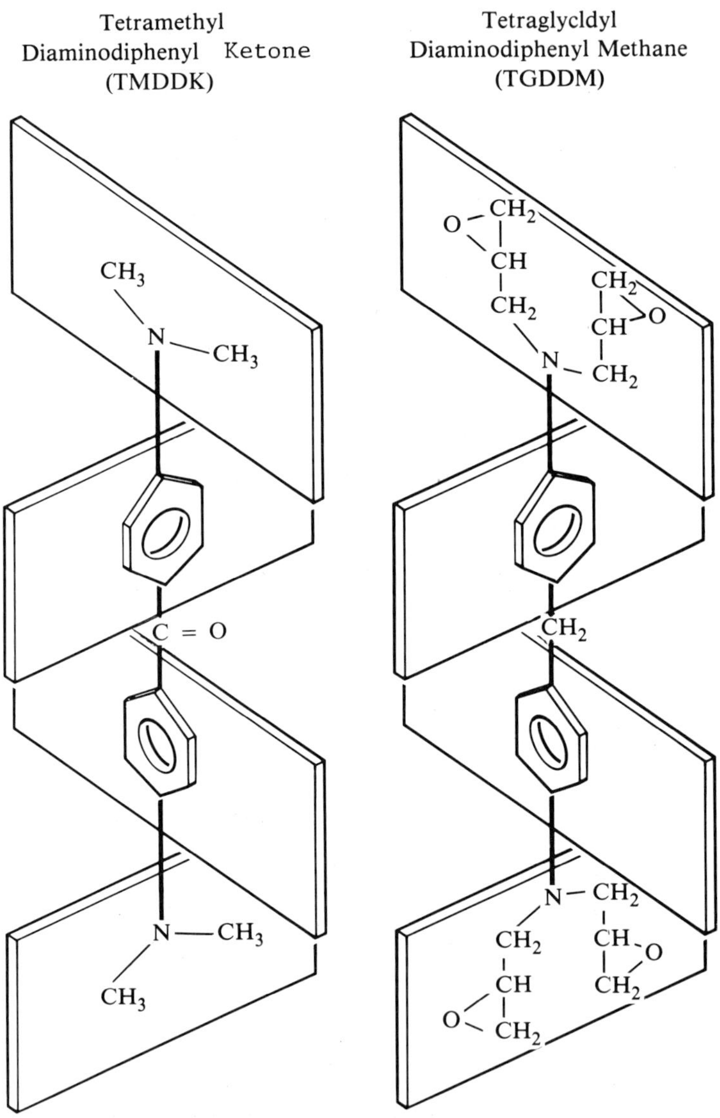

Fig. 4. Comparison of torsional relaxation modes of TMDDK and TGDDM.

REFERENCES

1. C.A. May, M.R. Dusi, J.S. Fritzen, D.K. Hada, M.G. Maximovich, and K.S. Thrasher, Org. Coat. Appl. Polym. Sci. Proc., 47, 419 (1982).

2. J.M. Barton, Thermochim. Acta, 30, 153 (1979).

3. P. Hedrig, Dielectric Spectroscopy of Polymers, Halsted Press, (Wiley) NY, 1977, p. 332.

4. N.F. Sheppard, S.L. Garverick, D.R. Day, and S.D. Senturi, Proc. 26th SAMPE Symposium, (1981), p. 65.

5. J. Stark and P. Lipp, Z. Physik Chem., 86, 36 (1913).

6. G. Oster and Y. Nishijima, J. Am. Chem. Soc., 78, 1581 (1956).

7. Th. Forster and G. Hoffmann, Z. Physik. Chem. (n.F.), 75, 63 (1971).

8. S. Sharafy and K.A. Muskat, J. Am. Chem. Soc., 93, 4119 (1971).

9. G. Jones, W.R. Jackson, and A.M. Halpern, Chem. Phys. Letters, 72 (2), 391 (1980).

10. C.J. Tredwell and A.D. Osborne, J. Chem. Soc. Faraday Trans., 2, 76, 1627 (1980).

11. A.D. Osborne, J. Chem. Soc. Faraday Trans., 2, 76, 1938 (1980).

12. A.T. Eske and K.R. Nagvi, Chem. Phys. Letters, 63, (1), 128 (1979).

13. R.O. Loutfy, Macromolecules, 14, 270 (1981).

14. R.O. Loutfy, J. Polym. Sci. Polym., Phys. Ed., 20, 825 (1982).

15. R.O. Loutfy and K.Y. Law, J. Phys. Chem., 84, 2803 (1980).

16. K.Y. Law and R.O. Loutfy, Macromolecules, 14, 587 (1981).

17. K.Y. Law, Chem. Phys. Letters, 75, 545 (1980).

18. R.O. Loutfy and D.M. Teegarden, Macromolecules, 16, 452 (1983).

19. G.L. Hagnauer, Ind. Res. and Dev., April 1981, p. 128.

20. E.S. Swinbourne, "The Analysis of Kinetic Data," Nelson, London, 1971, Chap. 4.

21. E.A. Guggenheim, Phil. Mag., 2, 538 (1926).

Dynamic Mechanical Properties of Silicone Pressure-Sensitive Adhesives

B.C. Copley

3M Center 230-1S-10
St. Paul, MN 55144

ABSTRACT

The chemistry of silicone resins and siloxane gums used to prepare silicone pressure sensitive adhesives is briefly reviewed. The thermal, dynamic mechanical and X-ray scattering properties of the silicone adhesives is presented. A specific morphological model for the silicone pressure sensitive adhesives is proposed based on the characterization data. The results for adhesives prepared from polydimethyl siloxane gums is compared to adhesive prepared from polydimethyl-co-diphenyl siloxane gums.

INTRODUCTION

The many novel properties and applications of silicone pressure sensitive adhesives (psas) have been discussed in numerous publications (1-5). Silicone adhesives exhibit excellent high temperature stability, good low-temperature properties, and good adhesion to low-energy surfaces such as fluorocarbons and silicones. Most of the major applications utilize one or more of these novel properties.

The preparation of silicone psas was first disclosed in the mid 1950s when it was observed that blends of silicone resins and polydimethylsiloxane gums (PDMS) produced a material with pressure sensitive adhesive properties (i.e., tack and adhesion under low application pressure). By 1956 a versatile process for the preparation of silicone resins suitable for use in silicone psas was described (6). Rapid progress in the preparation and application of silicone psas was observed from this point to the present.

The resins utilized in silicone psas are typically prepared from tetrafunctional (Q) and monofunctional (M) siloxanes in a two-phase, multistep reaction sequence. A typical synthesis consists of the following steps:

(1) A tetrafunctional sodium silicate in a carefully pH-controlled aqueous solution is warmed to 50°-90°C. The resulting hydrosol is a highly complex mixture of low molecular wt. oligomers of Q units (see the following).

(2) Addition of chlorotrimethylsilane in xylene and the capping of the reactive silanol groups results in a MQ resin with low residual silanol functionality which is soluble in xylene.

(3) Separation of the xylene phase, which contains the MQ resin, followed by evaporative concentration yields a high solids resin solution.

Blends of the MQ resin with high molecular weight siloxane gum which may be cross-linked by the addition of benzoyl peroxide result in silicone pressure sensitive adhesives. The purpose of this work is to examine the dynamic mechanical properties of the formulated adhesives in an effort to understand how the silicone resins and gums interact to form a pressure sensitive adhesive. A specific morphological model as suggested by experimental results will be discussed.

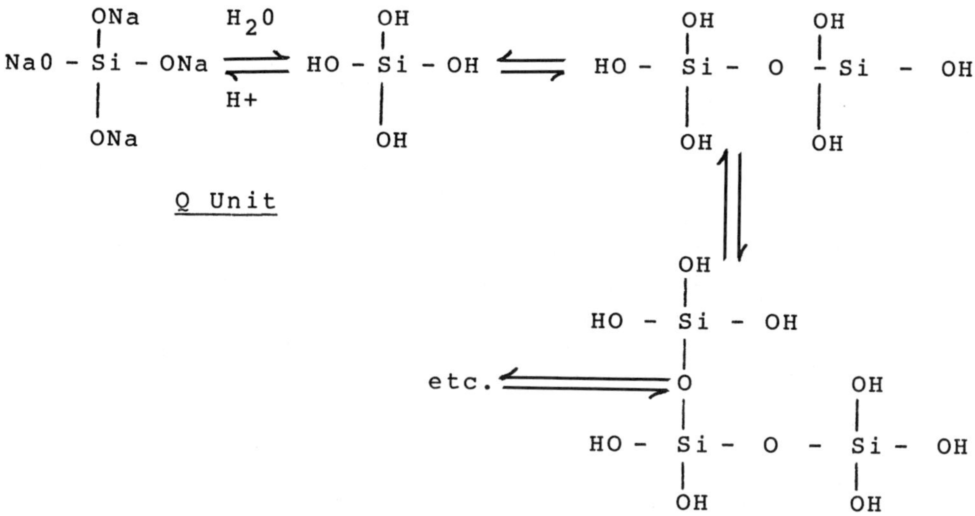

Q Unit

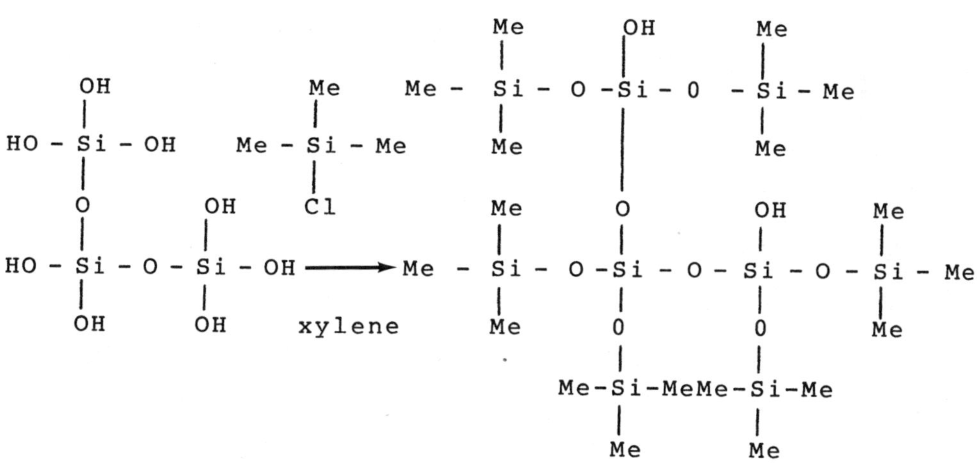

Hydrosol MQ Resin

EXPERIMENTAL

The polydimethyl siloxane adhesives were prepared by dissolving a polydimethylsiloxane (PDMS) gum ($\overline{M}_n$ = 357,000 $\overline{M}_w$ = 697,000 by GPC) in toluene and admixing a xylene solution of MQ resin (M"Q = 0.8:1, 70% solids solution). Film samples for measurement were prepared by solution casting onto Teflon film, air dried for 15 hrs., followed by 24 hrs. in a vacuum oven.

The phenyl containing siloxane adhesives were prepared by a similar procedure using the same MQ resin solution. In place of the polydimethyl siloxane gum was substituted a polydimethyl-co-diphenyl siloxane (PDMDPS) gum ($\overline{M}_n$ = 250,000, $\overline{M}_w$ = 830,000, 86:14 = dimethyl:diphenyl siloxane ratio).

Thermal analysis was performed with a Perkin-Elmer DSC-2 at 20°C/min. scan rate. The glass transition temperature and change in specific heat was calculated using the Perkin-Elmer software.

The dynamic mechanical data were obtained in two different modes. The low temperature data (-140°C to 20°C) were obtained in the torsion mode on a Rheometrics System IV at 10 rad/sec. The high temperature data were obtained in the parallel plate mode on a Rheometrics Mechanical Spectrometer at 10 rad/sec.

RESULTS

PDMS Adhesives

The dynamic storage modulus (G') vs. temperature data are displayed in Figs. 1 - 4. The low temperature torsion data (Figs. 1 and 2) show a distinct transition zone at approximately -115°C. This is the glass transition temperature (T_g) of PDMS. The T_g of the adhesive does not appear to change as the R/G ratio increases. Also, the adhesives of high R/G ratio have a relatively high modulus plateau zone, the level of which is proportional to the R/G ratio (i.e., higher R/G samples have higher modulus in this zone). At low R/G ratios, the modulus appears to drop sharply without any evidence of a plateau zone.

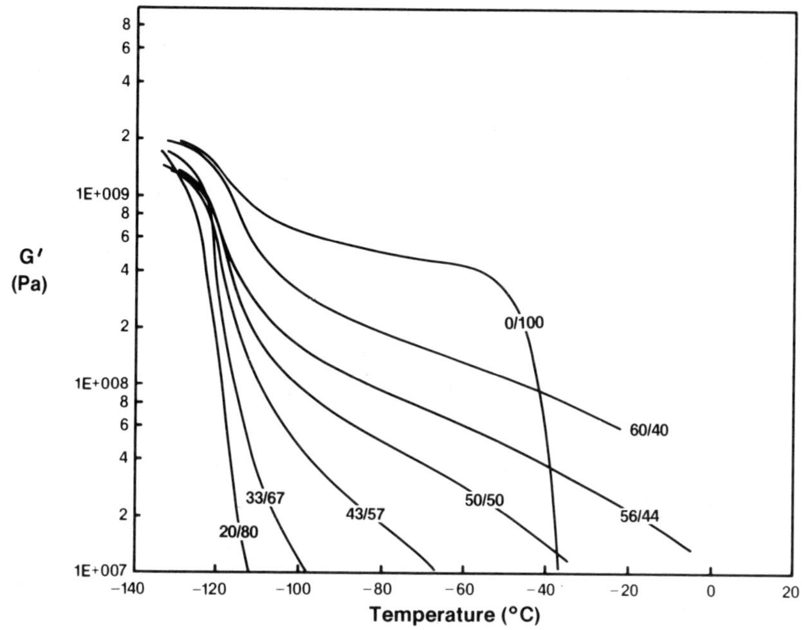

Fig. 1. PDMS Adhesives - Dynamic Mechanical Torsion

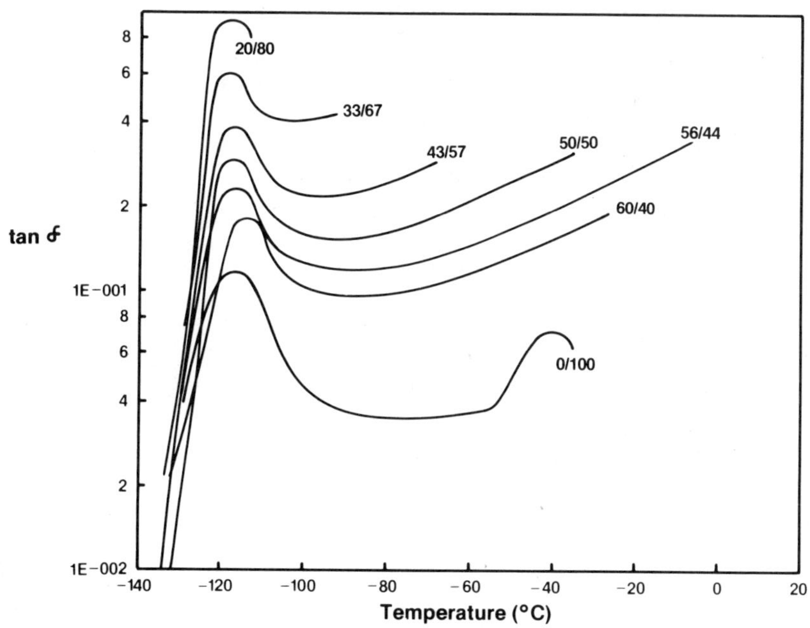

Fig. 2. PDMS Adhesives - Dynamic Mechanical Parallel Plate

The data for the pure PDMS exhibit a glass transition at -115°C, but the modulus remains high after passing through the T_g. The presence of crystalinity reinforces the polymer up to the melting point of -45°C. The modulus then drops sharply at~-45°C to the level expected for pure PDMS in the amorphous state. DSC data confirm the presence of crystals below -45°C.

The parallel plate data (Figs. 3 and 4) show the presence of a second transition zone. The higher temperature transition is directly proportional to the R/G ratio. This is quite different from the low temperature transition, which does not shift as the R/G ratio changes.

The DSC data in Table I confirm the torsional dynamic mechanical data. As the R/G ratio increases, the T_g remains at~-109°C. In addition, as the R/G ratio increases, the change in specific heat decreases. This is explained since, as R/G ratio increases, the concentration of the gum decreases and thus the ΔC_p decreases since it is proportional to the weight fraction of gum present.

No T_g was observed for pure MQ by DSC measurements and no other transitions were observed by DSC in the adhesives. The fact that two transitions are observed in the dynamic mechanical data suggests the presence of two material phases. The low temperature transition appears to be relatively pure PDMS whereas the high temperature may be due to a homogeneous blend of PDMS and MQ resin. This model is consistent with the observed shift of the transition zone to higher temperatures as the R/G ratio increases.

PDMDPS Adhesives

The low temperature dynamic mechanical data for the PDMDPS adhesives (Figs. 5 and 6) are similar to those for the PDMS adhesive. The T_g of the PDMDPS is higher (-85°C) than for PDMS (-110°C) and the transition zones are shifted to the higher temperatures. A contrasting feature of the PDMDPS adhesives is that a small but distinct increase in the T_g is observed as the R/G ratio increases. This suggests some solubility of a high T_g resin component in the gum. The plateau zone in the PDMDPS adhesive is much less pronounced, even at high R/G ratios.

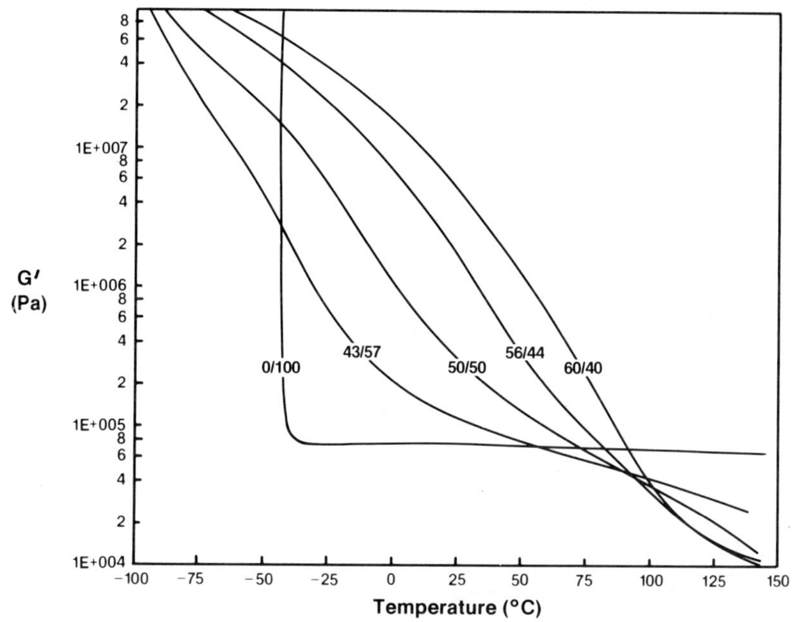

Fig. 3. PDMDPS Adhesives - Dynamic Mechanical Torsion

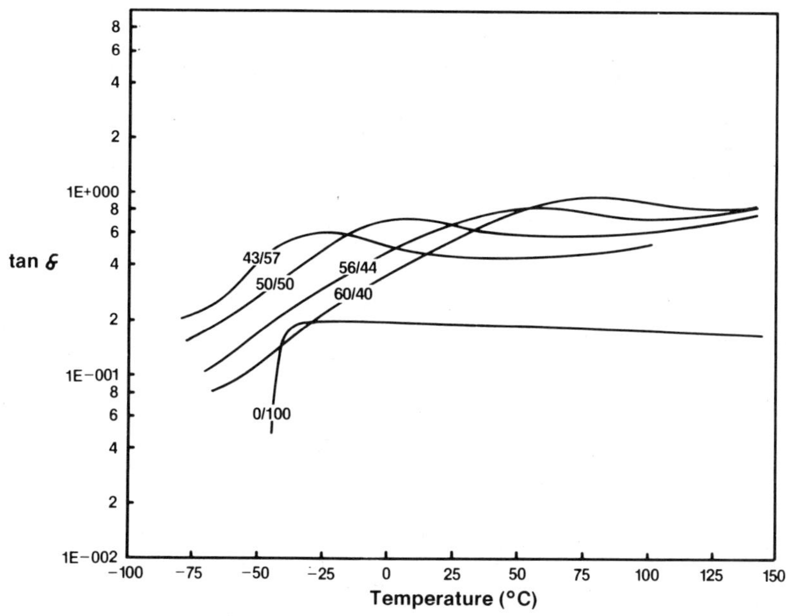

Fig. 4. PDMDPS Adhesives - Dynamic Mechanical Parallel Plate

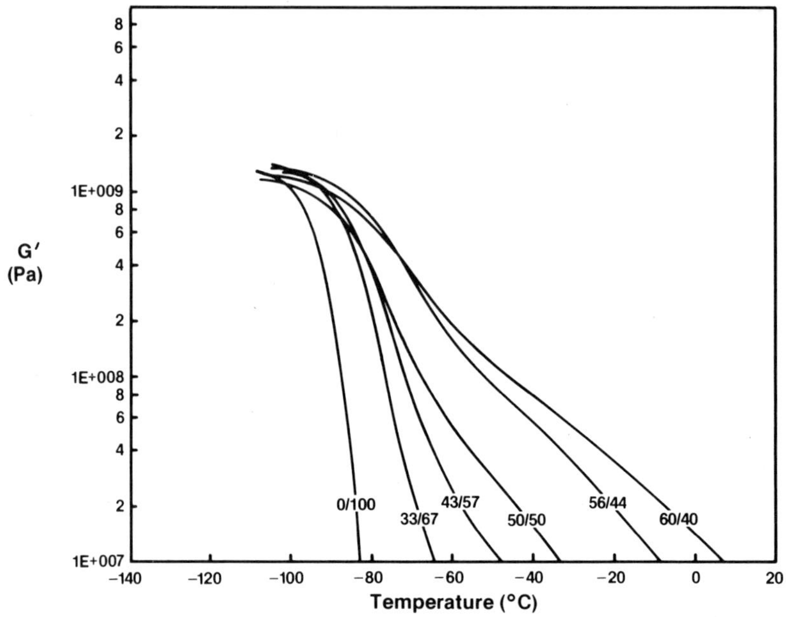

Fig. 5. PDMDPS Adhesives - Dynamic Mechanical Rectangular Bar Torsion

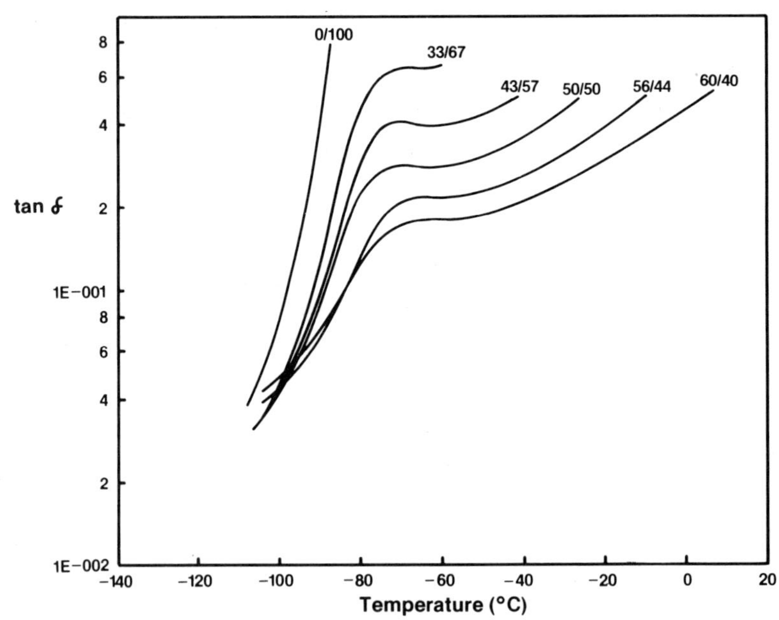

Fig. 6. PDMDPS Adhesives - Dynamic Mechanical Rectangular Bar Torsion

Table 1. DSC Thermal Analysis

POLYDIMETHYL SILOXANE ADHESIVES

R/G	T_g (°C)	ΔC_p (cal/g-°K)
0/100	-110.2	0.010
20/80	-109.3	0.071
33/67	-110.9	0.037
43/57	-110.3	0.021
50/50	-109.7	0.015
56/44	-109.2	0.014
60/40	-108.8	0.010
100/0	--	--

POLYDIMETHYL-CO-DIPHENYL SILOXANE ADHESIVES

R/G	T_g (°C)	ΔC_p (cal/g-°K)
0/100	-82.3	0.055
20/80	-79.3	0.031
33/67	-75.3	0.029
43/57	-72.8	0.020
50/50	-72.6	0.018
56/44	-70.2	0.013
60/40	-67.3	0.004
100/0	--	--

The high temperature parallel plate data (Figs. 7 and 8) show the same increase in the temperature of the transition zone as was observed for the PDMS adhesives. A morphology similar to the PDMS adhesives is considered for the PDMDPS adhesive. It appears that these adhesives are also two-phase, but there appears to be some solubility of the resin in the gum phase, resulting in a shift of both transition zones.

The DSC data in Table I confirm the shift in the low temperature Tg of the PDMDPS adhesives. This shift can be attributed to the presence of a PDMDPS gum phase containing enough resin to increase the Tg. Here again, in no case was any other transition observed by DSC.

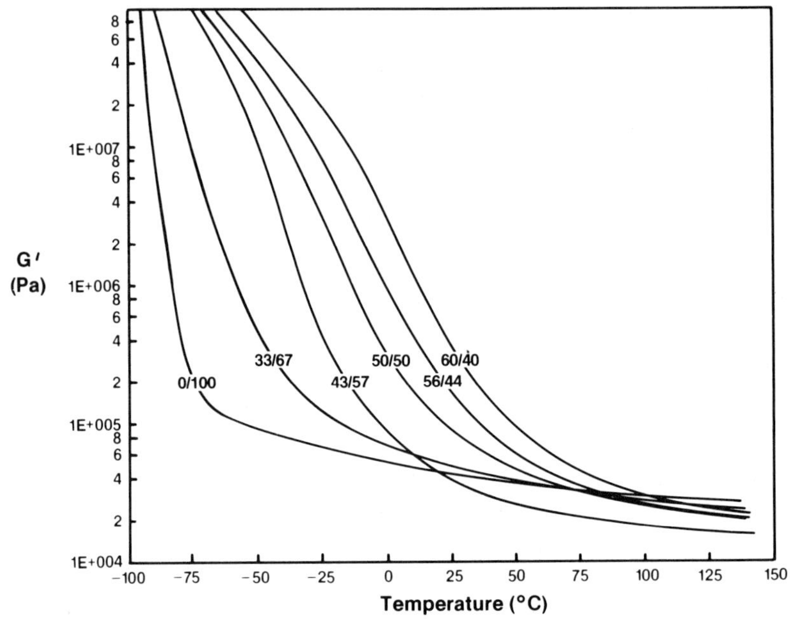

Fig. 7. PDMDPS Adhesives - Dynamic Mechanical Parallel Plate

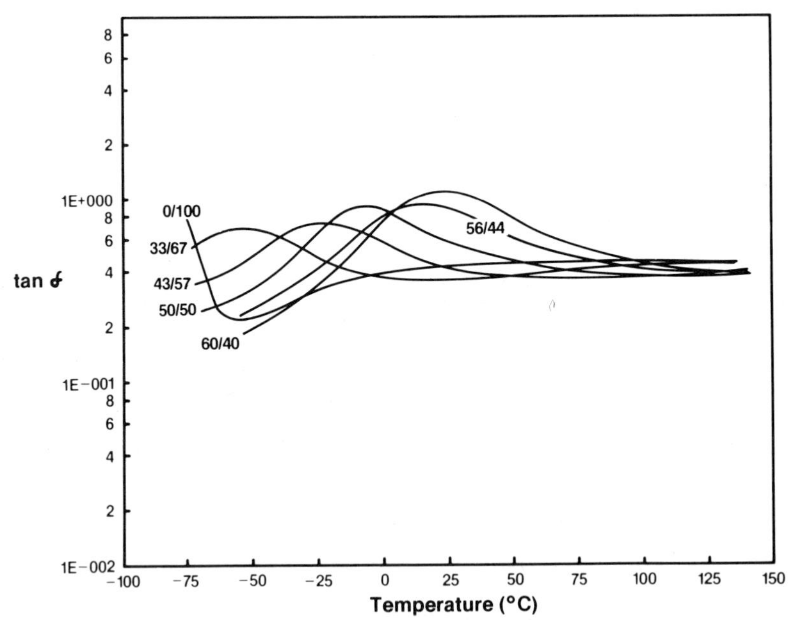

Fig. 8. PDMDPS Adhesives - Dynamic Mechanical Parallel Plate

CONCLUSIONS

Silicone pressure sensitive adhesives appear to be heterogeneous blends of MQ resin and a gum. A gum phase containing little if any dissolved resin is observed in the PDMS adhesive. A resin phase, probably containing low molecular weight PDMS is detected by dynamic mechanical analysis. The PDMDPS adhesives are similar, except that the solubility of the resin in the gum phase is higher. The result is an impure, gum-rich phase which coexists with a resin-rich phase of resin and gum.

REFERENCES

1. D.F. Merrill, Adhesives Age, 22, 39 (March 1979).
2. W.J. O'Malley, Adhesives Age, 18, 17 (June 1975).
3. J.C. Smith, Adhesives Age, 17, 27 (June 1974).
4. S.J. Price and J.B. Nathan, Adhesives Age, 17, 37 (Sept. 1974).
5. F.J. Modic, Adhesives Age, 5, 36 (Dec. 1962).
6. J.F. Dexter, U.S. Patent 2,736,721, Feb. 28, 1956.

The Viscoelastic Properties of Pressure-Sensitive Adhesives

J.B. Class and S.G. Chu

Hercules Incorporated
Hercules Research Center
Wilmington, DE 19899

ABSTRACT

It is generally accepted that the viscoelastic properties of a pressure sensitive adhesive control its performance. These adhesives are usually blends of rubber with low molecular weight resins. The significance of the resin type and concentration on the viscoelastic properties of the composition is not well understood. Low molecular weight resins of varying structures and molecular weights are examined in natural rubber and SBR at various concentrations. G' and tan δ of the systems depend upon the selection of resin, its molecular weight, and its concentration in the formulation. The shift in T_g resulting from the resin can be estimated using the Fox equation. Plateau modulus, identified as G' at the minimum in tan δ in the plateau region, is proportional to the volume fraction of the polymer raised to about the 2.3-2.5 power.

INTRODUCTION

Testing of a material for performance as a pressure sensitive adhesive involves both a bonding and debonding step. For satisfactory performance in each of these steps, the material must respond to a deforming force in a prescribed manner. It must adhere to the substrate under bonding conditions and resist separation at debonding conditions. Both the rate and extent of deformation are important in each of these steps. Bonding is a low rate process at low deformation, while debonding is a high rate process at high deformation. Thus, the viscoelastic behavior of the adhesive is a major factor in its performance. The specific viscoelastic requirements for performance of a material as a pressure sensitive adhesive are not well understood.

Pressure sensitive adhesives are typically blends of a rubber, such as natural rubber or styrene-butadiene rubber, with low molecular weight resins. Effective resins have weight average molecular weights, $\bar{M}_w$, in the range of 300 to 2,000. They are usually oligomers of cyclic monomers or of monomers that form cyclic structures during polymerization. The resins may also be esters of cyclic organic acids such as rosin. These materials are usually brittle glasses at room temperature and have glass transition temperatures from about 0° to about 70°C. A resin is described as a tackifier if by adding it to a rubber the resulting composition has the properties of a pressure sensitive adhesive. The significance of resin type and concentration on the viscoelastic properties of the composition has not been reported.

There have been very few studies reported on the viscoelastic properties of rubber-resin pressure sensitive adhesive systems. In 1973, M. Sherriff and co-workers (1) reported on the effect of adding poly(β-pinene) resin to natural rubber. Based on a G' master curve, they showed that the resin shifted the entry to the transition zone to a lower frequency and reduced the modulus in the rubbery plateau. G. Kraus and K.W. Rollman (2) reported in 1977 on their study of resins blended with styrene-isoprene-styrene block copolymers. They showed that the addition of a resin increased the glass transition temperature of the rubbery mid-block and decreased the plateau modulus. Accordingly, a satisfactory tackifying resin should produce these changes.

EXPERIMENTAL

The resins examined in this work are described in Table 1. The rosin-based products were prepared by esterification of hydrogenated rosin. The terpene, aromatic, and C_5-stream resins were prepared by cationic polymerization. Cycloaliphatic resins were prepared by complete hydrogenation of selected aromatic resins. Molecular weight data were obtained by size exclusion chromatography, using a Styragel column set with THF as the mobile phase. Calculations were based on a polystyrene calibration. Glass transition temperatures were obtained by differential scanning calorimetry from the first break on the second heat.

The natural rubber and styrene-butadiene rubber are described in Table 2. Natural rubber was shear degraded on a two-roll mill to reduce the Mooney viscosity to 53. The styrene-butadiene rubber was SBR 1011, containing 25% styrene.

Rubber and resins were individually dissolved in toluene and then blended to selected ratios of solids. The solutions were cast onto a release surface and allowed to evaporate at room temperature. The last trace of solvent was removed by drying for three days in a vacuum oven at 40°C. Absence of solvent was confirmed by thermogravimetric analysis. Photomicrographs of the blends were made using phase contrast on a Zeiss light microscope at 490X. Resolution limits of the microscopic system was about 0.5 μ.

The rubber-resin blends were examined on a mechanical spectrometer (Rheometrics, Inc.) in the parallel plate mode. Small-diameter (8 mm) plates were used with a 2 mm gap. G', G", and tan δ were plotted from the glassy state to 120°C at 10 rad/sec.

Frequency scans were run from 0.1 to 100 rad/sec at the temperature at which tan δ was at a minimum in the plateau zone of the temperature scan. Plateau modulus, G_N^o, was identified as G' at the minimum in tan δ on the frequency scans.

Table 1. Identification of Resins

	$\bar{M}_w$[a]	$\bar{M}_w/\bar{M}_n$[a]	T_g, °C[b]
ROSIN BASED			
Glycerol Ester of Highly Stabilized Rosin	550	1.2	32
Pentaerythritol Ester of Highly Stabilized Rosin	650	1.3	48
Pentaerythritol Ester of Stabilized Rosin	650	1.3	48
TERPENE			
Poly α-Pinene	650	1.4	68
Poly β-Pinene	1600	2.6	62
PURE MONOMER			
Polystyrene	990	1.6	32
Poly (vinyl cyclohexane)	650	1.4	38
Poly (*tert*-butyl styrene)	850	1.3	59
C_5 STREAM			
Polymerized Petroleum Stream, Low MW	1350	1.8	45
Polymerized Petroleum Stream, High MW	2850	2.8	41

a. Apparent molecular weight data based on polystyrene standards.

b. DSC data; first break on second heat.

Table 2. Description of Polymers

	Mooney[a] Viscosity	$\bar{M}_w$[b]	$\bar{M}_w/\bar{M}_n$[b]	T_g, °C[c]
Natural Rubber	53	266,000	2.53	-66
SBR	53	--	--	-57

a. ML(1 + 4) at 100°C.

b. Apparent molecular weight data based on polystyrene standards.

c. DSC data; first break on second heat.

EFFECT OF RESIN STRUCTURE

An examination of rubber-resin blends shows that compatibility is very sensitive to structure, even with resins that have apparent molecular weights below 1,000. Polystyrene resin ($\overline{M}_w$ 900, $\overline{M}_w/\overline{M}_n$ 1.6) was found to be incompatible with natural rubber based on an examination of G' and tan δ scans vs temperature for 1:1 systems as illustrated in Fig. 1. The temperature peak in tanδ did not shift significantly when the natural rubber was diluted with the resin and G' showed an increase in the plateau compared with that observed for the unmodified natural rubber. Also a second high temperature peak in tan δ was seen, presumably due to a resin-rich separate phase. On the other hand, a poly(vinyl cyclohexane) resin ($\overline{M}_w$ 650, $\overline{M}_w/\overline{M}_n$ 1.4), prepared by hydrogenating the low molecular weight polystyrene resin, was compatible with natural rubber as illustrated in Fig. 2. The temperature of the major peak in tan δ shifted significantly, and G' was depressed in the plateau, as predicted by previous workers.

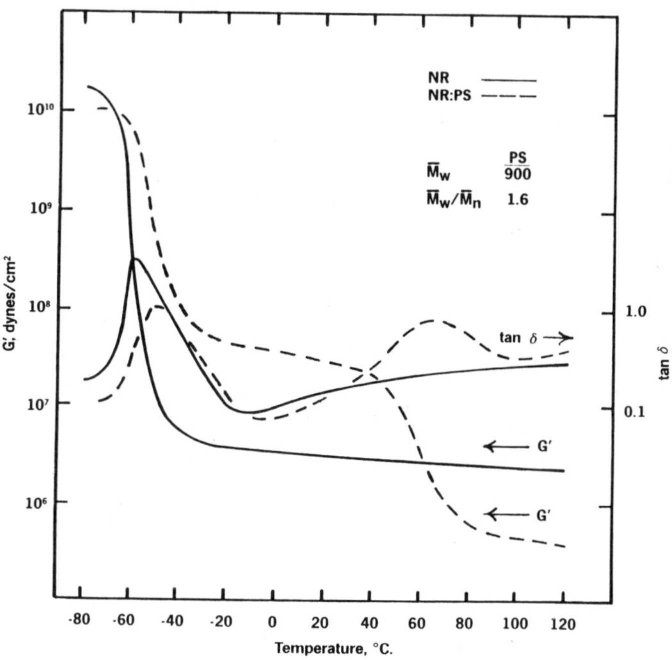

Fig. 1. G' and tan δ vs temperature of a 1:1 blend of natural rubber with polystyrene resin.

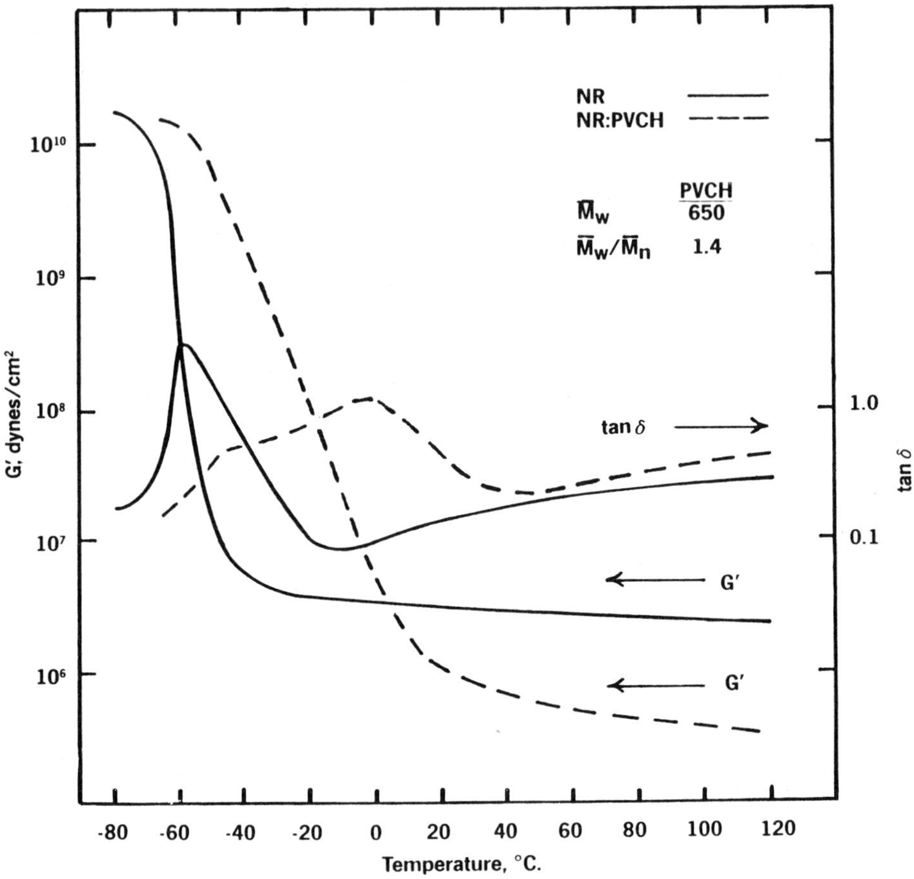

Fig. 2. G' and tan δ temperature of a 1:1 blend of natural rubber with poly(vinyl cyclohexane) resin.

Photomicrographs of these compositions at the 25% resin concentration confirm the incompatibility of polystyrene and the compatibility of poly(vinyl cyclohexane) with natural rubber. These are shown in Figs. 3 and 4. A dispersed resin about 2-4 µ in size, phase is clearly seen in the polystyrene example, while the blend containing poly(vinyl cyclohexane) has no visible structure within the limits of resolution of the equipment. The circles in the center of Fig. 4 are artifacts resulting from preparation of the sample for microscopy.

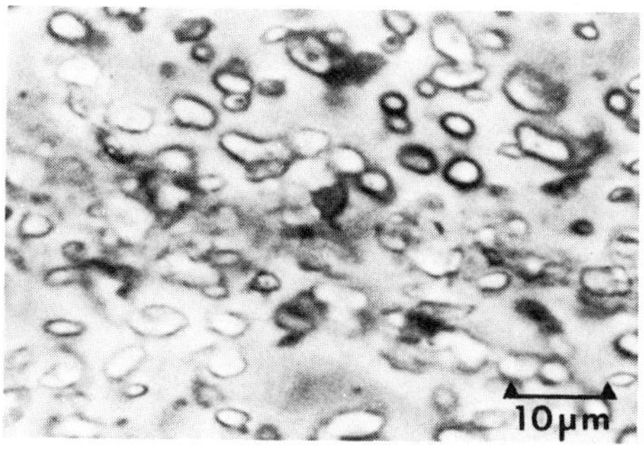

Fig 3. Photomicrograph of a blend of natural rubber (75%) with polystyrene resin (25%)

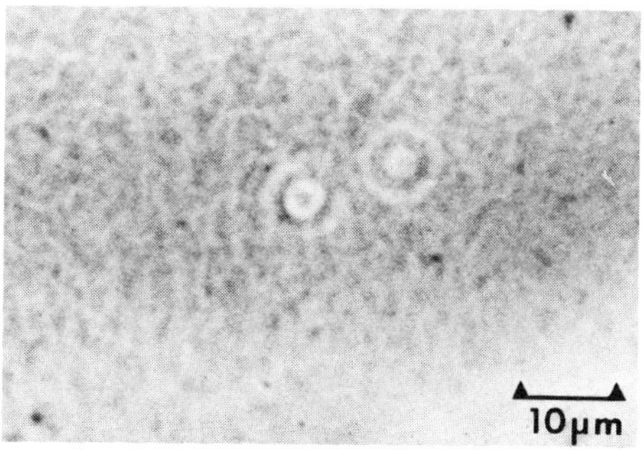

Fig. 4. Photomicrograph of a blend of natural rubber (75%) with poly-(vinyl cyclohexane) (25%)

An examination of the same two resins in styrene-butadiene rubber shows a similar performance, but the compatibility is reversed. The aromatic polystyrene resin is mostly compatible with styrene-butadiene rubber as illustrated in Fig. 5, although the beginning of a small second peak in tan δ is evident at about 80°C. The cycloaliphatic poly(vinyl cyclohexane) resin appears to be incompatible with styrene-butadiene rubber. The blend shows a significant second tan δ peak at about 80°C and an elevation in the plateau modulus, as illustrated in Fig. 6.

After determining the compatibility of an aromatic and a cycloaliphatic resin, it is of interest to examine an alkyl aromatic resin. Poly(*tert*-butyl styrene) $\bar{M}_w$ 850, $\bar{M}_w/\bar{M}_n$ 1.3) was found to be compatible with both natural rubber, as shown in Fig. 7, and with styrene-butadiene rubber as shown in Fig. 8. The *tert*-butyl group on the polystyrene resin altered the compatibility of the resin in natural rubber, but did not detract from compatibility with styrene-butadiene rubber.

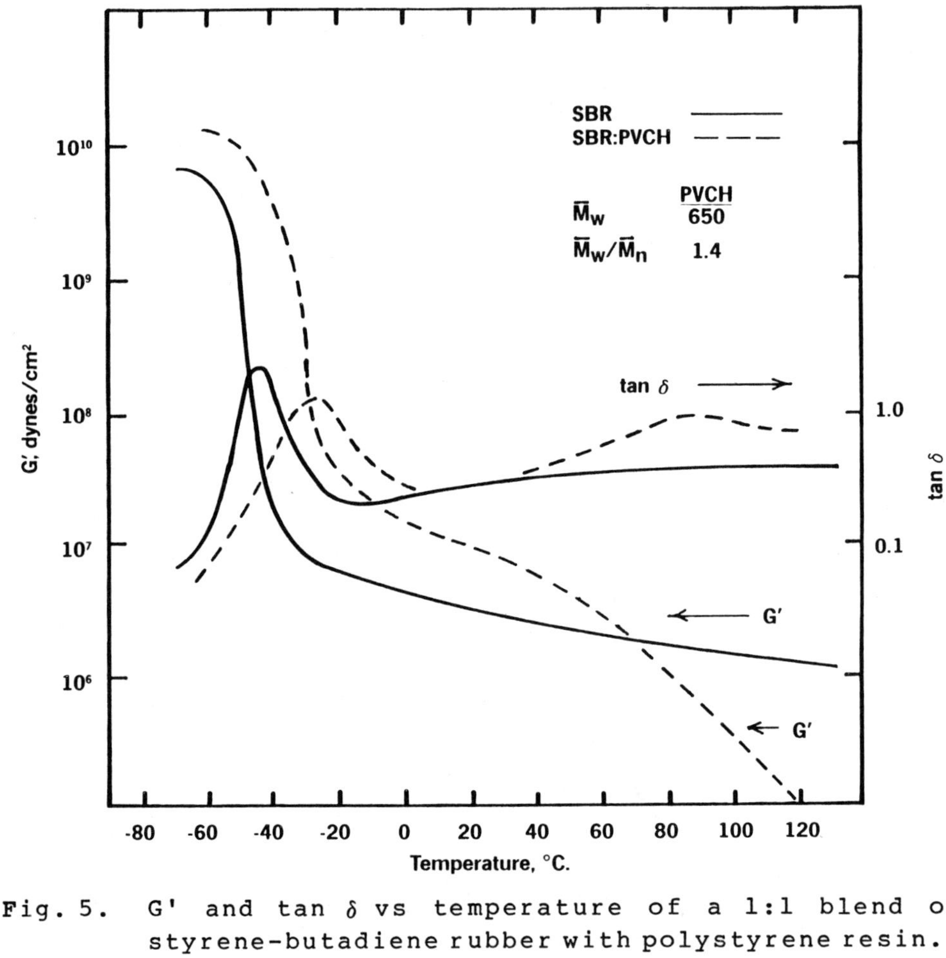

Fig. 5. G' and tan δ vs temperature of a 1:1 blend of styrene-butadiene rubber with polystyrene resin.

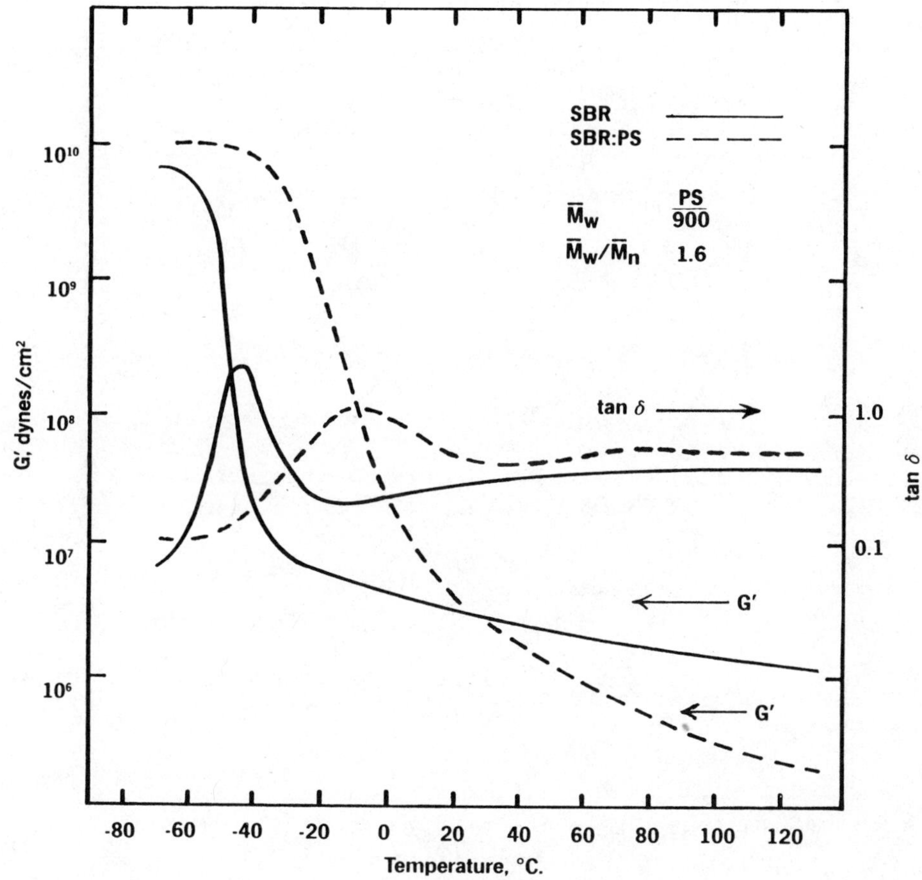

Fig. 6. G' and tan δ vs temperature of a 1:1 blend of styrene-butadiene rubber with poly(vinyl cyclohexane) resin.

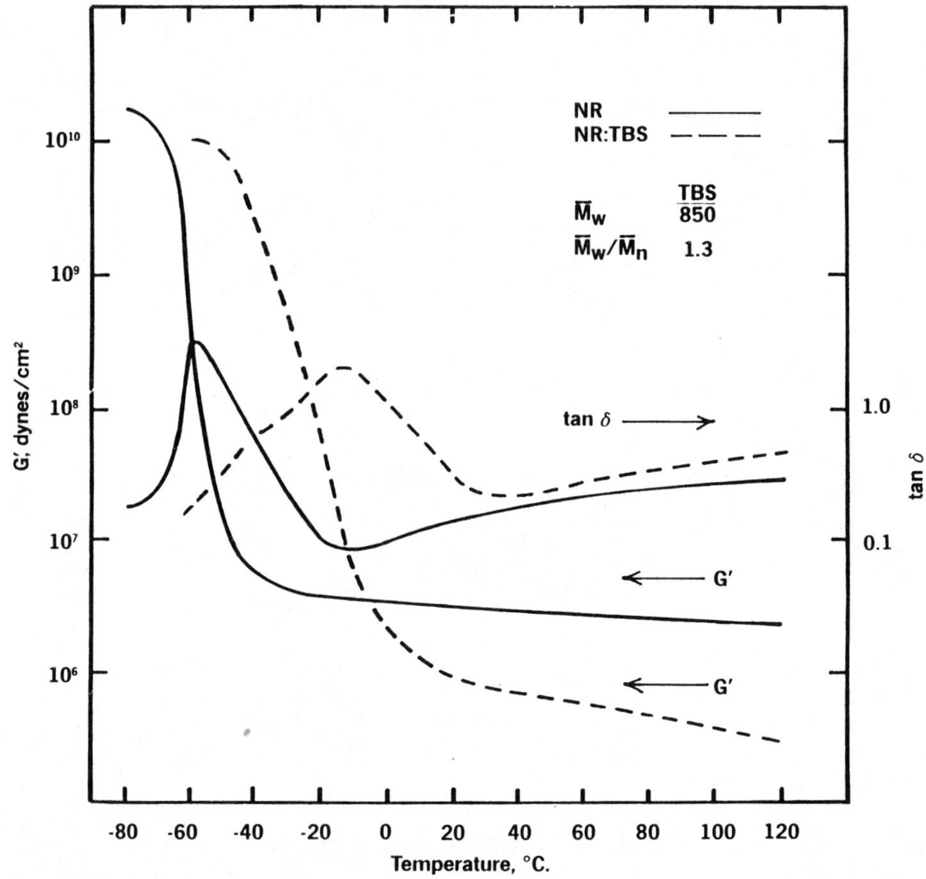

Fig. 7. G' and tan δ vs temperature of a 1:1 blend of natural rubber with poly(tert-butyl styrene) resin.

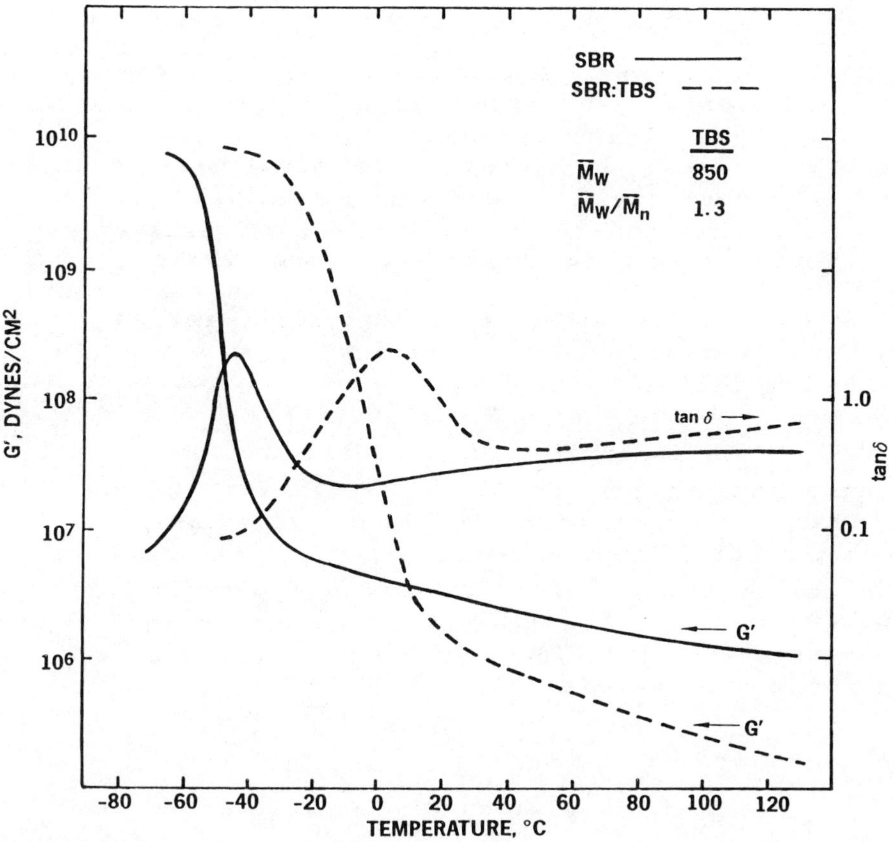

Fig. 8. G' and tan δ vs temperature of a 1:1 blend of styrene-butadiene rubber with poly(tert-butyl styrene) resin.

These studies demonstrate that in selecting resins to tackify either natural rubber or styrene-butadiene rubber, consideration must be given to the resin structure. An aliphatic resin is more likely to be compatible with natural rubber, while an aromatic resin would be compatible with styrene-butadiene rubber.

EFFECT OF RESIN MOLECULAR WEIGHT

The molecular weight of a resin is also very critical to its performance. A series of polystyrene resins were prepared by cationic polymerization. These resins are described in Table 3. Blends of the polystyrene resins with natural rubber (1:1) showed incompatibility throughout, with a large secondary higher temperature peak evident in the blends containing the higher molecular weight resin.

Hydrogenation of the polystyrene resins gave a corresponding series of poly(vinyl cyclohexane) resins which are described in Table 4. In addition, a higher molecular weight poly(vinyl cyclohexane) resin ($\bar{M}_w$ 1,800, $\bar{M}_w/\bar{M}_n$ 2.2) was included. The properties of these resins in natural rubber (1:1) are shown in Fig. 9 for tanδ and in Fig. 10 for G'. Molecular weight is shown on the Z-axis, but there is no significance to the scale. Data in the glassy region have been extrapolated to the -80°C limit to clarify the presentation. At the highest molecular weight, incompatibility is evident from the location of the major tan δ peak and the elevation of G' in the plateau.

Table 3. Description of Polystyrene Resins

$\bar{M}_w$ [a]	$\bar{M}_w/\bar{M}_n$ [a]	T_g, ^{o}C [b]
325	1.29	-42
350	1.26	-14
600	1.63	8
900	1.60	32

a Apparent molecular weight data based on polystyrene standards.

b. DSC data; first break on second heat.

Table 4. Description of Poly(Vinyl Cyclohexane) Resins

$\bar{M}_w$ [a]	$\bar{M}_w/\bar{M}_n$ [b]	T_g, °C [b]
300	1.22	-25
325	1.21	-8
375	1.17	13
650	1.37	38
1800	2.22	76

a. Apparent molecular weight data based on polystyrene standards.

b. DSC data; first break on second heat.

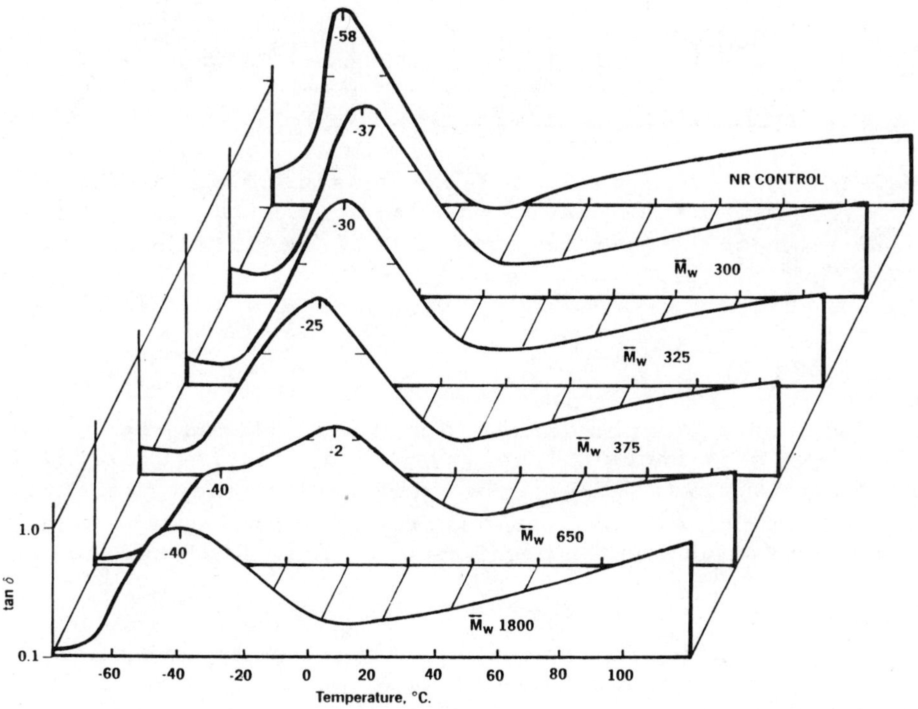

Fig. 9. Effect of resin molecular weight on tan δ plots of 1:1 blends of natural rubber with poly (vinyl cyhclohexane) resins. Resin $\bar{M}_w$ as shown.

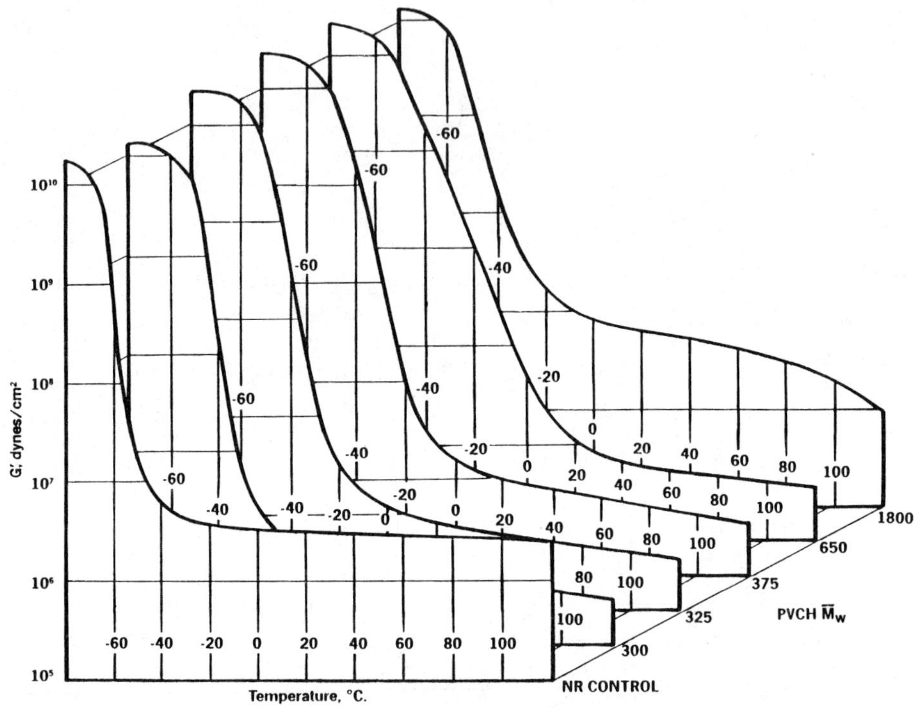

Fig. 10. Effect of resin molecular weight on G' plots of 1:1 blends of natural rubber with poly (vinyl cyclohexane) resins. Resin $\overline{M}_w$ as shown.

Photomicrographs obtained on these blends with natural rubber confirm incompatibility at higher molecular weights as shown in Figs. 11-14. Two distinct phases are clearly seen in Fig. 14, the composition containing the highest molecular weight resin ($\overline{M}_w$ 1,800). The dispersed phase is about 3-5 μ in size.

As expected from compatibility, blends with styrene-butadiene rubber show the reverse performance. Based on tan δ and G'plots, the polystyrene resins are compatible with styrene-butadiene rubber, while the poly(vinyl cyclohexane) resins are not. In all cases, the highest molecular weight resin is incompatible.

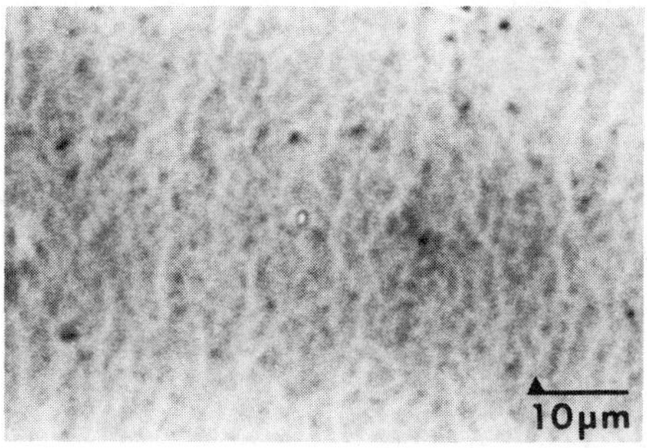

Fig. 11. Resin $\overline{M}_w$ 300

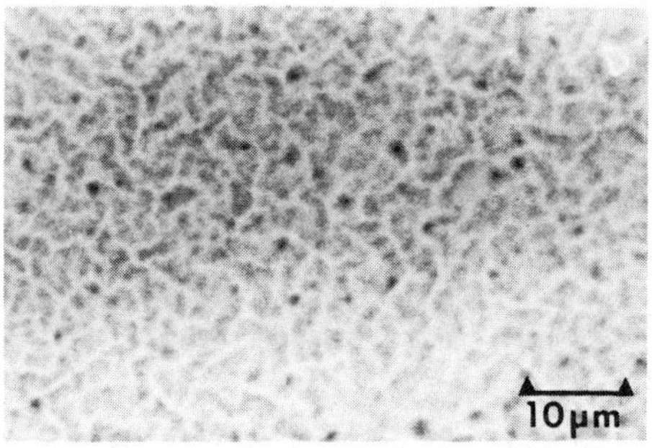

Fig. 12. Resin $\overline{M}_w$ 325

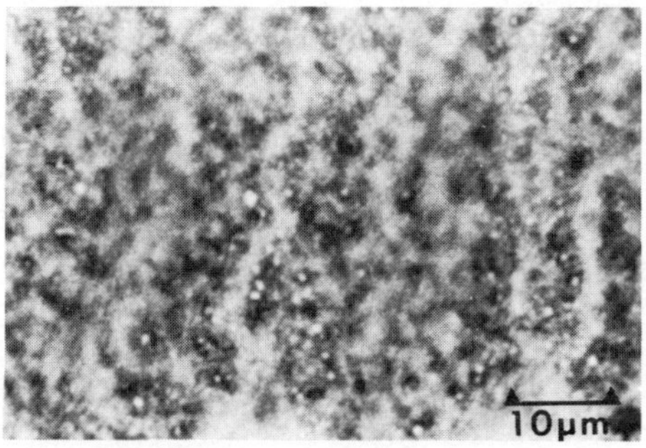

Fig. 13. Resin $\overline{M}_w$ 650

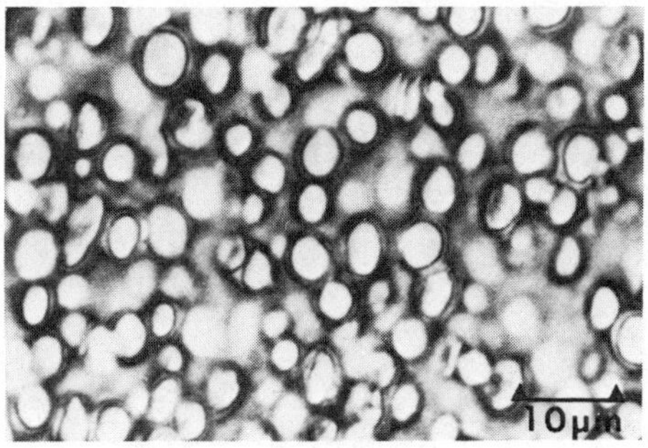

Fig. 14. Resin $\overline{M}_w$ 1,800

Figs. 11-14. Photomicrographs of 1:1 blends of natural rubber with poly(vinyl cyclohexane) resins.

This study clearly shows the significance of the molecular weight of the resin to its compatibility with an elastomer. At a molecular weight greater than 1,000, incompatibility begins to appear, even with a resin which is usually considered to be compatible with either natural rubber or styrene-butadiene rubber.

EFFECT OF RESIN CONCENTRATION

As the concentration of a normally compatible resin is increased in a rubber-resin blend, a level is reached where two phases appear. Natural rubber blended with a poly(vinyl cyclohexane) resin and styrene-butadiene rubber blended with a polystyrene resin, systems which were compatible at the 50% resin level, appear to have two phases at the 75% resin level.

Plots of tan δ and G' vs temperature were made for each series of resins with the Z-axis proportional to the concentration. The tan δ curves for the natural rubber-poly(vinyl cyclohexane) resin blends, shown in Fig. 15, indicate incompatibility at the low rubber (25%), high resin level (75%). Two distinct peaks, -40° and $+49^{\circ}C$, are seen. Tan δ remains high between these peaks, suggesting a complex system. The G' curves, shown in Fig. 16, confirm the effect of resin concentration. At the 75% poly(vinyl cyclohexane) level, the G' plot indicates a single broad transition region extending for about 120°, from $-60^{\circ}C$ to $60^{\circ}C$.

The blends of styrene-butadiene rubber with polystyrene resin perform similarly to the natural rubber-poly(vinyl cyclohexane) resin blends. Based on temperature scans of tan δ as shown in Fig. 17 and G' as shown in Fig. 18, a second incompatible phase is apparent at the 75% polystyrene resin concentration.

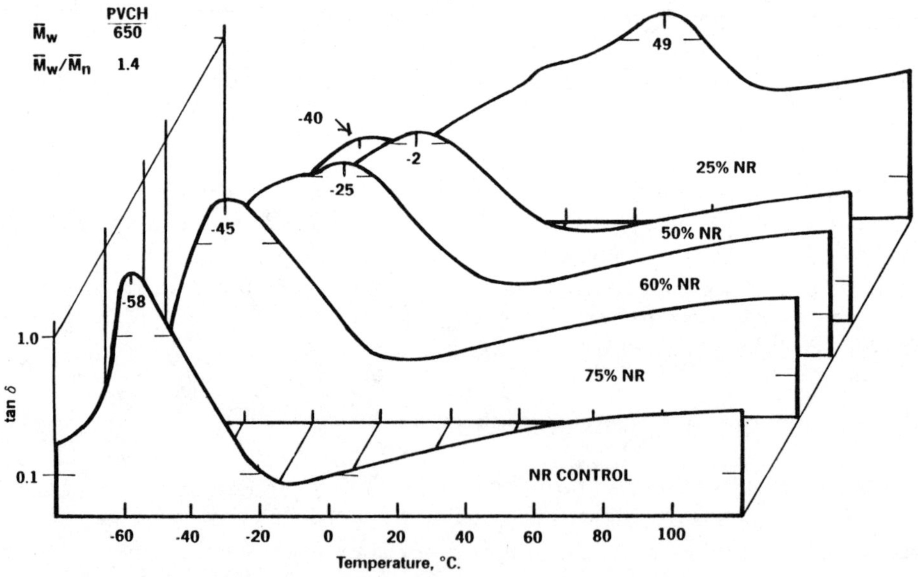

Fig. 15. Effect of concentration on tan δ plots of blends of natural rubber with poly(vinyl cyclohexane) resin.

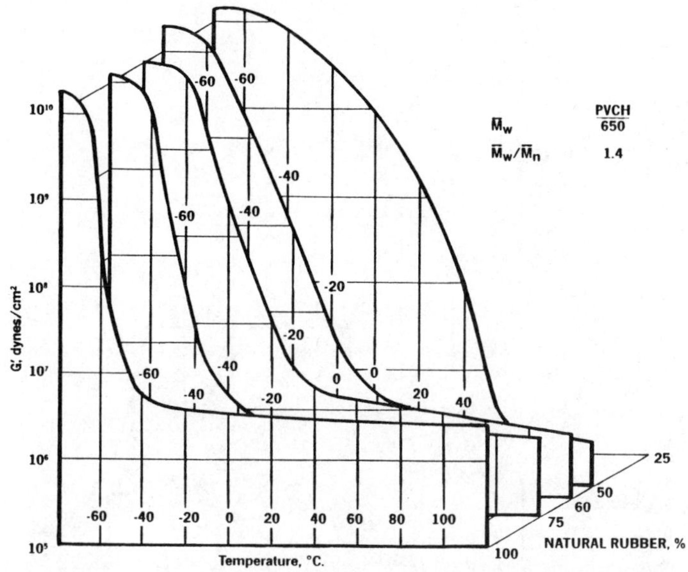

Fig. 16. Effect of concentration on G' plots of blends of natural rubber with poly(vinyl cyclohexane) resin.

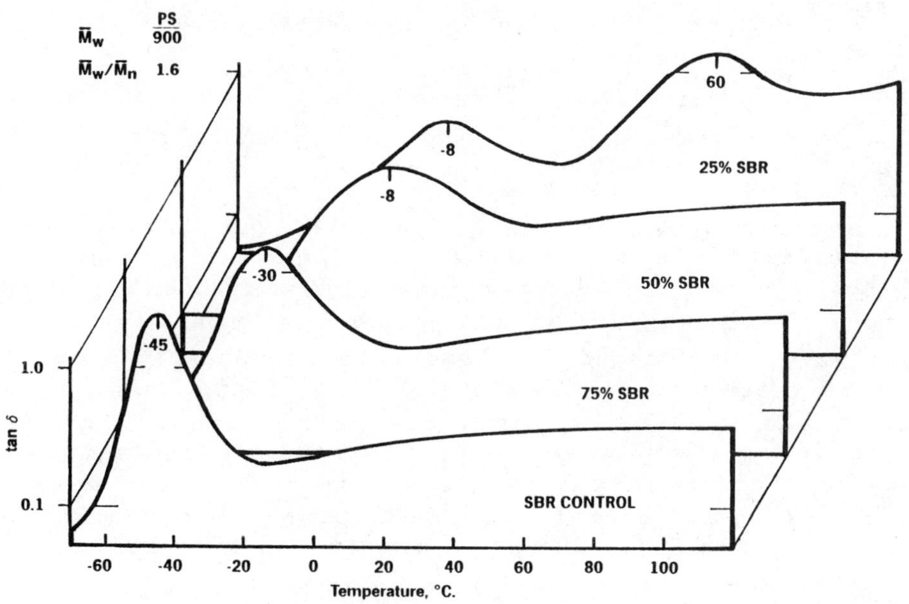

Fig. 17. Effect of concentration on tan δ plots of blends of styrene-butadiene rubber with poly(vinyl cyclohexane) resin.

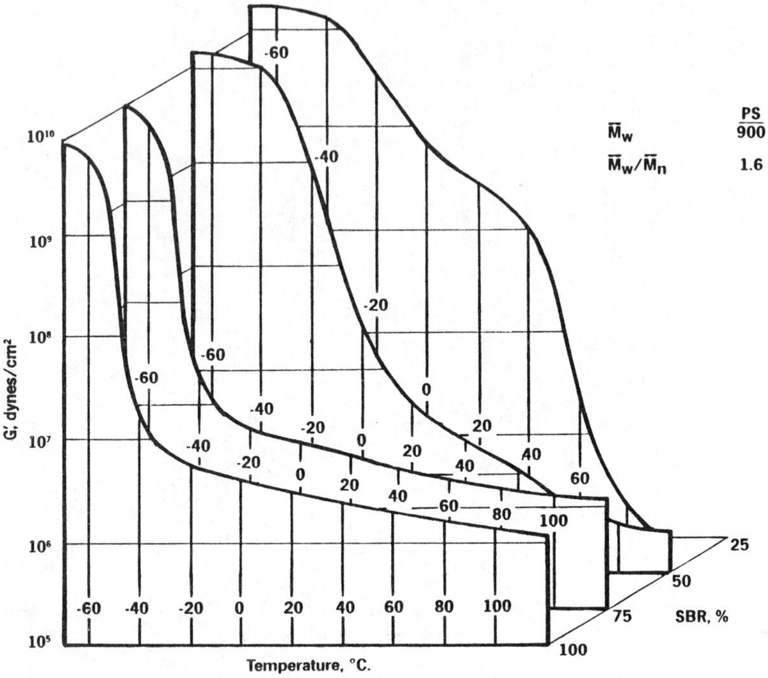

Fig. 18. Effect of concentration on G' plots of blends of styrene-butadiene rubber with poly(vinyl cyclohexane) resin.

Systems known to yield pressure sensitive adhesive compositions at specific ratios show an orderly progression of the tan δ peak temperature and the G' value in the plateau. For example, natural rubber-C_5 stream resin gives good pressure sensitive adhesive performance at about 50-60% resin concentration. The tan δ plots vs concentration, illustrated in Fig. 19, show a stepwise movement in the peak temperature as the resin concentration is increased. Also note that the tan δ in the plateau is at a higher value at higher resin concentration. The G' plots vs concentration, illustrated in Fig. 20, show a depression in the G' plateau as the resin concentration is increased. These uniform shifts for a system which includes pressure sensitive adhesive compositions suggest that it should be possible to predict tan δ and G' based on the resin used and its concentration.

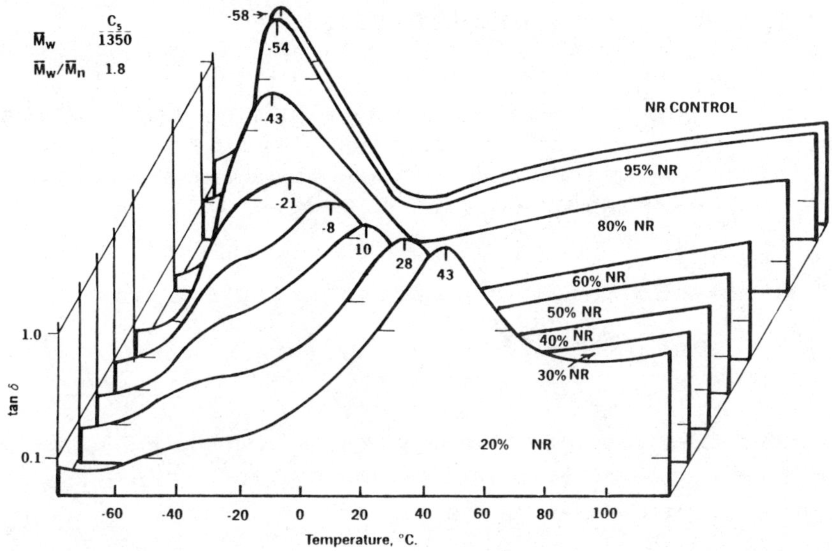

Fig. 19. Effect of concentration on tan δ plots of blends of natural rubber with a C_5-stream resin.

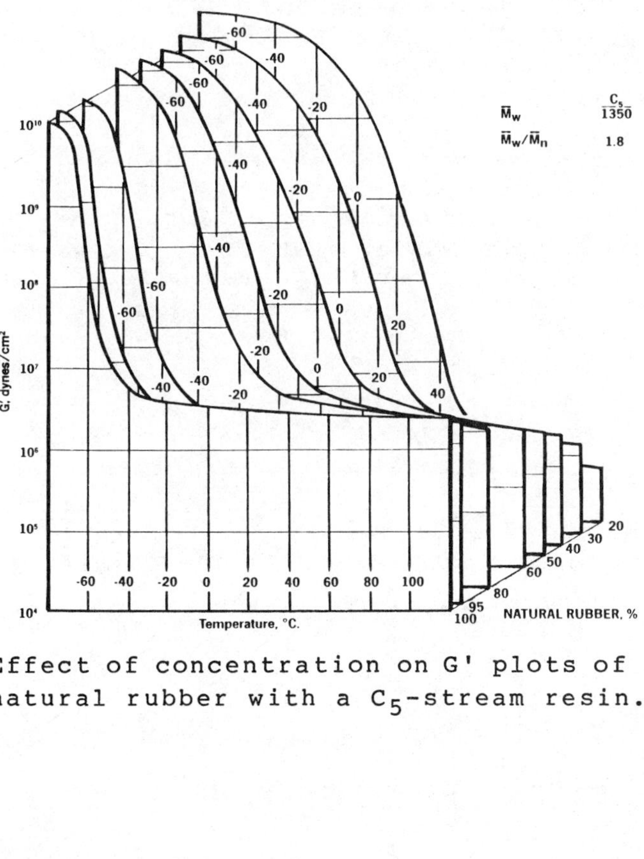

Fig. 20. Effect of concentration on G' plots of blends of natural rubber with a C_5-stream resin.

QUANTITATIVE EFFECTS OF CONCENTRATION

To study the effects of concentration on tan δ peak temperature and plateau modulus, a number of compositions known to give pressure sensitive adhesive performance at some concentration were examined at varying rubber-resin ratios. The resins examined are described in Table 5.

Table 5. Resins Used for Concentration Studies

Resin[a]	Figure Symbol
Glycerol Ester of Highly Stabilized Rosin	R-GLY, R-EST
Pentaerythritol Ester of Stabilized Rosin	R-PE-1
Pentaerythritol Ester of Highly Stabilized Rosin	R-PE-2
Poly α-Pinene	-PIN
Poly (vinyl cyclohexane)	PVCH
Poly (*tert*-butyl styrene)	TBS
Polymerized Petroleum Stream, Low MW	C_5-LMW
Polymerized Petroleum Stream, High MW	C_5-HMW

a. Molecular weight and T_g data are presented in Table 1.

Tan δ Peak Temperature

Of the proposed mixing laws, the reciprocal relationship suggested by T.G. Fox (3) gives the most satisfactory agreement with observations of the tan δ peak temperature.

$$T_g^{-1} = w_1 T_{g1}^{-1} + w_2 T_{g2}^{-1}$$

In this study, T_g, T_{g1} and T_{g2} refer to the tan δ peak temperature for the blend and pure components in ^oK; w_1 and w_2 are the weight fractions of the components.

Plots of the tan δ peak temperature vs concentration, according to the Fox equation, show a straight line relationship. The natural rubber blends are shown on Fig. 21 and the styrene-butadiene blends are on Fig. 22. The resin symbols are identified in Table 5. On these figures, the plot for each resin is shifted by 0.2 weight fraction unit to simplify comparison among the systems. The slopes of the lines are a function of the difference between the tan δ peak temperature for the polymer and the resin. The greater the difference between these temperatures, the greater is the slope.

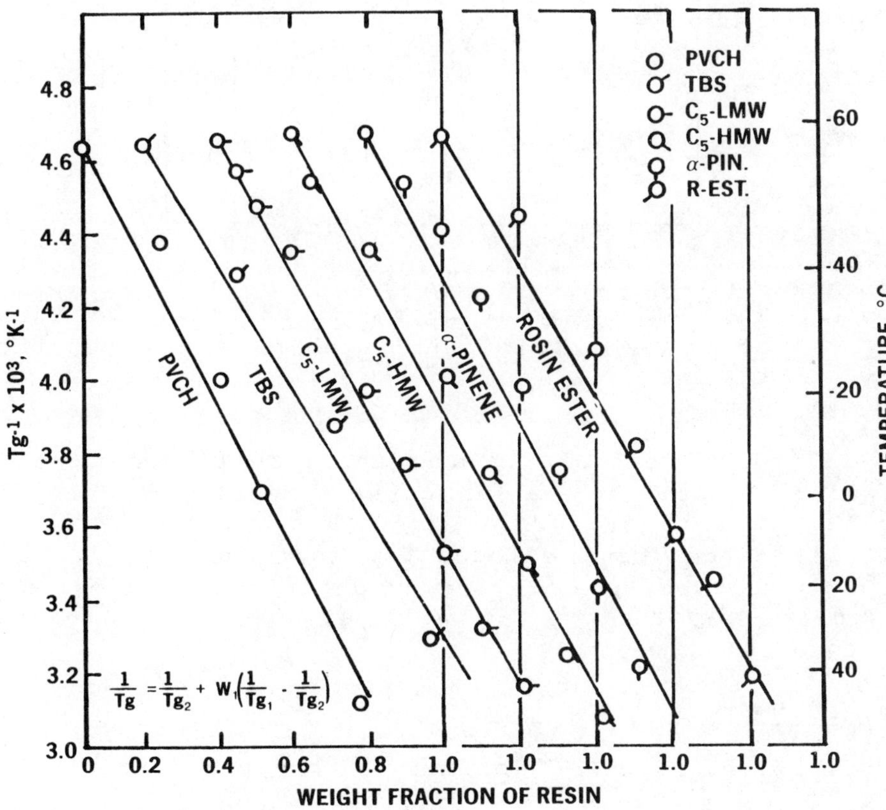

Fig. 21. Tan δ peak temperature vs concentration for blends of natural rubber with compatible resins.

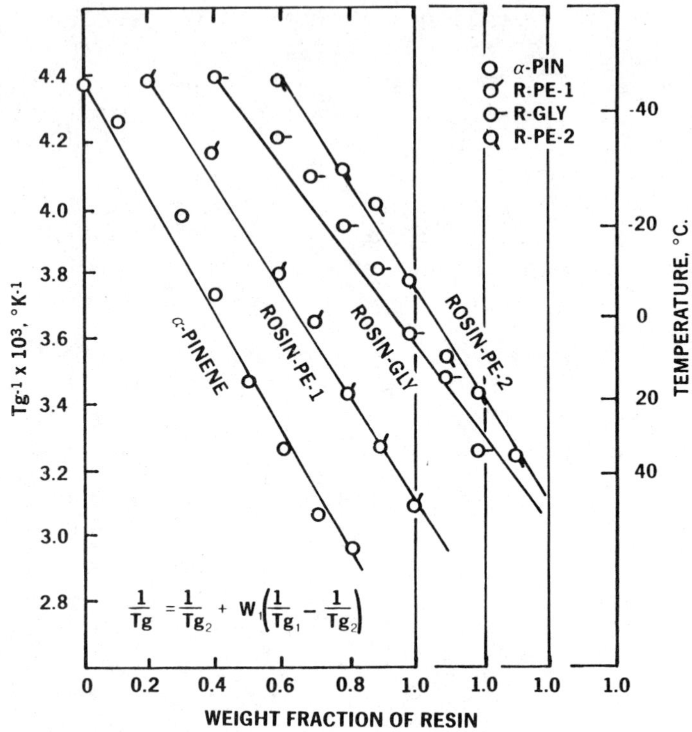

Fig. 22. Tan δ peak temperature vs concentration for blends of styrene-butadiene rubber with compatible resins.

The plots demonstrate the applicability of the Fox equation to estimate the tan δ peak temperature of rubber-resin systems. Using the Fox equation, the tan δ peak temperature can be predicted for any concentration of a compatible rubber-resin system by determining the value for a single composition, as well as for the unmodified elastomer.

Plateau Modulus

Classical rubber elasticity theory predicts that the plateau modulus, G_N^o, for a diluted polymer, is proportional to the square of the volume fraction of the polymer. Scaling law interpretations for dilute solutions by P.G. DeGennes (4) suggest the G_N^o should be proportional to the 2.25 power of the polymer volume fraction.

Studies of polymers dissolved in various oils indicate general agreement, with the volume fraction exponent reported to be 2 to 2.3 (5,6). In a recent study, W.W. Graessley and co-workers (7) reported exponents of 2.22 and 2.26 for polybutadiene and hydrogenated polybutadiene in various diluents. G. Kraus and K.W. Rollmann (8) reported on the only study of resins as the polymer diluent. Using a styrene-isoprene-styrene block copolymer, they found exponents of 2.28 for the glycerin ester of stabilized rosin and 2.62 for a C_5-stream resin. Plateau modulus was identified as G' in the rubbery plateau where tan δ was a minimum in a temperature scan of the systems.

A similar technique was used in this work. A series of compositions were prepared in a range of concentrations of resin with both natural rubber and styrene-butadiene rubber. Six resins known to be compatible with natural rubber were examined and four with styrene-butadiene rubber. A temperature scan of G' and tan δ was run at 10 rad/sec on each composition to identify the temperature at which tan δ was a minimum. G' was determined with greater accuracy by running a frequency scan from 0.1-100 rad/sec at this temperature. An example of this approach is shown in Fig. 23.

Log G_N^o values are plotted versus log v_2, the log of the volume fraction of the polymer, for natural rubber blended with various resins. This is illustrated in Fig. 24. The resin symbols are identified in Table 5. The plot for each resin is shifted by 0.2 unit on the log v_2 scale to simplify comparison among the systems. Each of the plots shows a satisfactory straight line relationship. The slopes give exponents for the polymer volume fraction of about 2.25-2.40, in reasonable agreement with the values predicted by DeGennes.

Plots of the styrene-butadiene rubber blends give higher values of 2.45 to 2.65 for the exponent, as shown in Fig. 25. A possible explanation is that the styrene-butadiene rubber contains gels that do not participate in the dilution by the resin. Therefore, the resin concentration in the amorphous phase is higher than calculated which would reduce the modulus more than expected. This would result in an apparent higher power for the polymer volume fraction.

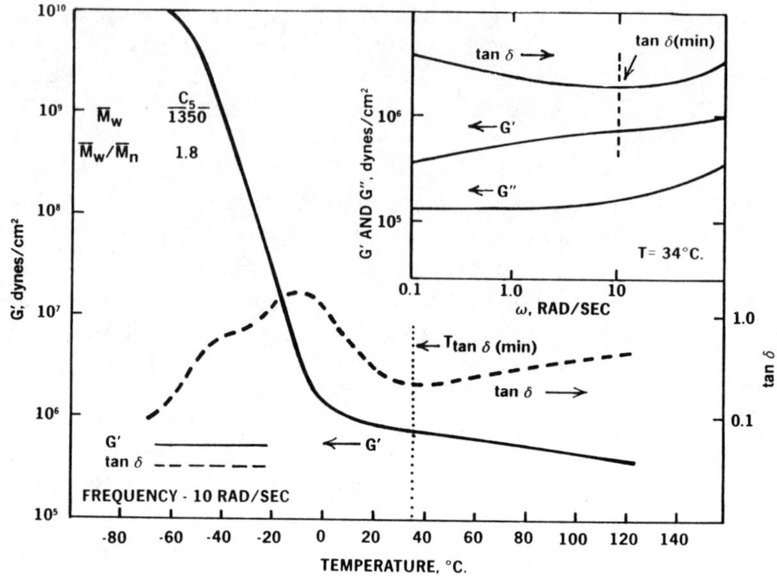

Fig. 23. Determination of G_N^o from G' at the tan δ minimum in the rubbery plateau. The sample is a 1:1 blend of natural rubber with C_5-stream resin.

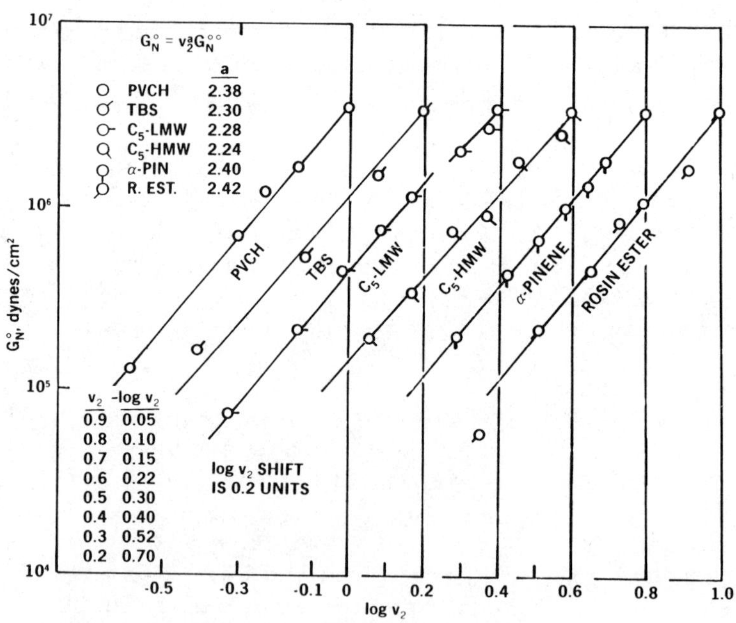

Fig. 24. G_N^o vs log v_2 for blends of natural rubber with compatible resins.

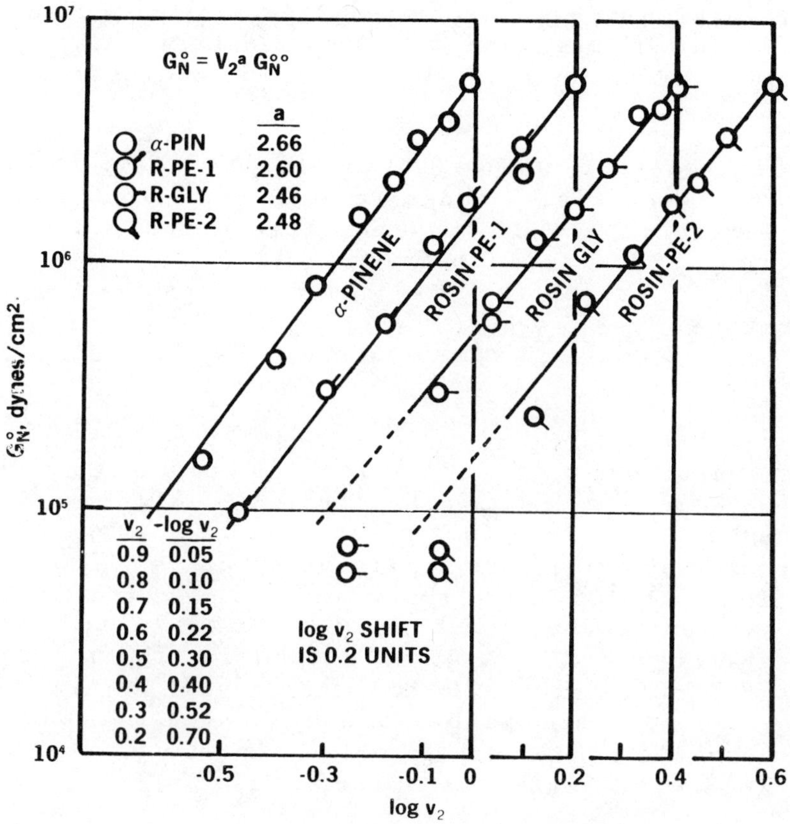

Fig. 25. G_N^o vs log v_2 for blends of styrene-butadiene rubber with compatible resins.

RELATIONSHIP BETWEEN VISCOELASTIC PROPERTIES AND PRESSURE SENSITIVE ADHESIVE PERFORMANCE

It has been shown that the plateau modulus of rubber-resin systems is a function of the volume fraction of the polymer, while the tan δ peak temperature is related to both the concentration and glass transition temperature of the resin additive. Pressure sensitive adhesive performance of rubber-resin systems is also related to the concentration of the resin. The optimum balance of tack, peel strength, and shear strength for typical systems can be found at a specific resin concentration. Therefore, performance zones can be identified on plots of tan δ peak temperature and plateau modulus. This is significant because identification of these zones will provide guidance in the formulation of pressure sensitive adhesives. Further definition of these zones will require additional study.

Plateau modulus and tan δ peak temperature may not be the most relevant viscoelastic properties to relate to pressure sensitive adhesive performance. Other viscoelastic parameters may be better predictors. This will be determined in continuing work to relate viscoelastic properties to pressure sensitive adhesive performance.

CONCLUSIONS

The structure of the low molecular weight resin is very important to its compatibility with elastomers and, consequently, to its effect on viscoelastic properties and performance as a pressure sensitive adhesive. A completely aromatic resin such as polystyrene has poor compatibility with natural rubber, but is compatible with styrene-butadiene rubber. A cycloaliphatic resin such as poly(vinyl cyclohexane) is compatible with natural rubber and is incompatible with styrene-butadiene rubber. An alkyl aromatic resin such as poly-(tert-butyl styrene) is compatible with both elastomers.

The molecular weight of the resin is significant to compatibility. At an apparent $\bar{M}_w$ of 1,000, the beginning of an incompatible phase is observed. For optimum effect on viscoelastic properties $\bar{M}_w$ should be less than about 1,500, and the resin should have a narrow molecular weight distribution.

Resins of appropriate structure are compatible but only to some limiting concentration. At 75% resin, two phases are observed in blends of natural rubber with poly(vinyl cyclohexane) resin ($\bar{M}_w$ 650) and styrene-butadiene rubber with polystyrene resin ($\bar{M}_w$ 900).

The tan δ peak temperature for compatible rubber-resin blends can be estimated using the reciprocal relationship proposed by T.G. Fox.

The plateau modulus of compatible rubber-resin blends can be calculated from the volume fraction of the polymer.

For natural rubber, the volume fraction exponent is 2.25-2.40, close to the value of 2.25 predicted by DeGennes. For styrene-butadiene rubber, the exponent was found to be somewhat higher at 2.45-2.65. A gel fraction in the styrene-butadiene rubber may explain these high values.

Identification of pressure sensitive performance zones on plots of tan δ peak temperature and plateau modulus will provide guidance in the formulation of pressure sensitive adhesives.

REFERENCES

1. M. Sherriff, R.W. Knibbs, and P.G. Langley, "Mechanism for the Action of Tackifying Resins in Pressure-Sensitive Adhesives," J. Appl. Polym. Sci., 17, 3423 (1974).

2. G. Kraus and K.W. Rollmann, "The Entanglement Plateau in the Dymanic Modulus of Rubber Styrene-Diene Block Copolymers," J. Appl. Polym. Sci., 21, 3311 (1977).

3. T.G. Fox, "Influence of Diluent and of Copolymer Composition on the Glass Temperature of a Polymer System," Bull. Ann. Phys. Soc. (Series 2), 1, 123 (1956).

4. P.G. DeGennes, "Dynamics of Entangled Polymer Solutions I. The Rouse Model," Macromolecules, 9, 587 (1976).

5. Y. Isono, T. Fujimoto, N. Takeno, H. Kjiuara, and M. Nagasawa, "Viscoelastic Properties of Linear Polymers with High Molecular Weights and Sharp Molecular Weight Distributions," Macromolecules, 11, 888 (1978).

6. E. Riande, H. Markovitz, D.J. Plazek, and N. Raghupathi, "Viscoelastic Behavior of Polystyrene-Tricresyl Phosphate Solutions," J. Poly. Sci. Polym. Symp., 50, 405 (1975).

7. V.R. Raju, E.V. Menezes, G. Marin, W.W. Graessley, and L.J. Fetters, "Concentration and Molecular Weight Dependence of Viscoelastic Properties in Linear and Star Polymers," Macromolecules, 14, 1668 (1981).

8. G. Kraus, K.W. Rollmann, and R.A. Gray, "Tack and Viscoelasticity of Block Copolymer Based Adhesives," J. Adhesion, 10, 221 (1979).

Block Copolymer Adhesive Studies

T.C. Ward, Anne Wood Brinkley,
Margaret Sheridan and Paul Koning

Department of Chemistry and Polymer Materials and Interfaces Laboratory
Virginia Polytechnic Institute and State University
Blacksburg, VA 24061

ABSTRACT

The thermal and dynamic mechanical behaviors of triblock copolymers with a styrene/isoprene/styrene architecture were investigated in order to understand their adhesive properties. Both copolymer free films and films bonding together two titanium alloy plates were found to have thermal and mechanical response that was strongly dependent on joint preparation. Microphase separation in the melts of these triblock materials was felt to contribute to the observed phenomena; namely, the presence of residual stresses in thin films which had been cooled while under high pressure.

INTRODUCTION

It is well known that viscoelastic stress fields are developed in the flow of molten polymer. When such a flowing polymer is cooled a molecular orientation process may lead to anisotropic mechanical properties. Studies relating residual stresses and impact failure behavior of ABS resins strongly point out the possible subtlety of structure-property relationships in oriented polymeric solids (1). Work in our laboratory has shown that styrene/isoprene/styrene triblock copolymers may also be prepared which show anisotropic behavior with respect to fracture in adhesively bonded lap shear joints made with rigid substrates (2). Furthermore, extruded films of similar triblock materials used in the fracture study showed different free film storage moduli in the extruded and transverse directions (3).

Recent theories (4,5) indicate that a sharp boundary does not exist in polystyrene-polydiene block copolymers, but rather, partial mixing exists in the interfacial region between the two thermodynamically incompatible phases. The thickness of the interface is temperature dependent. With increasing temperature, intermixing of the two phases increases at the expense of the pure phases, but the copolymers remain phase segregated. Indeed, structural changes continue to exist well into the melt, but the two phase structure of styrene-butadiene block copolymers persists to nearly 180°C (6). At a critical temperature T_c, complete mixing of both phases occurs; the system may be considered as thermodynamically homogeneous. Block copolymers, as a result of the existence of microphase separated morphologies in the melt, exhibit complex rheological properties. Sivashinsky, Moon, and Soong (7) have recently studied the time-dependent stress growth and relaxation behavior of SBS systems as a function of casting solvent, temperature, and shear history. Their objective was to provide insight into the mechanism of structure breakdown and reformation in SBS melts. The underlying principle in their investigation was that the rheology of SBS triblock copolymers is highly sensitive to the prevailing morphology. The morphology of the SBS systems, in turn, was found to be highly sensitive to many factors, such as casting solvent, temperature, shear history, and overall sample history. Therefore, rheological measurements may be very difficult to reproduce. Further complexities were found in studies of the stress relaxation behavior, where the morphology of the fluid evolves with time as stresses relax (7-9). Rheology work in our laboratories confirms their observations. Since molten adhesives of the microphase-separated type flow into bonding positions and rapidly cool, variable morphologies may exist in almost identically prepared joints.

J. M. Widmaier and G. C. Meyer (10) have previously demonstrated that only SIS copolymers with a segregated structure exhibit noticeable shear adhesive strength. The joint strength was found to pass through a maximum on increase in temperature. Such behavior has been interpreted as depending on the morphology of the copolymers, as discussed above. The maximum corresponds to the temperature just below T_c (dependent on the molecular weight and composition of the copolymers) where the morphology of the sample is favorable for strong bonding.

The morphology of the adhesive may be altered by kinetic effects; thus, different cooling conditions (i.e., quenched or annealed) results in bonds of differing lap shear strengths (10). Studies on SIS copolymers in our laboratory also indicate that the presence or lack of pressure upon cooling of a bonded joint produces structures which exhibit different dynamic mechanical damping patterns as will be presented below. Clearly, the history of the adhesive was very important in determining its behavior, pressure included. Some background in this area is appropriate.

Densified glasses, prepared by cooling amorphous polymers from the liquid state under elevated pressures, have been a topic of much investigation for the past decade. Currently, three types of properties are usually measured and available in the literature concerning densified glasses: specific volume data, density fluctuation as measured by SAXS, and enthalpy data from DSC measurements. Two different effects appear to result from the application of pressure during glass formation depending upon whether the formation pressure is above or below 1.5 kbar. For pressures less than 1.5 kbar all three of the above properties decrease with pressure, which is indicative of more efficient molecular packing. However, for pressures greater than 1.5 kbar a decrease in specific volume and a relatively constant fluctuation density is observed, accompanied by an increase in enthalpy. Weitz and Wunderlich (11) accounted for the increase in enthalpy at elevated pressures by noting that T_g increased with pressure; therefore, the glass prepared under elevated pressure contains chains of higher conformational energy. Prest (12), studying PVC densified under different pressures, at constant temperature, observed an increase in enthalpy with pressure; however, he attributed the higher enthalpy of a pressure-densified glass not to high energy conformational states but rather to frozen in stored energy, either in terms of bond angle distortions or increased intermolecular potential energy. Curro and Roe (13), who have recently done SAXS studies of density fluctuation in pressure-densified polystyrene glasses, concur with Prest. They feel the results at higher pressures indicate that molecular packing can no longer be made more efficient, and any further volume reduction is obtained by forcing polymer segments into positions of local energy minimum, requiring distortion of chain bonds and higher overall energy.

Wunderlich (11) observed, by means of DSC studies on pressure densified PS glasses, an additional endothermic event at temperatures far below the glass transition temperature. This is consistent with the suggestion that strain energy is trapped in glasses prepared under elevated pressure. The strain energy, stored in the form of distorted chain bonds or unfavorable intersegmental interactions, could provide a driving force to overcome the local energy barriers; the chains, on a local level, could undergo cooperative rearrangements, even at temperatures well below the glass transition temperature. Also, as an interesting aside, there is preliminary evidence that such low temperature rearrangements, occuring on a local scale, may lead to the formation of microcavities in such pressure densified glasses (13,15).

In our work presented below, free and bonded films of variable thicknesses were investigated using triblock copolymers. We were interested in this because other workers have cited data to indicate that adhesive joints often do not perform as predicted from properties of the bulk polymer adhesive. However, some researchers (17) have reported good correlation between in situ and bulk properties. In our present investigation, the observations on different bonded and free, oriented polymer adhesives with (molecular) orientation both parallel and transverse to the testing direction are summarized in an effort to clarify the effect of a rigid substrate on bulk polymer properties. We want, additionally, to gain further insight into the role of molecular orientation of adhesive films in bonded joint properties.

Even though a "systems approach" has been utilized throughout our work, a specific set of measurements was necessary for analysis. Our exploration primarily involved the linear viscoelastic quantity tan δ, which is widely used by polymer researchers to characterize primary and secondary relaxations in bulk polymers, most noticeably the glass transition, T_g (3).

EXPERIMENTAL

The fracture experiments were conducted on single lap shear joints with an overlap length of 1.27 cm, bond area of 3.22 cm^2, and adhesive layer thickness averaging 0.38 mm. The adherends were titanium 6-4 alloy. The bonds were formed under conditions of minimal temperature, pressure, and time for retention of orientation within the adhesive. The oriented films were extruded on a CSI mini-max extruder and have been labeled "parallel" and "perpendicular" with respect to the machine direction of extrusion.

The joints were fractured at room temperature on an Instron Table Model 1130 with a crosshead speed of 1 cm/min. Joint strength was measured as the maximum stress applied, and the total fracture energy was obtained by integrating over the fracture curve.

DSC thermograms were collected on the Perkin-Elmer DSC-2. The heating rate was $10^{\circ}K$/min.

The relaxation patterns for polymers in the bulk (free) form were obtained from experimentation conducted on a Rheovibron DDV-II-C in the usual Rheovibron tensile geometry. The heating rate was $1^{\circ}C$/min.

In the Rheovibron work on bonded films a test design which is shown in Fig. 1 was developed. Each end of the joint is fitted into new clamps on the Rheovibron having slots offset as indicated in the figure. Shims were used to control bond thickness. The entire assembly is rigidly screwed together. Titanium 6-4 alloy (phosphate fluoride etch) was used for the bonded substrates (3).

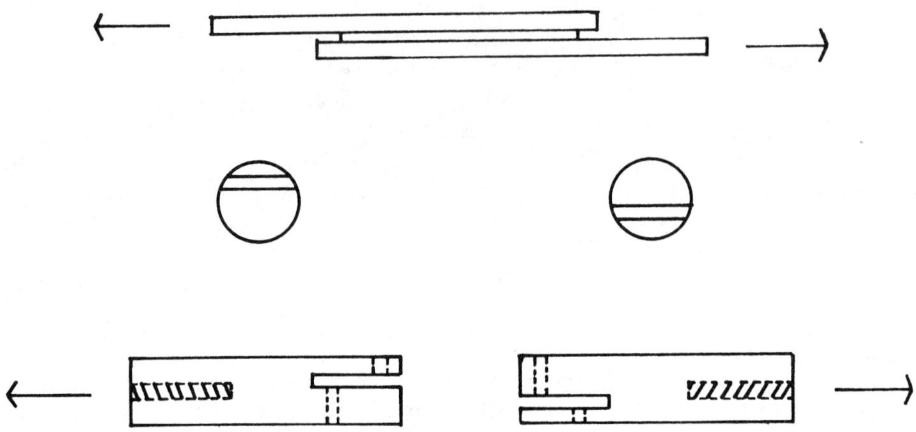

Fig. 1. Modified Lap Shear Assembly for Mounting in Rheovibron. Middle and Bottom Views are of New Clamps for Rheovibron (Ref. 3).

The weight percent styrene, and values of M_w/M_n and M_n of the styrene and isoprene blocks of the SIS copolymers used in the studies are listed in Table 1.

RESULTS AND DISCUSSION

We have determined that S/I/S triblock copolymers may be prepared to show anisotropic behavior with respect to fracture in adhesively bonded lap shear joints made with rigid substrates. Table 2 summarizes joint strength and fracture energy data for lap shear joints prepared with the orientation (machine) direction of the adhesive layer either parallel or perpendicular to the fracture axis of the joint. The important comparison is that of the fracture energies of the different orientations for the 50% styrene S/I/S copolymer (E2) sample. The high fracture energy for the parallel orientation corresponded exclusively to ductile drawing failure, the perpendicular orientation

Table 1

MATERIALS (Ref. 2)

SAMPLE	WEIGHT PERCENT STYRENE	M_w/M_n (10^{-3})	S/I/S M_n (10^{-3})
C2	43	98/93	20/53/20
D3	50	65/63	16/31/16
E1	60	81/77	23/31/23
E2	59	64/61	18/25/18

Table 2

ORIENTATION STUDY (Ref. 2)

SAMPLE	% STYRENE	FRACTURE AXIS ORIENTATION	AVERAGE JOINT STRENGTH (N/cm^2)	AVERAGE FRACTURE STRENGTH (J)
E2	59	$\parallel$	430 ± 70	3.9 ± 0.8
		$\perp$	380 ± 40	1.3 ± 0.1
D3	50	$\parallel$	290 ± 40	3.0 ± 1.0
		$\perp$	320 ± 60	2.6 ± 0.6
C2	43	$\parallel$	$210 + 40$	2.6 ± 0.7
		$\perp$	180 ± 10	2.1 ± 0.9

produced only a brittle failure and low fracture energies (2).

Another important contribution to the orientation study came from the data on the elastic moduli for the two orthogonal orientations of triblock materials similar to those used in the fracture study. Figs. 2 and 3 show different free film storage moduli in the parallel and perpendicular direction for a S/I/S triblock material and for a second triblock material also based on styrene hard blocks, respectively.

Results of experiments on S/I/S free films and on S/I/S bonded joints with thermoplastic glue line thickness of 0.3, 0.5 and 0.7 mm appear together in Fig. 4, previously published (3). All preparations were made with the 50% styrene content polymer. The preparation temperature was 150°C and at least 5000 psi was applied in the molding. Pressure was maintained during cooling in all cases. As the bond thickness increased the tan δ behavior for joints approached that of the bulk, but was shifted to higher temperatures. Identical experiments on the 40% styrene copolymer gave essentially the same conclusions (not shown). The curves were quite reproducible.

Neither just pressure applied during cooling, nor the S/I/S being contained in a bonded joint were sufficient to produce the "fine structure" in the tan δ vs. temperature experimental results shown in the figures. It was the combination of a thin bond, or free film, along with high pressure maintained during cooling that produced the details seen in Fig. 4. This was confirmed by repeated experiments on free and bonded, thick and thin films, prepared with and without pressure. As an example, in Fig. 5, two free films behave quite differently; both were cooled with pressure applied in the molding operation; however, one was 0.1 and the other 0.6 mm thick.

In a further exploration of the conclusions above we used the oriented films to prepare bonded joints, with thin bondlines, and cooled with pressure. The results are shown in Fig. 6. While there are some similarities, the parallel and perpendicular (machine direction of film with respect to Rheovibron strain direction) responses are shifted from one another and not identical. These curves are reproducible in all major features.

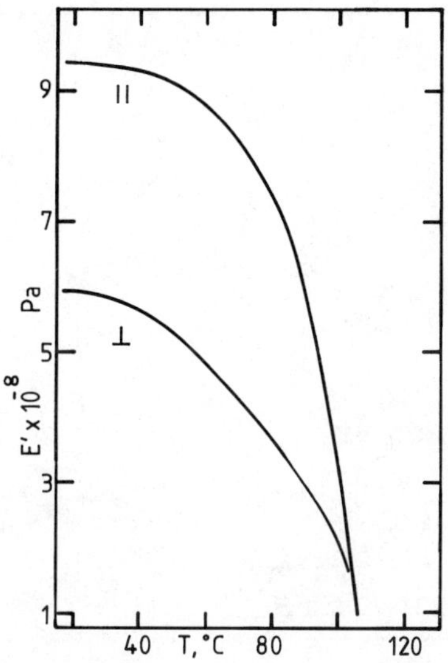

Fig. 2. Dynamic Mechanical Storage Modulus for S/I/S Free Films with Molecular Orientation Parallel (||) and Perpendicular (⊥) to the Testing Direction (at 11 Hz).

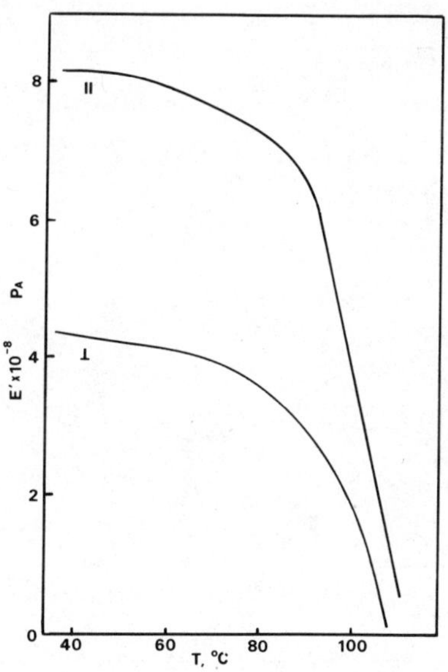

Fig. 3. Dynamic Mechanical Storage Modulus Free Film with Molecular Orientation Parallel (||) and Perpendicular (⊥) to the Testing Direction (at 110 Hz).

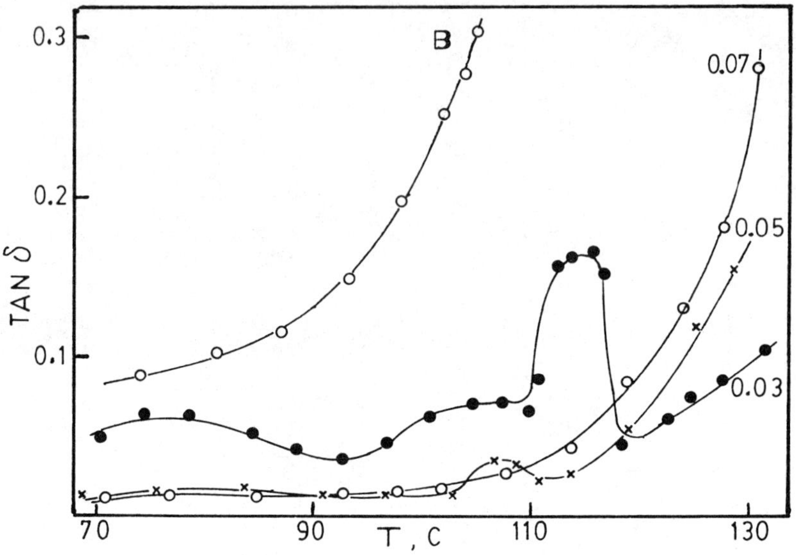

Fig. 4. Dynamic Mechanical Damping of 50% Styrene S/I/S Bonded Joint and Free Film at 110 Hz. Bond Thicknesses Shown on Curves in CM for Joints. Free Film = B.

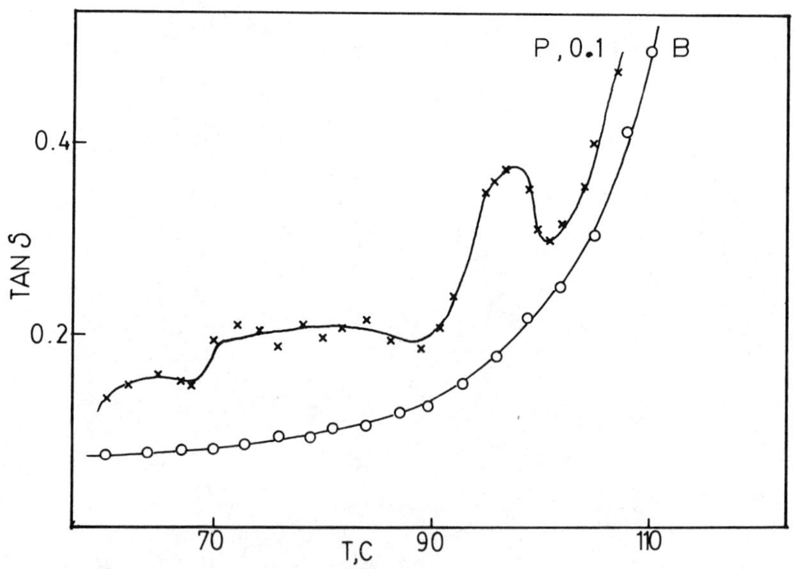

Fig. 5. Dynamic Mechanical Damping of 50% Styrene S/I/S (D3) Free Films, Cooled with >5000 PSI Pressure. B = 0.6 mm; P = 0.1 mm Thick (Ref. 3).

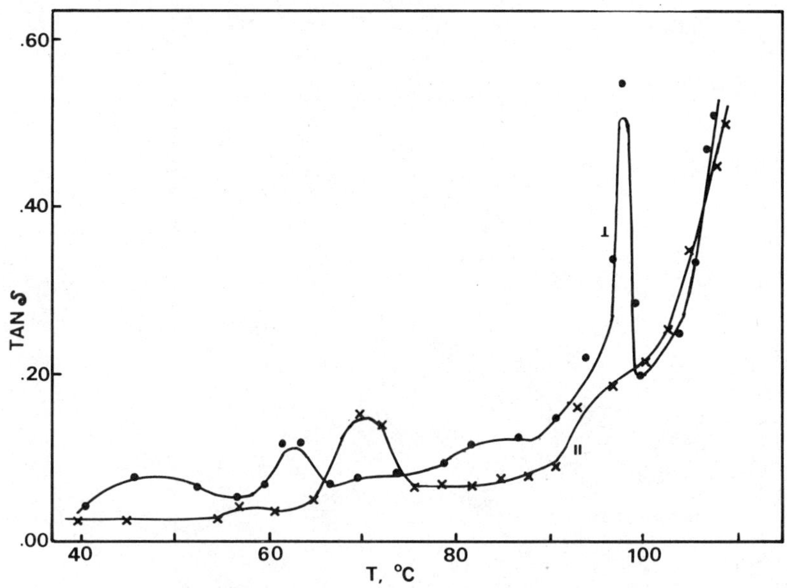

Fig. 6. Dynamic Mechanical Damping of a Block Copolymer Film with Molecular Orientation Parallel (X) and Perpendicular (•) to the Testing Direction at 110 Hz.

Earlier studies conducted in our laboratory dealing with isotropic adhesive films demonstrated that differences in the compression molding operation could be associated with either the presence or the absence of stress release events (2). Using DSC we sometimes noted unexpected thermal response of S/I/S copolymers about 20°C below the glass transition temperature which was attributed to a "stress release phenomena" (3). Fig. 7 shows the DSC curves for a 60% styrene S/I/S copolymer (E1). Note that in curve II, for the compression-molded film cooled under pressure, a baseline shift appears at about 70°C. This event was completely reproducible when similar thermal and pressure histories were established. No shift was observed on repeat runs on the same sample cooled from the melt in the DSC instrument.

In conclusion we have followed a path of analysis which includes the following facts:

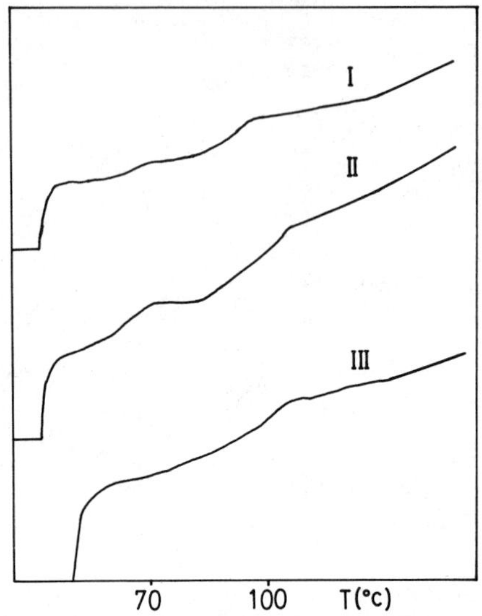

Fig. 7. First Run DSC Curves for Sample E1 (I) Powder, (II) Compression Molded Film Cooled Under Pressure. (III) Compression Molded Film Cooled Without Pressure (Ref. 2).

1) Microphase separated triblock thermoplastic materials have complex melt morphologies when flowing, which may be trapped by rapid cooling in a variety of states.

2) Cooling microphase separated triblock copolymer melts of S/I/S under pressure in situations where large amounts of flow occur (thin bonds) and allowing them to solidify may produce behavior that is related to that observed in pressure densified homopolymer glasses. Thermal analysis supports this conclusion.

3) Compression molding produced thin films that were basically isotropic with respect to "frozen stress" release (as measured by tan δ) on heating. Extrusion, in contrast, could produce S/I/S polymer films which exhibited anisotropic mechanical response in free films and in bonded joints.

A summary of the importance of these reported conclusions would certainly emphasize extreme care that should be observed in measuring either ultimate (adhesive failure) or small displacement (linear viscoelastic) properties of block copolymers which have been extruded or molded.

ACKNOWLEDGMENTS

Support from this research came from the Army Research Office.

REFERENCES

1. J. H. Daane and S. Matsuoka, Polym. Eng. Sci. 8 (4), 246 (1968).

2. A. Wood, M.S. Thesis, Virginia Polytechnic Institute and State University, Blacksburg, VA 24061 (1982).

3. T. C. Ward, Margaret Sheridan, D. L. Kotzev, "Nondestructive Evaluation of Some Bonded Joints", in Adhesive Joints: Formation, Characteristics, and Testing, K. L. Mittal (ed.), Plenum Press, New York, August (1984).

4. D. J. Meier, J. Polym. Sci., Part C, 26, 81 (1969).

5. D. F. Leary and M. C. Williams, J. Polym. Sci., Polym. Phys. Ed., 11, 345 (1973).

6. G. Kraus and T. Hashimoto, J. Appl. Polym. Sci., 27, 1745 (1982).

7. N. Sivashinsky, T. J. Moon and D. S. Soong, J. Macromol. Sci.-Phys., B22 (2), 213 (1983).

8. A. Ghijsels and J. Raadsen, Pure Appl. Chem., 52, 1361 (1980).

9. J. Kelterborn and D. S. Soong, Polym. Eng. Sci., 19, 1140 (1979).

10. J. M. Widmaier and G. C. Meyer, J. Appl. Polym. Sci., 28, 1429-1437 (1974).

11. A. Weitz and B. Wunderlich, J. Polym. Sci., Polym. Phys. Ed., 12, 2473 (1974).

12. W. M. Prest, Jr. and F. J. Roberts, Jr., Structure and Mobility in Molecular and Atomic Glasses, Vol. 371, 67 (1981).

13. J. J. Curro and R. J. Roe, J. Polym. Sci., Polym. Phys. Ed., 21, 1785 (1983).

14. H. J. Oils and G. Rehage, Macromolecules, 10, 1036 (1977).

15. J. B. Yourtee and S. L. Cooper, J. Appl. Polym. Sci., 18, 897 (1974).

16. D. W. Dwight, E. Sancaktar and H. F. Brinson, Adhesion and Adsorption of Polymers, Part A, 141, L. H. Lee (ed.) Plenum (New York), 1980.

17. G. Dolev and O. Ishai, J. Adhesion, 12, 283 (1981).

Discussion

On the Paper by J. T. Dickinson

D. P. Stengle (Monsanto Polymer Products): Based on your results for FE in studying modes of failure, to what do you attribute the difference in EE for adhesive mode versus the mixed adhesive/cohesive mode which you presented today? That is, what mechanisms occur on a molecular level?

J. T. Dickinson (Washington State University, Pullman, WA): The strength of charge separation between patches on the fracture surface is significantly higher for adhesive failure vs. cohesive failure due to the contact charging that occurs between dissimilar materials. We believe the resulting intense electrical fields lead to a gaseous discharge in the crack tip causing considerable bombardment of the fracture surfaces. We have shown recently* that electron bombardment alone of polybutadiene is followed by sustained electron emission from the surface with similar decay kinetics. Therefore in cohesive failure, where the charge separation may be very small, we expect not only lower emission intensities, but also lower energy emission since surface charging is responsible for the high energies observed in adhesive failure.

*J. T. Dickinson, L. C. Jensen, A. Jahan-Latibari, submitted to J. Vac. Sci. and Technol.

L. H. Lee (Xerox Corporation): Dr. Dickinson should be congratulated for developing this new technique of fracto-emission to explore the fracture phenomena at the molecular level. By this technique, not only have many energetic species been identified during the fracture process but the energy distribution of each species has also been accurately measured.

On the Paper by R. L. Levy and D. P. Ames

J. A. Simms (DuPont, Wilmington, DE): (1) Will probe work in acidic systems? (2) Can cure be followed to completion?

R. L. Levy (McDonnell Douglas Research Laboratories): (1) Yes. (2) Probably, but a series of probes with progressively smaller rotating groups might be needed for monitoring advanced stages of cure.

J. T. Dickinson (Washington State University): Have you considered using a pulsed excitation source and measuring the decay time of the fluorescence? This decay time should depend on viscosity.

R. L. Levy: Yes, this would be a good idea. It would, however, require a different source and a more complex experimental setup.

C. Byrne (Army Materials and Mechanics Research Center, Watertown, MA): How would lack of solubility of the fluorescent probe in resin affect the fluorescence?

R. L. Levy: It is not a problem for self-probe TGDDM. For solute dye probe, sometimes one has to heat the resin to dissolve the dye. This, however, means that the cure has already started.

S-D. Hong (Jet Propulsion Lab, Pasadena, CA): We have observed for the same system (TGDDM/DDS) a shift of wavelength of fluorescence at the peak intensity. In addition, we have also observed that there are actually two peaks (bands) of the fluorescence; the relative intensities of the two peaks change as a function of curing.

D. R. Day (MIT, Cambridge, MA): Have you been able to measure Tg's of cured materials by monitoring fluorescence as a function of temperature?

R. L. Levy: We have not done it but yes, this is definitely a potential application and we want to do it.

On the Paper by J. B. Class and S. G. Chu

L. H. Sharpe (Bell Laboratories, Murray Hill, NJ): I'd always thought that to be pressure-sensitive, a material had to be 2-phase (the Wetzel model). Apparently, that's not so. Is a pressure-sensitive material, then, just a high-viscosity, viscoelastic fluid? Or is this too naive a view?

J. B. Class (Hercules Research Center): A pressure-sensitive adhesive will have the properties of a viscoelastic fluid in the sense that its performance during bonding and debonding depend upon the rate and extent of deformation of the adhesive layer. However, there are upper and lower limits to its viscosity or modulus. If the modulus is too high, bonding cannot take place during the bonding step. If the modulus is too low, the adhesive will be weak during debonding. Surface energies also are important to bond formation and pressure-sensitive performance.

PART THREE:

RADIATION-CURABLE ADHESIVES

Introductory Remarks

V. D. McGinniss

Battelle-Columbus Laboratories

505 King Avenue

Columbus, OH 43201

The use of electromagnetic radiation induced polymerization reactions to produce rapid setting coating and adhesive systems has increased dramatically in the time period between 1979 to 1983. There are, of course, many alternatives to radiation-curable adhesives such as 100% solids reactive hot melt materials; water and solvent base adhesives (low and high solids content); two component-single component room temperature-elevated temperature cure adhesives and even gas-phase initiated curing of blocked isocyanate adhesive formulations. But even with these competing technologies, radiation-curable adhesive systems have several distinct advantages over conventional methods of applying, drying and setting historical adhesive formulations.

The papers presented in the following chapters represent advances in pressure sensitive adhesives (ultraviolet light activated acrylate monomer - low T_g polyether formulations); photoinitiated cationic polymerization (light activated aryliodonium and arylsulfonium salts of lewis acids in epoxy resin formulations); polymer and formulation design criteria for radiation curable adhesives; radiation curable composites (dynamic thermal analysis characterization of electron beam cured adhesives) and advances in reactive liquid or solid elastomers for use in radiation-curable adhesives.

Ultraviolet Cured Pressure-Sensitive Adhesives

Kenneth C. Stueben

Union Carbide Corporation
Specialty Chemicals Division
Bound Brook, New Jersey 08805

ABSTRACT

A variety of pressure-sensitive adhesives were prepared by photocuring mono- and multifunctional acrylate/polyether/benzophenone blends in air. Among the polyethers which have been employed successfully are poly (vinyl ethyl ether), poly(ethylene oxide), poly(ethylene oxide-propylene oxide), and poly(butylene oxide) while representative acrylates include 2-(N-methylcarbamoyloxy) ethyl acrylate, isobornyl acrylate, and neopentyl glycol diacrylate. Peel strengths of approximately 2 pli (adhesive failure) to 9 pli (cohesive failure) were achieved, often in combination with good shear resistance. The photocured products, which are believed to contain acrylates grafted onto low T_g polyethers, contain considerable cross-linked material. A number of important relationships between the properties of these adhesives and both cure and formulating parameters have been uncovered. Use of these relationships can significantly enhance the chances for designing and optimizing superior photocure pressure-sensitive adhesives.

BACKGROUND

The term "pressure-sensitive adhesive" (PSA) refers to a permanently tacky composition which will adhere to a variety of surfaces merely by application of light hand pressure. Such materials find widespread application in tapes, labels, wall coverings, floor tiles, and protective maskings (1). Typical property requirements for various pressure-sensitive products are shown in Table 1. For decades, such products have been manufactured by the deposition of preformed polymers from solution. However, as concern over energy and environmental problems began to surface, the pressure to find alternate methods of manufacture intensified and the use of solvents declined. The use of radiation to cure such materials in place is but one alternate to the conventional solvent-based approach.

In making a pressure-sensitive adhesive by this new process, a formulated fluid composition is applied to an appropriate substrate, polymerized by a few seconds exposure to a radiation source, and thereby converted from a low strength material to a tacky, higher strength adhesive with little loss of volatiles. In the majority of cases, acrylates, in the form of monomers or polymers are included in such compositions. Systems that are to be photocured also require a photoinitiator or photosensitizer to absorb the impinging light and induce polymerization while electron beam cured formulations do not require such initiators. Because irradiation usually results in a low degree of polymerization due to the simultaneous generation of numerous initiating radicals, other components are utilized to provide increased molecular weight or sites for grafting or cross-linking. Thus, most radiation curable adhesive systems are made up of photoactivators, monofunctional monomers, polyfunctional monomers, oligomers, and reactive or nonreactive resins which, upon irradiation, collectively develop desirable adhesive properties through polymerization. The radiation source employed in most of our experimental work was a Linde medium pressure mercury unit.

Brief reflection on the simplicity of this process reveals the many advantages of photocure over other techniques (Table 2). Of course, at the present state of development, a number of shortcomings can be cited as well and these will be discussed in a later section.

Table 1. Typical PSA Performance Requirements

Product	Peel Adhesion (pli)	Quick Tack (pli)	Shear (Hours)
General Purpose Masking Tape	2.3	1.1	>167[a]
Aluminum Foil Duct Tape	> 3.8	1.8	> 6[b]
Wall Covering	0.3-1.3	0.5	0.5-0.2
Paper Labels	1.1	-	0.6

[a] 1 in.^{2}/1000 gm.

[b] 1 in.^{2}/4540 gm.

Literature on Radiation-Curable Pressure-Sensitive Adhesives

Most of the information which has been published in this field is in the form of patents and, consequently, the importance of compositional changes and cure parameters on final properties and product structure are only vaguely understood in most instances. Nonetheless, some discussion of the prior art is desirable for the reader in order to be able to view our current results in the proper perspective.

Table 2. Advantages of Radiation Curing

No Disposal of Solvents Into The Atmosphere
Extremely Low Energy Requirements
High Line Speeds
Compact Operation
Heat-Sensitive Substrates May Be Used
Improved Adhesion to Substrate Due To Grafting

What follows are selected highlights taken from the attached literature review of this field (both UV and EB) published in 1977 (2). A list of the publications appearing since then is appended as an addition to that review (3,4, 8-10, 13-15, and Table 10).

For convenience, the components used in radiation-curable PSAs can be divided into three groups; i.e., monomer alone, polymers in monomers, and polymers alone (Table 3). The simplest means of preparing a PSA, by irradiating monomer alone, is rarely employed due to coating problems resulting from the low viscosity of the system, incomplete polymerization, and inadequate properties. However, one example of this type can be cited in which a blend of butyl acrylate containing 9 percent acrylic acid was applied at a thickness of 50 microns to an aluminum foil and subjected to 5 Mrads of EB radiation. The resulting pressure-sensitive adhesive displayed a 180° peel strength of 4.2 pounds/inch and a dead load shear time of 10 hours.

More commonly, acrylic polymers have been utilized as the sole or major components of radiation-curable adhesive systems. These acrylics have included a homopolymer and many examples of copolymers. One UV-cured acrylic ester copolymer system cited was claimed to have 180° peel strength of 4.8 to 5.5 pounds/inch. Radiation-curable systems based on copolymers of vinyl acetate and 2-ethylhexyl acrylate have been reported more frequently than other compositions and sometimes have been applied by hot melt techniques. These pressure-sensitive adhesives have displayed 180° peel strengths ranging from about 1.5 to 3 pounds/inch when cured by either UV or EB radiation. Not surprisingly, solutions of acrylic polymers in acrylic and vinyl monomers have also been used as radiation-curable pressure-sensitive adhesives.

Normally, a small amount of photoinitiator is required to cure UV systems efficiently. Several patents have described the use of specially prepared monomers or polymers which contain built-in photoinitiators.

Although acrylates have received the most attention, a number of other polymer types have been reported in

Table 3. Approaches to Radiation-Curable PSAs

<u>Monomer Alone</u>

Butyl Acrylate/Acrylic Acid

<u>Polymers Dissolved in Monomers</u>

Acrylic (Co)Polymers
Vinyl Acetate/2-Ethylhexyl Acrylate Copolymers
Ethylene Vinyl Acetate Copolymers
Polybutadiene, SBR
Polyvinylalkyl Ethers
Polyalkylene Oxides

<u>Oligomers/Polymers Alone</u>

Vinyl Acetate/2-Ethylhexyl Acrylate
Polyacrylates
Polyethers
Silicones
Polyisobutylene Rubber
Thermoplastic ABA

radiation-curable pressure-sensitive adhesives. Several systems have contained polybutadiene and its copolymers, particularly styrene-butadiene rubber. Other rubber bases utilized have included those derived from natural sources, polyisobutylene and silicones. In addition, components such as tackifiers, fillers, and (meth)acrylate monomers have been employed in these rubber-based systems. Ethylene copolymers, specifically ethylene-vinyl acetate and ethylene-5-ethylidene-2-norbornene-propylene, have been used with tackifiers.

Pressure-sensitive adhesives based on an entirely different type of chemistry, namely polythiol-ene adducts, have also been patented. A UV-cured system of this type has been reported to exhibit a 180° peel strength of 3.5 pounds/inch.

Radiation curing is also being used increasingly to upgrade the pressure-sensitive adhesive properties of compositions which contain substantial amounts of polymer (2-4). This technique is of particular interest for correcting some of the deficiencies of hot melt adhesives based on the thermoplastic A-B-A copolymers. In contrast to other types of pressure sensitive adhesives which can be chemically cross-linked by post heating, hot melt products must be largely free of such bonds in order to be melt processable. However, the hot melt processing of an adhesive which contains a radiation initiated cross-linking component would be a desirable approach to this problem. Because radiation curable cross-linking agents may also gel when heated, it is necessary to minimize the thermal abuse of such compositions during processing. A successful demonstration of such a process was recently described (3).

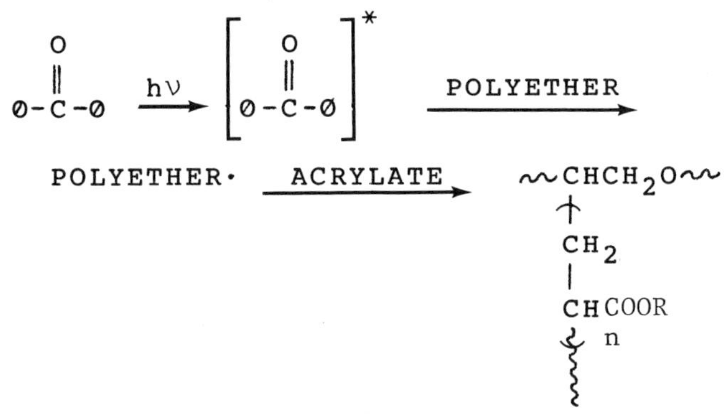

Fig. 1. Mechanism of Benzophenone Assisted Grafting

POLYETHER-BASED UV CURABLE PSAs

Choice of Components

As already mentioned, the majority of compositions utilized for radiation-curable pressure-sensitive adhesives have contained acrylates in the form of monomers or polymers reflecting not only the rapid cure capability of such materials but also the excellent properties of acrylate-based pressure-sensitive adhesives in general. Because only low degrees of polymerization are usually achieved in photocured systems, we sought to extend the molecular weight by grafting onto an existing polymer such as a polyether (Fig. 1.). The excited state of benzophenone is known to be effective at generating radicals via hydrogen abstraction from positions alpha to nitrogen and oxygen (5-7). The polyethers selected for study were poly(vinyl alkyl ethers) (often recommended as components of conventional pressure-sensitive adhesive formulations in the past), alkylene oxide polymers such as poly(ethylene oxide) (8), ethylene oxide-propylene oxide copolymers and a polytetramethylene oxide. Such polymers should be suitable backbones for photografting and, because of their low glass transition temperatures, offer the possibility of generating useful pressure-sensitive adhesives.

The structures of these components and the multifunctional acrylates studied are summarized in Table 4. Although many acrylates are known which could graft effectively, few have sufficiently low volatility to avoid substantial losses by evaporation during cure. As a result, two high boiling acrylates were selected for investigation. The first, 2-(N-methylcarbamoyloxy)ethyl acrylate (MCEA) is a polar urethane derivative while the second, isobornyl acrylate (IBA) is a relatively nonpolar, cycloaliphatic based monomer.

Multifunctional Acrylate System

A large number of compositions containing poly(vinyl alkyl ethers), multifunctional acrylates and benzophenone were irradiated and examined for their pressure-sensitive adhesive properties. Among the components which were found to confer useful properties on the films were the methyl, ethyl, and isobutyl poly(vinyl alkyl ethers) and the multifunctional acrylate derivatives of neopentyl glycol, pentaerythritol, certain epoxidized oils and an aliphatic polyurethane (9). For our present discussion, however, we will focus on the behavior of the poly(vinyl ethyl ether) (PVEE/neopentyl glycol diacrylate (NPGDA) system.

Table 4. Components Used in Photocure Pressure-Sensitive Adhesive Systems

ACRYLATES	POLYETHERS
$CH_2{=}CHCOOCH_2CH_2OC(=O)NHCH_3$ (MCEA)	$\text{-}(CH_2CH(OR))_n\text{-}$ R = CH_3, C_2H_5, ISO-C_4H_9
ISOBORNYL ACRYLATE	$\text{-}(CH_2CH_2O)_n\text{-}$
$HOCH_2C(CH_2OC(=O)CH{=}CH_2)_3$ (PETA)	$[\text{-}(CH_2CH_2O)_x\text{-} \text{-}(CH_2CH(CH_3)O)\text{-}]_y$
$(CH_3)_2C(CH_2OC(=O)CH{=}CH_2)_2$ (NPGDA)	$\text{-}([CH_2]_4O)_n\text{-}$

Urethane (isophorone-caprolactone) diacrylates
Acrylated soybean or linseed oils

The results of our initial experiments indicated that the best properties would be obtained with systems containing 80 percent or more of poly(vinyl alkyl ethers). One of the more promising compositions containing approximately 10 percent of neopentyl glycol diacrylate and 2 percent of benzophenone dissolved in poly(vinyl ethyl ether) was found to be very sensitive to changes in the concentration of the components. Fig. 2 shows the combined effects of changing the photoinitiator and multifunctional acrylate levels on the peel strength of the adhesive produced by a four-second exposure to ultraviolet light in air. The data suggests that an optimum degree of cross-linking is attained at a concentration of approximately 13 percent NPGDA.

In an attempt to simplify the system further, both exposure time and photoinitiator level were fixed and the concentration of NPGDA varied with the results illustrated in Fig. 3. Two well defined maxima were observed at 5

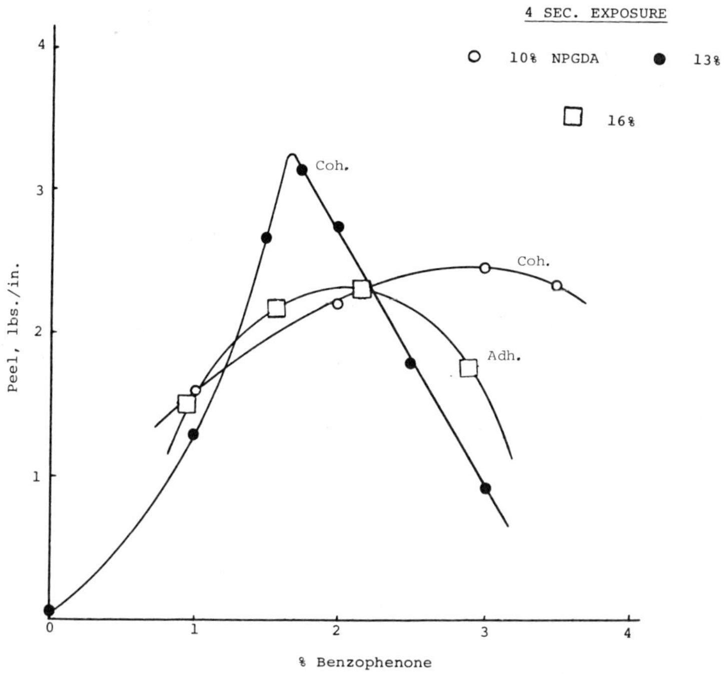

Fig. 2. Effect of Changing Photoinitiator and NPGDA Levels on Peel

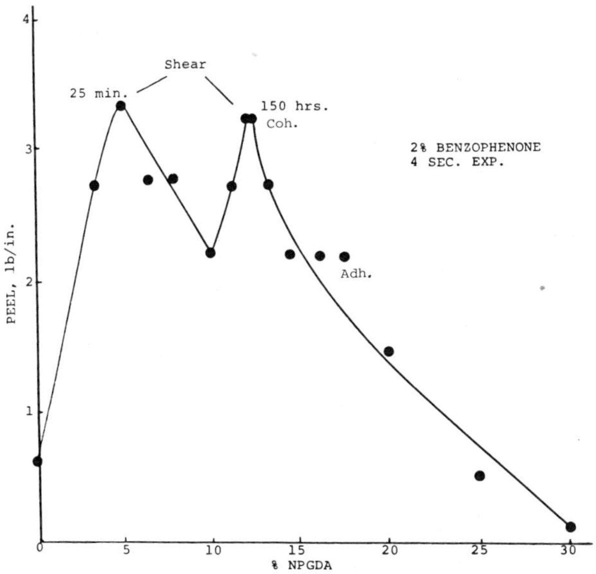

Fig. 3. Effect of Varying NPGDA Levels on Peel

percent and 12.5 percent NPGDA levels suggesting that at least two factors are responsible for developing peel in this system. The composition representing the first maximum has a peel strength of 3.3 pli coupled with a low shear while the second, with a similar peel strength, has excellent resistance to shear. The quick stick or tack of these systems was very high, being 3.3 pli and 2.0 pli, respectively. One can speculate that the peak seen at 5 percent is due to a lightly cross-linked PVEE-NPGDA graft copolymer while that at 12.5 percent NPGDA is due to an optimum level of cross-link density of a richer NPGDA copolymer. At still higher concentrations of the multifunctional acrylate, the cured composition becomes hard and tack is lost.

Some insight into the composition of the irradiated pressure-sensitive adhesives was gained by separating the materials into soluble and insoluble fragments via extraction. Thus, the pressure-sensitive adhesive obtained by irradiating a composition containing 13 percent NPGDA for four seconds was found to consist of 44 percent gel (Table 5). The gel was a rubbery, nontacky material which had an infrared spectrum with absorption characteristics of both poly(vinyl ethyl ether) and polyacrylate. The soluble portion, however, had an infrared spectrum which was equivalent to the starting polyether as did an irradiated formulation composed of only PVEE and benzophenone. The latter system also remained completely soluble, another indication that no cross-linked material was present. As a result of these findings, the pressure-sensitive adhesive itself is believed to consist of a cross-linked PVEE-NPGDA matrix which is plasticized/tackified by unreacted polyether. The presence of such a large fraction of unreacted polyether indicates that the benzophenone assisted photografting is not a highly efficient process under the conditions used. Although benzophenone is able to function in an air environment with reactive hydrogen donors, some quenching of its photoactivated state by oxygen is likely to reduce its efficiency. Similarly, the propagation of growing chains would also be inhibited which would lead to a reduction in average chain length. Nonetheless, sufficient reaction does take place in air to develop a good balance of pressure-sensitive properties.

A measure of the level of unreacted NPGDA was also obtained via extraction coupled with gas chromatography. In this way, the level of residual acrylate monomer was

Table 5. Extraction and IR Data on Poly(vinyl alkyl ether) Systems

Composition[a]	% Soluble	% Gel
Poly(vinyl ethyl ether), NPGDA (85/13)	56, No C=O	44, Strong C=O
Poly(vinyl methyl ether), NPGDA (85/13)	64, Some C=O	36, Strong C=O
Poly(vinyl isobutyl ether), NPGDA, CELLOSOLVE® acrylate (68/15/15)	18, No C=O	82, Strong C=O
Poly(vinyl ethyl ether), NPGDA (93/5)	73	27

[a] 2% Benzophenone, cured 4 sec.

found to be 0.75 percent. In regard to the peel data, it should be noted that when adhesive failure was observed, it was found to occur at the original supporting substrate, that is beneath the irradiated coating as compared to the surface closest to the ultraviolet source. This effect will be discussed shortly. The shear resistance of these adhesives at elevated temperature was outstanding. No failure was noted with the 12.5 percent NPGDA system in 100 hours when a 1 inch x 1 inch Mylar/steel bonded surface was tested at 70°C with a 500 gm. weight.

Poly(vinyl alkyl ethers) other than PVEE were also found to react similarly when irradiated in the presence of benzophenone and NPGDA. Substantial carbonyl containing gel fractions were obtained in this way from both poly(vinyl methyl ether) and poly(vinyl isobutyl ether), Table 5. In addition, multifunctional acrylates other than NPGDA can be employed including those based on pentaerythritol and various acrylated urethane oligomers. As expected, higher concentrations of neopentyl glycol diacrylate resulted in higher degrees of cross-linking as evidenced by increased gel content and lower swelling indices.

Monofunctional Acrylate Systems

Additional compositions containing polyethers and monoacrylates were prepared and irradiated in the form of thin films on Mylar (10). In the case of the PVEE/MCEA system, exposures of 4-6 seconds (20-30 feet/minute) were required to complete the cure in air when a 2 percent level of benzophenone was present. With such exposures, the level of unreacted monomer was established at less than 0.1 percent by gas chromatography (Table 6).

Despite the lack of any multifunctional acrylate, the cured PVEE-MCEA compositions were found to contain a large amount of gel (53-64 percent) and similar observations were also made with an isobornyl acrylate-based adhesive system. Presumably, the large instantaneous concentration of polymer radicals generated during UV exposure results in radical-radical coupling. The higher degree of cross-linking resulting from extended exposures is reflected in higher gel contents and lower swell indices.

The presence of a cross-linked network again led to the development of good pressure-sensitive properties. A plot of peel strength vs. composition for the cured PVEE-MCEA system is shown in Fig. 4. In subsequent experiments, the pressure-sensitive properties were maximized at 3.8 pli/48 hour shear for the 55/45 ratio.

Table 6. Cure of Poly(vinylether ether)/ 2-(N-Methylcarbamoyloxy) Ethyl Acrylate System[a]

Ratio PVEE/MCEA	Exposure Sec.	Loss On Cure, %	Unreacted MCEA By G.C., %
35/65	2	0.9	19
	4	1.0	<0.1
	6	1.3	<0.1
	10	1.5	<0.1
55/45	2	0.8	5.8
	6	1.2	<0.1
	10	1.3	<0.1

[a] Containing 2% Benzophenone

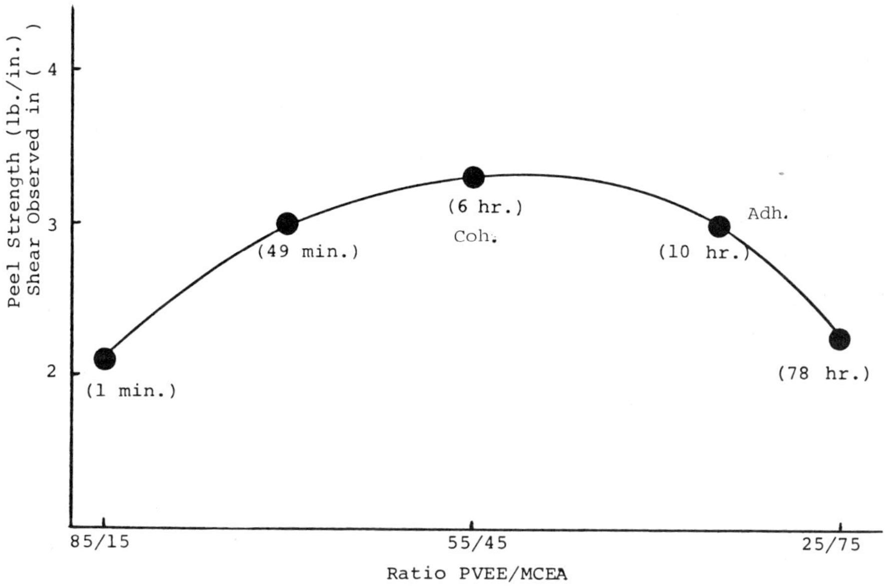

Fig. 4. Effect of PVEE/MCEA Ratio on Pressure-Sensitive Adhesive Properties (6 Sec. Cure, 2 % Benzophenone)

When isobornyl acrylate was used in place of MCEA, cured films with good pressure-sensitive adhesive properties were again obtained although somewhat longer irradiation times were required. For example, blends of poly(vinyl ethyl ether) in IBA displayed properties ranging from 2 pli with adhesive failure and over 90 hours of shear to nearly 9 pli (cohesive failure) and 1 hour of shear (Table 7). The latter product was examined closely for complete cure and found to have 1.6 percent unreacted monomer at 6 sec. radiation and 0.23 percent at 9 sec. Again, a large gel fraction was present in the cured adhesive as noted earlier.

Although the gel content indicated the presence of substantial cross-linking, the pressure-sensitive adhesive properties could be improved further by incorporating a small amount (1 percent) of pentaerythritol triacrylate (PETA). This system developed a peel value of 5.6 pli with adhesive failure and 22 hours of shear with only 6 seconds of cure time in air (Table 7).

Table 7. Photocured Pressure-Sensitive Adhesives Based on PVEE and Isobornyl Acrylate[a]

Composition	Exposure (sec.)	Peel (pli)	Shear (hrs.)
PVEE/IBA (60/40)	9	6.8 Coh.	0.6
A-50/A-25/IBA (36/24/40)	9	8.9 Coh.	1.0
Above With 1% PETA	6	5.6 Adh.	22
A-50/IBA (50/50)	6	2.0 Adh.	>90

[a] 2% Benzophenone. A-25 and A-50 are BASF PVEE products.

Based on our results thus far, it should also be possible to prepare pressure-sensitive adhesives by grafting onto other structurally similar polymer backbones such as poly(alkylene oxides). Some encouraging confirmation of this was obtained by curing solutions of poly(ethylene oxide) (Union Carbide CARBOWAX® polyethylene glycols, molecular weight 6,000) in 2-N-methylcarbamoyloxy) ethyl acrylate. In this fashion, PSAs with peel strengths in excess of 3 pli (adhesive failure) and shear resistance of over 50 hours were obtained (8) (Fig. 5). Useful properties were also obtained with lower molecular weight polymer (Union Carbide CARBOWAX polyethylene glycols 4000). Again, as in the case of the poly(vinyl alkyl ethers), a significant gel fraction was obtained (Table 8). The extracted gel was rubbery and displayed little tack.

Reasonably attractive properties were also observed with ethylene oxide-propylene oxide copolymers. A blend of 53 percent Union Carbide UCON 75-H fluids 380,000 (molecular weight 21,000; 75 percent ethylene oxide) in MCEA was irradiated for 9 sec. and found to have peels of 3.5 pli (adhesive failure) and shears of 31 hours.

Finally, a composition consisting of 29 percent Polymeg-2000 ($\text{-[}(CH_2)_4O\text{]-})_n$ and 2.9 percent benzophenone in MCEA was found to develop a modest pressure-sensitive adhesive when photocured for 9 sec. (0.4 pli, adhesive failure/40 min. shear).

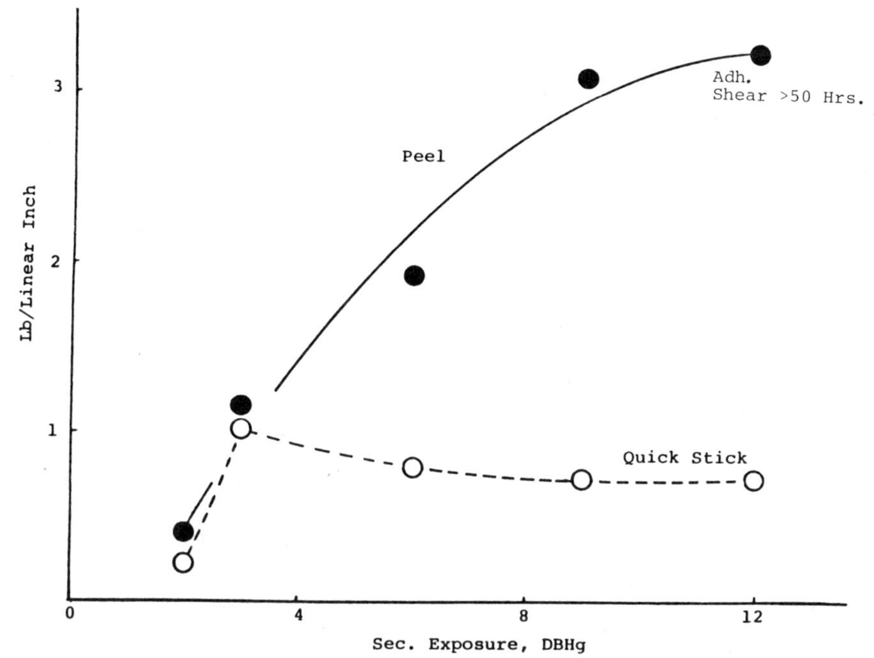

Fig. 5. Effect of Cure Time on PSA Properties of a CARBOWAX®/MCEA System

Table 8. Cure of Poly(ethylene oxide)/2-(N-Methylcarbamoyloxy) Ethyl Acrylate System[a]

Exposure, sec.	Loss on Cure, %	% Unreacted Monomer by G.C.	% Gel	Shear, hrs.
2	0.8	-	49	12 Min.
3	1.3	36	56	17
6	1.6	11	65	>50
9	2.0	1.8	68	>50
12	1.9	0.8	67	>50

[a] Union Carbide's CARBOWAX® polyethylene glycols 6000 (25)/MCEA(72)/water(2.5)/benzophenone(2). Cured in air with Linde DBHg unit.

CORRELATIONS BETWEEN CURING/FORMULATING PARAMETERS AND PROPERTIES

Those attempting to design photocure pressure-sensitive adhesives are faced with the fact that the majority of initial attempts yield little in the way of the desired properties. Progressing from this point has long been a difficult problem because the required properties are a complex interaction of the molecular parameters (polar character of repeat units, extent of branching and cross-linking, molecular weight, and molecular weight distribution) of such an adhesive and the curing process itself (intensity of radiation, presence of oxygen, type and amount of photoinitiator system).

Frequently, it will be found that the irradiated material displays little tack, being either rigid or ranging from soft to fluid. Such systems require extensive reformulating or changes in cure time. In the intermediate, tacky condition, it will be found that the key properties of peel strength and shear resistance will develop according to different patterns. With the polyethers we have studied and presumably with other systems as well, as optimum conditions are approached, shear resistance will usually rise abruptly and then decline even more rapidly as a function of increasing cure, cross-link density or other formulating parameters. Likewise, a plot of peel strength will also show a maximum but the curve will be of a gradual form. Such profiles are characteristic of pressure sensitive adhesives in general and are normally observed to be in close spatial relation to one another (11). However, when the system is other than ideal, we have observed that the maximum in the shear plot may precede or follow the maximum in the peel (Fig. 6) depending on whether the system is "hard or soft". A hard system may be defined as one in which shear values have begun to decline before peel values have maximized while in a soft system the reverse is true. The plots in Fig. 6 (A-D) are idealized in the sense tht complete experimental verification has not been obtained for all cases. However, a relatively complete study has been done on the progression from B-D. In reference to these figures, it should be pointed out that while plateaus for shear resistance are sometimes observed in practice, during our experimentation, we often discontinued tests after the arbitrary time value of 80 hours was reached. The plateaus, then, should be viewed as the limit of our testing.

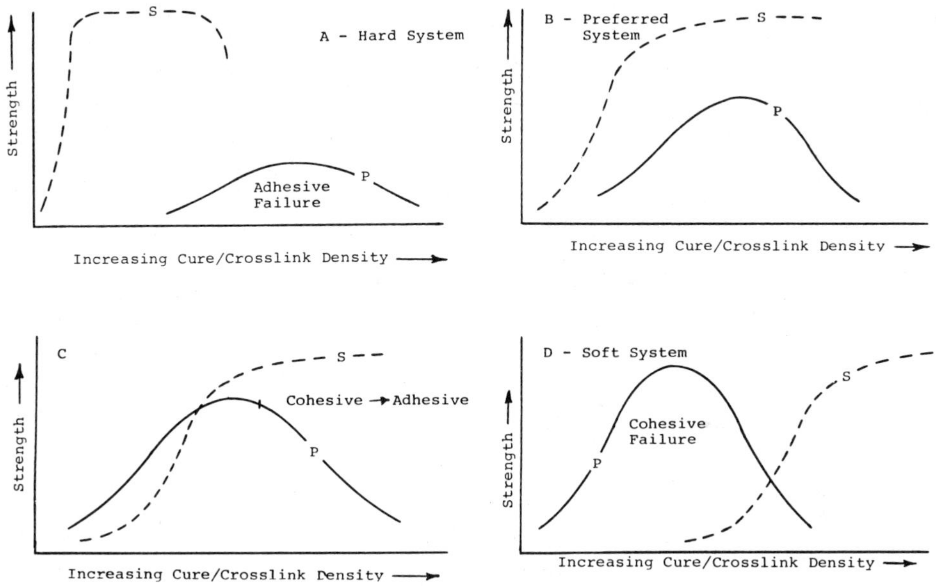

Fig. 6. Idealized Relationship of Shear and Peel in Hard and Soft Systems

Another important consideration is the mode of failure experienced during the 180° peel test. Pressure-sensitive adhesives with internal strength relatively low in comparison to their adhesion to a test substrate will fail cohesively while the same type of adhesive with improved internal strength will often fail adhesively. This adhesive failure can be important for temporary applications but is of less importance for high strength, permanent uses. Some systems change from the cohesive to the adhesive failure mode as the amount of cure is increased, a point which will be illustrated later.

As stated before, when cured, each adhesive will display some combination of peel (with adhesive or cohesive failure) and shear resistance as represented by a point on the shear and peel plots. The relative position of the plots to one another determines what kind of an adhesive has been formed and how amenable to improvement it is. Thus, a relatively hard PSA will display a combination of high shear resistance and lower peel strength, probably with adhesive failure (hypothetically Fig. 6-A). Continued cure or additional cross-linking with this type of system is unlikely to generate desirable properties. At the opposite extreme, a soft system (e.g., Fig. 6-D) may show moderate-high peels (cohesive failure), but shear values will be low. High shear may develop with additional cure, but the peel value will have already begun to decline.

Further modifications to the system to increase "hardness" can result in additional shifting of the shear plot to the left. Systems can then be found in which the shear rises abruptly to a maximum value at a point close to where the maximum peel is reached. While this type of system (Fig. 6-C) yields high values, exposure times must be controlled very carefully as small variations will have a large effect upon performance. A more preferable system is shown in Fig. 6-B. Here, the shear is maximized early and peel then increases to its maximum value with more exposure. In this way, only the peel value is varying and exposure time is less critical.

The shifting of the peel and shear relationships with formulation and cure changes can be illustrated with the poly(vinyl ethyl ether)/MCEA/benzophenone system (Fig. 7). At low MCEA levels (0-15 percent) increasing exposure time causes an increase in peel but the film does not have any

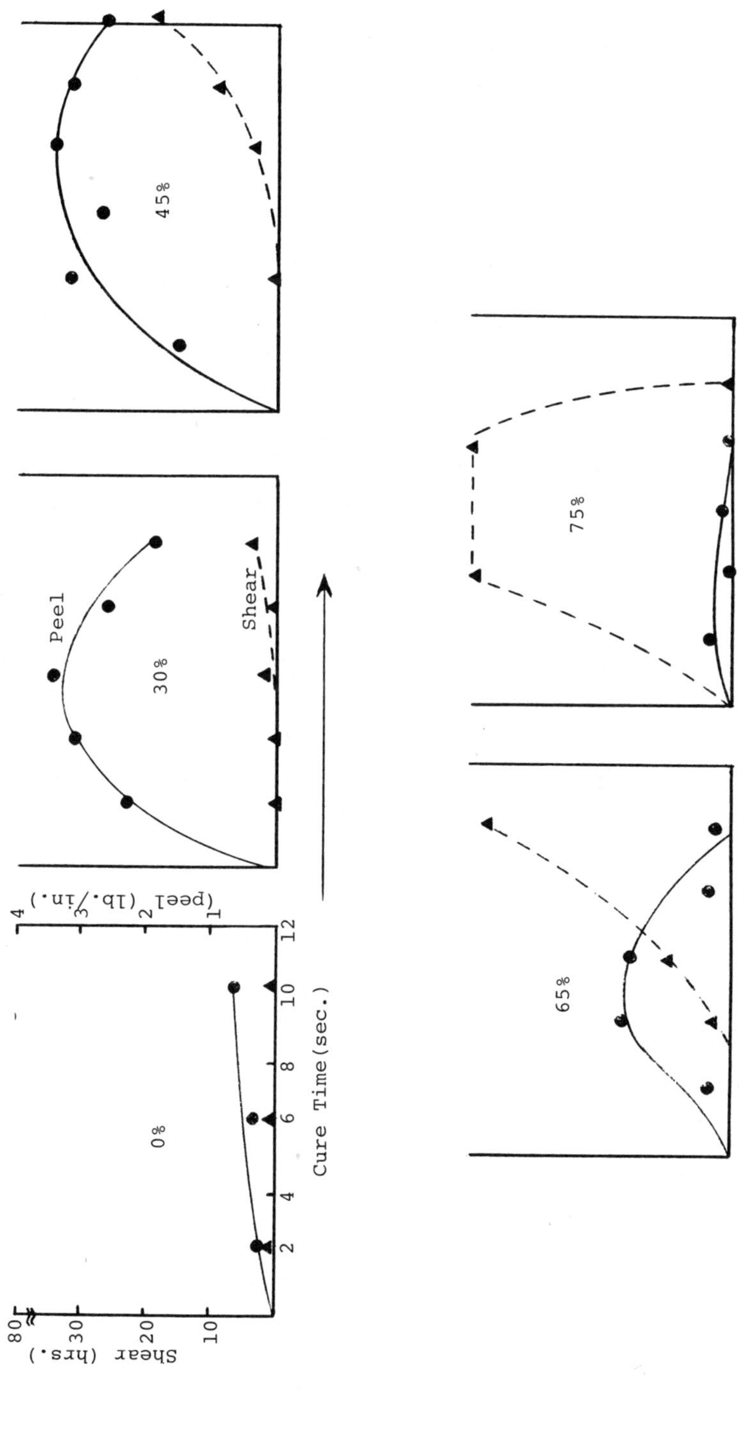

Fig. 7. Pressure Sensitive Properties as a Function of Cure Time and Composition

appreciable shear strength. When the acrylate level is increased to 30 percent, shear resistance begins to build slowly with additional exposure while at the 45 percent level, shear is further enhanced at longer irradiation times. At 65 percent MCEA, shear is approaching a measured value of 74 hours while peel has declined substantially. The shear plateau is fully developed and maximized with 75 percent MCEA present, even at low exposures, but peel has now declined to near zero. Adhesive failure was noted in the 65 percent MCEA composition at six or more seconds exposure along with the various exposures at the 75 percent level. Increased benzophenone levels did not increase the maximum peel strength but the systems did develop good shear values at shorter exposures.

While cure time is of substantial importance, other variables have also been found to result in major property changes. A semiquantitative comparison of the importance of three factors - multifunctional acrylate concentration, photoinitiator concentration, and cure time - is given in Table 9 for peel strength, shear, and the adhesive/cohesive peel mode. We have assigned relative values of high, medium, and low to each of the independent variables for their effect on the dependent properties.

Table 9. Importance of Cure/Formulating Factors on Properties

	Effect of Small Changes[c]		
	Multifunctional Acrylate	Photoinitiator	MPHg Cure Time
Peel Strength	Medium	Medium	Medium
Shear	High	Low[a]	Low[b]
Adhesive/Cohesive Peel Mode	High	Medium	Low[b]

[a] Assuming a minimum critical amount is present.

[b] There is a critical point where major effects are seen.

[c] Defined as 1 phr for multifunctional acrylate; 0.5 phr photoinitiator; 1 sec. change for cure time.

The amount of cross-linker present has the strongest effect upon properties of any of the variables shown in Table 9. An example of this effect is a blend of A-50, A-25, isobornyl acrylate, and benzophenone; 36/24/40/2. At 6-sec. exposure, this formulation has a peel of 7.8 pounds cohesive failure and a 0.6 hour shear. By adding 1 percent PETA the peel is now 5.6 pounds adhesive failure and a 22 hour shear. The peel has not only dropped over two pounds, but has also changed over to adhesive failure.

Changes in the photoinitiator concentration generally have moderate effects upon properties. Blends of PVEE and isobornyl acrylate (70/30) containing 2 and 3 percent benzophenone afforded values of 4.6 pli (cohesive failure) with 0.43 hour shear, respectively, using six seconds irradiation. Initiator concentration usually has its greatest effect on the time of exposure needed to achieve the maximum property values, but does not change the maximum obtainable for that formulation.

Cure time, as discussed in the previous section, has relatively little effect over most of the practicable exposure range. It is primarily at the critical points where peel changes its failure mode and shear increases rapidly, that a small exposure change will have a major effect. An example which illustrates this critical point can be seen with a blend of poly(vinyl ethyl ether) (A-50), 2-(N-methylcarbamoyloxy) ethyl acrylate, Union Carbide CELLOSOLVE® acrylate, and benzophenone (40/60/5/2). When irradiated for 6.5 seconds in air, a peel value of 5.5 pli is observed with cohesive failure while at 7.0 seconds it has changed to adhesive failure (4.0 pli). The maximum peel value for these formulations occurs just before the adhesive failure mode begins (point C→A, Figure 6-C).

Effect of Direction of Radiation

Many of the cured PSA films exhibit cohesive failure upon curing. This failure seems to occur close to the substrate upon which it was cured. Thus, when a formulation is coated on a rigid substrate, cured, overlaid with a flexible film such as Mylar and tested to failure, cohesive failure occurs toward the rigid substrate. If the procedure is reversed and the coating is initially done on the flexible film, failure will occur at that surface.

It was presumed that since the ultraviolet light enters the adhesive formulation from the top, this upper portion utilizes most of the energy therefore curing more completely. To gain some additional insight into this behavior, various portions of the adhesive were examined by attenuated total reflectance spectroscopy (ATR). In this technique, the infrared radiation penetrates into the surface approximately 1 to 5μ depending on the reflection conditions and the wavelength of the light; a spectrum is obtained which is similar to a normal transmission infrared spectrum. A comparison of the spectra of the uncured formulation (PVEE/MCEA/benzophenone--61/37/2), with that of the portion of the adhesive closest to the UV source and that furthest away from the source reveals that appreciable differences in composition exist through the cross section of the film which are not due to completeness of cure.

THE FUTURE

The many advantages of using radiation curing for the manufacture of pressure-sensitive adhesives are undeniable. Despite these advantages, however, only a small percentage of such products are being produced in this way. A recent study (12) indicated that the total sales of radiation curable PSAs totaled only $400,000 in 1979. The reasons why radiation curing has not captured a larger share of this market are complex but involve the following elements:

Technology less well developed than with alternate techniques - a broad array of materials/properties not available.

Raw materials costs tend to be high.

Many of the acrylates employed in these systems may cause skin burns and other health problems.

Nonetheless, interest remains high and, in time, solutions to these problems will be found.

EXPERIMENTAL

Materials

Most of the materials employed in this study were obtained from commercial sources and were used as received. The acrylates were stored at 5^oC until needed. Exceptions are noted.

2-(N-Methylcarbamoyloxy) ethyl acrylate - MCEA was prepared in accordance with Ref. 3; isobornyl acrylate - Rohm and Haas; neopentyl glycol diacrylate - Union Carbide Corp.; pentaerythritol triacrylate - Union Carbide Corp.

Poly(vinyl alkyl) ethers were obtained as follows: Poly(vinyl ethyl ether) - EDBC from Union Carbide. This material is relatively low in molecular weight (reduced viscosity 0.3 ± 0.1 in benzene, 0.1 gm/100 ml at 20^oC) and contains 0.5% monotertiary butyl hydroquinone stabilizer. It consists of 98% nonvolatiles. In some cases, PVEE, manufactured by BASF-Wyandotte, was used (A-25 and A-50 with reduced viscosities of 0.07 and 0.66 respectively in benzene at 20^oC - 0.2 gm/100 ml.); poly(vinyl methyl ether) - (M-574) from General Aniline and Film Corp.; poly(vinyl isobutyl ether) - (Lutanol J-60) from BASF - Wyandotte Corp.; poly(ethylene oxide) - CARBOWAX 4000 and 6000) from Union Carbide; ethylene oxide-propylene oxide copolymer-UCON 75-H 380,000 random copolymer containing 75% ethylene oxide - from Union Carbide; poly(butylene oxide) - Polymeg 2000 - Quaker Oats.

Benzophenone - Matheson, Coleman and Bell.

Blends

Blends of these components were usually prepared at least 24 hours in advance of curing by thorough mixing. The compositions were then applied to type "A" Mylar using a wire rod so as to give a 1-2 mil film. The thickness of the cured films were estimated with a type "B" G.E. thickness gauge.

In order to prepare stable solutions of polyethylene oxide-acrylate blends, it was necessary to incorporate small amounts of water. These blends were prepared by adding together the CARBOWAX, acrylate and benzophenone, heating to 60°C with agitation and then adding the water after the CARBOWAX has dissolved. Without the solubilizing effect of the water, the polyether usually precipitated from solution a short time after preparing the blend by warming to ca. 60°C.

UV Equipment

Most compositions were cured in air with a Linde MPHg unit employing three 2.2 KW, medium pressure mercury arc lamps, each 24 inches in length positioned parallel to the travel of the moving belt. The total delivered flux was estimated at 500 Watts/sq.ft.

As noted, some cures were effected with a Linde DBHg PS-2000 unit employing three medium-pressure mercury arc lamps positioned parallel to the travel of the moving belt. The total delivered flux was estimated at 500 Watts/sq.ft. The equipment was run for twenty minutes before samples were cured to minimize temperature variations. All cures were done in air.

Testing Procedures

After curing, two or more samples were tested by Pressure Sensitive Tape Council (PSTC) procedures using minor variations.

Peel Test (PSTC-1) - Cold rolled steel "Q-Panels" (Q-Panel Co.) were used. A one-inch wide strip of coated Mylar is bonded to the panel under pressure of a 4 1/2-lb. roller. An Instron is employed to test 180° peel at ca. 76°F at a crosshead speed of 12"/minute within one minute of assembly.

Quick Stick (PSTC-5) - A one-inch wide specimen is allowed to bond to the metal panel without any external pressure. The Instron immediately is employed to measure a 90° peel value at a removal rate of 12"/minute.

Shear Test (PSTC-7) - Dead load shear test; 1/4 sq. in. (1/2 in. x 1/2 in.) overlap of Mylar backed adhesive is bonded to the panel under 4 1/2-lb. roller pressure. Sample is vertically mounted at 77°F and a 500-gram weight is suspended from Mylar; time to failure is noted.

Gel Fractions - Weighed samples of film were shaken with twenty-fold amount of toluene for 16 hours at room temperature followed by similar treatment in dimethylformamide. The insoluble fraction was dried 16 hrs. @ 50°C in a circulating air oven followed by 16 hrs. in vacuum at 80°C.

Residual Monomers - Compositions to be checked for unreacted monomer were prepared with 1% of hexadecane as an internal standard prior to curing. Following cure, the film was extracted with CCl_4 for two hours in the presence of 0.5% of phenothiazine in a rolling bottle. Gas chromagraphic analysis is run as quickly as possible. Samples not immediately analyzed were refrigerated until used.

REFERENCES

1. J.W. Hagan and K.C. Stueben, "Pressure-Sensitive Adhesives" in Adhesives in Manufacturing, Marcel Dekker, in press.

2. K.C. Stueben, "Radiation Curing of Pressure-Sensitive Adhesives: A Literature Review," Adhesives Age, 20, No. 6, 16 (1977).

3. D.J. St. Clair, "Radiation Curing of PSAs Based on Thermoplastic Rubbers," Adhesives Age, 23, No. 3, 30 (1980).

4. S.H. Ganslaw, "Hot Melt Pressure-Sensitive Adhesives, Alternatives to Rubber-Resin Systems," in 1979 Course Notes, "Hot Melts--An Overview for Management" (TAPPI, South Carolina, June 4-7, 1979).

5. J. Hutchinson and A. Ledwith, Advances in Polymer Sci., 14, No. 1, 49 (1974).

6. J.F. Kinstle and S.L. Watson, Jr., J. Radiation Curing, 2, 7, (1975).

7. S.P. Pappas, Progress in Organic Coatings, 2, 333 (1973/74).

8. K.C. Stueben, U.S. 4,111,769 (September 5, 1978).

9. K.C. Stueben, R.G. Azrak, and M.F. Patrylow, U.S. 4,151,055 (April 24, 1979).

10. K.C. Stueben, R.G. Azrak, and M.F. Patrylow, U.S. 4,165,266 (August 21, 1979).

11. J.W. Hagan, C.B. Mallon, and M.R. Rifi, Adhesives Age, 22, 29 (1979).

12. Chemical Marketing Reporter, p. 5, March 10, 1980.

13. W.C. Perkins, "Radiation Curable Pressure-Sensitive Adhesives," Radiation Curing, 7, 4 (August 1980).

14. T. Okada, "Review of UV and EB PSAs," (in Japanese), Chem. Abstr. 90, 72690.

15. G. Hausman, Proc. Adhes. Coat. Technol., "Pressure-Sensitive Tape Ind. 80's," Tech. Semin. 1980, 183-91, Pressure-Sensitive Tape Council, Glenview, IL.

Reprinted from JOURNAL OF RADIATION CURING® Volume 9/Number 2 dated April 1982, published by Technology Marketing Corporation, 17 Park Street, Norwalk, CT 06851 USA.

Table 10. Radiation Curable PSAs - Literature Update

U.S. Patent	Date	Company	Comments
4,022,926	1977	Dennison Mfg.	A method for making a novel assembly of laid-on labels without die cutting.
4,052,527	1977	National Starch & Chemical	UV radiation of hot melt comprised of at least one 3-(chlorinated aryloxy)-2-hydroxypropyl ester of an alpha, beta unsaturated carboxylic acid with acrylate based copolymerizable monomers. PSA properties.
4,069,123	1978	National Starch & Chemical	Improved PSA compositions based on acrylate ester polymers. UV radiation PSA properties reported.
4,111,769	1978	Union Carbide	Polyoxyalkylene homo- or copolymer, carbamyloxy alkylacrylate, and optionally a photoinitiator. UV.
4,133,731	1979	Shell Oil Co.	EB/UV radiation cure of composition comprising a monoalkenyl arene/conjugated diene block copolymer, tackifying resin, and a di-tetra functional acrylate or methacrylate selected from the group consisting of acrylic and methacrylic acid esters of polyols. Improved high-temperature properties and solvent resistance. PSA properties.
4,150,170	1979	Celanese	Curable compositions containing butyl acrylate, vinyl acetate, acrylic acid, and benzoin alkyl ethers irradiated with UV, some PSA properties given.

U.S. Patent	Date	Company	Comments
4,151,055	1979	Union Carbide	Polyoxylalkyl ether, multifunctional acrylates and optionally monoacrylates cured by UV or EB.
4,141,057	1979	Shell Oil Co.	EB cure of composition comprising selectively hydrogenated monoalkylenyl arene/conjugated diene block copolymer/tackifier and a di-to-tetra-functional acrylate or methacrylate selected from the group consisting of acrylic and methacrylic acid esters of polyols. PSA properties given. High temperature properties/solvent resistance improved.
4,152,231	1979	Shell Oil Co.	UV or EB cure of system containing a linear or radial polydiene, tackifier, and a di-to-tetra-functional acrylate or methacrylate selected from the group consisting of acrylic and methacrylic acid esters and polyols. PSA properties given. High temperature properties/solvent resistance improved.
4,165,266	1979	Union Carbide	Polyvinyl alkyl ether, monofunctional acrylate monomer, photoinitiator blends cured by UV.
4,181,752	1980	3M Co.	Acrylate ester, acrylic acid, benzoin ethyl ether blends cured by UV. PSA properties given. Some blends with acrylamides and N-vinylpyrrolidone and other ingredients studied.

U.S. Patent	Date	Company	Comments
4,234,662	1980	National Starch & Chemical	Hot melt copolymer containing acrylic, allyl acrylate or methacrylate, polymerized by EB. PSA properties given.
4,243,500	1981	International Coatings	Monofunctional acrylate ester, tackifying resin, elastomer, initiator cured by UV or other radiation. Properties given.
4,305,854	1981	Polychrome Corp.	Castor Oil - hydroxyethyl acrylate - toluene diisocyanate reaction product and trimethylol propane triacrylate cured using DEAP and UV. 180° peel (1.5 mil polyester substrate) of 25-30 oz. Shear $\leq$8 hr. (500g.)

Foreign Patent	Date	Company	CA Ref.	Comments
Brit. 1,267,798	1972	Beiersdorf	78:112382	Bu Acrylate-bis(2-ethylhexyl)fumarate-N-vinyl-2-pyrrolidone copolymer.
Jap. Kokai 77 78,239	1977	Showa Elec. Wire & Cable	87:169058	EPT - tackifier - EB.
Jap. Kokai 77 92,249	1977	Showa Elec. Wire & Cable	88:7942	Ethylene vinyl acetate - EB.
Jap. 78 31,898	1978	Nitto Elec. Ind.	89:198769	Bu Acrylate - acrylic acid copolymer, rosin ester, tetramethylene glycol dimethacrylate - UV.
Jap. 78 09,839	1978	Furakawa Electric	89:7414	EPDM or ethylene propylene rubber and polyisobutene - EB.
Brit. 1,546,753	1979	Letraset	91:202236	UV.
Ger. 2743-979	1979	Beiersdorf	Derwent	Acrylates copolymerized with benzoin acrylate photoinitiator - UV.
Jap. Kokai 80,127,481	1980	Hitachi	94:48516	UV, acrylate, benzophenone.
Jap. Kokai 80 98,271	1980	Furakawa Electric	93:240811	Ethylene vinyl acetate copolymer or SBR block hot melt - EB.
Jap. 5 5071-774	1980	Furakawa Electric	Derwent	SB Block copolymer - EB.
Ger. Offen. 3,033,359	1981	Hitachi	94:209711	Acrylic polymer with high boiling monomer such as propylene glycol monomethacrylate and cross-linking agents - UV.

Foreign Patent	Date	Company	CA Ref.	Comments
DE 3036-425	1981	Polychrome Corp.	Derwent	Water insoluble monomer, tackifier.
Jap. Kokai 81 14,574	1981	Hitachi	95:8425	Epoxy containing acrylic polymers which form Lewis acids on UV irradiations. Properties given.
Belg. BE 890,425	1982	General Electric		UV cured silicone PSA's. Siloxanes containing epoxy groups cured using UV and catalysts such as bis(dodecylphenyl) iodonium hexafluoroantimonate.
Jap. Kokai Tokkyo Koho JP 81 149,480	1981	Hitachi		Void free coatings on polyethylene film from EHA/acrylic acid and glycidyl methacrylate using 2-ethylanthriaquinone.
Jap. Kokai Tokkyo Koho JP 82 10,667	1982	Hitachi		Acrylates with chain transfer agent.
Jap. Kokai Tokkyo Koho JP 82 14,672	1982	Hitachi		Acrylic prepolymers containing 5-40% aminoalkyl(meth)acrylate units and having 1000-50,000 m.wt. are treated with functional group containing unsaturated compounds to give photocurable PSA's.
Jap. Kokai Tokkyo Koho JP 82 14,671	1982	Hitachi		Acrylic polymers (m.wt. 1000-50,000) containing 5-40% O- $CH_2=CRCO_2CH_2CH_2-O_2CC_6H_4CO_2H$ are treated with functional groups - containing unsaturated compounds to give photocurable PSA's.

Foreign Patent	Date	Company	Comments
Jap. Kokai Tokkyo Koho JP 82 14,670	1982	Hitachi	Similar to above but 5-40% tetrahydrofurfuryl acrylate or methacrylate.
Jap. Kokai Tokkyo Koho JP 82 10,668	1982	Hitachi	Acrylic copolymers containing units of a (phosphonoxy) alkyl(meth)acrylate are photocurable to PSA's.
Jap. Kokai Tokkyo Koho JP 82 051-780	1982	Hitachi	Radiation curable adhesive tape production from (meth)acrylate prepolymer.
Jap. Kokai Tokkyo Koho J 56120-786	1981	Hitachi	Photocuring PSA sheet comprises tacky layer containing (meth)acrylic ester photopolymerization initiator and accelerator.
EP 56-526	1982	Mobil Oil Corp.	Radiation curable coating composition comprising an acrylic ester monomer and polyfunctional acrylate crosslinking agents. The use of the adduct of (meth) acrylic acid or its hydroxy-alkyl ester and bisphenol A diglycidyl ether is claimed.

Recent Progress in Photoinitiated Cationic Polymerization

James V. Crivello

General Electric
Corporate Research & Development Center
Schenectady, New York 12301

ABSTRACT

Over the past several years, there have been developed several new classes of onium salt photoinitiators capable of initiating cationic polymerization. The most significant of these are aryldiazonium salts, diaryliodonium salts, triarylsulfonium salts, and dialkylphenacylsulfonium salts. The mechanisms involved in the photolysis of these compounds have been elucidated and will be discussed. In general, on irradiation acidic species are generated which interact with the monomer to initiate polymerization. Using photosensitive onium salts, it is possible to carry out the polymerization of virtually all known cationically polymerizable monomers. A discussion of the various structurally related and experimental parameters will be presented and illustrated with several monomer systems. Lastly, some new developments which make possible the combined radical and cationic polymerization to generate interpenetrating networks will be described.

I. INTRODUCTION

In recent years, there has been an increasing interest in the use of photopolymerizations for many industrial coating and printing processes. Because of their high cure and application speeds, essentially pollution-free operation, very low energy requirements and generally excellent properties, coatings prepared by photopolymerizations have made a substantial impact on the wood coating, metal decorating and printing industries. While the bulk of the current research effort continues to be directed toward photoinduced radical polymerizations, it is well recognized that ionic photopolymerizations also hold considerable promise in many application areas. Photoinduced cationic polymerizations are particularly attractive because of the wide breadth of chemical and physical properties which can potentially be realized through the polymerization of such monomeric substrates as electron-rich olefins, lactones, epoxides, cyclic ethers, sulfides, silicones, and acetals as well as numerous other heterocyclic compounds. Further, photoinitiated cationic polymerizations have the important commercial advantage that they are not inhibited by oxygen and thus may be carried out in air without the need for blanketing with an inert atmosphere to achieve rapid and complete polymerization.

II. RESULTS AND DISCUSSION

Our research group at General Electric, along with similar groups at 3M Company and ICI, have been successful in developing several new types of very active photoinitiators for cationic polymerization, i.e., diaryliodonium (1-5), triarylsulfonium (6-9), triarylselenonium (10), dialkylphenacylsulfonium (11), and dialkylhydroxyphenylsulfonium salts (12). The most practical of these photoinitiators are diaryliodonium salts, I, and triarylsulfonium salts, II.

$$\mathrm{Ar{-}\overset{+}{I}{-}Ar'}\quad X^- \qquad\qquad \mathrm{Ar{-}\overset{\displaystyle Ar'}{\underset{\displaystyle Ar''}{S^+}}}\quad X^-$$

I II

These latter compounds are readily prepared using conventional synthetic techniques, have good photosensitivity and excellent dark stability. Only those compounds bearing weak nucleophilic anions of the type $X^- = ClO_4^-$, BF_4^-, PF_6^-, AsF_6^-, SbF_6^- are useful as photoinitiators of cationic polymerization.

Mechanism of Photolysis

Although diaryliodonium and triarylsulfonium salts are thermally stable, they undergo rapid photolysis when irradiated at wavelengths from 200-300 nm. Under these conditions, facile homolytic rupture of one of the carbon-iodine or carbon-sulfur bonds results as shown in equations 1 and 2.

$$Ar_2I^+X^- \xrightarrow{h\nu} Ar^{\cdot} + ArI^{\cdot +}X^- \qquad (1)$$

$$Ar_3S^+X^- \xrightarrow{h\nu} Ar^{\cdot} + Ar_2S^{\cdot +}X^- \qquad (2)$$

Evidence for the above reactions consists of direct observation in certain cases of the aryliodinium (13) and diarylsulfinium (14) cation-radicals as well as the isolation of aromatic hydrocarbons and biaryls from the aryl free radicals.

Interaction of the respective cations radical with the solvent or monomer, R-H, results in the release of a proton and the resultant formation of strong protonic acids as depicted in equations 3-7:

$$ArI^{\cdot +} + R\text{-}H \longrightarrow ArI^+\text{-}H\ X^- + R^{\cdot} \qquad (3)$$

$$ArI^+\text{-}H\ X^- \longrightarrow ArI + HX \qquad (4)$$

$$Ar_2S^{\cdot +}\ X^- + R\text{-}H \longrightarrow Ar_2S^+\text{-}H\ X^- + R^{\cdot} \qquad (5)$$

$$Ar_2S^+\text{-}H\ X^- \longrightarrow Ar_2S + HX \qquad (6)$$

$$2Ar_2S^{\cdot +}\ X^- + H_2O \longrightarrow Ar_2SO + Ar_2S + 2HX \qquad (7)$$

Protonic acids such as HBF_4, HPF_6, $HSbF_6$, etc., derived from the photolysis of diaryliodonium and triarylsulfonium salt photoinitiators are well known initiators of cationic polymerization (15).

In addition to the above reactions which generate acids, cationic polymerization can probably be initiated by mechanisms involving direct electrophilic attack of the aryliodinium and diarylsulfinium cation-radicals on the monomer as suggested by Ledwith and shown in equation 8 (16).

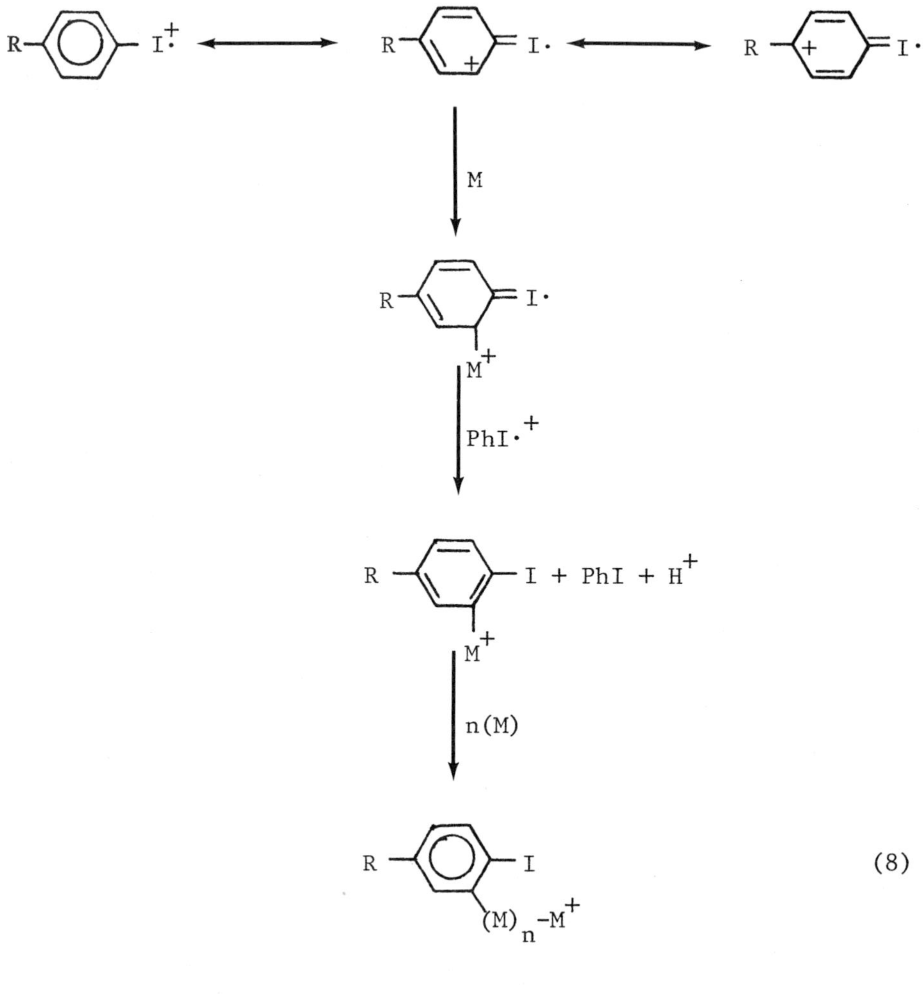

(8)

Attempts to quench the photolysis of diaryliodonium and triarylsulfonium salts using various triplet quenchers have failed, suggesting that cleavage of these compounds occurs from the excited singlet state. The photolysis rate of diaryliodonium salts has been shown to be solvent dependent (17,18). In alcohols and ethers, higher quantum yields have been reported than in acetonitrile. A free radical chain induced decomposition involving participation by the solvent has been proposed.

$$Ar_2I^+ \xrightarrow{h\nu} Ar^{\cdot} + ArI^{\cdot +} \tag{9}$$

$$Ar^{\cdot} + RCH_2OH \longrightarrow ArH + R\dot{C}HOH \tag{10}$$

$$R\dot{C}HOH + Ar_2I^+ \longrightarrow RCHO + H^+ + ArI + Ar^{\cdot} \tag{11}$$

The rate at which various onium salts undergo photolysis is a function of the cationic portion of the molecule. Manipulation of the absorption characteristics of these compounds through the introduction of various substituents can exert a profound effect on their apparent photosensitivity. These effects can be clearly seen in

Table 1. Structure and Spectral Properties of Triarylsulfonium Salts

Cation[a]	$\lambda_{max}(\varepsilon)$[b]
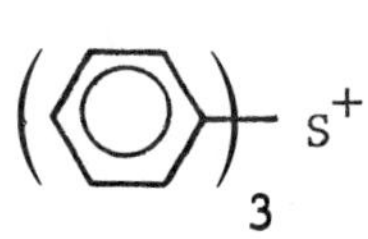	227(21,000)
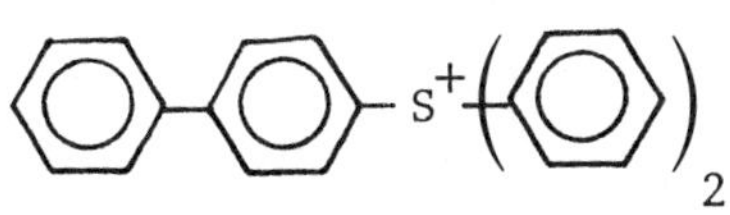	280(25,000)
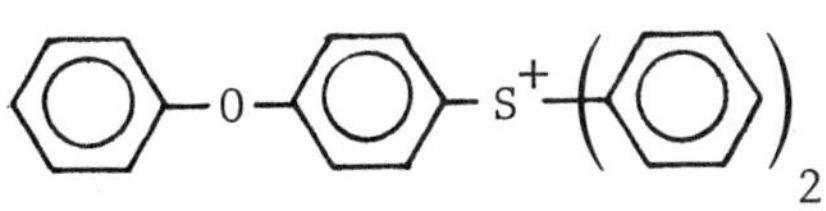	263(14,000)
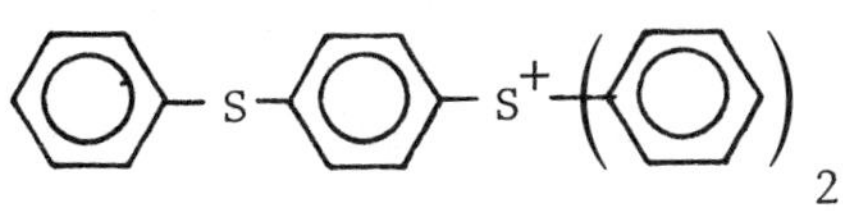	227(22,200) 300(18,000)
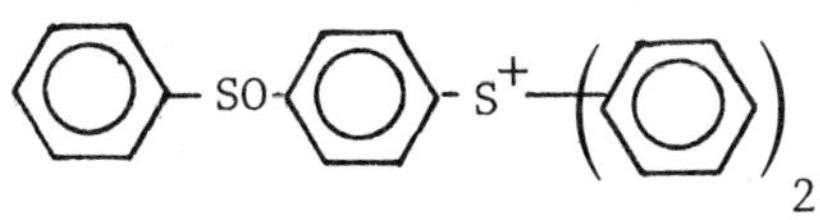	230(22,300) 300(19,300)

[a]All salts contained the AsF_6^- counterion. [b]UV spectra were recorded in methanol.

Table 1 in which resonance extending chromophores have been introduced into the triphenylsulfonium cation to produce a bathochromic shift in the absorption maxima. The compounds in Table 1 bearing these aromatic chromophores are much more efficient photoinitiators than the unsubstituted triphenylsulfonium salt when irradiated at wavelengths longer than 300 nm. The structure of the anions associated with these compounds has no effect on their rates of photolysis.

While the photolysis of diaryliodonium and triarylsulfonium salts has been observed to be independent of temperature, there is a marked dependence of the rate of photolysis on the intensity and wavelength of UV radiation. As the light intensity is increased, the photolysis rate also increases proportionally, indicating a first power relationship between these two factors.

Photosensitization

The spectral response of both diaryliodonium and triarylsulfonium salts can be broadened in both the long wavelength UV and visible regions through the use of photosensitizers. The function of these photosensitizers is to absorb light at the desired wavelengths and then to efficiently transfer that energy to the photoinitiator, causing its photolysis. The photosensitization of diaryliodonium salts is particularly facile, and a large number of condensed ring aromatic hydrocarbons, diaryl ketones (19), and dyes (20) are active photosensitizers. Triarylsulfonium salts are generally less responsive to photosensitization than diaryliodonium salts; however, they can be photosensitized by condensed ring aromatic hydrocarbons (19,21). In both cases, photosensitization appears to occur through an electron transfer mechanism as depicted in equations 12-15. The prime evidence for an electron transfer mechanism consists of a good correlation between the magnitude of the reduction potential of the excited photosensitizer, P*, and the oxidation potential of the onium salt. Further, the formation of the photosensitizer cation-radical, P. X-, has been observed spectroscopically in our laboratories as well as in those of Ledwith (16).

$$P \xrightarrow{h\nu} P^* \qquad (12)$$

$$P^* + Ar_2I^+ X^- \longrightarrow [P \cdots Ar_2I^+ X^-]^* \qquad (13)$$

$$[P \cdots Ar_2I^+ X^-]^* \longrightarrow P^{+\cdot} X^- + Ar_2I\cdot \qquad (14)$$

$$Ar_2I\cdot \longrightarrow Ar\cdot + ArI \qquad (15)$$

Using photosensitizers, the efficiency of photolysis of triarylsulfonium and diaryliodonium salt photoinitiators employing broad-band light sources can be significantly increased. Further increases in efficiency can be achieved through the use of multiple photosensitizers absorbing at different wavelengths. Photosensitization of diaryliodonium and triarylsulfonium salts makes it possible to carry out photoinitiated cationic polymerization efficiently in the long wavelength UV and even in the visible region using daylight.

Photoinitiated Cationic Polymerization

When photoinitiated cationic polymerizations employing diaryliodonium and triarylsulfonium salts are carried out under the usual laboratory "dry" conditions, the chief initiators of polymerization are protonic acids. The general mechanism by which these salts initiate polymerization is shown as follows using triarylsulfonium salts as an example:

Photolysis

$$Ar_3S^+ X^- \xrightarrow{h\nu} Ar_2S + Ar\cdot + HX \qquad (16)$$

Initiation

$$M + HX \longrightarrow HM^+ X^- \qquad (17)$$

Propagation

$$HM^+ X^- + nM \longrightarrow H(M)_nM^+ X^- \qquad (18)$$

The first step involves the photogeneration of the strong acid, HX, by reaction of the photoinitiator fragments. Initiation then proceeds from the direct protonation of the monomer, M, to form a carbon, oxygen, sulfur or nitrogen centered cationic species. In the last step, propagation takes place by the step-wise addition of monomer molecules to the growing cationic chain end.

Although the character of the anion in a cationic photoinitiator generally plays no role in determining its photochemical behavior, it does exert a major influence on the rate and extent of the polymerization of a cationically polymerizable monomer. At one extreme, photoinitiator salts bearing nucleophilic anions such as I^-, Cl^-, Br^-, and $CH_3CO_2^-$ compete so successfully with all cationically polymerizable monomers for the active cationic species that polymerization invariably fails. When anions of the intermediate nucleophilicity such as HSO_4^-, FSO_3^-, NO_3^-, and $CF_3CO_2^-$ are employed, successful polymerization of certain very reactive monomers, e.g., vinyl ethers, is observed while still others fail to polymerize at measurable rates. On the other hand, photoinitiator salts containing such non-nucleophilic anions as ClO_4^-, $CF_3SO_3^-$, BF_4^-, AsF_6^-, PF_6^-, and SbF_6^- are capable of polymerizing virtually all known types of cationically polymerizable monomers.

It has long been recognized in certain polymerizations proceeding by a cationic mechanism that polymerization will continue for very long times in the absence of terminating impurities. Such polymerizations have been termed "living" polymerizations. These effects have also been noted in photoinitiated cationic ring opening polymerizations in general and epoxide polymerizations in particular, and are manifested in the substantial "post cure" which has been observed in these systems. Other factors which affect the rates of the cationic polymerization of a particular substrate in the presence of diaryliodonium and triarylsulfonium salt photoinitiators are the intensity and wavelength of the irradiating light, the temperature, and the absence or presence of inhibitors or chain transfer agents. Appropriate measures must be taken to exclude basic materials in general and to limit the quantity of water and hydroxyl containing compounds present in the monomers.

Applications of Photoinitiated Cationic Polymerization

Photoinitiated cationic polymerizations are especially attractive for coating applications due to the wide breadth of chemical and physical properties which can potentially be realized through the polymerization of such monomeric substrates as electron-rich olefins, lactones, epoxides, cyclic ethers, sulfides, silicones, and acetals as well as numerous other heterocyclic compounds. Further, photoinitiated cationic polymerizations have the important commercial advantage that they are not inhibited by oxygen and thus may be carried out in air without the need for blanketing with an inert atmosphere to achieve rapid and complete polymerization. An additional advantage as mentioned before is the ability of these polymerizations to continue even after they are removed from the light. In practice, only enough light is required to produce sufficient handling characteristics. The final properties of the coating are obtained on standing. The substrates which have received the most attention for these coatings are epoxy resins. Such coatings are known for their excellent solvent and chemical resistance and have found major applications as insulating coatings, varnishes for wood products and protective overprint coatings for beverage cans.

Photopolymerizable epoxies using onium salt photoinitiators also show considerable promise for use in high performance composite applications. Traditional thermally cured epoxy resins are already well entrenched in these applications; however, the use of the recently developed photocurable epoxy materials offers considerable potential for rapid fabrication of fiber-reinforced composites without the need for cumbersome ovens and long cure times. Photopolymerized epoxy laminates and filament wound pipe have already been demonstrated in our laboratory.

Photoinitiated cationic polymerizations have numerous applications involving photoimaging. Epoxy-based photoresists with high resolution derived from commercially available resins have been developed (22) and the use of these same materials in plastic flexographic printing plates has been shown in our laboratory.

III. CONCLUSIONS

Over the past ten years, the development of onium salt and other cationic photoinitiators has moved from the realm of speculative investigation to the point today at which they are being employed in numerous commercial applications. Much work still needs to be done in this field, particularly to improve our understanding of the relationship between the structure and the photosensitivity of these photoinitiators. As the field advances, one can expect still other new classes of onium salt photoinitiators to be developed as well as continued improvements to be made in the efficiency of the present systems. An understanding of the mechanism of photosensitization should lead to discovery of more efficient photosensitizers and a further broadening of their spectral response in photoinitiated cationic polymerization.

Lastly, the process and polymers produced by onium salt photoinitiated cationic polymerizations are eminently useful. Like the corresponding photoinitiated radical polymerizations which they complement, these cationic systems will find a wide range of applications where polymerization speed and economy of energy utilization are of prime concern.

REFERENCES

1. J.V. Crivello and J.H.W. Lam, J. Polym. Sci., Symp. No. 56, 1-11 (1976).

2. J.V. Crivello and J.H.W. Lam, Macromolecules, 10, 1307 (1977).

3. J.V. Crivello, U.S. Patent 3,981,897, September 21, 1976.

4. G.H. Smith, Belg. Patent 828,841, November 7, 1975.

5. J. Nemcek, Belg. Patent 837,782, June 22, 1976.

6. J.V. Crivello and J.H.W. Lam, J. Polym. Sci., Polym. Chem. Ed., 17, 977 (1979); U.S. Patent 4,058,401, November 15, 1977.

7. J.V. Crivello and J.H.W. Lam, J. Rad. Curing, 5 (1), 2 (1978).

8. G.H. Smith, U.S. Patent 4,069,054, January 17, 1978.

9. J. Nemcek, Belg. Patent 833,372, March 16, 1976.

10. J.V. Crivello and J.H.W. Lam, J. Polym. Sci., Polym. Chem. Ed., 17, 1049 (1979).

11. J.V. Crivello and J.H.W. Lam, J. Polym. Sci., Polym. Chem. Ed., 17, 2877 (1979).

12. J.V. Crivello and J.H.W. Lam, J. Polym. Sci., Polym. Chem. Ed., 18, 1021 (1980).

13. S.P. Pappas, L.R. Gatechair, B.C. Pappas and W. Schnabel, Proc. Soc. Photogr. Sci. and Eng., 22nd Fall Symp., Washington, D.C., Nov. 15-18, 1982, p. 46.

14. J.V. Crivello and J.L. Lee, unpublished results.

15. J.P. Kennedy and E. Marechal, Carbocationic Polymerization, John Wiley & Sons, New York, 1982, p. 56.

16. A. Ledwith, Polymer Preprints, 23 (1), 323 (1982).

17. A. Ledwith, Makromol. Chem. Suppl., 3, 348 (1979).

18. L.R. Gatechair and S.P. Pappas, Org. Ctgs. & Appl. Polym. Sci. Proc., 46, 701 (1982).

19. S.P. Pappas and J.H. Jilek, Photogr. Sci. and Eng., 23 (3), 140 (1979).

20. J.V. Crivello and J.H.W. Lam, J. Polym. Sci., Polym. Chem. Ed., 16, 2441 (1977).

21. J.V. Crivello and J.H.W. Lam, J. Polym. Sci., Polym. Chem. Ed., 17, 1059 (1978).

22. S.J. Schlessinger, Photogro. Sci. and Eng., 18 (4), 385 (1974).

Polymer and Formulation Design Characteristics for Developing Bonding Capabilities of Radiation-Curable Coatings and Adhesive Systems

Vincent D. McGinniss

Battelle's Columbus Laboratories
505 King Avenue
Columbus, OH 43201

ABSTRACT

Development of adhesive bonding capabilities for rapid high energy electron or photoinduced polymerization reactions involving free radical intermediates requires knowledge of the following concepts.

1) Nature of the substrate/substrate preparation.

2) Monomer design/surface tension, solubility parameters, viscosity, polar-nonpolar, shrinkage during polymerization.

3) Cross-linking oligomer design/MW between cross-link sites or number of reactive sites, polarity, viscosity, shrinkage factor.

4) Polymer design/reactive-nonreactive T_g, MW, MW between cross-link sites, functionality, surface energies.

In this paper, several of the above concepts and their interrelationships directed towards achieving adhesive bonding of radiation curable coatings/adhesives to a wide variety of substrates will be discussed.

INTRODUCTION

Radiation curable coating and adhesive systems involve complex interactions of a large number of variables in order to achieve final film properties useful for commercial applications.

The development of satisfactory adhesive bonding capabilities for rapid high-energy electron or photo-induced polymerization reactions involving free radical intermediates is of prime importance for commercial utility and requires knowledge of the following concepts:

Polymer Design / reactive-nonreactive, T_g, MW, MW between cross-link sites, ionic functionality.

Monomer, Cross-linking Oligomer Design / number of reactive sites, surface tension values, solubility parameters, viscosity, polar-nonpolar, shrinkage during polymerization.

Formulation Guidelines and Constraints / viscosity, cross-link density.

Nature of the Substrate and Substrate Preparation / surface energies, solubility parameters (plastic substrates).

Each of the above concepts, as well as other factors such as energy requirements for cure and post-cure effects, are major contributors toward developing radiation curable coating systems that have bonding capabilities to a wide variety of substrate materials. The major emphasis of this paper will center on polymer design for improved adhesion.

POLYMER DESIGN IN RADIATION CURABLE COATINGS FOR IMPROVED ADHESION

A novel concept has been found that allows radiation curable coating systems to develop excellent adhesion to metal surfaces. This concept utilizes certain specially designed high T_g (glass transition temperature) polymeric additive materials in the formulation which is then cured by light energy or high-energy electron processing equipment. The only requirement is that the polymeric additives must have glass transition temperatures (T_g) above the ambient application and reaction zone curing temperatures in order to affect adhesive bonding capability in the radiation curing coating system. Another design criteria is that if the high T_g polymer contains unsaturation sites then these unsaturation sites should be placed between certain specific critical distances (effective chain length for adhesion) in order to achieve optimum bonding capability (Fig. 1) (1,2).

EXPERIMENTAL

The coating compositions used in this study consisted of 10-40% of a high T_g polymeric additive; 60-90% unsaturated acrylic monomer-multifunctional acrylate oligomer mixture; 1-10% of a photoinitiator when the coating system was cured through the use of ultraviolet light energy processing equipment. All adhesion promoting polymeric additives were prepared by conventional polymer synthesis techniques.

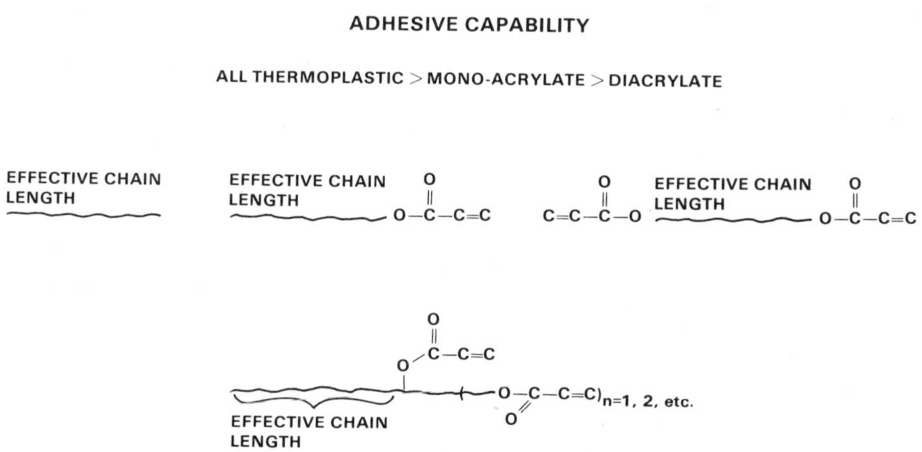

Fig. 1. Adhesive Bonding Capabilities of Radiation Curable Coating Systems Containing Polymer Additives as a Function of Effective Chain Length and Specific Placement of Unsaturation Sites

Coatings were applied to flat sheet metal substrates (cold rolled steel) and cured with ultraviolet light. After application, the coated metal was placed onto a moving conveyer line and exposed to U.V. radiation from an Ashdee Unit source (2 lamps of 2,000 Watts each) placed 5 to 7 inches from the coating. Exposure time was 9 seconds. (Line speeds of 50 to 100 ft. per minute.) The coating compositions were also applied onto a metal sheet substrate and subjected to electron beam from an Energy Science Electrocurtain (150 KV terminal voltage and 3.5×10^{-6} torr vacuum). Each metal sheet was mounted on a moving line of speeds varying from 10 to 40 ft. per minute. The dosage was 10 to 3 Mrads at 10 Mamps current.

Adhesion tests vary considerably depending on the type of coating being tested. By and large, the cellophane tape test is quite common, wherein a piece of adhesive cellophane tape is pressed on an area of the coating which had been scratched, scored or otherwise cut with a sharp edge.

RESULTS AND DISCUSSION

An ethylenically unsaturated binder coating was prepared from the following components: 23 weight parts pentaerythritol triacrylate (PETA), 19 weight parts phenyl Cellosolve acrylate (PCA); 1 weight part photosensitizer (2-chlorothioxanthone) and 3 weight parts methyl diethanolamine. (Photosensitizer is optional when used with high energy electron processing equipment.) To this binder composition was added approximately 16 weight parts of a polymeric additive as described in Tables 1 and 2. This binder system was then further pigmented with 38 parts of rutile TiO_2.

In Table 1, a wide variety of polymer systems were examined for their ability to promote adhesive bonding of a radiation curable coating formulation to a metal substrate. In all cases the high T_g values (greater than room temperature and 105°C) exhibited 100% adhesion while those polymeric additives having low T_g values (-55 and -5C°) did not exhibit adhesive bonding capabilities. Similar results were observed for a series of specially designed saturated and unsaturated high T_g polymer systems (Table 2). In

Table 1. Effect of T_g of Saturated Polymeric Additives on Adhesion of Radiation Curable Coating Formulations

Polymer	T_g (C^o)	Adhesion (%)
Monsanto RJ100 (styrene/allyl alcohol copolymer)	greater than room temperature	100
Polymethlymethacrylate	105	100
Polybutyl acrylate	-55	0
DOW epoxy resin DER 664	greater than room temperature	100
Polyester (4 moles hexa-hydrophthalic anhydride, 0.8 moles adipic acid, 5 moles cyclohexane di-methanol, 0.8 moles neopentyl glycol)	-5	0

Table 2. Effect of Polymeric Chain Length and Placement of Unsaturation Sites on Adhesive Bonding Capability For Radiation Curable Coating Formulations

Polymer/Additive	Chain Length (Å)	T_g (C^o)	Adhesion (%)
Tetrabenzoate ester of penta erythritol	30	greater than room temper-ature	0
Dow epoxy resin-DER 661 (diglycidyl ether of bisphenol-A)	40-60	ditto	100
Monsanto RJ100 (styrene allyl alcohol copolymer)	40-60	"	100
Monoacrylate of DER 661	40-60	"	100
Diacrylate of DER 661	40-60	"	30

certain cases where the effective chain length for adhesion is less than 30Å adhesive bonding of the coating to the substrate is not achieved. It is only possible to achieve bonding when the effective chain length of the high T_g polymeric additive is greater than 40Å (Fig. 2). Placement of unsaturation sites is also another critical factor in that both the non-acrylated and monoacrylate adduct of DER 661 (digylcidyl ether of bisphenol-A) demonstrated excellent adhesion capabilities but the diacrylate adduct of DER 661 showed only 30% adhesion (Table 2 and Fig 1).

The ability of the T_g value of a polymer additive to control adhesion of a radiation curable coating system is thought to be due to the fact that the high T_g polymer is compatible and soluble within the resinous binder prior to curing but becomes incompatible during the curing process. This incompatibility of the polymeric additive forces it to partition between the metal and cured coating network thus promoting physical adhesion to the metal surface due to the solid high T_g nature of the polymer reverting back to a solid state after curing. In contrast, low T_g polymers which are

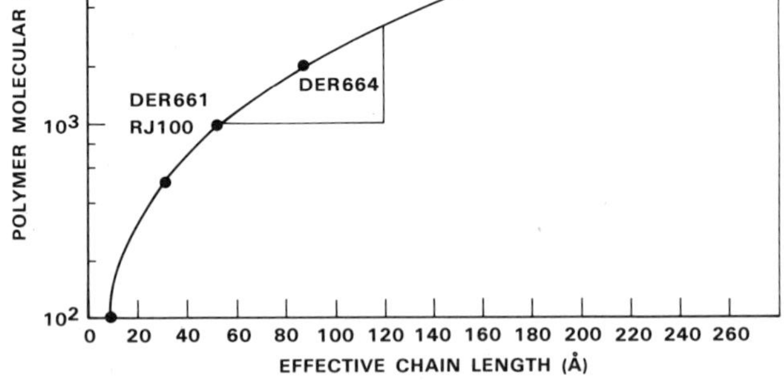

Fig. 2. Relationship Between Molecular Weight and Chain Length (Å) for Selected High T_g Polymer Additives

liquid or mobile at the ambient curing temperatures are encapsulated within the cross-linked binder matrix or are fluid at the metal-cured coating interface and do not provide mechanical adhesion to the metal substrate.

The practical aspect of the high T_g polymeric additive concept for adhesion becomes apparent when one considers certain temperature effects associated with a multi-lamp UV-radiation curable coating processing unit. If a radiation curing reaction zone temperature is 30°C or greater, then UV-curable coating formulations containing high T_g polymeric additives of about 40 to 100°C will demonstrate excellent adhesion to certain metal substrates when tested for adhesion immediately after the curing process. If the T_g of the polymeric additives are below 30°C, then adhesion will fall off correspondingly (Fig. 3). By lowering the reaction curing zone temperature adhesion can be developed with low T_g polymeric additives and by raising

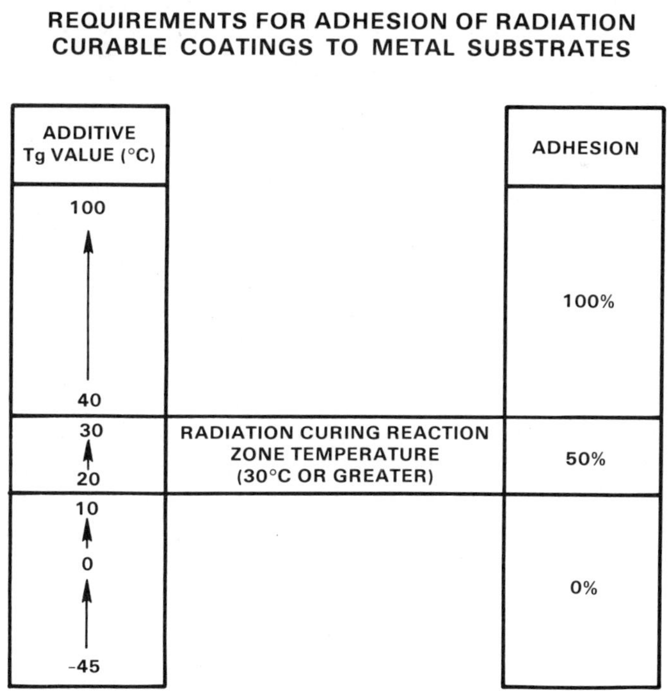

Fig. 3. Effect of Radiation Curing Reaction Zone Temperature on Polymer T_g and Adhesion

the reaction zone temperature the adhesion can be reduced with high T_g additives.

Identical results were obtained when the radiation curable coating formulation was cured on an Electrocurtain processing unit in that only the high T_g polymeric additives in the formulation caused adhesive bonding.

Monomer/Oligomer and Formulation Design Criteria

The coating systems used in radiation-curable technologies are a balance of three major variables; single vinyl unsaturated monomers (M), multifunctional unsaturated oligomers or cross-linking agents (C), and saturated or unsaturated polymer components (P). The primary function of the single vinyl unsaturated monomer component in the formulation is to reduce the viscosity of the coating system (Fig. 4) but the choice of monomer can strongly influence coating adhesion, surface energy of the formulation and the kinetic cure-rate response capability for the entire system. Some typical physical properties of common reactive monomers that can be used in radiation curable adhesive formulations are shown in Table 3.

In many situations it is not possible to readily determine certain physical properties of liquid or polymeric systems, thus a simple predictive model would be useful especially in a relatively new technology involving reactive solvent chemistry. A novel set of mathematical predictive relationships can be used to correlate and predict various physical properties of both liquid and polymeric materials (3,4). These simple predictive relationships for solvents and polymers have variables which are easily determined such as refractive index and molecular structural composition of the solvent or polymer. Application of these variables leads to a unique set of linear equations that take the general form:

Response =

$$C_1 n \pm C_2 \chi \pm C_3 Z' \pm C_4 n\chi \pm C_5 nZ' \pm C_6 \chi Z' \pm 1 \qquad (1)$$

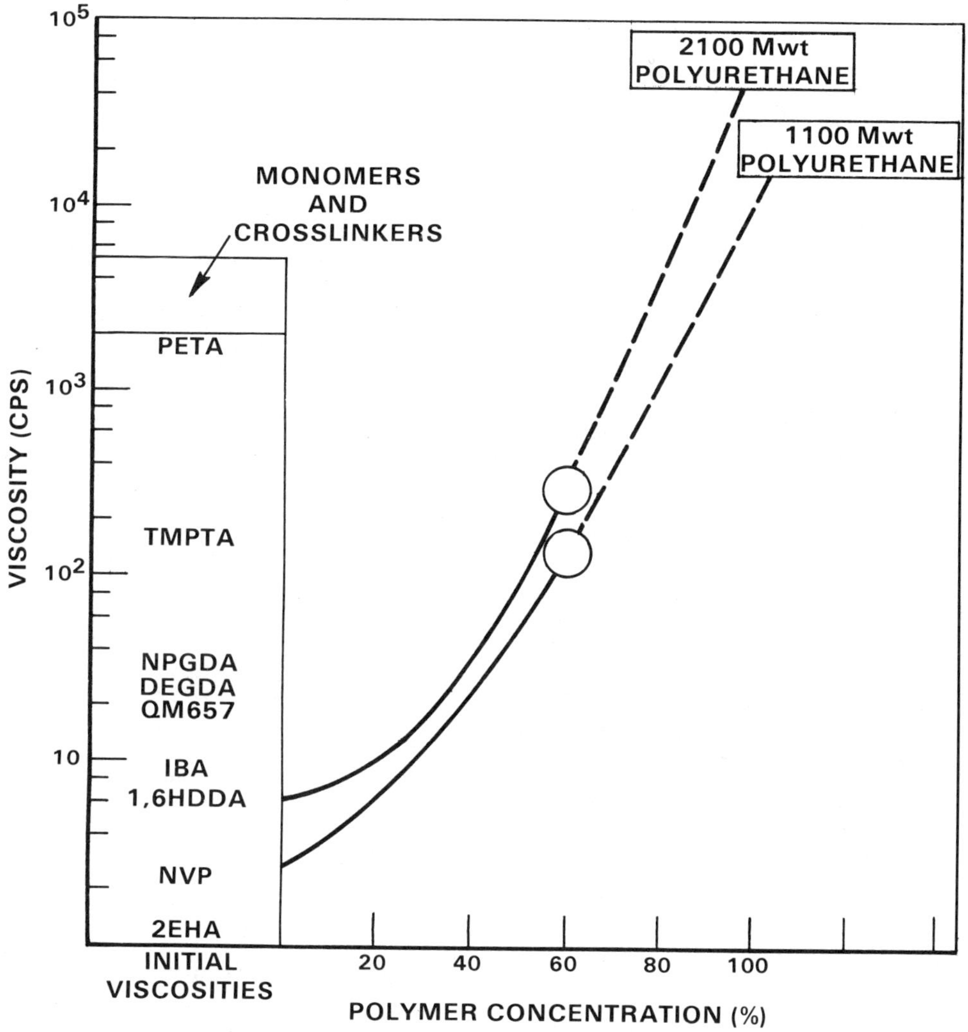

Fig. 4. Monomer/Cross-linker Viscosities and Solution Viscosities of Urethane Polymers in 1,6HDDA or NVP as a Function of Polymer Concentration (PETA - pentaerythritol triacrylate, TMPTA - trimethylol propane triacrylate, NPGDG - neopentyl glycol diacrylate, DEGDA - diethylene glycol diacrylate, QM657 - Rohm and Haas - dicyclo- pentadiene modified acrylate, IBA - isobornyl acrylate, 1,6HDDA - 1,6 hexanediol diacrylate, NVP - N-vinylpyrrolidone, 2EHA-2-ethylhexyl acrylate)

where

C_1-C_6 are the multiple linear regression coefficients.

n is the refractive index of the molecule or polymer.

χ is the weight or volume fraction of heteroatoms contained in the molecule or monomer repeat unit of the polymer.

Z' is the weight or volume fraction of electron density contained in the molecule or monomer repeat unit of the polymer.

1 is the regression equation intercept value, and $n\chi$, nZ', $\chi Z'$ are interactive terms.

The specific equations used to calculate the physical properties of reactive monomers in Table 3 are as follows:

$$\text{(dynes/cm)} = 167.49n + 35.73\chi - 180.06Z' - 218.84 \quad (2)$$

$$T = 53.70n + 18.67\chi - 162.50Z' - 71.73 \quad (3)$$

$$T = 0.19\gamma + 7.6\chi + 1.79 \quad (4)$$

Shrinkage of the reactive coating formulation is also of concern in developing radiation curable coating or adhesive systems (1).

$$\text{Monomer Shrinkage} = \frac{P_{density} - M_{density}}{P_{density}} \quad (5)$$

(P: polymer; M: monomer)

The addition of certain high T_g polymeric additives tend to reduce shrinkage and actually undergo phase separation in a similar manner as observed in styrene-unsaturated polyester low-profile molding resin technologies (5).

An idealized formulation guideline showing possible interactions of single vinyl unsaturated monomers, multifunctional vinyl cross-linking molecules and polymeric materials is represented by Fig. 5.

Table 3. Physical Properties of Reactive Monomers

	IDA	1,6HDDA	NPGDA	DCPA	PETA	TMPTA
Molecular weight	212.33	226	212.24	204.26	298	338
Density	0.8848	1.01	1.0104	1.0661	1.203	1.10
Refractive index	1.4416	1.4553	1.4522	1.5077	1.4866	1.4700
δ_T	8.24 (8.00)	9.8(10)	8.26(8.6)	9.18(9.5)	11.08(12)	-(9.2)
Viscosity (cps)	3	5-8	30	12	600-1800	30-150
γ(dynes/cm)	27(26.2)	35(31)	31.5(31.5)	35(35.7)	41.2(41.7)	36(34)
"McGinniss Equation" Parameters X	0.151	0.28	0.30	0.157	0.376	0.28
"McGinniss Equation" Parameters Z'	0.01	0.02	0.02	.01	0.02	

values in ( ) were calculated using equations 2, 3 and 4.

IDA - isodecylacrylate
1,6HDDA - 1,6-hexanedioldiacrylate
NPGDA - neopentylglycol diacrylate
DCPA - dicyclopentenyldiacrylate
PETA - pentaerythritoltriacrylate
TMPTA - trimethylolpropane triacrylate

Substrate/Substrate Preparation

Radiation curable coatings are almost exclusively applied to four general substrate classes 1) metals, 2) plastics, 3) cellulosics (wood, paper) and 4) glass. For adhesive bonding to occur between the radiation curable coating and the substrate it is important that the coating wets the surface of the substrate so that optimum chemical or physical bonding interactions can occur. Radiation curable coatings applied to metal substrates offers a special problem in that the cold application and rapid cure cycles inhibits wetting and spreading of the coating thus strongly influencing the possibility of forming intimate contact or adhesive bonding capabilities. The radiation curable coating systems generally are of a polar nature and have surface tension (γ) ranges between between 25 and 40 dynes/cm. Some of the common metal substrates used in the coatings industry have oil surfaces (dioctyl sebacate γ= 34 dynes/cm) or are clean and range in surface tension from 30 to 60 dynes/cm (Table 4).

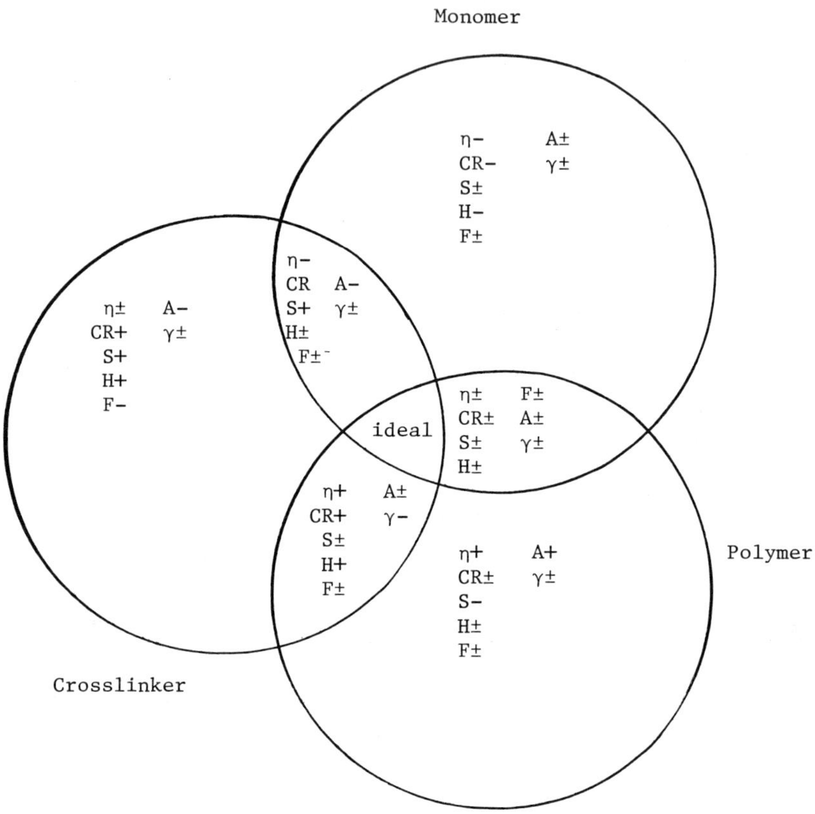

Viscosity	η	legend
Cure Rate	CR	
Shrinkage	S	
Hardness	H	
Flexibility	F	
Adhesion	A	
Surface Energy	γ	
Increase	+	
Decrease	−	

Fig. 5. Generalized Formulation Design Outline for Radiation Curable Coatings and Adhesive Systems

Table 4. Typical Surface Energies of Common Metals Used in the Coatings Industry

Substrate	γ (dynes/cm)
Tin plate	28-31.53
Tin-free steel	35.65
Bonderite 721	38.81
Aluminum	45
Cold-rolled steel	50-65

The ability of a radiation curable coating to adhere to a metal substrate is related to the surface energy values for the substrate in that adhesion is generally better and easier to obtain with higher energy surfaces such as rolled steel rather than lower surface energy substrates such as tin plate or tin free steel. Both tin plates and tin free steel are usually covered with a protective oil (DOS 34 dynes/cm) which is the reason for their low values in surface energy.

Surface preparation or pretreatment processes can have a very beneficial effect on developing improved bonding capabilities for radiation curable coatings on metal surfaces (Table 5).

Table 5. Effect of Surface Pretreatment on Adhesion of Radiation Curable Coatings to Metal Substrates

Substrate	Treatment	Surface Energy (dynes/cm)	Adhesion
Tin plate	none	28-31	Poor
Tin plate	1 min at 480°C	35-36	Excellent
Tin-free steel	none	35-36	Poor
Tin-free steel	6 min at 450°F	50	Excellent
Tin plate	none	29	Poor
Tin plate	polished	34-35	Excellent

Table 6. Radiation Curable Coating Formulation Concepts for Development of Adhesive Bonding to Various Classes of Substrate Materials

Substrate	Adhesive Bonding Components to be Contained in the Radiation Curable Coating Formulation
Aromatic	Aromatic monomers/polyesters
Vinyl substrate	N-vinyl pyrrolidone
Cellulose	Amine/acid/hydroxyl monomers
Glass	Silane/titanate coupling agents

Substrates other than metals (plastics, wood, paper, glass) all require individual consideration in developing bonding capabilities to radiation curable coating systems. Some general formulation considerations for radiation curable coatings in relationship to bonding to different substrates are outlined in Table 6.

SUMMARY

In this paper we have tried to examine several important variables associated with developing adhesive bonding of radiation curable coating and adhesive systems to different substrate materials. Specific adhesion development for a radiation (UV or EB) curable coating formulation was obtained on pretreated tin-free steel (baked for surface oil removal) or untreated cold-rolled steel substrates whenever a high T_g polymeric additive was present in the formulation. Other factors to consider in analyzing final performance parameters of the cured coating or adhesive are the physical properties associated with the individual liquid and polymeric components and their relationship to one another in the formulation before, during and after this curing process. A series of simple mathematical models have been proposed which enable one to readily calculate solubility parameters (δ_T), density, shrinkage and surface tension values (γ) for a wide variety of liquid and polymeric materials useful in radiation curable coating and adhesive technologies.

REFERENCES

1. V.D. McGinniss and A. Kah, Polym. Eng. and Sci., 17, 498 (1977).

2. V.D. McGinniss, V.W. Ting, and A.F. Kah, U.S. Patent 4,163,809 (1979).

3. V.D. McGinniss, Preprints of Organic Coatings and Plastics Chemistry, 39, 529 (1978).

4. V.D. McGinniss, Preprints of Organic Coatings and Applied Polymer Science, 46, 214 (1982).

5. E.J. Bartkus and C.H. Kroekel, Polyblends and Composites, B.F. Bruino, ed., Applied Polymer Symposia, No. 15, 113, Interscience, N.Y. (1970).

Dynamic Thermal Analysis Characterizations of Electron-Beam Cured Adhesives

F.J. Campbell

Naval Research Laboratory
Washington, DC 20375

ABSTRACT

Continuous scans of modulus versus temperature utilizing the DuPont Dynamic Mechanical Analyzer (DMA) has provided a comparison of the high temperature service capabilities of radiation-cured experimental formulations of a vinyl-modified epoxy resin, Shell Epocryl-12. These scans were compared to data obtained when the same materials were applied as adhesives on aluminum test panels, radiation-cured with an electron beam, and lap shear strength tested at discrete temperatures. The DMA instrument utilizes a thin rectangular specimen for the analysis, so specimens can be cut from blocks or from flat sheets. In this case the specimens were cured as sheets of resin-saturated graphite-fibers. The same order of high temperature stability was obtained by each method. However, the DMA method provided a more complete characterization of temperature performance in a much shorter test time and thus, it can be very useful for quick analyses of formulation and processing variables in many types of materials optimization studies. The paper will present details of this study with illustrations of the comparisons.

INTRODUCTION

This study was conducted as one of the approaches investigated in a larger Navy program to develop advanced processing technologies and characterize high performance materials leading to improved performance capabilities of future V/STOL aircraft (1).

There is a concerted effort being made in the Navy to increase the utilization of adhesive bonded joints to replace rivets and welds for joining sheets and structural supports in aircraft, missiles and high-performance ships. Light-weight fiber-reinforced plastic composites are also replacing metals in many places on these vehicles. The results are lighter weight, smoother skin surfaces, higher fatigue resistance and lower fabrication costs. The emerging technology of curing these high-performance adhesive and composite resins with electron-beam radiation energy, which can be performed at ambient temperature, offers additional advantages over conventional processing methods requiring autoclaving and heat-cure cycling. One of these advantages is that the many materials used in these complex structures can be joined together without sustaining the residual heat-shrink stresses that are inherent when bonds formed at high temperature are cooled to room temperature. These internal stresses will weaken the fatigue endurance and increase the susceptibility to stress corrosion of metallic adherends at the bondline. From the processing standpoint, the cure can be completed in a much shorter time by electron-beam equipment, and the amount of energy expended to accomplish it is much less. Also, radiation-curable resins do not contain solvents that must be removed in the processing, since the low-viscosity components of the formulation contribute to the cross-linking reaction.

During the course of this program, experimental studies were conducted to identify a number of polymer types in formulations which can be cured via radiation for adhesive applications. Performance studies were also made to determine endurance to exposures of various environmental conditions including moisture, solvents and high temperature aging. Data have been developed on a number of good candidate materials from both commercially available products and laboratory experimental formulations in which maximized properties were achieved by developing the optimum formulation and by controlling the irradiation conditions. These polymer types include various compositions of polyesters, vinyl esters, acrylics, and modified polyimides (2,3).

Electron-Beam Curing

Electron-beam curing has become a well established process in other industries such as protective coatings, wire insulation and heat shrinkable tubing so that industrial equipment is available that can readily be adapted to component assembly bonding, and the technology is available for assembling larger structures. Since material penetration depth is a function of its density and the energy of the incident electrons, the emphasis is on the accelerators with higher energy capabilities in order to reach the bonding interfaces of the assemblies. For example, continuous process accelerators with 4 MeV electron beams are commercially available for high throughput curing. This energy will effectively penetrate and cure to a depth of 10 mm in graphite/epoxy composites with a density of 1.5 g/cm^3. For reaching greater depths it will be feasible to utilize a portable linac, such as are being used in X-ray radiographic inspections. These are available with electron energies up to 15 MeV, having practical ranges of 29 mm in aluminum and approximately 52 mm in composites. Linacs with remotely controlled suspension systems can be pointed to designated spots or made to scan a bondline in the assembling or repairing of aircraft or ship components (3).

Dynamic Mechanical Analyzer

The measurements utilized in this study were obtained on a modified DuPont 980 Dynamic Mechanical Analyzer (DMA), coupled to a DuPont 990 Programmer/Recorder (4). With this patented system it is possible to obtain Young's modulus as a function of temperature of a wide variety of viscoelastic materials. Modulus is related to the resonant frequency by the sample geometry as the material is physically deformed by a flexing stress and then released. When stress is removed, the deformation energy is converted to mechanical oscillation at its resonant frequency. Energy dissipation, or damping, is a function of the restoring force needed to drive the system at a constant oscillation amplitude as a portion of the energy is dissipated in other forms such as heat.

The heart of this system is a pair of parallel, balanced sample support arms which oscillate freely around flexture pivots. Designed for low friction and precise balance, the natural frequency of the sample support system is less than 3 Hz, minimizing system contributions to damping. A schematic of this device is shown in Fig. 1. To make a measurement, a material of known dimensions is clamped between the two sample arms. The sample-arm-pivot system is oscillated at its resonant frequency by an electro-mechanical transducer. Frequency and amplitude of this oscillation are detected by a linear variable differential transformer (LVDT) positioned at the opposite end of the active arm. The LVDT provides a signal to an electromechanical transducer, which in turn keeps the sample oscillating at constant amplitude.

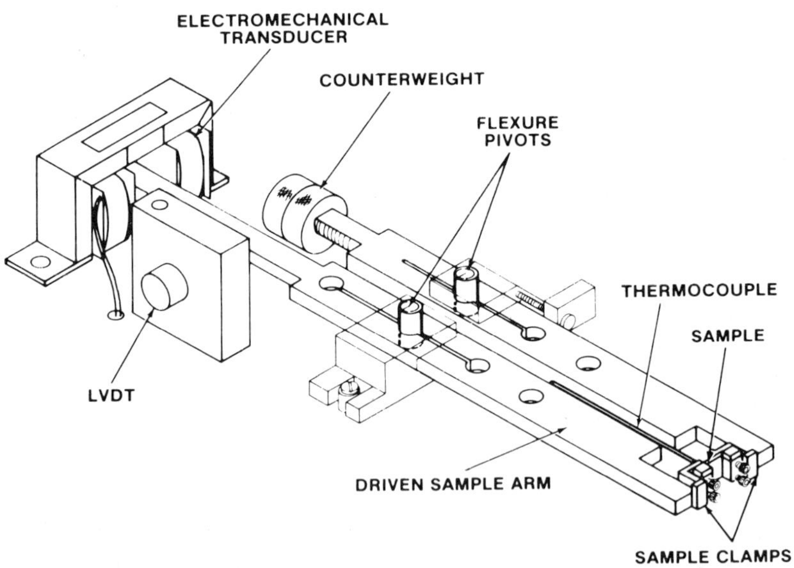

Fig.1. A sketch of the main components of the DuPont 980 DMA showing how the composite sample is clamped into position between the parallel arms.

Sample resonant frequency, and damping, are digitally displayed on the front panel of the 980. The same resonant frequency and damping signals are supplied to the 990 Programmer/Recorder where they are graphically recorded as a function of the measured sample temperature or time. Young's modulus for the sample can be obtained from resonant frequency by using the relationship

$$E = \frac{(4\pi^2 f^2 J - K)}{2W\left[L/2 + D\right]^2}\left[L/T\right]^3$$

where E is Young's modulus in pascals (Pa); f is the DMA oscillating frequency in hertz (Hz); J is the moment of inertia of the arm in kg-m^2; K is the spring constant of the pivot in Nm/rad; D is the clamping distance in meters; W is the sample width in meters; T is sample thickness in meters and L is sample length in meters.

EXPERIMENTAL PROGRAM

A commercially available vinyl ester was selected as the neat resin for a comprehensive study of the performance variations that can be achieved by variations in formulation and in the radiation dose required to cure them. In an initial study the adhesive materials were fabricated into lap shear specimens and tested at selected high temperatures in order to obtain an estimation of the limiting temperature stabilities and a comparison of high temperature strengths of the various formulations. At a later time the same formulations were fabricated into graphite-fiber supported composite sheets, and modulus versus temperature scans were performed by the DMA method. This method has provided a continuous curve over the entire temperature range of interest that gives not only a temperature stability comparison but also the exact limiting service temperature values for each radiation cured resin formulation.

Adhesive Formulations

The base resin used in the formulations of this study was Epocryl-12, made by Shell Chemical Company. It is typical of the vinyl ester resins produced by the reaction of an acrylic monomer with a bisphenol-A/epichlorohydrin epoxy which proceeds as follows:

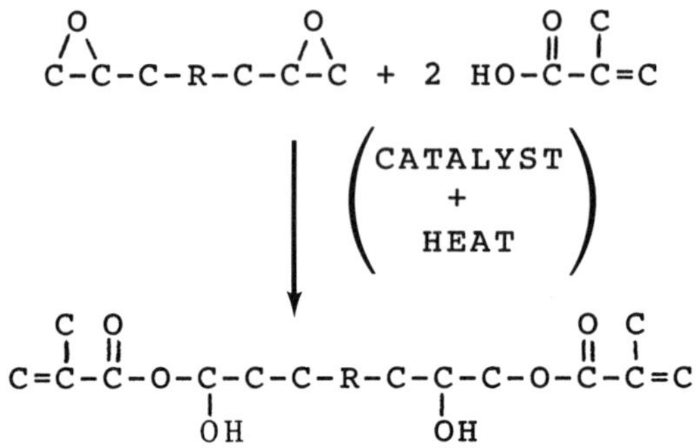

AND WHEN R IS THE BISPHENOL-A GROUP, WE HAVE

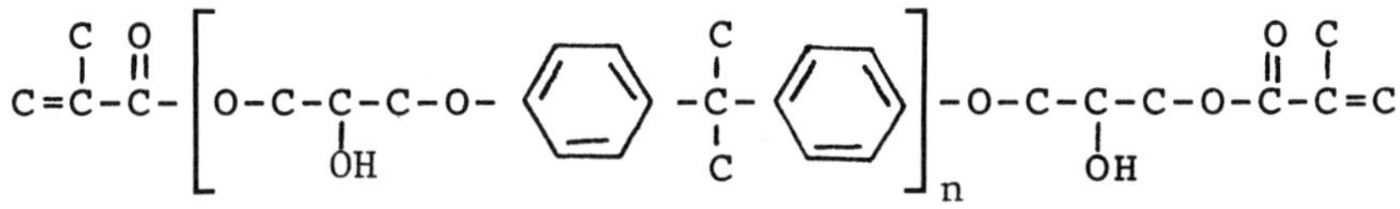

Depending on the size of n, the resulting linear molecules are either very high viscosity liquids or solids at room temperature. This makes it difficult or impossible to spread the resin over the areas to be bonded, and therefore, in this form, use as a room temperature assembled, radiation-curable adhesive is not possible without a reactive diluent. It is also desirable for some applications to enhance the reactivity and the degree of cross-linking by the selection of the diluent. Therefore, by formulating with vinyl functional monomers which are fluid at, or near, room temperature, solutions can be obtained with viscosities that are appropriate for adhesive applications and with more unsaturated sites for reaching higher cross-link densities than the base Epocryl resin itself. This resin was therefore modified with three different vinyl-functional monomers (wt ratio: 100 pt, Epocryl 12/30 pt monomer) to prepare the adhesive compositions which were used in the radiation curing of aluminum lap shear test assemblies.

The three monomers selected for these experiments were divinyl benzene (DVB), triallyl cyanurate (TAC) and styrene (STY). The addition of such monomers to Epocryl 12 resin makes possible not only viscosity adjustments for greater processing ease, but also the modification of such important properties of the cured resin as degree and type of cross-linking, elevated temperature performance, solvent resistance and physical strength. It was considered that the divinyl benzene (two double bonds per molecule) and triallyl cyanurate (three double bonds per molecule) modifications of Epocryl 12 would result in a higher level of cross-linking and provide superior ambient and elevated temperature strength properties compared to the styrene monomer (one double bond per molecule) blend. The structural formulas for each of these monomers are as follows:

STYRENE

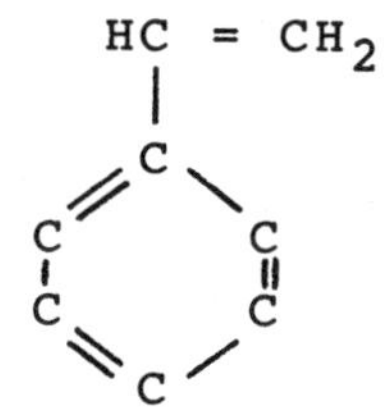

DIVINYL BENZENE

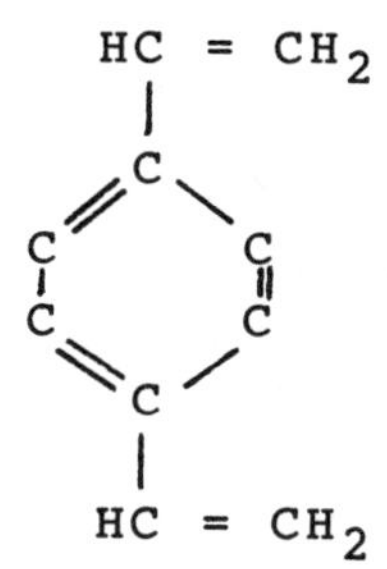

TRIALLYL CYANURATE

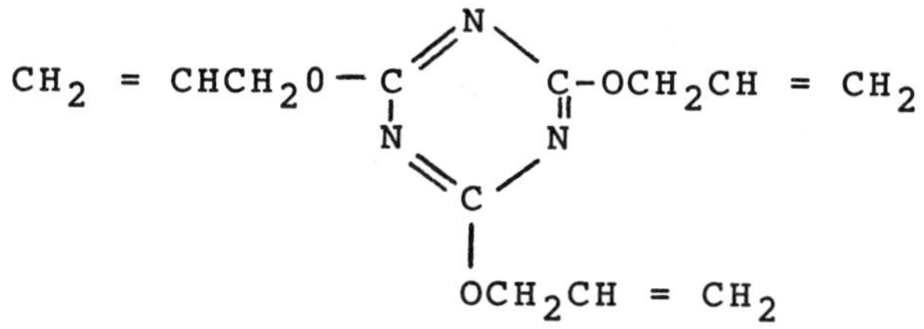

Lap Shear Strength Tests

Formulations of each of these monomers with the Epocryl 12 resin were prepared and evaluated as adhesives by bonding aluminum to aluminum in single lap shear tests. Strips of 7075-T6 aluminum alloy were prepared for bonding by the standard FPL acid-dichromate etching procedure. To assemble the specimens, a few drops of the adhesive composition were distributed on the bonding area of one aluminum strip and a second aluminum strip was positioned above the first so as to produce a 25 mm x 25 mm square lap joint. The assemblies were then tightly wrapped with glass-fiber reinforced strapping tape to maintain their configuration during handling and curing. No additional pressure was necessary during the curing process. The lap shear strength was then determined by the force required to break the adhesive bond in tensile shear with an Instron Tester at a jaw-separation rate of 1.25 mm per minute. An initial study was conducted to determine the effect of radiation dose on bond strengths of the different formulations. These specimens were prepared and electron-beam irradiated at room temperature to nominal doses of 75, 100, 200, and 300 kilograys (kGy). The test results at room temperature, 115°C, and 150°C showed that the values of lap shear strength reached a maximum at a dose of 100 kGy and then were dose-independent over the 100 to 300 kGy range. Therefore, the 100 kGy dose was adopted as the optimum and was used exclusively throughout the high-temperature test series which followed.

The lap shear strengths of the irradiated aluminum assemblies were compared with those values obtained with specimens conventionally cured with heat and pressure. The same vinyl ester resin formulations were employed, but 0.5 wt percent benzoyl peroxide was added to each composition to catalyze the thermal cure. These assemblies were cured in accordance with the manufacturer's recommendation of 2 hours at 80°C under pressure in a laminating press.

Both the irradiated and the conventionally heat-cured sets of specimens were subjected to the same test procedures. Lap shear strengths were tested at room temperature (25°C), 115°C, 150°C, 200°C, and 260°C. The high-temperature lap shear strength properties were measured at a given temperature after a two-hour heat soak at that temperature. The data from these tests are plotted in Fig. 2 (5). Each point was averaged from five individual specimens.

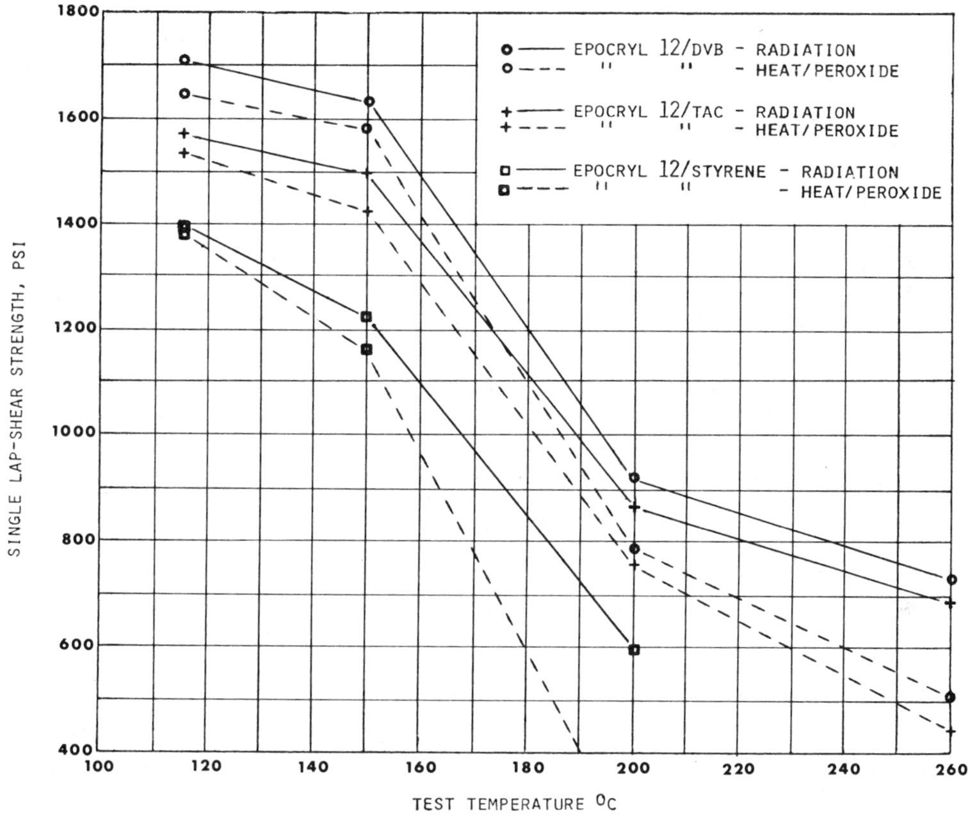

Fig. 2. Radiation-cured versus heat-cured formulations of Epocryl 12 (Shell). Comparisons of single lap shear strength at elevated temperatures.

The curves of each formulation on the graph of lap shear strength versus test temperature provides a visual analysis of the experimental results. In each of the three basic formulations the radiation cured specimens have greater strength than their peroxide/heat-cured counterpart at each test temperature. This comparison also shows that heat stability is a function of the reactive monomer used in the formulation, and therefore it is necessary to examine these variables experimentally in order to optimize heat stability of a basic resin system. In the order of their stabilizing characteristics the monomers appear in the following order:

divinyl benzene > triallyl cyanurate > styrene.

The plots show that the first two can be used for service at a temperature in the 150°C to 200°C range, whereas the third will only be serviceable up to 150°C.

DMA Experiments

The specimens for the DMA measurements were prepared as graphite-fiber reinforced resin composite sheets in which the fibers were in the form of a woven cloth, designated as T300-3000, 13 x 13 weave (Union Carbide Thornel 300 yarn, woven by Hexcel). To prepare a specimen, four thicknesses of cloth, in a bidirectional fiber lay, were saturated with resin, vacuum-pumped to remove trapped air, laid smoothly between two sheets of 0.025 mm polyethylene and irradiated to a total dose of 100 kGy. The irradiation was performed at the NRL Van de Graaff accelerator facility with electrons having an energy of 2 MeV. The composites, thus made were approximately 1 mm thick and contained approximately 60 wt percent resin. The specimens cut from these sheets were approximately 10 x 12.5 mm and were measured to the nearest 0.1 mm before inserting into the clamps of the DMA.

The DMA run was made over the temperature span of -100° to +400°C in a nitrogen atmosphere, with the oscillating frequency and relative damping being plotted vs temperature on the DuPont 990 X-Y, Y' Recorder. The rate of temperature rise was controlled at a programmed rate of 5°C/min. The graph plotted for the DMA run of the Epocryl 12/DVB composite is shown in Fig. 3. From this frequency curve the Young's modulus was calculated from the formula and replotted on a graph of Young's modulus vs temperature. The curves of all three resin formulations were overlaid on one graph as shown in Fig. 4. From this figure it is evident that the formulation with the highest temperature capability for strength retention is the one containing DVB. This material does not begin to lose strength until reaching a temperature of approximately 170°C. This temperature is even more

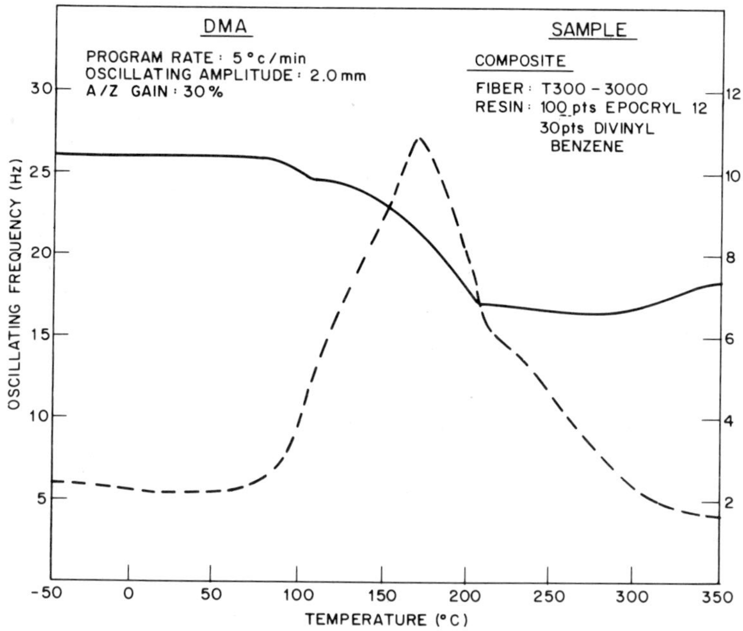

Fig. 3. DMA run of the composite of graphite-fiber/Epocryl 12-DVB formulation.

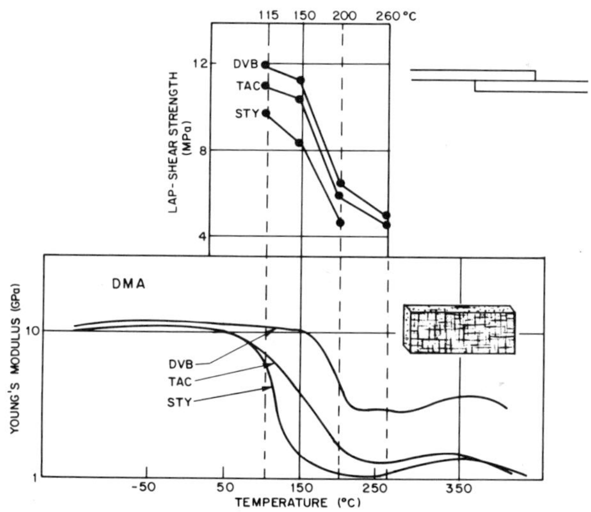

Fig. 4. Correlation of temperature stability determinations by lap shear strength and dynamic mechanical analysis (DMA) of electron-beam cured formulations of Epocryl-12/vinyl-reactive monomers.

clearly designated by the peak in the damping curve in Fig. 3. The temperature at which this peak occurs in a cross-linked polymeric material is generally considered to be the glass transition temperature T_g, and the upper limit of use. Curves of the other two formulations show modulus decreases beginning at lower temperatures, with the TAC having more retention of strength over the 150°-$200^{o}C$ range than the STY formulation.

Fig. 4 also has the graph of the radiation-cured lap-shear strength data, showing the strength data at the specific temperatures at which specimens were tested. By superpositioning these curves over the Young's modulus curves on the same temperature scale it can be seen that the same order of temperature stability was obtained by the DMA method on composite specimens as was determined on the adhesive specimens. That is, in order of temperature stability, the monomers to be used for maximum temperature service are: DVB, TAC, and STY. In addition, the DMA method of analysis provides more information since it shows the temperature at which the cured adhesive or composite material begins to lose strength and, thus, gives an actual value for the limiting service temperature.

CONCLUSIONS

These results have demonstrated that by utilizing a dynamic analysis system to obtain continuous scans of modulus vs temperature, such as can be performed with a DuPont DMA instrument, more exact measurements of the high temperature limits of cured resin systems can be obtained. As determined in this study, these limits correlate with the approximate ranges of temperature stability of the same electron-beam cured formulations of Epocryl 12/vinyl-reactive monomers as determined by a discrete temperature, lap shear strength measurement process. Thus, the utilization of the DMA instrument has facilitated the more rapid evaluation of variables in formulation studies on the ultimate high temperature capabilities of adhesives and composites resins in the continuing research effort to obtain high performance materials by electron-beam curing technology.

ACKNOWLEDGMENT

This work was supported as a portion of a program to develop high-performance adhesives and composites for further Navy V/STOL aircraft which was controlled by the Naval Air Systems Command, Washington, DC 20361.

REFERENCES

1. L.B. Lockhart, Jr., editor, "High Performance Composites and Adhesives for VISTOL Aircraft", Third Annual Report, NRL Memorandum Report 4005, May 1979.

2. F.J. Campbell, B.A. Rugg, R.P. Kumar, J. Arnon, and W. Brenner, Radiation Phys. and Chem., 14, 699 (1979).

3. F.J. Campbell and W. Brenner, Nav. Engs. J., 94, 160 (1982).

4. DuPont Co., Scientific and Process Instrumentation Division, 981 Dynamic Mechanical Analysis System (DMA), Bulletin E-08434.

5. F.J. Campbell, B.A. Rugg, and W. Brenner, "Diversity-Technology Explosion," SAMPE SERIES, 22, 59 (1977).

Reactive Butadiene/Acrylonitrile Liquid and Solid Elastomers: Formulating Acrylic, Anaerobic and Radiation-Curable Adhesives

R. S. Drake and A. R. Siebert

B. F. Goodrich Chemical Group
6100 Oak Tree Blvd.
Cleveland, OH 44131

ABSTRACT

Structural as well as pressure-sensitive adhesives having a high degree of permanence, which are based on various kinds of mono-, di-, and tri-functional monomers, usually employ an elastomer species for providing balance in performance properties associated with shear strength, peel strength and environmental durability. Thus, reactive butadiene/acrylonitrile liquid and solid elastomers have found usage among the growing number of products comprising the adhesive categories of acrylic (or liquid reaction), anaerobic and radiation-curable types. Reactive chemistries related to carboxyl-, hydroxyl-, and acrylated vinyl moieties account largely for the reactive elastomer types employed. It is the purpose of this paper to highlight the role that reactive butadiene/acrylonitrile elastomer modification plays in formulating these adhesives. The format is a systematic review of pertinent journal and patent literature.

INTRODUCTION

Newer adhesives of the acrylic, anaerobic or radiation-curable types must, if they are structural, have a relatively high degree of toughness and durability if they are to compete with or challenge epoxy adhesive systems. Likewise, newer radiation-curable, pressure-sensitive adhesive systems must exhibit the properties of permanence largely associated with cross-linked adhesive masses deposited from an acrylic solution polymer base. Epoxy resin structural adhesives largely define the existing area on the one hand; the cross-linked acrylics deposited on plastic or metallic films the other.

Toughness in structural adhesives of nearly all varieties, where a set of balanced properties is preeminent, is usually accomplished through the addition of an elastomeric, or at the very least, a flexibilizing modifier. Many of these types of adhesives have been in the marketplace for as long as 15 years with many newer developments occurring within the last five years.

Lees (1) reported as recently as late 1980 that during a lengthy period of 'controlled evaluation' the toughened epoxies proved more versatile than acrylic and anaerobic adhesive types. The reference is to single-component, heat-cured epoxy-nitrile paste adhesives in comparison to similarly toughened acrylics based either on methyl methacrylate or tetrahydrofurfuryl methacrylate. The comparisons are made with aluminum adherends where lap shear strength is examined over a temperature range of -60 to +120°C or against bondline thickness from 0.1 to 1.0 mm. This does not mean to imply, however, that the non-epoxy resin adhesives do not have their own assets and advantages. Comparisons are inevitable.

The use of elastomeric or flexibilizing modifiers occurred and grew with epoxy resins first. Various aspects of toughened epoxy adhesives have been covered in reviews by the present authors (2,3), where the elastomeric modifiers have essentially been carboxylic, liquid and solid butadiene/acrylonitrile polymers. There has not been a systematic review, however, of these and other reactive liquid polybutadiene/acrylonitriles in the burgeoning areas of acrylic, anaerobic and radiation-curable systems. Thus, this paper's intent.

Reactive Butadiene/Acrylonitrile Elastomers

Solid elastomers of the butadiene/acrylonitrile variety are relatively old synthetic polymeric materials. Critical reviews of Hofmann (4) and Brown (5), although 20 years old, offer excellent sources to their preparation, characterization, chemistry and applications. The newer reactive polybutadiene/acrylonitrile liquids are telechelic polymers having either carboxyl-, amine, acrylated vinyl or hydroxyl- terminal moieties (Table 1). Selected references for each of these materials are cited (6,7,8,9). Specific chemical reactivities of these telechelic polybutadiene/acrylonitrile liquids will permit (a) modified resinous oligomer preparations having elastomeric segments or (b) direct admixture to a system where end-group chemistry will permit the liquid elastomer to enter specific thermosetting networks.

In many instances, low-level additions of liquid or solid reactive butadiene/acrylonitrile elastomers serve to toughen free-radical curing resinous systems which are otherwise extremely brittle. Tables 2 and 3 show this for popular corrosion-resistant polyesters. Such toughened systems, when optimized, normally are opaque and are characterized by particulate second-phases existing at reasonably high volume fractions in the fully cured products. Low rubber levels, in these cases a liquid elastomer curing into the network, can provide impact and fracture surface energy enhancement with general improvements in tensile and flexural properties.

Table 1

	CTBN (1300X8)	CTBN (1300X13)	VTBN (1300X22)	ATBN (1300X16)	HTBN (1300X17)
Reactivity	Carboxyl	Carboxyl	Acrylated	2° Amine	1° Hydroxyl
B.V.cP at 27°C	150,000	570,000	225,000	235,000	140,000
Acrylonitrile,%	16	26	16	16	16
Sp.Gr.$25^{o}/25^{o}$C	0.948	0.960	0.984	0.956	0.960
Equiv.Wt.*	1920	1750	1470	900	1880

*unit weight per single equivalent unit

Table 2

Vinyl Ester Resins

(Epoxy-Methacrylates)

	A	B
Epocryl DRH 322 (Shell)	100	100
Hycar VTBN (1300X22)	-	10
t-butyl perbenzoate	1.5	1.5
Exotherm, ^{o}C/min	185/38	161/26
Cured 2 hrs/90°C + 2 hrs 120°C		
Shrinkage, %	6.4	1.2
Gardner Impact, J	2.4	4.5
Fracture Surface Energy J/m^2	542	1870

Table 3

Brominated Polyester Resins

	A	B
Atlac 711-05A (ICI-America)	100	100
Styrene	3	3
Hycar VTBNX (1300X23)	-	3
MEK Peroxide	1	1
CoNaphthenate (6%)	1	1
Cure Schedule:		
Gel - 15 min at R.T.		
Postcure - 2 hrs/150°C		
Barcol Hardness	41-42	38-40
Tensile Strength, mPA	42.83	68.27
Elongation, %	2.4	5.6
Tensile Energy, kN-cm	8.8	341.4
Flexural Strength, mPA	59.68	112.52
Izod Impact, J/cm (unnotched)	0.80	1.16
Gardner Impact, J	1.14	1.55
FSE, J/m^2	338	425

On the other hand, it is also possible to flexibilize a common polymerizing monomer such as methyl methacrylate. Table 4 demonstrates such systems at a relatively high level of vinyl-terminated polybutadiene/acrylonitrile liquid (VTBN). One finds, as would be expected, glass transition temperatures associated with rubber and with polymethyl methacrylate. Such monomers are excellent diluents for rubber where, as an example, the following relationship exists for VTBN (1300X22):

Percent MMA	B.V. cP at 27°C
25	8,500
50	600
75	100

Further examples of rubber inclusion will be couched in adhesive usage; however, bulk properties from casting compositions serve to illustrate the effects of rubber inclusion in free-radical polymerizing systems.

Table 4

Castable Systems Based on
Hycar VTBN and Methyl Methacrylate
With 100 parts Methyl Methacrylate
(Cumene hydroperoxide/Cobalt naphthenate)

Percent VTBN	40	60
Cured Schedule - 24 hrs/RT + 10'/120°C		
DTA, Tg, °C	-39.5	-43
	93	91
Tear, Die C (kN/m)	82.8	48.1
Original Properties		
Tensile Strength, mPa	24.1	12.4
Elongation, %	100	130
Shore D Hardness	75	57
After 70 hrs Water (100°C)		
Tensile Strength, mPa	22.7	9.3
Elongation, %	50	70
Shore D Hardness	67	44

Acrylic Adhesives

It's more than five years since toughened acrylic adhesives became routinely available. These materials provided a generous increase in adhesive performance over predecessors. Reliable joints and strong bonds can be obtained. And these materials perform well on unprepared (oily) metals. However, these newer acrylics did not catch on as expected. There were problems with handling (odor), accelerated aging (humidity) and withstanding temperatures associated with paint bake-cycles. Some of these objections have been overcome. Numerous articles review or detail specific instances of acrylic adhesive performance; and patent literature has grown accordingly. Green and Toy (10) dwell on improved performance of newer acrylics citing applications for metals, glass and various plastics (both thermoplastic and thermoset). The old and new are compared with some outstanding differences in lap shear values.

Prane (11) reviews the field and concentrates on history citing pivotal patents related to acrylics (or 'reactive') adhesives. These patents show that numerous materials have been utilized in toughening acrylic adhesives, some nonreactive in the systems, others reacting in, while still others using combinations of each general type modifier. Thus, we find examples for polychloroprene, thermoplastic urethanes, urethane adducts, acrylic rubber, chlorosulfonated polyethylene in addition to butadiene/acrylonitrile rubber (or its carboxylic version).

Lees (12) continues with toughened epoxy and toughened acrylic adhesive comparisons in a paper designed largely for the transportation engineer and the bonding of lightweight vehicles. It is interesting that each adhesive has excellent T-peel (aluminum) and impact resistance (steel assemblies - ASTM-D950/72). However, the single component epoxy outperforms the acrylic in shear (steel: 41.5 vs 22.7 N mm^{-2} respectively). And Charnock and Martin (13) have recently described the structure-property relationships of rubber-toughened acrylic adhesives showing that the continuous matrix polymer contains small rubber particles of 0.1-0.15 μm diameters.

Table 5 shows data from cures of VTBN (1300X22) with higher boiling monomers replacing a portion of methyl methacrylate. In these two instances, stearyl methacrylate

Table 5

	1	2
Hycar VTBN (1300X22)	45	45
Stearyl Methacrylate	15	-
Isodecyl Methacrylate	-	15
Methyl Methacrylate	40	40
t-Butyl perbenzoate	2	2
B.Visc. cps @ 27°C	480	800
Cure Schedule: 1 hr/100°C + 0.5 hr/120°C		
Tensile Strength, mPa		
Elongation, %	61	95
Duro A Hardness	61	75
Die C Tear, kN/m	20.5	29.2
Tabor Abrasion - % loss	0.37	0.09

(bp = 181°C) and isodecyl methacrylate (bp = 158°C) produce rubbery, cured materials having relatively high elongations. However, the isodecyl methacrylate compound has superior tensile strength.

Owston (14) has shown that fast curing, flexible industrial adhesives are capable of being prepared from solid butadiene/acrylonitrile or carboxylic butadiene/acrylonitrile elastomers in combination with monomers and catalysts of choice. The systems discussed are at about 30% rubber level and provide 3,300-4,000 psi lap shear strengths with aluminum adherends utilizing hardening times as low as five minutes.

One may combine solid and liquid butadiene/acrylonitrile elastomers in one acrylic adhesive formulation. There exists an exacting parallel to this in commercially available epoxy-nitrile film or prepreg adhesive systems. Such epoxy systems have been shown to be tougher than most.

Utilizing the vinyl-terminated liquid polymers constitutes one approach to this type of formulating. Another, and perhaps more nearly versatile, would be to synthesize epoxy-acrylates or epoxy-methacrylates utilizing the carboxyl-terminated polybutadiene/acrylonitrile (CTBN) liquids in the preparation scheme. Thus, one would have a butadiene/acrylonitrile rubber segment built into the oligomer types finding increasing use in performance applications. Both Navjar (15) and Waters (16) have described such materials primarily for the purposes of corrosion-resistant structural applications. Table 6 summarizes an example from the Waters invention. As polymer syrups in monomeric styrene, there is a useful low viscosity. In the cured state, there's appreciable tensile strength coupled with a useful elongation. Impact improvement is significant without generally prohibitive loss of corrosion resistance.

Anaerobic Adhesives

Numerous publications concerning anaerobic adhesives exist which delineate both performance properties and the healthy patent literature surrounding them. Since they are

Table 6

Elastomer-Modified Epoxy-Methacrylate

Synthesis and Typical Properties

	Grams
DGEBA Liquid (EEW = 182)	181
CTBN (1300X8)	180
Methacrylic Acid	77.6
Pyridine	1.4
Hydroquinone	0.26
Acid Number	3.5
Styrene, Blend	440
Viscosity at 25°C	137 mPa s
Tensile Strength*, mPa	346
Elongation, %	29.2

*Cure: 1 hr/93°C - 0.7% BPO and 0.2 tBp

W. WATERS, US 4,097,569

Table 7

	1	2
Hycar VTBN (1300X22)	60	40
Methyl Methacrylate	40	60
Cumene hydroperoxide	1.0	1.0
Cobalt naphthalenate (6% Co)	0.2	0.2
Cure Schedule: 16 hrs/RT + 10'/120°C		
Tensile Strength, mPa	91.4	205
Elongation, %	110	120
Duro D Hardness	50	72
Tear (Graves) - kN/m	46.4	117.3
Tabor Abrasion, % loss	0.19	0.20

close cousins to acrylics, Hauser (17) has discussed them together in terms of their structural bonding capability and performance. From the balances cited in peel, impact and shear strengths, it is obvious these models are rubber-toughened systems.

Table 7 illustrates the mechanical property behavior of VTBN in a simple hydroperoxide model. Recipe 2 with an appreciable tear strength broke in a tensile-yielding mode similar to toughened thermoplastics. VTBN (as well as CTBN and HTBN) have found utility in the newer anaerobic adhesive systems.

Baldwin and Lees (18) disclosed that VTBN with THF methacrylate can serve to optimize a set of properties based on shear (collar/pin and lap joint) and impact strengths. This balance appears to occur at reactive liquid rubber levels of 7.5-10.0%. These improvements occurred without detriment to storage stability and subsequent curability of the composition. Catena (19) essentially verifies that comparable activity is obtained when VTBN serves as a 10% monomer replacement in anaerobic formulations based on tetraethylene glycol dimethacrylate.

Hargreaves et al. (20) were able to show substantial improvement in shear strength (pin pressure) after elevated temperature aging (204°C) when CTBN is used to modify an anaerobic model based on ethoxylated bisphenol-A dimethacrylate and derivatives of amino alcohols. CTBN inclusion offers more than 3-fold shear improvement after aging 72 hrs at 204°C.

Baccei (21) discloses polymerizable adhesive and sealant compositions having as an important ingredient urethane-acrylate-capped oligomers based on liquid hydroxyl-terminated polybutadiene or polybutadiene/acrylonitrile copolymers. Properties associated with impact and thermal improvements are noted as well as the ability to cure adequately through the bondline gap. Model systems are largely triethylene glycol dimethacrylate based. Impressive adhesive performance properties are recorded. (Cf. Paper by Malofsky and Baccei in this volume.)

Radiation Curable Adhesives

In this burgeoning and rapidly changing area of adhesive and coating systems there exists no systematic review of all adhesive developments utilizing radiation processes - least of all an attempt to single out and demonstrate the systematic growth and use of oligomeric materials such as telechelic polybutadienes and copolymers. This reporting purports to cover both electron-beam and photocurable (free-radical and cationic) processes for these products.

Reasonably early, however, in the development of telechelic polybutadiene/acrylonitrile, Brenner and Drake (22) showed that mercaptan- and carboxyl-terminated liquid polybutadiene/acrylonitrile did respond to electron-beam (1-5 megarads) and cure (2-3 seconds/pass) from the liquid to the solid state. The mercaptan polymer (3.1% RSH, 23% bound acrylonitrile, Mn 1700) cured more readily. CTBN as expected required 50-100% higher irradiation dosage levels. With both MTBN and CTBN-type products of higher acid content (6% vs 3%), gum rubber properties of 13.8-34.5 mPa were obtained with 60-100% elongation. The E-beam cures were carried out in air at ambient temperature. Thus, unformulated telechelic polybutadienes were shown to be substantive to cure in the presence of ionizing radiation.

Table 8 examines simple photocurable models based on epoxy-acrylate oligomers. Two recipes represent two modification approaches: (a) Synthesized CTBN-modified epoxy-acrylates, or (b) VTBN admixed with an existing epoxy-acrylate. Each system shows free-radical cure response to actinic radiation. Ariga (23) discloses finished coatings developments utilizing CTBN-modified epoxy acrylates (concentrating on bisphenol-F based diepoxides and CTBN at 26% bound acrylonitrile). In syntheses with other lower molecular weight dicarboxylic acids (having at least 4 methylene groups interposed between 2 carboxyl groups), there existed lower-than-expected viscosities for uncured resins. This processing advantage is coupled with cured coatings possessing excellent scratch and crosshatch adhesions, good direct and reverse impacts (duPont impact tester) and bending diameter resistance (2 mm).

Table 8

Rubber-Modified Photocurable Systems

	1	2
CTBN-Modified Epoxy-Acrylate	75	-
Dow 8008.04	-	44
Hycar VTBN (1300X22)	-	31
Trimethylolpropane triacrylate	10	10
2-Ethylhexyl Acrylate	15	15
Vicure 10 (benzoin ether - Stauffer Chemical)	2	2
15 mil films under nitrogen blanket		
Observations:		
After 15 seconds		
Duro D Hardness	17	13
Condition	very slightly tacky	
After 30 seconds		
Duro D Hardness	27	32
Condition	- dry -	
After 60 seconds		
Duro D Hardness	31	30
	- dry -	
	slightly opaque	

Structural adhesives for glass/glass, glass/metal or glass/plastic are possible with photocurable adhesive masses. Albrecht (24) describes a rubber-modified cationic system based on selected diepoxides and triphenylsulfonium hexafluoroantimonate. 10-mil films curing in 40 secs (200 watt/inch; med. pressure Hg lamp) gave tensiles ranging from 186-269 mPa with bulk elongations of 12-25%. The systems passed low temperature (-20°F) testing and have good impact resistance. VTBN-type polymers are employed at about 7-8% of the adhesive weight.

Dudgeon (25) describes systems suitable for curing in deep sections (e.g., as auto body solders) which may be adapted for heat or cationic UV curing (or a combination of both). Fast curing (2-3 min) systems with full depth cure are based on CTBN or HTBN modification of an epoxycyclohexyl ether and mixtures of triphenylsulfonium hexafluoroantimonate (50% in propylene carbonate) and diphenyliodonium hexafluoroarsenate (50% in MEK) with low levels of copper naphthenate. These systems operate at high talc loadings to form deep-section curable putties. Examples suffice where a cluster of two infrared lamps and two ultraviolet lamps operate efficiently to effect cure.

Pressure-sensitive adhesive masses, in one sense the opposite of structural adhesives (but also requiring permanence within their own performance definition), have been reported for both UV and E-beam processes where telechelic polybutadiene/acrylonitrile liquids are concerned.

Glennon (26) has been able to modify a resin-tackified linear block copolymer of butadiene/styrene dissolved in n-butyl acrylate monomer with a CTBN-type polymer. This easily E-beam cures on Mylar at 3 megarads irradiation dosage to obtain 700 N/m Mylar to painted steel adhesion (180° peel). This system suggests a move from similar solvent-borne or hot-melt approaches to a polymer syrup/monomer mixture that is chain-extendable and cross-linkable with ionizing radiation.

Perkins (27) published extensively on radiation curable pressure sensitive adhesives showing that permanent systems can be obtained if liquid hydroxyl-terminated polybutadiene/acrylonitrile were used with a selected acrylic interpolymer (e.g., 2-ethylhexyl-acrylate/acrylic acid/vinylacetate - I.V.=0.35) in conjunction with the monomers 2-ethylhexyl acrylate and tetraethylene glycol diacrylate. This system photocures with the preferred benzoin ethyl ether initiator to give at cure speeds of 10 ft/min.

Peel Adhesion (N/m)	432
Type PS Adhesion	permanent
Creep Resistance - hrs (500g. weight)	>24
Tack	high

Other references (28,29) show the use of telechelic polybutadiene liquids in photocurable printing or flexographic printing plates. While these are not adhesives, they aim for softer cured polymer masses which, if tackified, would most likely have pressure-sensitive adhesive properties.

SUMMARY

Selected examples have been cited to demonstrate that areas of acrylic, anaerobic and radiation-curable adhesive developments have begun to consider telechelic polybutadiene/acrylonitrile liquids as useful formulating ingredients either in direct admixture or in the preparation of rubber-modified oligomers serving as an adhesive formulating base. These cited adhesive examples cover structural, semi-structural, gap-filling and pressure-sensitive types.

REFERENCES

1. W.A. Lees, "Modified Epoxides-Practical Aspects of Toughening", Adhesion and Adhesives (PRI) Sept. 3, 1980, Durham, U.K.

2. R.S. Drake, "Elastomer-Modified Epoxy Resins for Structural Applications", SAMPE Quarterly, 6, (4), July 1975.

3. R.S. Drake and A.R. Siebert, "Reactive Butadiene Acrylonitrile Liquid and Solid Elastomers: Formulating Epoxy Structural Adhesives", ACS-OCPL, Symposium on Recent Developments in Adhesive Chemistry, March 21-

23, 1983, Seattle, Wash., and this volume.

4. W. Hofmann, "Nitrile Rubber", RC&T Technical Review Supplement, 1963.

5. H.P. Brown, "Cross-linking Reactions of Carboxylic Elastomers", RC&T, 36, (4) p 931, Oct. 11, 1963.

6. R.S. Drake and W.J. McCarthy, "Liquid Butadiene/-Acrylonitrile Polymers with Reactive Terminals", Rubber World, Oct. 1968.

7. C.K. Riew, "Amine-Terminated Reactive Reactive Liquid Polymers: Modification of Thermoset Resins", 54, (2) May 6, 1975.

8. D. Skillicorn, US Patent 3,910,992, "Vinylidene-Terminated Liquid Polymers", Oct. 7, 1975.

9. A.R. Siebert, US Patents 3,551,471 and 3,551,472, "Liquid Hydroxyl-Terminated Polymers, Dec. 29, 1970.

10. D.H. Green and L.E. Toy, "Acrylic Structural Adhesives: Technology in Rapid Evolution", Technology Conferences (SPE) - ADHESIVES FOR INDUSTRY, Aug. 24, 1980, El Segundo, CA.

11. J.W. Prane, "Reactive Adhesives", Adhesives Age, 23, No. 8, 35 (1980).

12. W.A. Lees, "New Adhesives and the Structural Engineer", SITEV Conference, May 1981, Geneva.

13. R.S. Charnock and F.R. Martin, "Structure Property Relationships of a Rubber-Modified Acrylic Adhesive", PRI Conference: Adhesion and Adhesives, Sept. 3-5, 1980, Durham, U.K.

14. W.J. Owston, US Patent 3,832,274, "Fast Curing Adhesives", Aug. 27, 1974.

15. D. Navjar, US Patent 3,892,819, "Impact Resistant Vinyl Ester Resin and Process for Making Same", July 1, 1975.

16. W. Waters, US Patent 3,928,491, "Modified Vinyl Ester Resin...", Dec. 23, 1975.

17. M. Hauser, "Anaerobic and Modified Acrylic Engineering Adhesives: Technological Innovations in Structural Bonding, Technology Conference (SPE), ADHESIVES FOR INDUSTRY, June 24-25, 1980, El Segundo, CA.

18. T.R. Baldwin and W.A. Lees, US Patent 4,138,449, "Anaerobic Curing Compositions", Feb. 6, 1979.

19. W.J. Catena, US Patent 4,235,986, "Anaerobic Curing Adhesive Composition and Process", Nov. 25, 1980.

20. K. Hargreaves et al., US Patent 4,243,791, "Anaerobic Adhesives", Jan. 6, 1981.

21. L.J. Baccei, US Patent 4,295,909, "Curable Polybutadiene-Based Resins", Oct. 20, 1981.

22. W. Brenner and R. Drake, "Electron Beam Curing of Liquid Mercaptan-Terminated Butadiene/Acrylonitrile Elastomers", SAMPE, 14th Nat. Symp.

23. N. Ariga et al., US Patent 4,085,018, April 1978.

24. D.E. Albrecht, US Patent 4,219,377, "Photocurable Epoxy Composition", Aug. 26, 1980.

25. C.D. Dudgeon, US Patent 4,308,118, "Deep Section Curable Epoxy Resin Putty", Dec. 29, 1981.

26. A.E. Glennon, US Patent 4,243,500, "Pressure Sensitive Adhesives", Jan. 6, 1981.

27. W.C. Perkins, "Radiation Curable Pressure Sensitive Adhesives", Radiation Curing, Aug. 1980.

28. S. Proskow, US Patent 4,177,074, "Photosensitive Elastomeric Polymer Compositions", Dec. 4, 1979.

29. R.L. Pohl, US Patent 4,174,218, "Release Plates from Polymer with Terminal Unsaturation", Nov. 13, 1979.

Photoinitiators: A Review of Mechanisims and Applications

L.R. Gatechair and D. Wostratzky

CIBA-GEIGY Corporation
Ardsley, NY 10502

ABSTRACT

The commercialization of many new applications for UV-cured adhesives has created a need for new classes of photoinitiators having greater curing efficiency, improved solubility, lower odor, and resistance to yellowing. The objective of this paper is to review the current major classes of commercial photoinitiators, describe their mechanisms and discuss key properties which should be considered when choosing a photoinitiator for various applications.

INTRODUCTION

The broad field of adhesives has provided many new opportunities for using UV-curing technology. Increasingly, printers, coaters, and converters are learning that UV-curing can make important contributions to greater profitability. Initially, the conversion to UV-curing was motivated by the desire for reduced emissions or possibly, lower energy consumption. Those industrial leaders who adapted UV-curing are now finding the added benefits of reduced cure cycles resulting in greater throughput, coupled with improved properties in the final product. For example, a product which required a time consuming lamination to provide a protective barrier coating may now be laminated with a UV-curable resin. The increased demand for UV-curable formulations which are pigmented or used in exterior applications requiring photostability has prompted photoinitiator suppliers to introduce new products which can meet these new challenges. UV-cured adhesives have many advantages including reduced solvent emissions, high cure rates and excellent solvent resistance. However, obtaining the correct balance of physical properties requires an understanding of how the various components of a formulation interact.

The six classes of photoinitiators which have evolved having commercial success in the various fields of UV-curing are:

- Benzoin ethers
- Benzophenones
- Thioxanthones
- Ketals
- Acetophenones
- Onium salts

Each of these classes has certain performance characteristics which account for their commercial acceptance. It is the objective of this paper to review 11 commercial photoinitiators; by presenting an overview of their use in the UV-curing industry, their properties, and the mechanisms by which they cure coatings. It is hoped that a greater understanding of the characteristics of each class will provide the coating chemist with a basis on which to choose the appropriate photoinitiator for a specific application.

In the process of UV-curing, the photoinitiator acts as the curing catalyst. As shown in Fig. 1, it is the role of the photoinitiator to absorb the available radiation and produce an active initiating species resulting in polymerization of the available monomer (1,2). Polymerization of olefins may be anionic, free radical or cationic in nature. This refers to the charge on the growing end of the polymer during the polymerization, being negative, neutral or positive, respectively. In general, anionic polymerization of olefins is too easily terminated by contaminants, water, or oxygen to be of commercial interest for the curing of coatings. This paper will mainly discuss photoinitiators for free radical polymerization where the growing end of the polymer carries a neutral charge (1 unpaired electron).

Recently, photoinitiators have been developed which cationically polymerize vinyl ether and epoxy functional coatings (3,21). The last section of this paper will briefly introduce this new technology.

Photoinitiators for Free Radical Polymerization

The process of free radical polymerization can be divided into three steps: initiation, propagation, and termination. Fig. 1 shows a simplified scheme of the photoinitiation process. The photoinitiator (PI) absorbs radiation over the range of 250-450 nm, resulting in the formation of the electronically excited state photoinitiator (PI*). As most photoinitiators are derivatives of aromatic ketones, the active excited state is generally the triplet state. The details of this photochemistry will not be reviewed in this paper as Pappas and McGinniss have described the nature of the excited states involved during photoinitiation (2).

In general, excited state photoinitiators have two mechanisms by which they can form the free radical species which initiate polymerization. Type I photoinitiators undergo homolytic bond cleavage resulting in the formation of two free radical units (P. and I.). When absorbed radiation causes the bond between a carbonyl groups and an adjacent carbon to break, the reaction is called Norrish Type I Cleavage. Type II photoinitiators undergo hydrogen abstraction from a second molecule in the formulation (RH) resulting in the formation of two different free radical

PI —RADIATION (hν)→ PI*

TYPE 1: (ALPHA CLEAVAGE)

PI* ⟶ P· + I·

TYPE 2: (HYDROGEN ABSTRACTION)

PI* + RH ⟶ ·PIH + R·

INITIATION STEP:

P·		PM·
I·	⟶	IM·
HIP·	+ MONOMER (M)	HIPM·
R·		RM·

PROPAGATION STEPS:

PM· + M ⟶ POLYMER

TERMINATION STEPS:

(COMBINATION) POLYMER· + POLYMER· ⟶ POLYMER - POLYMER

(CHAIN TRANSFER) POLYMER· + RH· ⟶ POLYMER-H + R·

(CHAIN TRANSFER) POLYMER-OO· + RH ⟶ POLYMER-OOH + R·

(REACTION WITH OXYGEN) POLYMER· + O_2 ⟶ POLYMER-OO·

Fig. 1. Photoinitiated Free Radical Polymerization Steps

species (·PIH and R·). The resulting radical species react with styrene, unsaturated polyester, or acrylate functional resins initiating the polymerization process with varying efficiency. Propagation results in the formation of polymer by the addition of successive monomer units. Termination may occur via combination which results in the increase in molecular weight or chain transfer which results in a decrease in molecular weight. Termination may also result via reaction with oxygen which results in reduced molecular weight coupled with the formation of peroxide and hydroperoxide by-products. These oxidation products may reduce the performance life of the coating.

Table 1 shows a comparison of physical properties for the 11 free radical photoinitiators reviewed in this paper. This comparison includes the physical state, color and comments on solubility in typical multifunctional acrylate diluents used in UV-curing. Also included is a list of photoinitiator abbreviations used throughout this report.

Table 2 shows a comparison of UV-cured coating market characteristics for 7 diverse U.S. industries. This table compares film thickness, pigments, sensitivity to yellowing, comments on curing conditions, and typical resins used in various applications. It is clear that the wide range of these performance requirements results in a need for different types of photoinitiators. The challenge confronting the formulating chemist is to maximize curing efficiency while attaining the other desired properties of the cured coating or film. The following 4 considerations are presented as guidelines in order to maximize the efficiency of the UV-curing process. The initiator must be chosen in order to:

1. Absorb radiation efficiently. Avoid contaminants, pigments, additives, or resinous materials which compete for the available radiation. Choose a photoinitiator which has the optimum balance of extinction coefficients for the given application. In pigmented systems, choose photoinitiators which absorb strongly at wavelengths where the pigment absorbs weakly. In many cases, a blend of two photoinitiators will make better use of the available radiation, as will be discussed later.

Table 1. Photoinitiator Properties

CLASS	EXAMPLE PHOTOINITIATORS	ABBREVIATION	PHYSICAL STATE	COLOR	SOLUBILITY
BENZOIN ETHERS:					
	ISOPROPYL BENZOIN ETHERS	IPBE	SOLID	WHITE	GOOD
	ISOBUTYL BENZOIN ETHER (MIXTURE)	IBBE	LIQUID	COLORLESS	GOOD
BENZOPHENONES:					
	BENZOPHENONE	BP	SOLID	WHITE-AMBER	GOOD
	MICHLER'S KETONE	MK	SOLID	YELLOW-BLUE	FAIR
THIOXANTHONES:					
	CHLOROTHIOXANTHONE	CTX	SOLID	YELLOW	POOR
	ISOPROPYLTHIOXANTHONE	IPX	SOLID	YELLOW	FAIR
	DODECYLTHIOXANTHONE	DTX	LIQUID	BROWN	GOOD
KETALS:					
	BENZIL DIMETHYL KETAL	BDMK	SOLID	WHITE	GOOD
	ACETOPHENONE DIETHYL KETAL	ADEK	LIQUID	COLORLESS	GOOD
ACETOPHENONES:					
	α-HYDROXY, CYCLOHEXYL PHENYL KETONE	HCPK	SOLID	WHITE	GOOD
	2-HYDROXY, 2-METHYL PHENYL PROPANONE	HMPP	LIQUID	COLORLESS	GOOD

Table 2. Characteristics of UV-Curing Market Segments

	FILM THICKNESS (MILS)	PIGMENT	RESISTANCE TO YELLOWING	CURING ATMOSPHERE	CURING SPEED	TYPICAL COMPOSITION
WOOD FILLERS	1.-5.	FILLER	YELLOWING ACCEPTABLE	AIR/N_2	LOW	POLYESTER STYRENE
PAPER COATINGS (CLEAR)	.1-.5	NONE	SOME APPLICATIONS	AIR/N_2	HIGH	EPOXY ACRYLATE
INKS	.1	ORGANIC	NOT APPLICABLE	AIR	HIGH	EPOXY ACRYLATE, AROMATIC URETHANE ACRYLATE
WHITE PIGMENTED INKS & COATINGS	.3-1.	TiO_2	SOME APPLICATIONS	AIR	LOW	EPOXY ACRYLATE, AROMATIC URETHANE ACRYLATE
FLOOR COATINGS	2.-10.	NONE	CRITICAL	AIR/N_2	LOW	ALIPHATIC URETHANE ACRYLATE
CLEAR EXTERIOR PROTECTIVE COATINGS	.5-3.	NONE	CRITICAL	AIR	MEDIUM	ALIPHATIC URETHANE ACRYLATE
ADHESIVES	.5-3.0	NONE	SOME APPLICATIONS	AIR	MEDIUM	URETHANE ACRYLATE + TACKIFIER

2. Use the absorbed radiation efficiently to form free radicals. Avoid components which might quench (deactivate the excited state of) the photoinitiator, such as styrene or conjugated double bonds (5). The ideal photoinitiator will not undergo side reactions which produce inactive free radicals.

3. Form reactive free radicals which diffuse rapidly to react with vinyl groups. A knowledge of mechanisms will aid in choosing a photoinitiator which produces small energetic free radicals having high diffusion coefficients. This diffusibility becomes more important in later stages of the curing process because large, bulky free radicals will not reach the residual reactive sites resulting in incomplete conversion of monomer to polymer.

4. Reduce the inhibition of curing caused by the presence of oxygen. Oxygen has two effects on the UV-curing process: it may quench the excited state photoinitiator, and it may retard the free radical polymerization. Some applications lend themselves to curing in an inert atmosphere such as nitrogen or argon, however this may be economically impractical for most large scale facilities. In thinner coatings where there is more contact with oxygen, a photoinitiator or synergist may be necessary which has been designed to reduce oxygen inhibition.

Using these guidelines, we will discuss the various classes of photoinitiators and show why they have advantages for some of the applications listed in Table 2. Table 3 shows a selection chart for the use of various photoinitiators in the applications of: woodfillers, clear overprint varnishes for paper, inks, white pigmented coatings, floor coatings, exterior coatings, and electronics (photoresists and potting compounds). The authors are fully aware that to some degree, almost all photoinitiators may be used in any of these applications. In Table 3 we have attempted to highlight the major U.S. markets for photoinitiators based on available information. Naturally, any suggestions are system dependent and the formulator must test a range of products and concentrations in order to find the best photoinitiator for the system to be studied.

Table 3. Photoinitiator Selection Chart (For Major UV-Curing Applications in the USA)

CLASS / EXAMPLE PHOTOINITIATORS	ADHESIVES	POLYESTER-STYRENE WOOD FILLER, COMPOSITES	PAPER VARNISHES	INKS	WHITE PIGMENTED	FLOOR	EXTERIOR
BENZOIN ETHERS:							
ISOPROPYL BENZOIN ETHERS		*					
ISOBUTYL BENZOIN ETHER (MIXTURE)		*					
BENZOPHENONES: (+ AMINE)							
BENZOPHENONE			*	*			
MICHLER'S KETONE				*	*		
THIOXANTHONES: (+ AMINE)							
CHLOROTHIOXANTHONE				*	*		
ISOPROPYLTHIOXANTHONE				*	*		
DODECYLTHIOXANTHONE				*	*		
KETALS:							
BENZIL DIMETHYL KETAL	*	*		*	*		
ACETOPHENONE DIETHYL KETAL	*		*			*	*
ACETOPHENONES:							
α-HYDROXY, CYCLOHEXYL PHENYL KETONE			*			*	*
2-HYDROXY, 2-METHYL PHENYL PROPANONE			*			*	*

It is clear from Table 3 that certain trends exist within the UV-curing industry in the United States. For instance, benzophenones and thioxanthones dominate the pigmented applications, while the ketals and benzoin ethers dominate the polyester-styrene applications. Finally, acetophenones and ketals tend to dominate the applications which require excellent through-cure and resistance to yellowing.

The Effects of the Radiation Source

There are many radiation sources used in the UV-curing industry (4). Most utilize the emission of electronically excited (hot) gases, predominantly mercury. In principle, such lamps operate by passing an electronic current through the gas promoting electrons from the ground (lowest) electronic state to higher excited states. The atoms release this stored energy by emitting radiation as the electron returns to the ground state. In the simplest case, i.e., mercury lamps, only certain electronic transitions are allowed by the laws of quantum mechanics. Thus, the radiation from excited gases is concentrated in a few discrete lines (narrow wavelengths range). The wavelength of the emission line is determined by the energy of the electronic transition (5), and can be calculated using the following formula for the relationship between wavelength and energy:

$$\lambda = \frac{hc}{E} = \frac{28591}{E}$$

λ = wavelength (nm) h = Planck's Constant
E = Energy (Kcal/mole) c = Velocity of light

One of the most important factors during the curing process, is the efficient use of the available radiation. Although most photoinitiator suppliers provide the extinction coefficients for the initiators at their absorption maxima, the only wavelengths of importance are those emitted by the radiation source. A typical medium-pressure mercury arc lamp which is commonly used for industrial photocuring (operating at 200 Watts per inch), has significant emissions clustered near the 260, 313, 360, and 405 nanometer ranges. Fig. 2 shows the wavelengths and

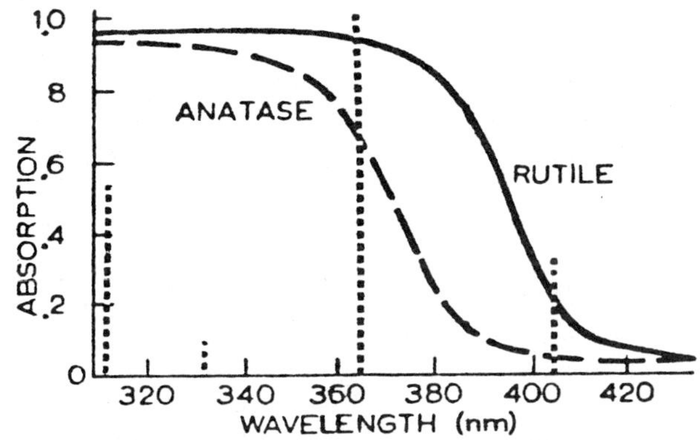

Anatase Absorption — — — — —
Rutile Absorption ————————
Mercury Emission Lines -----------

Fig. 2. TiO_2 Absorption Spectra[a] and Mercury Emission Wavelengths[b]

a) From Ref. 6.

b) Relative intensity of emission from a PPG 200 W/inch medium pressure mercury arc lamp.

relative intensity of these emissions superimposed over the absorption spectra of TiO_2 (6). From this figure, it is apparent that in order to have a UV-curable white pigmented coating, the initiators must absorb at wavelengths greater than 360 nanometers. This is also true in principle, for colored organic pigments which are used in inks, however, the spectra are very much dependent on the specific pigment. A generalization can be made that photoinitiators for such pigmented systems, must have significant absorptivity at long wavelengths.

In the absence of such pigments, the formulating chemist has more flexibility to choose one or more photoinitiators to make the best possible use of the total range of wavelengths available from the radiation source. Since by definition, photoinitiators which have absorptivity at wavelengths greater than 400 nanometers will be colored, their use may not be practical in many applications such as clear overprint varnishes used over white substrates. It should be noted that some lamps are available with significantly increased emissions at longer wavelengths which may result in improved curing efficiency for pigmented coatings (4). Wicks and Pappas have discussed theoretical considerations for UV-curing of pigmented coatings (7). For practical suggestions in formulating white pigmented UV-cured coatings, the authors recommend an article by Hulme (22). Table 4 compares molar extinction coefficients and molecular weights of the eleven photoinitiators in this review. Following Beer's Law (5), the absorptivity (A) is proportional to the molar extinction coefficient (ε) x molar concentration (C) x path length (L).

$$A = \varepsilon CL$$

A high extinction coefficient translates to more efficient absorption of radiation at a given wavelength and concentration. An optimum concentration of a given photoinitiator exists, where the best combination surface-cure and through-cure will be obtained. Lower concentrations will be needed for photoinitiators having significantly higher extinction coefficients. The lower regions of the coating will not be sufficiently cured if too much radiation is absorbed near the surface (i.e., concentration is too high). This may be indicated by poor adhesion, wrinkling, or the formation of a cured skin over uncured resin. For the same reason, thicker applications (>3 mil) need a relatively <u>lower</u> concentration of photoinitiator in order to maintain the correct balance of surface and body cure. In contrast, thin applications (<0.5 mil) may need higher photoinitiator concentrations in order to overcome the inhibition due to the diffusion of oxygen.

The optimum balance of surface and body cure may be obtained by using a combination of two or more photoinitiators. While such blends are often described as sensitizers, co-sensitizers, or unique synergistic

Table 4. Photoinitiator Molar Extinction Coefficients at Selected Wavelengths

CLASS	EXAMPLE PHOTOINITIATORS	WAVELENGTH (NANOMETERS) 260 NM	360 NM	405 NM	MOLECULAR WEIGHT
BENZOIN ETHERS:					
	ISOPROPYL BENZOIN ETHERS	11,379.	50.	0	254.
	ISOBUTYL BENZOIN ETHER (MIXTURE)	*	*	*	268.
BENZOPHENONES:					
	BENZOPHENONE	14,922.	51.	0	182.
	MICHLER'S KETONE	8,040.	37,500.	1,340.	268.
THIOXANTHONES:					
	CHLOROTHIOXANTHONE	*	3,944.	197.	247.
	ISOPROPYLTHIOXANTHONE	*	5,182.	102.	254.
	DODECYLTHIOXANTHONE	*	3,620.	127.	381.
KETALS:					
	BENZIL DIMETHYL KETAL	9,740.	97.	0	256.
	ACETOPHENONE DIETHYL KETAL	5,775.	19.	0	210.
ACETOPHENONES:					
	α -HYDROXY, CYCLOHEXYL PHENYL KETONE	3,170.	18.3	0	204.
	2-HYDROXY, 2-METHYL PHENYL PROPANONE	2,710.	9.	0	164.

* NOT MEASURED

combinations, the enhanced curing efficiency may best be explained simply by more efficient use of available radiation. For example, the relatively high triplet energies of the benzoin ethers and ketals (~70 Kcal/mole) make triplet sensitization by benzophenones or thioxanthones doubtful (triplet energies <70 Kcal/mole).

Comparing Tables 1, 3, and 4 shows that the photoinitiators which dominate the pigmented applications have high extinction coefficients at visible wavelengths (>400 nm). Because these initiators are generally colored, the resulting white pigmented coatings may have a slight amber appearance. BDMK is not colored, but in pigmented applications, it is generally used in combination with MK or a thioxanthone.

Photoinitiator Mechanisms

BENZOINETHERS (IBBE AND IPBE): Benzoinethers were one of the first classes of photoinitiators developed for the UV-curing industry and were used successfully in both polyester-styrene and acrylate based formulations. Various substitutions for R such as propyl, isopropyl (IPBE), butyl, and mixtures of butyl isomers (IBBE) results in photoinitiators which all undergo Norrish Type I photocleavage resulting in the formation of the benzylic and the benzylether radicals as shown in Fig. 3. Both radicals are known to initiate polymerization, although the relative reactivity appears to be somewhat system and condition dependent (2).

As is shown in Table 5, the (dark) package stability of acrylate and polyester formulations proved to be poor for this class of photoinitiators compared to those developed subsequently. It has been suggested that the instability of benzoin ethers as a class is due to the presence of the readily abstractable benzylic H-atom (2).

Oxygen has a significant retarding effect on the UV-curing process. This is believed to originate from two interactions as described in Fig. 4. Quenching of the initiator is the process by which a molecule of oxygen (which normally exists as a ground state triplet $^3O_2$) interacts with the excited state triplet initiator ($^3S^*$) in such a way as to form the deactivated photoinitiator (ground

Table 5. Comparison of Package Stability

ACCELERATED PACKAGE STABILITY (DAYS AT 60°C)

PHOTOINITIATOR	EPOXY ACRYLATE	POLYESTER STYRENE[2]
(NONE)	> 40	35
BDMK	> 40	35
HCPK	> 40	-
IPBE	3	14
IBBE[2]	1	25
BP (+ 3%) + MDEOA (5%)	1	-

PHOTOINITIATORS TESTED AT 3% EXCEPT AS NOTED.

2) TESTED AT 2%.

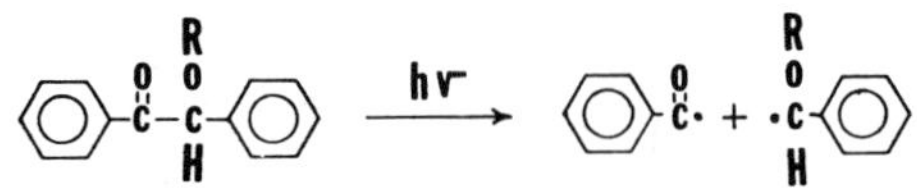

Fig. 3. Photochemistry of Benzoid Ethers (Norrish Type 1 cleavage)

QUENCHING THE EXCITED STATE PHOTOINITIATOR

$$^{3}PI^{*} + {}^{3}O_2 \longrightarrow PI + {}^{1}O_2^{*}$$

RETARDING THE POLYMERIZATION:

$$M' + M \longrightarrow MM' \quad \text{(PROPAGATION)}$$

$$M' + O_2 \longrightarrow MOO'$$

(TERMINATION)

$$MOO' + M \not\longrightarrow MOOM'$$

$$MOO' + RH \xrightarrow{\text{SLOW}} MOOH + R' \quad \text{(HYDROGEN ABSTRACTION)}$$

Fig. 4. Effects of Oxygen on UV Curing

BENZOPHENONE (BP): Because of the sensitivity of benzoinethers to oxygen inhibition, a great deal of effort was made to find new photoinitiator classes which would be less sensitive to the presence of oxygen. Osborne at Union Carbide (9) and separately, Gaske at DeSoto, Inc. (10-12) discovered the synergistic interaction of benzophenone and various alkyl amines for overcoming oxygen inhibition resulting in superior surface-cure (see Fig. 5). This technology is particularly effective in extremely thin coatings such as inks or thin clear overprint varnishes where a great deal of contact with oxygen may occur during the application and curing processes.

The interaction of benzophenone and alkyl amines is truly synergistic in that neither alone is an effective photoinitiator, yet the two used in combination presents one of the most efficient systems for the curing of thin coatings. Fig. 5 shows that benzophenone (I) is a Type II photoinitiator in that following excitation, it abstracts a hydrogen atom from the amine (II) resulting in the formation of the benzhydril free radical (III) and the alkyl-amino free radical (IV). Both free radicals may react as initiators for free radical polymerization. However, the alpha—amino free radical has the ability to react preferentially with oxygen resulting in the formation of peroxy radicals (V). These peroxy radicals further react by abstracting hydrogen from a second amine molecule resulting in the formation of a hydroperoxide (VI) and a new alkyl amino free radical (IV). These latter reactions form a chain sequence which results in the efficient consumption of oxygen. In this manner, the formation of one alkyl amino radical may remove as many as 12 molecules of oxygen from the formulation (17). The result is that the benzophenone-amine photoinitiator system which is much less sensitive to the presence of oxygen than most other photoinitiator mechanisms.

In order to function as a curing synergist, the alkyl amine must have hydrogens alpha to the nitrogen, on one or more of the substituent groups. Typical synergists include methyldiethanolamine, and dialkylamino alkylbenzoates. The aminobenzoates may have some advantages in being less basic and having a lower tendency to form yellow photoproducts which may discolor the cured coating. The addition of alkyl amines (19) generally reduces dark package stability (see Table 5). The aminobenzoates are reported to have less stability problems.

PHOTOINITIATION VIA HYDROGEN ABSTRACTION

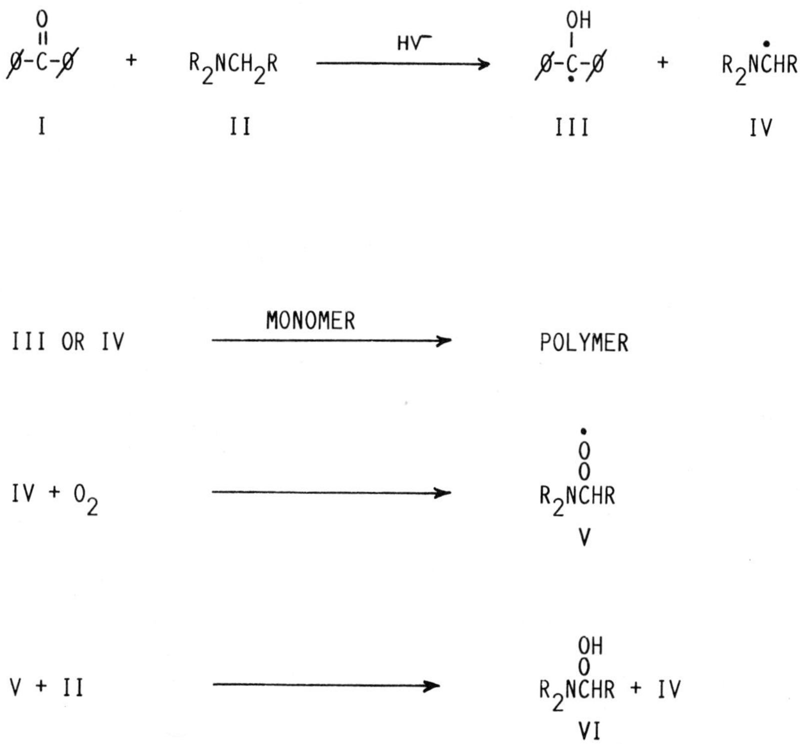

Fig. 5. Benzophenone + Amines

MICHLER'S KETONE (MK): Michler's Ketone is a unique member of the benzophenone group in that it possesses both the benzophenone chromophore and tertiary amine group. The effects of this molecular combination photoinitiator are two-fold: first, the amine group is immediately available for the abstraction of hydrogen atoms and second, the amine substituents on the benzophenone chromophore result in a greatly enhanced UV absorption spectrum, especially at longer wavelengths. In printing ink applications, Michler's Ketone is generally used in combination with benzophenone. Although it is generally agreed that the combination of these two photoinitiators is better than either used alone in pigmented coatings, the mechanism for interaction is not clear (2,18). Structures of benzophenone and Michler's Ketone are shown below:

BENZOPHENONE

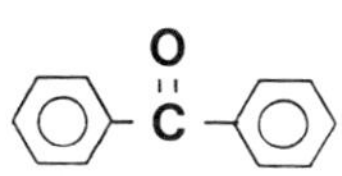

MICHLER'S KETONE

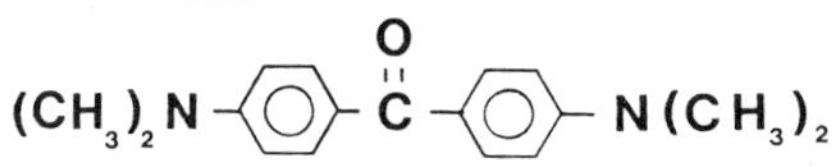

THIOXANTHONES (CTX, DTX, ITX): Thioxanthones are a subclass of the benzophenone chromophore modified by the presence of a bridging sulfur atom, resulting in a greatly enhanced UV absorption spectra as is shown in Table 4. Like benzophenone, these aromatic ketones require an amine synergist for efficient curing. Fig. 6 shows 3 examples of this class of photoinitiators: chlorothioxanthone (CTX),

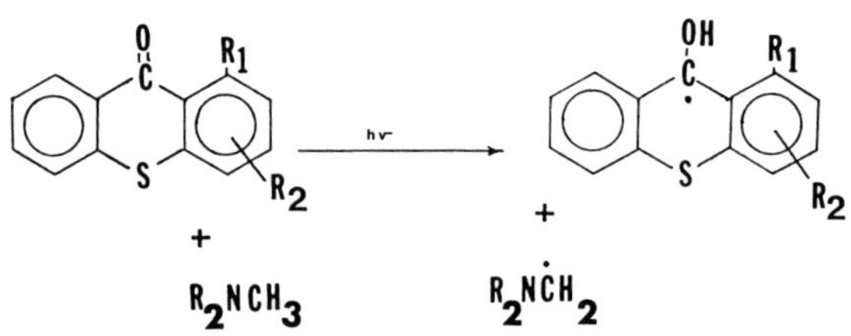

PHOTOINITIATOR	R_1	R_2
CTX	CL	H
DTX	H	DODECYL
ITX	H	ISOPROPYL

Fig. 6. Thioxanthone Photochemistry

dodecyl thioxanthone (DTX), and isopropyl thioxanthone (ITX). Although the photolysis of these molecules results in the formation of large, sterically hindered free radicals, they nonetheless are efficient for the curing of pigmented coatings. Evidently, the increased absorptivity resulting in more efficient use of the available radiation at longer wavelengths, more than offsets the reduced diffusion coefficient of the resulting free radical.

KETALS (ADEK, BDMK): More recently, two photoinitiators from the ketal class were commercialized. Acetophenone diethylketal (ADEK), also called diethoxy acetophenone, is a Type II photoinitiator (see Fig. 7). Unlike the benzophenone class, ADEK, undergoes internal hydrogen abstraction resulting in the formation of a diradical (Norrish Type II cleavage) (15,16). There is also some indication at ADEK undergoes Norrish Type I cleavage (Fig. 8). Although the Norrish Type II cleavage is believed to be the predominant mechanism, it has been suggested that Norrish Type I free radical products are the true initiating species (2). Osborne demonstrated the addition of acrylate monomer during the photolysis of ADEK has no effect on the formation of the oxytanol product of Norrish Type II cleavage but prevented the formation of benzaldehyde and ethyl formate, the recombination products of Norrish Type I cleavage (14).

ADEK has been widely recognized as an efficient photoinitiator for many applications. The absence of a benzylic hydrogen makes the ketal class relatively free from package stability problems (19). However, it is clear that if the minor mechanism accounts for the formation of the initiating species, there is room for further improvements in the efficiency of initiation. Norrish Type I reactions can be promoted by either making it more difficult to undergo intramolecular hydrogen abstraction, or by facilitating (stabilizing) the formation of the Norrish Type I free radical products (or the transition state leading to them) (13). BDMK is an example of a molecule which uses both concepts described above to facilitate Norrish Type I cleavage reactions.

BENZYL DIMETHYLKETAL (23) (BDMK): As is shown in Fig. 8, BDMK reacts primarily by a Norrish Type I cleavage efficiently generating a benzoyl radical and a substituted benzylic radical. The substituted benzylic radical re-

A) BIRADICALS VIA INTERNAL HYDROGEN ABSTRACTION (NORRISH TYPE II CLEAVAGE)

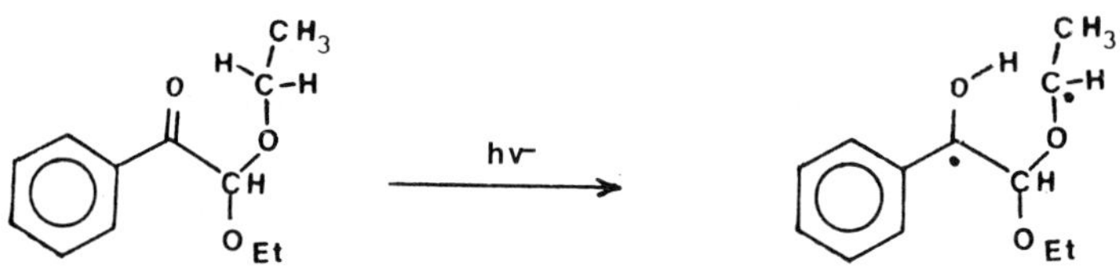

B) FREE RADICALS VIA ALPHA-CLEAVAGE (NORRISH TYPE I)

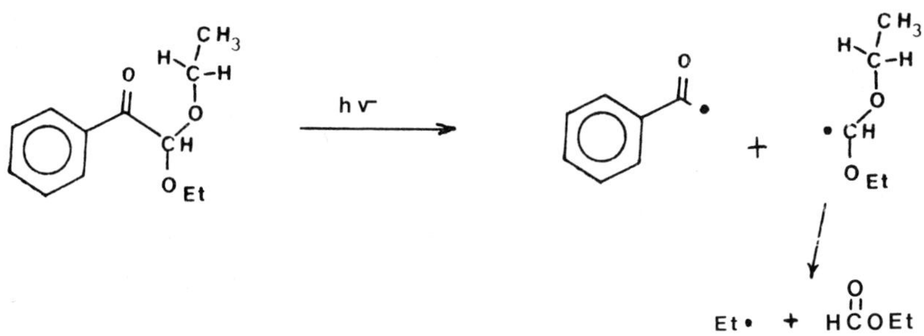

G. BERNER, J. PUGLISI, R. KIRCHMAYER, G. RIST, J. RADIATION CURING, 6 (1979) PP. 2-9

Fig. 7. Photochemistry of Acetophenone Diethylketal (ADEK)

arranges into a highly reactive methyl radical and the stable methylbenzoate molecule (20). The methyl radical is a highly desirable initiating species as it is very non-selective when seeking a monomer to initiate polymerization. The small size of the methyl radical results in high diffusibility in more viscous formulations or during later stages of the polymerization. The Norrish Type I reaction is facilitated in preference to the Norrish Type II for two reasons. The benzylic radical formed by BDMK is much more stable than the ketal radical formed by ADEK. In addition, hydrogen abstraction from the methyl carbon on BDMK is not as favorable as from the secondary carbon (in the case of ADEK).

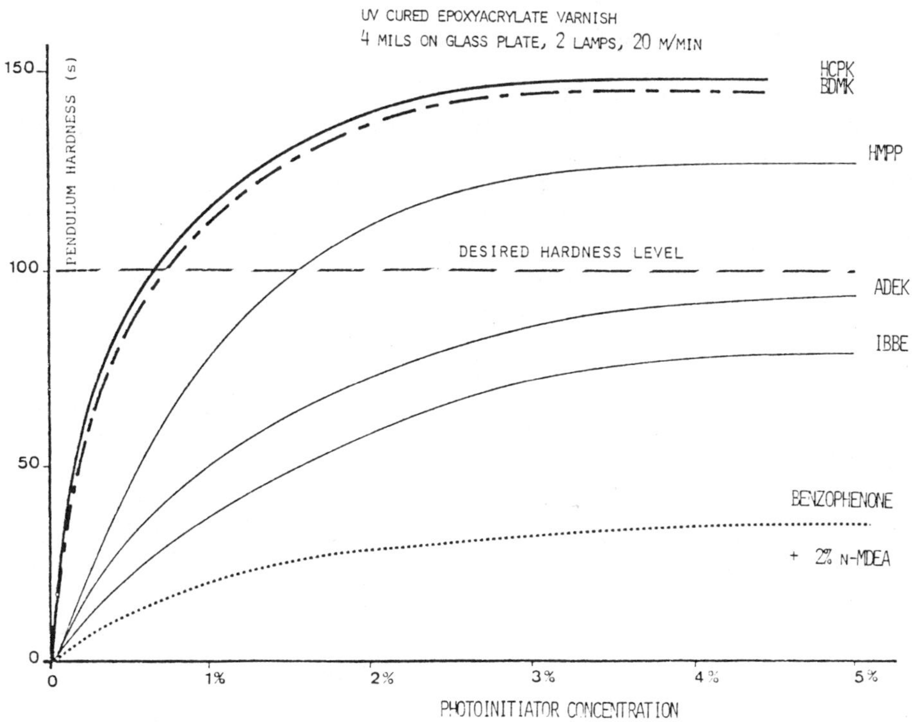

Fig. 8. Curing Efficiency as a Function of Photoiniator Concentration

UV Cured Epoxyacrylate Varnish
4 mils on Glass Plate, 2 Lamps, 20 m/mm

Although similar in structure, the different mechanisms of ADEK and BDMK may, in part, explain the differences in reactivity and the applications in which these two photoinitiators tend to be used (see Table 3). ADEK tends to find use in the adhesive, flooring and possibly exterior varnish applications, which are mainly thinner coatings. In addition to varnishes, BDMK finds applications in adhesives and polyester-styrene applications, both of which tend to be thicker coatings with high demands for through-cure (body-cure) efficiency. The fact that BDMK finds wide use in the unsaturated polyester/styrene formulations may be due to the efficient formation of the highly energetic methyl radical which may be a much more efficient initiating species for these less reactive olefins than the more stabilized free radical products of the other photoinitiator classes (see Fig. 8). Table 5 shows the improved package stability of formulations containing BDMK compared to the class of benzoin ethers.

ACETOPHENONES (HCPK and HMPP): Although unsubstituted acetophenone does not undergo efficient photocleavage and thus, is not commercially used as a photoinitiator, the addition of alpha substituents facilitates Norrish Type I cleavage resulting in the efficient formation of free radical products as shown in Fig. 10. Both hydroxycyclohexyl phenylketone (23) (HCPK) and 2-hydroxy-2-methylphenylpropanone (HMPP) are examples of alpha-substituted acetophenone which are highly efficient photoinitiators for acrylate functional coatings. This new class is also reported as having excellent package stability (see Table 5) (19). The clear varnish applications in which this class of photoinitiators tend to be used require efficient curing of thin films. In addition to curing efficiency, a critical requirement is the need for resistance to yellowing, both immediately after cure as well as following exposure to sunlight or fluorescent light, i.e., coatings for floor tiles and exterior applications. Fig. 11 shows a comparison of acetophenone and ketal type photoinitiators for resistance to yellowing during QUV exposure in an aliphatic urethane acrylate based formulation cured into 20 mil thick plaques. Formulations containing HCPK show superior resistance to yellowing as compared to the other photoinitiators.

NORRISH TYPE I CLEAVAGE

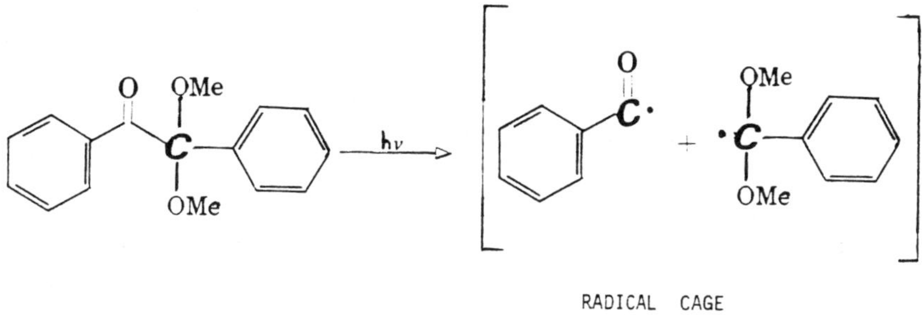

SUBSTITUTED BENZIL-RADICAL REARRANGEMENT

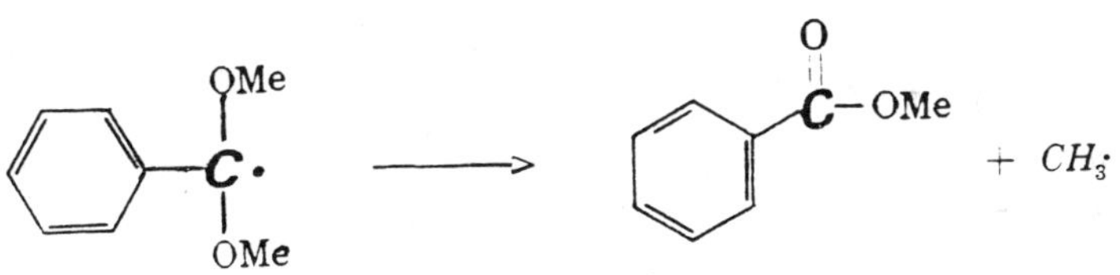

G. BERNER, J. PUGLISI, R. KIRCHMAYER, G. RIST,
J. RADIATION CURING, 6 (1979) PP. 2-9

Fig. 9. Photochemistry of Benzil Dimethylketal (BDMK)

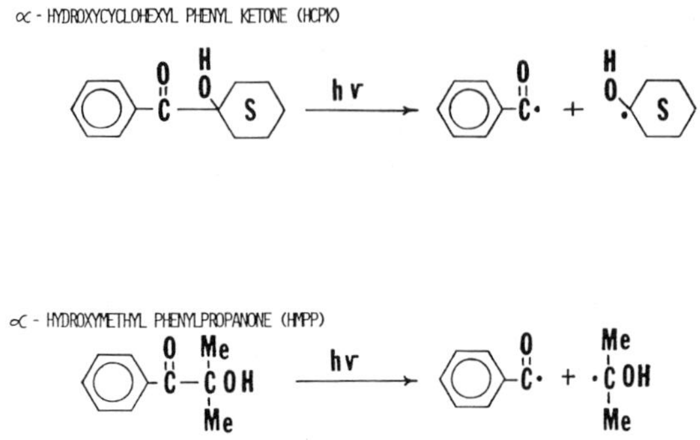

Fig. 10. Photochemistry of Alpha-Substituted Acetophenones (Norrish Type I Cleavage)

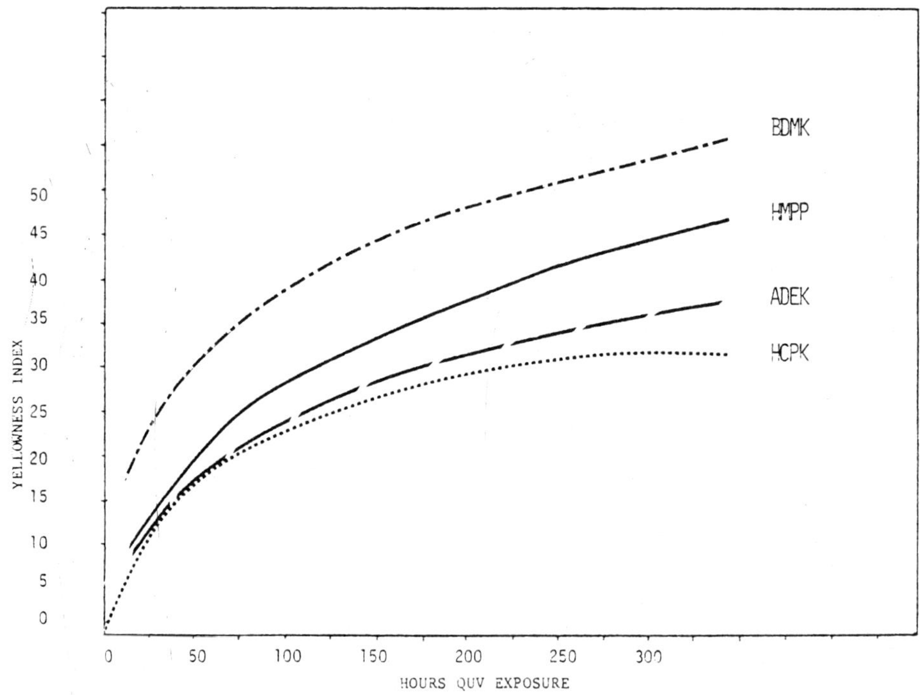

Fig. 11. Comparison of Commercial Photoinitiators in Aliphatic Urethane Acrylate Formula

Yellowing during QUV Exposure of 20 mil Plaques

Fig. 8 compares representatives of four classes of photoinitiators in an epoxy acrylate formulation for curing efficiency as measured by pendulum hardness and photoinitiator concentration. The ketal and acetophenone classes appear to represent a significant advance in the state of the art for curing efficiency in unpigmented UV-cured coatings. Using a pendulum hardness of 100 seconds as a desired level of cure for this formulation, only 0.75% of either HCPK or BDMK would be required. In contrast, almost twice this concentration of HMPP would be required to reach this same level of hardness. Under the curing conditions used in Fig. 12, over 5% of ADEK, IBBE, or BP (+2% - MDEA) would be required to achieve this degree of cure. For applications where both curing efficiency and resistance to yellowing are important, Figs. 10 and 11 demonstrate that alpha-hydroxy cyclohexyl phenyl ketone (HCPK) will provide the optimum balance of properties.

Photoinitiated Cationic Polymerization of Adhesives

Many recent publications and patents give evidence of the growing interest in photoinitiated cationic polymerization (see the accompanying chapter by J.V. Crivello). In general, aryliodonium and arylsulfonium salts are utilized as photoinitiators for formulations containing vinyl ether or cyclic ether cross-linking agents.

Since epoxy functional resins are widely available and contribute many desirable characteristics to coatings such as adhesion, toughness and abrasion resistance, it may be desirable to have UV-curable adhesives based on epoxy systems. In conventional UV-curable coatings, the photoinitiator absorbs light and produces free radical species generally resulting in addition polymerization of acrylate or unsaturated polyester components. Epoxy resins, however, are not curable by free radical polymerization (except epoxyacrylates). In order to cure an epoxy resin based coating, the photoinitiator must produce an acidic species capable of catalyzing the cationic polymerization of the epoxy resin.

$$\text{Photoinitiator} \xrightarrow{\text{Light } (h\nu)} H^+$$

$$H^+ + \text{Epoxy Resin} \longrightarrow \text{Crosslinked Polymer}$$

The cure rates of cationic formulations based on epoxy resins are usually much slower than the acrylates. Studies of photoinitiation efficiency have shown that polymerization rates may be enhanced by the appropriate choice of sensitizers (3,21,24).

One useful property of cationic polymerization is that it is not air inhibited. In the absence of nucleophilic impurities, there are no inherent modes of termination. Thus, the polymerization may continue for long periods (hours to days) after the light is turned off, in contrast to photoinitiated free radical polymerization (24). This post-cure should be useful for laminating two opaque materials, in that irradiation of the adhesive followed by lamination should result in a well cured adhesive, although it may take 1-2 days to form maximum bond strength.

Post-Cure in Free Radical UV-Cured Adhesives

Post-cure is also observed in free radical UV-cured formulations. Although no detailed studies are known to the author, this change in properties is often observed over a period of up to 60 minutes following the curing process. This crosslinking may be caused by diffusion controlled recombination of trapped free radicals, possibly peroxy radicals, in the cured coating.

One property common to UV-cured acrylate formulations as was discussed earlier is relatively poor surface-cure due to oxygen inhibition (see Fig. 1). This effect is most predominant at low photoinitiator concentrations, in thin coatings. Adhesive formulations can benefit from this property by enhancing surface tack via the presence of residual uncured resin. Post-cure, from residual crosslinking, or other anaerobic chemistry, can later result in enhanced bond strength.

CONCLUSIONS

Each UV-curing application has a unique combination of requirements. In order to meet these requirements, the formulating chemist must carefully optimize his choice of many components including one or more photoinitiators, and their concentrations. Many properties such as surface or through-curing efficiency, yellowing, adhesion, hardness, abrasion resistance, durability, and package stability depend directly on the correct choice and concentration of photoinitiator. The authors hope that this paper has provided some practical guidelines to aid in the correct use of photoinitiators when formulating UV-curable adhesives.

ACKNOWLEDGMENTS

The figures and portions of this text were preprinted with the permission of Technology Marketing Corporation and the Society of Manufacturing Engineers. The authors also acknowledge the support of the CIBA-GEIGY Corporation during the preparation of this manuscript.

REFERENCES

1. S.P. Pappas, Prog. Org. Coatings, 2, (1974) 333.

2. S.P. Pappas and V.D. McGuinniss, U.V. Curing: Science and Technology (S.P. Pappas, editor), Technology Marketing Corp., Norwalk, Connecticut (1978), pp. 1-22.

3. J.V. Crivello, ibid, pp. 23-77.

4. V.D. McGinniss, ibid, pp. 97-132.

5. J.G. Calvent and J.N. Pitts Jr., Photochemistry, J. Wiley and Sons, Inc., New York (1966).

6. M.A. Parrish, Oil and Col. Chem. Assoc., 60, 474-478 (1977).

7. Z.W. Wicks and S.P. Pappas, Ref. 2, pp. 79-96.

8. C.T. Niessner, Radiation Curing, 8, 26 (1981).

9. C.L. Osborne, U.S. Patent 3,759,807 (assigned to Union Carbide Corp., Sept. 10, 1973).

10. J.E. Gaske, U.S. Patent 3,844,916 (assigned to DeSoto Inc., Oct. 29, 1974).

11. J.E. Gaske, U.S. Patent 3,914,165 (assigned to DeSoto Inc., Oct. 21, 1975).

12. J.E. Gaske, U.S. Patent 3,925,349 (assigned to DeSoto Inc., Dec. 9, 1975).

13. J.E. Christensen, et al. (Ref. 8) pp. 4-12.

14. C.L. Osborne and S.L. Watson, Abstracts, 9th Central Regional Meeting of the Amer. Chem. Soc., Charleston, West Virginia, POLY 21, p. 75, October 12-14, 1977.

15. C.L. Osborne, Preprints, Am. Chem. Soc., Div. Org. Coatings & Plastics Chem. 34, (1) 660 (1974).

16. M.R. Sandner and C.L. Osborne, Tetrahedron Letters, 30, (5) 415 (1974).

17. R.F. Bartholomew and R.S. Davidson, J. Chem. Soc. (C), 2342 (1971).

18. G.S. Hammond, C.C. Wamser, C.T. Chang, and C. Baylor, J. Amer. Chem. Soc., 92, 6362 (1970).

19. R.A. Leiberman, (Ref. 8) pp. 13-24.

20. G. Berner, J. Puglisi, R. Kirchmayer, and G. Rist, J. Radiation Curing, 6, April 1979, 2-9.

21. L.R. Gatechair and S.P. Pappas, presented at the 183rd Meeting of the American Chemical Society, March 25-April 2, 1982 Macromolecular Secretariat. (Preprints pp. 701-707.) ACS Symposium Series: Initiation of Polymerization (F.E. Bailer, ed.), 173-190 (1983).

22. B.E. Hulme, J. Oil Col. Chem. Assoc., 59, 245-252 (1976).

23. Irgacure® Photoinitiators are products of the CIBA-GEIGY Corporation.

24. J.V. Crivello and J.L. Lee, Macromolecules, 14, (5), 1141-1147 (1981). Note: Post-cure was not observed with dialkylphenacyl sulfonium and dialkyl(4-hydroxyphenyl)sulfonium salts.

Discussion

On the Paper by K. C. Stueben

J. B. Class (Hercules Research Center): Did you examine PSA properties as a function of the time elapsed between curing of the film and assembling the test piece?

K. C. Stueben (Union Carbide Corporation): No, although most samples were tested about 24 hours after curing under PSTC conditions.

C. Bluestein (Captan Associates, Inc.): For pressure sensitive adhesives for film substrates, Mylar, current users are seeking 5 lbs. peel minimum both before and after aging.

M. Zeldin (Purdue University, Indiana): Have you applied statistical (chemometric) analysis to your multivariable problem?

K. C. Stueben: Yes, but it did not give unambiguous answers.

On the Paper by J. V. Crivello

W. J. Bailey (University of Maryland): How thick a sample can be cured by your photoinitiated cationic system? The composites that you showed appeared to be quite thick.

J. V. Crivello (GE Development and Research Center): The depth of cure that can be achieved by our system depends on a number of factors. Visible light penetrates much deeper than UV. A monomer that does not absorb the light in competition with the initiator will allow deeper cure. Under ideal conditions we have been able to cure films in excess of 100 mils.

L. H. Lee (Xerox Corporation): Is there a problem of the catalyst residue which can lower the resistivity of the dielectric coating?

J. V. Crivello: There does not appear to be a problem of poor dielectric properties in films produced by photoinitiated cationic polymerization of epoxies which is related to catalyst residues. In fact, films with excellent dielectric properties have been obtained particularly with photoinitiators which have SBF as their anions.

L. S. Corley (Shell Development Company): A number of your cationic initiators were used in the presence of photosensitizers containing amino groups. Were the photosensitizer levels much smaller than the initiator levels? Was any inhibition of cationic polymerization by the amino groups observed?

J. V. Crivello: The amino groups present in the photosensitizers do not substantially inhibit polymerization for two reasons. First, their concentration is very low, approximately 0.1 mol photosensitizer per mol photoinitiator is necessary in such systems. Second, the mechanism of photosensitization as discussed in my talk involves electron transfer in which the photosensitizer cation-radical is the initiator of cationic polymerization. The aminium cation-radicals derived from the dyes containing dimethylamino groups are, thus initiators and not inhibitors.

J. R. Gardon (Sherwin Williams): Has there been any advance in the "dark reaction?" In my experience, reaction post irradiation occurred only at exposed sites.

J. V. Crivello: In photoinitiated cationic polymerizations one does observe considerable dark polymerization after irradiation has ceased. This polymerization will basically be limited by diffusion rates of the monomer and the growing chain ends. When multifunctional monomers are used, network polymers are formed. In these systems since the growing chain end is attached to the matrix, it cannot diffuse and the rate of polymerization is dependent only on the rate of diffusion of the monomer into the swollen network. These diffusion-controlled processes are slow processes and are confined to very small distances away from the site of initiation. Therefore, in practice, one is limited with respect to the depth of cure in such systems to the depth of penetration of the irradiating light employed.

P. H. Scott (W. R. Grace): Would you please comment on the thermal sensitivity of these photoinitiated cationic systems?

J. V. Crivello: The thermal stability of diaryliodonium and triarylsulfonium salts themselves and in the presence of polymerizable monomers is excellent. The salts are shelf stable for an indefinite period of time as long as they are protected from the light. Similarly, solutions of these salts in the presence of highly reactive monomers remain latent even when heated to temperatures as high as 100°C. Shelf life studies of such solutions have been conducted for as long as one year at 50°C without evidence of polymerization. Yet, these solutions remain highly photosensitive when exposed to UV light.

On the Paper by F. J. Campbell

L. S. Corley (Shell Development Company): (1) Did you use purified divinylbenzene, or did you use the commercial 55% material containing substantial amounts of monofunctional and nonpolymerizable materials? (2) Did you perform any fracture toughness measurements on your laminates?

F. J. Campbell (Naval Research Laboratory): (1) The divinylbenzene used in these formulations was obtained from a specialty chemical supply catalog company which listed it as 55-60% active. (2) No.

PART FOUR:

HIGH TEMPERATURE ADHESIVES

Introductory Remarks

K. C. Stueben

Union Carbide Corporation

Bound Brook, New Jersey 08805

The vast majority of adhesives consumed throughout the world are used in applications requiring but little resistance to elevated temperature. Some notable exceptions to this include foundry molds, abrasives and brake linings, all of which require adhesive binders having good resistance to high temperatures. Ordinary phenolic and furan based adhesives have proven to be excellent and cost effective for such applications and are used in substantial quantities. However, there are numerous other highly specialized applications in the microelectronic, aircraft and aerospace industries where conventional adhesives are not satisfactory. Epoxy resins find some use here, but again, they are extremely limited in temperature resistance. The needs in these industries are for materials which can perform over a broad spectrum of conditions; namely, thousands of hours at 232°C, hundreds of hours at 316°C, and/or minutes at 538°C.

As a result, extensive research has been underway for over two decades in search of structural adhesives which have the requisite balance of processability, properties and cost effectiveness. The most promising materials have been derived from polymers containing various aromatic and heteroaromatic rings including, among others, various polyimides, polybenzimidazoles and polyquinoxalines.

A major problem with many of these polymers has been the difficulty in balancing the needs for high temperature performance with those of processbility. High temperature resistance is often closely allied with chain rigidity and thus, low solubility and/or extremely high softening points - factors which make it difficult to properly wet the substrate. Unlike most other heteroaromatic polymers, the high molecular weight polyquinoxalines are soluble and more readily processable. In those cases where processability is a problem, a strategy involving a two-step synthesis of the adhesive has been used with success. With polyimides, for example, a first-stage polymer containing the poly(amic acid) precursor to the final heterocycle is prepared and applied to the material to be bonded. Further heating completes the formation of the polyimide heterocycle in situ. A shortcoming of this approach is that solvents and/or by-products of the reaction are difficult to remove in many constructions and will create void-like imperfections. The formation of volatiles during cure can be avoided through the use of oligomers terminated with various unsaturated groups which cure by free radical means rather than condensation polymerization. End groups consisting of acetylenic moieties and "nadic" (norbornyl) residues have been successfully used in this fashion.

Even when processability problems have been overcome, it is often found that high temperature adhesives lack the toughness required for long-term use. Some improvement in this property has been achieved by incorporating elastomers. Finally, to achieve maximum performance from these adhesives, it is most important that we increase our understanding of the molecular factors taking place at the interface as well as the test results obtained during failure. This session will present the most recent advances in regard to the synthesis and modification of such adhesives as well as the testing and enhancement (through adhesion promoters) of structural high temperature adhesive joints.

Status of High Temperature Adhesives

P. M. Hergenrother

NASA Langley Research Center
Hampton, VA 23665

ABSTRACT

The status of high temperature organic adhesives is reviewed. These materials are defined by having usable adhesive properties at temperatures ranging from 232°C (after hundreds of hours at 232°C) to 538°C (after a few minutes at 538°C). The search for high temperature adhesives began in the 1960s, primarily under government funding. Since then, many high temperature polymers have undergone evaluation as adhesives. In spite of this effort, only a few are commercially available because of an elusive market and an unfavorable combination of price, processability and performance. The principal adhesives to be discussed will be addition and condensation polyimides, polyquinoxalines, polybenzimidazoles, and addition, condensation, and cross-linked polyphenylquinoxalines. The synthesis, processability, adhesive performance, and availability of these materials are addressed.

I. INTRODUCTION

High temperature adhesives are required in many different applications in advanced aircraft, space vehicles, missiles, ground vehicles, and electronics. Their use involves the joining of metals, composites, ceramics, films and combinations thereof. In addition, insulators and coatings with high temperature stability are needed where adherence to the substrate is of prime concern. Although each application may have a unique set of requirements, the common criterion is high temperature stability. The adhesive must withstand the processing temperature and must be able to perform (resist deformation and environmental degradation under load) at the use temperature for the required service period. The degree of high temperature stability needed depends upon the specific application. In microelectronics, the polymer must adhere to the substrate during various processing cycles where the temperature may reach 400°C in an inert environment. Tens of thousands of hours at temperatures to 230°C will be required for use in supersonic aircraft structures. Application in missiles requires mechanical performance for less than a minute at temperature $>540°C$. Non-stick coatings and decorative coatings on cookware also require good adhesive properties and stability at high temperatures.

The need for structural adhesives with performance at elevated temperatures superior to that of the modified epoxy phenolics was recognized in the late 1950's. A major effort was initiated in the early 1960's, primarily under government funding, to develop high temperature adhesives. In these early years, there was no real appreciation for the thermal limitation of organic polymers and the performance requirements were frequently in excess of the potential of the polymer. Organic polymers with remarkable thermal stability but little or no processbility were reported in the late 1950 - early 1960 era. Work then shifted to the development of these thermally stable polymers into a processable form for evaluation as adhesives and composite matrices. Since then, many promising polymers have been evaluated as high temperature adhesives.

This paper presents a status review of high temperature structural adhesives. For the purpose of this paper, high temperature adhesives are classified as those which retain usable strength at temperatures ranging from 232°C after hundreds of hours at 232°C to 538°C after minutes at 538°C. Many of the polymers discussed are not commercially available but were selected to demonstrate the adhesive properties of that particular family of materials.

In most cases, tensile shear specimens (TSS, Lap shear) were used for preliminary evaluation. It must be recognized that the performance of an adhesive as judged by TSS can be misleading due to the ease of fabrication (small bond area, volatiles tolerated) and the nature of TSS failure mode (peel plus shear). Although many polymers exhibit good adhesive performance in TSS, more comprehensive evaluation using other specimens such as flatwise tensile, climbing drum peel, edgewise compression, and wedge opening often reveal shortcomings. Heat transfer problems are often encountered in the fabrication of large bonded structures. The larger and more complex the bonded structure, the more forgiving an adhesive must be.

Adhesives are designed for specific applications. For example, a honeycomb structure adhesive (requiring good filleting around the cells) would have different flow characteristics than one used for metal-to-metal joining. In addition, the type of stress (e.g., shear or peel) on the bonded area varies and will depend upon the design of the structure.

Discretion is advised when comparing the performance of high temperature adhesives where the evaluations were performed in different laboratories. Important parameters critical to the performance of the adhesive, such as quality of the polymer, adhesive type, adherend, surface treatment, bonding conditions, aging and test conditions, often vary between one investigation and another.

DISCUSSION

Polybenzimidazoles

One of the first all-aromatic heterocyclic polymers to undergo development as a high temperature adhesive was a polybenzimidazole (PBI). In 1961, Vogel and Marvel (1) reported the synthesis of high molecular weight PBI from the melt polycondensation of 3,3',4,4'-tetraaminobiphenyl (TAB) and diphenyl isophthalate (eq. 1). In high molecular weight form, the early polymer was infusible and insoluble.

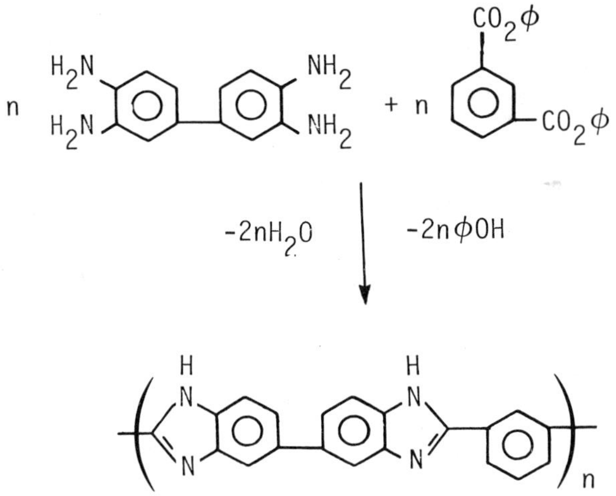

(1)

To obtain processable material, capable of wetting and flowing, the prepolymer concept was originated. This involved the preparation of a low molecular weight oligomer (dimer to trimer) from the partial reaction of a stoichiometric mixture of the monomers.

A commercial form of the processable PBI, Imidite 850 (formerly available from Narmco Materials, Inc.), was extensively evaluated as a high temperature adhesive. Although severe processing problems were encountered with Imidite 850 due primarily to the evolution of a large amount of volatiles (phenol and water), PBI provided respectable adhesive performance (2,3). Typical TSS on PH 15-7 molybdenum stainless steel adherends were 4000 psi at room temperature (RT), 2500 and 1100 psi at 300°C after 100 and 200 hr at 300°C in air and 1100 psi at 538°C after 10 min. Perhaps the most impressive behavior of PBI as an adhesive was in the cryogenic temperature region where TSS of 4600 psi were obtained at -196°C. In addition, fatigue (10^6 cycles) resistance was better at -196°C than at RT (2200 psi vs 1500 psi).

To achieve maximum adhesive properties at the higher temperatures (e.g., 538°C), PBI, like other high temperature adhesives, must be postcured. The final postcure temperature depends upon the maximum use temperature. Postcure temperatures as high as 454°C have been employed. The high temperature postcure induces cross-linking in the polymer which is generally reflected in lower room temperature TSS (more brittle) but improved short term performance at elevated temperatures. A PBI similar to Imidite 850 is available in experimental quantities and is still under evaluation as an adhesive (4).

Other benzimidazole polymers where the imidazole hydrogen has been substituted with a phenyl (5) or arylene group (6) have also undergone preliminary adhesive evaluation. Although these polymers are more thermo-oxidatively stable than the conventional PBI, elimination of the hydrogen bonding resulted in lower glass transition temperatures (T_gs) and accordingly, pronounced thermoplasticity at elevated temperatures. Although postcure studies were conducted, only moderate improvement in high temperature performance was attained. This was generally accompanied by poorer thermo-oxidative stability than exhibited in the linear polymer.

Condensation Polyimides

Concurrent with the development of PBI as a high temperature adhesive, work was also performed on polyimides (PI). The early synthesis of PI utilized the method of Edwards (7) and Endrey (8) where a soluble processable polyamic acid was formed from the stoichiometric reaction of an aromatic diamine with an aromatic tetracarboxylic acid dianhydride (eq. 2). The polyamic acid was subsequently cyclodehydrated to the PI. In adhesive work, this ring closure was accomplished thermally.

Early PI were insoluble and infusible. The polyamic acid form was the processable precursor for PI. This form, however, has two major shortcomings, hydrolytic instability (9) and volatiles (H_2O). The evolution of volatiles causes pronounced processing difficulties, particularly in large area bonding, and porosity in the bondline. In spite of this, certain PI adhesives are still processed in the polyamic acid form. Early adhesive work

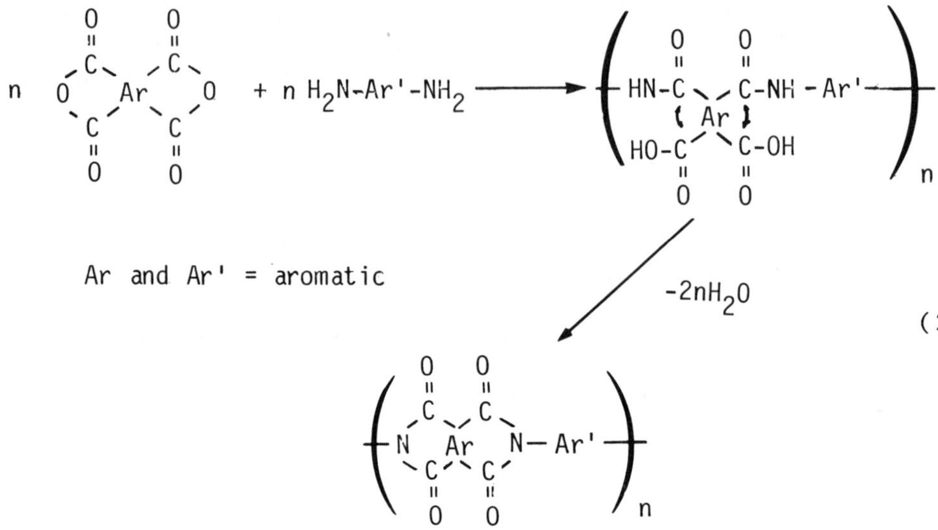

(2)

on PI with systems such as PI-3301 and PI-1101 (DuPont) provided TSS of 3000 psi at RT and 1900 and 1400 psi at 300°C after 100 and 1000 hr, respectively. More recent adhesive properties of a commercially available PI adhesive (FM 34B-18)* are stainless steel TSS of 4050 psi at RT, 2250 psi at 260°C after 750 hr at 260°C, and 1281 psi at 260°C after 192 hr at 316°C.

In an attempt to circumvent the volatile problem, Freeman and coworkers (10) were the first to process polyimides as thermoplastics, above the T_gs. Generally, pressures ~200 psi and temperatures ~399°C were employed. Representative TSS on titanium adherent (Ti/Ti) of the PI from the reaction of 3,3',4,4'benzophenonetetracarboxylic acid dianhydride (BTDA) and 1,3-phenylenediamine (eq. 3) were ~3000 psi at RT and ~2000 psi at 288°C after 100 hr at 288°C in air.

Other workers have since processed PI by virtue of their thermoplastic nature. The polymer must have the

*Bloomingdale Products Brochure BPT-217, American Cyanamid Co.

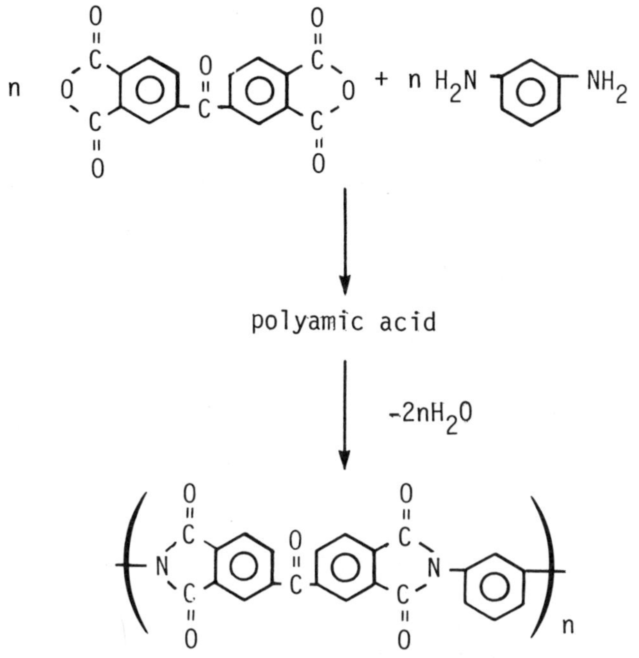

(3)

proper rheological behavior above the T_g to allow good flow and wetting in a bond. It should be noted that too much flow in an adhesive is as detrimental as too little flow. To obtain the proper combination of properties, there is generally a compromise between processability and high temperature performance, as governed by the T_g. New monomers, primarily diamines, became available and were used to make PI with better processability. An example was the work of Bell and coworkers (11,12) who studied structure/property relationship of PI using isomeric diamines. One polymer, from the reaction of BTDA and 3,3'-diaminobenzophenone (DABP) (eq. 4), had a favorable combination of attractive properties. This polymer has been referred to as LARC-2 and LARC-TPI.

St. Clair pioneered the evaluation of the amic acid and imidized form of this polymer as a high temperature adhesive. Bis(2-methoxyethyl)ether (diglyme) was found to be an excellent solvent for the polyamic acid (13) and when used in adhesive work, Ti/Ti TSS as high as 6180 psi at RT, 2600 psi at 225°C, and 950 psi at 250°C were obtained (14). The T_g of the BTDA/DABP polyimide has been reported to be 256 (14), 264 (11), and 275°C (15). Hence it was not surprising to observe thermoplastic failure of adhesive specimens at

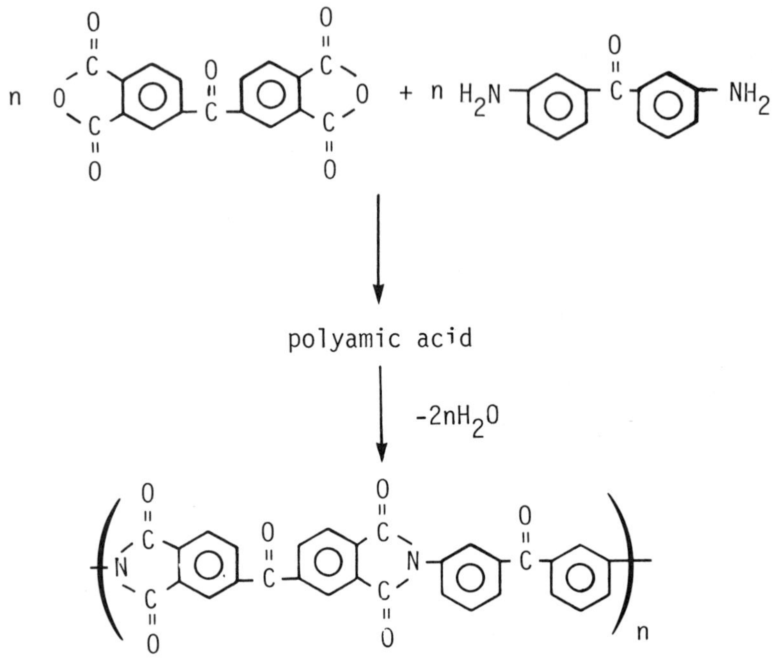

(4)

250°C. More recent adhesive work has shown that LARC-TPI has good resistance to aircraft solvents, moisture, and creep at 232°C. Ti/Ti TSS of 3140 psi at 232°C after 5000 hours at 232°C were obtained (16). LARC-2 has also performed well as a high temperature adhesive for bonding Kapton® (a polyimide film) (17). Experimental samples of LARC-TPI are available (18).

Many other linear condensation PI have undergone preliminary high temperature adhesive evaluation (e.g., DuPont's NR-150 A2 and B2). A modification of NR-150 PI is NR-056X, made from the reaction of 2,2'-bis(3,4-dicarboxyphenyl)hexafluoropropane, 75 mole % of 1,4-phenylenediamine and 25 mole % of 4,4'-oxydianiline in diglyme (eq. 5). Ti/Ti TSS (65 phr Al filler) were 3500 psi at RT and 2200 at 316°C (19). After aging at 232°C for 5000 hr and testing at 232°C, TSS of 2000 psi were obtained (20).

Of the many PI evaluated as high temperature adhesives, only a few are commercially available, such as American Cyanamid's FM34B and FM36. Both are PI adhesive tapes with high volatiles, due to the condensation by-products and residual solvent. The role of the residual solvent in the adhesive tape is to provide flow, tack and drape. Without

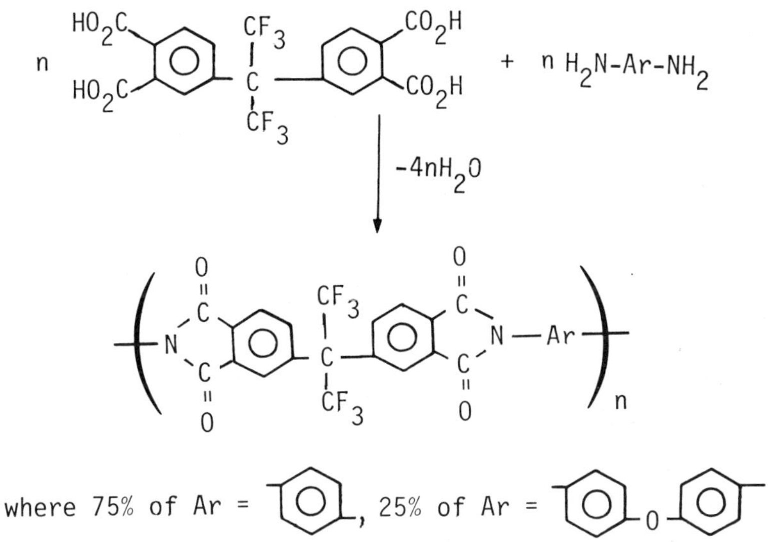

(5)

it, the adhesive tape would be extremely boardy and difficult to handle, especially in complex adhesive layups.

Addition PI

In an attempt to develop high temperature PI adhesives which are essentially free of volatiles, yet retain good processability, ethynyl (acetylenic) and norbornene (nadic) groups have been used to end-cap amic acid oligomers. Nadic end-capped amic acid oligomers can be thermally converted to the imide oligomer apparently without inducing reaction of the nadic end-groups. However, the ethynyl-terminated amic acid oligomer must be chemically imidized since thermal imidization induces reaction of the ethynyl end-groups. The resultant acetylene-terminated imide (ATI) has poor solubility and when free of solvent, poor flow. To improve the solubility and flow characteristics, recent work has involved the chemical conversion of the ethynyl-terminated amic acid oligomer to isoimide oligomer (21). The isoimide form is reported to exhibit better solubility and flow properties than the imide. To date, no adhesive work on the ethynyl-terminated isoimides has been reported. It is worthy to note that polyisoimides (iminolactones) were initially reported in 1963 (22). The isoimide rearranges to the imide at high temperatures (~300-350°C) (23). The ethynyl-terminated imides evolved from the effort of Bilow and coworkers (24) while the nadic-terminated imides (NTI) emanated from the early work of Lubowitz (25).

The principal problems with using the ethynyl group to terminate high temperature oligomers is that the ethynyl group often begins to react prior to the formation of a melt, thereby inhibiting flow and wetting. Hydroquinone has been used to retard the reaction of the ethynyl groups (and possibly serve as a plasticizer) to improve the processability of ATI (26). The product formed from the thermally induced reaction of the acetylenic group is less thermally and oxidatively stable than the parent polymer. In spite of these drawbacks, various ATI's such as that in Eq. 6 have been evaluated as high temperature adhesives, providing Ti/Ti TSS as high as 3200 psi at RT, 1900 psi at 232°C after 1000 hr at 232°C, and 1200 psi at 260°C after 1000 hr at 260°C (27).

NTI with a molecular weight of ~1300 g/mole, prepared from the reaction of appropriate molar amounts of BTDA, 3,3'-diaminodiphenylmethane, and nadic anhydride in N,N-dimethylformamide, was evaluated as a high temperature adhesive (eq. 7). This material has been referred to as LARC-13. The nadic group on NTI undergoes a thermally

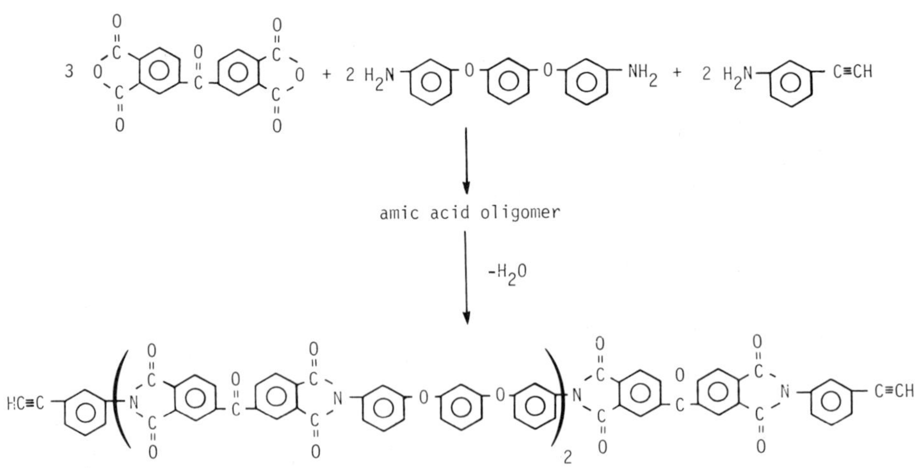

(6)

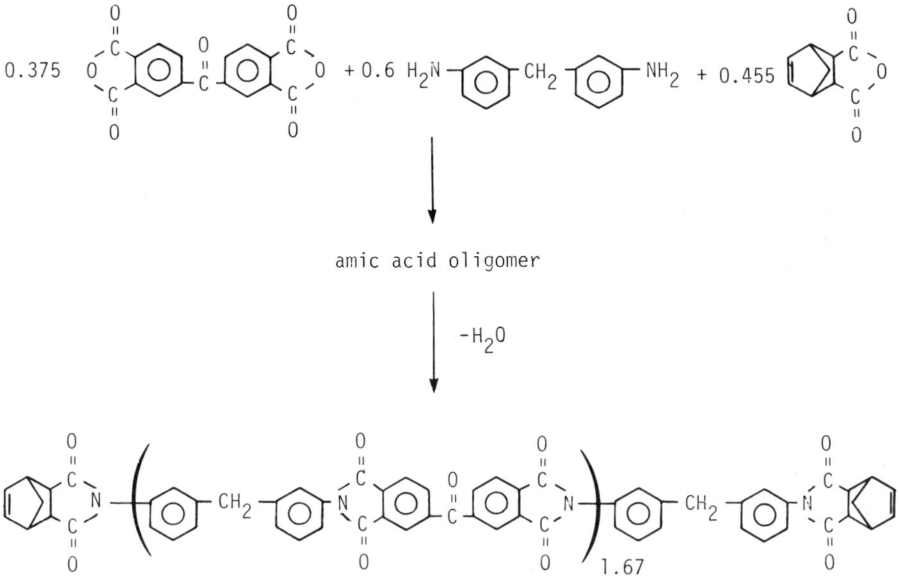

(7)

induced reaction with the evolution of a small amount of cyclopentadiene to yield a brittle resin. Ti/Ti TSS (Aℓ filled) as high as 3300 at RT and 2800 psi at 260°C have been reported (28). LARC-13 exhibited a drop in TSS to 1200 psi at 232°C after aging at 232°C for 2000 hr (20). To toughen LARC-13, the technology used in toughening epoxies by the addition of rubbers (29) was employed. Various elastomers (e.g., 15% Dow Corning Sylgard 184) incorporated within LARC-13 improved the T-peel strength (toughness) and provided surprisingly respectable Ti/Ti TSS (2200 psi) at 232°C after 500 hr at 232°C (30). Other work has been done with LARC-13 such as blending with a polyamideimide (AI 1130L, Amoco) to form A7F to control the flow and improve filleting in sandwich structure (31). An addition PI adhesive (BXR10214-151C) based upon NTI chemistry is available from American Cyanamid. Typical Ti/Ti TSS are 3200 psi at RT and 2100 psi at 260°C.

Polyquinoxalines

Polyquinoxalines (PQ) from the reaction of aromatic bis(o-diamines) with aromatic bis(glyoxals) were first reported in 1964 (32,33). The reaction of bisglyoxals with aromatic bis(o-diamines) yields a configurationally unordered polymer, containing three isomers along the chain. PQs from aromatic bis(o-diamines) are totally amorphous and unlike most PIs, are soluble in the ring-closed high molcular weight form in solvents such as m-cresol. This permits solvent impregnation of the carrier and subsequent drying to low level of volatiles. A PQ from the reaction of TAB with 4,4'-oxybis(phenylgloyoxal hydrate) (eq. 8) filled with amorphous boron provided the highest 371°C TSS of any organic adhesive after 371°C aging. Values of 3350 psi at RT, 2280 at 316°C after 200 hr at 316°C, 2540 psi at 371°C after 50 hr at 371°C, and 1325 psi at 538°C were reported (34). The bonds were fabricated under the following conditions; 1 hr each at 344, 426, and 455°C under 200 psi.

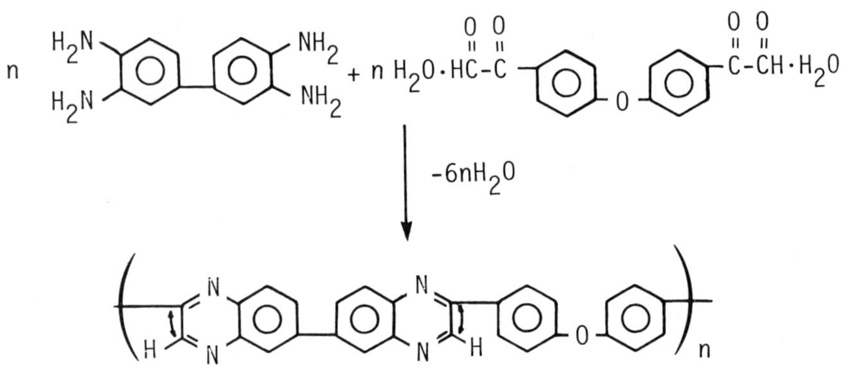

(8)

The temperature and pressure employed in the fabrication of the bonds were beyond the capability of conventional processing equipment. Conventional autoclaves are defined as those with maximum operating temperatures of 177°C even though there are relatively large autoclaves with operating temperatures of 316°C. Smaller laboratory autoclaves are available with 427°C capability. In bonding at the higher temperatures (e.g., >316°C), severe problems are encountered with bagging materials and sealants, and autoclave malfunctions are frequent. In addition, surface degradation of the adherend, residual stresses in the bondline, and thermal and/or oxidative degradation of the adhesives are typical of the detrimental effects of processing at high temperatures.

<u>Condensation Polyphenylquinoxalines</u>

Polyphenylquinoxalines (PPQ) are prepared from the reaction of aromatic bis(*o*-diamines) and bis(phenyl α-diketones) as initially reported in 1967 (35). The pendant phenyl group and the isomeric units in the polymer backbone impart good solubility and thermoplasticity to the relatively high molecular weight PPQ. Utilizing the thermoplastic nature of PPQ, large area (4 ft. x 4 ft.) metal-to-metal bonded panels have been successfully fabricated using adhesive tape containing low volatiles (<1%). The properties (processability and mechanical performance) of PPQ as with other polymers can be controlled by altering the chemical backbone and molecular weight.

Adhesive work on the Supersonic Transport (SST) Program at Boeing involved the polymer from the reaction of TAB and 4,4'-oxydibenzil (ODB) (eq. 9). Ti/Ti (phosphate

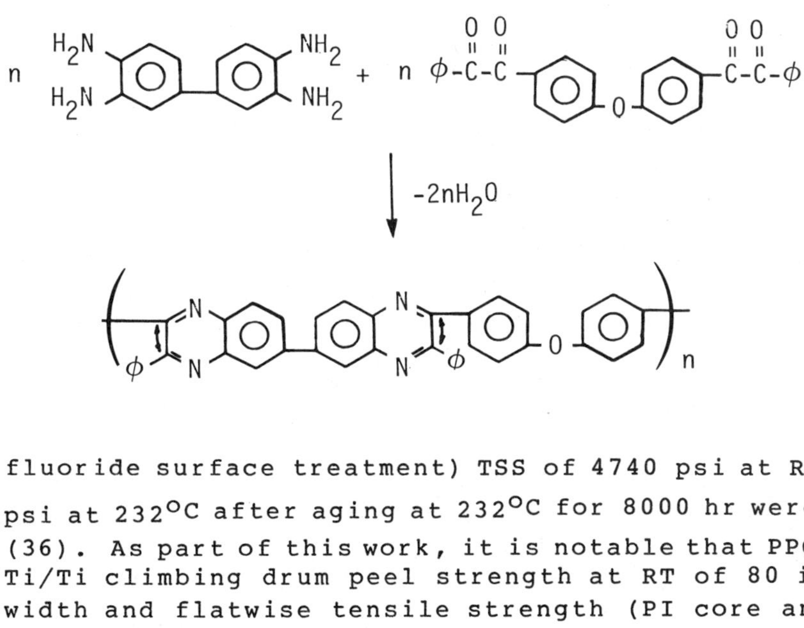

(9)

fluoride surface treatment) TSS of 4740 psi at RT and 3350 psi at 232°C after aging at 232°C for 8000 hr were reported (36). As part of this work, it is notable that PPQ provided Ti/Ti climbing drum peel strength at RT of 80 in-lb/3 in width and flatwise tensile strength (PI core and Ti face sheets) of 950 psi at RT and 825 psi at 232°C (36). More recent adhesive work with PPQ involved a study of Ti surface treatment (20). Chromic acid anodized surface treatment of Ti adherends provided TSS of 5000 psi at RT, 2000 psi at 232°C after 5000 hr at 232°C, and low values at 232°C after

10,000 hr at 232°C. Analysis of the failed specimens after 10,000 hr at 232°C revealed failure in the oxide layer (20).

An interesting phenomenon was observed in adhesive work with the PPQ from the reaction of TAB with 1,3-bis(phenylglyoxalyl)benzene. The polymer exhibited an apparent T_g of 318°C. Ti/Ti TSS were 4400 psi at RT, 2500 psi at 288°C, 300 and 500 psi at 316°C after 10 min and 300 hr at 316°C respectively. Thermoplastic failure occurred at 316°C. Using the same bonding conditions, adhesive bonds were fabricated using composite adherends (matrix was DuPont's NR 150B2 polimide with Hercules HTS carbon/graphite reinforcement). TSS were 6000 psi at RT, 3700 psi at 288°C, 2800 and 2500 psi at 316°C after 10 min and 300 hr respectively at 316°C in air (36). The reason for the improvement in the 316°C strength is unknown although it is thought that mechanical interlocking between the adhesive and the composite occurred. PPQ are unavailable as adhesive tapes although a PPQ solution is available in experimental quantities (38).

Addition PPQ

Addition PPQ work primarily involved the end-capping of phenylquinoxaline oligomers with acetylenic groups. In the initial report by Kovar and Arnold (39) in 1975, α-diketone-terminated phenylquinoxaline oligomers were end-capped with 3(3,4-diaminophenoxy)phenylacetylene. An alternate and improved approach was reported in 1976 (40) where o-diamino-terminated phenylquinoxaline oligomers were end-capped with 4(4-ethynylphenoxy)benzil. Acetylene-terminated phenylquinoxalines (ATPQ) exhibit shortcomings similar to those of the ATI. The thermal reaction of the ethynyl group inhibits flow and the cured ATPQ are less thermo-oxidatively stable than a comparable PPQ (41). Representative Ti/Ti TSS of the ATPQ shown in Eq. 10 were 4730 psi at RT, 1350 psi at 260°C after 500 hr at 260°C, and 1525 psi at 316°C (41).

Cross-linked PPQ

In 1981, PPQ containing various amounts of pendant phenylethynyl groups were reported (42). These groups undergo a thermally induced reaction to cross-link the PPQ

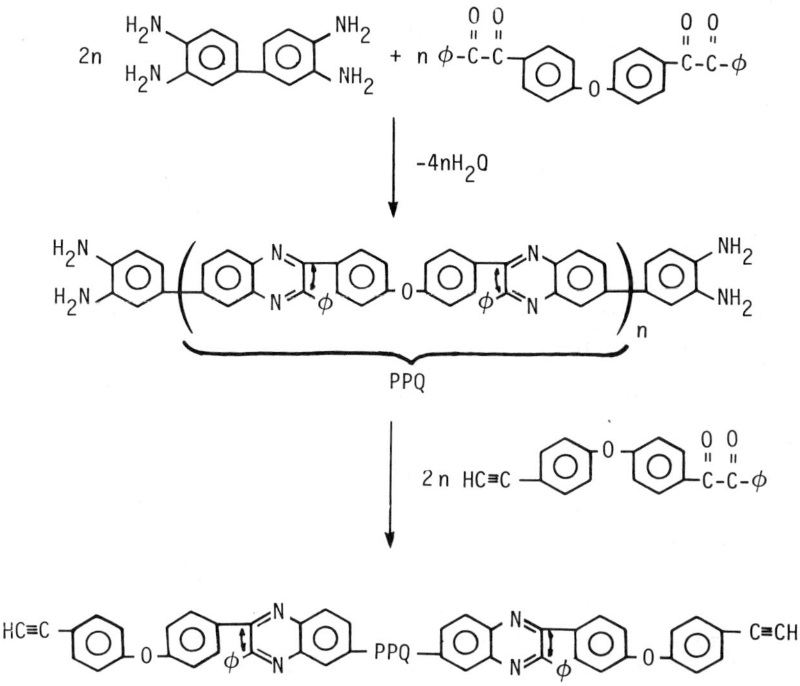

(10)

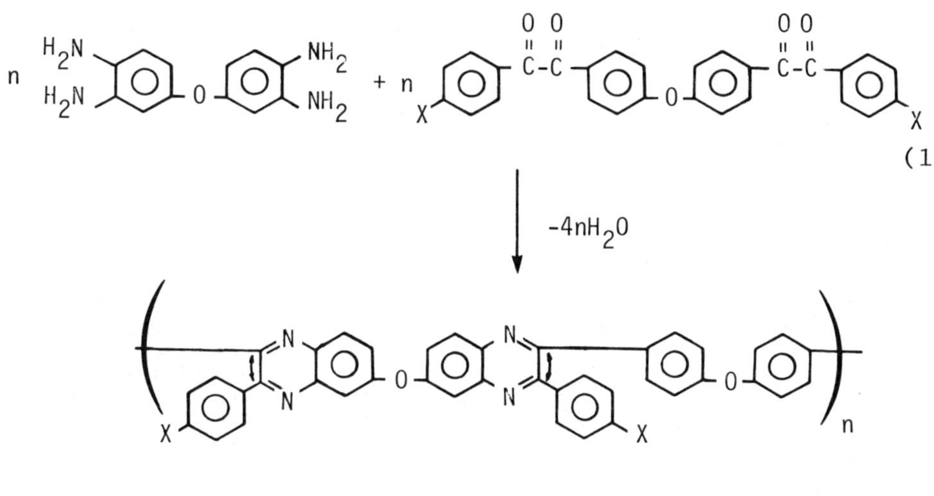

(11)

Where 10 mole % of X = C≡C-ϕ and 90 mole % of X = H

Where 10 mole % of X = C≡C-ϕ and 90 mole % of X = H.

and substantially increase the T_g. Preliminary adhesive evaluation of a phenylethynyl containing PPQ from the reaction of 3,3',4,4'-tetraaminodiphenyl ether (TADE), ODB (90 mole%), and 4,4'-oxybis(4"-phenylethynylbenzil) (10 mole %) (eq. 11) provided Ti/Ti TSS of 4520 psi at RT and 2700 psi at 232°C after 1000 hr at 232°C (43). TSS of a comparable PPQ from the reaction of TADE and ODB was 5150 at RT and 1510 at 232°C after 1000 hr at 232°C. The difference in the 232°C strength between the two PPQ is due to cross-linking. The PPQ void of phenylethynyl groups had a Tg of 256°C whereas the PPQ containing 10 mole % of phenylethynyl groups had a T_g of 281°C.

Other Polymers

Many other polymers have undergone evaluation as high temperature adhesives such as Upjohn 2080 polyimide, polyamideimides (e.g., Torlon), polybenzimidazoquinazolines, polybenzothiazoles, polybenzoxazoles, polyoxadiazoles, and polyphenyl-*as*-triazines. A few of these materials such as Upjohn 2080 polyimide and polybenzothiazoles showed good TSS at elevated temperatures. However, the problems which have plagued other high temperature adhesives are also present in these materials. Such problems include inadequate flow, volatile evolution, high processing temperature and pressures, and thermoplasticity at elevated temperatures.

CONCLUDING REMARKS

Although several polymers have shown promise as high temperature adhesives, they are relatively expensive and/or demand processing temperatures >316°C. Few companies have production equipment capable of operating at 316°C except for small area bonding. In addition, flexible bagging material and sealants to operate at these temperatures are a problem. Those companies that have a special need for a high temperature adhesive often develop a material for that particular application and do not rely upon a commercial source. These special needs vary from the joining of electronic elements to the bonding of ceramic components in a missile. It is unlikely that a single polymer will meet the requirements of all the applications.

Very little high temperature adhesive development has been sponsored by companies who supply adhesives, primarily because of a questionable market. Although research and development work is continuing on high temperature adhesives to meet future needs, there is no single application that demands a sizable quantity of a high temperature adhesive at this time. As improvements are made in the processing and performance of high temperature adhesives, a larger market will develop which should lower the cost of adhesive bonding for high temperature applications.

The use of trade names of manufacturers does not constitute an official endorsement of such products or manufacturers, either expressed or implied, by the National Aeronautics and Space Administration.

REFERENCES

1. H. Vogel and C.S. Marvel, J. Polym. Sci., 50, 511 (1961).

2. H.H. Levine, in Encycl. Polym. Sci. and Technol., (H.F. Mark, N.G. Gaylor, and N.W. Bikales, eds.) Vol. 11, Intersciences Pub., Inc., New York, 1969, p. 188.

3. H.H. Levine et al. AFML-TR-64-365, Pt. 1, Vol. 1, Dec. 1963.

4. Chad Delano, Aerotherm Div./Acurex Corp., private communication.

5. H.H. Levine, N.P. Loire, and C.B. Delano, AFML-TR-67-63, 1967.

6. A.A.R. Sayigh, B.W. Tucker, and H. Ulrich, U.S. Pat. 3,708,439 (1973) (To Upjohn Co.).

7. W.M. Edwards, U.S. Pat. 3,179,614 and 3,179,634 (1965) (To DuPont).

8. A.L. Endrey, U.S. Pat. 3,179,631 and 3,179,633 (1965) (To DuPont).

9. L.W. Frost and I. Kesse, J. Appl. Polym. Sci., 8, 1039 (1964).

10. H.A. Burgman, J.H. Freeman, L.W. Frost, G.M. Bower, E.J. Traynor and C.R. Ruffing, J. Appl. Polym. Sci., 12, 805 (1968).

11. V.L. Bell, B.L. Stump, and H. Gager, J. Polym. Sci., Polym. Chem. Ed. 14, 2275 (1976).

12. V.L. Bell, U.S. Pat. 4,094,862 (1978) (To NASA).

13. D.J. Progar, V.L. Bell, and T.L. St. Clair, U.S. Pat. 4,065,345 (1977) (To NASA).

14. D.J. Progar and T. St. Clair, 7th Nat'l. SAMPE Tech. Conf. Series, 7, 53 (1975).

15. A.K. St. Clair and T.L. St. Clair, Sci. Adv. Mat'l. and Proc. Eng. Series, 26 (165 (1981).

16. S.G. Hill, P.D. Peters, and C.L. Hendricks, NASA Contractor Report 165944 (1982).

17. A.K. St. Clair, W.S. Slemp, and T.L. St. Clair, Sci. Adv. Mat'l. Proc. Eng. Series, 23, 113 (1978).

18. Kiran Baksi, Gulf Research and Development Co., private communication.

19. P.S. Blatz, Adhesives Age, 21, No. 9, 39 (1978).

20. C.L. Hendricks and S.G. Hill, SAMPE Quart., 12, 32 (1981).

21. A.L. Landis and A.B. Naselow, 14th Nat'l. SAMPE Tech. Conf. Series, 14, 236 (1982).

22. R.J. Angelo, U.S. Pat. 3,345,342 (1967) (To DuPont).

23. F.P. Gay and C.E. Berr, J. Poly. Sci. A-1, 6, 1935 (1968).

24. N. Bilow, A.L. Landis, and L.J. Miller, U.S. Pat. 3,845,018 (1974) (To Hughes Aircraft Co.).

25. H.R. Lubowitz, U.S. Pat. 3,528,950 (1970) (To TRW, Inc.).

26. R.J. Kuhbander and T.J. Aponyi, 11th Nat'l. SAMPE Tech. Conf. Series, 11, 295 (1979).

27. N. Bilow, A.L. Landis, R.H. Boschan, and J.G. Fasold, SAMPE Journal, 18 (1), 8 (1982).

28. T.L. St. Clair and D.J. Progar, Sci. Adv. Mat'l. Proc. Eng. Series, 24 (2), 1081 (1979).

29. J.N. Sultan and F.J. McGarry, Polym. Eng. Sci., 13 (1), 29 (1973).

30. A.K. St. Clair and T.L. St. Clair, Int. J. Adhesion and Adhesives, 5, 249 (1981).

31. S.G. Hill and C.H. Sheppard, 12th Nat'l. SAMPE Tech. Conf. Series, 12, 1040 (1980).

32. G.P. deGaudemaris and B.J. Sillion, J. Polym. Sci., B-2, 203 (1964).

33. J.K. Stille and J.R. Williams, J. Polym. Sci., B-2, 209 (1964).

34. P.M. Hergenrother and H.H. Levine, J. Appl. Polym. Sci., 14, 1037 (1970).

35. P.M. Hergenrother and H.H. Levine, J. Polym. Sci., A-1, 5, 1453 (1967).

36. P.M. Hergenrother, SAMPE Quart., 3 (1), 1 (1971).

37. P.M. Hergenrother and D.J. Progar, Adhesives Age, 20, No. 12, 38 (1977).

38. R.B. Gosnell, King-Mar Laboratories, private communication.

39. R.F. Kovar, G.F.L. Ehlers, and F.E. Arnold, Am. Chem. Soc. Div. Polym. Chem. Prepr., 15, 1081 (1977); R.F. Kovar and F.E. Arnold, U.S. Pat. 3,966,729 (1976) (To U.S. Dept. of the Air Force).

40. P.M. Hergenrother, Am. Chem. Soc. Div. Org. Coat. Plast. Chem. Prepr., 36 (2), 264 (1976); Chemistry and Prop. of Cross-linked Polymers, (S.S. Labana, ed.), Academic, New York, 1977, p. 107.

41. P.M. Hergenrother, Poly. Eng. & Sci., 21 (16), 1072 (1981).

42. P.M. Hergenrother, Am. Chem. Soc. Div. Polym. Chem. Prepr., 22 (2) 102 (1981; Macromolecules, 14, 898 (1981).

43. P.M. Hergenrother, J. Appl. Polym. Sci., 28, 355 (1983).

Polyimide Adhesives: Modified with ATBN and Silicone Elastomers

Anne K. St. Clair, Terry L. St. Clair and Stephen A. Ezzell

NASA Langely Research Center
Hampton, VA 23665

ABSTRACT

A series of studies evaluating the effects of added elastomers on the properties of a high-temperature addition polyimide adhesive is reviewed. First, thermoset polyimides containing various butadiene/acrylonitrile and silicone elastomers were synthesized either as physical polyblends or by chemically reacting the elastomers directly onto the polymer backbone. The modified adhesive resins were characterized for thermomechanical properties, fracture toughness and adhesive strength. A second series of elastomer-containing polyimides was also prepared in order to study the effects of the elastomer chain length on polymer properties. Aromatic amine-terminated silicone rubbers with repeat units varying from n=10 to 105 were reacted onto the polyimide backbone, and the resulting polymers were characterized for their adhesive properties.

INTRODUCTION

Because of their thermal stability and easy processability, thermoset addition polyimides are candidate high-temperature adhesives for bonding composite materials and metals on future aircraft and spacecraft. Addition polyimides are cured via reactive unsaturated end groups causing them to be highly cross-linked, insoluble and extremely brittle. A mechanism for toughening addition polyimides is needed to provide these otherwise excellent structural materials with such properties as improved peel strength and resistance to adhesive fracture.

The rubber-toughening process has been one of the most successful methods for modifying polymer toughness. The incorporation of small amounts of rubber into polymer matrices has resulted in significantly improved fracture resistance (1). Particularly interesting improvements in the toughness of epoxies have been accomplished by the addition of carboxyl-terminated butadiene/acrylonitrile (CTBN) elastomers (2,3).

Recent studies at NASA Langley Research Center have been directed towards the toughening of higher temperature thermoset addition polyimides (4,5). ATBN and silicone elastomers have been incorporated into an addition polyimide adhesive, LARC-13 (6,7), which is a candidate for bonding metals such as titanium and graphite/polyimide

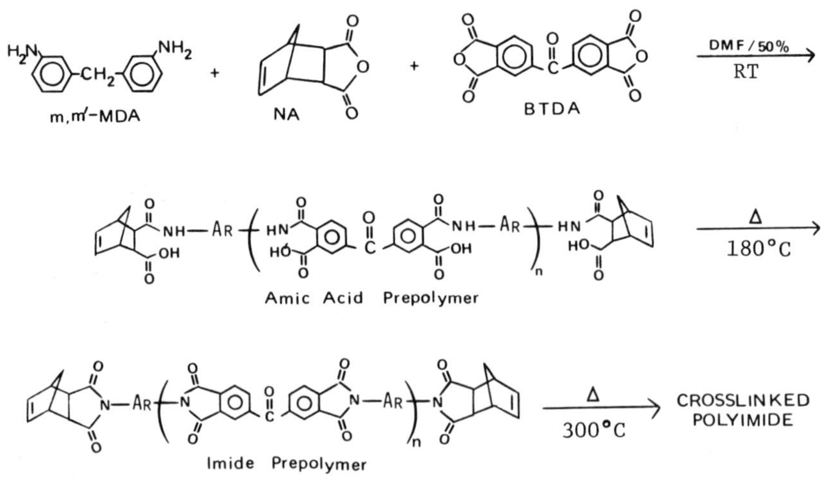

Fig. 1. LARC-13 adhesive chemistry

composites for applications on advanced aircraft. Some of the results from these studies evaluating the effects of added elastomers and the variation of elastomer chain length on the adhesive properties of LARC-13 will be reviewed.

ADHESIVE SYNTHESIS AND CHARACTERIZATION

Preparation of LARC-13 Adhesive

The synthetic route for preparing the addition polyimide adhesive LARC-13 is displayed in Fig. 1. Appropriate quantities of 3,3'-methylenedianiline (m,m'-MDA), nadic anhydride (NA), and 3,3',4,4'-benzophenonetetracarboxylic acid dianhydride (BTDA) are combined in an amide solvent at a concentration of 50% solids by weight to form a 1300 molecular weight amic acid. This amic acid prepolymer is used as the adhesive resin and is later cured to the cross-linked polyimide by the removal of solvent, imidization, and bonding under pressure at 300°C.

Preparation of Elastomer-Containing Adhesives

Structures of the various butadiene/acrylonitrile and silicone elastomers used in preparing these adhesives are represented in Fig. 2. Silastic LS-420 and Sylgard 184 elastomers were obtained from Dow Corning. The aromatic amine-terminated butadiene/acrylonitrile (ATBN) is a highly viscous experimental elastomer and was contributed by B. F. Goodrich Research Center, Brecksville, Ohio. The aromatic amine-terminated liquid silicone elastomers were prepared by M & T Chemicals, Rahway, NJ.

Two approaches were used to incorporate these elastomers into the base LARC-13 resin at a concentration of 15% by weight. The first method was to physically blend the Silastic and Sylgard silicone rubbers into the LARC-13 amic acid prepolymer solution. The second method was to chemically react the rubber directly onto the prepolymer backbone by replacing a portion of the MDA diamine with an aromatic amine-terminated elastomer, ATBN or ATS, so that the elastomer was equivalent to 15% by weight of the total solution solids. The addition of the elastomeric materials to LARC-13 caused a reduction in the viscosity of the amic acids in all cases.

SILASTIC

SYLGARD

ATBN

ATS

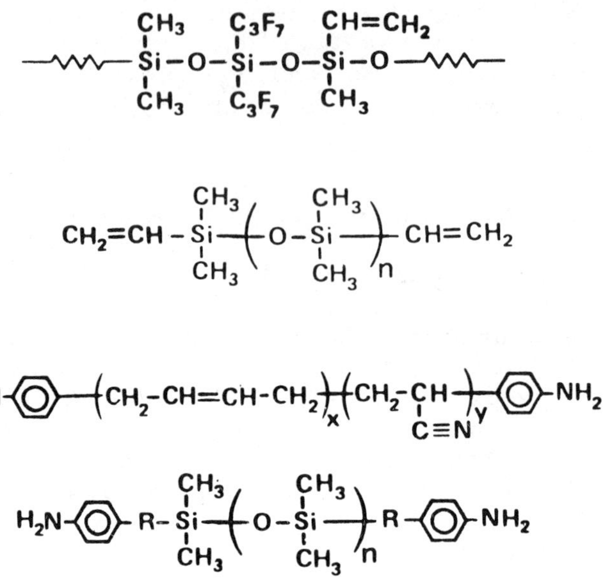

Fig. 2. Elastomer structures

Compatibility and Phase Separation

Varying degrees of compatibility were achieved between resin and elastomer phases. The physically blended solutions were transparent with rubber particles dispersed throughout. After curing at 300°C, however, the polymer blends appeared homogeneous. The LARC-13/ATBN was both translucent and homogeneous before and after cure. The LARC-13/ATS solution became turbid as molecular weight build-up was achieved and was opaque upon curing.

Two separate phase transitions were displayed for each of the cured elastomer-containing resins as evidenced by torsional braid analysis (TBA). A distinct glass transition temperature (T_g) of the LARC-13 phase was represented by a high-temperature peak between 277° and 300°C; and a distinct T_g was displayed by the elastomers at low temperature from -115° to -65°C. TBA spectra of the LARC-13 control and ATS-modified resin are reproduced in Fig. 3. Further evidence of microphase separation between

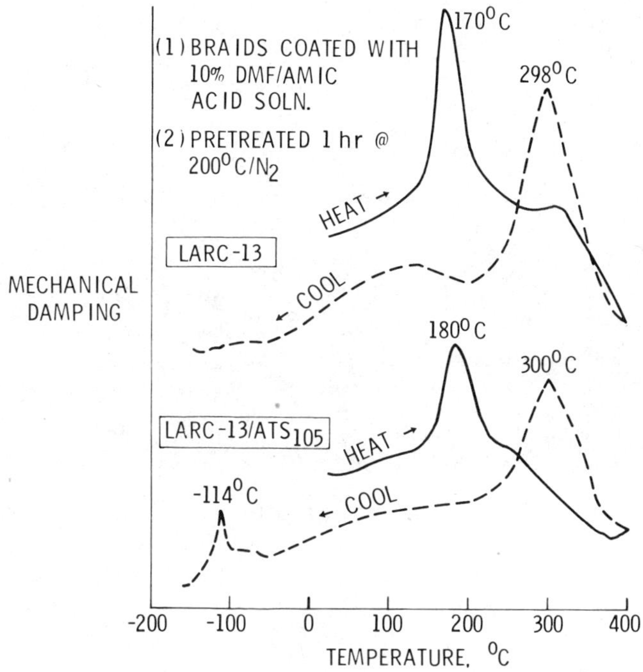

Fig. 3. Torsional braid analysis spectra of LARC-13 and LARC-13/ATS resins

elastomer and the matrix is shown in the scanning electron photomicrograph of LARC-13/ATS resin, Fig. 4. After a five-minute etch in chromic and phosphoric acids, the adhesive surface shows elastomer particles (cavities remaining after removal of particles) to be relatively uniform in distribution and size having an average diameter on the order of 0.5 μm. Micrographs of the other elastomer-modified adhesives were unobtainable due to failures at staining the elastomer particles.

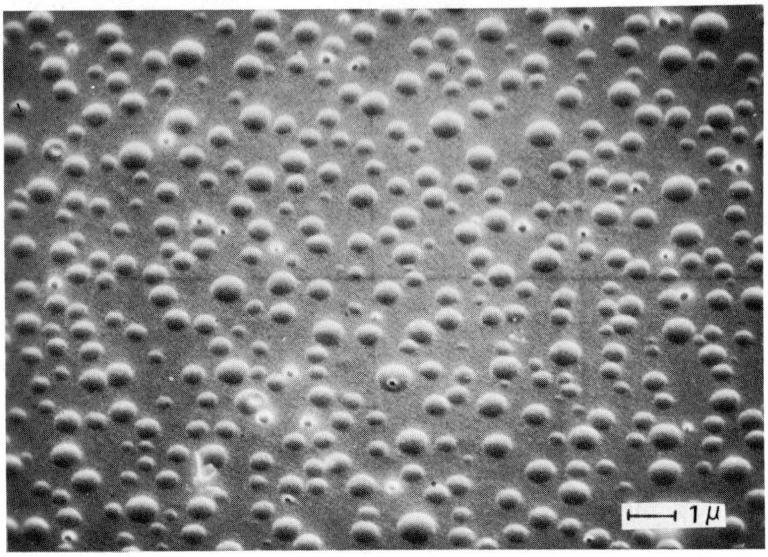

Fig. 4. Photomicrograph of LARC-13/ATS

EFFECT OF ELASTOMERS ON ADHESIVE PROPERTIES

Single Lap Shear Bond Testing

The adhesive resins containing 30% (w/w) aluminum powder were applied to a 112 E-glass carrier cloth and staged to 180°C prior to bonding. Simultaneously, 6Al-4V titanium adherends (2.54 cm wide, 1.27 mm thick) were prepared for bonding using the Pasa Jel 107 surface treatment. Adherends were primed with adhesive and heated for one hour at 100°C. Adhesive-coated scrim cloths were sandwiched between primed adherends with a 1.27 cm overlap and bonded as follows:

(1) RT to 300°C at 5°C/min, hold 30 min;

(2) Apply 50 psi, hold 30 min;

(3) Cool at 5°C/min to 150°C or below.

Lap shear tests were performed according to ASTM D-1002 at a cross-head speed of 0.127 cm/min. Elevated temperature tests were conducted after soaking the specimens for 10 min at test temperature.

Results from room temperature and elevated temperature lap shear testing of the elastomer-modified LARC-13/titanium bonds are presented in Table 1. After 500 hr at elevated temperature, all four adhesive formulations were reduced in strength compared to LARC-13 control specimen. However, it should be noted that the silicon-containing adhesives showed improved strengths at 260°C after aging at elevated temperature. Such behavior is probably due to additional cross-linking.

T-Peel Testing

T-peel specimens were bonded by a similar method to the one described previously using 0.25 mm grit-blasted titanium sheets. Tests were conducted according to ASTM D-1876 at a crosshead speed of 25.4 cm/min. Results from room temperature and 232°C testing of unaged and aged specimens are shown in Table 2. All four of the elastomer-modified

Table 1. Lap Shear Strengths (LSS) of Adhesive/Titanium Bonds

Adhesive	LSS,psi Samples Unaged			LSS,psi Aged 500 hr @ 232°C		
	RT	232°C	260°C	RT	232°C	260°C
LARC-13	2900	2800	2100	2900	2900	2500
LARC-13/ Silastic	2200	1800	1100	1600	1800	1500
LARC-13 Sylgard	3300	2400	1500	2500	2300	2100
LARC-13/ ATBN	3700	2800	2200	2000	1900	1900
LARC-13/ ATS	2600	1800	1100	1700	1600	1600

resins showed improved peel strengths compared to LARC-13 indicating an increase in toughness. The Silastic and ATS-modified adhesives displayed a 6- to 7-fold increase in peel strength at room temperature.

Adhesive Fracture Energy

A further indication that an increase in toughness has been achieved for these systems is evidenced by an improvement in adhesive fracture energy. Three of the modified resins were tested for fracture toughness at the Naval Research Laboratory, Washington, DC, using the aluminum tapered double-cantilever beam specimen (8,9). The fracture energies (opening mode strain energy release rates) are listed in Table 3. Tests were performed within 2 weeks of receipt of the resins and were repeated 4 months later. In general, failures were cohesive, and crack propagation continuous. The in-chain ATBN and ATS elastomers produced a definite toughening effect on the LARC-13. The LARC-13/ATBN retained its strength at elevated temperature.

Table 2. T-Peel Strengths of Adhesive/Titanium Bonds

Adhesive	T-Peel Strength, lbs/in		T-Peel Strength, lbs/in after 500 hr @ 232°C	
	RT	232°C	RT	232°C
LARC-13	1.3	2.9	1.9	---
LARC-13/Silastic	8.7	5.2	6.5	---
LARC-13/Sylgard	3.1	5.5	2.8	2.5
LARC-13/ATBN	5.5	3.9	3.0	2.4
LARC-13/ATS	8.5	6.2	4.5	5.8

Table 3. Adhesive Fracture Energy Data on Elastomer-Modified Polyimides

Adhesive	G_{I_C} (J/m^2) Initial Test	G_{I_C} (J/m^2) after 4 months	
	RT	RT	225°C
LARC-13	117	70	---
LARC-13/Sylgard	146	110	---
LARC-13/ATBN	393	371	399
LARC-13/ATS	---	317	---

EFFECT OF ELASTOMER CHAIN LENGTH ON ADHESIVE PROPERTIES

Adhesive Formulations

In order to study the effects of varying the length of the elastomer chain on the properties of LARC-13 adhesive, a series of polymers were prepared using the ATS elastomer (Fig. 2). Resins were prepared with a concentration of 15% (w/w) elastomer using four ATS elastomers containing chain repeat units of 10, 41, 63 and 105. (ATS with repeat unit of 105 was used in studies previous to this.) Incorporation of these four elastomers into the LARC-13 backbone caused a reduction in the amic acid viscosity as shown in Table 4. This phenomenon was observed previously and was especially pronounced for the polymer containing 10 repeat units in the elastomer (ATS_{10}).

Three additional LARC-13/ATS formulations were prepared using a 50:50 combination of elastomers having repeat units of 105 and 10. These resins containing a combination of two elastomers were made at three elastomer concentrations of 8, 15 and 25% w/w solids. Bimodal formulations were of particular interest because of a previous study on epoxies resulting in the enhancement of properties when the resins were modified using bimodally distributed elastomer particles (10). A significant drop in the Brookfield viscosity of these resins was observed as the elastomer content was increased from 8 to 25%. This may be due to an offset in the monomer stoichiometry due to elastomer impurity or the presence of small amounts of water in the elastomers.

Table 4. ATS-Modified LARC-13 Adhesive Formulations

ADHESIVE	ELASTOMER STRUCTURE (ATS) 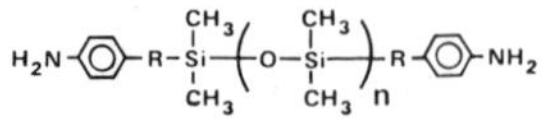	ELASTOMER CONCENTRATION %	BROOKFIELD VISCOSITY (cps)
LARC-13	NONE	0	2300
LARC-13/ATS_{105}	n = 105	15	1675
LARC-13/ATS_{63}	n = 63	15	1900
LARC-13/ATS_{41}	n = 41	15	1750
LARC-13/ATS_{10}	n = 10	15	1350
LARC-13/ATS_{105} + ATS_{10}	n = 105, n = 10 (50:50)	8	1850
LARC-13/ATS_{105} + ATS_{10}	n = 105, n = 10	15	680
LARC-13/ATS_{105} + ATS_{10}	n = 105, n = 10	25	490

Lap Shear Bond Testing

Adhesive lap shear strengths of the ATS-modified LARC-13/titanium bonds tested at ambient and elevated temperatures are presented in Table 5. As was found in previous studies (Table 1), toughness has been achieved at a sacrifice to the elevated temperature strengths of the adhesives containing individual elastomers ATS_{10} to ATS_{105}. Elevated temperature lap shear strengths of all the ATS-modified adhesives were consistently lower than for the LARC-13 control resin. Strengths did improve significantly, however, upon aging at 232°C as was observed in the control. The most surprising result of this study was the improvement in strength displayed by the LARC-13 formulation containing the 50:50 bimodal distribution of elastomers ATS_{105} + ATS_{10}. Lap shear strengths of the unaged polymer were equivalent to those of the LARC-13 control at RT and 288°C. After aging at 232°C, the LARC-13/ATS_{105} + ATS_{10} bonds exhibited improved strengths over those adhesives containing a single elastomer, but lower than those of LARC-13.

Table 5. Adhesive Strengths of ATS-Modified LARC-13/-Titanium Bonds

ADHESIVE	LAP SHEAR STRENGTH, psi UNAGED SAMPLES				LAP SHEAR STRENGTH, psi AGED 500 hrs @ 232°C			
	RT	232°C	260°C	288°C	RT	232°C	260°C	288°C
LARC-13	2850	2800	2130	1280	2920	2910	2480	2160
LARC-13/ATS_{105}	2550	1800	1130	870	1710	1630	1620	1390
LARC-13/ATS_{63}	2550	1380	1030	700	1580	1550	1640	1320
LARC-13/ATS_{41}	2480	1770	1430	805	1710	1630	1600	1410
LARC-13/ATS_{10}	2910	1850	1150	690	1810	1670	1650	1430
LARC-13/ ATS_{105} + ATS_{10}	3000	2070	1850	1300	2300	1930	1910	1720

Effect of Elastomer Concentration on Adhesive Strength

Because in the previous study the LARC-13/ATS_{105} + ATS_{10} polymer was found to have the best lap shear strengths, this bimodally distributed elastomer-containing adhesive was placed under further investigation. The concentration of the elastomer in the polymer matrix was varied while leaving the ratio of ATS_{105} to ATS_{10} constant in order to find that elastomer concentration which would maximize the adhesive properties of this material.

As shown in Table 6, the elastomers were incorporated into LARC-13 at concentrations of 8, 15, and 25% solids by weight. The adhesive prepared with 15% elastomer content exhibited the best lap shear strengths before and after aging. This is an interesting finding in view of a study on epoxies which proved that a 15-20% concentration of elastomer particles provided maximum properties (1).

Table 6. Adhesive Strengths of LARC-13/ATS_{105} + AT_{10}

ELASTOMER CONCENTRATION %	INITIAL LAP SHEAR STRENGTH, psi		LAP SHEAR STRENGTH, psi AFTER 500 hrs@ 232°C	
	RT	288°C	RT	288°C
0	2850	1280	2920	2160
8	2500	950	1670	1450
15	3000	1300	2300	1720
25	2890	830	1840	1340

SUMMARY

Silicone and butadiene/acrylonitrile elastomers have been successfully incorporated into an addition polyimide adhesive, LARC-13. Two approaches were used to accomplish this task. Fluorosilicone (Silastic) and vinyl-terminated (Sylgard) silicone rubbers were physically blended into the adhesive amic acid; and aromatic ATBN and aromatic amine-terminated silicone elastomers were chemically reacted into the prepolymer backbone. Each of the four resins displayed a distinct and separate T_g for the resin and elastomer phases. Incorporation of elastomer particles at a concentration of 15% by weight solids resulted in a 6- to 7-fold increase in peel strength and a 3- to 5-fold increase in the fracture toughness of LARC-13. This improvement in toughness was accomplished at a sacrifice in the elevated temperature adhesive strength of the material.

A second series of LARC-13 adhesives was synthesized containing aromatic amine-terminated silicone elastomers of varying chain lengths. Changing the elastomer repeat unit from n = 10 to n = 105 had very little effect on polymer properties. However, when a 50:50 combination of long and short-chained elastomers were reacted into the LARC-13 amic acid, adhesive strengths were enhanced. A concentration of 15% by weight of elastomers was found to maximize adhesive properties for this system.

According to the results of these combined studies, the incorporation of elastomers into a high-temperature polyimide thermoset, LARC-13, shows much potential for broadening the applicability of this polymer system as an adhesive.

REFERENCES

1. C.B. Bucknall, Toughened Plastics, Materials Science Series, edited by L. Holiday and A. Kelley, Applied Science Publishers, London, pp. 1-56 (1977).
2. J.N. Sultan and F.J. McGarry, "Effect of Rubber Particle Size on Deformation Mechanisms in Glassy Epoxy," Polym. Eng. Sci., 13 (1) 29 (1973).
3. E.H. Rowe, A.R. Siebert, and R.S. Drake, "Toughening Thermosets With Liquid Butadiene/acrylonitrile Polymers," Mod. Plastics, 49, 110 (1970).
4. A.K. St. Clair and T.L. St. Clair, "Addition Polyimide Adhesives Containing ATBN and Silicone Elastomers," Int. J. Adhesion and Adhesives, 249, July (1981).
5. A.K. St. Clair, T.L. St. Clair, and S.A. Ezzell, "The Effect of Elastomer Chain Length on Properties of Silicone-Modified Polyimide Adhesives," NASA TM-83172 (1981).
6. A.K. St. Clair and T.L. St. Clair, "Structure-Property Relationships of Isomeric Addition Polyimides Containing Nadimide End Groups," Polym. Eng. Sci., 16, (5) 314 (1976).
7. T.L. St. Clair and D.J. Progar, "LARC-13 Polyimide Adhesive Bonding," Preprints 24th National SAMPE Symposium, 24 (2) 1081 (1979).
8. R.Y. Ting and R.L. Cottington, "A Comparison of Test Methods in Polymer Fracture Evaluation," ACS Org. Coat. and Plast. Preprints, 41, 405 (1979).
9. W.D. Bascom and S. Mostovoy, "Adhesive Fatigue Failure of an Elastomer-Modified Epoxy," ACS Org. Coat. and Plast. Preprints, 38, 152 (1978).
10. W.D. Bascom and D.L. Hunston, "Toughening of Plastics," Preprints Plastics and Rubber Institute, London (1978).

The Application of Thermodynamic and Spectroscopic Techniques to Adhesion in the Polyimide/Ti 6-4 and Polyphenylquinoxaline/Ti 6-4 Systems

S. Dias and J.P. Wightman

Chemistry Department
Virginia Polytechnic Institute &
State University, Blacksburg, VA 24061

ABSTRACT

A variety of specialized surface pretreatments have been developed for joining of titanium and its alloys by adhesive bonding. A surface oxide film is invariably formed yet the role of this oxide film in determining bond durability is not established. The objective of the present work was to characterize Ti (6% Al - 4% V) adherend surfaces after several chemical pretreatments and after fracture of lap shear samples aged at elevated temperatures. The primary spectroscopic techniques were XPS and specular reflectance FT-IR. The adhesives used were polyimides (PI) and polyphenylquinoxalines (PPQ). Each chemical pretreatment gave rise to a unique XPS fingerprint. SPS spectra characteristic of the adhesives were obtained following cohesive failure. Energetics of the interaction between primer solutions and the Ti 6-4 adherend were assessed by microcalorimetry. Changes in the structure of the surface oxide layer on heating of the Ti 6-4 adherend at elevated temperatures were inferred from heats of immersion in PI and PPQ solutions. [Work supported under NASA Grants NAG-1-127 and NAG-1-248, and ONR Center Grant.]

INTRODUCTION

There is a continuing and demonstrated. need for a detailed molecular understanding of the role which the surface oxide layer of the adherend plays in titanium bonding. We have been involved in a research effort aimed at establishing the baseline data necessary in such an understanding. The research has utilized principally pretreated Ti 6-4 adherends. The effects of adherend pretreatment, bonding conditions, and thermal aging of lap shear specimens have been studied. Our primary emphasis has focused on the use of the SEM/EDX and XPS techniques to study surface morphology and surface composition. In addition, contact angles and both infrared and visible reflection spectroscopy have been used in ancillary studies. The results have been summarized recently (1).

Venables and co-workers have reported (2) results of depth profiling of the surface oxide layer on Ti 6-4 adherends by AES. Acidic etches lead to rather thin (20 nm) oxide layers whereas anodized oxide films are considerably thicker (40-80 nm). Venables has further reported (3) that initially an amorphous oxide layer forms on Ti 6-4 adherends which can be converted to a crystalline layer at elevated temperatures.

One phase of our most recent work has focused on calorimetric studies using Ti 6-4 powders (4). The higher surface area of the powder compared to the standard lap shear coupon permits heats of immersion to be measured. The heats of immersion of Ti 6-4 have been measured in water, in PPQ and LARC-13 primer solutions and their respective solvents. The Ti 6-4 powder was heated to different temperatures in vacuum for varying time periods prior to immersion. The heats of immersion sharply increased between 300° and $400^{o}C$ which was attributed to a chemical reaction of the wetting liquid with elemental titanium. The calorimetric results support a model of thermal cracking of the surface oxide layer on Ti 6-4 adherends. Venables' results (3) appear to be a plausible mechanism for this cracking phenomenon. There is a clear need then to pursue further research in this critical area. The objective of this research is the characterization by SEM and XPS of the surfaces resulting from fracture of adhesively bonded Ti 6-4 lap shear samples which had been thermally aged at $450^{o}F$ for up to 10,000 hours.

EXPERIMENTAL

Samples

Fractured Ti 6-4 lap shear samples were sent from the Boeing Aerospace Company and these panels were used as received. A detailed microscopic/spectroscopic study of similar samples including crack extension measurements has been reported (5). The samples used in this study are described in Table 1. It is noted that these samples are quite interesting because few bond durability studies have extended to 10,000 hr at 450°F. The Ti 6-4 adherends were pretreated by 10V chromic acid anodize. The Ti 6-4 panels were bonded with LARC-13 (polyimide) and PPQ (polyphenylquinoxaline). The fractured Ti 6-4 surfaces were categorized as cohesive, interfacial and mixed mode failure [see Table 1]. A 3/8" diameter sample was punched after visual examination of both fracture surfaces. These punched samples were then assigned to the following groups: metal failure surface (MFS), adhesive failure surface (AFS), metal substrate surface (MSS), and adhesive substrate surface (ASS). A Bausch and Lomb stereo-zoom optical microscope (OM) was used to photograph each punched sample at 20X.

Table 1. Description of Boeing Ti 6-4 Fractured Lap Shear Samples.

Sample No.	Time (hrs) [Temp]	Lap Shear Strength (psi)	Failure Mode
L13-10-21	5000 [450°F]	1180	Interfacial
L13-10-50	10000 [450°F]	300	Interfacial
L13-10-53	500 [120°F]	3000	Cohesive
L13-10-60	500 [450°F]	2280	Cohesive
PPQ-10-36	10000 [450°F]	910	Interfacial
PPQ-10-46	500 [450°F]	2560	Cohesive
PPQ-10-67	500 [120°F]	2850	Cohesive

Scanning Electron Microscopy (SEM)

The samples were mounted with copper tape on an aluminum stub. One set of samples was coated with a gold/palladium alloy and photomicrographs at various magnifications were obtained on an AMR (Advanced Metal Research Corporation: Model 900) scanning electron microscope. A second set of samples was gold-coated and photomicrographs of these specimens were obtained using a JEOL JFM 35c scanning electron microscope. The surface of some of the samples were coated with carbon and were examined by EDX (energy dispersive X-ray analysis) in order to identify the elements present in that particular surface.

X-ray Photoelectron Spectroscopy (XPS)

The XPS studies of the fractured samples were done using a Physical Electronics ESCA/SAM Model 550 electron spectrometer. Data acquisition was accomplished using a SAM 550 data system and a Digital PDP-1104 computer. The punched samples were mounted with double sided stick tape. The binding energies were referenced to the C 1s photopeak at 284.6 eV. Wide scan (0 to 1100 eV) spectra were used to identify the major elements present on the surface of the samples. Samples were scanned repetitively to obtain the atomic fractions of elements present in the sample surface.

RESULTS AND DISCUSSION

Scanning Electron Microscopy (SEM)

The SEM photomicrographs in general supported the failure mode assignments based on optical microscopy. However, detailed features of the Ti 6-4 adherend surfaces are clearly delineated in the SEM photomicrographs.

X-Ray Photoelectron Spectroscopy (XPS)

The XPS results are summarized in Tables 2 and 3 and are discussed below. There were no detectable quantities of fluorine, chromium, calcium or phosphorus on the metal failure surface or metal substrate surface of any of the samples which fractured interfacially. This result is in contrast to the results of the XPS analysis of 10V chromic

Table 2. XPS Analysis of Fractured Boeing Ti 6-4 Lap Shear Samples Bonded with L13

Sample No.	L13-10-21(AFS)		L13-10-21(MFS)		L13-10-50(AFS)	
Photopeak	B.E.	A.F.	B.E.	A.F.	B.E.	A.F.
F 1s				NSP		NSP
Cr 2p						NSP
O 1s	532.0	0.19	531.6$_D$	0.24	531.8	0.24
Ti 2p3	458.8	0.004	458.2	0.036	458.2	0.009
N 1s	400.4	0.046	400.0	0.037	399.8	0.032
Ca 2p						NSP
C 1s	284.6	0.72	284.6	0.67	284.6	0.70
P 2p						NSP
Si 2p	102.0	0.035	101.8	0.023	102.0	0.004
Ti/C		0.0056		0.054		0.013

Sample No.	L13-10-50(MFS)		L13-10-53		L13-10-60	
Photopeak	B.E.	A.F.	B.E.	A.F.	B.E.	A.F.
F 1s		NSP				NSP
Cr 2p		NSP				NSP
O 1s	531.2		531.9	0.17	531.8	0.22
		0.33				
	529.8					
Ti 2p3	458.0	0.08	458.2	0.001		NSP
N 1s	399.6	0.018	400.2	0.020	399.8	0.041
Ca 2p		NSP				NSP
C 1s	284.6	0.54	284.6	0.77	284.6	0.70
P 2p		NSP				NSP
Si 2p	101.6	0.014	102.6	0.036	102.6	0.020
Ti/C		0.15		0.0013		--

B.E. - Binding energy (eV)
A.F. - atomic fraction
NSP - no significant peak
D - doublet

Table 3. XPS Analysis of Fractured Boeing Ti 6-4 Lap Shear Samples Bonded with PPQ

Sample No.	PPQ-10-46		PPQ-10-67		PPQ-10-36(AFS)	
Photopeak	B.E.	A.F.	B.E.	A.F.	B.E.	A.F.
F 1s						NSP
Cr 2p						NSP
O 1s	532.4	0.18	532.4	0.13	532.2	0.15
Ti 2p3	458.4	0.001	458.6	0.001	458.6	trace
N 1s	399.0	0.02	399.0	0.030	398.8	0.028
Ca 2p						NSP
C 1s	284.6	0.75	284.6	0.80	284.6	0.73
Pb 4f7					138.6	0.005
Si 2p	102.8	0.046	102.4	0.040	102.0	0.005
Ti/C		0.0013			0.0012	--

Sample No.	PPQ-10-36(MFS)		PPQ-10-36(ASS)		PPQ-10-36(MSS)	
Photopeak	B.E.	A.F.	B.E.	A.F.	B.E.	A.F.
F 1s		NSP		NSP		NSP
Cr 2p		NSP		NSP		NSP
O 1s	531.8		532.4	0.14	531.9	
		0.33				0.27
	529.9				530.0	
Ti 2p3	458.2	0.028		NSP		
N 1s	399.2	0.02	398.8	0.021	399.2	0.016
Ca 2p		NSP		NSP		
C 1s	284.6	0.59	284.6	0.74	284.6	0.71
Pb 4f7	138.1	0.017	138.8	0.005	138.4	0.02
P 2p						NSP
Si 2p	102.0	0.014	102.2	0.011		NSP
Ti/C		0.047		--		--

B.E. - Binding energy (eV)
A.F. - atomic fraction
NSP - no significant peak

PPQ-10-46 and PPQ-10-67. - Both of these high strength (2560 and 2850 psi) lap shear samples showed minimal Ti photopeaks. The O 1s photopeaks both at 532.4 eV are due only to the PPQ adhesive.

A review of the XPS results in Tables 2 and 3 suggests that for the 10V chromic acid anodized Ti 6-4 adherend, the high Ti surface concentration on the L13-10-50 adhesive failure surface (AFS) thermally aged for 10,000 hr is associated with the lowest strength. It is proposed that long-term thermal aging of bonded lap shear samples weakens the surface oxide. The high Ti concentration observed on the L13-10-50 adhesive failure surface (AFS) results from the fracture of the surface oxide layer.

The fact that the adhesive failure surface (AFS) of Sample No. PPQ-10-36 which had been aged similarly to Sample No. L13-10-50 does not give a significant Ti peak suggests a difference in the interaction of L13 and PPQ with the surface oxide layer. We have shown (7) such a difference in the heats of immersion of L13 and PPQ primer solutions with anatase titanium dioxide.

Supporting evidence for the cracking of the surface oxide layer on Ti 6-4 has been obtained from microcalorimetry. We have observed (4) that the average heat of immersion of Ti 6-4 powder evacuated in vacuo for 2 hr at 300° and $400^{o}C$ in primer solutions of L13 and PPQ increased from 8.0 to 40 J/m^2. This five-fold increase in heat is attributed to a cracking of the surface oxide layer at elevated temperatures and subsequent penetration and reaction of the liquids with elemental titanium in the substrate. A similar process could be occurring in the thermally aged Ti 6-4 lap shear samples. The fact that reduced lap shear strengths were not observed until 10,000 hr at $450^{o}F$ ($232^{o}C$) could be due to an activation energy for the cracking process and/or to an inhibiting effect of the L13 and PPQ adhesives.

ACKNOWLEDGMENT

Work supported under NASA Grants NAG-1-127 and NAG-1-248 and by the Office of Naval Research.

REFERENCES

1. J.P. Wightman, SAMPE Quarterly, 13, 1-8 (1981).

2. B.M. Ditchek, K.R. Breen, T.S. Sun and J.D. Venables, in Proc. 25th Natl. SAMPE Symp., M. Smith, Ed., pp. 13-24, SAMPE, Azusa, CA (1980).

3. M. Natan, J.D. Venables and K.R. Breen, Proc. 27th Natl. SAMPE Symp., M.Smith, Ed., pp. 178-191, SAMPE, Azusa, CA (1982).

4. R.V. Siriwardane and J.P. Wightman, J. Adhesion, 15, 225-239 (1983).

5. S.G. Hill, P.D. Peters and C.L. Hendricks, NASA Contractor Report 165944, NASA-Langley Research Center, Hampton, VA, July, 1982.

6. J.G. Mason, R. Siriwardane and J.P. Wightman, J. Adhesion, 11, 315-378 (1981).

7. R. Siriwardane and J.P. Wightman, J. Colloid Interface Sci., 92, 256-259 (1983).

Evaluation of High Temperature Structural Adhesives for Extended Service

C.L. Hendricks and S.G. Hill

Boeing Aerospace Company
P.O. Box 3999
Seattle, WA 98124

ABSTRACT

Polyphenylquinoxaline (PPQ) and Polyimide (LARC-TPI) adhesive systems were evaluated for titanium metal bond characteristics in various environments. Each polymer was chemically analyzed prior to fabrication into glass reinforced films and bonded into titanium test coupons. Effects of 450°F continuous exposure upon selected adhesives demonstrated that LARC-TPI bonds were stable to at least 10,000 hours. PPQ in the same high temperature environment exhibited significant loss of bond strength between 5,000 and 10,000 hours exposure. Failures were attributed to disbond between PPQ primer and the oxide layer of the titanium surface. Exposure of both adhesive systems to various aircraft fluids showed very little effect upon bond strength except for minor degradation in the presence of Skydrol. Large area bond titanium metal-to-metal and metal sandwich panels were fabricated using processes developed on the program.

INTRODUCTION

High temperature stable adhesive systems were evaluated for potential Supersonic Cruise Research (SCR) vehicle applications. The program was divided into two (2) major phases: Phase I "Adhesive Screening" evaluated eleven (11) selected polyimide (PI) and polyphenylquinoxaline (PPQ) adhesive resins using eight (8) different titanium (6Al-4V) adherend surface preparations; Phase II "Adhesive Optimization and Characterization" extensively evaluated two (2) adhesive systems, selected from Phase I studies, for chemical characterization and environmental durability. The adhesive systems which exhibited superior thermal and environmental bond properties were LARC-TPI polyimide and polyphenlyquinoxaline both developed at the NASA Langley Research Center. The latter adhesive system did develop bond failures at extended thermal aging due primarily to incompatibility between the surface preparation and the polymer. However, this study did demonstrtate that suitable adhesive systems are available for extended supersonic cruise vehicle design applications.

The increased demands upon aerospace vehicles for improvements in structural efficiency, performance, and durability have pushed to the limit the capabilities of conventional materials and designs. One area of materials and design that can provide significant improvements in airframe structures is the use of structural bonding. This method of joining is highly resistant to fatigue damage and offers the potential for lower production costs. For supersonic cruise vehicles that experience elevated temperatures to 450°F for extended times, bonding has lacked the materials and the data to adequately demonstrate its feasibility.

The program described in this paper, funded by the NASA Langley Research Center (1), extensively evaluated all available high temperature stable resins as adhesives and all production practical titanium surface preparations to advance the state-of-knowledge and technology of high temperature bonding. Previously developed adhesive systems and studies failed to solve a combination of problems which included lack of adequate thermal stability, impractical process requirements and/or lack of ability to be processed into large area bond assemblies. This program has as its primary objective the evaluation, test, and demonstration of high temperature structure adhesives for supersonic vehicle applications. Efforts were concentrated into two (2) phases: Phase I, "Adhesive Screening," and Phase II, "Adhesive Optimization and Characterization." Phase I evaluated eleven (11) adhesive resins and eight (8) Titanium (6Al-4V) surface preparations. The candidate adhesives studied were LARC-13 (NASA Langley), NR150A2 (DuPont), NR150B2 (DuPont), NR056X (DuPont), HR-602 (Hughes), LARC-13 with amide-imide modification (Boeing), LARC-13 with methylnadic capping and m-phenylenediamine modification (Boeing), Polyphenylquinoxaline (NASA Langley), PPQ with boron-powder modification (Boeing), LARC-TPI (NASA Langley), and FM-34 (American Cyanamid) as baseline.

EXPERIMENTAL

LARC-13 polyimide adhesive was selected as a candidate based on preliminary data obtained by Boeing which demonstrated good elevated temperature thermal stability and desirable failure modes. The base resin synthesis and cure chemical reactions are illustrated below:

LARC-13 base resin formulated with 30 weight percent aluminum powder (Alcoa 101) was supplied by NASA Langley for this program. The adhesive was subsequently impregnated by Boeing on Style 112, A-1100 silane finished glass fabric and B-staged to a low flow state. The control of B-staging just prior to cross-linking reaction is critical to low or no volatile release during final cure.

LARC-13 Modification I - LARC-13 does demonstrate good adhesive properties; however, it also exhibits undesirable processing characteristics. This modification consisted of Amoco Amide-imide AI-1130 mixed at 50 phr with the base LARC-13 formulation plus addition of Alcoa 101 aluminum powder to 60 phr. This adhesive formulation provides controlled flow and honeycomb filleting for structural bonding. Style 112, A-1100 glass fabric films were prepared for bonding operations.

LARC-13 Modification II - In addition to copolymer (Amide-imide) studies performed on Modification I, a second modification was selected/evaluated which involves methyl

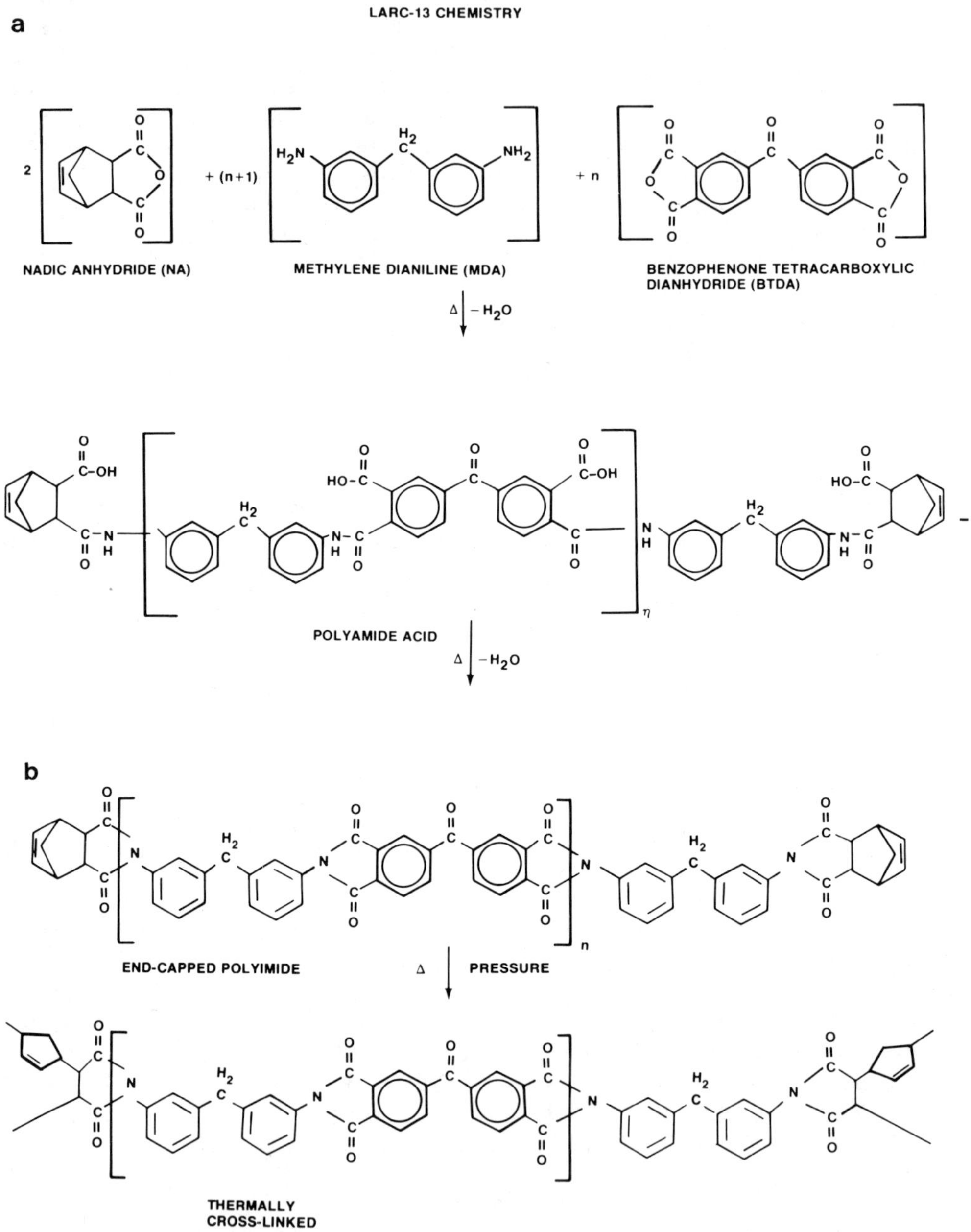
a
LARC-13 CHEMISTRY
NADIC ANHYDRIDE (NA)
METHYLENE DIANILINE (MDA)
BENZOPHENONE TETRACARBOXYLIC DIANHYDRIDE (BTDA)
POLYAMIDE ACID
b
END-CAPPED POLYIMIDE
PRESSURE
THERMALLY CROSS-LINKED

nadic-capped polymer and addition of 20 mole percent of m-phenylenediamine as co-diamines. This results in a nominal polymer molecular weight of 1,300. The modification procedures just described were performed by Boeing personnel using identical synthesis procedures as those used by NASA Langley in making the LARC-13 resin.

NR150B2 - This resin supplied by DuPont was selected and evaluated because of excellent strength retention in 588K (600°F) environments. The chemical reaction is shown below:

NR150B2 CHEMISTRY

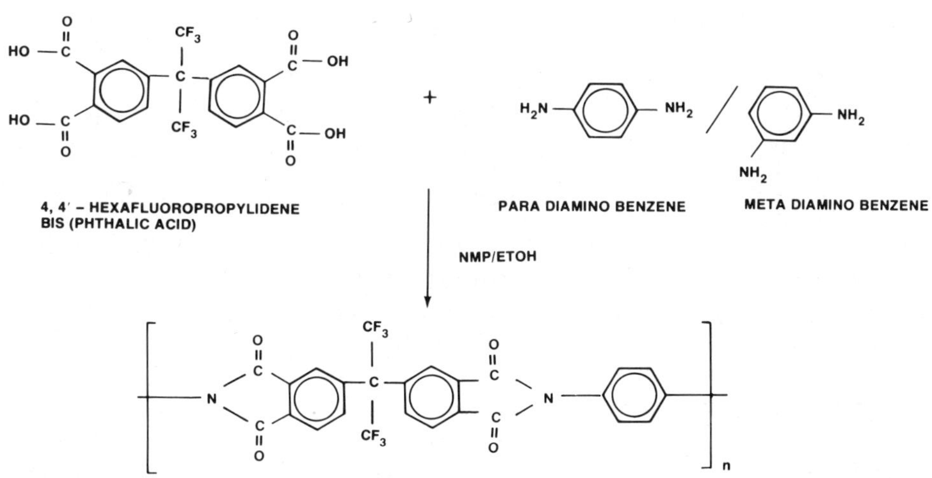

NR150A2 - This polyimide resin, also supplied by DuPont, is stable at elevated temperatures, exhibits good processing characteristics, and is tougher, but possesses lower T_g than NR150B2. The base resin was evaluated unfilled and with 30 phr aluminum powder (Alcoa 101). Both formulations were applied to 112 A-1100 glass fabric for preparation of adhesive films. Chemicals used for synthesis and cured polymer structure is shown below:

NR150A2 CHEMISTRY

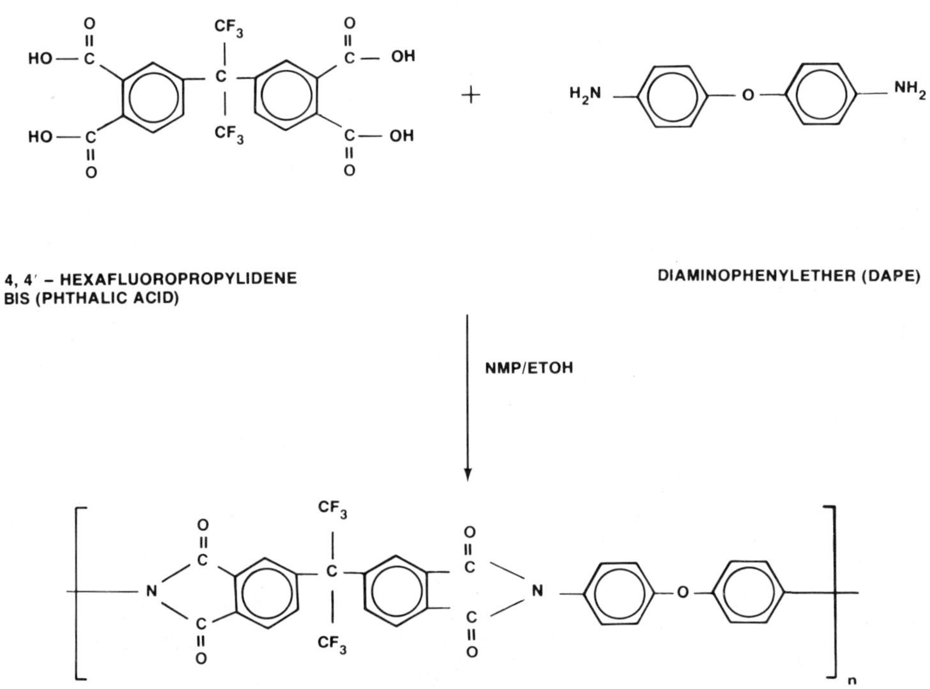

NR056X - This resin represents a modification of the NR150 series synthesized from a monomeric solution of 4,4'-hexafluoropropylidene-bis-phthalic acid, diaminobenzene, and diaminophenyl ether (mole ration 1.0/0.75/0.25) in diglyme solvent. This resin formulation in diglyme offers potential for a tough adhesive bond and improvement in processibility by removing the undesirable polar solvent NMP. The chemistry of this system is very similar to the NR150B2 system.

NR056X CHEMISTRY

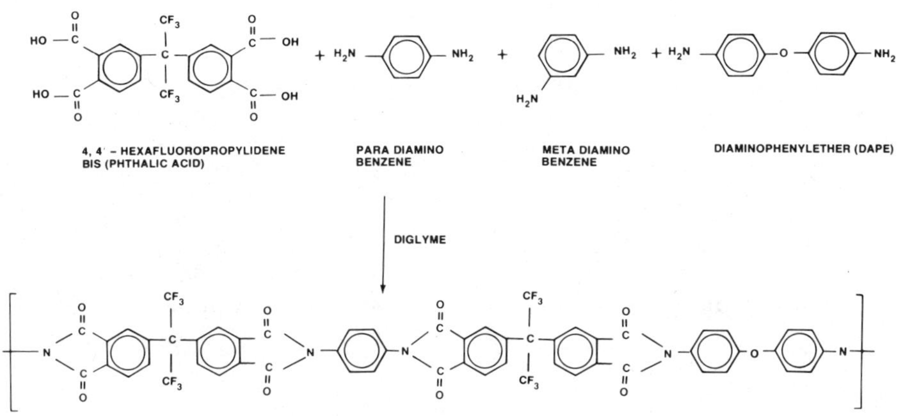

Polyphenylquinoxaline (PPQ) - This resin is supplied by NASA Langley as monoether polyphenylquinoxaline in a solvent mixture 1:1 of practical grade m-cresol and mixed xylenes. The solid content is about 16.6% based upon final polymer weight. PPQ has demonstrated excellent elevated temperature stability and represents a different family of polymers for evaluation in this program. The chemistry of this polymer is shown below.

PPQ MOD I - This modification consists of the addition of Boron powder at 30 phr to PPQ resin solution. This modification was an attempt to modify the adhesive's coefficient of thermal expansion to the coefficient of thermal of expansion of the metal substrate. This modification was deemed desirable for adapting the system for large-area bonding.

POLYPHENYLQUINOXALINE CHEMISTRY

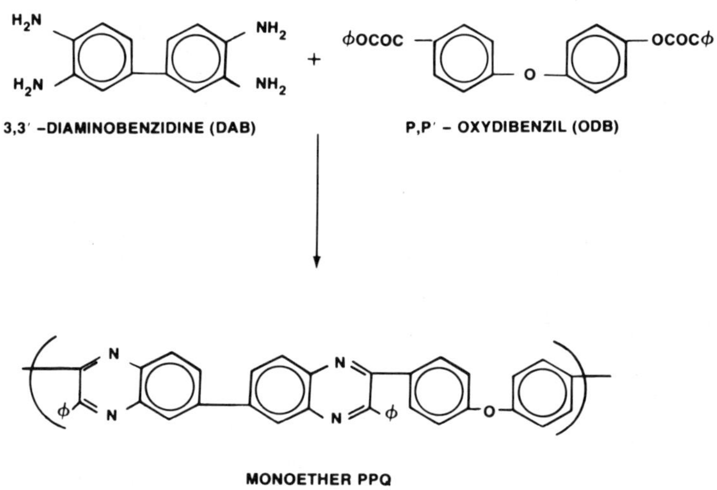

HR-602 - This system represents a different resin chemistry through use of acetylene-terminated polyimide structure. The cure mechanism consists of an addition reaction with no release of volatiles during cure. The intermediate acetylene-terminated polyimide structure is shown below. Reaction to the final cured structure is proposed as a trimerization of the acetylene groups to form an aromatic structure.

HR-602 CHEMISTRY

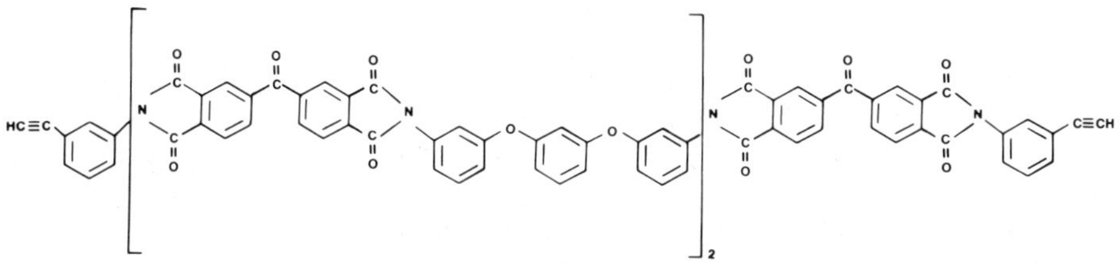

LARC-2 (LARC-TPI) - This is a thermoplastic polyimide supplied by NASA Langley which was added to Phase II portion of the program. All other adhesive system candidates, except PPQ, were not suitable for full evaluations of Phase II, and to maintain two systems in this study LARC-TPI was screened to Phase I requirements then added to Phase II. This polymer is the reaction product of pyromellitic dianhydride (PMDA) and 3,3', 4,4'-benzophenone tetracarboxylic dianhydride (BTDA) using reagent grade bis(2-methoxyether) ether (diglyme) solvent. The synthesis and polymer structure are illustrated below.

LARC-TPI CHEMISTRY

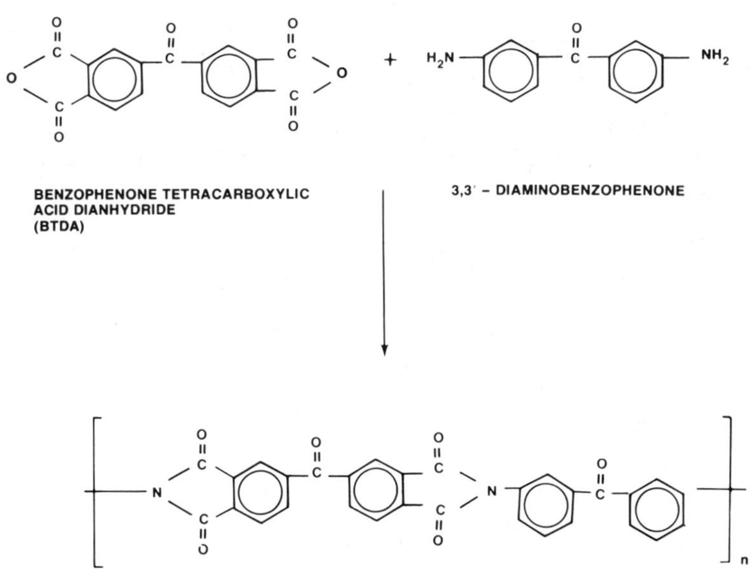

FM-34 - This polyimide system was selected as a baseline control. The chemistry and thermal stability are well known. Because it is not suitable for large area bonding this system was limited to the initial evaluation phase only.

The titanium surface preparations evaluated were 5-Volt chromic-acid anodize (Boeing BAC 5890), 10-Volt chromic-acid anodize (Boeing BAC 5890), phosphoric-acid anodize (Boeing BAC 5555), Pasa-Jell 107, phosphate fluoride (Boeing BAC 5514), phosphate fluoride (Picatiny Modified), phosphate fluoride with grit blast (Boeing BAC 5514 modified), Turco 5578 alkaline etch, and the British RAE H_2O_2/NaOH process. Lap shear strength and crack extension were the primary tests used to judge adhesive bond properties. In addition, titanium surface preparation characterization and test coupon adhesive failure analyses were conducted using SEM, STEM, and Auger techniques. These analyses were performed to more fully understand the mechanisms for bonding or debonding.

RESULTS AND DISCUSSION

Fig. 1 illustrates the lap shear strengths obtained with LARC-13 using the various surface preparations. Both chromic-acid anodize (CAA) and phosphoric-acid anodize (PAA) exhibit superior bond properties.

Fig. 2 illustrates lap shear strengths obtained with PPQ in this portion of the study. Both CAA and PAA produced exceptional values at ambient.

Surface preparation evaluations for NR056X, shown in Fig. 3, also demonstrate superior properties for CAA and PAA.

FM-34 adhesive appeared to be unaffected by changes in surface preparation as seen in Fig. 4.

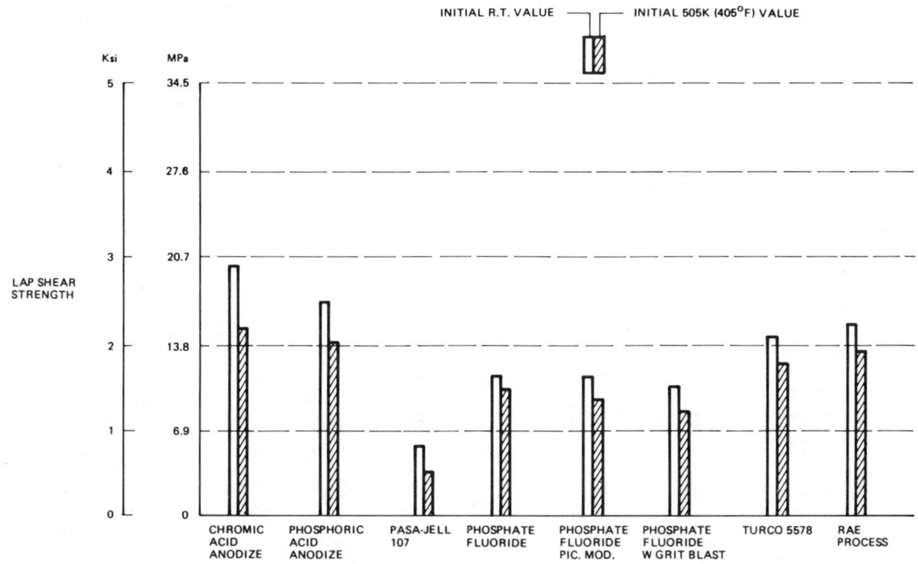

Fig. 1. Adhesive Evaluation Lap Shear Strength Test Results - LARC-13

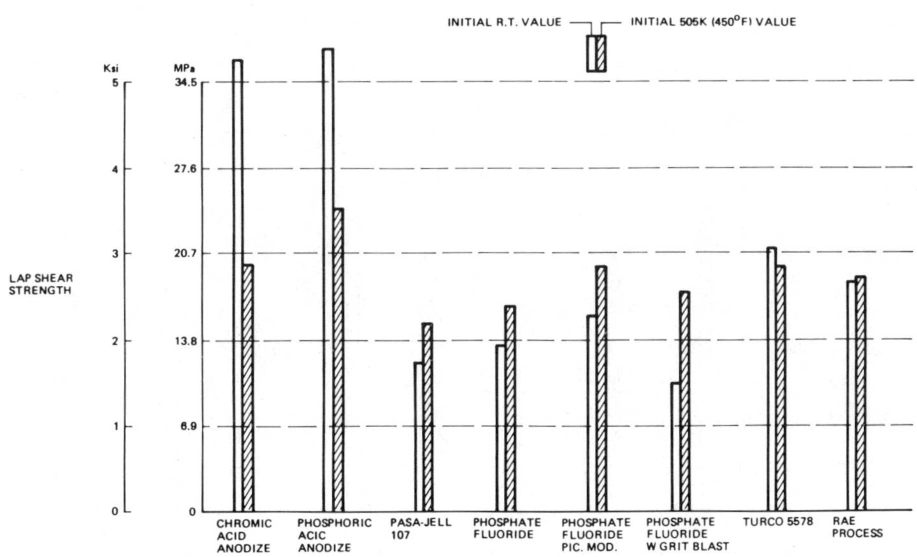

Fig. 2. Adhesive Evaluation Lap Shear Strength Test Results - PPQ

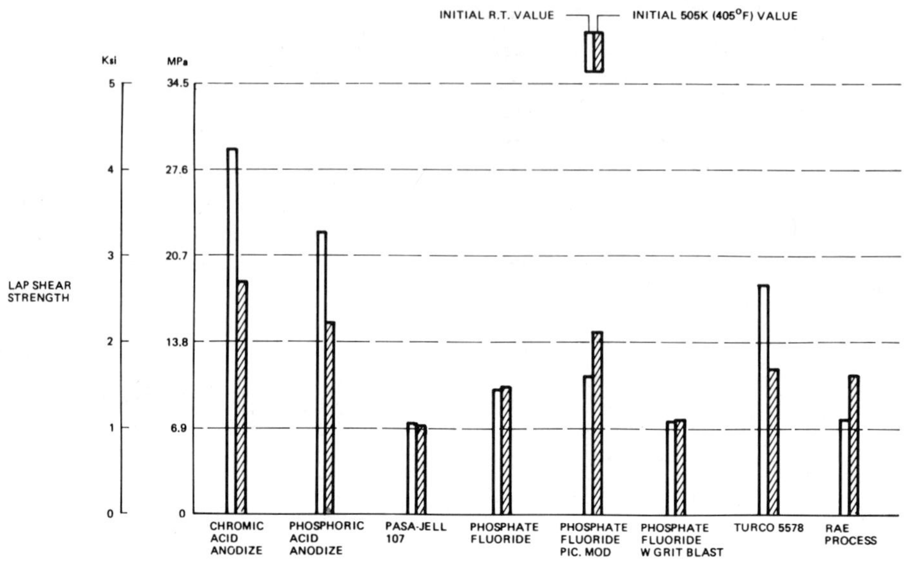

Fig. 3. Adhesive Evaluation Lap Shear Strength Test Results - NR056X

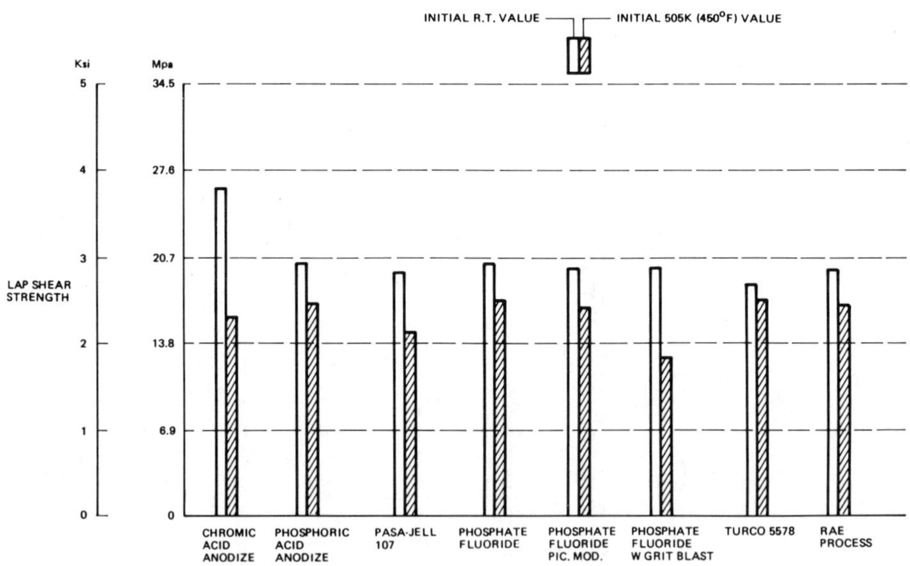

Fig. 4. Adhesive Evaluation Lap Shear Strength Test Results - FM-34

Crack extension test results are shown in Figs. 5-8 for four of the adhesives evaluated. These test data correlate very well with lap shear data demonstrating the superior characteristics of CAA and PAA with the candidate adhesives except for FM-34 where crack growth appears independent of surface preparation.

Based on the data shown in Figs. 1-8 and similar results obtained on the other candidate adhesives CAA (10 Volts) was selected as the surface preparation for use in the remaining studies of this program. These data were also used to select the adhesive systems for continued evaluation.

LARC-13 demonstrated superior adhesive properties when used in combination with dilute LARC-13 primer and chromic-acid anodized titanium surface treatment. The primary disadvantage with LARC-13 is the high flow exhibited during its curing cycle. This would tend to produce sandwich panels with marginal flatwise tensile properties. The data gathered on Boeing and the Phase I studies indicate no significant difference in the properties obtained with BR34 primed substrates when compared to LARC-13 primed substrates. The LARC-13 polyimide system is less thermally stable than the NR056X system. However, its advantage is that it possesses the capability of complete devolatilization prior to the bonding cycle.

The PPQ adhesive system made up of PPQ adhesive resin, dilute PPQ resin primer, and chromic-acid anodized surface treatment was selected based upon the data generated from this study, as well as from internal Boeing research programs. Lap shear data in Fig. 2 shows the phosphoric-acid anodized surface treatment superior to the chromic-acid anodized. However, Boeing data on both these surface treatments have shown that phosphoric-acid anodize is not stable in long-term (125 hr) elevated temperature (600°F) environments. Chromic-acid anodize exhibited good thermal stability at 450°F. PPQ demonstrated excellent lap shear values. Adhesive film producibility and quality were excellent. Primary disadvantage with PPQ was the high 644K (700°F) process temperature. However, the overall process time was shorter than for the other candidate systems. The combined advantages of good mechanical properties, ease of film fabrication, and potential thermal stability were the main reasons for its selection for further evaluation.

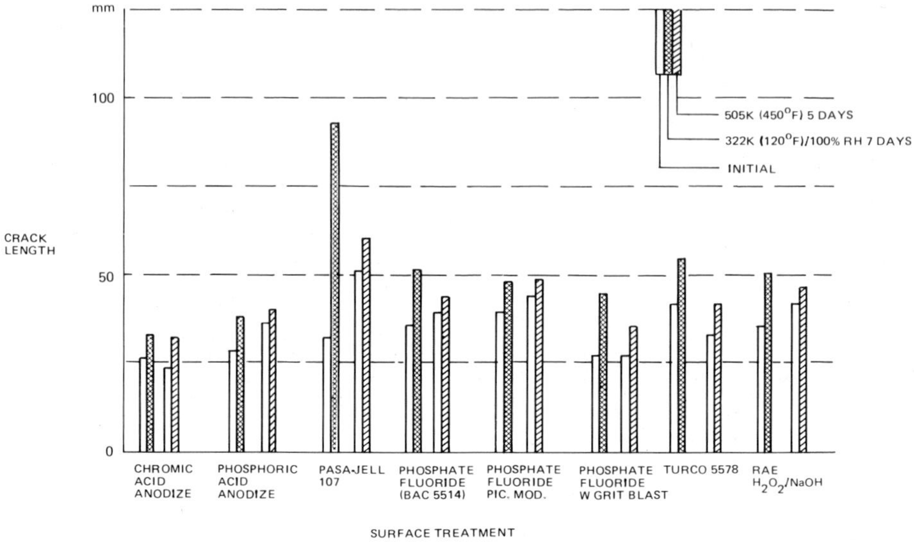

Fig. 5. Crack EXtension Environmental Exposure Results - LARC-13

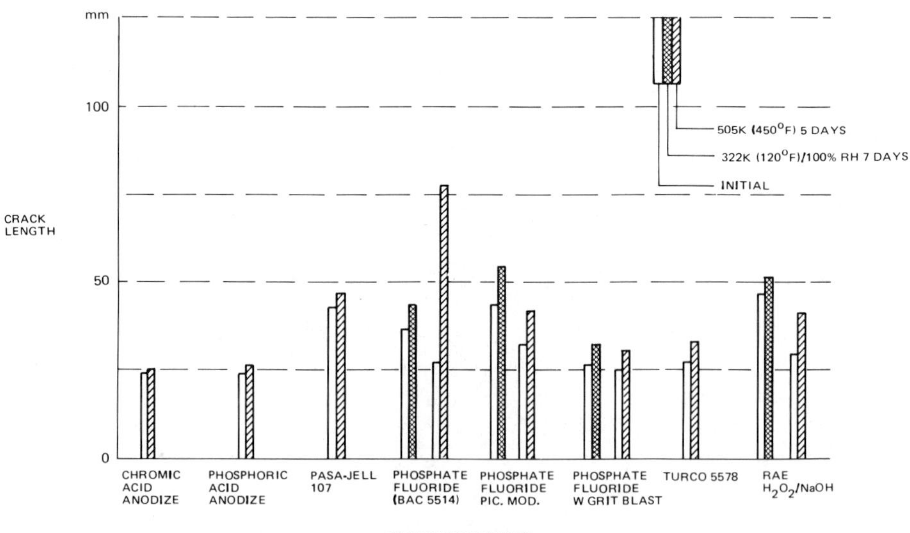

Fig. 6. PPQ Crack Extension Environmental Exposure Results

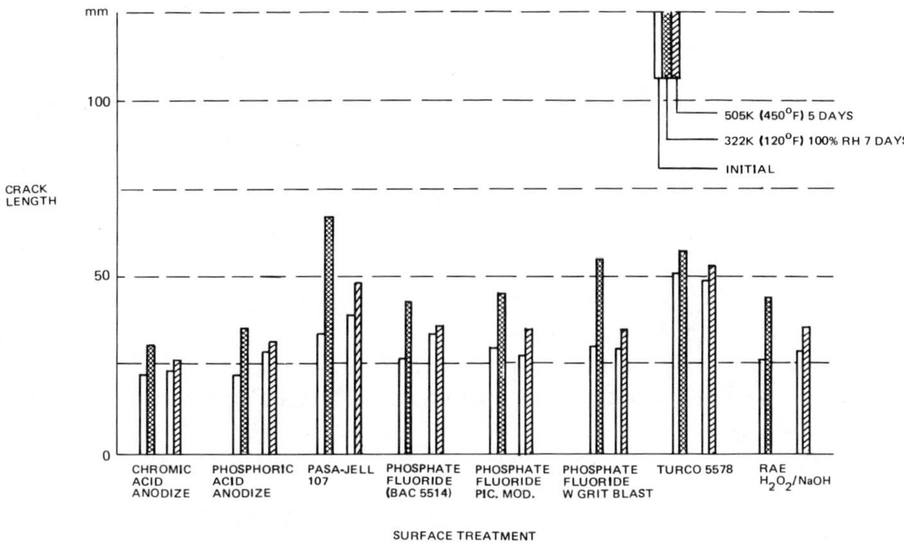

Fig. 7. Crack Extension Environmental Exposure Results - NR150B2

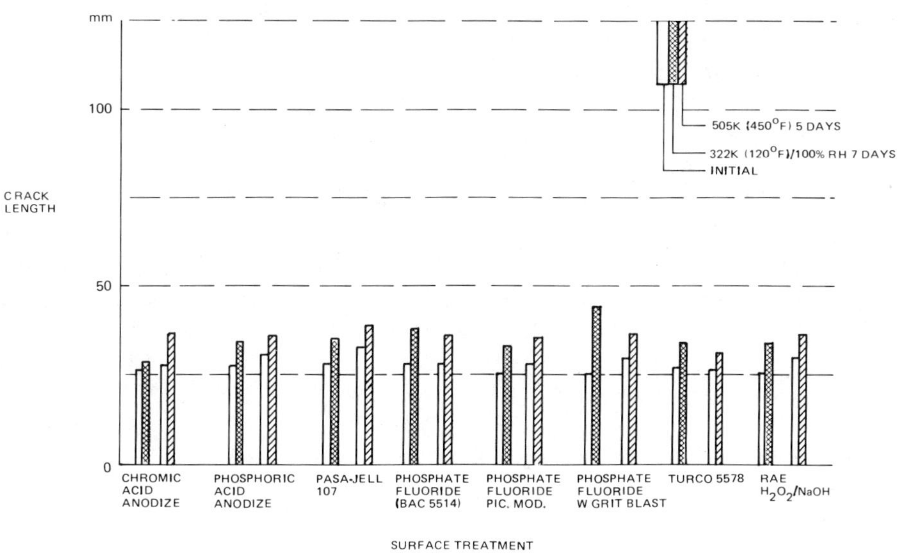

Fig. 8. Crack Extension Environmental Exposure Results - FM-34

NR056X - This polyimide system possesses excellent strength retention at 450°F because of its chemical structure. Even though its cured chemical structure is similar to NR150B2 it has the added advantage of using diglyme solvent rather than NMP. This enhances solvent removal in adhesive tape manufacture at lower temperatures than necessary for NR150B2. The major deficiency is that it is a condensation polymer which causes volatile release problems during bonding using large area substrates.

The modifications to PPQ and LARC-13 did not exhibit any significant improvement over the base formulations as seen from the data previously listed. PPQ Mod 1 with boron powder apparently reduced the flow of the system to such an extent that processing conditions used in panel preparation resulted in properties being somewhat lower than expected.

With respect to the LARC-13 modifications, the following reasoning was used to eliminate them from further evaluation. LARC-13 Mod 1 simply replaced 50 phr LARC-13 resin with AMOCO AI 1130 polyamide-imide resin. Based on results of Phase I, the properties were essentially identical with LARC-13. However, utilizing data obtained on other Boeing programs, it was found that the AI 1130 polyamide-imide made it more difficult to bond large areas because of the NMP solvent that it contained. Since the properties required in this program (SCR) are based on solid metallic substrates, it was determined that the basic LARC-13 system would provide data for the addition type polyimides. With respect to the LARC-13 Mod II, the chemistry is essentially the same as LARC-13. The variation of the ingredients would only cause the Mod II to possess a slightly higher T_g. The data from this program was available before Phase II of this program started, leaving the option open for its inclusion in the program if desired.

FM-34 was eliminated at this stage of the program since it cannot be reliably processed in large area bond structures. This adhesive did function successfully as a baseline control to compare with the newer adhesive resin formulations.

HR602 ws not selected because of its low lap shear properties with all metal surface treatments. Data generated on this adhesive for this program was very similar to the test data submitted by Hughes on the test sample material. All process procedures used for HR602 were performed as specified by Hughes. Subsequent evaluation of this adhesive by Boeing (independent of this contract) did produce much improved bond properties. These improvements were primarily the result of changes in the cure cycle.

LARC-TPI was added to the program early in Phase II in anticipation of possible failure in some or all of the other three adhesive systems. Test data confirming LARC-TPI as a suitable candidate adhesive were generated on Boeing Research.

Test results from Phase I based upon a combination of bond strength, processibility, and availability were used to select LARC-13, PPQ, NR056X, and LARC-TPI for additional evaluaiton in Phase II.

Phase II initial activities were to chemically characterize each of the adhesive resins. Three (3) different batches of resin for LARC-13, PPQ, and NR056X were analyzed using high-pressure liquid chromatography(HPLC) and infrared techniques. In general, all the adhesive resins exhibited relatively low variability from batch-to-batch.

To further characterize each adhesive resin lap shear bonds were prepared using three different cure cycles. These coupons were tested at initial ambient and at 505°K after extended (to 3000 hours) exposure to 505°K. The cure-cycles selected concentrated on varying upper dwell temperature. For this study the dwell temperatures were 588°K, 602°K, and 616°K.

At this point in the program both LARC-13 and NR056X were eliminated for any future study. LARC-13 was dropped because Phase I long-term 505°K exposure data showed a steady reduction in lap shear strength attributed to thermal degradation. NR056X was withdrawn from the market by DuPont and this made further evaluation questionable since it could not support future potential production demand. The 602°K cure-cycle was selected as most optimum for PPQ (highest strength values) and the 616°K cure-cycle was selected for LARC-TPI (also highest strength values).

The next major effort in Phase II was to determine the environmental durability of the remaining PPQ and LARC-TPI systems.

Lap shear and crack-extension coupons were prepared for these systems using their respective selected cure-cycles and subjected to environmental exposure and test.

Unstressed thermal exposure was conducted on the two remaining systems. For this study lap shear coupons were exposed to 505°K for up to 10,000 hours. The PPQ adhesive system was tested after 100, 500, 1000, and 5000 hours. A separate set of coupons (5 each set) was tested at 219°K, ambient, 422°K and 505°K after exposure. The 10,000 hour coupons were too weak to test.

All test values showed a relative rapid decline to 5,000 hours (Fig. 9). The horizontal dashed line in all plots represents the minimum acceptable structure bond strength for design (14 MPa).

The PPQ failures were characterized as gradually changing from cohesive to adhesive. This type of thermal failure also appeared in Phase I exposure and test. Surface failure analysis had shown that an incompatibility between PPQ resin and the titanium oxide layer had occurred. The PPQ resin was pulling clear from the oxide. Conclusions at this point were that the polymer was thermally stable but an unknown factor was creating an incompatibility with the anodized, prepared oxide.

The results for unstressed thermal (505°K) aging for LARC-TPI were much different (Fig. 10). This adhesive exhibits stability to 9500 hours exposure. The data for expected 20,000 hours exposure have not been completed at the publication of this paper. All LARC-TPI lap bond failures have been cohesive.

This adhesive system is considered to be stable at extended 505°K exposure conditions based on these test results.

Stressed bond exposure testing was performed on each system. PPQ (Fig. 11) and LARC-TPI (Fig. 13) lap shear coupons were subjected to 219°K, ambient, and 505°K while stressed at 25% of ultimate. Additional coupons were

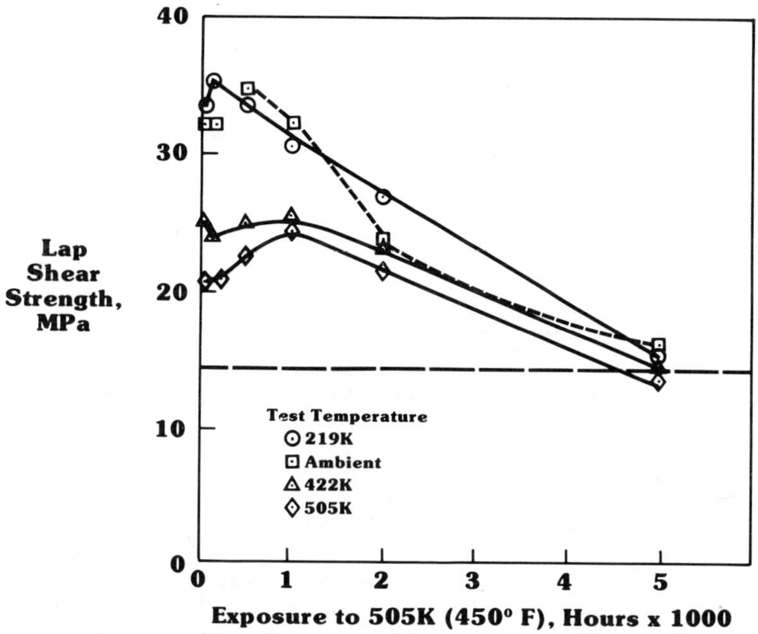

Fig. 9. Unstressed Thermal Exposure, PPQ

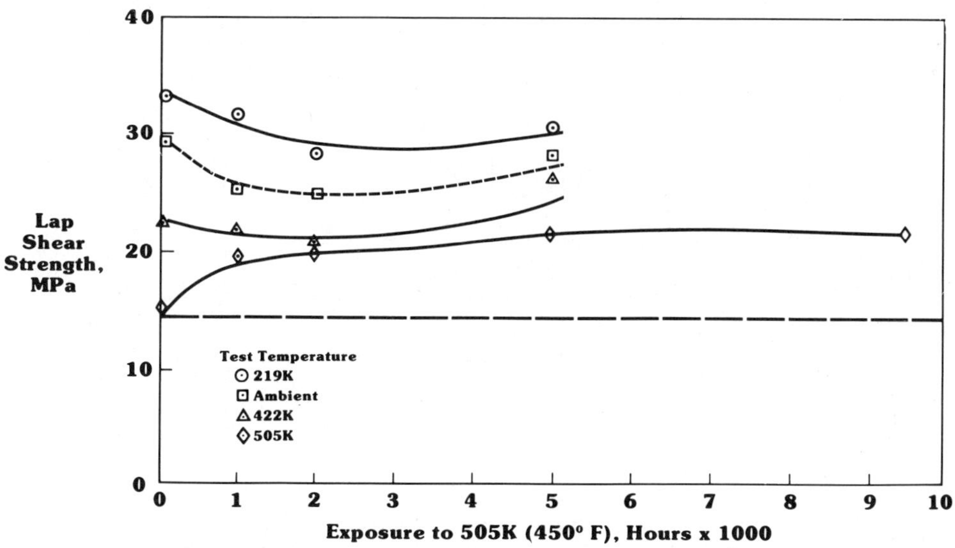

Fig. 10. Unstressed Thermal Exposure, LARC-TPI

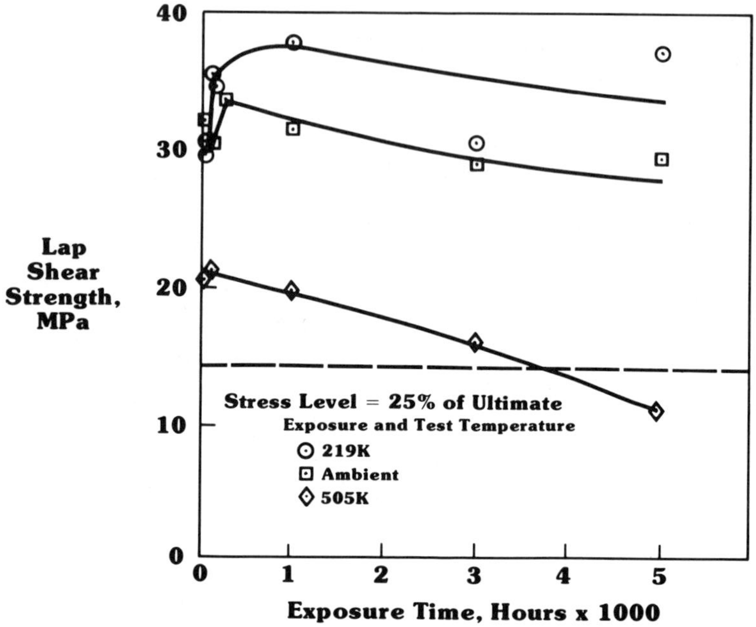

Fig. 11. Stressed Thermal Exposure, PPQ at 25% of Ultimate

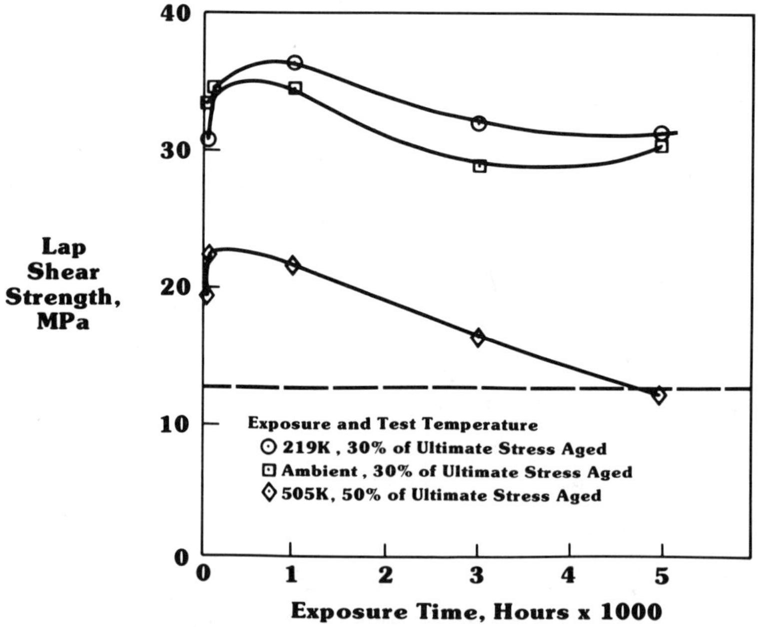

Fig. 12. Stressed Thermal Exposure, PPQ at 30 and 50% of Ultimate

stressed at 30% of ultimate at 219°K and ambient temperatures, and at 50% while exposed to 505°K (Figs. 12 and 14). Stressed coupons from each group were tested to failure at their exposure temperature after 100, 1000, 3000, and 5000 hours. For PPQ the 219°K and ambient exposed shear strengths were relatively unchanged from initial test values. However, the 505°K bond strengths exhibited a steady decline in strength and at the 5000 hour test had dropped to about 11 MPa (below structural 14.5 MPa requirements).

The test results for LARC-TPI were comparable to PPQ at the 219°K and ambient exposure conditions. At 505°K the lap shear strengths at all stress levels remained significantly above 14.5 MPa.

Conclusions from these tests were that 219°K and ambient exposure has no significant effect upon PPQ and LARC-TPI bond strengths. PPQ is affected by the 505°K exposure after 5000 hours exposure whether stressed or unstressed. LARC-TPI exhibited bond stability at all temperatures and stress levels.

Humidity exposure was conducted also in this study. The adhesive systems, PPQ and LARC-TPI, were subjected to 332°K/95% relative humidity for up to 2000 hours using lap shear (Figs. 15 and 16) and crack extension coupons (Fig. 17). The coupons were tested (lap shear) or measured (crack extension) after 1000 and 2000 hours exposure. Both PPQ and LARC-TPI do not show any shear strength degradation when tested at 219°K. LARC-TPI strengths appear to drop more rapidly than PPQ when tested at ambient after exposure. Both systems drop in strength to the same degree when tested at 505°K after exposure. Crack extension results appear to show about the same crack growth rate for PPQ and LARC-TPI at the 1000- to 2000-hour exposure times. General conclusion from these test results is that both adhesive systems show moderate bond strength loss when tested at 505°K after high humidity exposure.

To further establish the durability of PPQ and LARC-TPI adhesive bonds both systems were exposed to various fuels and fluids using lap shear and crack extension coupons. The

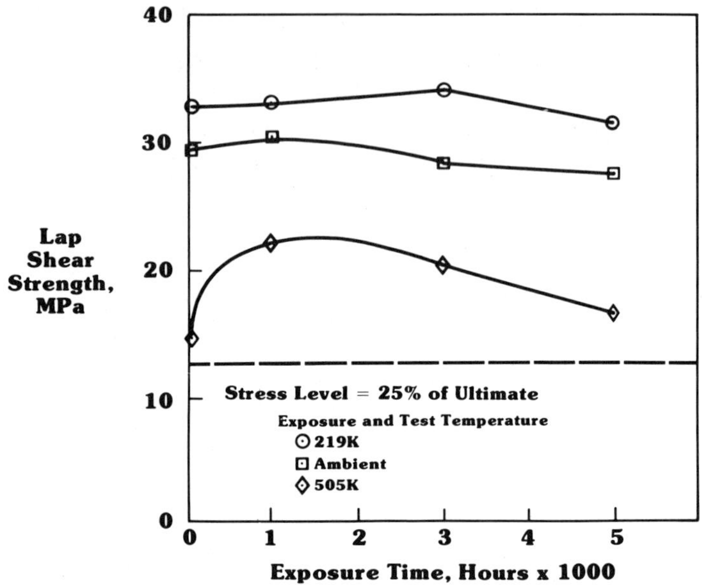

Fig. 13. Stressed Thermal Exposure, LARC-TPI at 25% of Ultimate

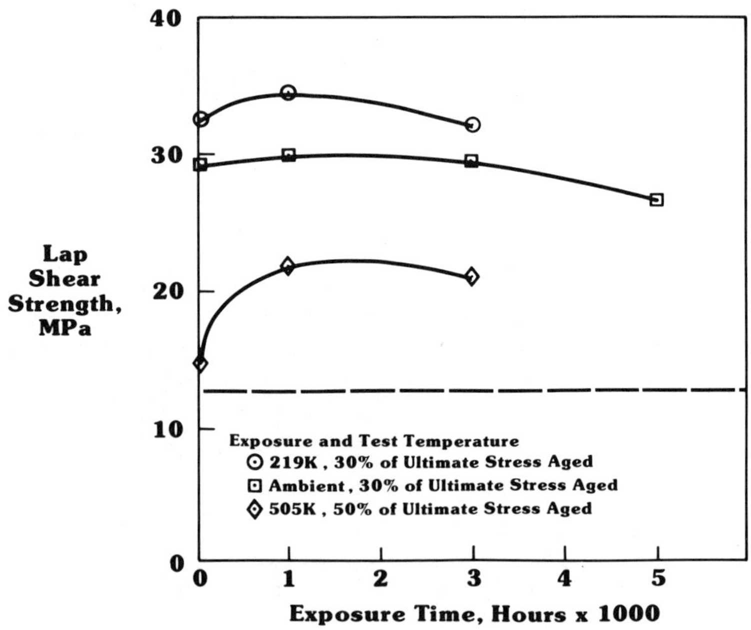

Fig. 14. Stressed Thermal Exposure LARC-TPI at 30 and 50% of Ultimate

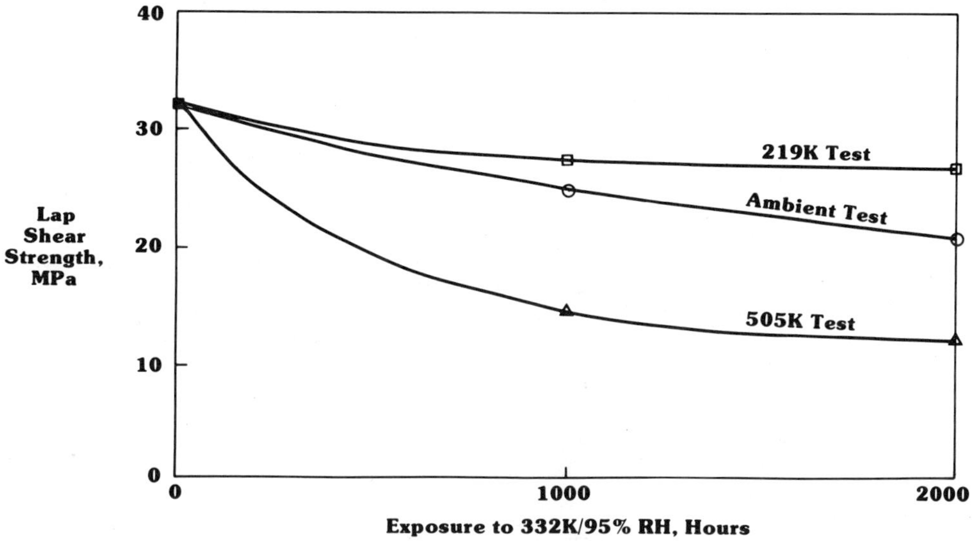

Fig. 15. Humidity Exposure - PPQ Lap Shear

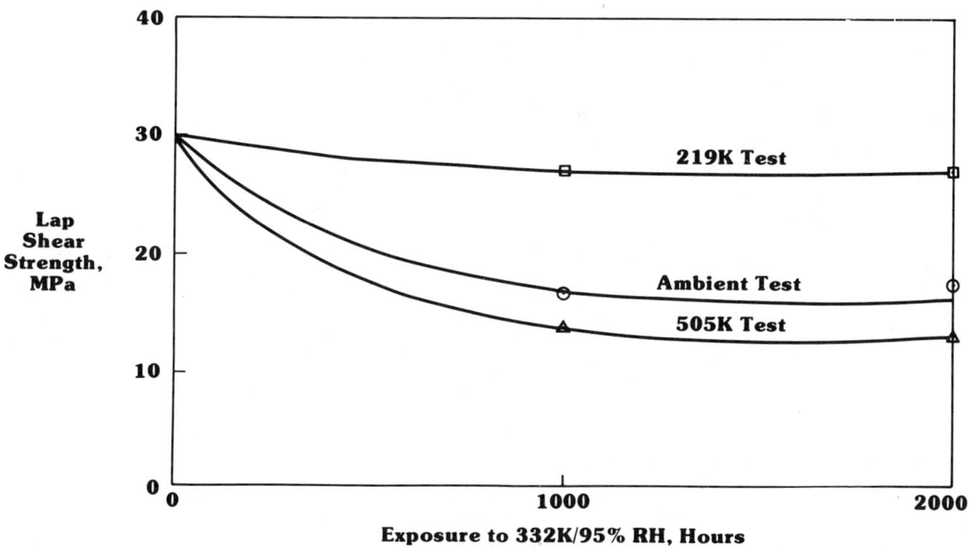

Fig. 16. Humidity Exposure - LARC-TPI, Lap Shear

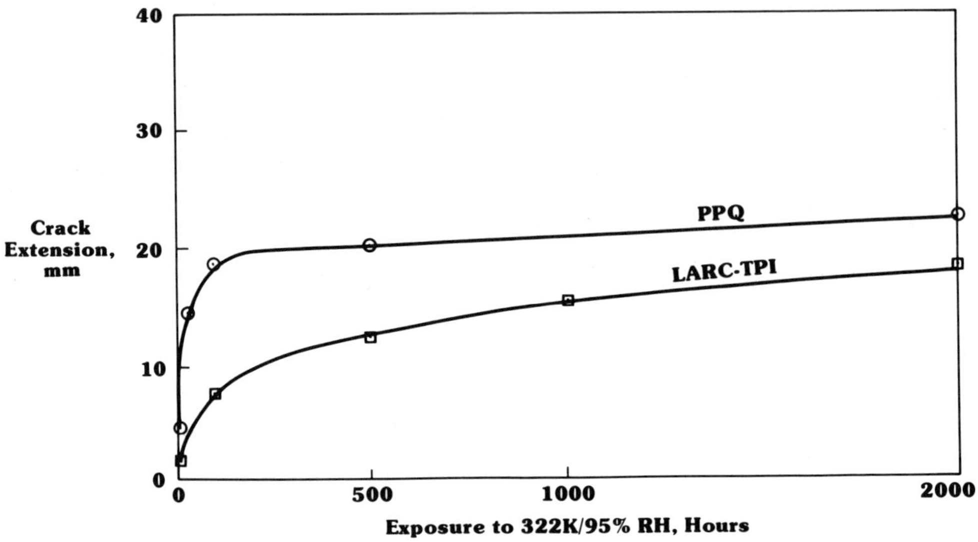

Fig. 17. Humidity Exposure, Crack Extension

fuels and fluids used were JP-4 (ambient exposure), MIL-H-5606 (ambient and 334°K exposure), Skydrol (ambient and 334°K exposure), MIL-H-7807 (ambient exposure), and de-icing fluid (ambient exposure). Test coupons were tested (lap shear) Figs. 18 and 19) or measured (crack extension) (Figs. 20 and 21) after 30-days, 60-days, and 5000-hours exposure.

Lap shear strength values (Figs. 18 and 19) were relatively unaffected by the various fuels and fluids. The difference between crack extension and lap-shear results is attributed to the fact that the former was a stressed bond condition while the latter was exposed while unstressed.

As shown the only significant crack growth occurred with exposure to Skydrol. This is not unexpected since this fluid is known to attack a wide variety of polymers. PPQ was especially sensitive to hot Skydrol (Fig. 20). LARC-TPI did show some crack growth in hot Skydrol but at much lower levels than for PPQ (Fig. 21).

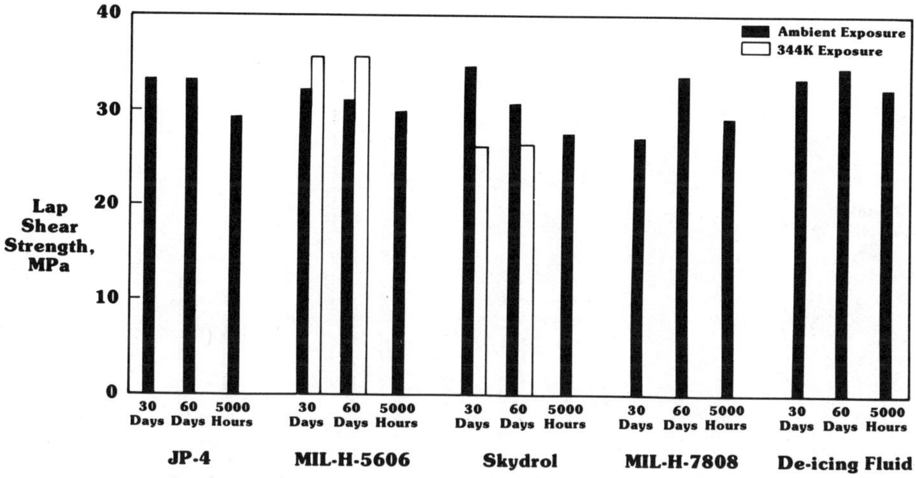

Fig. 18. Aircraft Fluid Exposure - PPQ, Lap Shear

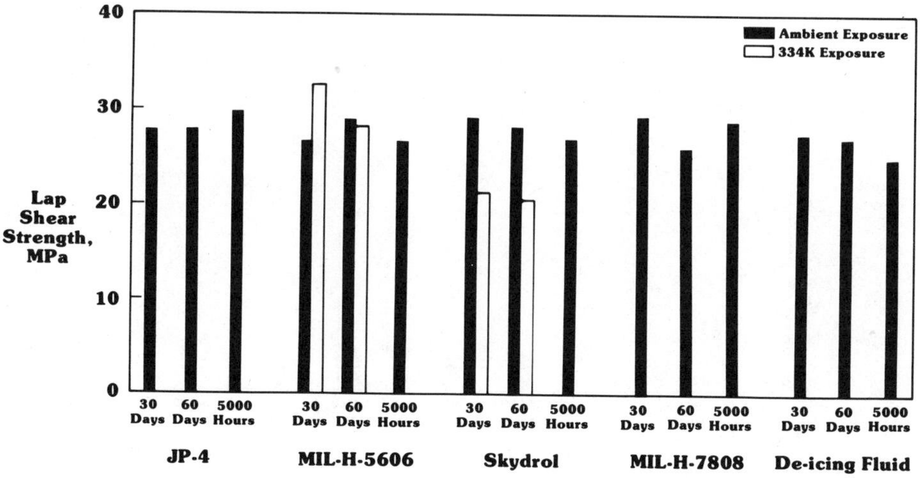

Fig. 19. Aircraft Fluid Exposure - LARC-TPI, Lap Shear

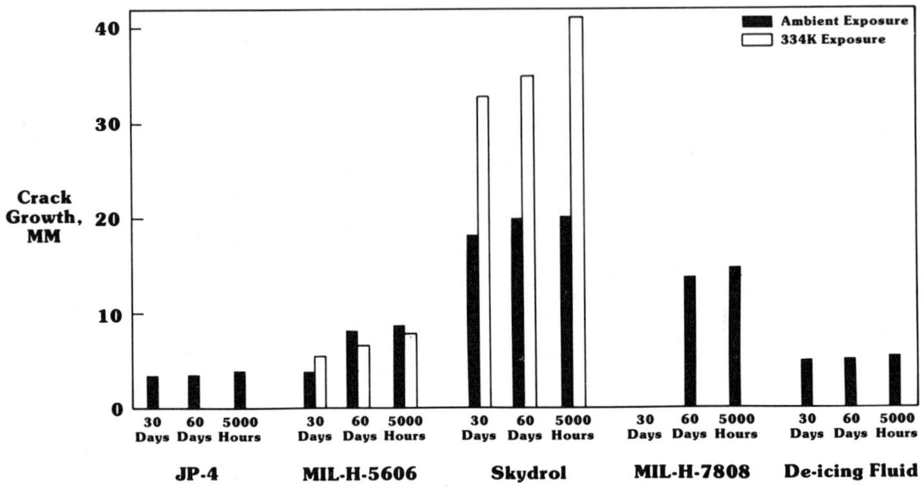

Fig. 20. Aircraft Fluid Exposure - PPQ, Crack Extension

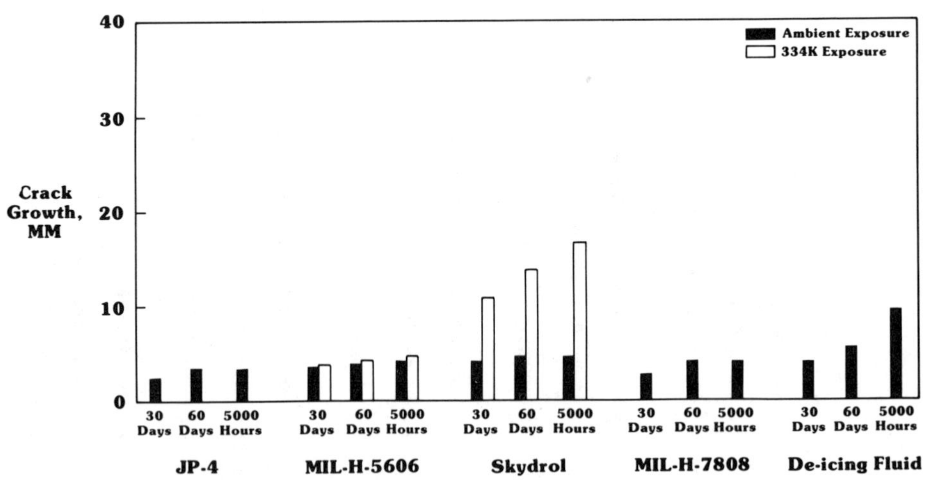

Fig. 21. Aircraft Fluid Exposure - LARC-TPI, Crack Extension

CONCLUSIONS

Conclusions from this part of the program are that PPQ is affected by hot Skydrol; LARC-TPI to a lesser extent. Stressed bonds such as that produced by crack extension specimens are more sensitive to fluid exposure conditions.

The overall program results indicate that good quality structural bonds (titanium-to-titanium) can be produced which exhibit durability to supersonic cruise vehicle environments. LARC-TPI demonstrated the best overall properties of all systems evaluated. PPQ with proper metal surface preparation could possibly also perform as desired in these environments.

ACKNOWLEDGMENTS

Support of this work by the NASA Langley Research Center is gratefully acknowledged.

REFERENCES

1. NASA Contract NAS1-15605: "Evaluation of High Temperature Structural Adhesives for Extended Service," J.B. Nelson, Contract Monitor, initiated Nov. 1978.

2. C.L. Hendricks and S.G. Hill, "Evaluation of High Temperature Structural Adhesives for Extended Service," SAMPE National Symposium, 25, 39-55 (1980).

Aminosilane and Metal Modifications of Polyimide Surfaces

H. J. Leary, Jr. and D. S. Campbell*

Department A-46
IBM Corporation
Poughkeepsie, NY 12602

ABSTRACT

Alterations in polyamic acid-polyimide surfaces produced by the use of different methods of curing (oven vs. hot plate), the application of a silane adhesion promotor (γ-aminopropyltriethoxysilane), and the deposition of copper have been studied using ESCA. Our objective was to develop chemical state information (elemental composition, chemical bonding features) about the modifications occurring in the outermost few tens of angstroms of polyimide films. The results show different C 1s spectral features for hot plate vs. oven cured films but identical stoichiometries. The silane treated amic acid surface showed an additional N 1s component at 402.0 eV due to NH^+. Heat treatment of this film resulted in the formation of a silicon-polyimide. Grazing angle ESCA results showed that the silane diffused into the sub-surface of the amic-acid film. Analysis of cured polyimide films after the etch ($FeCl_3$) removal of a copper film indicates that unzipping of the polymer may have occurred. A disrupted polyimide matrix that is vertically inhomogeneous in elemental composition was detectable after the sputter deposition of ~1.5 nm of Cu onto a cured polyimide film.

*IBM Corporation, Essex Junction, VT 05452

INTRODUCTION

The interfacial chemistries occurring in polyimide-inorganic and in Cu-polyimide systems are of concern to us. Our approach has been to use ESCA (or XPS) in conjunction with the "separation of variables" experimental method. To initiate studies of the interfacial chemistry occurring when polyimide layers are joined to inorganic substrates when an interlayer of the adhesion promotor Union Carbide A-1100 (γ-aminopropyltriethoxysilane, γ-APS) is used, we have investigated the A-1100-polyimide chemistry that occurs within the outermost 30-100Å of the polymer surface. For insights about what occurs at Cu-polyimide joints, we have investigated the interface formed by the deposition of thin metal films (~1.5 nm) onto cured polyimide substrates.

Our studies had one objective, to define the basic surface interactions occurring when thin layers of organic films (i.e.,γ-APS), and thin layers of inorganic films (i.e., Cu and Ni) are adsorbed or deposited onto organic substrates.

EXPERIMENTAL

The organic substrates were resin layers formed by the spin coating of PI-5878 (8μ thick) onto Si wafers. The resin was heated to approximately 100°C on a hot plate for 20-30 minutes to remove excess solvent (i.e.,N-methylpyrrolidone) so as to form a rigid condensed-phase polyamic acid (PAA) layer. A 0.1% aqueous solution of γ-APS was then applied to the PAA film layer by use of a puddle-spin application method. These specimens were examined by XPS before and after <u>in situ</u> thermal treatment. The results were compared with those obtained for non-silane treated PAA films before and after curing.

PI-5878 films cured to 380°C in an oven were the substrates onto which 1.5 nm Cu and 1.5 nm Ni were deposited. XPS characterizations were performed before and after temperature-humidity (80°C-80%) exposure for 500 hours. In addition, a thick layer of Cu was evaporated onto <u>one-half</u>

of a cured PI-5878 substrate. The complete substrate was subsequently exposed to a $FeCl_3$ solution so that the chemical etch removal of Cu would occur. Each half of the substrate was analyzed by ESCA in order to determine if the Cu induced changes in the polymer matrix.

The XPS measurements were performed using the Hewlett-Packard 5950B ESCA Spectrometer System. The procedures used are outlined elsewhere (1,2).

SELECTED RESULTS AND DISCUSSION

Figure 1 shows the photoelectron spectral data for γ-APS coated PAA films before and after a 200°C *in situ* heat treatment. The binding energy values enclosed in parentheses are those from non-γ-APS coated substrates. The N 1s component at binding energy 402.0 eV (Spectrum B) is due to NH^+ in agreement with literature data (3). Evidence for imidization in the surface region after the thermal treatment is shown by the existence of the single component carbonyl functionality at 289.0 eV due to imide (1,2). The reactions occurring, as detectable by our ESCA results, are summarized in Fig. 2 and are consistent with results and conclusions obtained by other analytical methods. Grazing angle ESCA measurements produced results indicating that penetration of the γ-APS into the polyamic acid matrix does take place.

Table 1 contains the elemental compositional results obtained by ESCA for Cu coated polyimide films before and after T & H exposure. As shown, the metal diffuses into the polymer matrix after the T & H exposure. The same result was obtained for Ni. The diffusion of Cu into PI-5878 surfaces detected in our ESCA results correlates with the 4X increase in peel strength (after 500 hours at T & H - 85°C 80%) measured (4) for thick Cu films deposited onto cured PI-5878 substrates that were held at 25°C during the metal deposition. Diffusion is also thought to be responsible for the *larger* increase in peel strengths measured (4) for Cu when the polyimide substrate temperature was purposefully maintained at 200°C during the metal deposition.

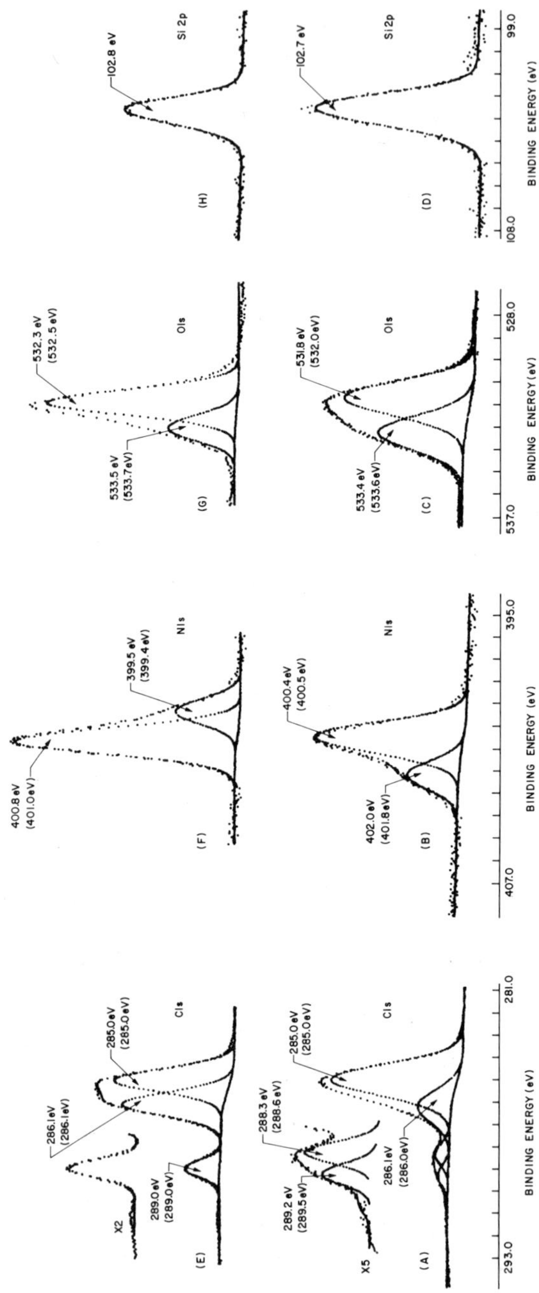

Fig. 1. ESCA Spectral Data for Polyamic Acid (PI-5878)-A-1100 Surfaces

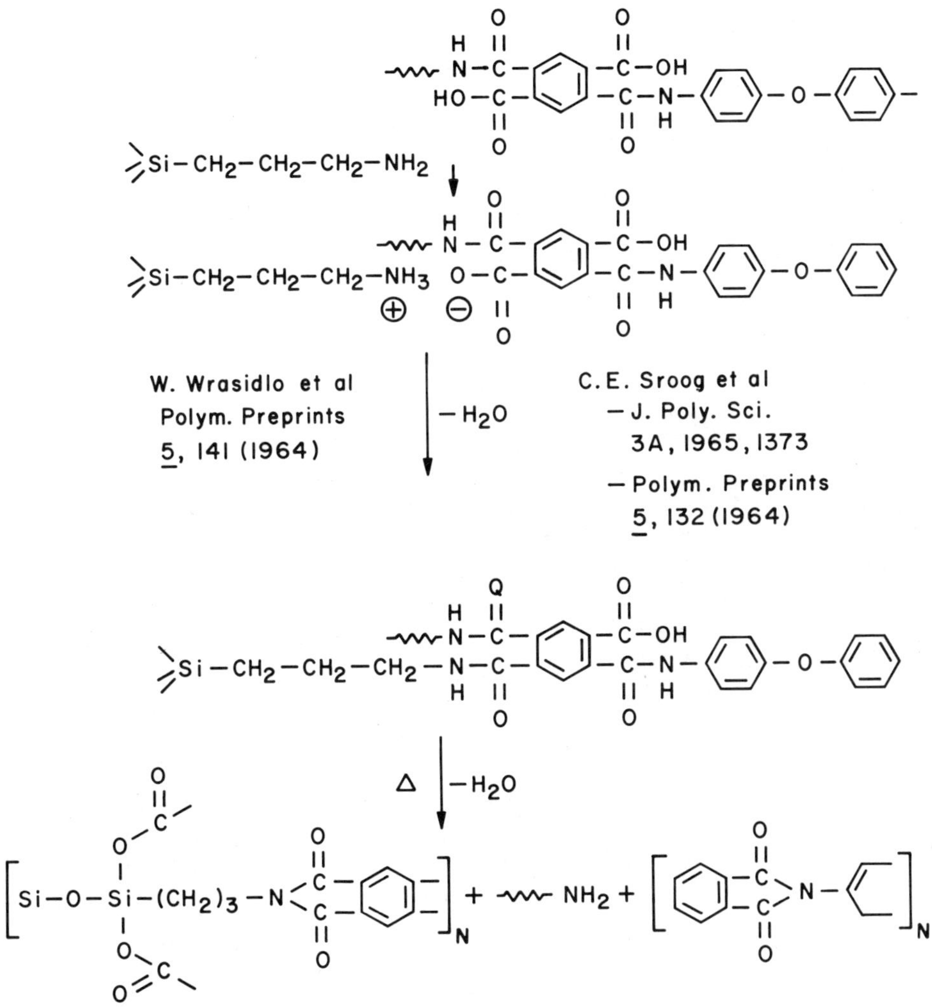

Fig. 2. Silicon Polyimide Resin

E. Bouldebuck et al., US Patent 3,303,157, Feb. 1967.
F. Holub, US Patent 3,392,144, July 1968.

Table 1

15Å Copper (Evaporated) Coated onto Oven-Cured (380°C) PI-5878

Elemental Composition Results (θ = 38.5Å)

	Mean Atom %			
Sample	C 1s	Cu 2P 3/2	N 1s	O 1s
Cu-Coated Surface (No T & H Exposure)	50.63 ± 0.14	19.35 ± 0.14	2.00 ± 0.94	28.02 ± 0.23
Cu-Coated Surface (After 500 hrs at 80°C-80%)	67.97 ± 0.33	4.34 ± 0.04	2.89 ± 0.03	24.80 ± 0.33

CONCLUSIONS

ESCA was found to be useful in studying the effects of different methods of curing, the adhesion promoter and the deposition of copper on polyamic acid-polyimide surfaces. The results show different C 1s spectral features for hot plate vs. oven cured films but identical stoichiometries. The silane treated amic acid surface showed an additional N 1s component at 402.0 eV due to NH^+. Heat treatment of this film resulted in the formation of a silicon-polyimide. Grazing angle ESCA results showed that the silane diffused into the sub-surface of the amic-acid film. Analysis of cured polyimide films after the etch ($FeCl_3$) removal of a copper film indicates that unzipping of the polymer may have occurred. A disrupted polyimide matrix that is vertically inhomogeneous in elemental composition was detectable after the sputter deposition of ~1.5 nm of Cu onto a cured polyimide film.

ACKNOWLEDGMENTS

The authors express their gratitude to E. Sacher, R. White and R. Ludwig for providing the samples used in this study.

REFERENCES

1. H. J. Leary, Jr. and D. S. Campbell, Surface and Interface Analysis, 1, 75 (1979).

2. H. J. Leary, Jr. and D. S. Campbell, "ESCA Studies of Polyimide and Modified Polyimide Surfaces", in *Photon, Electron and Ion Probes of Polymer Structure and Properties*, D. W. Dwight et al., Eds., American Chemical Society, Washington, D.C. (1981).

3. P. Moses et al., Anal. Chem., 47, 1882 (1975).

4. B. Fontaine, unpublished results.

Influence of Molecular Weight on Fracture Behavior of Polyphenylquinoxaline Thermosets

T.E. Helminiak
AF Wright Aeronautical Labs
W-PAFB, Ohio 45433

and

W.B. Jones
Texas Tech University
Lubbock, TX 79409

ABSTRACT

The fracture behavior of a family of polyphenylquinoxaline thermosets has been experimentally investigated, along with a polyphenylquinoxaline thermoplastic. Engineering fracture methods were adapted and applied to small quantities of materials. Miniature compact tension (CT) fracture specimens were fabricated using highly purified material samples, and were tested. The room temperature test results show the samples having low monomeric molecular weight (between reactive end groups) to be comparatively brittle, and samples having progressively higher (monomeric) molecular weight displayed progressively higher fracture toughness. The fracture toughness was found to increase monotonically with molecular weight between reactive end groups or entanglements.

INTRODUCTION

Polymers are being more frequently placed in structural roles as, e.g., matrix resin in particulate and fiber-reinforced composites and adhesives. This places requirements on polymers that heretofore have received little attention. One important property required of structural materials is resistance to fracture.

As a class of materials, polymers are usually considered to be comparatively tougher, or more resistant to fracture, than glass and other ceramics; and more brittle than metals. On closer examination, it is noted that various polymers may range in fracture toughness from that displayed by ceramics to that displayed by metals. Some polymers are even tougher than many metals.

Bearing in mind that the molecular arrangement possibilities (through synthesis) are almost infinite, one inquires whether a molecular structure can be adjusted for high fracture toughness while maintaining other material requirements such as high–temperature dimensional and thermo-oxidative stability. At present, it does not appear that this can be done with accuracy for very many materials, yet the process of inquiry produces beneficial results, and improves prospects for the future.

In recent years, substantial effort in industrial, governmental, and university laboratories has been directed to develop high temperature adhesives and matrix resins for composites. Many of the materials developed are referred to as "organic glasses," probably since their fracture behavior tends to remind one of the brittleness of glass. Upon thorough characterization and examination of these materials, it turns out most have distinguishable fracture behavior, and most are tougher than inorganic glasses. One such investigation was conducted at the Materials Lab (of the Air Force Wright-Aeronautical Laboratories) on a family of acetylene-terminated polyphenylquinoxaline thermosets.

MATERIALS

During the course of generating the science base necessary to develop the new acetylene addition-type cure chemistry, polymer synthesists in the Materials Lab prepared limited quantities (50 gms) of four oligomers. The samples were prepared according to the procedures developed by Arnold and Hedberg (1), and were purified by chromatographic column separation, and precipitation into a non-solvent (which was easily removed) in order to have a narrow range of molecular weight, and very little retained solvents (less than 0.01%). The molecular structure of the monomeric units is shown in Fig. 1. It is also noted from this figure that there are "building blocks" in common. One can see that the material called BADABA is essentially a dimer of the material called SBQ. And ATQ is essentially a tetramer of SBQ. BATQ is essentially a dimer of BADABA with a phenyl ether insert, and has a chain length only slightly longer than ATQ. A fifth material, PPQ-401, (a high molecular weight thermoplastic version of the polyphenylquinoxaline molecular structure, which was commercially available earlier from Whittaker Corp.), was also tested.

SAMPLE TESTING

Specimen preparation included drying the materials in a vacuum chamber at 40 to 100°C for several days to remove solvents and absorbed moisture, then introducing into preheated molds to form bars 2.5mm thick x 12.5mm wide by 75mm long. Only the low molecular weight materials, SBQ became fluid enough at the preheat temperature to facilitate casting by pouring into a closed silicone rubber mold. The other materials were formed to shape at 220°C in metal molds at 100 psi in a hydraulic press. After being partially cured for an hour at 235°C, the specimens were placed in an inert gas (N_2) oven and cure continued at 372°C for 1 hour. The oven was then turned off and allowed to slowly cool to about 50°C. Specimens were then removed from the oven and machined to the miniature compact tension fracture test specimen configuration shown in Fig. 2. Fracture specimens were then stored in a desiccator, to avoid moisture absorption, until testing.

Specimens were fracture tested at constant cross-head rate in a tensile testing machine. The specimens were initially precracked using a razor blade as a wedge. Specimens were then mounted in the test machine using

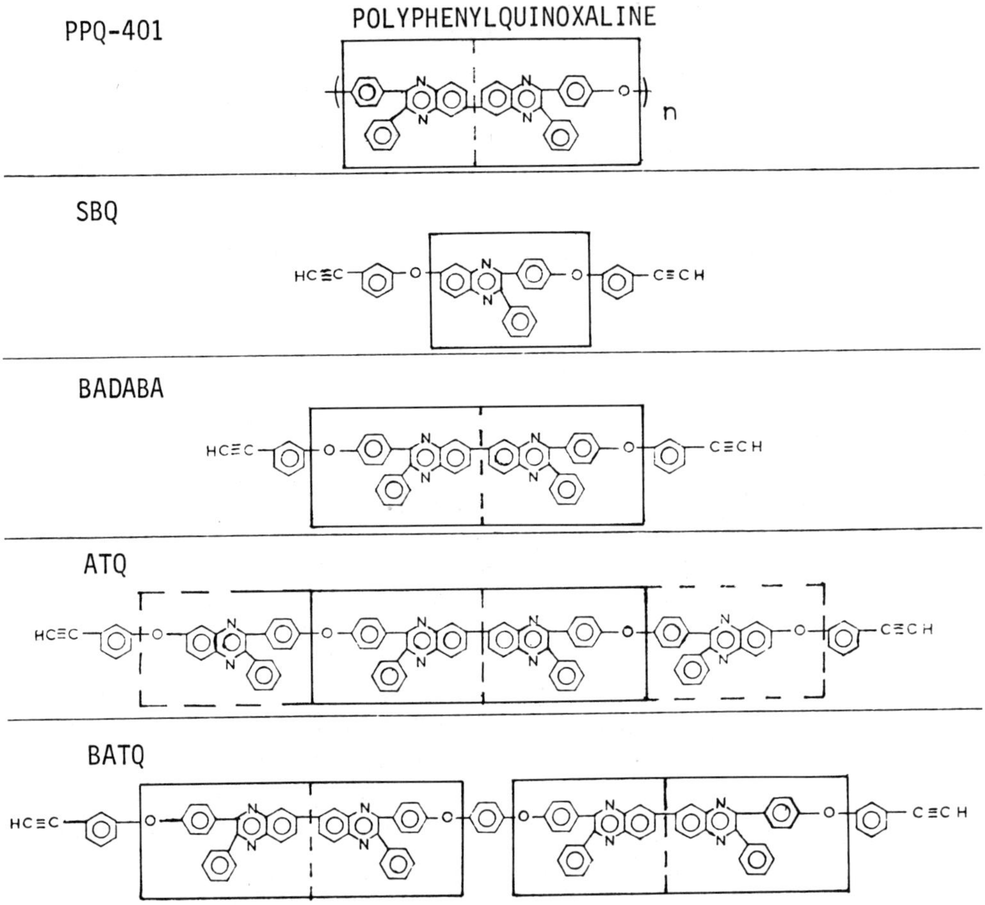

Fig. 1. Family of Polyphenylquinoxaline Thermosets and a Thermoplastic

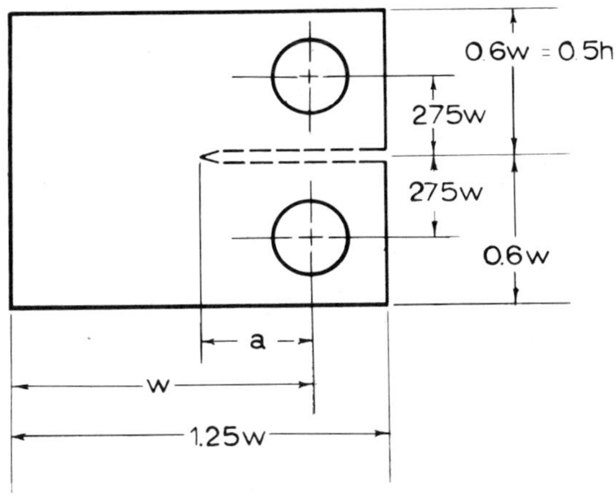

Fig. 2. Plan View of Compact Tension Specimen

specially designed crevices, and were pulled to fracture at constant cross-head rate of 8.47×10^{-4}mm/s. Often, upon fracture the load would drop sufficiently for the crack to arrest without completely parting the specimen. Loading was continued on these specimens, and multiple data points were collected. Data were recorded as load versus cross-head displacement such as shown in Fig. 3. The sharp drop in load, as well as the arrest, can be noted on these graphs.

RESULTS

The data were reduced by associating a crack length with each load-at-fracture, and calculating the fracture toughness K_c, according to the formula:

$$K_c = P_c/Bw^{1/2} f(a/w)$$

where the function f(a/w) is given in ASTM-E399 (2). Here B is the specimen thickness and w is defined in Fig. 2. The fracture load, P_c, is the maximum load prior to crack advance. The crack lengths, a, were measured after test using an optical microscope.

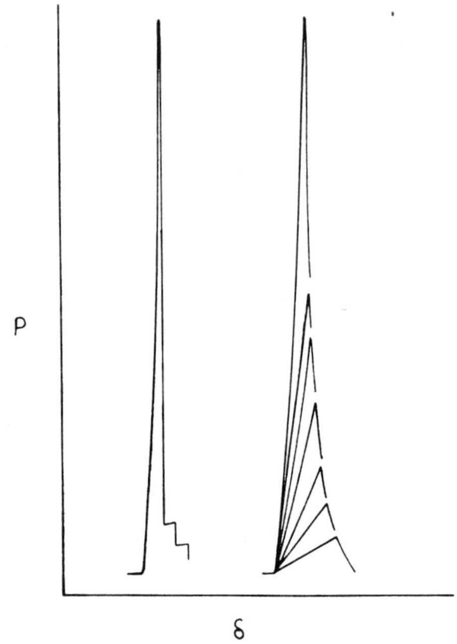

Fig. 3. Force-displacement (P- δ) Records

The reduced data for the five materials tested are summarized in Fig. 4. The fracture toughness data are plotted versus the normalized effective chain "length." The "length" for the thermosetting materials was determined by counting bonds along the chain, in the shortest distance between the reactive acetylene end terminations. These "lengths" were then normalized by dividing by the "length" of SBQ. For the thermoplastic, PPQ-401, the same procedure was used, for the molecule having the entanglement length. The range of entanglement length was estimated on the basis of viscosity data measured using fractionated samples of PPQ-401 (3).

DISCUSSION

The data shown in Fig. 4 clearly indicate a trend of higher fracture toughness for longer molecules (or higher molecular weight between cross-links or entanglements). Intuitively, this is an acceptable result. We know of no exceptions, and would expect additional fracture data for other polyphenylquinoxalines to reasonably fall between existing data points. As yet we have no reason to believe data obtained from a different polymer to fall on the same curve, but we would expect the same trend of higher fracture toughness for increasing molecular weight.

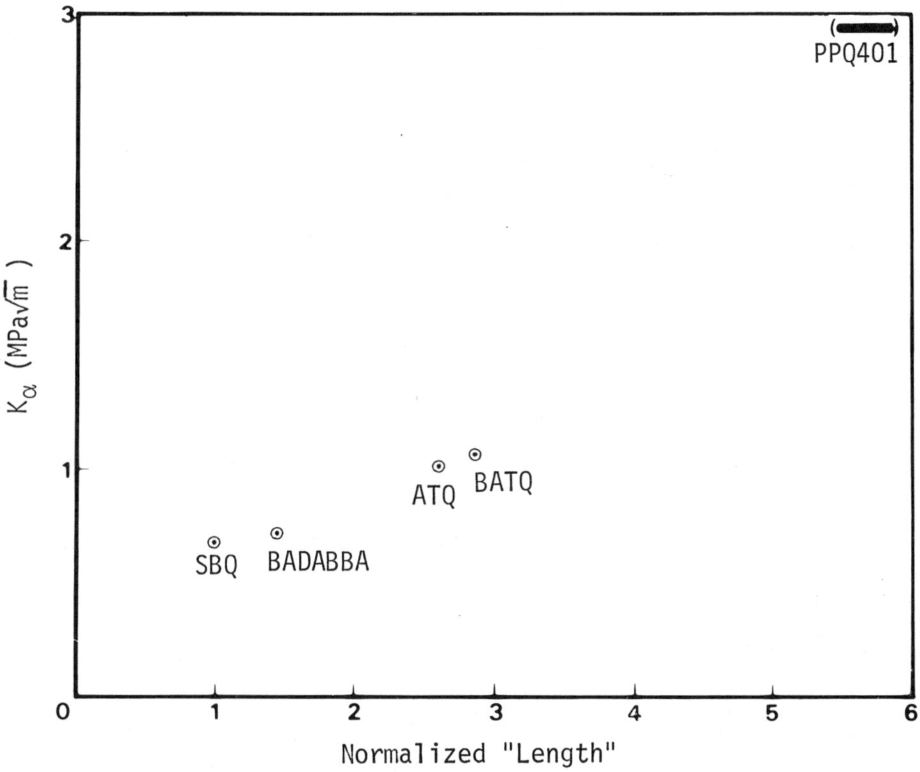

Fig. 4. Fracture Toughness vs. Normalized Chain Length for Acetylene-terminated Quinoxalines

In conclusion, we wish to emphasize that fracture toughness is an especially meaningful structural parameter, which can be used as a figure of merit for evaluating polymers for potential structural applications. Recognizing there are a number of requirements a structural resin must meet, we recommend all polymers be systematically characterized as early as possible in their development, and certainly that the most important structural property, fracture toughness, should be included in that characterization.

REFERENCES

1. F.L. Hedberg and F.E. Arnold, J. Appl. Polym. Sci., 24, 763 (1979).

2. F.L. Hedberg and F.E. Arnold, "Plane Strain Fracture Toughness of Metallic Materials," ASTM E399-78a, 1978.

3. Private communications with Dr. Joe S. Ham, Department of Physics, Texas A & M University, College Station, Texas 77841.

Discussion

On the Paper by P. M. Hergenrother

H. J. Leary, Jr. (IBM Corporation):

What evidence, and from whom, do you have that isoimides exist at a 3-6% level in thermally cured polyimides? Does that evidence extend to systems other than PMDA-ODA structures? Why have we not had isoimide reported in the numerous IR results generated in the literature of the 1970's and early 1980's?

P. M. Hergenrother (NASA Langley Research Center):

Researchers at IBM Research Center, San Jose, CA, particularly Dr. D. Y. Yoon, have done extensive work with the polyamic acid and polyimide from the reaction of pyromellitic dianhydride and 4,4'-oxydianiline. In private communication, Dr. Yoon reported that the thermally cyclodehydrated polyimide contained ~4% isoimide as evidenced by ESCA (Electron Spectroscopy for Chemical Analysis). Dr. Fred Hedberg, AFWAL, Dayton, OH, is using Fourier Transform Infrared Spectroscopy (FTIR) to determine residual isoimide content in the thermally cured product from an acetylene-terminated isoimide oligomer. Dr. Hedberg reported in private communication that the major isoimide band is in the 929-936 cm^{-1} region. Unfortunately, there are other weak bands in the IR spectra of polyimides which interfere with the isoimide band. The IR spectra of thermally cured polyimides from 3,3',4,4'-benzophenonetetracarboxylic acid dianhydride exhibited bands in the 929-936 cm^{-1} region, presumably due to isoimide, which decreases in intensity after further heating at 300-350°C in N_2. The isoimide rearranges to imide. The reason isoimides have not been reported in the polyimide IR literature of the 1970's and 1980's can only be answered by the researchers who did the work. Presumably, bands due to small amounts of isoimides are masked by other bands.

D. A. Scola (United Technologies Research Center):

How would you rate the thermooxidative stability of the high temperature polymers? Recent work showed PMR-15/graphite fiber composites to have good thermal stability, but poor thermooxidative stability. The composite is protected by the graphite fiber.

P. M. Hergenrother: In neat resin forms such as films, moldings, or powders, linear polyimides and polyphenylquinoxalines exhibit the highest thermo-oxidative stability at 316°C and 371°C of all the processable high temperature polymers. As reactive terminal groups (e.g., ethynyl, norborene, phenylethynyl, etc.) and reactive pendant groups (e.g., ethynyl and phenylethynyl) are introduced into the molecule, the thermooxidative stability of the cured polymer is lowered, relative to a comparable linear polymer. In comparing the thermooxidative stability of polymers, it is important that factors which influence the test be held constant. These include items such as oven design, air flow, temperature control, thermal history of sample, and sample form. In regards to the performance of PMR-15/graphite fiber composites under thermooxidative conditions, the specimens erode on the surface. Although the performance of PMR-15/Gr composites have varied from one laboratory to another, usable mechanical properties have been measured at 232°C after ~20,000 hours at 232°C and at 316°C after ~500 hours at 316°C.

J. Koutshy (University of Wisconsin): Where was the locus of failure for the lap shear fracture data reported?

P. M. Hergenrother: At RT, the high temperature adhesives exhibited predominantly cohesive failure in the lap shear specimens. As the test temperature was increased, the lap shear specimens generally exhibited mixed adhesive and cohesive failures. At test temperatures close to the Tg's of the adhesives, pronounced thermoplastic failures occurred. Chromic or phosphoric acid anodized titanium surface treatment generally exhibited excellent long term stability at 232°C but after ~300 hours at 316°C, the titanium oxide surface degrades to give adhesive-type failure. Most articles on the adhesive performance of high temperature polymers reports specific details including failure modes.

On the Paper by J. P. Wightman and S. Dias

D. Briggs (ICI, England): How does the Ti:0 stoichiometry work out when you analyze the Ti2p signal and the low binding energy 01S signal for fracture surfaces which show the peaks? Does it back up the theory that failure would take place partially in the TiO_2 surface layer?

J. P. Wightman (VPI, Blacksburg, VA): Total oxygen atomic fractions are reported in the tables. However, as Dr. Briggs points out, it is the lower binding energy oxygen photopeak which is assigned to TiO_2. A calculated value for the 0/Ti ratio of 2.7 based on curve fitting is in agreement with the value for TiO_2 powders as reported by Egerton et al. in Colloids and Surfaces.

On the Paper by C. L. Hendricks and S. G. Hill

J. P. Wightman (VPI, Blacksburg, VA): Why do you think anodization (phosphoric acid or chromic acid) of Ti 6-4 adherends gives better bond properties when compared to other pretreatments?

C. L. Hendricks (Boeing Aerospace Company): Surface characterization and failure analysis studies have shown that anodization of Ti 6Al-4V produces relatively deep oxide structures. This structure is penetrated by adhesive resin to the oxide base forming a strong bond attributed to mechanical locking to the large oxide surface area. Variation of the anodize voltage to change the oxide depth has shown that deeper oxides produce higher bond shear strengths. The other surface treatments studied have relatively shallow oxide layers which contribute to lower shear properties.

PART FIVE:

ANAEROBIC AND STRUCTURAL ADHESIVES

Introductory Remarks

James P. Wightman

Department of Chemistry

Virginia Polytechnic Institute
& State University

Blacksburg, VA 24061

Buchanan from ISI, describes the basic principles of scanning electron microscopy (SEM) and has emphasized a review of recent developments which has resulted in improvements in SEM performance and flexibility. Modifications in the secondary electron detector/objective lens design are discussed. Resolution of 2 nm can be achieved using the so-called top stage configuration. Large specimens such as silicon wafers can be examined at high resolution using a dual detector system. Several advantages of low voltage operation are pointed out and a comparison of three electron sources is made including a double anode LaB_6 electron gun.

Pike and Lamm, in a contribution from United Technologies, describe the preparation of acrylic modified polyester adhesives which exhibited anaerobic cure behavior. The hydroxyl containing polyester was synthesized from dimethyl terephthalate, isophthalic acid, trimellitic acid and diethylene glycol or ethylene glycol using a butyl titanate catalyst. A series of model compound reactions was carried out to establish optimum conditions for the acid-catalyzed addition of glycidyl methacrylate with the hydroxyl containing polyester. Sulfuric acid and boron trifluoride etherate were the most effective catalysts at reaction temperatures of 75° and 40°C, respectively.

Malofsky and Baccei, from the Loctite Corporation, report on the significant mechanical property improvements obtained with maleimide and nadic capped monomers and prepolymers as additives in anaerobic methacrylate/acrylate adhesive systems. Significant improvement in thermal resistance was noted in a PEGMA (polyethyleneglycol dimethacrylate) system in which a polymerizable maleimide addition was incorporated. A "neat" polymerized urethane methacrylate failed catastrophically within 1000 hrs. at 450°F (232°C) whereas with the additive m-phenylene dimaleimide, the co-monomer system gave nearly 1000 psi tensile shear strength after 2700 hrs. aging.

Humphreys, of the Loctite Corporation, reported on the chemistry of accelerators for curing anaerobic adhesives. He showed that the reaction between N,N-dimethylaniline derivatives and cumene hydroperoxide is relatively slow even at 100°C. He concluded that the accelerated polymerization of anaerobic adhesives of ambient temperatures caused by cure systems containing combinations of tertiary aromatic amines, hydroperoxides, and sulfonimides does not result from the hydroperoxide-amine reaction.

Pocius, Wangsness, Almer and McKown, at 3M, have reported quite interesting results of long term (up to 10 years) bond durability studies on a variety of adhesive/adherent system. The authors have used the sustained bond overlap shear test at elevated temperatures (up to 140°F) and at 100% relative humidity to make qualitative correlations of bond durability with a number of variables. Combinations of three aluminum alloys and one steel adherend pretreated in three ways and twenty adhesives were analyzed. Variables examined included adhesion type, cure temperature, adherent type, surface preparation and load placed on the specimen. It is significant that the failure mode of samples exposed for long times is gradual undercutting of the adhesive bond at the metal oxide/adhesive (primer) interface. This undercutting continues until the remaining bonded surface is not large enough to support the applied load.

In a contribution from B. F. Goodrich, Drake and Siebert extensively review the journal and patent literature since 1975 on reactive butadiene/acrylonitrile liquid and solid elastomers used in formulating epoxy structural adhesives. Areas reviewed include the preparation of elastomer-modified epoxy resins, the characterization of rubber-toughened epoxy resins, fracture mechanics and adhesive formulation and testing.

Some Recent Developments in Scanning Electron Microscopy

R. Buchanan

International Scientific Instrument, Inc.
Santa Clara, CA 95051

ABSTRACT

During the past few years a number of SEM instrumentation developments have taken place which have resulted in an improvement in image resolution, both under high and low voltage operating conditions. The areas where the most significant advances have been made are in electron gun, objective lens, and secondary electron detection system design. Secondary electron image resolution of 2nm at 30kV accelerating voltage, and 15nm at 1kV can now be realized. This opens up a number of new application areas for the SEM, such as semiconductor wafer inspection and the examination of delicate biological materials. A number of techniques are now available which allow the examination of uncoated insulating materials. These include low voltage operation, the use of a high efficiency backscattered electron detector, and observation of the specimen in a low vacuum environment.

INTRODUCTION

The scanning electron microscope (SEM) has been available in commercial form since 1965, and is now well-established as an indispensable tool in a very wide range of applications. In addition to providing a unique imaging capability, it can, with suitable accessories, perform microanalysis, reveal electrical faults in integrated circuits, and measure some fundamental properties of semiconductor materials. Until recently, the design concept, based on that developed at the Engineering Department of Cambridge University in England between 1948 and 1965, has changed little.

Several important advances have been made during the last few years, however, which have resulted in image resolution improvement, both at high and low operating voltages, the development of techniques for examining uncoated insulating materials, and the ability to produce images with enhanced image surface sensitivity. The subject of this paper is a review of some of these new developments.

Operating Principles

An image is formed in a SEM by scanning, in a raster fashion, a focused beam of electrons over the surface of a solid specimen and detecting, at least in its most common operating mode, low energy secondary electrons (SE) which are ejected from the specimen by the incident beam. This signal is detected and used to modulate the brightness of a CRT which is scanned in synchronism with the beam striking the specimen.

As the SE intensity varies across the surface of the specimen as a function of topography, and to some extent composition, an image of the scanned area is produced. Magnification is determined by the ratio of the CRT display area to the area scanned on the specimen. Reducing the scanned area on the specimen increases the magnification and vice versa. A schematic diagram of a SEM is shown in Fig. 1. For a detailed treatment of the subject of scanning electron microscopy, and an in-depth description of the operating principles of the instrumentation, the reader is referred to Refs. 1 and 2.

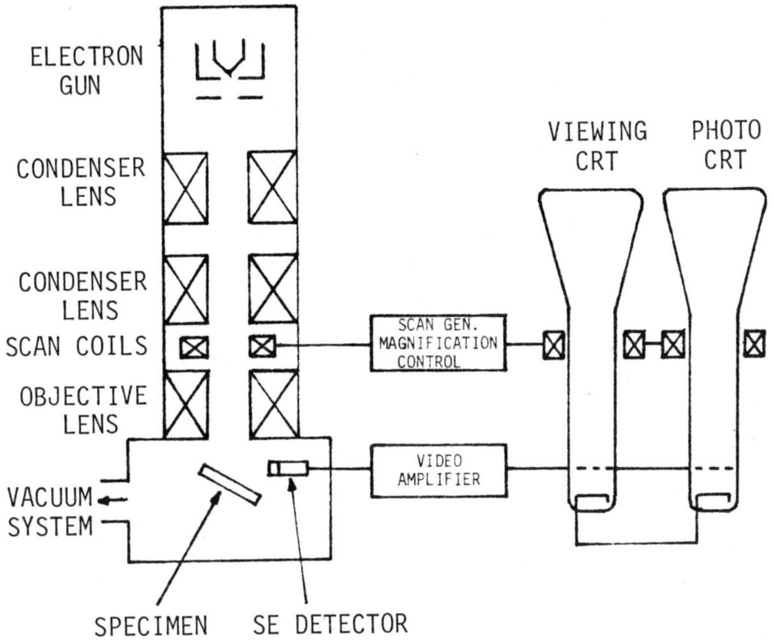

Fig. 1. Schematic diagram of a typical SEM.

Ignoring electron diffusion effects within the specimen, SEM resolution is determined by the diameter of the electron beam which scans the specimen. The minimum useable beam diameter, and therefore the optimum resolution, is determined by a number of instrument parameters, which include the brightness of the electron gun, the SE collection efficiency, and the aberrations of the final focusing lens in the electron optical column. This very critical lens is normally referred to as the objective lens. While aberrations are associated with the condenser lenses, their magnitude is much smaller than those of the objective lens, and they can be ignored.

The electron gun should provide a source brightness as high as possible and the energy spread in the emitted beam should be as low as possible. The two principal aberrations of concern in a SEM are spherical and chromatic aberration, and they should be as low as possible.

Spherical aberration dominates when operating at accelerating voltages above about 10kV. This results from electrons which pass through the objective lens far from the electron optic axis being brought to a focus closer to the lens than near axis electrons. Chromatic aberration, which dominates at low accelerating voltages, results from electrons which have slightly different energies being brought to a focus at different points in the image plane.

The aberration characteristics of the objective lens are determined by the polepiece geometry and magnetic field strength. The diameters of the discs of confusion associated with these aberrations, i.e. the magnitude of the aberration and the associated image resolution limitation, are directly proportional to two lens parameters called the spherical aberration coefficient (C_s) and the chromatic aberration coefficient (C_c).

Most currently available SEMs are capable of resolving better than 10nm, with the most sophisticated instruments achieving 2nm resolutions. This is considerably better than that possible with an optical microscope.

Yet another important advantage of the SEM over the light microscope is the much larger depth of field of the electron optical instrument. It is typically about X200 greater at any given magnification. This is due to the very small convergence angle of the electrons striking the specimen surface (typically a few milliradians)--equivalent to an extremely small numerical aperture in light optical terms.

Signal Generation in the SEM

We have thus far reported only to the SE signal as a means of generating an image in the SEM. In fact, several other signals are generated as a result of the interaction of the incident electron beam with the specimen. These include backscattered electrons (BSEs), X-rays, light, heat, specimen current, electron beam induced conductivity, transmitted electrons (if the specimen is thin enough), to name the most commonly used.

SEs have energies less than about 20eV, while BSEs have energies ranging up to that of the incident electrons. Due to the wide energy difference between the SEs and BSEs (usually several thousand eV), it is possible to easily separate the two signals. A typical Everhart/Thornley SE detector (3) is shown in Fig. 2. This consists of a mesh, to which a positive potential of a few hundred volts is

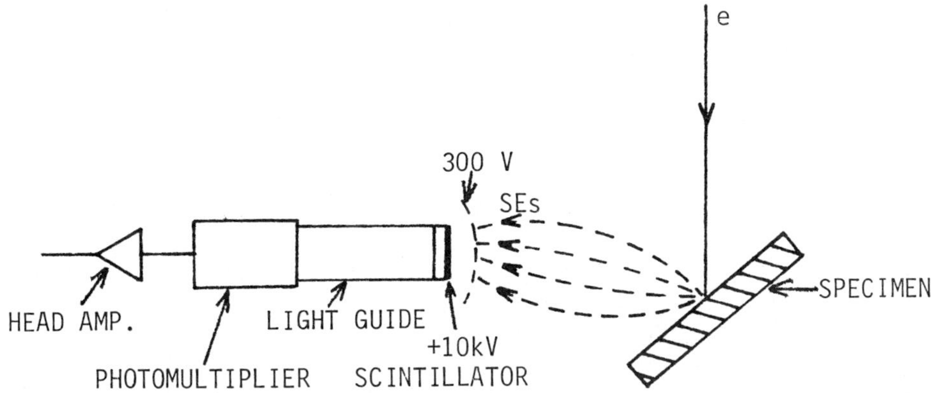

Fig. 2. Everhart/Thornley SE Detector.

applied, located in front of a scintillator, which is coupled via a light guide to a photomultiplier. SEs are attracted to the detector mesh and accelerated through the mesh onto the scintillator, which is biased with a positive voltage, typically around 10kV. The scintillator is coated with a thin film of phosphor which produces light when struck by the electrons. This is carried down a light guide to a photomultiplier, which converts it to an electrical signal suitable for image generation.

This type of detector can also detect BSEs. Due to the high energy of BSEs, they are not attracted to the SE detector by the relatively low grid potential. Only those which by chance travel from the specimen towards the detector, or enter the detector after backscattering from surfaces inside the specimen chamber, contribute to the output signal. As the solid angle subtended by the scintillator is very small, however, the contribution of BSEs directly to the SE signal is normally very small. As discussed in a subsequent section of this paper, there is in addition a very substantial indirect contribution from BSEs to the detected SE signal.

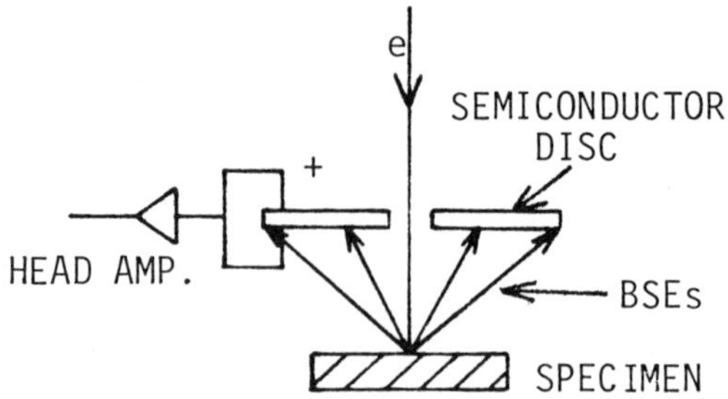

a) Solid State BSE Detector

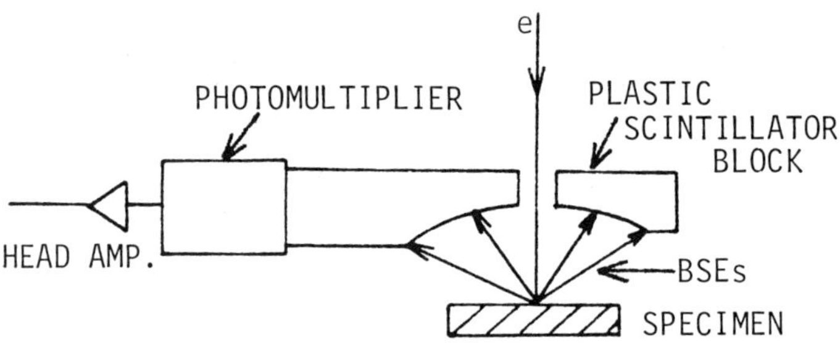

b) Robinson BSE Detectors

Fig. 3. Backscattered Electron Detectors.

Efficient BSE collection clearly requires that the detector subtend a large solid angle, and be insensitive to low energy SEs. Both of these requirements are satisfied by the solid state disc, and Robinson detector (4) shown in Fig. 3. In the case of the solid state detector, the incident electrons require to have energies greater than approximately 10keV, to generate hole/electron pairs in the semiconductor disc, which results in the signal used for image formation. The Robinson detector, on the other hand, will operate down to 5keV and exhibits a higher signal to noise characteristic, allowing the generation of images with 10nm resolution. The two main advantages of BSE images are that they exhibit enhanced elemental contrast, which is especially useful in conjunction with an X-ray analysis system, and the images are considerably less sensitive to degradation due to specimen charging.

X-rays are emitted from a small near-surface region, of the order of a few cubic microns, when a specimen is bombarded by the electron probe in a SEM. This volume is somewhat larger than that within which the incident electrons dissipate their energy, as a result of the production of X-rays outside this volume due to fluorescence. As the radiation is characteristic of the material within the excitation volume, microanalysis, both qualitative and quantitative, of selected regions of the specimen may be performed. Alternatively, the distribution of one particular element within the area scanned by the beam can be shown.

There are currently two types of X-rays analysis system used with SEMs. By far the most commonly employed is the energy dispersive system. Here X-rays are detected by a liquid-nitrogen cooled Li-drifted Si-disc enclosed in a vacuum housing with a thin Be entrance window through which the X-rays pass. This is illustrated in Fig. 4. An output pulse is generated by individual incident X-ray quanta, the amplitude of which is proportional to the energy of the

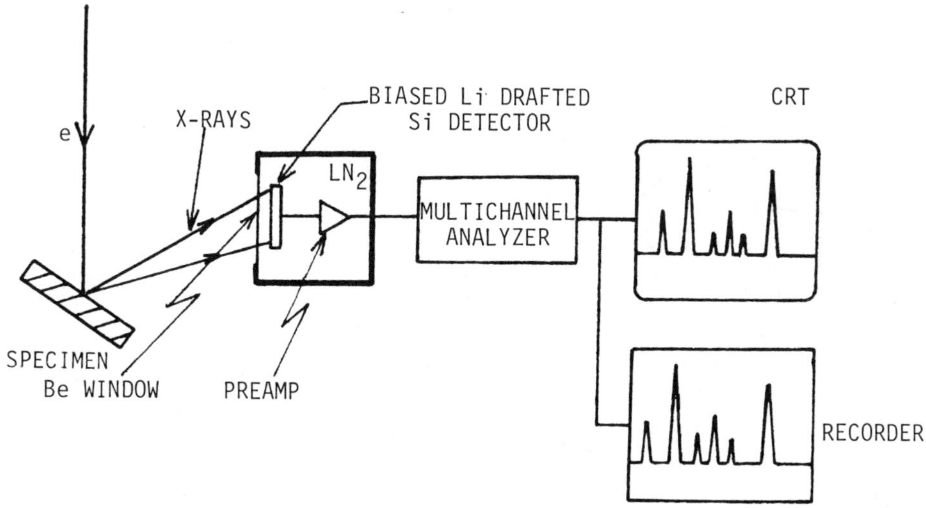

Fig. 4. Energy Dispersive X-Ray Analysis System.

quanta. By sorting and storing the output pulses according to their amplitude, a spectrum can be generated. The elemental detection range is from Na to U, although F can be detected with some difficulty. By removing the Be entrance window, the light element limit can, however, be lowered to C. Windowless EDX detectors are a relatively recent development, but are now available from a number of suppliers.

The other type of X-ray analysis system used with SEMs today is the wavelength spectrometer, illustrated in Fig. 5. Here the selective diffraction properties of a crystal in conjunction with a proportional counter detector is used to sort out the radiation according to wavelength. With suitable crystals, this system allows the analysis of all elements from Be-U. In addition to allowing the detection of light element radiation, the WDX spectrometer provides superior elemental discrimination compared to an EDX system.

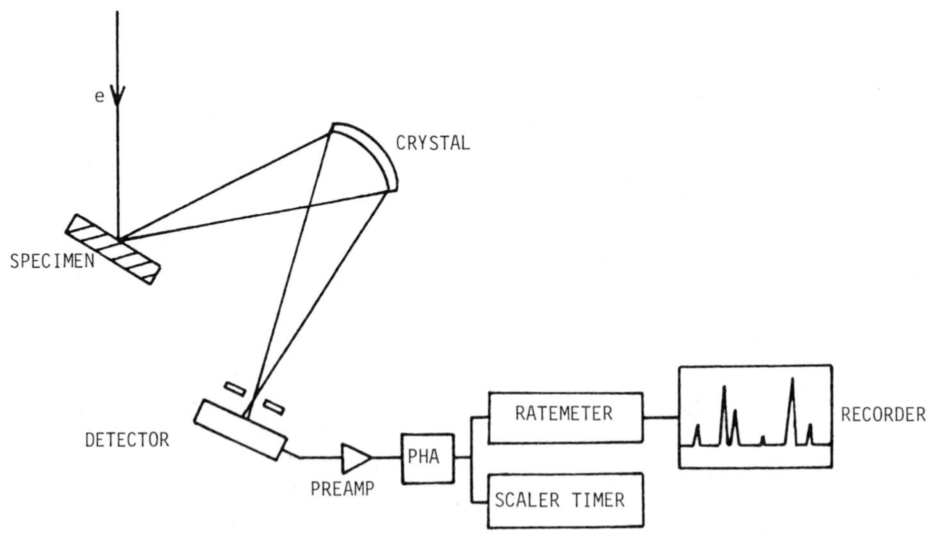

Fig. 5. Wavelength Dispersive X-Ray Analysis System.

There are, however, many advantages of the EDX equipment over WDX spectrometers. These include analysis speed, ease of spectra interpretation, and compatibility with the SEM. An overview of all the elements present, within the elemental detection range of the EDX spectrometer, is obtained almost instantaneously, while the WDX spectrometer requires to be driven from one setting limit to the other with each of the normally four crystals provided. The interpretation of WDX spectra is often complicated by the presence of higher order diffraction, which does not occur in EDX spectra. Due to the inefficiences involved in the diffraction of X-rays from a crystal surface and the fact that the percentage of generated radiation incident on the crystal can be very small, it is necessary to operate the SEM with higher beam currents with a WDX system than with an EDX spectrometer. This leads to accelerated contamination of the electron optical column which results in degraded image resolution.

Recent developments, such as the windowless EDX detector, have allowed the light element range to be extended down to C. Automated WDX spectrometers with computer control of the operating parameters are now available, such as the Microspec WDX-2A system. This considerably simplifies WDX analysis. The ideal system, however, requires both an EDX and WDX system mounted on the microscope simultaneously. This would permit the rapid determination of the elements present with the EDX system, and a detailed analysis of these elements using the WDX spectrometer. Combined EDX/WDX systems have already been developed where components of the hardware are shared, such as a computer to perform corrections on the measured data for atomic number, absorption, and fluorescence effects. These corrections are necessary when performing quantitative analysis.

SE Detector/Objective Lens Design

Until recently, all SEMs used a "pinhole" objective lens and single SE detector located below the lens, as shown in Fig. 6(a). A characteristic of the "pinhole" lens, in which the polepiece bore diameter closest to the specimen is smaller than the other bore, is that the stray magnetic field below the lens is very low. With the specimen in field-free space, the low energy SEs are readily collected by a SE detector located in the vicinity of the specimen. The other reason for selecting this particular lens design

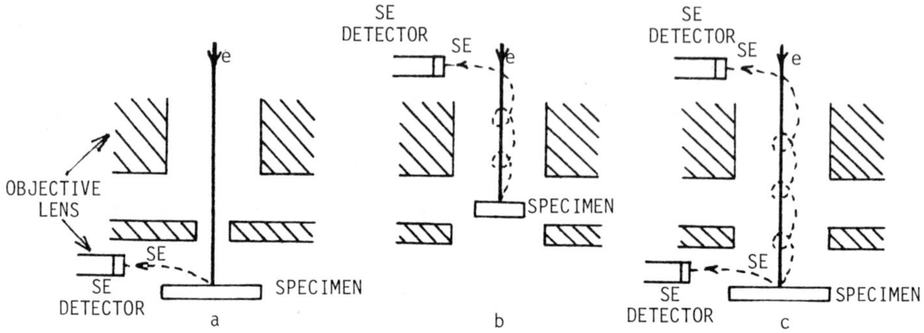

Fig. 6(a,b,c). Objective lens polepiece and secondary-electron detector systems. (a) Conventional "pinhole" lens with bottom SE detector. (b) Low-aberration configuration employing top SE detector. (c) Open-bore configuration with dual SE detectors.

is to avoid magnetising the specimen. A limitation of the "pinhole" lens, however, is that the aberration characteristics are poorer than with some other lens designs.

A departure from convention appeared in 1978 with the introduction of the ISI DS-130 SEM. In this instrument, two objective lenses, two specimen stages, and two SE detectors are used. Of particular interest is the top stage and associated objective lens. Here the specimen is introduced from the side into the gap between the polepieces of a high field strength objective lens with very low aberration characteristics. Refer to Table 1. SEs are trapped by the magnetic field and spiral back through the lens, where they are picked up by a detector above the lens. This is illustrated in Fig. 6(b). Resolution of 2nm is achieved. Figure 7 illustrates the high resolution performance obtained.

Table 1. Objective Lens Aberration Coefficients

	Dual-Stage SEM		SS-Series/ IC-130	
Aberration Coefficients	Top Stage	Bottom Stage "pinhole"	At 3mm Working Distance	With polepiece insert - "pinhole"
C_s (mm)	3.3	24	5.2	24
C_c (mm)	3.3	14	5.6	14

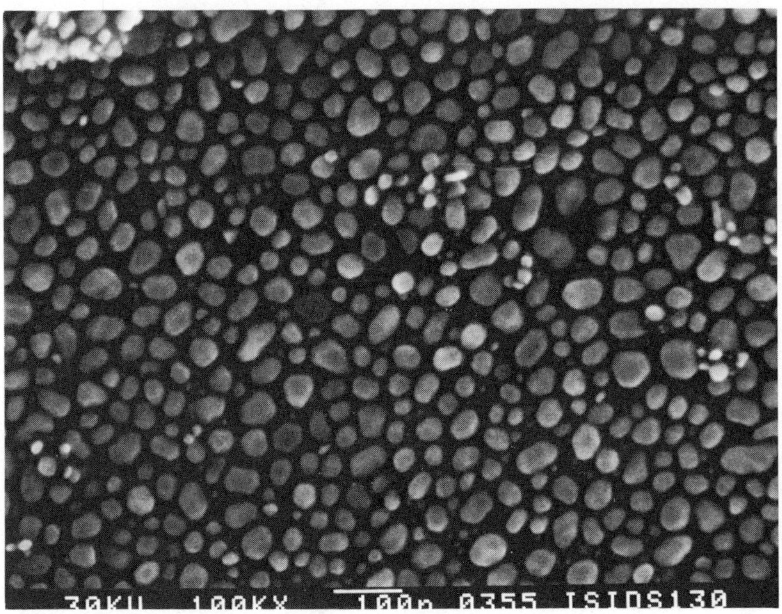

Fig. 7. Gold particles on carbon taken in top stage DS-130.

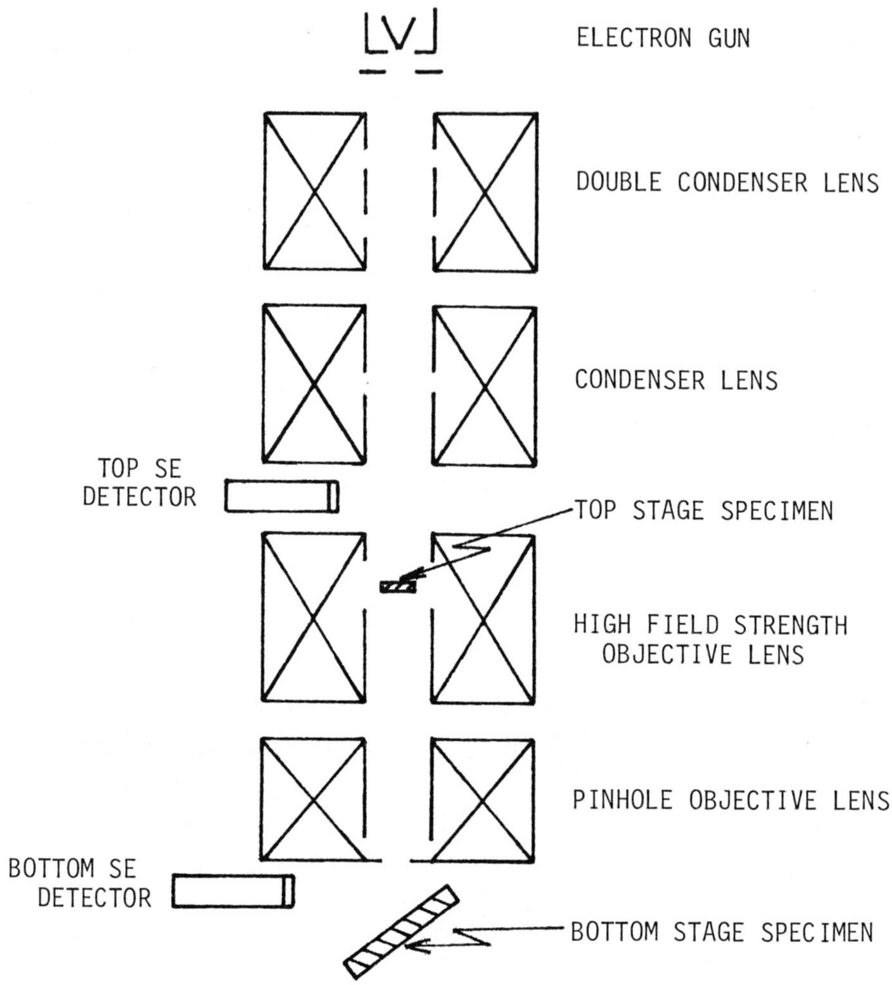

Fig. 8. DS-130 Schematic Diagram.

Magnetic materials and samples too large to fit in the top stage of the instrument can be examined in the bottom chamber, where a conventional "pinhole" lens and SE detector arrangement is used. A schematic diagram of the complete electron optical column is shown in Fig. 8.

In order to simplify the instrument design, while still retaining the high resolution and flexible specimen handling capabilities of the DS-130, a new series of microscopes was introduced in 1981 by ISI--the SS-series. This is based on a design concept described by Koike et al. (5). The instruments incorporate a single objective lens with three operating modes, a single specimen stage, and dual SE detectors which are used in combination.

In the high resolution (HR) mode, the specimen is elevated through the bore in the bottom polepiece of the objective lens and positioned in the gap between the polepieces. High field strength conditions are then set up which results in low aberration characteristics. SEs spiral back through the lens and are collected by a detector located above the lens. This is comparable to the arrangement used in the DS-130 top stage, as illustrated in Fig. 6(b). As can be seen from Table 1, the C_s and C_c values, while not as good as the DS-130 top objective lens, are a considerable improvement over the typical "pinhole" lens. 3nm resolution is readily obtained.

Large specimens are examined below the lens, and here the signals from both SE detectors are either summed or used individually, and the resultant signal used for image formation. At a short working distance most of the signal is contributed by the top detector. The contribution from the bottom detector increases with increasing working distance. This operating mode is referred to as Mode 1 and is illustrated in Fig. 6(c). Magnetic materials are examined after screwing a polepiece insert into the bottom of the lens, thus converting it to a "pinhole" configuration. This is referred to as Mode 2, and here the bottom detector is used alone.

In order to avoid the specimen size limitations imposed by both the DS-130 top stage and SS-series HR operating mode (8mm diameter X 5mm high), a new objective lens, in which the magnetic center lies outside the polepiece gap, has been developed. C_s and C_c values obtained at a working distance of 3mm are similar to those of the SS-series in the HR mode. This lens is incorporated in the ISI IC-130 semiconductor wafer inspection SEM, which has recently been introduced.

A practical limitation of conventional SEMs which is often overlooked, is that high resolution cannot normally be obtained when examining a large specimen. This is due to the fact that the SE collection efficiency with large specimens at short working distances is very poor. In order to achieve an acceptable signal level, it is necesary to use a relatively long working distance. This allows the field from the SE detector to penetrate to the point on the specimen impacted by the incident electron beam. This is necessary for efficient SE collection. Extending the working distance, however, degrades the resolution. The dual detector system eliminates the need to work at long working distance due to the high collection efficiency of the top detector with the specimen close to the lens. Thus a large specimen, such as a semiconductor wafer, can be examined under high resolution conditions.

SE Signal Sources

We have thus far referred to the SE signal as low energy electrons ejected from the specimen as a result of the interaction of the incident electron beam with the specimen. There is, however, another source of SEs in a SEM, and there are two mechanisms for generating SEs in the specimen. We will refer to these electrons as Type I, II, and III, as defined below:

Type I: SEs produced by the primary electron beam as it enters the specimen. Only those produced close to the surface escape and contribute to the detected SE signal. The escape depth depends on the composition of the specimen. It is typically in the range from 2-20nm for metals and 5-50nm for insulators.

Type II: SEs produced by BSEs as they leave the specimen. Again only those produced near the surface of the specimen escape.

Type III: SEs produced by BSEs colliding with a surface in the vicinity of the specimen, such as the bottom of the objective lens and the inside surfaces of the specimen chamber.

The sources of Type I, II, and III secondaries are illustrated in Fig. 9.

As most SEMs collect all three types of SEs simultaneously, and offer no means of separating the signals, the images produced are influenced by both the SE

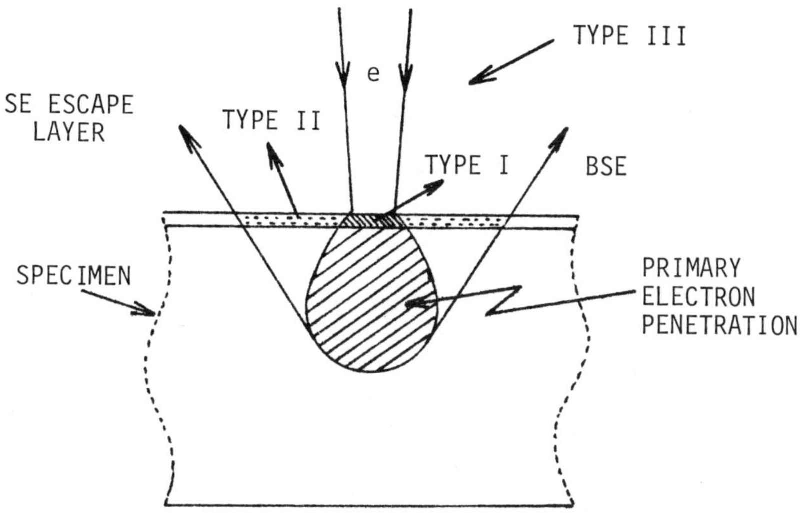

Fig. 9. Sources of Types I, II, and III SEs in a SEM.

and BSE emission characteristics of the specimen. The relative contribution of each type of SE signal to the image is a function of the energy of the incident electron beam, the composition and topography of the specimen, the geometrical relations between the specimen and its surroundings, as well as the materials present in the vicinity of the specimen. This subject has been discussed extensively by Peters (6,7).

SE Production

The relative contributions of Type I and II SEs have been investigated experimentally by Bronshtein and Fraiman (8), and Kanter (9) for the incident electron beam energy ranges from 1-4kV and 2-20kV respectively. From this work it can be concluded that the total number of Type II SEs leaving the specimen is quite significant. Values varying from 25% to 75% of the total SE emission from the specimen (Type I plus II) are typical.

The area density for Type I SEs is, however, higher than for Type IIs, which come from a larger surface area. Image resolution is therefore normally determined by the Type I SEs, with a lower resolution background contribution from Type II and IIIs.

The Type III SE signal level is dependent on the BSE characteristics of the specimen and the BSE/SE conversion characteristics of the specimen surroundings. Measured data (9) shows that the contribution of Type III SEs compared with the total SE production (Type I plus II plus III) can be as high as 80%. Suffice to say that the number of Type III SEs produced is relatively high.

As the BSE/SE conversion characteristics of the surroundings remain approximately constant during image formation, there is no signal modulation imposed on the Type III SEs other than by the specimen. Type III SEs, therefore contribute BSE information to the SE image.

True Surface Imaging

As explained previously, most SEMs employ a "pinhole" objective lens and single SE detector. In this arrangement, there is no discrimination between the various types of SEs detected, and all contribute towards image formation.

With the dual detector and open bore lens design described previously, some separation of the various types of SEs is possible.

At a short working distance, the majority of Type I and II SEs are trapped by the stray magnetic field which extends below the open bore objective lens. They spiral back through the lens, and are picked up by the top detector. There is a contribution to the top detector signal as a result of BSEs producing SEs from the lens bore, but this is less than in the "pinhole" arrangement. It has been determined experimentally (10) that the percentage contribution of Type I and II SEs to the detected signal is more than 2X higher in the arrangement shown in Fig. 6(b) using the top detector only, than in the conventional arrangement depicted in Fig. 6(a).

The bottom detector in Fig. 6(b), on the other hand, collects predominantly Type III SEs at a short working distance, and therefore allows the production of a BSE image. A signal mixing unit allows the signals from the two detectors to be used individually or mixed.

Comparing a top detector image with a bottom detector image allows surface and sub-surface contrast effects to be easily separated. The left side of Fig. 10 is a bottom detector image, of a molybdenum foil, and the right side is a micrograph of the same area of the foil using the top detector only. The dark islands surrounding groups of particulates are due to surface contamination, which cannot be detected in the left image.

Figure 11 shows bottom and top detector images, respectively, of a gold coated IC. Again, the information content of both images is dramatically different, allowing the extraction of much more information than is possible with a single combined image as obtained in conventional SEMs.

Figure 12 shows bottom and top detector images of an uncoated IC. Note the circular black surface contaminants visible only in the right image. Another interesting phenomenon is also illustrated in this micrograph, namely the high sensitivity of the top detector image to voltage contrast, induced, in this case, by the incident electron beam.

SE images produced by most SEMs are strongly influenced by the BSE characteristics of the specimen. This results in the production of composite SE/BSE images. The use of dual SE detectors allows the partial separation of the SE and BSE information, thereby facilitating the identification of surface related phenomena.

The images produced by the SE detector above the objective lens are strongly influenced by specimen surface characteristics such as topography, composition, and voltage differences. The BSE information content is considerably less than in the case of a conventional SEM.

The bottom SE detector, on the other hand, produces BSE images at a short working distance, which are much less sensitive to specimen surface phenomena. The selection of this detector allows adverse specimen charging effects to be minimized, and can be used to enhance bulk elemental contrast.

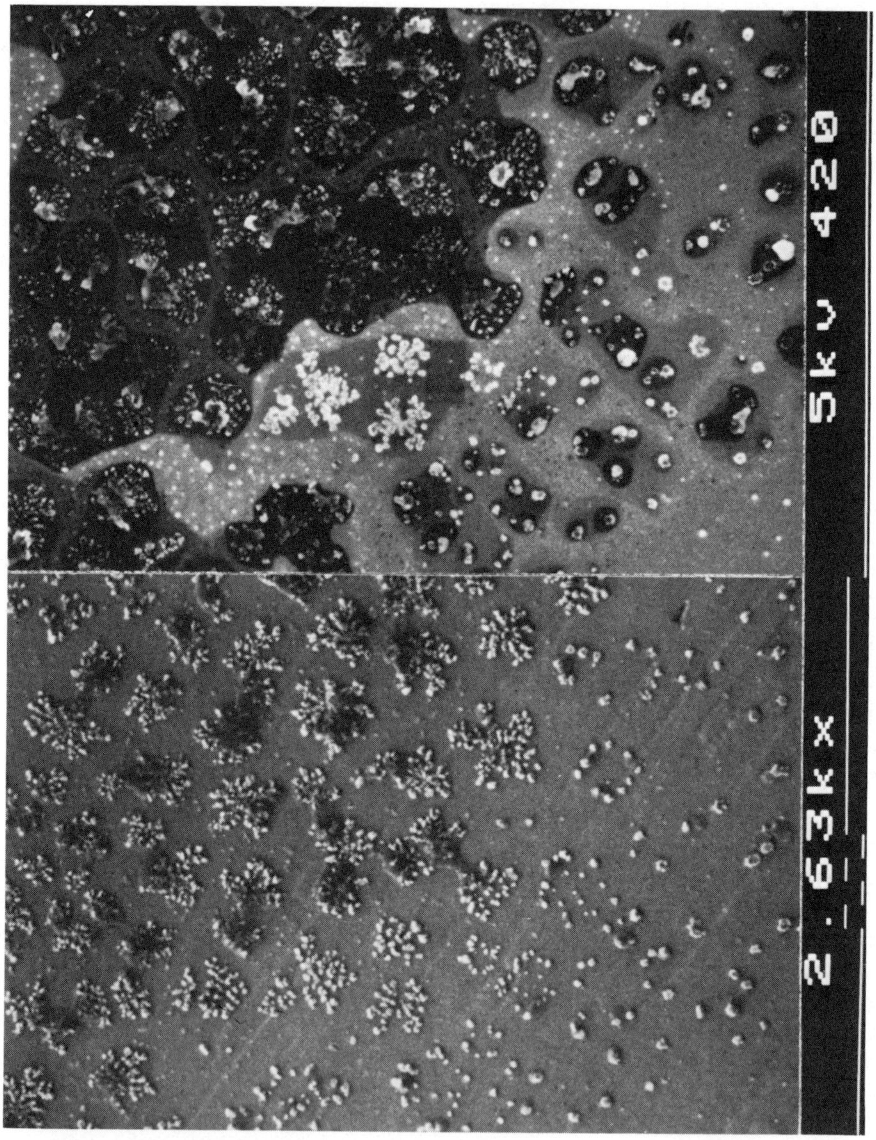

Fig. 10. Molybednum foil image produced with bottom SE detector (left), and top SE detector (right).

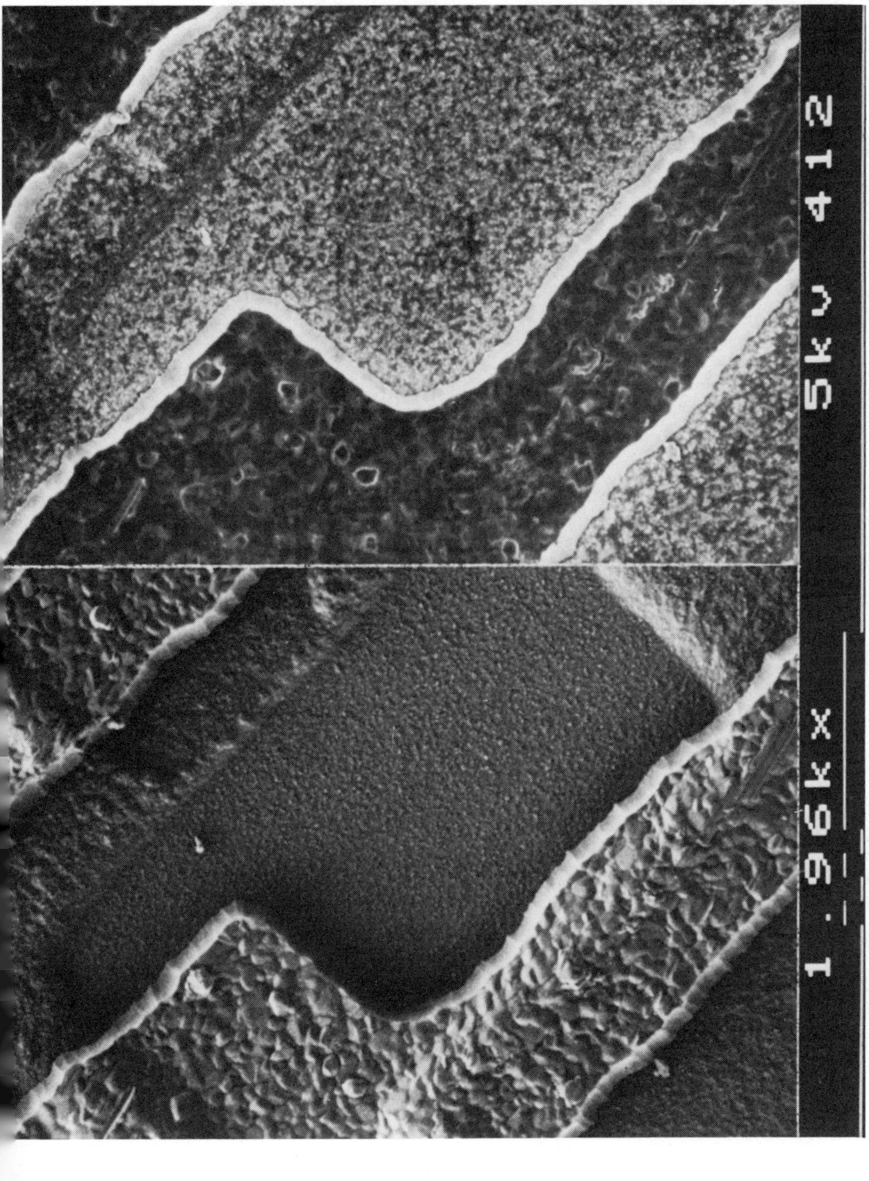

Fig. 11. Gold coated IC images produced with bottom SE detector (left), and top SE detector (right).

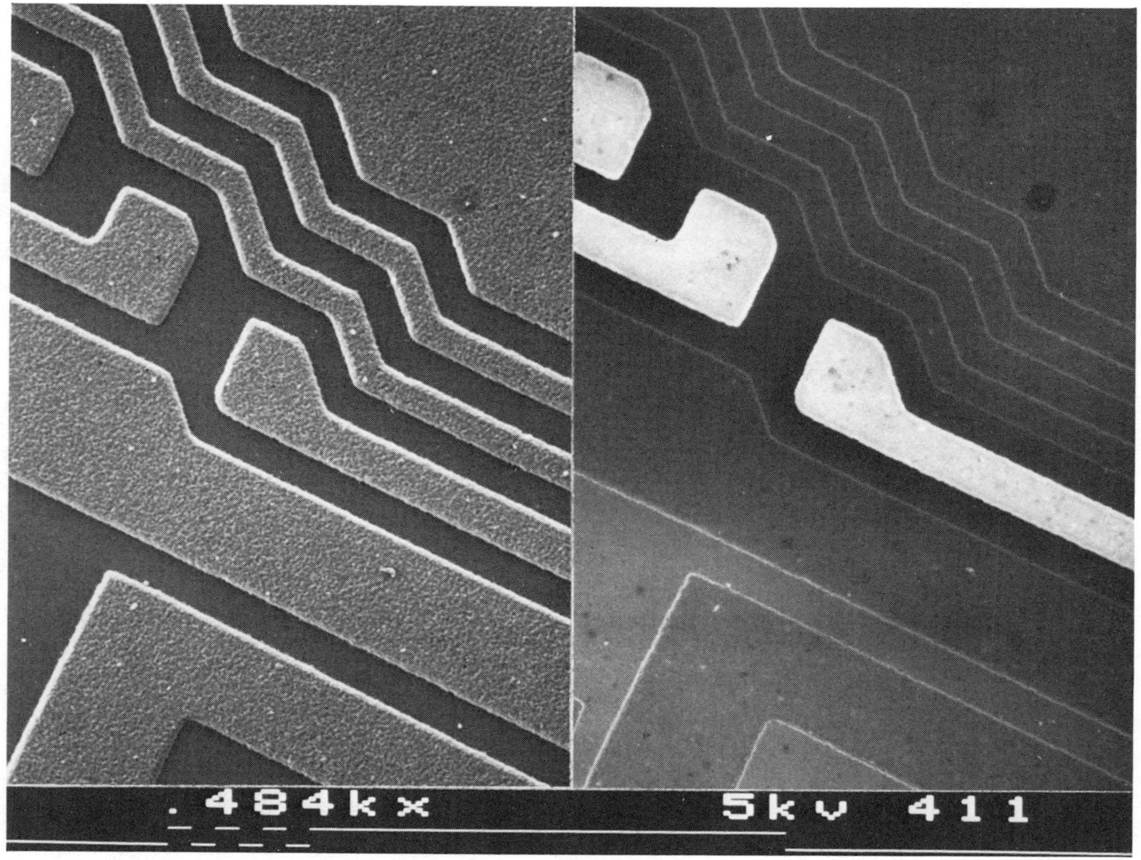

Fig. 12. Uncoated IC images produced with bottom SE detector (left), and top SE detector (right).

An in-depth discussion of secondary electron production with specific application to the double SE detector system referred to here can be found in Ref. (11).

Low Voltage Operation

Most scanning electron microscopy is performed using a beam accelerating voltage in the range from 10-30kV. This is due to the better resolution and signal to noise levels achieved under such conditions. It is also mandatory in some applications, such as X-ray analysis, where higher voltages are required to excite the characteristic X-ray lines of interest. The resolution of most SEMs degrades rapidly as the beam voltage is decreased below about 10kV, due to increasing chromatic aberration. There are, however, several advantages to low voltage SEM operation.

Beam damage, which is a concern in many biological and semiconductor applications, is much reduced, or effectively eliminated. The images produced are much more sensitive to surface effects, both topographical and elemental, due to the decreased penetration of the incident electron beam.

Yet another important advantage of low voltage operation is that specimen charging problems can be eliminated. This is due to the fact that as the beam energy decreases, the secondary electron emission, i.e. the number of SEs emitted per incident electron, increases, and in fact exceeds unity in the range from about 1-3kV.

The magnitude of the chromatic aberration is proportional to the factor $\Delta V\, C_c$, where ΔV is the energy spread in the electron beam, as determined by the type and operating conditions of the electron gun. C_c is the chromatic aberration coefficient of the objective lens, as explained previously.

Table 2 summarizes some important characteristics of three electron sources currently used in SEMs. Clearly the tungsten field emitter is most attractive, due to its high brightness and low energy spread. There are, however, a number of drawbacks associated with this type of source. In the first place, a gun vacuum of 10^{-10} torr is required, which can result in problems when examining a specimen which outgasses. The emission current is unstable. The maximum

Table 2. Electron Source Characteristics

Source Type	Brightness (A/cm^2 str)	Temperature (^{o}k)	Source Size (μm)	Energy Spread (eV)
W-Hairpin	1×10^5	2,800	30	~3
LaB_6	5×10^6	1,800	5	~1
W-F.E.	$\sim 1 \times 10^8$	295 Room Temp.	0.01	~0.2

specimen current is only on the order of 10^{-8} Amps, which is about three orders of magnitude less than can be achieved with the other two sources listed. These limitations prohibit the field emission SEM from being used for a number of applications such as quantitative X-ray analysis, cathodoluminescence, channeling pattern generation, etc. In spite of these limitations, excellent low voltage performance can be achieved with a field emission SEM.

The single-crystal LaB_6 source suffers from none of the practical problems associated with field emitters, offers a good compromise in terms of performance characteristics and ease of operation, and is clearly superior to the most commonly used tungsten hairpin filament. The LaB_6 brightness listed in Table 2 is for a triode gun operated at an accelerating voltage of about 30kV. If, however, the voltage is dropped to 1kV, the brightness drops to a value considerably below that shown in the table.

In order to compensate for this loss of brightness at low voltage, a new double-anode LaB_6 gun has recently been developed, and is used at present on the ISI IC-130 SEM. This is illustrated schematically in Fig. 13. In the example shown, the gun is operated at 1kV. The electron beam is first accelerated to 8kV by the first anode, to which a fixed 7kV potential is applied. A crossover is formed between the two anodes with a brightness equivalent to a triode gun operated at 8kV. The beam is then decelerated to 1kV on passing through the grounded second anode, while

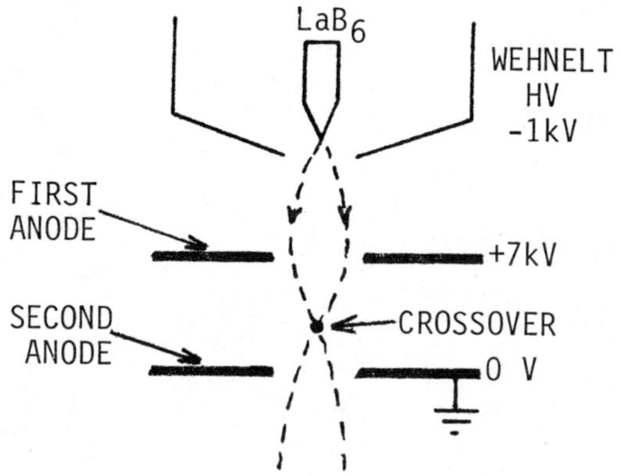

Fig. 13. Dual anode LaB_6 electron gun.

still maintaining the same source brightness. A further gain in brightness at low voltage is obtained by shortening the Wehnelt to first anode spacing.

The combination of the high brightness double anode LaB_6 electron gun, low C_c objective lens, and high SE collection efficiency double SE detector system results in the attainment of 15nm resolution at 1kV beam energy. This is an improvement of about an order of magnitude over conventional SEMs, and opens up a multitude of new applications. An example of an SE image obtained using an accelerating voltage of 1kV is shown in Fig. 14.

Charge-free Anti-contamination System

In many SEM applications, electrically non-conducting specimens, and samples containing moisture and other liquids such as oil, require to be examined. Conventional metal coating techniques to render the sample conductive are not always tolerable, and the normal SEM is too sensitive to vacuum degradation and contamination to allow

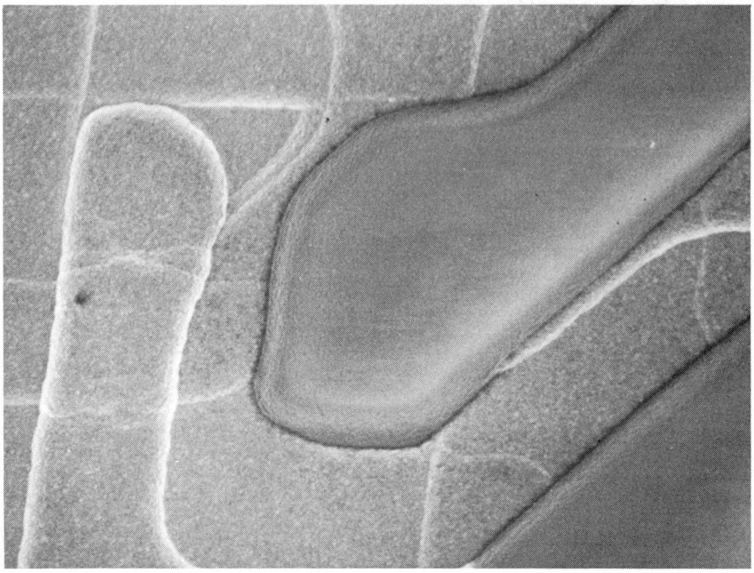

Fig. 14. Uncoated photoresist on polysilicon. 1kV, 5,000X magnification.

the examination of wet or oily samples. There is therefore a requirement to developing methods of examining such specimens in their natural state. A system has been developed by Robinson (12), which permits this to be accomplished, and still allows the microscope to be easily and quickly returned to its normal operating configuration. Interest in this technique is currently on the increase.

The method relies on the ability to provide a vacuum differential across the objective aperture in the electron optical column. This is in principle possible in all SEMs, but in practice easier to achieve in some microscopes than others. Image formation requires a BSE detector rather than the more commonly used SE detector. While any backscattered electron detector can be used, the high resolution potential of the SEM can only be maintained by employing a high efficiency detector, such as the Robinson detector used here.

Fig. 15. SE image of uncoated paper exhibiting severe charging effect.

As the sample in a SEM is irradiated by an electron beam, there is a negative charge injection into the specimen during observation. If the specimen is electrically conductive, charge build-up is easily avoided by connecting it to ground. If it is an insulator, or poor conductor; however, charge accumulation will take place. This can seriously affect the low energy SEs detected, and also displace and distort the incident electron beam.

Severe image degradation therefore results. Figure 15 shows a typical SE image exhibiting severe charging. In order to prevent charge accumulation on poorly conducting specimens, it is a common practice to coat the surface of the specimen with a conducting film, such as Au, which is then grounded. The coating is usually performed using a sputter coater. There are, however, situations where this technique is either unacceptable or doesn't work. Metal coating in many instances is destructive, e.g., when it is required to examine a semiconductor wafer which must be returned to the production line, or a specimen which must be subjected to further tests which would be adversely affected by the metal coating. It is often impossible to lay down a thin continuous film on a complex surface structure. This results in regions being isolated from ground, which then charge. A further objection to coating is that it can introduce artifacts which complicate image interpretation. Clearly, if coating can be avoided, this is desirable.

One technique to avoid charging has already been referred to, namely to operate at low incident electron beam energies. This allows severe image distortion of the type shown in Fig. 15 to be avoided. The resolution of most SEMs degrades severely under such low voltage operating conditions, however, and many microscopes have insufficient control in this low voltage range. A further limitation is that X-ray analysis, which requires high beam energies, cannot be performed.

An alternative technique is offered by the Charge-Free Anti-contamination System (CFAS) which was used to produce the series of micrographs shown in Figs. 16-21. In this system, the specimen chamber of the SEM is isolated from the main instrument vacuum pumps. A mechanical roughing pump with controlled leak is used to evaluate the specimen chamber. The objective aperture requires to be mounted in the electron optical column, in such a way that a pressure differential can be maintained across it. Thus, the electron gun and path between the electron source and the aperture can be maintained at about 10^{-4} torr, while allowing the specimen chamber to be held at about 10^{-1} torr. In the SEM used for the work reported here, a 200 micron aperture was screwed into the bore of the objective lens bottom polepiece, thus allowing the required differential pumping. A further essential component of the system is an efficient BSE detector. In this case, a Robinson BSE detector was used. A schematic diagram of the system is shown in Fig. 22.

When examining an insulating specimen in the approximately 10^{-1} torr specimen chamber vacuum environment, charge accumulation is approximately equal to gas ionization in the vicinity of the specimen, so that charging is automatically eliminated. The correct balance is struck by observing the image for charging effects, and adjusting the variable leak until they disappear. Any small residual charge does not significantly affect either the incident

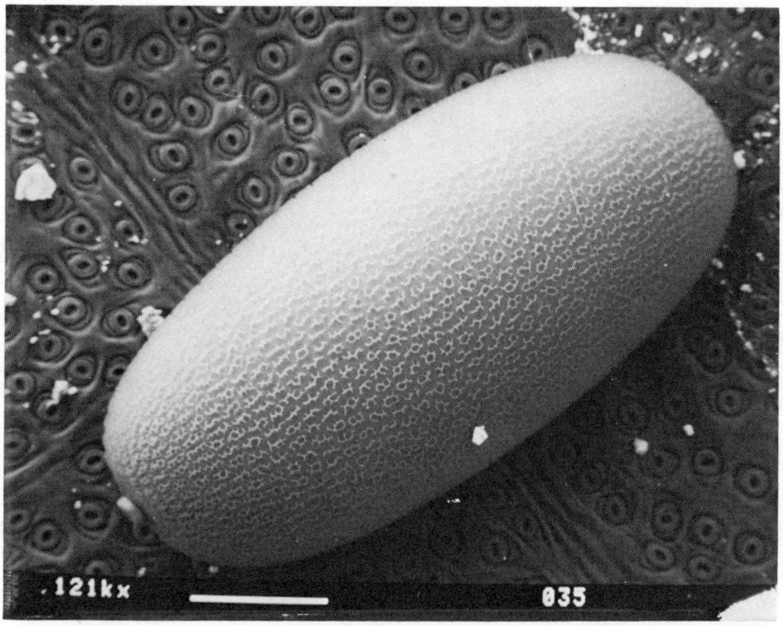

Fig. 16. Uncoated insect larva.

Fig. 17. Uncoated household sugar.

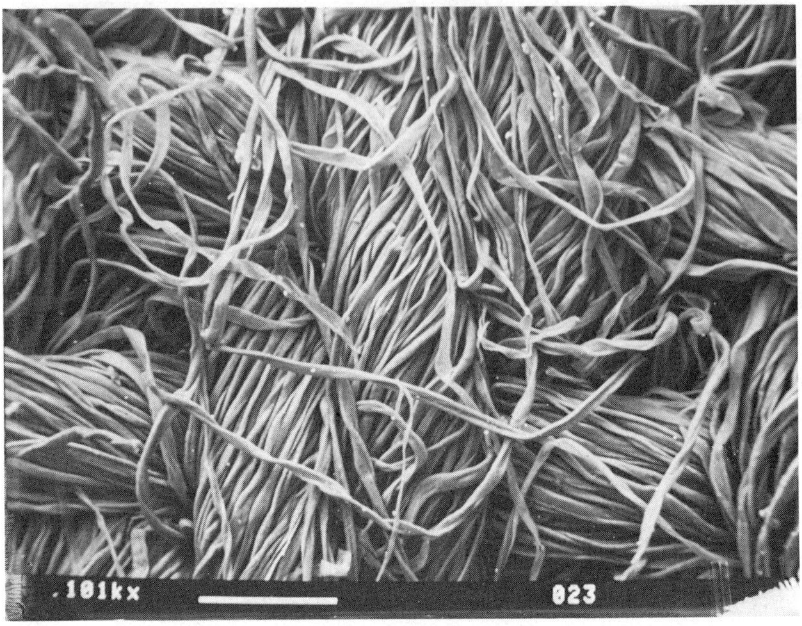

Fig. 18. Uncoated cotton cloth.

Fig. 19. Uncoated gypsum.

Fig. 20. Uncoated household salt.

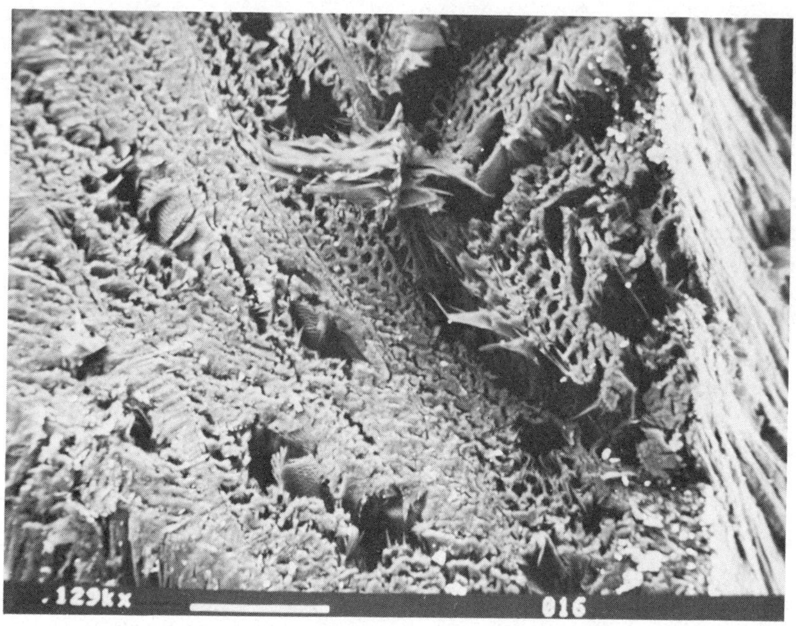

Fig. 21. Uncoated wood.

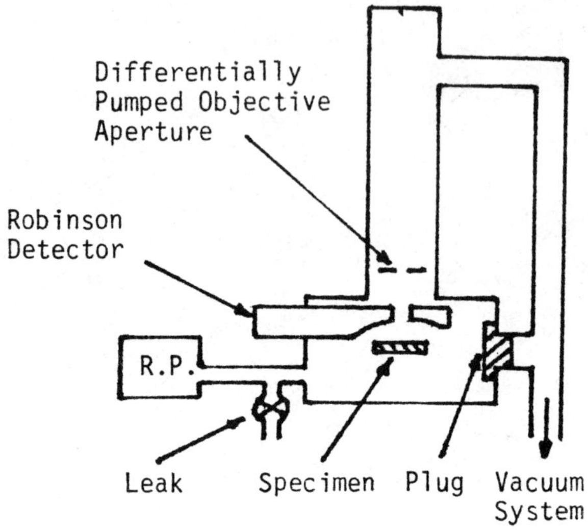

Fig. 22. CFAS System.

electron beam, or the detected BSEs which have energies close to that of the incident electrons.

There is some resolution degradation due to interaction of the incident electron beam with gas molecules in the specimen chamber. The magnitude of the degradation is a function of the chamber vacuum level, the energy of the incident electrons, and the working distance at which the specimen is examined. Resolution of a few tens of nanometers is, however, readily obtained.

A further benefit of the set-up is that system contamination, which normally results when examining oily samples, is avoided. This is due to the presence of significantly higher levels of oxygen in the specimen area than in normal SEM operation. This results in carbonaceous contamination being oxidized and removed by the roughing pump as a gas. This technique has been used in the past (13) in X-ray microprobe analyzers to prevent or remove specimen contamination. In that case, a jet of oxygen or air was blown across the surface of the specimen during analysis.

A series of micrographs of various uncoated insulating materials is shown in Figs. 16-21. Beyond attaching the specimen to the sample stub, no special procedures were undertaken. Virtually any specimen, even those containing substantial amounts of moisture, can be examined with a minimum of specimen preparation.

CONCLUSIONS

When the SEM was first introduced in 1965, it was generally accepted that the ultimate resolution of SE images would be limited to about 10nm by electron optical design, and electron beam/specimen interaction effects. Resolutions of better than 2nm have been achieved as a result of recent instrumentation improvements. Future resolution improvements can certainly be expected, especially under low-voltage operating conditions. The impetus to provide good low voltage performance has come mainly from the semiconductor industry, where specimen damage must be avoided, and coating the specimen to eliminate charging cannot be tolerated. It is certain, however, that high-resolution low-voltage capability will prove of benefit in many other areas.

The CFAS system allows the examination of virtually all specimens with a minimum of specimen preparation. It couples the high resolution and large depth of field of the SEM with the ease of specimen preparation of the optical microscope.

One of the most significant recent developments is the ability to enhance specimen surface contrast effects by using a SE detector above the objective lens and placing the specimen in the lens field. This is further augmented by the ability to produce high quality BSE images under the same operating conditions with a second SE detector in the specimen chamber, thus aiding in image interpretation.

The application of microprocessors to SEMs has already started. This trend will continue. In current applications, the microprocessors are used to control a number of simple switching functions. More extensive use in this manner could prove useful in simplifying operation and increasing system reliability. Further possible applications include image enhancement and image analysis.

REFERENCES

1. O.C. Wells, Scanning Electron Microscopy, McGraw-Hill, New York (1974).

2. J.I. Goldstein, et al., Scanning Electron Microscopy and X-Ray Microanalysis, Plenum Press, New York (1974).

3. T.E. Everhart and R.F.M. Thornley, J. Sci. Instr., 37, 246 (1960).

4. V.N.E. Robinson, Scanning Electron Microscopy, IITRI, Chicago, IL, 1975, p. 51.

5. H. Koike, et al., U.K. Patent Application No. 8104621 (1981).

6. K-R. Peters, Scanning Electron Microscopy, IV, 1359 (1982).

7. K-R. Peters, Asilomar Conf., SEM, Inc. (To be published.)

8. J.M. Bronshtein and B.S. Fraiman, Rad. Eng. Elec. Phys., 7, 1530 (1962).

9. H. Kanter, Phys. Rev., 121, 689 (1961).

10. H. Kawamoto, et al., Scanning Electron Microscopy, SEM, Inc. (To be published.)

11. R. Buchanan, "Secondary Electron Image Formation in a Dual SE Detector SEM," ISI Technical Bulletin (1983).

12. V.N.E. Robinson and B.W. Robinson, Scanning Electron Microscopy, SEM, Inc., 595 (1978).

13. S.H. Moll and G.W. Bruno, "Second National Conference on Electron Microprobe Analysis," Boston, MA (1967).

Potential Anaerobic and Structural Acrylic Modified Polyester Adhesives

R.A. Pike and F.P. Lamm

United Technologies Research Center
East Hartford, CT 06108

ABSTRACT

The acid-catalyzed addition of a hydroxyl group across the oxirane ring of an epoxy acrylate to yield a beta-hydroxy ether has been demonstrated to be an acceptable procedure for introducing acrylate functionality into compounds containing free hydroxy groups. The reaction is readily monitored using liquid chromatography and goes to completion rapidly under mild conditions. Monomers and prepolymers acrylated by this technique exhibit anaerobic as well as thermal and transition metal/peroxide cure characteristics. Test data on bonded parts will be presented to show that hydroxy prepolymers, acrylated using the described chemistry, exhibit acceptable bond strengths when used as anaerobic adhesives under elevated temperature and high humidity exposure as well as ambient conditions.

INTRODUCTION

Anaerobic and modified acrylic adhesive systems have begun to achieve acceptance as substitutes for mechanical fastening methods. This is largely due to increasing energy and labor costs as well as the necessity of improving productivity. These two relatively new classes of adhesives have gained approval as assembly tools because of their ability to cure rapidly at room temperature without mixing or metering and demonstrated adherence to a variety of substrates.

Fundamental to the performance of engineering adhesive systems is the ability to provide both the rigidity required for high tensile strength and the toughness and flexibility needed for high peel and impact resistance. Although chemically different the anaerobic and modified acrylic adhesives are both capable with proper formulation of providing this combination.

Anaerobic adhesives were initially introduced as locking and sealing compounds. These products, although capable of improving the strength and reliability of threaded fasteners and sealing leaks, in general do not possess the flexibility and toughness of true structural adhesives. The modified acrylics are normally solutions of monomethacrylate monomers (methyl methacrylate) and polymeric rubbers. There are also systems available which are formulated using higher molecular weight monomethacrylates. Both systems are similar to general epoxies in that they consist of two parts. One part consists of polymer, monomer (to act as viscosity controllers) and initiator (peroxide or hydroperoxide). The other contains the activator, normally a transition metal (cobalt acid salt) or tertiary amine.

During the course of an investigation to synthesize radiation curable resins, materials were synthesized starting from high hydroxyl containing polyesters which exhibited anaerobic cure behavior. A limited study was initiated to determine if such materials could be formulated to provide adhesive characteristics similar to commercially available anaerobics or could be cured by other methods. The following sections of this paper present the results of this brief investigation.

EXPERIMENTAL

The hydroxyl containing polyester was synthesized from dimethyl terephthalate, isophthalic acid, trimellitic acid, and diethylene glycol or ethylene glycol using a butyl titanate catalyst. Ratios of reactants can be varied within limits to produce desired levels of hydroxyl content.

Glycidyl methacrylate was purchased from Polyscience Corp. All other chemicals were reagent grade and were used without further purification. Infrared spectra were taken on a Perkin-Elmer Model 298 Infrared Spectrometer.

The HPLC analysis of the alcohol epoxy reactions was carried out as follows regardless of the alcohol or catalyst:

Equipment:	Waters ALC/GPD-202/R401 W 6000 A Solvent Delivery System
Column:	C_{18} Reverse Phase 3.9 X 300 mm
Solvent:	Methanol/water 1:1
Sample Size:	4.0 μℓ Neat

The data was taken by measuring peak heights and normalizing to the alcohol peak. Peak areas could be used to obtain a more accurate analysis of the reaction. An example of changing peak heights as the reaction proceeded is illustrated by the chromatograms in Fig. 1.

The model compound studies were done by adding 5ml of the glycidyl methacrylate to a 10 mole excess of the alcohol with 0.01g hydroquinone. The reaction mixture was then warmed by oil bath to the final temperature and 8 drops of the appropriate catalyst were added to initiate the reaction. Aliquots were taken directly and analyzed by the LC technique to monitor the reaction.

A typical reaction sequence for the addition of the glycidyl methacrylate to the polyester prepolymer was as follows:

To a 500cc, 3-necked, r.b. flask equipped with a mechanical stirrer and thermometer was charged 39.4g of the polyester, 28.4g glycidyl methacrylate, 0.01g of hydroquinone. With stirring, 8 drops of conc. sulfuric acid was added. An exotherm occurred raising the temperature from

25° to 35°C. The temperature was raised to 75°C and held. In the course of 8 hours the relative viscosity (as measured by flow) had increased from a flow of 4.6mm to 0.1mm. An infrared spectrum of the adduct showed the presence of the methacrylate ester, hydroxyl groups and an absence of oxirane (epoxy) rings. The final kinematic viscosity of the fluid was 138,000 cps at 70°C.

RESULTS AND DISCUSSION

A. Addition Reactions

Examination of the literature confirmed that acid or basic type catalysts can catalyze the reaction of alcohols with oxirane (epoxy) groups. However, the reported studies dealt only with saturated epoxidized compounds and low molecular weight alcohols primarily in an attempt to define

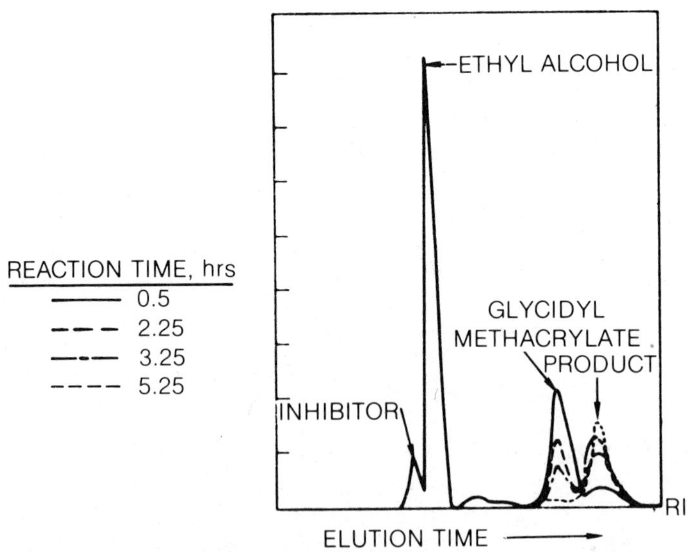

Fig. 1. Ethyl Alcohol + Glycidyl Methacrylate.

the mechanism of the reaction. Winstein and Henderson reported that the reaction of methanol with a saturated epoxy was catalyzed by sulfuric acid (1). A U.S. Patent described the reaction of starch with alkylene oxides to produce starch ethers using quaternary ammonium hydroxides, trialkyl amines or tri-(hydroxy alkyl) amines or mixtures thereof as catalysts (2). More extensive studies were carried out on a mechanistic basis by Chitwood et al., which defined the reaction of propylene oxide with alcohols using Lewis acids (BF_3) and metal hydroxides and alcoholates (3). Schechter and Wynstra, using the same types of catalysts, carried out an extensive study of glycidyl ether reactions with alcohols, phenols, carboxylic acids and anhydrides (4). The latter work showed that with acidic catalysts the course of the reaction became highly selective and produced only one of the two possible isomeric products. It was found, however, that sulfuric acid was ineffective even up to 100°C in catalyzing the reaction of dipropylene glycol and phenylglycidyl ether. Weaker acids were also inactive. Stannic chloride was found to be extremely effective even at room temperature in producing complete reaction. Thus, from these studies it was apparent that the course of the reaction was influenced by the type of epoxide, alcohol and catalyst employed. Since no definite conclusions could be made as to the optimum catalyst, reaction conditions, and structure of the product in the reaction of alcohols with unsaturated epoxides such as glycidyl methacrylate, where vinyl polymerization was also possible, a series of model compound reactions was carried out to establish the conditions for ultimate reaction of glycidyl methacrylate with the hydroxyl containing polymer described above. The desired addition reaction was shown below:

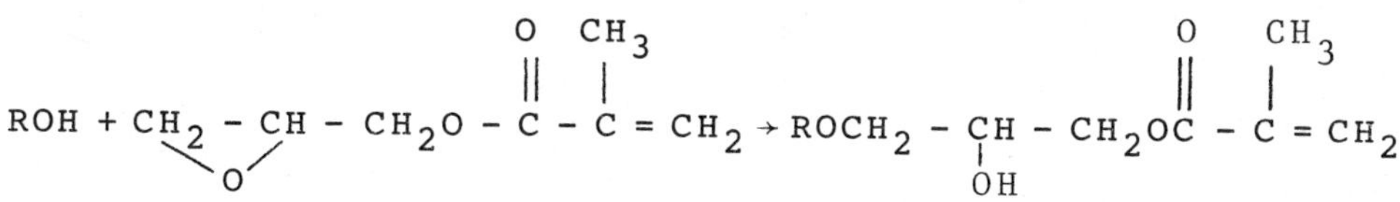

The initial studies were carried out where R was either ethyl or isopropyl. Conditions were varied to optimize the reaction. The results are tabulated in Table 1.

The results indicated sulfuric acid had the greatest effect on the reaction and did not produce detectable amounts of vinyl polymerization if run for limited times.

Table 1. Reaction of Alcohols with Glycidyl Methacrylate

Alcohol	Catalyst[a]	Reaction[b] Conditions, °C	Time (Hours)	Product
isopropyl	none	84°	8	no evidence of reaction
isopropyl	H_2SO_4	84°	1	polymer plus ~20 w/o adduct
isopropyl	acetic acid	84°	1	low amount of polymer formed
isopropyl	TBT	84°	6	no evidence of reaction
ethyl	H_2SO_4	82°	1	sweet smelling, viscous liquid, η_D^{25} 1.4538
ethyl	H_2SO_4 (1/2 concn)	82°	1/2	~50% reaction
ethyl	TBT or acetic acid	82°	2	no evidence of reaction
ethyl	isophthalic acid (not complete solubility)	82°	4	some polymer ~30% reaction
2-methoxy ethanol	H_2SO_4	82°	1	viscous, sweet smelling liquid, some polymer evident.

[a] Acid concentration, 0.25 w/o (total reactants).

[b] All reactions carried out using 10 molar excess of alcohol.

This is in direct contrast to the ineffective reaction of phenyl glycidyl ether with alcohol (4).

To more accurately determine the course of the reactions involved, additional runs were made using liquid chromatography to follow the reaction sequence. The results indicated that nearly complete reaction of the glycidyl methacrylate was achieved at 80°C with H_2SO_4 as the catalyst (Fig. 1).

The results indicated that although sulfuric acid was an effective catalyst at the temperatures used (without added inhibitor), the desired addition reaction was accompanied by formation of a small amount of vinyl polymerization product. A lower reaction temperature together with a small amount of added hydroquinone inhibitor eliminated this undesirable side effect. The run using the BF_3 etherate catalyst gave no evidence of the polymer formation and the results showed, as expected, a temperature and catalyst concentration dependence.

To expand the number of possible catalysts for the addition reaction, tetrabutyl ammonium hydroxide, ferric chloride, stannous and stannic chloride were evaluated. Only the latter compound proved to be an effective catalyst, producing the desired adduct with isopropyl alcohol at 40°C reaction temperature. The $SnCl_4$ had been reported by Schecter to be an effective catalyst for the reaction of alcohols with epoxy rings (4).

To verify the structure of the adduct formed in these reactions, several runs using 2:1 and 1:1 mole ratios of alcohol were carried out with the BF_3-etherate catalyst. The use of 1:1 ratios gave, with both isopropyl and ethyl alcohol, some gel formation in conjunction with the adduct. The unreacted alcohol was removed by vacuum (weight loss equal to the excess alcohol added) and IR was used to identify the resulting adduct.

The infrared spectra of the products obtained showed strong hydroxyl absorption at 3400 and 1110 cm^{-1} (see OH), unsaturation at 1630 and 830 cm^{-1} as well as the strong ester peak at 1710 cm^{-1}. The spectrum of the isopropyl alcohol adduct was slightly different in the 2900-3000 cm^{-1} range reflecting the effect of the additional methyl group. No evidence of epoxy at 910 cm^{-1} was found in the products.

Based on the results of this limited study it was concluded that sulfuric acid and boron trifluoride etherate were the most effective catalysts at reaction temperatures of 75°-$80^{o}C$ and $40^{o}C$, respectively. Use of hydroquinone inhibitor was required to eliminate any vinyl polymerization during reaction.

The three polyester prepolymers prepared having varying viscosity (different di/tri acid ratio) with essentially the same hydroxyl content were used to define optimum reaction conditions for introduction of unsaturated groups via addition of glycidyl methacrylate to the liable hydroxyls.

Prepolymer	Viscosity, CS@RT
1	9,600
2	7,960
3	18,000

To determine molecular changes which occurred in the reaction, liquid chromatographic curves of the starting prepolymers were obtained using tetrahydrofuran as the carrier solvent. The peak due to unreacted diethylene glycol used in the synthesis was also identified. These curves were used as standards to define the molecular weight changes which occurred on reaction with the glycidyl methacrylate.

Reaction of the prepolymers with glycidyl methacrylate with varying ratios of hydroxyl to epoxy were carried out at two different temperatures using the BF_3 catalyst. $SnCl_4$ was also tested as a catalyst. Several conclusions were drawn from these initial reactions.

1. The stannic chloride catalyst effective in the model compound studies was ineffective with the polymers. The catalyst was complexed and deactivated by the tetrabutyl titanate catalyst used in the resin preparation.

2. Nitrogen atmosphere (elimination of oxygen) caused gel formation. This was consistent with the anaerobic curing of the materials discussed below.

The relative increase in viscosity which occurred during the addition reactions was determined by using a flow technique on a glass slide similar to that used to determine the flow of phenolic resins. Comparison of the flow of the adducts to that of the three starting polyesters indicated a final viscosity range of 2.5-3.0 stokes.

The model compounds indicated that boron trifluoride and sulfuric acid were the best potential catalysts for addition of glycidyl methacrylate to hydroxyl. To determine the applicability of the H_2SO_4 catalyst in a polymer system, three adducts were prepared, with two of the polyester prepolymers (1) 0.75/1.00 epoxy/hydroxyl ratio, (2) and (1) both at a 1.0/1.0 epoxy/hydroxyl ratio. The reactions proceeded smoothly with 0.4 w/o catalyst at 75°C. The change in flow (relative viscosity) was used to follow the reaction as illustrated in Fig. 2. In the initial stages of the reaction, the glycidyl methacrylate acted as a solvent reducing the viscosity. As expected, (1) having the highest starting viscosity at 1.0/1.0 produced the highest final relative viscosity. Infrared spectra of the adducts showed loss of the epoxy peak at 910 cm^{-1} and was used to follow the course of the reaction.

Reactions with boron trifluoride etherate catalyst at 40-60°C temperatures gave similar products. It should be noted that it was not necessary to react all available hydroxyls with glycidyl methacrylate. Intermediate levels, i.e., 25, 50 and 75 percent may be reacted if desired to control the final properties of the cross-linked cured polymer.

B. Adhesion Studies

Two observations led to the conclusion that the subject compounds exhibited anaerobic polymerization behavior. These were the necessity of carrying out the addition reaction in an air atmosphere to prevent gel formation, particularly when using the BF_3 etherate catalyst, and the sticking of several syringes if they were not cleaned immediately after use. In addition, on heating with added peroxide, the adducts cured to clear films or with added acrylate monomers and photosensitizers ultraviolet exposure caused polymerization.

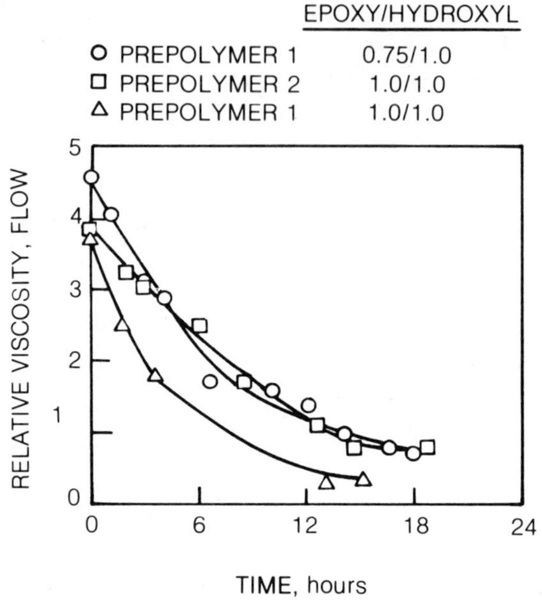

Fig. 2. Relative Viscosity Change of Prepolymer Reactions.

The acrylated polyester used in the adhesion studies described below was as the 1:1 adduct of the intermediate molecular weight polyester prepolymer (1).

To illustrate their effectiveness as anaerobic adhesives several combinations of the adduct, with or without added filler, reactive diluents, or peroxide, were screened using nut/bolt combinations. The results in terms of breakaway torque are listed in Table 2. The strengths obtained, although none of the systems were optimized, demonstrated the use of the methacrylated polyester compounds as anaerobic curing or thermal curing adhesives.

It was evident from some initial results that the compounds made using the sulfuric acid catalyst would not cure due to the inhibiting effect of the acid. Attempts to use organic amines to neutralize the acid as the nut/bolt combinations were made did not prove very successful. The addition of the filler, particularly alumina or calcium carbonate was the preferred method of catalyst neutralization. In addition, the filler could be used to control the viscosity of the final mixture which was of concern in the peroxide-cured adhesive systems.

It was of interest to find that at least with two of the peroxide-cured formulations boiling in water actually increased the bond strength. With an anaerobic cured formulation a decrease in bond strength was noted under the same conditions. With the alumina filled systems it appeared that the peroxide catalyst was not required to obtain excellent bond strength when curing thermally.

Table 2. Torque in/lb. Breakaway of Anaerobic Systems

Composition	Cure	Torque in/lb.
Polyester-100% acrylated		
1. + monomer.[a]	Peroxide, 180°C 1/2 hr.	5
2. + cabosil filler.	Peroxide, 180°C 1/2 hr.	37
3. + triallylcyanurate.	Peroxide, 170°C 1/2 hr.	19
4. + cabosil filler and triallylcyanurate.	Peroxide, 170°C 1/2 hr.	55
5. + triallylcyanurate.	Heat 2 hrs. boiling water after cure.	30
6. + cabosil filler and triallylcyanurate.	Heat 2 hrs. boiling water after cure.	75
7. + 50 w/o alumina.	No peroxide, 170°C 1/2 hr.	135
8. + 50 w/o alumina.	Peroxide, 170°C 1/2 hr.	130
9. + 50 w/o alumina.	No peroxide, RT cure	75
10.	No peroxide, RT cure	5-10
11. + monomer.[a]	No peroxide, RT cure	8
12. + monomer.[a]	Heat 2 hrs. boiling water after cure.	3

[a]Monomer used was adduct of isopropyl alcohol and glycidyl methacrylate.

No attempt has been made to date to optimize any of the possible combinations to give an acceptable adhesive system.

To obtain additional information as to the level of adhesion on aluminum adherends a series of anaerobically cured tensile lap shear specimens were fabricated and tested at room temperature. The results showed shear strengths of 925 $\pm$ 20 psi and indicated good bonding strengths could be obtained on aluminum with these materials. Using an optimized formulation, higher lap shear strength could be expected.

An alternate adhesive application area for the polyester acrylates described in this report was in the general field of two-part structural acrylics. Such systems generally cured at room temperature using transition metal or amine-activator and a hydroperoxide or peroxide radical source. A brief study was carried out to determine if the synthesized compounds had potential use as structural adhesives. As before, the nut/bolt combination was used as an indication of adhesion. The results of these limited tests are given in Table 3.

In general, the systems cured within one-half hour depending upon concentration of the catalyst and activator. No attempt has been made to optimize or further modify the systems. The results do indicate that with proper formulation the acrylated polyesters have potential as structural adhesives.

CONCLUSIONS

Based on the limited evaluation of the synthesized compounds, the following conclusions can be made: a) The synthetic process developed for introduction of methacrylate groups into hydroxyl containing polyester systems appears to be a viable route for modification of other hydroxyl containing monomers and polymers, b) Additional development work on selected systems to define optimum formulation recipes will be required, c) The materials evaluated have potential application in the areas of sealants, coatings and adhesives.

Table 3. Rapid R.T. Cure of Methacrylate Polyesters

Composition	Catalyst/activator[c]	Torque in/lb.
Polyester[a] + Al_2O_3 + Trimethylolpropane triacrylate (for activator).	Cumene hydroperoxide/ vanadium dodeconate	15
" " "[a]	Cumene hydroperoxide/rare earth octoate	5
" " "[a]	MEK peroxide/cobalt naphthenate	55
Polyester[b]	Cumene hydroperoxide/cobalt naphthenate	20

[a]Polyester adduct using sulfuric acid catalyst.

[b]Polyester adduct using BF_3 etherate catalyst.

[c]Peroxide monomer solvent Part A. Activator with or without added monomer solvent Part B. One part applied to bolt, other part applied to nut.

ACKNOWLEDGEMENTS

The authors would like to gratefully acknowledge J. Pinto (UTRC) for assistance with experimental work and C. McGregor, Essex Group UTC for preparation of the polyester prepolymers used in this work.

REFERENCES

1. S. Winstein and R.B. Henderson, J. Am. Chem. Soc., 65, 2196 (1943).

2. R.W. Kerr and W.A. Faucette, US 2,733,238, Jan. 31, 1956 (to Corn Products Refining Co.).

3. H.C. Chitwood and B.T. Frevre, J. Am. Chem. Soc., 68, 680 (1946).

4. L. Schechter and J. Wynstra, Ind. and Eng. Chem., 48, 86 (1956).

Anaerobic Adhesives Containing Maleimides Having Improved Thermal Resistance

L.J. Baccei and B.M. Malofsky

Loctite Corporation
North Mountain Road
Newington, CT 06111

ABSTRACT

The addition of maleimides as prepolymers to anaerobic adhesive systems which are composed primarily of methacrylate monomers has given materials which have improved thermal resistance. The systems, which are heterogeneous, polymerize at room temperature to give a mixture of maleimide in polymethacrylate matrix. When these materials are heated above 150°C, they become homogeneous as the maleimide polymerizes. The increase in thermal resistance was clearly demonstrated by various adhesive property measurements, infrared and instrumental thermal analysis. These adhesives exhibit significantly improved strengths when tested at actual elevated temperatures rather than ambient conditions.

INTRODUCTION

A significant number of reports have appeared in the literature on the uses of maleimide or polymaleimide additives to obtain polymers with increased thermal and mechanical properties. For example, various rubbers with residual unsaturation have been effectively vulcanized with these additives. Though the maleimides can be cured via an amine Michael-type addition reaction (1) or with certain alkali metal salts (2), the most common method appears to be thermal polymerization with or without the use of peroxide initiator (3,4).

The use of maleimides in adhesives is believed to have been first reported by Kehn (4). Meta-phenylene bis-maleimide was added to a halogenated rubber adhesive formulation and the system was cured at 150°C to yield improved adhesion to various highly unsaturated rubbers. N,N'-(methylene di-p-phenylene) dimaleimide was copolymerized with ethyl α-cyanoacrylate at 200°C to yield a cured adhesive with good strength at temperatures up to 200°C (5). We have reported previously on the use of maleimide and nadic capped monomers and prepolymers as adhesive and thermal resistance additives in anaerobic methacrylate/acrylate adhesive systems (6,7). Figure 1 illustrates the structures of three of the polymerizable maleimides used in our studies. We wish to report extensively, at this time, on the significant mechanical property improvements which can be obtained with these additives and the necessary curing conditions.

EXPERIMENTAL

Materials

The base methacrylate monomers utilized were a polyethyleneglycol dimethacrylate (PEGMA - produced by Loctite), an ethoxylated bisphenol—A dimethacrylate (EBIPMA - produced by AKZO), a urethane methacrylated capped poly(butadiene acrylonitrile) polyol, and hydroxy-propyl methacrylate (HPMA - Rohm and Haas). The following maleimide capped monomer/prepolymers were examined: N-phenyl maleimide (NPM), meta-phenylene dimaleimide (m-PDM), and a reaction product of methylenedi-aniline and excess methylenedianiline bismaleimide (P-MDA-2MDABM), sold by Rhodia Corp. under the tradename of Keramide

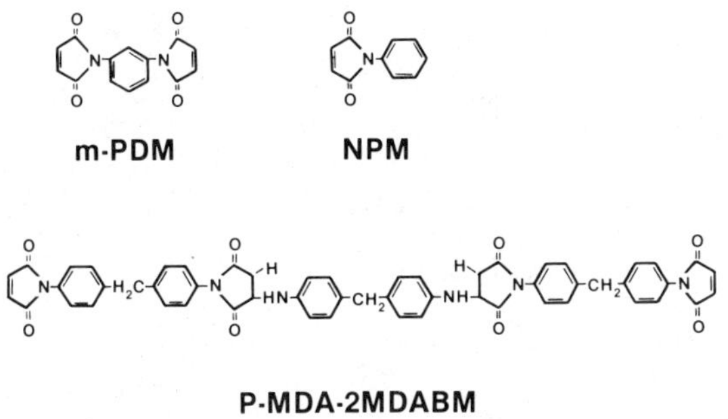

Fig. 1. Structures of Three Polymerizable Maleimides.

601). A nadic capped (capped with an endo-methylene ring) prepolymer of approximately 1050 molecular weight, which is a random copolymer of a 80/20 methylene dianiline/thio dianiline mix condensed with benzophenone tetracarboxylic acid dianhydride, was also studied (M,SDA-BTDA, formerly P105A from TRW). A "non-reactive" random polyimide copolymer of toluene diisocyanate and methylene diphenydiisocyanate reacted directly with benzophenone tetracarboxylic acid dianhydride (PI - available as Upjohn "2080") was used as a non-curable polyimide control.

Normally, the base methacrylate monomer was mixed with the maleimide, approximately two weight percent cumene hydroperoxide, HPMA and assorted accelerators, stabilizers and adhesive aids to form a fluid formulation.

Characterization

Adhesive strength tests were conducted such as a tensile lap shear test according to ASTM D-1002-65 (psi), a compressive shear according to military specification MIL-

R-46082A(MR) which measures the force required to press a cylindrical "pin" bonded within the bore of a mating collar through the collar (psi), and a "nuts/bolts" test in which bonded 3/8 - 24 steel nuts and bolts were aged at 232°C (450°F) and tested at room temperature for their "break/prevail" strengths in inch-lbs. Break strength is the torque required to cause the first movement between nut and bolt, whereas, prevail strength is the torque required to unwind the nut 180° beyond the break point.

Some of the cured materials were examined with a Perkin-Elmer 621 Grating Infrared Spectrophotometer to monitor the curing characteristics of the maleimide/methacrylate monomer mixtures. Thermal analysis was also performed with a Perkin-Elmer DSC-2 Differential Scanning Calorimeter and a Perkin-Elmer TMS-2 Thermomechanical Analyzer.

A polynomial least-squares curve fitting program was used as a guide for the curves shown in Figs. 8 and 9.

RESULTS AND DISCUSSION

The maleimide double bond appears to cure sluggishly until a significant amount of energy is available, such as near 150°C environment. This is not surprising as these materials are 1,2-disubstituted ethylenes (8). Studies with maleimide capped materials as comonomers with various methacrylate monomers show the rather poor polymerizability of the maleimide moiety relative to the methacrylate functionality. Figure 2 demonstrates, via infrared spectroscopy, the cure characteristics of the m-PDM comonomer with a poly(butadiene-acrylonitrile) urethane methacrylate (9). We have keyed inon the 3100 cm^{-1} band which is typical of the C-H stretch of the maleimide and methacrylate double bonds and a strong 690 cm^{-1} band which is a C-H deformation of a di-substituted, cis double bond (maleimide). As can be seen, after more than 1300 hours at ambient conditions with a standard ambient temperature accelerator (tetramethylthiourea - TMTU) for cumene hydroperoxide based systems, a "cured" matrix forms in which most of the maleimide double bonds remain. This is evidenced by only partial disappearance of the 3100 cm^{-1} band, due to methacrylate polymerization, and virtually no disappear-

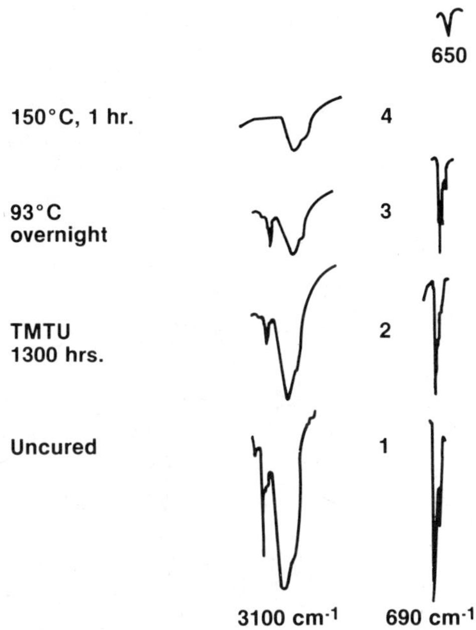

Fig. 2. IR Spectra of m-PDM/P (BD-CN) Urethane Methacrylate Before and After Various Cure Cycles.

ance of the maleimide 690 cm^{-1} band. If the comonomer mix is heated overnight instead at 93°C (200°F), no significant curing of the maleimide was observed. Curing at 150°C, however, for only one hour does reveal a high degree of polymerization of the maleimide as the 3100 cm^{-1} and the 690 cm^{-1} absorptions disappear.

This cure temperature region was independently confirmed by differential scanning calorimetry and thermomechanical analysis studies of another comonomer system, m-PDM in EBIPMA. First, Fig. 3 shows the DSC analysis of <u>neat</u> m-PDM which indicated the neat m-PDM appears to first melt and then spontaneously polymerize (large exotherm) at 200°C. This is shown by curve (1) and region (3) in the Figure. When the just-polymerized m-PDM is itself scanned, a smooth, "flat" curve (2) is obtained in the regions which earlier showed the endotherm melt and polymerization exotherm behavior of the neat monomer. This shows the material is now virtually fully cured.

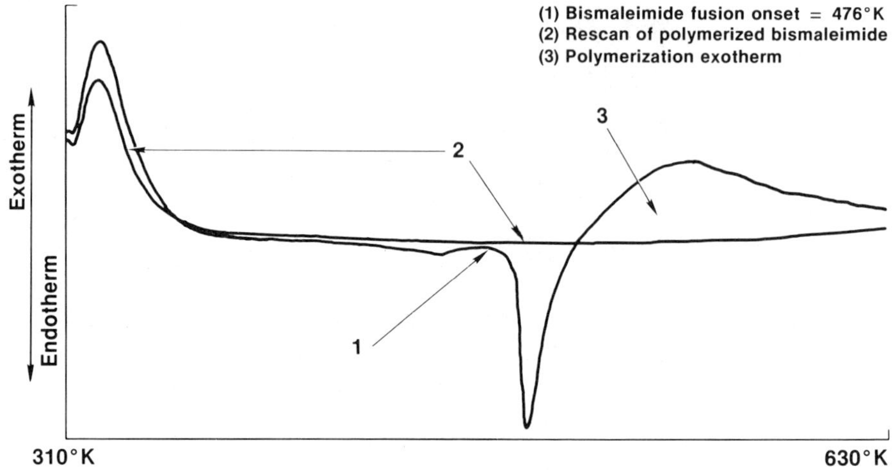

Fig. 3. Thermal Transition of m-PDM.

Whereas the neat m-PDM doesn't appear to thermally polymerize until near 200°C, the material as a comonomer, with liquid EBIPMA, will polymerize nearly fifty degrees lower. Figure 4 illustrates the DSC data obtained with the comonomer system which was pre-cured at 93°C. Curve 3 in this Figure clearly shows the onset of the polymerization of bismaleimide (m-PDM) component occurs near 150°C (see region 5). Curve 4 in the Figure shows the re-scan after this in situ polymerization and demonstrates that virtually all the material is fully cured. Thus, pre-curing at 93°C may polymerize a good portion of the EBIPMA comonomer but apparently little of the m-PDM itself.

A duplicate comonomer mixture, pre-cured at 150°C, demonstrated no exotherm with the DSC but rather a smooth baseline indicating an already apparently fully cured, cross-linked system (Fig. 5). Confirmation of the DSC data was essentially provided by Thermomechanical Analysis instrumentation and actual adhesive strength tests.

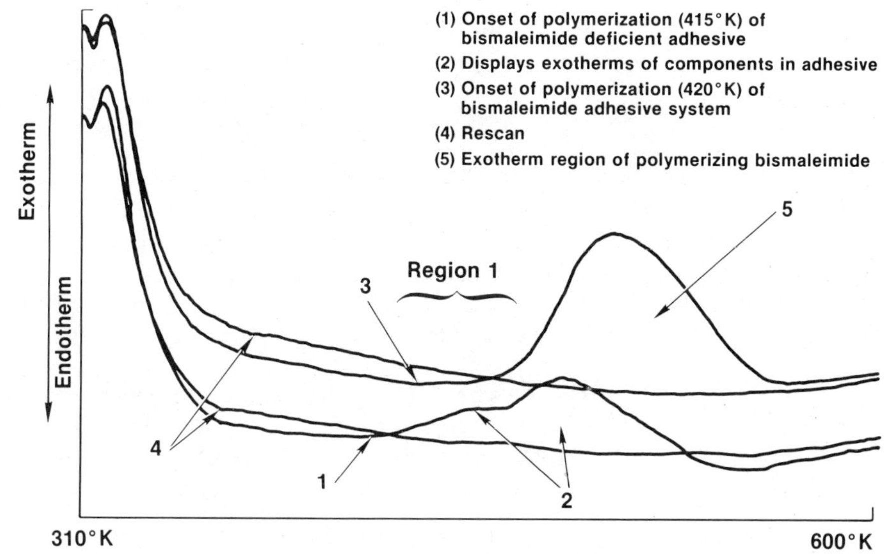

Fig. 4. Low Temperature Pre-Cured Adhesives.

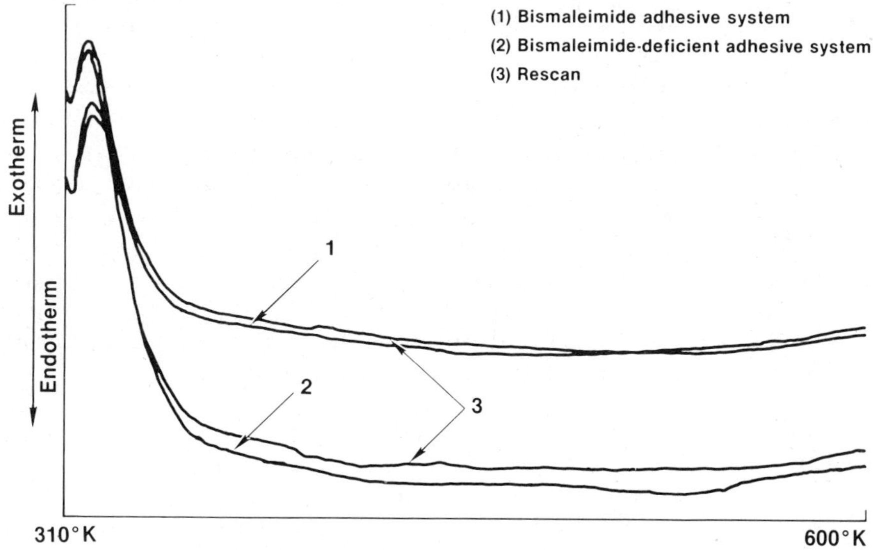

Fig. 5. High Temperature Pre-Cured Adhesives.

The TMA expansion cure data of a 93°C cured mixture showed a series of obvious transitions at the 100-250°C range, whereas the mixture initially cured at 150°C showed a steady expansion with no obvious transition up to 270°C where the scan was terminated. This would appear to also indicate the 150°C cured material was fully cured (Figs. 6 and 7). In the 110-170°C region of the 93°C pre-cured material (Fig. 6) there is a rapid change in the expansion rate followed by a volume contraction. This can be rationalized as the polymerization step with the expected shrinkage as the monomer forms the cross-linked polymer.

Impact resistance tests of a third comonomer system, a urethane methacrylate capped poly(butadiene acrylonitrile) polyol and m-PDM, yielded even further evidence of a more fully cured matrix at 150°C. An impact strength of 4.5 ft-lbs/in^2 was obtained for the material cured at 93°C, whereas a toughened material was formed at a cure temperature of 150°C as shown by the latter's impact strength of 11.5 ft-lbs/in^2.

A significant amount of further mechanical ensemble adhesive strength data was obtained which clearly indicated significant increases in the thermal oxidative resistance, hot strengths, and solvent resistances. For example, Fig. 8 illustrates the significant improvement in thermal resistance a PEGMA system can demonstrate with a polymerizable maleimide additive. Cured PEGMA, and PEGMA with a non-curable polyimide additive, PI, show virtually no residual strength after only 200 hours aging at 232°C. The mono-NPM, and the bismaleimide prepolymer, P-MDA-2MDABM, in PEGMA both show significantly higher strengths up to 500 hours. The nadic capped material, known to cure quite sluggishly, appears to indeed cure slowly and reaches its maximum strength around 200 hours and gradually falls off to approximately fifty percent of its initial strength after 670 hours and zero after 1300 hours.

Examination of pins and collars bonded with EBIPMA/maleimide formulations again show dramatically improved thermal resistance. Here, as shown by Fig. 9, one can clearly see the sequential improvement in going from neat monomer, or monomer with non-reactive PI, to a monomaleimide additive, to a cross-linkable bismaleimide, to

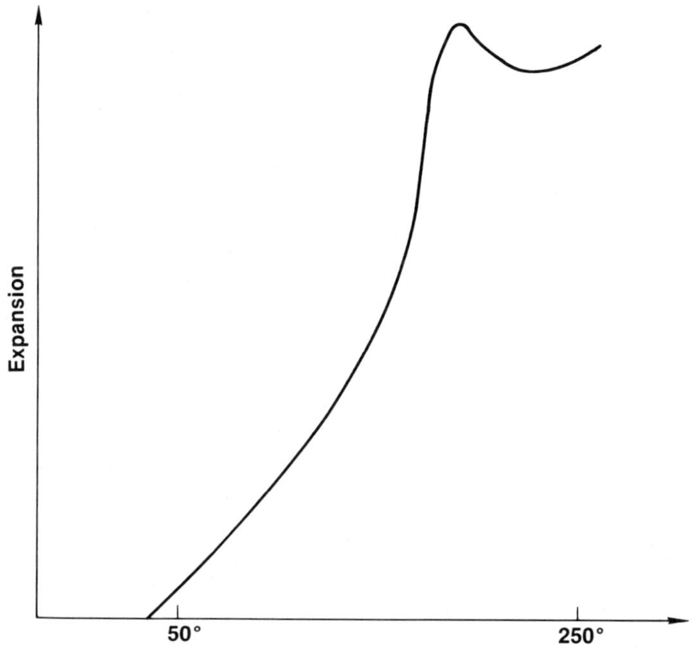

Fig. 6. Low Temperature Cured Maleimide Formulation.

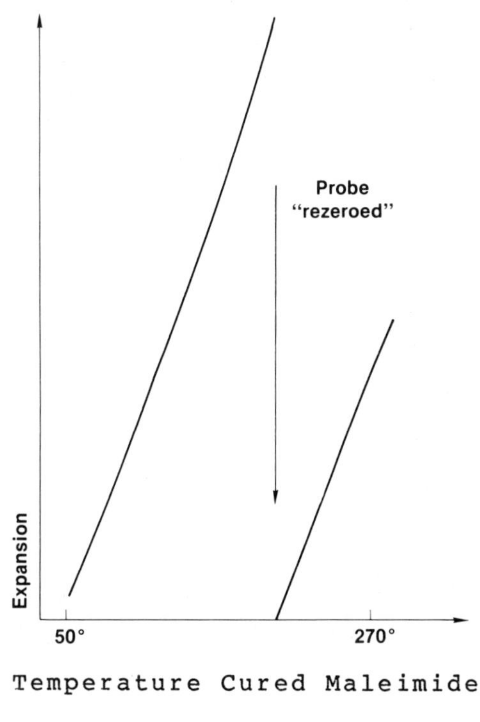

Fig. 7. High Temperature Cured Maleimide Formulation.

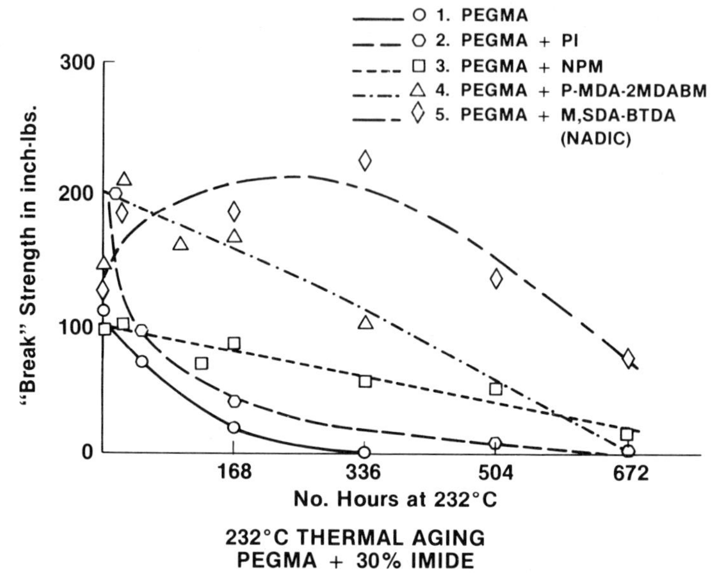

Fig. 8. Effect of Thermal Aging on Break Strengths Nuts/Bolts.

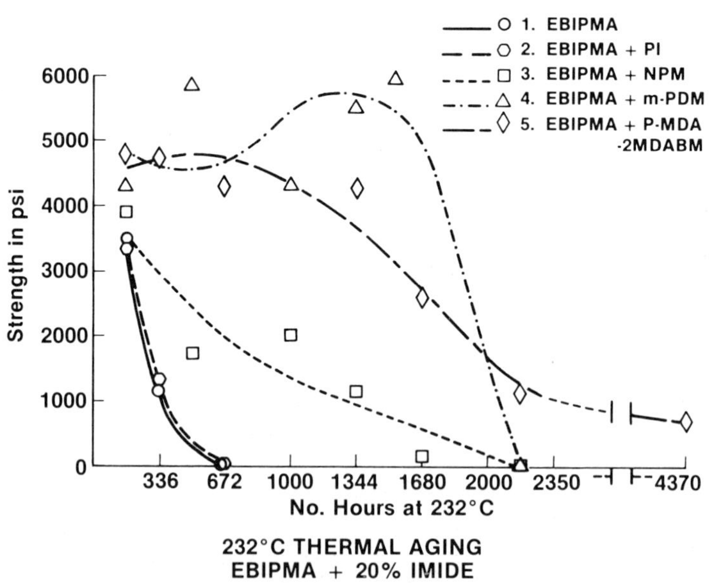

Fig. 9. Effect of Thermal Aging on Break Strengths Pins/Collars.

finally a bismaleimide capped polyimide (P-MDA-2MDABM). Also, whereas all the PEGMA based formulations indicated complete degradation by 8-weeks aging (Fig. 8), two of the EBIPMA/bismaleimide systems retained over 1000 psi strength after 2000 hours and one system still exhibited 700 psi strength after 4300 hours aging at 232°C.

Figure 10 demonstrates, once again, the outstanding resistance in thermal aging with the use of the meta-phenylene-dimaleimide (m-PDM) additive as a comonomer with a urethane methacrylate monomer. The "neat" polymerized urethane methacrylate failed catastrophically within 1000 hours at 232°C, whereas the m-PDM comonomer system yielded nearly 2000 psi tensile shear strength after 2700 hours aging of bonded, steel lap shears.

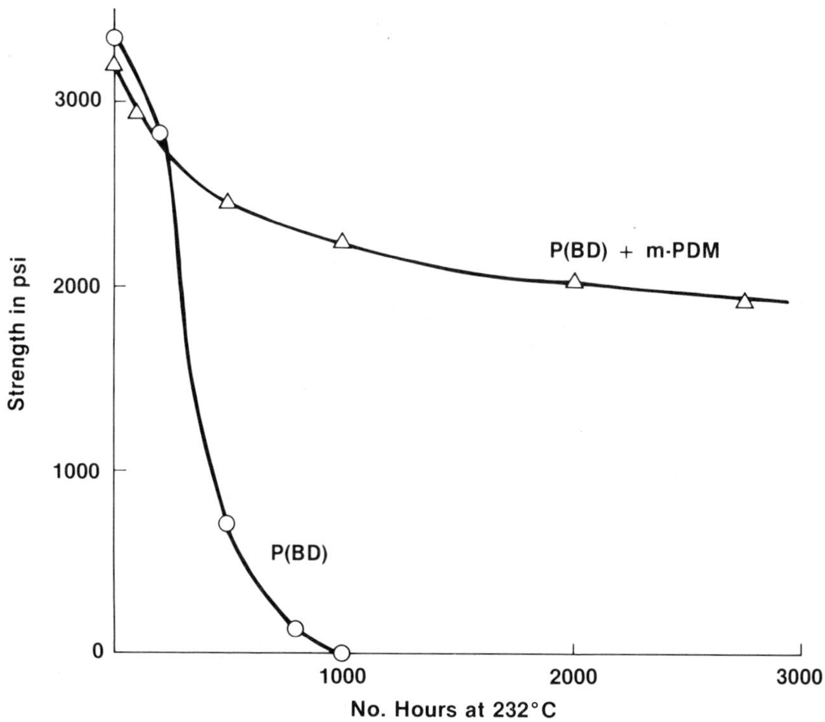

232°C THERMAL AGING
P(BD) URETHANE METHACRYLATE + 30% m-PDM

Fig. 10. Effect of Thermal Aging on Lap Shear Strengths (Steel Lap Shears).

The hot strengths of the above comonomer system showed superior values when compared to the neat cured urethane methacrylate (Table 1).

Significantly increased solvent resistance was also found with this maleimide comonomer system. For example, trichloroethane at 87°C, considered a relatively harsh organic environment, lowers the urethane methacrylate bonding strength to 1100 psi after 30 days, whereas the maleimide comonomer system yields over 3000 psi strength.

Though the latter cured comonomer system is of a highly cross-linked nature, it nevertheless demonstrates surprisingly good toughness as earlier evidenced by the impact strengths. Table 2 illustrates an attractive G_{I_c} fracture energy of 508 J/m^2 for the cured monomer system compared to two thermoset polyimide materials which yielded 104 and 175 J/m^2 respectively (10). Thus, not only can good thermal properties and solvent resistances be gained by the use of these polymerizable maleimides as additives/ comonomers, but also reasonably tough cured materials are produced.

Table 1. Hot Strengths (psi)

	Steel Lap Shears			Steel Pins/Collars	
Material	150°C	205°C	232°C	150°C	205°C
P[BD-CN]	1230	300	0	2350	1020
P[BD-CN] + m-PDM	2350	1770	1130	4880	4040

Table 2. Fracture Energy (G_{I_c})

P BD-CN + m-PDM	508 J/m^2
<u>Thermosets</u>	
Polyimide (addn) - Hexcel F-178	104
Polyimide (acetylenic) - Gulf T-600	175
<u>Thermoplastics</u>	
Polyimide - UpJohn 2080	920
Polyarylsulfone - Union Carbide, RADEL	1,300

*Naval Research Laboratory data

REFERENCES

1. T. Asahara and coworkers, Ger. Offen. 1,954,878, May 21, 1970.

2. G. D'Alelio, Ger. Offen. 2,425,260, December 19, 1974.

3a. T. Usamoto and T. Okita, U.S. Patent 3,641,215, February 8, 1972.

3b. Shin-Etsu Chemical Co., Japanese Patent 81,120,744, September 22, 1981.

4. J.T. Kehn, U.S. Patent 3,334,075, August 1, 1967.

5. K. Akiyama and coworkers, Ger. Offen. 2,419,882, November 7, 1974.

6. B. Malofsky, U.S. Patent 3,988,299, September 22, 1976.

7. L. Baccei and B. Malofsky, Gr. Br. Patent 1,546,815, May 31, 1979.

8. C.S. Cleaver and B.C. Pratt, J. Am. Chem. Soc., <u>77</u>, 1544 (1955).

9. L.J. Baccei, U.S. Patent 4,295,909, October 20, 1981.

10. Private communications from Naval Research Laboratory.

Chemistry of Accelerators for Curing Anaerobic Adhesives – Reaction of N, N-Dimethylaniline Derivatives with Cumene Hydroperoxide

R.W.R. Humphreys*

Loctite Corporation
North Mountain Road
Newington, CT 06111

ABSTRACT

Some of the chemistry of common accelerators used in curing anaerobic adhesives is discussed in detail. Emphasis is placed on the reactions of aromatic amines, particularly in the presence of hydroperoxides. Product studies are presented for the reaction of a series of amines with cumene hydroperoxide (CHP) and plausible mechanisms for product formation are proposed. Relationships are drawn between these results and the chemistry which occurs in anaerobic adhesive formulations.

*(New address: Lever Research Inc., 45 River Road, Edgewater, NJ 07020)

INTRODUCTION

Anaerobic adhesives are methacrylate containing fluids which remain liquid in the presence of oxygen but cure to rigid, cross-linked networks in its absence. The slow polymerization of anaerobic adhesives upon exclusion of oxygen was an early technological problem. One of the many developments which led to the commercialization of these materials was a so-called cure system, which gave a long storage life but rapid, reliable polymerization after proper application of the adhesive.

Tertiary aromatic amines, hydroperoxides and sulfonimides are important components of many of the common anaerobic adhesive cure systems. While various formulative aspects of these compounds are well understood, a detailed explanation for their dramatic effect on the rate of polymerization of the adhesive has been lacking. Our approach to the problem has been to study the chemistry of the isolated components of this cure system under well defined conditions and to apply the results to understanding the mechanism by which these compounds accelerate the polymerization of anaerobic adhesives. Herein, we report some of the results of our studies of the reactions of N,N-dimethylaniline derivatives, which are typical amines used in anaerobic formulations, with cumene hydroperoxide (CHP). Connections will be made between the chemistry of the isolated systems and that which occurs in anaerobic formulations, both during storage and cure.

EXPERIMENTAL

GC analyses were performed with a Perkin-Elmer 3920B FlD instrument modified with a capillary attachment. A 50-m OV-17 glass capillary column was used for all GC analyses. HPLC analyses were carried out with a Perkin-Elmer Series 3B LC equipped with both refractive index and UV detection. Most ^{1}H NMR spectra were recorded on a Varian EM-360 60 MHz Spectrometer. GC/MS data were obtained with a Hewlett-Packard 5985 GC/MS System. A Bruker WM-250 Spectrometer was used for some ^{1}H NMR and all ^{13}C NMR spectra (62.8 MHz).

Most compounds used in this study were commercial materials or were prepared by standard synthetic methods.

Dimethylaniline derivatives were purified by a procedure described elsewhere (1) and were >99% pure as determined by GC. HPLC grade acetonitrile was distilled from calcium hydride under argon. CHP was purified by very careful fractional vacuum distillation; the best fraction from a preliminary distillation was redistilled. The best fraction from the second distillation was usually>97% pure with cumyl alcohol as the only impurity. CHP purity was determined by HPLC with a Waters cyano-terminated column run in reverse phase with gradient elution (10-90% acetonitrile-water).

All glassware used for reagent distillations and in the thermolysis experiments was treated beforehand with 0.25 M aqueous Na_4EDTA, followed by repeated rinsing with deionized water and oven drying. In this way, interference from trace metal impurities was minimized.

Thermolysis experiments were carried out in sealed glass tubes in a thermostatically controlled oil bath at $100^{\circ}C$ for at least 48 hours. Thermolysis solutions were usually 0.2 M in amine and hydroperoxide and were purged with argon for 15 minutes before the tubes were sealed. Products were analyzed by quantitative GC using the internal standard technique with response factors calibrated against authentic reference compounds. For the reaction of N,N-dimethyl-p-toluidine with CHP, products were also identified spectroscopically (^1H NMR, IR, MS) after isolation by preparative GC.

For reactions monitored by NMR, thermolyses were carried out in acetonitrile or acetonitrile-d_3 in sealed NMR tubes in a thermostatically controlled bath, usually at $100^{\circ}C$. Spectra were recorded every 30 or 60 minutes throughout the thermolysis. Unstable peroxidic products formed during the thermolysis were identified unambiguously by stopping the thermolysis, adding the authentic compound to the reaction mixture and observing an increase in the intensity of the appropriate NMR signals.

Analysis of anaerobic adhesive formulations during aging studies was carried out by GC. Important peaks in the GC traces were identified by co-injection of authentic compounds.

RESULTS AND DISCUSSION

Reactions between N,N-dimethylaniline derivatives 1a-c and CHP were slow even at 100°C and, thus, thermolyses were carried out for 48 hours to ensure complete reaction. In every case, the results of product analysis could be described by Fig. 1. Product yields are listed in Table 1.

It is obvious from Table 1 that the material balance for CHP is excellent. CHP is essentially quantitatively accounted for as cumyl alcohol and acetophenone. These results agree well with those from previous studies of hydroperoxide reduction by amines (2); alcohols are usually obtained in greater than 80% yield. Alkoxy radical products are also formed, consistent with our observation of acetophenone from β-scission of the cumyloxy radical.

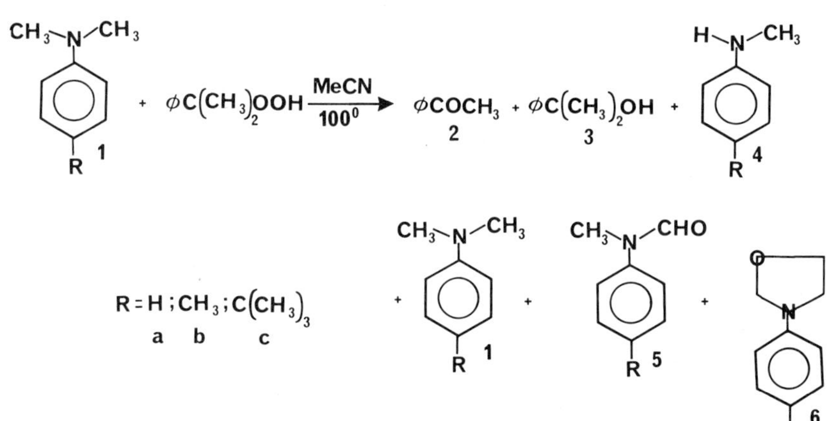

Fig. 1. Products of Reactions Between N,N-Dimethylaniline Derivatives and CHP

On the other hand, the material balance for 1a-c is much less impressive. About 20% of 1 could not be accounted for by GC analysis and may involve the formation of higher molecular weight products of very low volatility. The major amine derived product is the monomethylaniline, 4, formed by oxidative de-methylation, as shown in Fig. 2. This

Table 1. Product Yields (%) for Reaction of <u>1a</u>, <u>b</u>, and <u>c</u> with CHP

Starting Amine	1[a]	2	3	4	5	6
1a	45.8 ± 2.3[b]	19.1 ± 1.4	76.8 ± 3.0	19.7 ± 0.8	13.4 ± 1.0	4.6 ± 0.6
1b	36.2 ± 0.7	15.1 ± 1.3	85.3 ± 2.6	24.5 ± 1.4	10.9 ± 0.1	4.9 ± 1.0
1c	35.9	13.1	82.4	24.9	11.8	5.09

[a]Unreacted starting material

[b]Range of two or more individual experiments

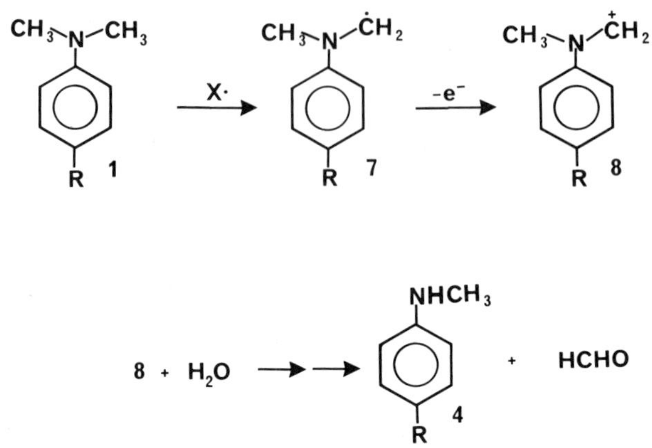

Fig. 2. Oxidative Demethylation of 1

sequence is consistent with the results of previous reports of oxidation and autoxidation of N,N-dialkylaniline derivatives. Amines such as 1 are excellent H-atom donors (3). Radicals such as 7 are expected from the reactions of tertiary amines with hydroperoxides, which clearly involve a radical chain (2c). The very low oxidation potential of α-aminomethyl radicals (4) should lead to the facile production of the N-methylanilinium ion in the presence of CHP, followed by reaction with water to give 4 and formaldehyde as a co-product.

The recovery of 35-40% unreacted 1 after complete consumption of CHP suggests the formation of amine products in higher oxidation states than 4. Two such products are formamide 5 and oxazolidine 6. The identification of significant amounts of oxazolidine is interesting in that its formation involves the addition of α-aminomethyl radical 7 to formaldehyde at the carbonyl carbon (Fig. 3), a reaction for which there is a clear precedent in the literature (5). The addition can occur in two ways, both of which are presumably reversible. Path A gives the thermodynamically more stable radical 9. However, alkoxy radical 10, generated in Path B by addition to the formaldehyde carbonyl carbon, can be trapped by facile intramolecular H-atom abstraction through a six-membered transition state. One-electron oxidation of the resulting radical, followed by cyclization, gives the oxazolidine directly.

This mechanism for oxazolidine formation is strongly supported by the following experimental evidence (6):

(1) The yield of 6 increases when the formaldehyde concentration is increased. When 1 and CHP were heated in acetonitrile in the presence of 0.2 M formaldehyde (added as paraformaldehyde) the oxazolidine yield increased to ca. 10% (see Table 2). An oxazolidine yield of 21% was obtained upon addition of paraformaldehyde equivalent to 0.4 M formaldehyde.

(2) Thermolysis of 1b-d_6 and CHP in the presence of excess protio-paraformaldehyde should give labelled 6 which is either 4-H_2 or 4-d_2 (the 4-d_2 product is produced from formaldehyde-d_2, a co-product of oxidative de-methylation of 1b-d_6), as shown in Fig. 4. This experiment was carried out and the oxazolidine was isolated by preparative GC. Analysis by ^1H NMR, ^{13}C NMR and GC/MS confirmed that 6 was either D_2 or H_2 at position 4; no evidence was found for mixed isotopic substitution.

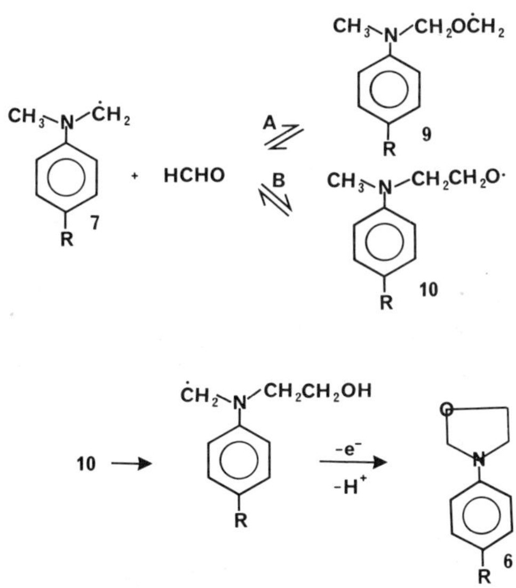

Fig. 3. Formation of Oxazolidine 6

Table 2. Product Yields (%) for Reaction of 1a, b, and c, with CHP, Paraformaldehyde Added

Starting Amine	1[a]	2	3	4	5	6
1a	53.5 ± 1.2[b]	19.9 ± 0.3	76.0 ± 0.1	3.2 ± 0.1	5.8 ± 0.3	11.0 ± 0.1
1b	45.3	13.9	77.5	5.96	5.82	10.2
1c	49.8	13.1	82.9	1.76	4.91	13.3

[a] Unreacted starting material

[b] Range of two or more individual experiments

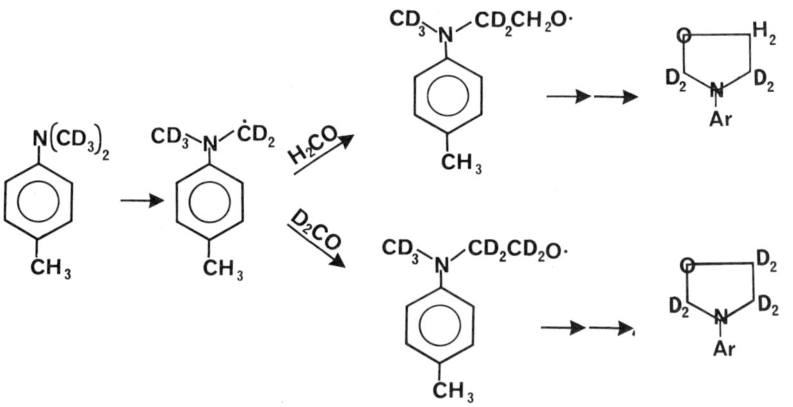

Fig. 4. Labelling Experiments

(3) When peroxide 11 (Fig. 5) was heated at 100°C in acetonitrile containing a good H-atom donor such as 1b or in toluene, oxazolidine 6a was produced in 14 and 18% yield, respectively. Peroxide 11 should give alkoxy radical 10a directly, thus verifying that the radical resulting from α-aminomethyl radical addition to the formaldehyde carbonyl carbon gives the oxazolidine.

Equally interesting is the formation of formamide 5. A plausible mechanism for the formation of 5 involves the thermal decomposition of peroxyamine 12 (Fig. 6). Indeed, previous studies of the chemistry of similar peroxyamines have shown that formamides are formed in significant yield (7). We have prepared 12b by a modification of a procedure described by Kharasch et al. (8) and formamide 5b is a major decomposition product in acetonitrile at 100°C.

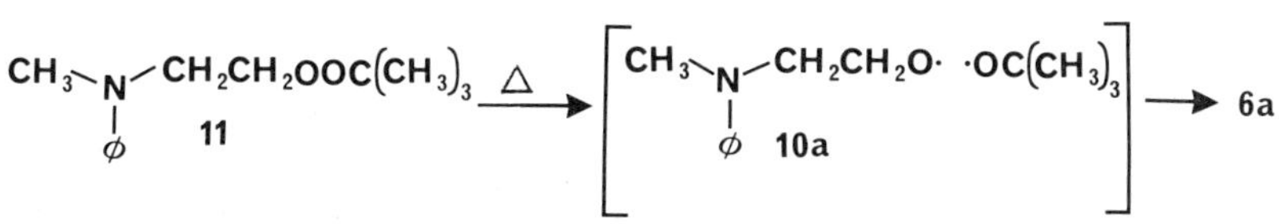

Fig. 5. Thermoloysis of Peroxide 11

Evidence for the formation and subsequent decomposition of peroxyamine 12 during the reaction of 1 with CHP was provided by NMR spectroscopy. Thus, when 1b and CHP were heated in acetonitrile at 100°C in a sealed NMR tube, the formation of 12b was clearly indicated by the appearance of a sharp singlet at δ5 ppm. Proof of the formation of 12b was provided by the addition of an authentic sample to the reaction mixture after partial thermolysis, leading to an increase in intensity for all appropriate NMR signals. We suggest that 12 is formed by addition of CHP to the formaldehyde N-methylanilinium ion, generated from α-aminomethyl radical 7 by one-electron oxidation.

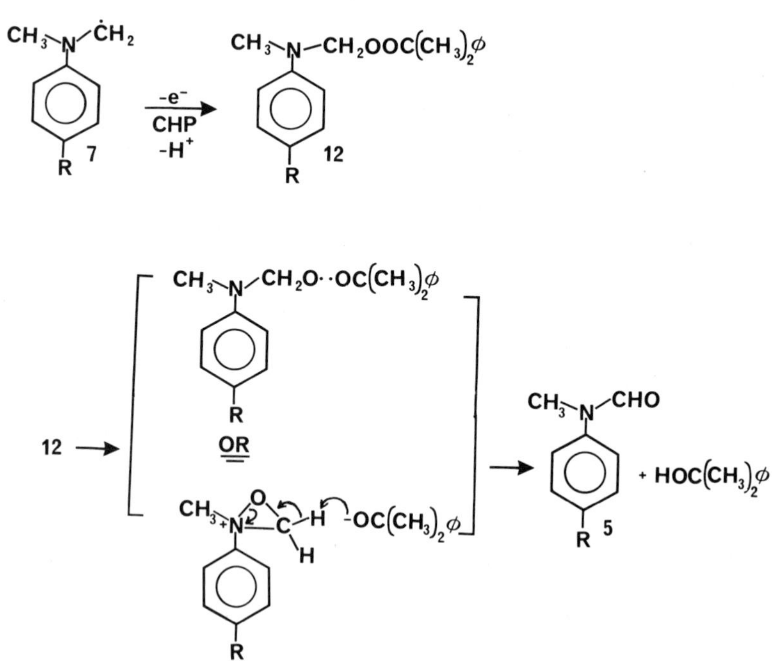

Fig. 6. Mechanism of Formamide Formation

How do these results relate to the chemistry which occurs in an anaerobic adhesive? We examined anaerobic adhesive formulations that had aged at room temperature in an air atmosphere for several months. A representative formulation consisted of polyethylene glycol dimethacrylate, N,N-dimethyl-p-toluidine (1b), saccharin, CHP and a small amount of naphthoquinone and Na_4EDTA. Gas chromatographic and GC/MS analysis of these formulations after aging clearly identified both the monomethylaniline 4b and formamide 5b. We suggest that these products are formed by the autoxidation of 1b. As shown in Fig. 7, 4b and 5b may be formed during the decomposition of hydroperoxyamine 13, generated from α-aminomethyl radical 7b by reaction with molecular oxygen and H-atom abstraction. Howard (3) identified the analogous hydroperoxyamine in autoxidations carried out in the presence of N,N-dimethylaniline and found the major decomposition product to be the corresponding formamide. This type of amine autoxidation may explain the decrease in polymerization rates of anaerobic adhesive formulations after prolonged storage. Autoxidation of the aniline derivative removes an important component of the cure system. We are currently seeking evidence for the hydroperoxyamine or its peroxy radical precursor in aging anaerobic formulations.

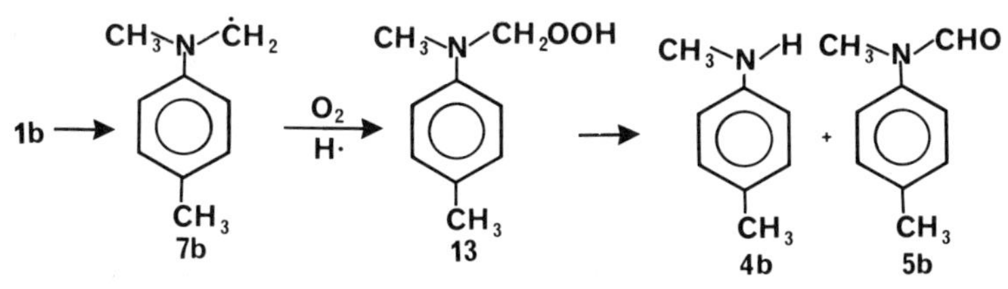

Fig. 7. Autoxidation of 1b in Anaerobic Adhesive Formulation

CONCLUSIONS

The reactions of tertiary amines with hydroperoxides involve radical chains and are capable of initiating the polymerization of vinyl monomers. However, we have demonstrated that the reaction between N,N-dimethylaniline derivatives and CHP is relatively slow even at 100°C. We conclude that the accelerated polymerization of anaerobic adhesives at ambient temperature caused by cure systems containing combinations of tertiary aromatic amines, hydroperoxides, and sulfonimides does not result from hydroperoxide-amine reaction. This reaction may become important at elevated temperatures, such as during 82°C, formulation stability testing.

REFERENCES

1. D.D. Perrin, W.L.F. Armarego and D.R. Perrin, Purification of Laboratory Chemicals, 2nd ed., Pergamon, New York (1980).

2. (a) C.W. Capp and E.G.E. Hawkins, Reactions of Organic Peroxides. VI. Reactions With Amines, J. Chem. Soc., 4106 (1953).

   (b) H.E. De La Mare, "Reaction of Some Aliphatic Amines with tert-Butyl Hydroperoxide. The Fate of the Amine," J. Org. Chem., 25, 2114 (1960).

   (c) R. Hiatt, Hydroperoxides, in Organic Peroxides , Vol II, D. Swern, ed., Wiley, New York (1971).

3. J.A. Howard and T. Yamada, "Absolute Rate Constants for Hydrocarbon Autoxidation, 31. Autoxidation of Cumene in the Presence of Tertiary Amines," J. Am. Chem. Soc., 103, 7102 (1981).

4. D. Griller and F.P. Lossing, "On the Thermochemistry of α-Aminoalkyl Radicals," J. Am. Chem. Soc., 103, 1586 (1981).

5. (a) F.F. Rust, F.H. Seubold, and W.E. Vaughan, "Formation of Secondary Alcohols in the Free Radical Decomposition of Aldehydes," J. Am. Chem. Soc., 70, 4253 (1948).

(b) G. Fuller and F.F. Rust, "Free-Radical Addition of Cyclopentane and Cyclohexane to Formaldehyde," J. Am. Chem. Soc., 80, 6148 (1958).

(c) M. Oyama, "A Free-Radical Reaction of Primary and Secondary Alcohols With Formaldehyde," J. Org. Chem., 30, 2429 (1965).

(d) W.H. Urry, A. Nishihara, and J.H.Y. Niu, "Free-Radical Chain Addition Reactions of Aldehydes With Perfluoro Ketones and Chloro Perfluoro Ketones," J. Org. Chem., 32, 347 (1967).

(e) W.G. Bentrude and K.R. Darnall, "A Free-Radical Acylation," J. Am. Chem. Soc., 90, 3588 (1968).

(f) K. Maruyama, M. Taniuchi, and S. Oka, "Radical Addition Reactions to the Carbonyl Group. I. The Reaction of Aliphatic Aldehydes and Di-t-Butyl Peroxide," Bull. Chem. Soc. Jap., 47, 712 (1974).

(g) E.P. Petryaev, O.I. Skadyro, and P.N. Davidovich, "Mechanism of Oxidation of α-Hydroxyalkyl Radicals by Formaldehyde," Zh. Org. Khim., 14, 2488 (1978).

6. R.W.R. Humphreys, "Reaction of N,N-Dimethylaniline Derivatives With Cumene Hydroperoxide. Oxazolidine Formation Via Addition of α-Aminomethyl Radicals to Formaldehyde," J. Org. Chem., 48, 1483 (1983).

7. E.G.E. Hawkins, "α-Peroxyamines," Angew. Chem. Int. Ed. Engl., 12, 783 (1973).

8. M.S. Karasch and A. Fono, "Metal Salt-Induced Homolytic Reactions. I. New Method of Introducing Peroxy Groups Into Organic Molecules," J. Org. Chem., 24, 72 (1959).

Chemistry, Physical Properties and Durability of Structural Adhesive Bonds

A.V. Pocius, D.A. Wangsness, C.J. Almer,
and A.G. McKown

Adhesives, Coatings, and Sealers Division
3M Company
St. Paul, MN 55144

ABSTRACT

Many types of tests for evaluating and/or predicting the durability of metal-to-metal adhesive bonds are known in the literature. In our laboratory we have made extensive use of the sustained load overlap shear test at high temperature and high humidity for this purpose. As a result of more than 10 years of evaluation we are able to make qualitative correlations of the durability of metal-to-metal adhesive bonds as a function of adhesive type, temperature of cure, type of adherend, type of surface preparation and the amount of load placed on the specimen.

The authors acknowledge the help of Mr. D.J. Thone in the preparation of this paper.

INTRODUCTION

Structural adhesives may be loosely defined as thermosetting resin systems which are used to bond high strength materials, usually metals, in such a way that the bonded structure is able to bear a permanent load. Commonly, structural adhesives are designed to withstand loads in lap shear of over 1000 lbs/in^2 (6.89 MPa) at room temperature. The load experienced by a bonded joint is likely to be applied over extended periods of time in possibly hostile environmental conditions. Hence, the durability of the structural adhesive bonded joint becomes as important as the initial strength of the bond. Many types of tests for evaluating and/or predicting the durability of metal-to-metal adhesive bonds are known in the literature. Minford (1) has recently published a review on this topic. In our laboratory, we have made extensive use of the sustained load lap shear test at high temperature and high humidity for this purpose. In the following, we will discuss the results of more than 10 years of evaluation. With our results, we can make qualitative correlations of the durability of metal-to-metal bonds as a function of adhesive chemistry, cure temperature, adherend type, surface preparation and load conditions.

CHEMISTRY AND PHYSICAL PROPERTIES

There are three main polymers presently used for structural adhesive bonding and they are phenolics, epoxies and urethanes. We can also include in our classification, the initial physical state of the uncured adhesive since it governs the type of application and curing conditions. Structural adhesives are manufactured in the form of films or pastes. The films are one-part adhesives, i.e., they contain a latent catalyst which requires heat for activation. Structural adhesive films are made in a number of different thicknesses and can either be supported (containing a scrim) or unsupported. In general, film adhesives require the application of pressure during cure in order to obtain ultimate properties. Paste adhesives are either one- or two-part materials. That is, pastes can either contain a latent heat activatable catalyst, or they can be a system which is separated into two parts, one of which contains the curative/catalyst. Two-part pastes cure at room temperature. In this section the chemistry of one-

and two-part adhesives will be discussed in a general sense, i.e., generic chemical formulas and reactions for each type of adhesive will be presented. The physical properties presented are those of structural adhesives manufactured by 3M.

Phenolic Based Adhesives

Both resole and novolac phenolics are used in structural adhesives, although novolac phenolics predominate. The resole phenolic structure is obtained by heating phenol in an excess of formaldehyde in the presence of base to obtain a structure similar to that shown below.

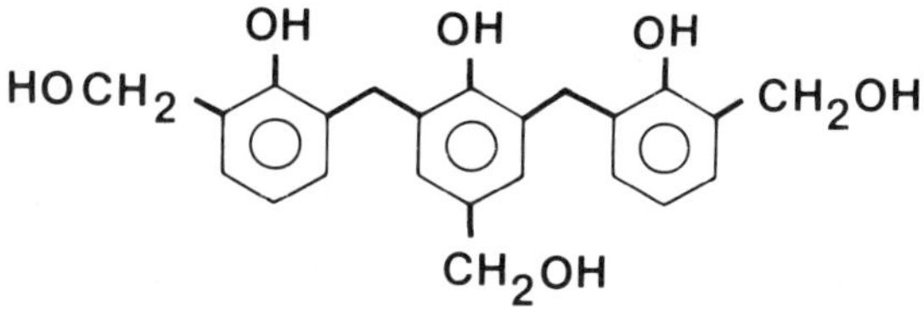

Further condensation can be obtained by heating the resole again to continue the reaction of the methylol groups. This further condensation, however, results in the evolution of gas which could result in a porous bondline. This evolution of small molecules limits the utility of resole phenolics in structural adhesives. Novolac phenolic resins have a structure similar to that shown in Fig. 1 and they are obtained by heating formaldehyde in the presence of excess phenol. As one can see, the novolac structure contains no residual methylol groups. Novolacs will not cure with themselves as do resoles, but they can be crosslinked by means of hexamethylene tetramine ("HEXA"). This is also shown in Fig. 1.

Phenolic based adhesives are outstanding in their wetting characteristics and strength. In the unmodified form, however, they tend to be brittle. To overcome this problem, phenolic based adhesives usually contain a rubber modifier to increase the crack resistance of the adhesive. The first modified phenolic contained polyvinyl butyral and was known as the "Redux" adhesive. Many modern phenolic adhesives contain nitrile rubbers (copolymers of acrylonitrile and butadiene) as the modifiers. An example of a phenolic based film adhesive is AF-30 which is listed in Table 1 along with its physical properties. This film adhesive cures at 170°C (350°F) with applied pressure of 100 psi, AF-30 has been in general use in the aircraft industry for metal-to-metal bonding of aluminum for many years.

Table 1. Physical Properties of Adhesive Bonds made with 171°C Curing Film Adhesives. The metal is FPL etched 2024T-3 clad aluminum. (Shear numbers in parnethesis are in psi, peel numbers in parenthesis are in pounds/inch width).

Adhesive	Chemistry	Lap Shear (MPa) -55°C	Lap Shear (MPa) 25°C	Lap Shear (MPa) 82°C	T-Peel (N/25 mm) 25°C	Floating Drum Peel (N/25 mm) 25°C	Control Strength (MPa) 60°C
AF-30	Nitrile/Phenolic	26.73[1] (3880)	22.6[1] (3280)	12.74[1] (1850)	151.24[1] (34)		15.49[1] (2248)
AF-143	Novalac Epoxy	18.32 (2660)	22.39 (3250)	21.36[2] (3100)	31.13 (7)	44.5 (10)	18.9 (2743)
AF-147	Novalac Epoxy	24.11 (3500)	31.0 (4500)	17.22[2] (2500)	75.62 (17)	155.7 (35)	2893 (4200)

[1] With EC-1593 primer.

[2] At 149°C.

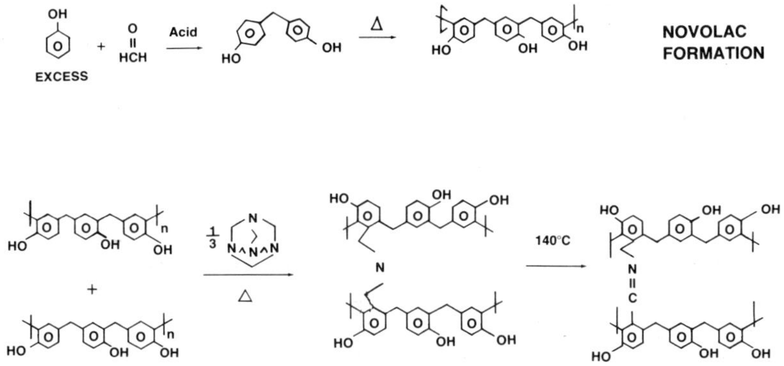

Fig. 1. Chemistry of Formation and Curing of Novolac Phenolic Resins

Epoxy Based Adhesives

The novolac phenolic resins described above can also be used to generate materials known as novolac epoxies by the reaction of the novolac with epichlorohydrin. These materials use latent catalyst systems based on aromatic amines. The reactions which take place will be discussed below. Table 1 also contains information about the physical properties of adhesive bonds made with adhesives based upon novolac epoxies, AF-143 and AF-147. These adhesives have similar chemistries, but AF-147 is a more flexible version of AF-143. This is, of course, to be ascertained by the higher peel strength of AF-147. The high strength values at 300°F for AF-143 and AF-147 indicate that they are targeted for high temperature applications such as aircraft engine nacelle structure.

The majority of structural adhesives are based upon epoxy resins whose molecular structure is as follows:

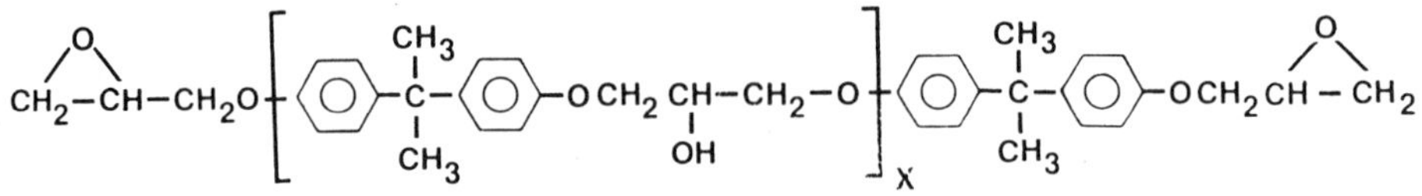

This material is the reaction product of epichlorohydrin and 4,4'-isopropylidene bisphenol. Epoxy resins of this sort are available in a large number of molecular weights and can vary from moderately viscous liquids to solids at room temperature. In general, epoxy based structural adhesives are cured by primary, secondary or tertiary amines. The catalysis of homopolymerization of epoxy resins by tertiary amines is thought to proceed by one of the routes (2) shown in Fig. 2. The second of these reaction mechanisms may be most likely since it does not require separation of charge in a low dielectric constant medium. The reaction sequence also indicates that a hydroxyl containing material is necessary for the progress of the reaction. The hydroxyl containing material may be aliphatic hydroxyls from the epoxy resin itself or it may be added as a separate reagent. Secondary amines react by the well known mechanism (2) shown in Fig. 3. If the R groups on the secondary amine do not cause steric hindrance, the reaction product of the epoxy and the secondary amine can also lead to tertiary-amine-catalyzed reactions. The primary amine reaction with epoxies is also shown in Fig. 3. In this case, since the tertiary amine is likely sterically hindered by the epoxy adducts, no tertiary amine catalysis occurs. The primary amine reaction occurs at room temperature and these materials are used as the curative for the two-part room temperature curing epoxy pastes.

As mentioned above, many structural adhesives require latent curatives which react in the system at an elevated temperature. Latent epoxy curing agents are usually materials which are insoluble in the resin at room temperature but become soluble at some high temperature. One of the most widely used latent curing agents for epoxy resins is dicyandiamide. The chemical structure of this material is

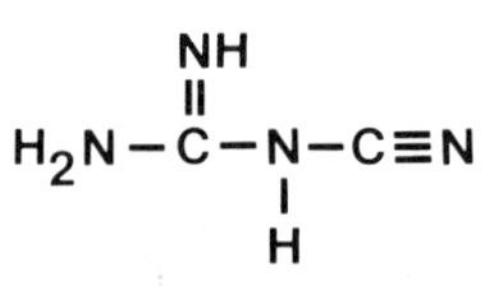

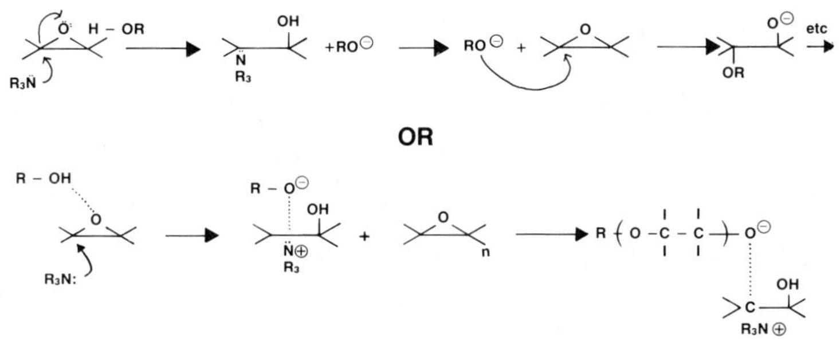

Fig. 2. Chemistry of Tertiary Amine Catalyzed Anionic Polymerization of Epoxies

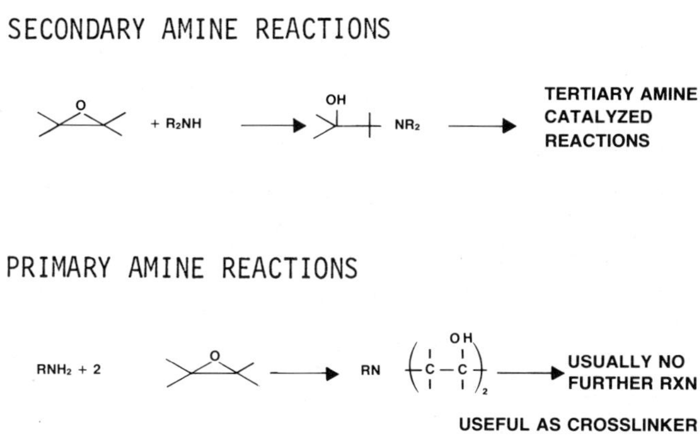

Fig. 3. Chemistry of Secondary and Primary Amine Cross-linking Reactions with Epoxies

It is water soluble and organic solvent insoluble at room temperature. Saunders, Levy and Serino (3) have performed detailed spectroscopic studies of cures by dicyandiamide and have proposed a mechanism yielding a five-fold cross-links. This occurs at temperatures in excess of 325°F (163°C). This would make dicyandiamide a material of only limited utility because there is a need for materials that cure at lower temperatures such as 250°F (121°C). The need was filled a number of years ago by materials such as the dimethylamine urea of p-chlorophenyl isocyanate which act as accelerators for the dicyandiamide cure by lowering the cure temperature to 121°C. Laliberte and Bornstein (4) have recently offered a mechanism for this reaction.

Adhesives based solely on epoxy resins and amine curing agents give cross-linked materials that have high modulus but are very brittle and have no resistance to peel forces. Epoxy resins became much more useful when it was found that rubber flexibilizers could be added to increase crack resistance. Further, it was found that under appropriate reaction conditions, these rubbers could act to "toughen" the epoxy system. That is, addition of rubber flexibilizers tends to reduce the T_g and shear properties of an epoxy. When an epoxy is "toughened" little or no change occurs in the T_g and shear properties while a substantial increase in crack resistance is obtained. For example, Riew, Rowe and Siebert (5) have shown that carboxy-terminated butadiene acrylonitrile rubbers will react with epoxy resins and bisphenols in such a manner as to chain extend the nitrile rubber from the carboxyl group to form an ABA block copolymer. At low conversions, the B segment is soluble in the epoxy resin. However, as the A segment grows in molecular weight, the B block phase separates in the form of rubber particles which are covalently bound in the interphase to the epoxy matrix. McGarry (6) and others have shown that this phase separation lends crack arrest properties to the overall resin system. Recently Gillham (7) has studied the cure mechanism and Hunston (8) has studied the viscoelastic properties of such resin systems.

The physical properties of aluminum-aluminum adhesive bonds made with 121°C curing epoxy film adhesives are shown in Table 2. All of these materials are either flexibilized or toughened. AF-44 is a nylon toughened film adhesive

Table 2. Physical Properties of Adhesive Bonds made with 120°C Curing Film Adhesives. The metal is FPL etched 2024T-3 Clad Aluminum. (Shear numbers in parentheses are in psi, peel numbers in parentheses are in pounds/inch width).

Adhesive	Chemistry	Lap Shear (MPa) -55°C	Lap Shear (MPa) 25°C	Lap Shear (MPa) 82°C	T-Peel (N/25mm) 25°C	Floating Drum Peel (N/25mm) 25°C	Control Strength (MPa) 60°C
AF-44	Nylon-Epoxy	44.78 (6500)	44.78 (6500)	20.67 (3000)	244.7 (55)	533.8 (120)	36.28 (5267)
AF-126-2	Nitrile/Epoxy	32.86 (4770)	32.79 (4760)	16.19 (2350)	182.4 (41)	355.9 (80)	28.41 (4124)
AF-163	Polyether/Epoxy	47.19 (6850)	41.64 6045)	24.8 (3600)	177.9 (40)	355.9 (80)	27.56 (4000)
AF-163-2	Polyether/Epoxy	45.47 (6600)	40.51 (5880)	26.73 (3880)	177.9 (40)	338.1 (76)	27.56 (4000)

which gives outstanding initial physical properties. This adhesive has been used to bond helicopter rotor blades and other metal-to-metal applications. AF-126-2 is a nitrile rubber flexibilized adhesive which has been used extensively for aircraft metal-to-metal and metal-honeycomb bonding. AF-163 and AF-163-2 are 3M's newest aerospace adhesives. They are based on a polyether-toughened epoxy system. Physical properties of these materials, especially as they relate to resistance to precure high-humidity exposure, are discussed in an earlier publication (10).

With the exception of EC-3455, bonds made with the epoxy based one-part paste adhesives shown in Table 3 do not have the physical properties of bonds made with adhesive films. This is due to the fact that the resins in the paste adhesives are lower molecular weight (to achieve handling properties) than those in the film adhesives. The 121°C curing pastes are targeted for general industrial applications. For example, EC-2214 HD has been used to bond halves of magnesium chain saw gas tanks while EC-2214 HT has been used to bond returns on heat exchanger coils. EC-3445, a material similar to EC-3455 has been used in landing gear steps in Cessna aircraft. XA-3430 was targeted for automotive applications.

The physical properties of some room-temperature cured epoxies are shown in Table 4. In general, all of the room-temperature curing materials result in adhesive bonds which have properties that are not as good as those for the higher-temperature cured adhesives. This could possibly be due to the overall lower molecular weight of the room-temperature curing materials. This lower molecular weight allows one to have a lower viscosity for mixing. EC-2158 B/A has been used for bonding floor tile while EC-2216 B/A has been used in a wide variety of applications from bonding of stainless steel sinks to plywood countertops to the bonding of air frames and missile components. EC-1838 B/A has been used for bonding balsa wood to aluminum in baggage and to bond bushings and sleeves in submarine engines.

Table 3. Physical Properties of Adhesive Bonds made with 120°C Curing Paste Adhesives. The Metal is FPL etched 2024T-3 clad aluminum. (Shear numbers in parenthesis are in psi, peel numbers are in pounds/inch width).

Adhesive	Chemistry	Lap Shear (MPa) -55°C	Lap Shear (MPa) 25°C	Lap Shear (MPa) 82°C	T-Peel (N/25mm) 25°C	Control Strength[1] MPa 38°C	Control Strength[2] MPa 38°C
C-2214 HT	Epoxy/Amine	13.78 (2000)	13.78 (2000)	17.22 (2500)	26.7 (6)	10.88 (1580)	
EC-2214 HD	Nitrile/Epoxy	27.96 (4058)	33.94 (4927)	30.10 (4370)	22.2 (5)	17.37 (2522)	17.1 (2482)
XA-3430	Vinyl/Epoxy	18.94 (2750)	19.08 (2770)	10.92 (1585)	48.9 (11)	14.05 (2040)	16.1 (2337)
EC-3455	Polyether/Epoxy	42.71 (6200)	39.27 (5700)	19.98 (2900)	355.9 (80)	24.5 (3556)	17.6 (2555)

[1] Measured on FPL etched 5052T-4 aluminum.

[2] Measured on solvent wiped 1010 cold rolled steel.

Table 4. Physical Properties of Adhesive Bonds made with 25°C Curing Epoxy Paste Adhesives. The metal is FPL etched 2024T-3 clad aluminum. (Shear numbers in parenthesis are in psi, peel numbers in parentheses are in pounds/inch width).

Adhesive	Chemistry	Lap Shear (MPa) -55°C	Lap Shear (MPa) 25°C	Lap Shear (MPa) 82°C	T-Peel (N/25mm) 25°C	Control Strength (MPa)[1] 38°C	
EC-2158 B/A	Filled Epoxy	11.71 (1700)	13.78 (2000)	4.13 (600)	13.3 (3)	10.86 (1577)	
EC-2216 B/A	Flexibilized Epoxy	13.78 (2000)	17.22 (2500)	2.76 (400)	111.2 (25)	15.29 (2220)	9.91 (1439)[2]
EC-2054 B/A	Aluminum Filled Epoxy	10.33 (1500)	13.78 (2000)	2.76 (400)	18 (4)	17.20 (2497)	
EC-1838 B/A	Cr_2O_3Filled Epoxy	10.47 (1520)	20.67 (3000)	5.51 (800)	9 (2)	16.69 (2422)	
EC-1751 B/A	Aluminum Filled Epoxy	9.64 (1400)	13.78 (2000)	3.44 (500)		14.71 (2135)	

[1] Measured on FPL etched 5052T-4 bare aluminum.

[2] Measured on EC-1945 B/A primed FPL etched 5052T-4 bare aluminum.

Urethane Based Adhesives

Although epoxies have become the work horse of structural adhesives, urethane-based materials have also come into significant usage. The reaction used in urethane adhesives, which are, in general, two-part room-temperature curing pastes, is shown in Fig. 4. In Fig. 4, the diol can be any diol but is usually prereacted to give a high molecular weight diisocyanate. This is done to reduce the toxicity of the isocyanates. Various accelerators for this reaction can also be included.

Table 5 presents the physical properties of some room-temperature curing two-part urethanes manufactured by 3M. As one can see, these materials are just on the edge of our definition of a structural adhesive. However, from the peel values shown, they are quite tough and crack resistant. Cooper and his colleagues at Wisconsin (10) have done extensive work on the toughening mechanism in urethanes and have indicated that the mechanism is not unlike that for rubber-toughened epoxies, but on a molecular scale, Cooper's contention is that urethanes phase-separate into hard (urethane) and soft (diol) segments and that this phase separation provides the crack arresting properties.

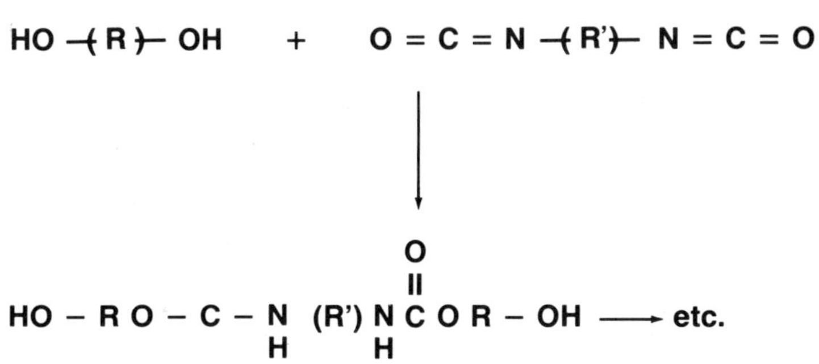

Fig. 4. Chemistry of Urethane Formation

Table 5. Physical Properties of Adhesive Bonds made with 25°C Curing Urethane Paste Adhesives. (Shear numbers in parenthesis are in psi, peel numbers in parenthesis are in pounds/inch width). All adhesive chemistries are isocyanate polyol.

Adhesive	Lap Shear (MPa) -55°C	Lap Shear (MPa) 25°C	Lap Shear (MPa) 82°C	T-Peel (N/25mm) 25°C	Control Strength[4] (MPa) 38°C
EC-3529 B/A	3.44[2] (500)	8.27[2] (1200)	1.38[2] (200)		8.78 (1275)
EC-3539 B/A	4.13[1] (600)	8.27[1] (1200)	2.0[1] (290)		8.49 (1232)
EC-3532 B/A	3.44[2] (500)	8.27[2] (1200)	1.38[2] (200)	169[3] (38)	13.0 (1893)
EC-3535 B/A	17.22[2] (2500)	17.22[3] (2500)	2.07[3] (300)	289[3] (65)	13.8 (2002)

1 Abraded aluminum and primed with EC-1945 B/A.

2 Polyester plastic to sand blasted steel.

3 Measured on FPL etched 2024T-3 alclad aluminum.

4 Measured on FPL etched 5052T-4 bare aluminum.

The adhesives in Table 5 have many industrial applications. For example, EC-3535 B/A has been used to bond aluminum extrusions into auto roofs while EC-3532 B/A has been used to bond vinyl clad wire to printed circuit board. Urethane adhesives have also been used to bond polyester molded resin to the steel frame in Corvette auto bodies.

SURFACE PREPARATION

In addition to the effect of the chemistry and the physical properties of the adhesives, the surface preparation of the adherend also has a marked effect. Aluminum surface treatments vary from simple solvent wiping to anodization. Steel treatments vary from abrasive grit blasting to acid etching. In general, it is found that the better the initial surface preparation, the more durable the bond.

Much work has been done concerning the surface preparation of aluminum for adhesive bonding in the aerospace industry. Bethune (11) described the evolution of surface treatments from the FPL-etch (H_2SO_4/CrO_3) to the "optimized" FPL-etch to H_3PO_4 anodization. McMillan (12) has indicated that proper surface pretreatment is the key to adhesive bond durability. Venables and co-workers (13) at Martin-Marietta have investigated the surface morphology of various aluminum surface pretreatments and conclude that the surface morphology and the hydratability of the aluminum surface controls bond durability. This is amply demonstrated in a U.S. patent (13c) covering a technology for making FPL-etch oxides less hydratable by reacting the surface with materials such as nitrilotris(methylene)-triphosphonic acid. Noland (14) has postulated that the failure mode with an FPL-etched surface is the conversion of the Al_2O_3 to boehmite. The boehmite is more crystalline and brittle and thus more apt to fail. According to Noland, phosphoric acid anodized aluminum is very resistant to hydration (conversion to boehmite) and hence is more stable than the FPL-etch under use conditions. Brockmann and co-workers (15) have also investigated the durability of chromic-acid-anodized surfaces.

Structural adhesives are not usually applied directly to the prepared metal. Rather, a primer is applied first. The primers are usually of the same generic chemical composition as the adhesive being used over them (i.e., epoxy primer to epoxy adhesive). The primers may or may not contain a corrosion inhibiting pigment and may or may not have to be cured before application of the adhesive. EC-1593 is a 350°F (121°C) curing nitrile-phenolic primer while EC-3917 is a chromated epoxy-phenolic primer curing at 250°F (121°C) but formulated for 171°C-performance. EC-3950 is a chromated epoxy primer curing at 121°C with 121°C-performance while EC-1945 B/A is a chromated room-temperature curing epoxy primer.

SUSTAINED LOAD DURABILITY TESTING

Experimental

The method used for testing durability of adhesive bonds was developed at 3M by W.D. Sell (16) and is called "sustained load stress durability". The metal substrates were 2024T-3 clad or bare aluminum alloy, 5052T-4 bare aluminum alloy or 1010 cold rolled steel. The surface preparation for the aluminum alloys was either the "optimized" FPL-etch" or the H_3PO_4-anodization process (12). The steel was solvent wiped. If a primer was used, it was cured before application of the adhesive. Film adhesives were cured in an autoclave at either 120°C or 177°C with appropriate applied pressures and cure schedules. Paste adhesives were cured in a press at room temperature or 120°C. The specimen used is known as a "blister detection" specimen and is shown schematically in Fig. 5. The specimens were loaded in series in the test fixture shown in Fig. 6. After applying a certain stress by means of a spring loading mechanism, the fixture is placed in one of two environments: 140°F (60°C)/100% RH or 100°F (38°C)/100% RH (RH means relative humidity). A photograph of one of the environmental cabinets is shown in Fig. 7. The fixture is monitored and the time to break for each of the specimens is determined. Each adhesive/primer/surface preparation/metal system can be exposed at various levels of stress and a durability bar-graph can be developed (Figs. 8-12).

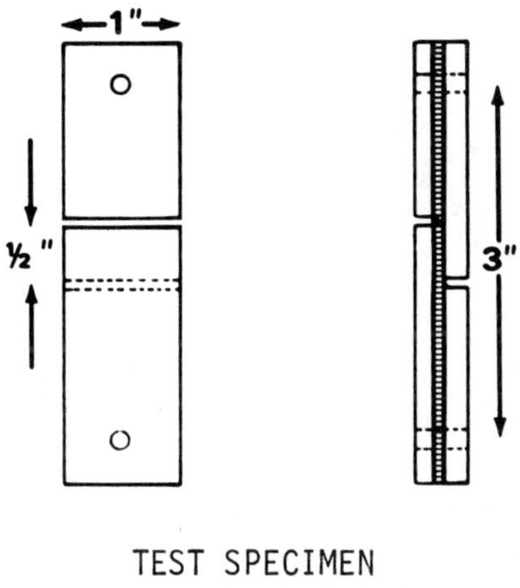

Fig. 5. Blister Detection Specimen

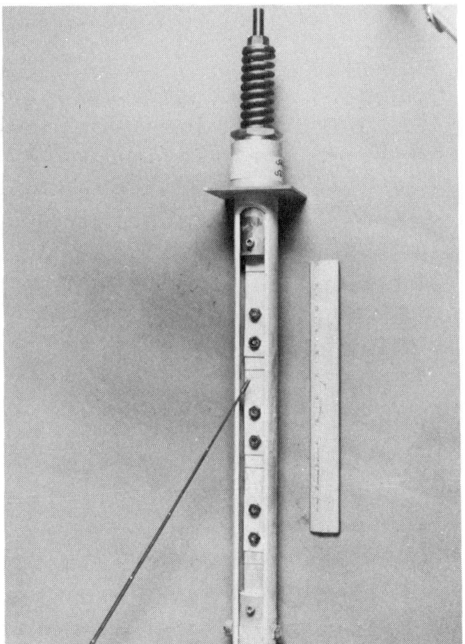

Fig. 6. Spring Loaded Dead Load Stress Fixture

Fig. 7. High Temperature-High Humidity Cabinet

Durability Test Results

In order to scale the test results, one also needs to know the control strength of adhesive bonds tested at the temperature of the high humidity chamber. One can then compare the lifetime of the adhesive bond under various load conditions against this control strength. These values are listed in Tables 1-5.

In all cases, more than one specimen was tested for sustained load durability and the legend for the bar-graphs is as follows. The ordinate shows the sustained load applied to the samples while the abscissa shows the days of exposure. A white bar with an arrow at the end indicates no failure of the specimens, while a white crosshatched bar indicates that at least one sample is still in test. A white bar with a blunt end indicates that all of the samples had failed. In this case, the average of the days to failure of all the tested specimens is given. When possible, the degree of cohesive failure is shown at the end of the failed specimen bar-graph. For example, 0.1 C means 10% cohesive failure and 90% apparent adhesion failure.

The durability test results in the 171°C-film-curing adhesives are shown in Fig. 8. In general, sustained loads of 50% or less of the control strength with these adhesives show them to be exceedingly durable systems. The materials shown in this figure have lives of over 6 years at 60°C/100% RH and at loads of 900-1200 psi. Even though the exact relationship between this sort of test and actual in service experience is not known, we do know that AF-30 has given excellent service experience with respect to disbonding for over 20 years. Applied loads over 50% of the control strength in general lead to early failure with a primarily cohesive failure mode.

The results for the 120°C-curing epoxy film adhesives are shown in Fig. 9. Marked differences are noted here. The AF-44 nylon-toughened epoxy has very poor durability at any load and adhesion failure predominates. AF-126-2, the first-generation nitrile-rubber-flexibilized epoxy has somewhat better durability and the failure mode is at least partially cohesive. AF-163 and AF-163-2, the second-generation-toughened epoxies, show a substantial improvement over AF-126-2 and AF-44. The durability of these materials seems to approach that of the 170°C-curing film adhesives.

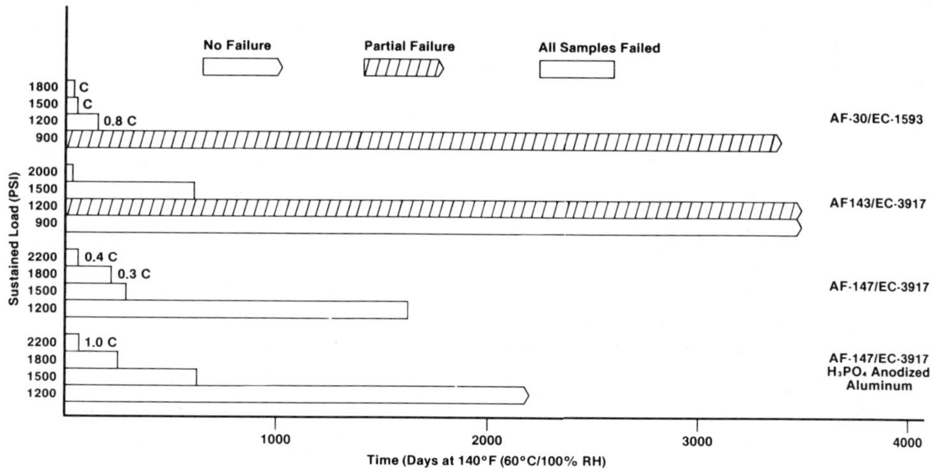

Fig. 8. Sustained Load Durability Results for 177°C Cured Film Adhesives on Clad 2024T-3 Aluminum. (The metal was FPL etched, unless otherwise noted.)

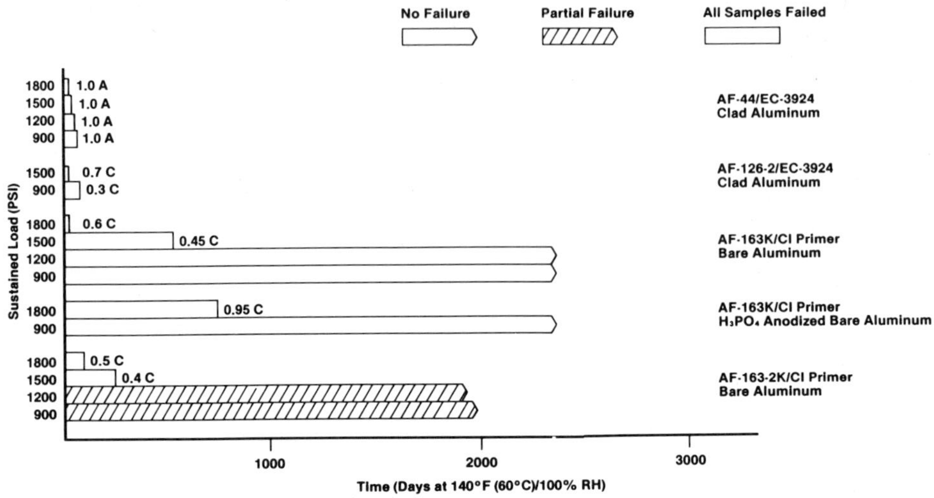

Fig. 9. Sustained Load Durability Results for 120°C Cured Film Adhesives on 2024T-3 Aluminum. (The metal was FPL etched unless otherwise noted.)

The 120°C-curing epoxy paste adhesives on aluminum are covered in Fig. 10. In all cases, the sustained load is substantially less than 50% of the control strength and most of the adhesives show excellent durability by sustaining loads for close to 8 years. One interesting phenomenon is noted. The toughened system, EC-3455 shows much improved durability over the less crack resistant EC-2214 HD. Three of the 120°C-curing pastes were also tested on solvent-wiped steel as shown in Fig. 11. The durability of these bonds is substantially less than those obtained with FPL-etched aluminum. The failure mode was primarily adhesion failure.

In contrast to all of the above data, the durability of room-temperature curing paste adhesives is shown in Fig. 12. These products show poorer durability than any of the higher-temperature curing materials. It is of interest to compare the data on EC-2216 B/A with and without primer. The substantially improved durability of this product when used in conjunction with a corrosion adhesion inhibiting primer is easily seen. Also, it can be noted that the amount of adhesion failure in these materials is greater than that observed with the elevated-temperature curing materials.

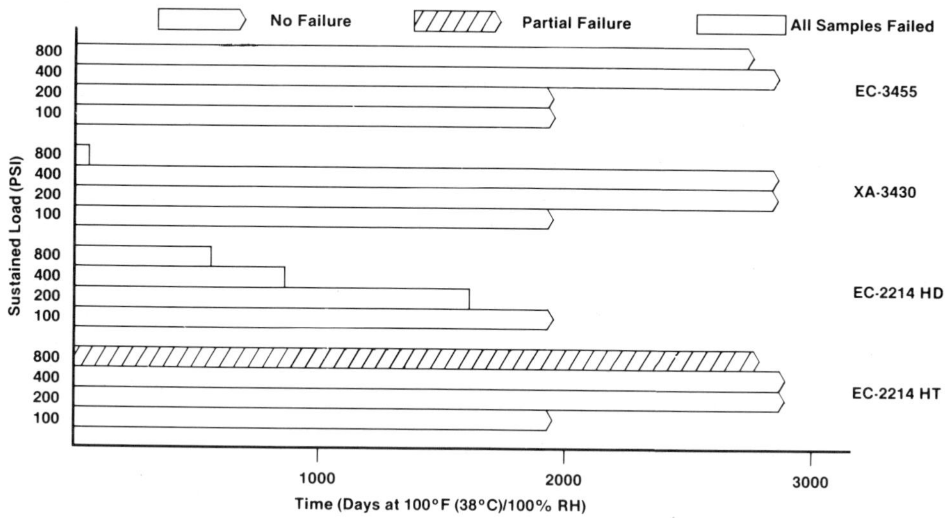

Fig. 10. Sustained Load Durability Results for 120° Cured Paste Adhesives on FPL Etched 5052T-4 Bare Aluminum

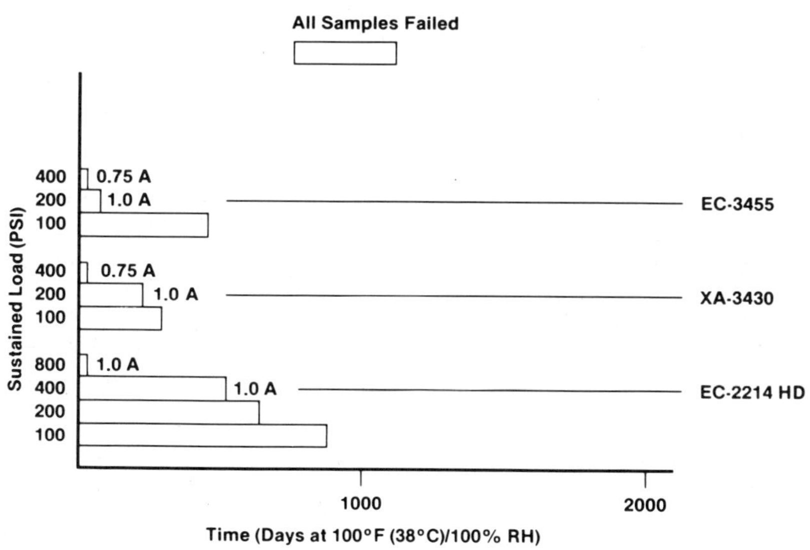

Fig. 11. Sustained Load Durability Results for 120°C Cured Paste Adhesives on Solvent Cleaned Cold Rolled Steel

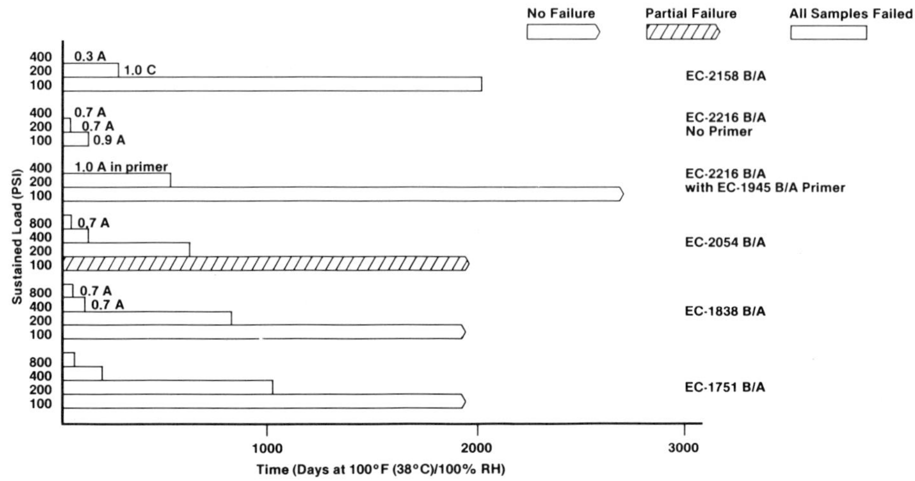

Fig. 12. Sustained Load Durability Results for Room Temperature Cured Epoxy Based Paste Adhesives on FPL Etched 5052T-4 Bare Aluminum

The data concerning room-temperature curing urethane adhesives are displayed in Fig. 13. The urethanes (even when used with a primer) are somewhat less durable than the room-temperature curing epoxies tested without a primer. We also see a somewhat loose relationship between applied load and control strength (i.e., products having the higher control strength are more durable than the materials having lower control strength).

Two data sets highlighting the effect of surface preparation are shown in Figs. 6 and 7. Here we have compared FPL-etched alloys to H_3PO_4-anodized aluminum alloys. For AF-147, the durability at each load level is substantially increased by using H_3PO_4-anodized adherends. For AF-163, at a load-level of 1800 psi (12.4 MPa), the durability is increased markedly for H_3PO_4-anodized adherends over FPL-etched adherends and the failure mode changes from 0.6C to 0.95C.

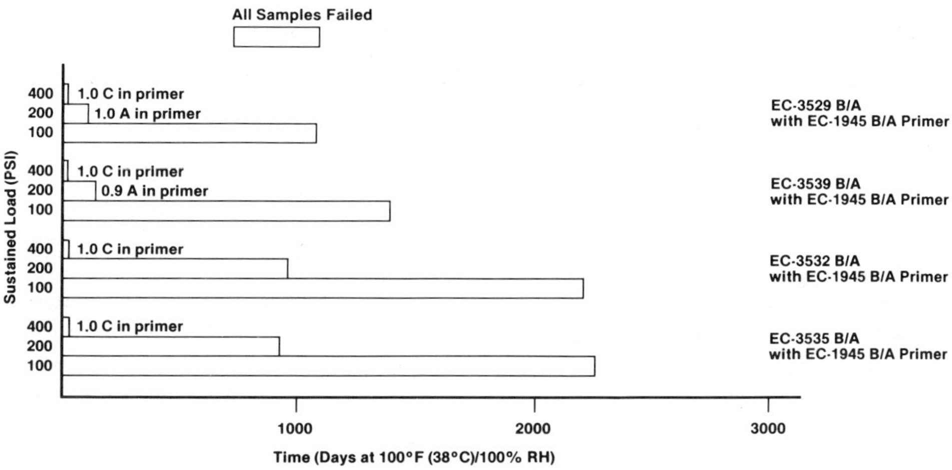

Fig. 13. Sustained Load Durability Results for Room Temperature Cured Urethane Based Paste Adhesives on FPL Etched 5052T-4 Bare Aluminum

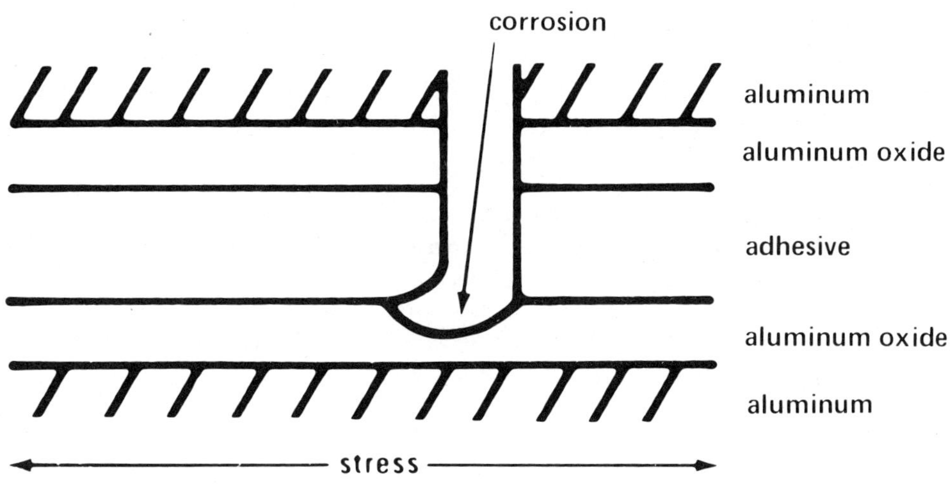

Fig. 14. Schematic Diagram Showing Observed Failure Mechanism

The failure mode of samples exposed for long times is gradual undercutting of the adhesive bond at the metal oxide/primer or metal oxide/adhesive interface. This is shown schematically in Fig. 14. This undercutting continues until the remaining bonded surface is not enough to support the load.

CONCLUSIONS

From the testing done at 3M we can conclude certain relationships between bond durability and the chemical structure of the adhesive. Certainly, the high-temperature curing adhesives have better durability than the room-temperature cures. This is likely due to a higher cross-link density for the higher temperature cures and the additional effect of toughening which would resist stress cracking. The phenolic adhesives seem to be especially good with respect to durability. Phenolics are known to bond very strongly to aluminum oxide. Indeed, Weinberg and his co-workers (19) have shown that phenol will dissociate on aluminum oxide to form aluminum phenoxide. Aliphatic alcohols such as those in epoxy resins are not acidic enough to react with aluminum oxide. In addition, cured phenolics have low water swell.

The primary problem with epoxy resins appears to be the swellability of the resin in moisture (18). This is likely due to the preponderance of hydrophilic groups in the cured epoxy-adhesive. Urethanes have the additional problem that metal ion catalyzed hydrolysis of the urethane bond can take place.

The nylon-toughened adhesive, AF-44, presents a special problem. The nylon-toughening material which gives this adhesive its exceptional properties is hydrophilic and its "toughening" is likely due to the presence of intermolecular hydrogen bonds. In the presence of moisture, these hydrogen bonds can weaken or break and drastically reduce the properties of this material. In contrast, the polyether-toughened materials such as AF-163-2 do not have this problem.

The variation in durability from one substrate to another (together with the surface preparation utilized in each case) can be dramatic. Bonding to solvent-wiped steel appears to give bonds of poor durability. The oxide which naturally forms on steel is not tenacious and will flake off in the presence of moisture and stress. Aluminum, in contrast, has a very tenacious oxide which can be further modified by chemical or electrochemical means to give chemical and surface morphological structures which can give bonds of excellent durability. In addition, primers, especially those with corrosion inhibiting pigments, seem to be quite useful in promoting durability on aluminum substrates.

The sustained load that is applied to the bond in relation to the bonds control strength is certainly a critical parameter. Whether or not the load has any effect on the adhesive itself (analogous to stress corrosion cracking of metals) cannot be ascertained from our data. The literature, especially that of Bascom (19) would indicate that it does. This work does indicate that designers of adhesively bonded structure should take durability testing of this sort into account in their design experimentation.

REFERENCES

1. J.D. Minford, "Aluminum Adhesive Bond Permanence," in Treatise on Adhesion and Adhesives, R.L. Patrick, ed, Vol. 5, Marcel Dekker, New York, 1981.

2. L. Schechter, J. Wynstra, and R.P. Kurkjy, Ind. Eng. Chem., 48, 94 (1956).

3. T.F. Saunders, M.F. Levy, and J.F. Serino, J. Polym. Sci. (A-1), 5, 1609 (1967).

4. B.R. LaLiberte and J. Bornstein, Govt. Report #AMMRC-TR-81-34, NTIS #AD-A 104658.

5. C.K. Riew, E.H. Rowe, and A.R. Siebert, in Toughness and Brittleness of Plastics, Adv. in Chemistry Series #154, R.D. Deanin and A.M. Crugnola, eds., American Chemical Society, Washington, D.C., 1976.

6. F.J. McGarry and A.H. Willner, Amer. Chem. Soc. Div. Org. Coatings, Plast. Chem Preprints, 28, 512 (1968).

7. L.T. Manzione and J.K. Gillham, Org. Coat. Plast. Chem., 41, 371 (1979).

8. D.L. Hunston, J.L. Bitner, J.L. Rushford, J. Oroshnik, and W.S. Rose, J. Elastomers and Plastics, 12, 133 (1980).

9. A.V. Pocius and C.J. Almer, 12th National SAMPE Technical Conf. Proc., p. 924, 1980.

10. G.M. Estes, S.L. Cooper sand A.V. Tobolsky, J. Macromol. Sci. Rev. Macromol. Chem., C4(2), 313 (1970).

11. A.W. Bethune, SAMPE J., 11, 4 (1975).

12. J.C. McMillan, in "Bonded Joints and Preparation for Bonding," AGARD Lecture Series No. 102, Harford House, London, 1979.

13. a. J.D. Venables, D.K. McNamara, J.M. Chen, T.S. Sun, and R.L. Hopping, Appl. Surf. Sci., 3, 88 (1979).

    b. G.D. Davis, T.S. Sun, J.S. Ahearn, and J.D. Venables, J. Materials Sci., 17, 1807 (1982).

    c. U.S. Patent 4,308,079.

14. J.S. Noland, in Adhesion Science and Technology, L.H. Lee, ed., Plenum Press, New York, 1975.

15. W. Brockmann, O-D. Hennemann, and H. Kollek, Int. J. Adhesion and Adhesives, 2,33 (1982).

16. W.D. Sell, Proc. 19th SAMPE Symposium, 1974.

17. B.F. Lewis, W.M. Bowser, J.L. Horn, T. Luu, and W.H. Weinberg, J. Vac. Sci. Tech., 11, 262 (1974).

18. a. J. Comyn, D.M. Brewis, R.J.A. Shalash, and J.L. Tegg, in Adhesion 3, K.W. Allen, ed, Applied Science, London, 1979.

    b. R.J. Morgan and E.T. Moves, Proc. 11th National SAMPE Tech. Conf., 1979.

19. W.D. Bascom, Adhesives Age, 22, 28, (1979).

Reactive Butadiene/Acrylonitrile Liquid and Solid Elastomers: A Bibliography for Formulating Epoxy Structural Adhesives

R.S. Drake and A.R. Siebert

BF Goodrich Chemical Group
6100 Oak Tree Blvd.
Cleveland, Ohio 44131

ABSTRACT

Epoxy structural adhesives using carboxylic polybutadiene/acrylonitrile solid and liquid elastomers as modifiers have increased in number and use since their introduction mid-1960. The authors reviewed this subject through early years of development and commercialization (SAMPE Quarterly, 6, No. 4, July, 1975). These 'epoxy-nitrile' adhesives were largely supported or unsupported films for metal/metal bonding in aircraft. Since 1974, elastomer toughening and flexibilization has benefited epoxy structural adhesive technology as it proliferated and broadened to embrace electrical/electronics, automotive and industrial applications. Each area requires structural adhesives with balanced shear and peel strengths, excellent vibration/impact resistances as well as environmental durability. 'Epoxy-nitrile' adhesives meet these demands while functioning with a wide variety of substrates including prepared and oily metals, rigid plastics, ceramics and glass. More recently, amine-reactive butadiene/acrylonitrile liquid polymers were developed for two-part epoxy structural adhesives curing under ambient or mildly elevated temperature conditions. The present paper pivots from the 1975 review article and proceeds to cover journal and patent literature since that time. Of concern are preparation of the 'epoxy-nitrile' formulating base, elevated- and room-temperature cure systems, structural adhesive testing highlights as well as characterization by morphology, fracture toughness, viscoelastic effects, durability and failure mechanisms. Data exist for both the adhesive bulk and the adhesive joint.

INTRODUCTION

Epoxy structural adhesives which employ carboxylic polybutadiene/acrylonitrile solid and liquid (CTBN) elastomers as modifiers have increased in number and proliferated in use since their introduction in the mid-'60's. Such adhesive systems are now used in aircraft, electronics, automotive and industrial bonding operations. In the mid-'70's, amine-reactive versions of the liquid polymers (ATBN) were issued, thereby offering another way to introduce rubber modification into a cured epoxy network. References are cited which provide detailed discussions of nitrile rubber, carboxylic nitrile rubber and both carboxyl- and amine-terminated nitrile liquid polymers (1-4).

The authors (5) have reviewed this subject along with reinforced epoxy resin composites through the early years of development and commercialization. Most of the available 'epoxy-nitrile' adhesives at that point in time were supported or unsupported films for aircraft metal/metal bonding. As recent as 1981, Drake (6) reviewed some of the more recent progress in this area before The Adhesion Society concentrating on the synthesis of elastomer-modified epoxy resins and their subsequent formulating into epoxy structural adhesives. These adhesives, films or pastes, are capable of providing high peeling resistance, high shear strength and good durability. The present paper pivots from the 1975 review article of the authors and proceeds to cover systematically the journal and patent literature since that time.

Clarke (7), as early as 1971, studied the behavior of a nitrile rubber modified, epoxy paste adhesive using the climbing drum peel strength test. Peeling resistance was measured at various temperatures from 77 to 300°F using three different Instron speeds for dicyandiamide-cured epoxy adhesives. Later, Clarke (8) looked at these and other adhesive developments in historical perspective. Progress in developing peeling resistance without loss (or with gain) in shear strengths was reported for epoxy-urethane, epoxy-nylon and the epoxy-nitrile alloys formulated into adhesives. However, it is noted that the epoxy-nitrile systems best survived environmental testing (bond degradation by moisture plus stress hastened by mildly elevated temperatures). The model epoxy-nitrile systems retained the largest percentage of original lap shear properties under these durability test conditions.

Preparation of Elastomer-Modified Epoxy Resins

Carboxylic nitrile liquid and solid elastomers are used to prepare elastomer-modified epoxy liquid and solid resins when it is desirable to have the modifier in the epoxy portion of the system. This is effected through alkylhydroxy esterification reactions. This is covered in the literature for uncatalyzed liquid resins (9), for tert-amine catalyzed systems (10), for tert-phosphine catalyzed systems (11), for low molecular weight solid resins advanced from the liquid state (12) and for liquid and solid resins where an additional rubber vulcanization step is carried out in addition to the alkylhydroxy ester adducting step (13). Such adduct preparations offer formulation stability with a wide range of hardener types: amines, anhydrides, catalytic, Lewis acids/bases.

The amine-reactive nitrile liquids are useful in formulating a modified hardener package through admixture with an amine, amidoamine or fatty poly-amide curing agent of choice (14). And novel elastomer-modified epoxy resin hardeners giving excellent reported adhesive properties have begun to appear in the marketplace (15).

Adhesive systems based on free-radical curing epoxy-methacrylates are cited (16). And Dudgeon (17) has covered cationic, heat-curable rubber-modified epoxy resin systems which very likely have latent-cure, structural adhesive capability.

Characterization

An intense interest, currently without abatement, in the characterization of rubber-toughened thermosets in general, and rubber-toughened epoxy resins in particular, has surfaced over the last several years. Riew (18) summarized work leading to the optimization of toughness in an epoxy/CTBN/piperidine model. Concurrently, Kaelble (19) reported on such systems from a block copolymer orientation and examined properties related to adhesive performance. Automated Torsional Braid Analysis offers a distinctly unique study tool for rubber-toughened systems as they progress through gel into vitrified thermoset networks. Comprehensive data exist for various liquid and solid epoxy models (20) and adhesive films (21). Crabtree (22) reported extensively on dynamic mechanical analyses of modified epoxy adhesives and composites.

Relationships between structure and mechanical properties have been reported (23,24), phase separation and phase structure studied (25,26) and equilibrium physical properties (27) as well as those resulting from sub-Tg annealing (28) determined.

Fracture Mechanics

With the development and optimization of toughened thermoset resins, fracture surface energy measurements served as one means of ascertaining the degree of improvement over brittle, highly crosslinked glassy polymers. Nearly all the earlier fracture mechanics evaluations were conducted on bulk specimens, i.e., unwarped, air-free, thick castings. The mid-70's saw a change, however, in these types of investigations. The question was asked: If discrete particle size rubber inclusions offer several-fold improvements in fracture surface energy in the bulk, to what degree is this improvement transferred to the adhesive joint or the confines of a fiber-reinforced composite? This question has largely been answered by Bascom, Hunston and co-workers (29-34) and Kinloch (35,36). Attention to adhesive joint fracture in terms of mixed mode fracture of model structural adhesives, bondline thickness effects, rate and temperature effects, bond angle effects and fracture reliability largely concerned these researchers.

Earlier Rowe and co-workers (37) had discussed improved toughness of cured epoxy polymers having a bimodal distribution of rubber particles embedded in the matrix--especially in light of failure mechanisms and what they suggest or mean. Similar models were addressed by Bascom and Hunston (38,39) and Ting (40,41) to show that such toughness improvements are carried over to the adhesive joint or to the fiber-reinforced composite.

Adhesive Formulating and Testing

The formulating and subsequent testing of various nitrile-rubber modified epoxy adhesive systems has become an impressive body of journal, patent and trade literature. It speaks practically, for the most part, about the introduction of peeling resistance into one- and two-component adhesive systems coupled with control over low-temperature properties and good-to-excellent durability in comparison to other modified epoxy resins. Importantly it demonstrates this for several epoxy curing agent or hardener types. With popular two-component systems it also shows the effects of rubber inclusion in either the epoxy or the hardener with studies embracing room-temperature or mildly elevated temperature cure conditions. One-component systems have been studied extensively as pastes or films (prepregs).

Modified polyfunctional aliphatic amines, amidoamines and fatty polyamides are covered for both CTBN/epoxy adducts or ATBN hardener systems (42,44). Examples of systems based on selected epoxy-anhydride (45), an amine-ether (46) and Lewis acid/amine complexes (47) augment this information base. Dicyandiamide (1-cyano guanidine) containing systems alone (48) and with melamine (49) and various proprietary accelerators (50-53) are shown to serve as latent, one-component film and paste adhesives with an excellent balance of adhesive properties.

Extensive adhesive testing has been reported for a variety of systems aimed at widely different bonding applications. Aircraft is represented with numerous studies of which selected references are representative for nitrile rubber modified adhesive systems (54-56). Underwater curing adhesives (57), construction adhesives (58), intumescent adhesives (59) and materials related to improved adhesion or adhesives in electrical/electronic circuitry fabrication (60-62) serve to show the wide latitude of nitrile rubber modified adhesive proliferation.

Durability (63,64) of bonded assemblies continues to receive the prime attention it deserves in the performance regime of structural adhesive testing. And aging studies of selected nitrile-epoxy structural adhesive models are being reported (65,66).

The last four years have witnessed an intense interest in designed epoxy adhesive systems for the transportation sectors embracing cars, buses, trucks and rail. Environmental durability, peel, vibration resistance and impact are of prime concern in the newer paste adhesive systems for the applications involving metal/metal, metal/plastic and plastic/plastic bonding. An impressive documentation has begun to emerge which addresses these newer areas for designed epoxy structural adhesives (67-70).

Related Areas

Application areas that involve adhesion and fracture toughness (but are not adhesives per se) exist for epoxy-nitrile systems in parallel to the structural adhesive developments. This growing body of literature is concerned with high performance materials and is related to organic body solders or deep-sectioned putties(71-73) and fiber-reinforced composites (74-76).

REFERENCES

1. W. Hofmann, "Nitrile Rubber," Rub. Chem. Techn. Technical Review Supplement, 1963.

2. H.P. Brown, "Crosslinking Reactions of Carboxylic Elastomers," Rub. Chem. Techn., 36, No. 4, pg. 931 (10/11/1963).

3. R.S. Drake and W.J. McCarthy, "Liquid Butadiene/Acrylonitrile Polymers with Reactive Terminals," Rubber World (10/1968).

4. C.K. Riew, "Amine-Terminated Reactive Liquid Polymers: Modification of Thermoset Resins," Rub. Chem. Techn., 54, No. 2 (5/6/1981).

5. R.S. Drake and A.R. Siebert, "Elastomer-Modified Epoxy Resins for Structural Applications," SAMPE Quarterly, 6, No. 4 (7/1975).

6. R.S. Drake, "Elastomer Toughened Epoxy Resins for Structural Adhesives," 4th Annual Meeting, The Adhesion Society, Savannah, GA (2/22/1981).

7. J.A. Clarke, "Polymer Viscoelasticity: Relation to Peel Strength in Structural Adhesives," in Recent Advances in Adhesion, L.H. Lee, Editor, 239, Gordon & Breach, New York (1973).

8. J.A. Clarke, "A Field Trip into the World of Adhesives," SPI-ERF 25th Anniversary Meeting (8/11-13/1982), Boston, MA.

9. BF Goodrich Chemical Group, Product Bulletin, RLP-2, "Toughen Epoxy Resins with Hycar RLP" (3/1980).

10. N.C. Paul, et al., "An Aliphatic Amine Cured Rubber Modified Epoxy Adhesive...," Polymer, 18, 945 (1977).

11. K. Dusek, et al., "Modification of Epoxy Resins with Liquid Rubbers Having Carboxyl Groups. Part I. Preparation of the Mixtures, Curing and Morphology of the Cured Resins," Chemicky Prumysl, 30/55, No. 11, 591-597 (1980).

12. W.A. Romanchick and J.F. Geibel, "Synthesis of Solid Rubber-Modified Epoxy Resins", ACS, ORPL Preprints, April, 1982, Las Vegas, Nevada.

13. R.R. Shelley and J.A. Clarke, U.S. Patent 3,947,552, "Epoxy Resin Compositions" (3/30/1976).

14. BF Goodrich Chemical Group, Product Bulletin RLP-3, "Improving Epoxy Resins - Amine Cure Study" (6/1982).

15. Tohto Chemical Co., Product Bulletin, "Goodmide GX-730" (1982).

16. K. Hawthorne, et al., "High Bond-Super Tough-CR Resin," 32nd Annual SPI RP/Composites Institute, Washington, D.C. (1977).

17. C.D. Dudgeon, U.S. Patent 4,275,190, "Latent Heat-Curable Iodonium/Copper Salt-Catalyzed Epoxy Resin Compositions" (6/23/1981).

18. C.K. Riew, et al., "Rubber-Toughened Thermosets," Advances in Chemistry, No. 154, Toughness and Brittleness in Plastics, ACS (1976).

19. D.H. Kaelble, "Block Copolymers as Adhesives," Adhesion Science and Technology, Editor, L.H. Lee, Plenum Press, New York (1975).

20. L.T. Manzione, J.K. Gillham, and C.A. McPherson, "Rubber Modified Epoxies. I. Transitions and Morphology, II. Morphology and Mechanical Properties," J. Appl. Polym. Sci., 26, 889-919 (1981).

21. M.J. Doyle, et al., "Time-Temperature Cure Behavior of Epoxy Based Structural Adhesives," Polym. Engr. & Sci., 19, No. 10, p. 687 (Mid-Aug. 1979).

22. D.J. Crabtree, "Dynamic Mechanical Analysis (DMA) for Quality Assurance of Epoxy-Based Adhesives and Composite Materials," 12th National SAMPE Tech. Conf., 10/7-9/1980.

23. C.B. Bucknall and T. Yoshii, "Relationship Between Structure and Mechanical Properties in Rubber-Toughened Epoxy Resins," British Polymer J., 10 (3/1978).

24. M.J. Doyle, "Molecular Structure and Cohesive Properties of Polymeric Adhesives," AFML-TR-78-117, Final Report, 11/1978.

25. T.T. Wang and H.M. Zupko, "Phase Separation Behavior of Rubber-Modified Epoxies," J. Appl. Polym. Sci., 26, 293-2401 (1981).

26. J.A. Sayre, et al., "Characterization of the Phase Structure of an Amine Cured Rubber Modified Epoxy," Polymer, 22, 87 (1981).

27. R.A. Pethrick, et al., "Rubber-Modified Epoxy Resins: I. Equilibrium Physical Properties, II. Dielectric and Ultrasonic Relaxation Studies," Polymer, 22, 28 (1981).

28. G. Wilkes, et al., "Sub-Tg Annealing Studies of Rubber-Modified and Unmodified Epoxy Systems, J. Appl. Phys., 49, 10 (10/1978).

29. W.D. Bascom and C.O. Timmons, "Mixed Mode Fracture of Structural Adhesives," ACS 169th Meeting, ORPL Preprints, 35, No. 1, 4/7-11/1975.

30. W.D. Bascom and R.L. Cottington, "Effect of Temperature on the Adhesive Fracture Behavior of an Elastomer-Epoxy Resin," J. Adhesion, 7, 333-346 (1976).

31. W.D. Bascom, et al., "Fracture Reliability of Structural Adhesives," J. Appl. Polym. Sci., Applied Polymer Symposium 32, 165-188 (1977).

32. W.D. Bascom and J. Oroshnik, "Effect of Bond Angle on Mixed-Mode Adhesive Fracture," J. Mats. Sci., 13, 1411-1418 (1978).

33. W.D. Bascom, et al., "The Fracture of an Epoxy Polymer Containing Elastomeric Modifiers," J. Mats. Sci., 16, 2657-2664 (1981).

34. D.L. Hunston, et al.,"Characterization of Fracture Behavior of Adhesive Joints," ACS, ORPL Preprints, Kansas City. MO, 9/1982.

35. A.J. Kinloch, "Crack Blunting Mechanisms in Polymers," J. Mats. Sci., 15, 997-996 (1980).

36. A.J. Kinloch and S.J. Shaw, "The Fracture Resistance of a Toughened Epoxy Adhesive," J. Adhesion, 12, 59-77 (1981).

37. E.H. Rowe, et al., "What Failure Mechanisms Tell About Toughened Epoxy Resins," Plastics Engr., 31, No. 3 (1975).

38. W.D. Bascom, et al., "Fracture Behavior of an Epoxy Polymer Having a Bimodal Distribution of Elastomeric Inclusions," ACS ORPL Preprints, Miami, FL, 9/1978.

39. D.L. Hunston, "Rate and Temperature Effects in the Failure of Adhesive Joints," ACS ORPL Preprints, 41, Washington, D.C., 9/1979.

40. R.Y. Ting and R.L. Cottington, "A Comparison of Test Methods in Polymer Fracture Evaluation," ACS ORPL Preprints, 41, Washington, D.C., 9/1979.

41. R.Y. Ting and R.J. Moulton, "Fracture Properties of Elastomer-Toughened Epoxies," 12th National SAMPE Tech. Conf., 265, 10/7-9/1980.

42. Spencer-Kellogg, Product Bulletin, TD-7922 (Revised 12/13/1979), "Kelpoxy G293-100, Elastomer-Modified Epoxy Resin for Adhesives."

43. BF Goodrich Chemical Group, Product Bulletin RLP-3, "Improving Epoxy Resins-Amine Cure Study," 6/1982.

44. BF Goodrich Chemical Group-Preliminary Data Sheets, (a) Hycar CTBN-Modified Epoxy Adhesives," (b) "Hycar ATBN-Modified Epoxy Adhesives," 8/1982.

45. W.P. Barie, U.S. Patent 3,948,849, "Adhesive Compositions," 4/6/1976.

46. Ibid (10). See also N.C. Paul, et al., "Synthesis of Flexibilized Epoxy Adhesives," Adhesion, 3, Appl. Sci., p. 65.

47. J.W. Hill, "Adhesive and Electrical Properties of BF_3 Cured Elastomer Modified Epoxies," SPE ANTEC, 5/1974, San Francisco, CA.

48. Ibid (44).

49. J.H. Flickinger, "Elastomer-Modified Epoxy Resins," Adhesives and Sealants Council, Spring Meeting, 1974, Philadelphia, PA.

50. U.K. Patent 1,389,493, "Epoxy Resin Compositions," 4/1975.

51. U.S. Patent 3,894,113, "Bonding Film," 6/1975.

52. J.P. 82-21, 450, "Epoxy Adhesives and Binders", 2/1982.

53. C.T. Mooney, et al., "Fracture Toughness of Elastomer-Modified Epoxy Adhesive," ORPL Preprints, ACS, Kansas City, MO, 9/1982.

54. M.E. Gross, et al., "Controlled Flow Structural Adhesives for Film Reticulation," SAMPE, "Materials Review-75", 7, 10/14-16/1975, Albuquerque, NM (Adhesives Age, 12/1976).

55. McDonnell Aircraft Company, "Acceptance Criteria for Aerospace Structural Adhesives," AFML-TR-78-53, Final Report, 5/1978.

56. J.P. 82-21, 427, "Prepregs", 2/1982.

57. T.R. Cassutt, U.K. Patent 1,545,807, "Underwater Curing of Epoxide Resin...," 5/16/1979.

58. J.P. 82-34, 151, "Epoxy Resin Adhesives," 2/1982.

59. J.P. 81-45, 927, "Intumescent Resin Compound...," 1981.

60. C.A. McPherson, U.S. Patent 4,121,015, "Curing Epoxy Polymer Having Improved Adhesive Properties," 10/17/1978.

61. U.K. Patent 1,531,826, "New Article Comprising a Resinous Surface...," 11/8/1978.

62. E.J. Leech, "Adhesive-Coated Laminates for Additive Circuits," Adhesives Age, 6/75.

63. Ibid (49).

64. J.D. Minford, "Joint Durability Studies with Abraded, Etched, Coated and Anodized Aluminum Adhesives," ACS, ORPL Preprints, Kansas City, MO, 9/1982.

65. C.E.M. Morris, et al., "Short-Term Aging Effects in a Nitrile-Epoxy Structural Adhesive," J. Macromol. Sci.-Chem, A 17(2), 227-242, 1982.

66. C.E.M. Morris, et al., "Aging of Structural Film Adhesives...," 9/1982, ACS, ORPL Preprints, Kansas City, MO, 9/1982.

67. W.A. Lees, "Modified Epoxides; Practical Aspects of Toughening," J. Adhesion, 12, 233-240, 1981 (PRI Conf. 9/3-5/1980, Durham, England).

68. W.A. Lees, "Use of Adhesives in Constructing Vehicles," Adhesives Age, 2/1981.

69. W.A. Lees, "New Adhesives and the Structural Engineer," SITEV Symposium, Geneva, 5/1981.

70. J.C. Bolger, "Epoxies for Manufacturing Cars, Buses and Trucks," Adhesives Age, 12/1980 (SAE-Paper #800209).

71. J.C. Bolger, "Epoxies and Urethanes-Changing Times Breed New Products," SAE, Automotive Engr. Congress, Detroit, MI, 2/24-28/1975.

72. U.S. Patent 4,273,686, Noland, et al., "Epoxy Based Synthetic Solder," 6/16/1981.

73. C.D. Dudgeon, U.S. Patent 4,308,118, "Deep Section Curable Epoxy Resin Putty," 12/29/1981.

74. A.R. Siebert and R.S. Drake, "Elastomer-Modified Epoxy Resins for Composite Applications," 27th National SAMPE Symposium Preprints, 5/4-6/1982, pg. 739.

75. W.D. Bascom, et al.,"The Interlaminar Fracture of Organic-Matrix, Woven Reinforcement Composites," Composites, 1/1980.

76. J. Harper-Tervet, "Impacting Testing of Toughened Epoxy Resin Systems," 27th National SAMPE Symposium, 5/4-6/1982.

Low Curing Temperature CTBN-Toughened Epoxy Adhesive

Nie Xuzong, Wang Lijuan, Xue Ruilan,
Li Yiming and Yu Yunchao

Institute of Chemistry
Academia Sinica
Beijing, China

ABSTRACT

A carboxyl-terminated butadiene-acrylonitrile copolymer (CTBN)-toughened epoxy adhesive cured with 2-ethyl-4-methyl-imidazole (EMI) is described. In the presence of EMI the reaction of carboxyl end-group of the liquid rubber with epoxy group is very rapid, it is practically completed before the appearance of phase separation or gelation. At low CTBN concentrations the in situ formed domain are discrete globules, which toughen the resin matrix. Aggregation of the dispersed phase begins at the CTBN concentration 30 parts per hundred parts resin (phr); at much higher CTBN concentrations the rubber phase becomes continuous. The EMI concentration is a critical factor for the toughening of the adhesive. A room-temperature lap shear strength maximum of adhesive joint was found at CTBN concentration in the range of 30 to 40 phr. The epoxy adhesive KH-225, toughened with CTBN, has superior fatigue strength to those modified with a non-functional liquid butadiene-acrylonitrile copolymer or a liquid statistical copolymer of butadiene-acrylonitrile-acrylic acid.

INTRODUCTION

Toughness of a glassy polymer can be improved by introducing an appropriate amount of elastomeric particles without remarkable loss in thermal resistance of the matrix polymer. The research on toughened thermosetting resins was so far concentrated on the carboxyl-terminated butadiene-acrylonitrile copolymer (CTBN)-epoxy resin system (1,2). In situ toughened epoxy adhesives have the following advantages such as high fluidity, good bond strength and superior bond durability (3). In our laboratory rubber-toughened epoxy adhesives with latent curing agent have been developed (4). Disadvantage of these single-part adhesives is the high cure temperature. In this article a CTBN-toughened adhesive cured with 2-ethyl-4-methyl-imidazole (EMI) is described. Imidazoles catalyze the polymerization of epoxy group at relatively low temperatures. The cured products have much higher heat distortion temperature (HDT) than those cured with aliphatic amines. By incorporation of a CTBN toughener low-temperature curing structural adhesives of high performance can be formulated.

EXPERIMENTAL

Materials

Liquid carboxyl-terminated butadiene-acrylonitrile copolymer was prepared in this laboratory (5). The number average molecular weight was 2500 and acrylonitrile content 30 wt %. The butadiene unit was mainly in 1,4-trans structure with 11% 1,2-structure. The carboxyl functionality was about 1.9. This product was designated as E-30.

Bisphenol-A type epoxy resin E-51 was produced by Wuxi Resin Factory. The epoxy value was 5.1 meq/g. 2-Ethyl-4-methyl-imidazole from Tienjin Second Chemicals Factory was a special reagent and used as received.

Analysis

The reaction of the carboxyl terminal group of the liquid rubber with epoxy group was followed by IR analysis; the disappearance of 1710 cm^{-1} absorption indicated the completion of the esterification. The conversion of the carboxyl group was determined quantitatively by titration in benzene-methanol (4:1) with 0.1 N sodium methanolate to the thymol blue end point.

Morphology and Dynamic Mechanical Test

The morphology of the cured resin was studied by electron micrography of OsO_4-stained microsections and by SEM observation of fracture surface and the surface etched with HNO_3. The dynamic mechanical test was conducted on a viscoelastometer (Kukusui Electron Corp.) at 3.3 Hz over a temperature range of -196 to 180° at 2°C/min.

Mechanical Testing of Adhesive Joints

For tensile shear tests, aluminum alloy LY-12-CZ and cold rolled steel pieces of 0.5mm x 20mm x 60mm were used. The aluminum was etched in a $K_2Cr_2O_7$-H_2SO_4 bath, and the steel panels were polished and degreased with acetone. The overlap was 15mm. Five specimens for each formulation were tested, and the average shear strength was calculated.

Fatigue tests were conducted at 100 Hz and a stress ratio 0.1. Aluminum alloy LY-12-CZ panels of 5mm x 30mm x 100mm etched in a $K_2Cr_2O_7$-H_2SO_4 bath were used for the preparation of double overlap joints. The overlap was 10mm for each side.

RESULTS AND DISCUSSION

Chemistry of Curing

In an epoxy resin-curing agent-CTBN system chemical reactions can be divided into two kinds: (i) the reaction between carboxyl group of the toughener and epoxy group forming block prepolymer (reaction (a) in Fig. 1), and (ii) cross-linking polymerization of the resin. The relative rate of these reactions is essential to phase separation and subsequently the morphology and mechanical performance of the end product.

When m-phenylenediamine (mPDA) or other aromatic primary amine is used as curing agent, the reaction (b), which leads to cross-linking of the resin, is catalyzed by carboxyl group as shown by the substantial reduction of the gel time (see Table 1). On the other hand, the esterification of carboxyl group is less accelerated by primary amines. As a result, only a small part of the

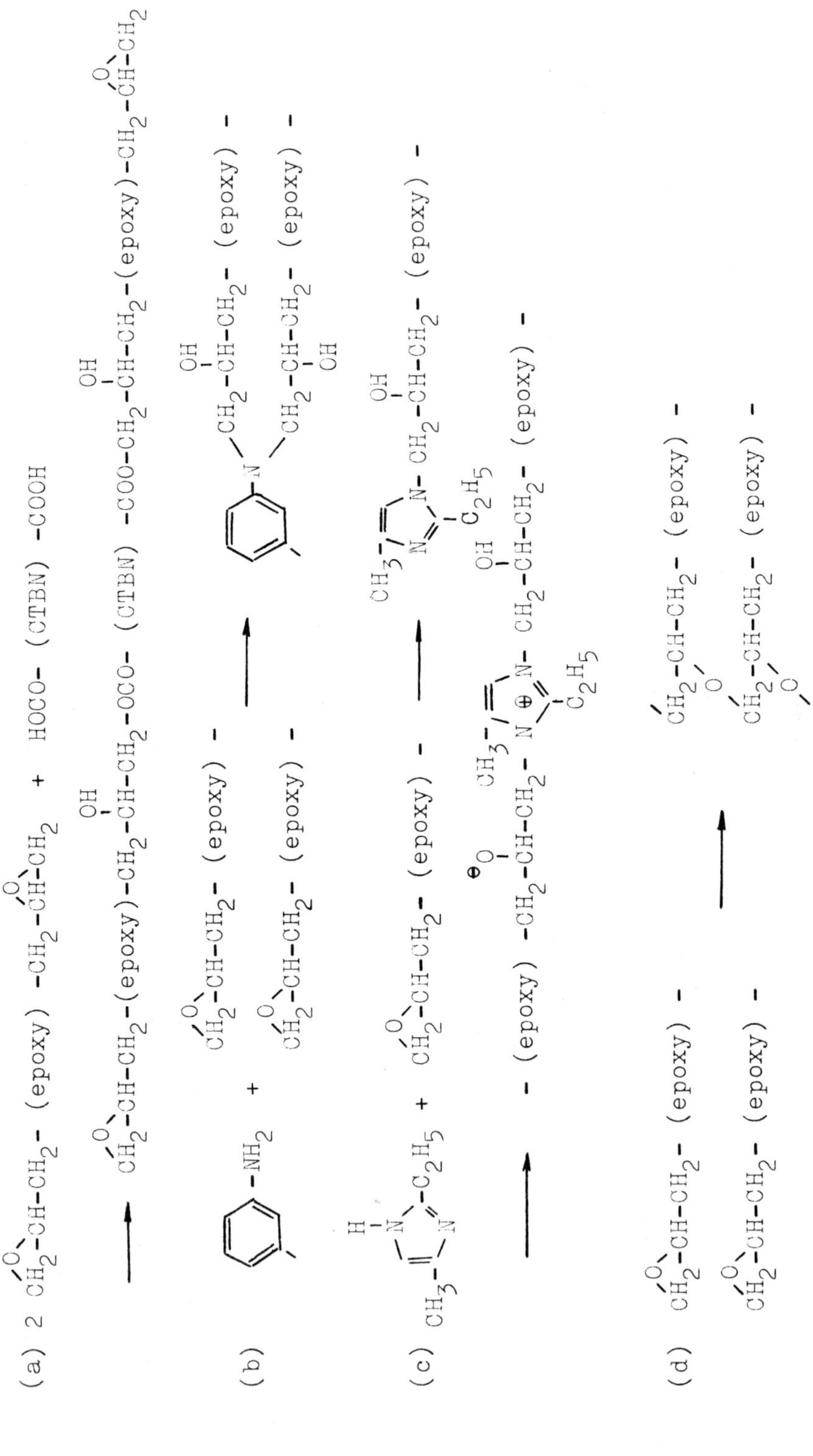

Fig. 1. Chemical Reactions During the Cure of CTBN-Toughened Epoxy Resin

toughener molecule is chemically bound onto the resin before gelation of the system. Free CTBN molecules may deteriorate the mechanical performance of the bulk resin; moreover, their collection at the interface would form a weak boundary layer if the resin was used as metal adhesive (6). For this reason the esterification should be carried out in advance, or alternatively, an effective catalyst for esterification must be added along with a primary amine curing agent.

Imidazoles are very effective catalysts for the esterification of carboxyl group. The reaction between carboxyl group and epoxy group can be effected before gelation of the system by incorporation of 2 phr EMI when epoxy resin is cured by mPDA, or by incorporation of 1 phr EMI when 4,4'-diaminodiphenylsulfone was used as curing agent.

By using EMI itself as curing agent at a level from 1 phr to 10 phr, the reaction leading to the formation of epoxy-CTBN-epoxy block copolymer was practically completed long before the gelation or the appearance of the phase separation as shown by disappearance of the 1710 cm^{-1} absorption in IR spectra and the decreasing of carboxyl value (Table 1). In this case, incorporation of CTBN per se or its pre-reaction product with epoxy resin into an adhesive formulation did not make much difference.

Epoxy resins were cured by EMI through reactions (c) and (d). Elevation of cure temperature usually speeds up the reaction and leads to the completion of the network structure. However, the linkage formed by reaction (c) may break down at high temperatures, forming products with carbonyl or unsaturated groups (7). The influence of cure cycle is thus complicated.

Morphology and Mechanical Performance

CTBN can be vulcanized with an appropriate amount of epoxy resin. For example, the elastomer from 100 g E-30 and 36 g E-51 catalyzed by 2 g DMP-30 has a tensile strength 79 Kg/cm^2 and an elongation 300%. The rubber forms the continuous phase in this case.

Table 1. Conversion of Carboxyl Group at Gel Point when mPDA was used as Cure Agent.

Curing agent	EMI		mPDA	
Level of curing agent (phr)	10	10	14	14
Level of liquid rubber (phr)	0	35	0	35
Gel time at 100°C (min.)	20	23	30	13
Conversion of carboxyl group at gel point (%)	-	100	-	20

If CTBN is the minor ingredient of a CTBN-epoxy resin mixture, it will be separated as dispersed phase during the cure. The distance between the rubber domains diminishes with the increase of the CTBN concentration. At 20 phr E-30 level the dispersed phase was discrete globules (Fig. 2). Aggregation of the particles was observed at 30 phr E-30. Above 40 phr E-30 level, the rubber domains had a complex form as a result of particle aggregation. Owing to the high acrylonitrile content, the separation of E-30 from epoxy resin was not complete as shown by the decrease in the glass transition temperature of the continuous phase (see Table 2).

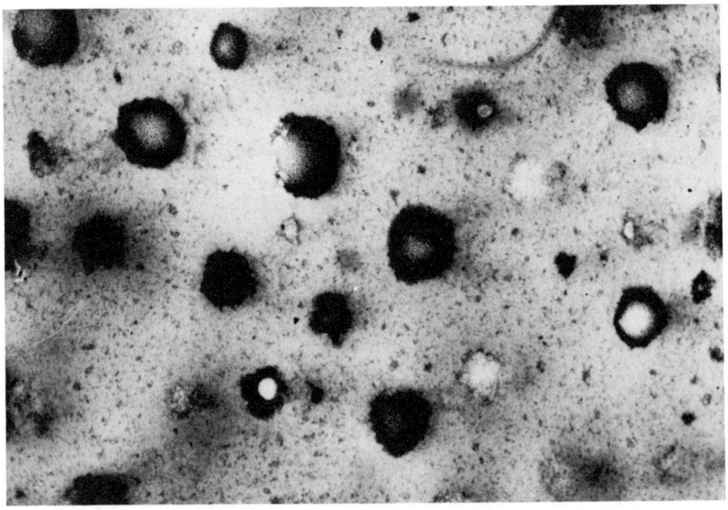

Fig. 2. Electron Micrograph of an OsO_4-Stained Microsection of Epoxy Resin Toughened with E-30

Table 2. Glass Transition Temperature of Epoxy Resin Toughened with E-30 (°C)

E-30 level (phr)	EMI level (phr) 4	6	10
0	162	142	118
10	149	127	107
20	144	125	107
30	137	117	100
40	137	114	104

The samples were cured at 80°C for 1 hr and 150°C for 4 hrs, T_g was determined by dynamic mechanical test.

Incorporation of the E-30 component usually improves the elongation at break but lowers the modulus and the strength of the resin. However, there was no simple relationship between adhesive performance and bulk properties of the resin. With respect to the tensile shear tests, a strength maximum at room temperature was found in the range of 30 to 40 phr E-30 (Fig. 3).

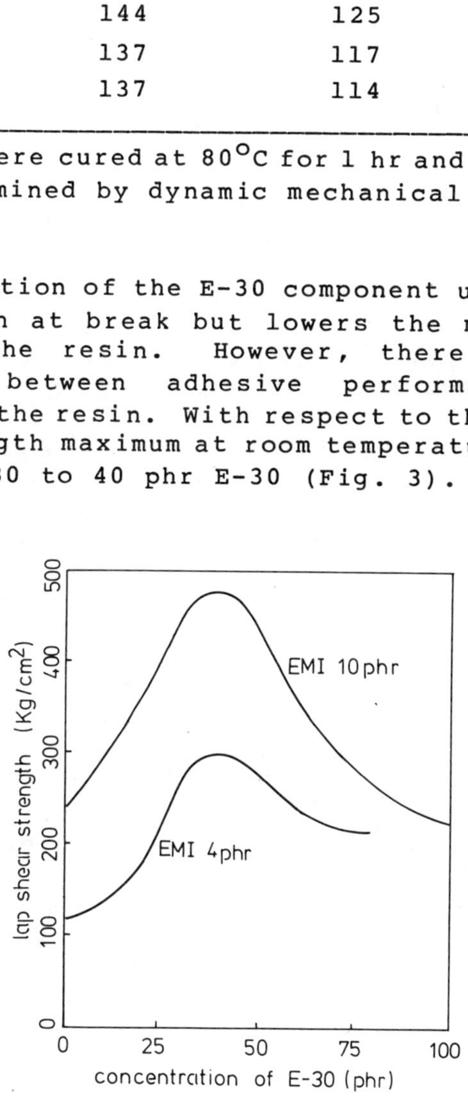

Fig. 3. Toughening of Epoxy Adhesive with CTBN at Different Level of Cure Agent (cold rolled steel, cured at 120°C for 3 hrs.)

The concentration of EMI was a critical factor to the adhesive performance. With 4 phr EMI as curing agent, only a moderate shear strength could be reached with an incorporation of E-30 as much as 40 phr. On the other hand, the shear strength increased sharply with the addition of E-30 at 10 phr EMI, as shown in Fig. 3. This is in agreement with the rule-of-thumb: a tougher resin is easier to be toughened. Epoxy resins cured with 10 phr EMI have a lower cross-link density and a lower glass transition temperature (see Table 2), they are tougher than those cured with 4 phr EMI.

Adhesive KH-225

Adhesive KH-225 is a paste structural adhesive based on CTBN-toughened epoxy resin. It can be cured with EMI at a temperature such as 60 to 80°C under contact pressure.

This adhesive has a very high bond strength and moderate thermal resistance (Table 3). It can be used at 120°C for prolonged time: no degradation was found after aging at that temperature for 400 hrs.

Because of the high polarity of E-30, Adhesive KH-225 has a high water affinity, so that the bond strength at elevated temperatures drops sharply after exposure to high humidity at elevated temperatures. Nevertheless, the water adsorption is reversible, the shear strength at elevated temperature could be recovered to some extent by baking at 120°C (see Table 4).

Being toughened with CTBN, the adhesive has good fatigue strength. Under 55% load coefficient the fatigue life of the adhesive toughened with E-30 was 4 times higher than those modified with a non-functional liquid butadiene-acrylonitrile copolymer, and 3 times higher than those modified with a liquid statistical copolymer of butadiene-acrylonitrile-acrylic acid. The limit fatigue ratio of double overlap joint with Adhesive KH-225 was 30%, thereby the maximum shear stress in the bonding area was 125 Kg/cm^2 (see Fig. 4).

Adhesive KH-225 has advantages of high toughness, high bond strength, good fluidity and low cure temperature. It has been used successfully for bonding metals, ceramics and reinforced plastics, especially for substrates of different thermal expansion coefficients.

Table 3. Bond Strength of the Adhesive KH-225 (cold rolled steel)

Cure Cycle		3 hrs at 120°C	4 hrs at 80°C
Lap shear strength	25°C	470	364
(Kg/cm^2)	100°C	180	165
	120°C	134	87

Table 4. Bond Strength of the Adhesive KH-225 after Soaking in Boiling Water (aluminum alloy)

Temperature	25°C	60°C
Lap shear strength before soaking (Kg/cm^2)	300	290
Lap shear strength after 510 hrs in boiling water (Kg/cm^2)	240	190
Lap shear strength after soaking and baking at 120°C for 3 hrs (Kg/cm^2)	240	225

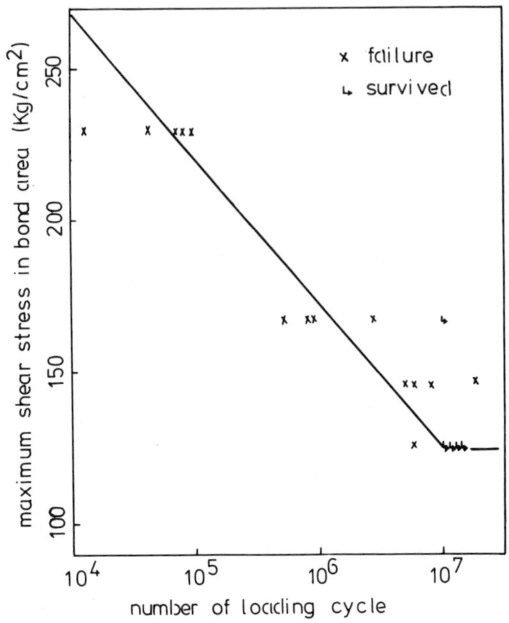

Fig. 4. S-N Diagram of Double Overlap Joint with Adhesive KH-225 at 100 Hz and R = 0.1.

REFERENCES

1. F.J. McGarry, Proc. Roy. Soc., A 319, 59 (1970).

2. A.R. Siebert and C.K. Riew, Preprints of Organic Coatings and Plastics Chemistry, 31, No. 1, 552 (1971).

3. C.D. Weber and M.E. Gross, Adhesives Age, 17(2), 18 (1974).

4. YANG Yukuen, LIAU Zengkuen, YU Yunchao and LU Fencai, Synthetic Adhesives, Chapter 6 (in Chinese), Science Press, Beijing, 1980.

5. YU Yunchao, LI Yiming, WANG Lijuan, GU Juying, JIN Shijiu, XUE Ruilan and TAN Kerong, "Synthetic Rubber Industry," (in Chinese), 1983, No. 1, 45.

6. JIN Shijiu and GU Juying, Polymer Communications (in Chinese), 1981, No. 4, 150.

7. F. Ricciardi, M.M. Joullie, W.A. Romanchick and A.A. Griscavage, J. Polym. Sci., Polym. Letter ed., 20, 127 (1982).

Summarizing Remarks

James P. Wightman

Department of Chemistry

Virginia Polytechnic Institute
& State University

Blacksburg, VA 24061

Recognition of Dr. L. H. Lee, Symposium Chairman

- Identification of relevant/timely Symposium topics: Recent Developments in Adhesive Chemistry.
- Solicitation of papers and the organization of contributed papers into a coherent program.
- Calibre of Plenary Lectures evidenced by high attendance.
- Encourage scientific exchange through oral/written questions.
- Insure rapid publication of Symposium papers.

Dr. Lee, the polymeric materials community is indebted to you for all your efforts which culminated in a highly successful Symposium, and I know those present will join with me in expressing our thanks to you.

Responses by Dr. L. H. Lee:

I would like to thank Professor Wightman, other Session chairmen, and all speakers for this Symposium for their excellent contributions.

Discussion

On the Paper by R. Buchanan

L. H. Lee (Xerox Corporation: We are really fortunate to have Dr. Buchanan here giving us a latest account about recent developments in scanning electron microscopy. SEM has been one of the useful tools for us to examine adhesive interfaces. New techniques described in this paper should enable us to observe the interfacial area with greater precision.

On the Paper by R. A. Pike and F. P. Lamm

D. Hicks (Celanese Corporation): Is H_2SO_4 any more selective for epoxy/OH reaction than BF_3?

F. P. Lamm (United Technologies Research Center): No difference was observed.

T. Rayman (Hanna Chemical Coatings Corp): Were the diacids and diols in the mixture in certain ratios?

F. P. Lamm: Yes, the formulations were done according to a definite hydroxyl number. Several OH/epoxy ratios were used to control final properties of the resin.

L. H. Lee (Xerox Corporation): (1) Does BF_3 affect the double bond? (2) Is the reaction between BF_3 and expoxide explosive?

F. P. Lamm: (1) No evidence of reaction was observed -

does not preclude some interaction. (2) No explosions. A small amount of BF_3 etherate causes the exothermic reaction to warm the overall reaction.

On the Paper by B. M. Malofsky and L. J. Baccei

T. Murdock (H. B. Fuller Co.): (1) How was formaldehyde detected in the N,N-dimethyl-p-toluidine-cumene hydroperoxide reaction mixture? (2) Please comment on the toxicology of the maleimides.

B. M. Malofsky (Loctite Corp.): (1) N.M.R. (2) Most maleimides are known respiratory irritants.

W. Brooks (Boeing-Renton Materials Technology): Are the materials discussed commercially available?

B. M. Malofsky: Yes, currently in production.

D. K. Ray-Chaudhuri (National Starch & Chemical Corp.): Is the lack of polymerization of bis-maleimide at RT due to insolubility of the monomer in the medium?

B. M. Malofsky: N-phenyl maleimide is soluble in the system. It also did not polymerize at RT.

On the Paper by A. V. Pocius, D. A. Wangsness, C. J. Almer and A. G. McKown

L. H. Lee (Xerox Corporation): What is the nature of the primer for polyurethane structural adhesives?

A. V. Pocius (3M Corporation): The primer EC-1945BIA is a 2-part solvent diluted amine-epoxy primer. The primer also contains chromates for corrosion inhibition.

J. P. Wightman (VPI, Blacksburg, VA): Why do you think H_3PO_4 anodization is more effective than the FPL etch in pretreating aluminum?

A. V. Pocius: In terms of initial bond strength, there is very little difference between the FPL etch and the H_3PO_4 anodized aluminum surface. The differences become apparent when bonds are exposed to hot-humid or corrosive environments. The H_3PO_4 anodized oxide is inherently resistant to hydration. This resistance is due to inclusion of $PO_4^{\equiv}$ in the oxide. The mechanism by which $PO_4^{\equiv}$ inhibits hydration of alumina is not entirely known but it could be due to both the insolubility of $AlPO_4$ and the rather exact crystal lattice matching of $PO_4^{\equiv}$ in the alumina lattice. This hydration resistance is amply demonstrated by the resistance of the H_3PO_4 anodized surface to sealing after anodizing. The FPL etch oxide converts to pseudo-boehmite rapidly under sealing conditions and other anodizes (e.g., CrO_3 anodizing) also seal well. In summary, initial bond strengths of aluminum prepared with "good" surface preparations do not vary widely. However, when the bond is stressed under hot-humid conditions, those surface oxides which are resistant to hydration are able to support the stress longer or retain strength longer than oxides which are prone to hydration.

K. K. Knock (Boeing, Seattle, WA): Do you have further comments on why the steel (non-CR) surface preparations which produce good paint adhesion don't produce durable adhesive bonds?

A. V. Pocius: I assume that the question concerns zinc phosphated steel surfaces. We can only theorize as to the reason for poor bond performance using zinc phosphated steel. The zinc phosphated steel surface consists of a myriad of Zn-Fe phosphate crystals. These crystallites are quite fragile and probably are unable to adequately transfer a load from adherend to adhesive. I would expect that the resistance of Zn-phosphated bonded steel to corrosion in an unstressed state to be similar to painted Zn-phosphated steel.

On the Paper by Nie Xuzong, et al.

Comments:

(1) I have been led to believe that the reaction selectivity of 2,4 EMI is as follows:

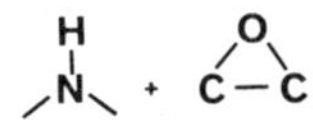

(i)

and then

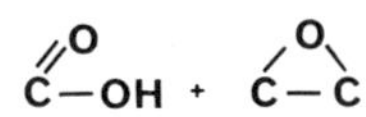

(ii)

as long as one is at epoxy excess, and that the only reason the second reaction (ii) occurs is due to the basicity of the system before, during, and after (i). I believe this needs commenting on as one moves from catalytic amounts of 2,4 EMI $\longrightarrow 4 \longrightarrow$ 10 phr.

(2) It is also not clear where CTBN is placed in this obvious 2-component adhesive.

- If it is placed with epoxy, then there's little one can do to prevent esterification (COOH disappearance, viscosity increase) upon normal storage under ambient. If the CTBN has the following conformation

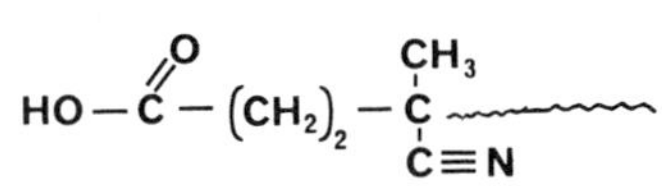

then the C N in the position greatly enhances the pKa value of the acid. Thus, if above, the pKa is 3.0. There's little to stop esterification at excess epoxy 3 mole/1 COOH. Even if the conformation is

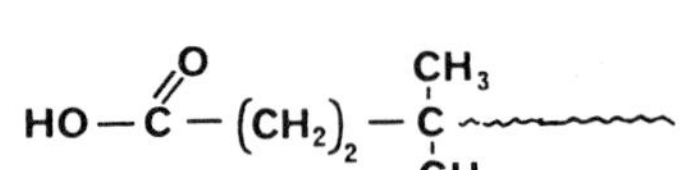

esterification under storage will occur.

- If it's placed with 2,4-EMI, then the substituted imidazole salt of the acid polymer is formed. This is clearly a different reacting species than pure COOH and pure 2,4-EMI.

These items need clarification, I believe.

(3) Pg. 4 - Para. 2. Is it true that the aromatic primary amines reaction with epoxy (the selective reaction in this instance) is catalyzed by acid? Salicylic acid in conjunction with an alcoholic hydroxyl carrier (benzyl alcohol, nonyl phenol, furfuryl alcohol, etc.) may do it. But I doubt it if CTBN will. Perhaps I'm wrong. But this needs explaining.

(4) Pg. 4 - Para 2. CTBN as free dibasic acid <u>does</u> deteriorate all amine, amidoamine or polyamide cured epoxy adhesives <u>if</u> it exists as free acid. This is why the overwhelming majority of instances <u>adducts</u> CTBN and epoxy first. No free acid.

(5) Pg. 6 - Para 3. Is the water affinity due to polarity only <u>or</u> possibly because of the high level of regenerated t-amine in the system?

Answers:

1. We have not evaluated the rate of the addition of secondary amino group to epoxy ring. This reaction does not lead to cross-linking and, therefore, is not so critical in respect to the adhesive performance.

2. CTBN was mixed with epoxy resin. The esterification does occur upon normal storage. However, the adhesive performance was not affected markedly.

3. The catalytic activity of carboxyl group on the cure of epoxy resin with aromatic primary amines is known in the literature (R. F. Gould, Adv. Chem. Ser. No. 92, 29, 1970), and it can be ascertained through the substantial reduction of the gel time.

4. We can demonstrate that free carboxyl acids do not deteriorate the aromatic amine cured epoxy adhesive if a powerful catalyst is present, so that the esterification can be completed before the gelation of the system. Good bond strength could be reached by the addition of 2 phr EMI when mPDA was used as curing agent for epoxy-CTBN mixture, or 1 phr EMI when DDS was used as curing agent.

5. Incorporation of E-30 does increase the water affinity of the cured resin. The water pick-up of the cured epoxy film of 0.5 mm thickness after 24 hrs in boiling water increased from 3.3% to 4.8% by incorporation of 20 phr E-30.

PART SIX:

TRENDS IN ADHESIVE RESEARCH

Developments in and Limitations of Adhesive Materials for Severe Environments and a Long Service Life

Lieng-Huang Lee

Webster Research Center
Xerox Corporation
Webster, New York 14580

ABSTRACT

Durability of adhesive materials is affected by environmental factors. The specific environments of concern are: (1) extreme high temperatures, (2) extreme low temperatures, (3) extreme high humidity, (4) salt water, (5) fire, (6) corrosive gases or liquids, and (7) external stresses. A detailed discussion of the first five factors is given. Future research needs about these factors are also suggested.

INTRODUCTION

Some new adhesive materials have been designed to meet the demands for use in a severe environment and for a long service life (1). The specific environments of concern are: (1) extreme high temperatures, (2) extreme low temperatures, (3) extreme high humidity, (4) salt water, (5) fire, (6) corrosive gases or liquids, and (7) external stresses. Although sealants may be mentioned, our major concern is structural adhesives.

EXTREME HIGH TEMPERATURES

During the last decade, many high-temperature adhesives have been synthesized for aerospace applications. Supersonic cruise aircraft adhesives must be durable for tens of thousands of cumulative hours at 232°C (500°F), while spacecraft adhesives must withstand hundreds of hours of 316°C (600°F) temperatures. Missile adhesives must survive for less than one minute at temperatures greater than 538°C (1000°F) (2). Besides thermal stability, high-temperature adhesive materials should exhibit processability at approximately 100°C above their Tg's. Yet many high-temperature polymers (polybenzoxazoles (PBO) or polybenzothiazoles (PBT) are difficult to process. The structures of phenyl-substituted PBO's and PBT's are shown as follows:

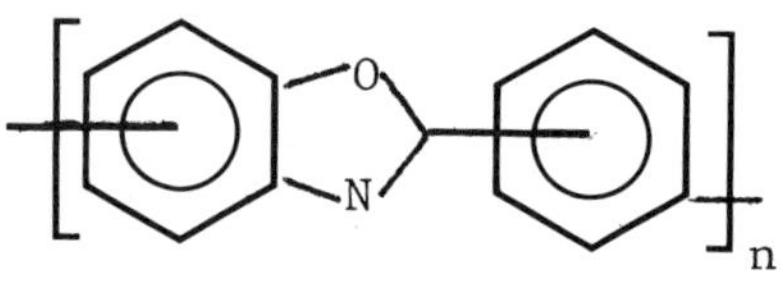

Polybenzoxazoles (PBO)

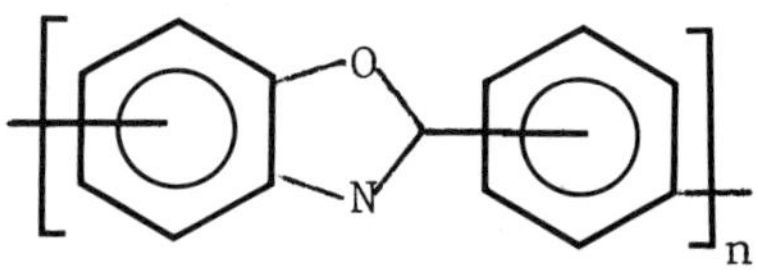

Polybenzothiazoles (PBT)

For durable bonds, adhesives should be tough. However, many high temperature adhesives (addition-type polyimides, bismaleimides (BMI), and acetylene-terminated resins) are brittle and need to be toughened by elastomers.

Several processable, thermally stable polymers have been developed, e.g., polyimides (PI), polyimidesulfone, polyphenylquinoxalines (PPQ), polyquinoxalines (PQ), and polyimidesulfide hot melt (3), and triaryl-s-triazine ring, polybenzimidazoles (PBI), have not been evaluated as structural adhesives. In an inert atmosphere, the thermal stability of the phenyl-substituted PBI appears to be better than that of the phenyl-substituted PBO or the PBT (Fig. 1) (5). PBI's are suitable for short-term use between 538°C and 760°C (1000-1400°F). However, we may have exceeded the ability of polymers alone to survive at such high temperatures.

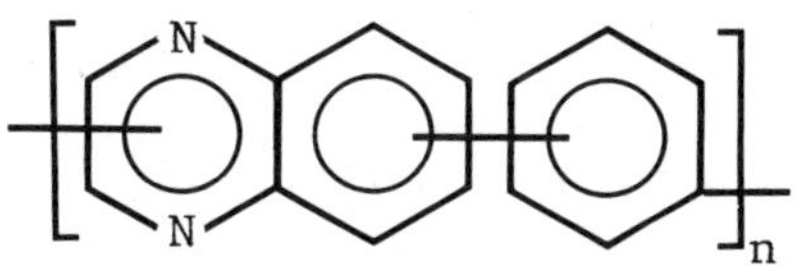

Polyquinoxalines (PQ)

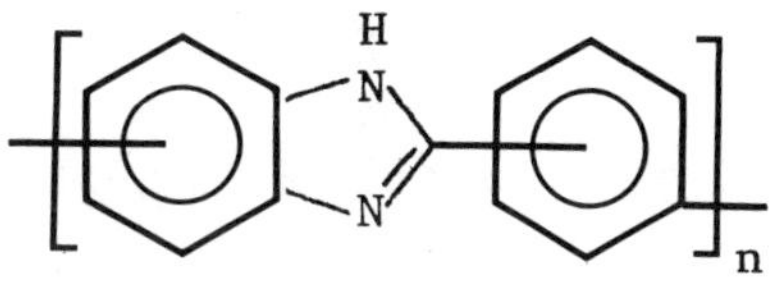

Polybenzimidazoles (PBI)

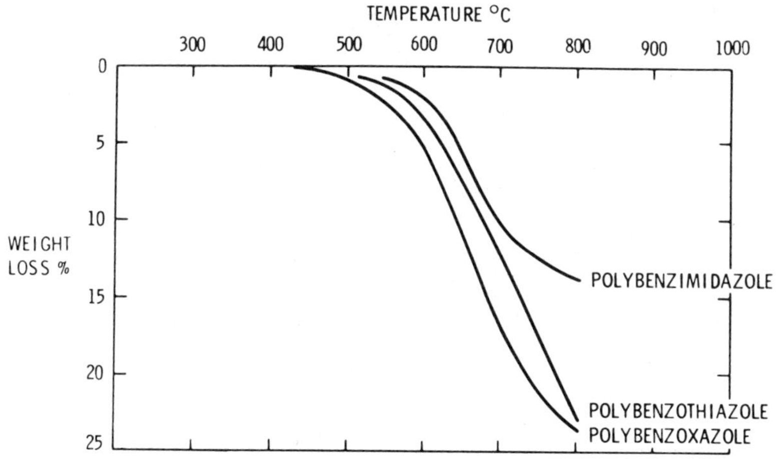

Figure 1. Comparison of the stabilities of 6/5 ring polymers in an inert atmosphere.

(Ref. C. Arnold, J. Polym. Sci., Macromole. Rev. 14, 265 (1979).)

The chemistry of TSTR is of considerable interest (4). TSTR is formed by the cyclotrimerization of benzonitriles:

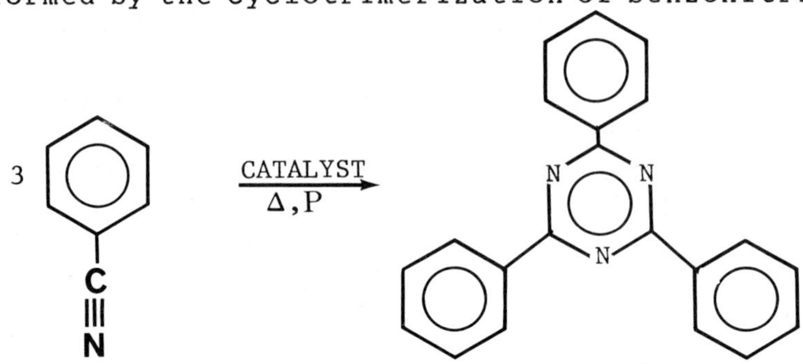

Cyclotrimerization of Benzonitriles	$t_d > 907^{o}F$ Triphenyl-S-Triazine
t_m -14°F	t_m 452-455°F

TSTR-type polymers are actually copolymers of TSTR and other heat-resistant polymers of aromatic or heterocyclic ring structures such as aromatic imides, phenylquinoxalines, etc.:

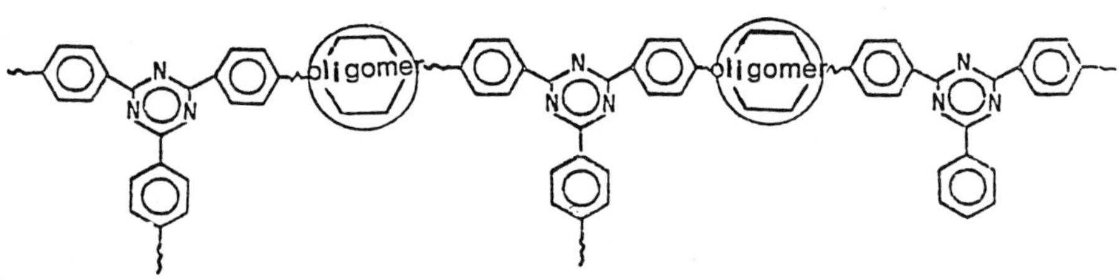

OLIGOMER	RING STRUCTURE
IMIDES	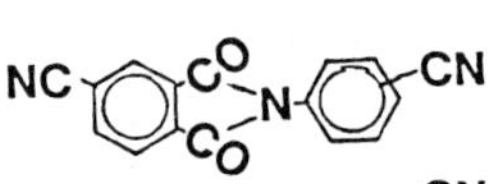
AZOLES	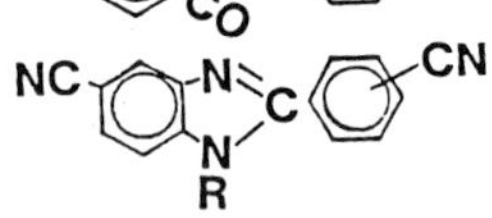
PYRROLONES	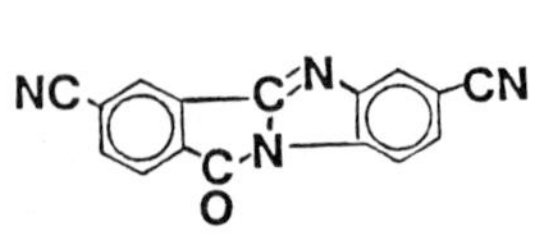
QUINOXALINES	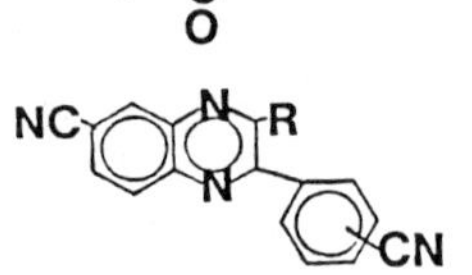

Though the thermal stability data for TSTR are yet unavailable, it is likely that the ultimate stability is determined by the stability of each oligomer. It is difficult to visualize how the TSTR ring can enhance the thermal stability above and beyond that of the oligomer. For mechanical properties, whether the TSTR ring has added to the rigidity or the flexibility (or toughness) of the copolymer still needs to be evaluated. Thus, the study of TSTR's as a high-temperature adhesive needs to be pursued.

Future needs for research on high-temperature polymers should be scrutinized. The following three areas are worth consideration:

1. Synthetic work to improve the processability via better flow prior to solidification.

2. Improved surface treatments of adherends to enhance stability upon high-temperature processing $>232^{o}C$ ($>450^{o}F$).

3. Generation of bond strength data at very high temperatures, e.g., 316°-$650^{o}C$ (600°-$1200^{o}F$) with appropriate composite and metal substrates.

EXTREME LOW TEMPERATURES

For aerospace adhesives and sealants, we need to consider not only extremely high temperatures, but also extremely low temperatures. Adhesives for cryogenic engines should maintain their mechanical performances between -184° and $-196^{o}C$ (-300° to $-320^{o}F$), but this seems to be beyond the limits of polymer capability. To meet this need, an experimental composite or a blend of adhesives in a multilayer form has been developed. Some work has been carried out at Hughes Aircraft, McDonnell Douglas, or NASA-Lewis, but is not very well known to the public.

Recently, an interesting article was published about the behavior of elastomers, such as cross-linked polybutadienes, at cryogenic temperatures (6). Their results are contrary to the common belief that the fracture energy would decrease to nothing at these temperatures. In fact, the fracture energy increases - presumably due to craze formation. If the same result holds true for the rubber-toughened systems, common structural adhesives reinforced with elastomers might survive the extreme low-temperature environments.

For intermediate low temperatures, e.g.,-70°C (-94°F), several polymers (7) have been developed for advanced aerospace sealants:

1. Cyanosilicones (8) ($T_g \sim -51^{\circ}C$ ($-60^{\circ}F$))
   (Products Research Chemical)
   Operative Range: $-51^{\circ}C$ ($-60^{\circ}F$) $\longrightarrow$ $232^{\circ}C$ ($450^{\circ}F$)

2. Fluororalkylarylenesiloxanylene (FASIL) (9)
   (Air Force Materials Laboratory)
   Operative Range: $-54^{\circ}C$ ($-66^{\circ}F$) $\longrightarrow$ $260^{\circ}C$ ($500^{\circ}F$)

3. Phosphonitrilic Fluorelastomer (PNF®) (10)
   (Firestone Rubber and Tire Co.)
   Operative Range: $-68^{\circ}C$ ($-91^{\circ}F$) $\longrightarrow$ $175^{\circ}C$ ($347^{\circ}F$)

4. Flexible polyimides (11)
   (TRW Defense and Space Systems (Troup))
   Operative Range: $-46^{\circ}C$ ($-50^{\circ}F$) $\longrightarrow$ $260^{\circ}C$ ($500^{\circ}F$)

Future research on intermediate low-temperature adhesives can be pursued along the following lines:

1. Phosphazene elastomers (12)

2. Poly(silarylene siloxanes) (13)

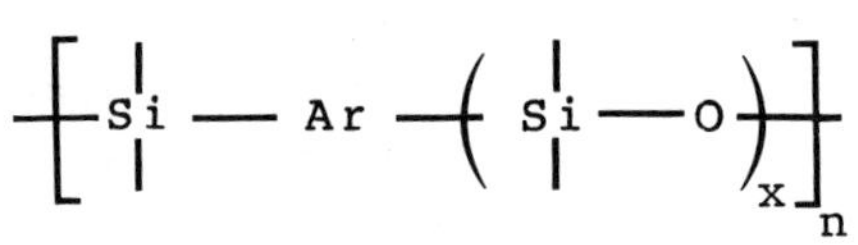

$T_g \simeq +1^{\circ}C$ ($33.8^{\circ}F$) $\longrightarrow$ $-102^{\circ}C$ ($-151.6^{\circ}F$)

3. Fluoroelastomers

EXTREME HIGH HUMIDITY

Besides temperature, extreme high humidity or liquid water immersion (14) is the chief environmental problem for all adhesive bonds. Water can attack three fronts: the adhesive, the interface (or the interphase), or the adherend. Since most adhesives are polar, water can diffuse into the bulk and cause the adhesive to swell and deform. At the interface, water can displace some of the bonds already formed between the adhesive and the adherend. On the adherend surface, water can corrode the metal or weaken the plastic or composite. All in all, water is the worst environmental enemy of adhesive bonds, especially at elevated temperatures. There is no clear-cut solution to the water problem.

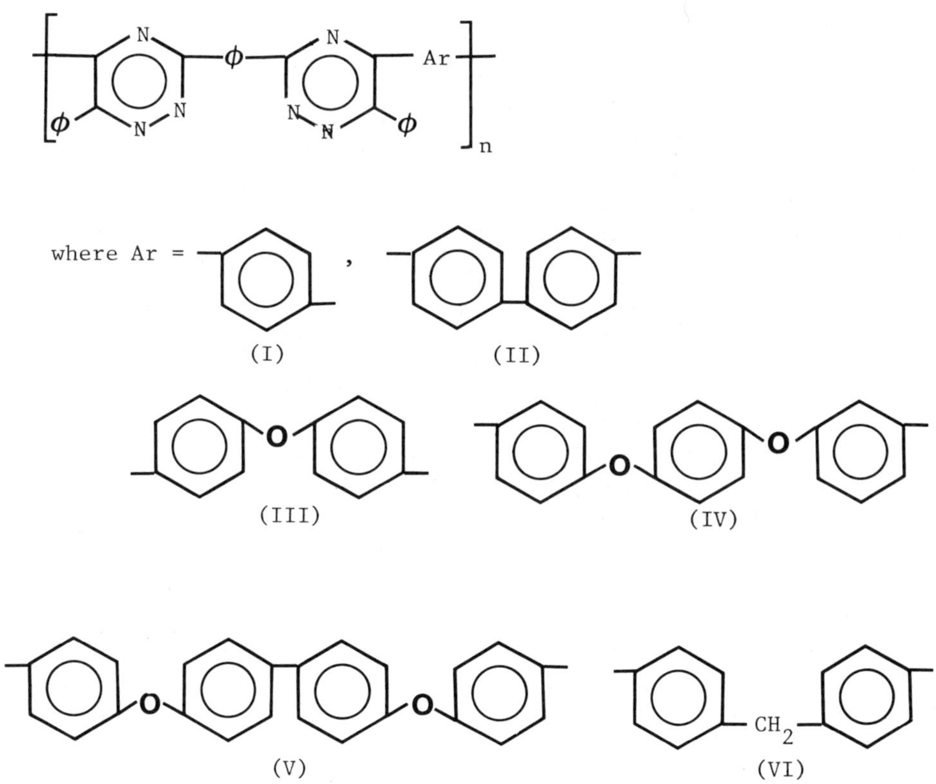

Fig. 2. Structures of Polyphenyl-as-Triazines (18).

However, despite the intensity of the water problem there are some improvements that can be made in each of the three fronts. For adhesives, e.g., epoxy resins or polyimides, the structural modification, such as fluorine-substitution (15) can certainly improve the water resistance. The use of fluoro compounds as curing agents (16,17) has also achieved the same purpose. For high-temperature polymers, some researchers (18) in China recently disclosed that polyphenyl-as-triazines (Fig. 2) from terephthalamidrazone had good water resistance. However, no work has yet been reported on the evaluation of those polymers as adhesives.

For interface, it is well known that silane coupling agents (19), if properly chosen, can generally increase the water resistance or wet strength of the adhesive bonding. For example, by using γ-aminopropyltriethoxysilane (20), a 25% increae in dry strength and a 50% increae in wet strength can be achieved between the 6AL-4V-Ti alloy and epoxy adhesives.

Recently, two coupling agents in the form of chelating agents have been shown to impart remarkable water resistance to the adhesive bonds. The first, Volan 82, a fumarato coordination compound, has been shown by Yang and Zhou (21) to provide unusual water resistance for a polyethylene and aluminum bond. Three years' immersion in water produced no separation of the bond. The second class of chelating agents is based on polyfunctional mercaptoester. These agents have been claimed by DeNicola and Bell (22) to improve water resistance of the epoxy-steel bond.

For adherends, primers can sometimes protect metal surfaces by inhibiting corrosion. For aluminum surfaces a hydration inhibitor (23) nitrilostris(methylene)phosphoric acid (NTMP or $N[CH_2P(O)(OH)_2]_3$), has been recently reported. The presence of NTMP resulted in a four-fold increase in bond durability of adherends etched in FPL over that of untreated adherends. Durability can also be improved by the incorporation in a primer of a corrosion-inhibitor such as strontium chromate. Corrosion inhibiting primers are generally used with bonded components in U.S. Air Force aircraft (24).

Future research to solve the water problem may have to be along some of the following leads:

1. New siloxane copolymers (25) (block or random).
2. New fluoropolymers (15).
3. Silane coupling agents (J. Koenig and H. Ishida, Case Western Reserve University, J. Boerio, University of Cincinnati, and C.S.P. Sung, MIT).
4. New chelating compounds (21,22).
5. Silarylene siloxanes (13).
6. New primers for various adherends.

SALT WATER

Because it is more corrosive than fresh water, salt water not only attacks metals but also degrades even the so-called inert polymers: polyethylene, polypropylene, polystyrene, polymethyl methacrylate, etc. (26). Indeed, salt water is very corrosive to adhesive joints. For example, it was reported (27) that exposure to a 5% salt spray for three months had a more severe effect on aluminum joints (stressed DCB) than exposure to a semi-tropical environment for three years. In Table 1, McMillan (27) compares the severity of salt spray with a natural, semi-tropical environment and various in-service conditions.

Table 1

Relative Joint Strength Loss
Caused by Different Environments

Mild flight-service environment	$<$	Semi-tropical natural weather exposure	$<$	Ground air-ground cycle	$<$	Condensing humidity at elevated temperatures	$<$	Severe flight service environment	$\ll$	5% Salt spray exposure

Deterioration in joint strength generally occurs much faster in salt water than in fresh water. It is likely that salt water attacks through the bondline instead of through diffusion alone as in the case of fresh water.

Minford (28) studied the effect of salt water on stressed aluminum joints. He concluded that:

1. For each alloy studied, the chromic-sulfuric acid etching, Alodining, and anodizing surface pretreatment produced the most durable joints, although the relative order of effectiveness could vary with different alloys.

2. Anodizing pretreatments on aluminum should be ranked as phosphoric acid $>$ chromic acid $>$ sulfuric acid with regards to their bond durability under simultaneous stressing and corrosive, saltwater exposure conditions.

3. Most etching, anodizing, or conversion coating surface preparation will produce joints of superior durability under stress and corrosive, saltwater exposure conditions. Mill-finish, vapor-degreased, or mechanically abraded surface joints performed poorly in comparison, regardless of alloy choice.

One of the most difficult adhesive problems involves the underwater sonar systems. Elastomers (29) (Neoprene rubbers) are adhered to metals with adhesives. The applications include cable sheath, vibration isolator and damper, acoustic absorber, sound reflector, sonar-dome window, and transducer boot. A poor bond at the Neoprene-steel interface could easily lead to saltwater penetration and corrosion, which eventually causes the failure of the sonar dome.

The lifetime for many sonar transducers is required to be ten to fifteen years. The device is constantly under environmental stresses due to saltwater attack, temperature and pressure cycling, plus mechanical stresses. Therefore, it has been very difficult so far to develop an adhesive system that can meet these requirements.

In the future, underwater acoustic system research should be designed to identify the mechanisms of bond degradation and to develop accelerated life testing methods and long-term predictive capability.

FIRE

Unlike water, fire is less a concern with adhesives than with composites. However, flame-resistant adhesives have potential applications in aircraft interior panel structure. In the last two years, some tragic civilian aircraft accidents have prompted the development of flame-resistant adhesives.

In the past, for structural materials, additives (phosphorous or halogen compounds or antimony oxide) were used to enhance flame-resistance. However, the dispersion of additives may be inhomogeneous, and some of these chemicals can generate toxic fumes. Thus, the alternative is to synthesize adhesives containing phosphorous or halogens. The process of incorporating phosphorous may be called phosphorylation. For example, a phosphorylated epoxide (30) (BGDPMPO) has the following structure:

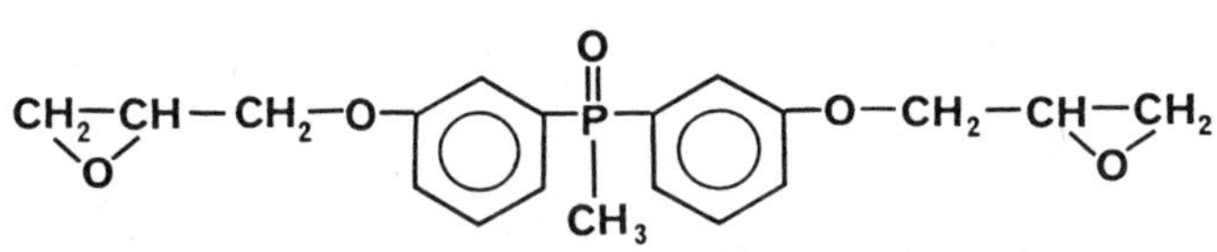

This epoxide has a higher oxygen index than the regular epoxide; thus, it is flame-resistant.

The work on phosphorylated polyimides carried out by Fohlen et al. (31) also appears to be very promising. For example, maleimide derivatives have been prepared from bis(m-aminophenyl) methyl phosphine oxide (I) and tri(m-aminophenyl) phosphine oxide (II).

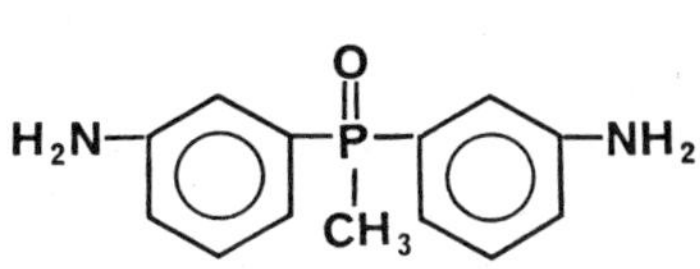

(I)

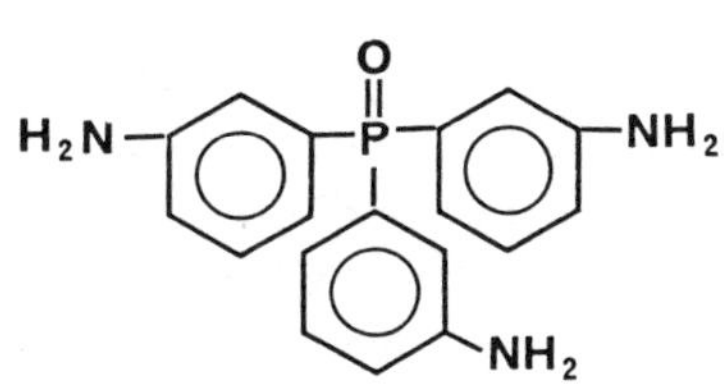

(II)

The maleimides can be cured by an addition reaction in which no volatiles are evolved. One of the remarkable features of the polymers is their high LOI's (limiting oxygen index) of 100. In other words, these materials do not burn in the presence of oxygen, even when the samples are preheated to 300°C.

Phosphazenes and cyclophosphazenes are another promising family of flame-resistant materials. Interestingly, cyclosphazenes (32) can be both fire- and heat-resistant. The polymer obtained from tris(aminophenoxy)-tris(maleimido-phenoxy)cyclotriphosphazene has good thermal stability and high char yield: 82% at 800°C in N_2 and 81% at 700°C in air. Moreover, according to Fohlen, a new synthetic route (31) has been found to make this material at a cost of only $20.00/lb.

Future research on fire-resistant adhesives should be aimed at: (1) phosphazenes and cyclophosphazenes, (2) phosphorous-containing epoxides and polyimides, and (3) nontoxic halogen-containing polymers.

For corrosive gases or liquids, no general rule exists. A chemical can be corrosive to a certain type of adhesive and, yet, may be harmless to another. We do not intend to discuss this topic further.

The last severe environment on our list is external stress. External stresses (14,33) especially affect the water uptake or saltwater penetration of adhesive bonds. However, some adhesives may appear insensitive to stresses. Presumably, the critical stress level (34) is so high that no environmental attacks can take place below that level. Thus, it is difficult to discuss the effect of external stresses without considering the fracture mechanics of adhesive joints. In general, external stresses accelerate bond degradation when the joint is exposed to severe environments.

Many publications have appeared during the last decade that discuss durability of adhesive joints (35,36). It would be inappropriate for us to discuss or predict a long service life for any adhesive in this short paper. However, the materials that we mention in the above sections are designed for long service life. An interesting report by Romanko, Leichti, and Knauss (37) illustrates the integrated methodology for predicting adhesive bond service life under various conditions. We shall not explore this subject any further.

CONCLUSIONS

In general, polymer properties have their limits. If these limitations are known, designers can design around them, and chemists can synthesize new monomers for better and stronger adhesives. Chemists have overcome some of the high temperature problems, but not many of the extreme low temperature ones. Water will remain the major problem, and breakthroughs are not in sight. Because a polymer is a viscoelastic material, its properties are time-dependent. External stresses can shorten their service lives. Thus, for a long service life, a polymeric adhesive should be toughened by rubber reinforcement, by an increase in molecular weight, or by other means.

ACKNOWLEDGEMENTS

The author would like to thank the many researchers who responded to the questionnaire for the National Materials Advisory Board Workshop on "Reliability of Adhesive Bonds Under Severe Environments." Some of their names appear in the references. A summary of this paper, presented to the Workshop held on May 3-4, 1984 at the National Academy of Sciences, will become a part of the Committee Report to be published separately.

REFERENCES

1. L.H. Lee, "Recent Developments in Adhesive and Sealant Chemistry," to be published in Adhesive Chemistry - Developments and Trends, Editor, L.H. Lee, Plenum, New York, 1984.

2. P.M. Hergenrother, "Status of High-Temperature Adhesives," in Ref. 1.

3. T.L. St. Clair, "Adhesive Development at NASA-Langley," presented to the Program Review/Workshop at the Virginia Polytechnic Institute and State University, May 1, 1984.

4. L.C. Hsu and W.H. Philipp, "Cyclopolymerization of Aromatic Nitriles and Polymers with Ring-Chain Structure Made Thereby," presented to the IUPAC Macro Conference, University of Massachusetts, Amherst, MA, July, 1982. Also L.C. Hsu, U.S. Patent 4,159,262 (1979).

5. C. Arnold, Jr., "Stability of High-Temperature Polymers," J. Polym. Sci. Macromole. Rev., 14, 265-378 (1979).

6. R.P. Burford, "Fracture Toughness of Polybutadiene at Cryogenic Temperatures," J. Mater. Sci., 18, 3756 (1983).

7. G.F. Schmitt, Jr., "Spacecraft, Aircraft, and Missiles: Pushing the Limits of Adhesives, Sealants, and Coatings," presented to Case Western Reserve University, January, 1983.

8. H. Singh, "Cyanosiloxane Sealants Used for Space Shuttle Orbiter," Adhesives Age, 21, No. 4, 24 (1978).

9. H. Rosenberg and E.W.Choe, "Synthesis and Properties of Fluoroalkylarylene Siloxanylene (FASIL) High Temperature Polymers," Org. Coatings and Plastics Chem., 40, 793 (1979).

10. "PNF® Elastomer by Firestone," Firestone Rubber and Tire Co., October, 1979.

11. R.J. Jones and H.N. Cassey, "Flexible Polyimide Fuel Tank Sealants," Adhesives Age, 22, No. 11, 33 (1979).

12. R.E. Singler, G.L. Hagnauer and R.W. Sicka, "Phosphazene Elastomers: Synthesis-Properties-Applications, Part II," to be published in the ACS Symposium Series, 1984.

13. N. Koide and R.W. Lenz, "Preparation and Properties of Poly(silarylene siloxane)s," J. Polym. Sci., Polym. Sym. 70, 91 (1983).

14. J. Comyn, "Kinetics and Mechanisms of Environmental Attack," in Durability of Structural Adhesives, Ed., A.J. Kinloch, Applied Science Publishers, London and New York (1983).

15. J.R. Griffith, "Epoxy Resins Containing Fluorine," CHEMTECH, 12, 290, 1982. Also J.R. Griffith, J.G. O'Rear, and J.P. Reardon, "A Fluoro Anhydride Curing Agent for Heavily Fluorinated Epoxy Resins," in Adhesion Science and Technology," L.H. Lee, Ed., 9A, 429, Plenum, NY (1975).

16. P. Johncock and G.F. Tudgey, "Epoxy Systems with Improved Water Resistance, and the Non-Fickian Behavior of Epoxy Systems During Water Ageing," Brit. Polym. J., 15, 14, 1983.

17. D.A. Scola and R.H. Pater, "Bisimide Amine Cured Epoxy Resins," Polym. Preprints, 22, No. 2, 228 (1981).

18. Feng-cai Lu, Lan-min Xing and Yu-lan Wang, "Heat and Water Resistant Polyphenyl-as-Triazines from Terephthalamidazone," Polym. Preprints, 22, No. 1, 62 (1981).

19. E.P. Plueddemann, Silane Coupling Agents, Plenum, New York, 1982.

20. M.E. Schrader and J.A. Cardamone, "Adhesion Promoters for Titanium-Resin Interface," J. Adhesion, 9, 305 (1978).

21. Qi-xiang Yang and Qin-Li Zhou, "The ESCA and AES Studies of the Interfacial Chemical Bonding Between Aluminum and Chromium (III) Fumarato-Coordination Compound," in Ref. 1.

22. J.A. DeNicola, Jr. and J.P. Bell, "Polyfunctional Chelating Agents for Improved Durability of Epoxy Adhesion to Steel," Org. Coatings and Appl. Polym. Sci. Proceedings, 46, 489 (1982).

23. D.A. Hardwick, J.S. Ahearn and J.D. Venables, "Environmental Durability of Aluminum Adhesive Joints Protected with Hydration Inhibitors," J. Mat. Sci., 19, 223 (1984).

24. T.J. Reinhart, Adhesion, II, Ed., K.W. Allen, p. 87, Applied Science Publishers, London (1978).

25. J.E. McGrath, Private Communication, April, 1984.

26. R.G. Donnelly and R.E. Cohen, "Enhancement of The Stability of Common Polymeric Materials Against Undersea Degradation," MTSG 81-14, Massachusetts Institute of Technology, Cambridge, Massachusetts, October, 1981.

27. J.C. McMillan, in Developments in Adhesives - 2, Ed., A.J. Kinloch, p. 243, Applied Science Publishers, London (1981).

28. J.D. Minford, "Effect of Surface Preparations on Stressed Aluminum Joints in Corrosive Saltwater Exposure," Adhesives Age, 23, No. 10, 36 (1980).

29. R.Y. Ting, "Elastomer-to-Metal Adhesion in Sonar Transducer," to be published in Polymer News, 1984.

30. D.A. Kourtides, J.A. Parker, T.W. Grants, N. Bilow and M.T. Hsu, "Thermophysical and Flammability Characterization of Phosphorylated Epoxy Adhesives," Adhesives for Industries, 92, Technology Conferences, El Segundo, California, June 25, 1980.

31. G.M. Fohlen, Private Communication, April 11, 1984.

32. D. Kumar, G.M. Fohlen and J.A. Parker, "Fire- and Heat-Resistant Laminating Resins Based on Maleimido-Substituted Aromatic Cyclotriphosphazenes," Macromolecules, 16, 1250 (1983).

33. J.D. Minford, "Adhesives" in Durability of Structural Adhesives, Ed., A.J. Kinloch, p. 135, Applied Science Publishers, London (1983).

34. B.W. Cherry ad K.W. Thomson, in Adhesion - 1, Ed., K.W. Allen, p. 251, Applied Science Publishers, London (1977).

35. G.P. Krueger and L. Bogne Sandberg, "Durability of Structural Adhesives for Use in the Manufacture of Mobile Homes," Report submitted to Dept. of HUD, March 1, 1979.

36. W. Althof and W. Brockmann, "New Test Methods for the Prediction of Environmental Resistance of Adhesive Bonded Joints," Adhesives Age, 20, No. 9, 27 (1977).

37. J. Romanko, K.M. Liechti and W.G. Knauss, "Integrated Methodology for Adhesive Bonded Joint Life Prediction," AFWAL-TR-82-4139, November, 1982.

Elastomers and Resin Modifiers for Water-Base Adhesives

Marlene J. Jones

Hercules Incorporated
Wilmington, DE 19899

ABSTRACT

A review of the traditional use of resins as tackifiers for elastomers in solvent-applied adhesives is presented. Following is a discussion of how this technology is extended to water-based systems. Data is discussed on the use of various backbone polymer latexes, modified with the proper resin emulsions to achieve pressure-sensitive adhesive properties. Recommendations are also given for the proper choice of resin type to achieve optimum properties with each specific class of elastomer.

INTRODUCTION

For many years pressure-sensitive adhesives have been formulated in and applied from solvent. In the early seventies, however, external pressures began a trend in the industry away from solvent-applied, pressure-sensitive adhesives. This trend is accelerating today. Initially, the reasons were primarily environmental. Solvents emitted to the air are pollutants, explosion hazards are great with many solvents and worker safety based on exposure to these toxic solvents is in doubt. Our local and federal governments began to restrict industry with regard to pollution and worker exposure forcing industry to change. Various choices evolved as potential routes to circumvent solvents. These choices have included: water-based systems, hot melts, 100% reactive systems, and solvent recovery. All of these systems have advantages and disadvantages. However, depending upon the cost and availability of solvents, the cost of energy and ultimately performance, some systems will be more practical than others.

Solvent recovery and recycle has limitations in that it still depends on solvents and requires expensive equipment. If complex mixtures of solvents are used, they must be separated and cleaned up before reuse. Workers may still be exposed to hazards, and insurance rates where solvents are involved are increasing. One-hundred percent solids systems may be the wave of the future, however, at this time the technology is limited. Current hot melt systems have thermal stability problems which inhibit their use in various applications. And though problems and disadvantages remain to be solved with water-based systems, the current technology is practical and presents advantages the other alternatives cannot achieve at this time.

An internally generated survey indicates that by 1987, 40% of the pressure-sensitive adhesive (PSA) market will be supplied by water-based products (Table 1). Solvent-applied systems will represent only a small fraction of the PSA volume growth (Fig. 1).

RESINS IN PRESSURE-SENSITIVE ADHESIVES

In order to see more clearly how the technology has

Table 1. Pressure-Sensitive Adhesives
1974 to 1987
By System - In %

1974	System	1987
95%	Solvent	30%
2%	Hot Melt	30%
3%	Water Based	40%

Total Pounds of Compounded PSA
MM Pounds

1974		1987
147	Polymer	--
98	Resin	--
12	Compounded Materials	--
257	Total Compound	550 to 650

Average G.R. 7-1/2%

changed and where we are today, I will first present a brief history of the traditional use of resins in pressure-sensitive adhesives.

As the PSA industry evolved, natural rubber (NR) and styrene-butadiene rubber (SBR) were the primary elastomers used. Other backbone polymers were available but were used to a lesser degree. These other elastomers include polychloroprene, butyl rubber and nitrile rubber. Traditionally, formulations containing natural rubber have made use of polyterpene resins as tackifiers, particularly beta-pinene resins. The probable structure of a beta-pinene resin is given as follows and represents the terpene class of resins.

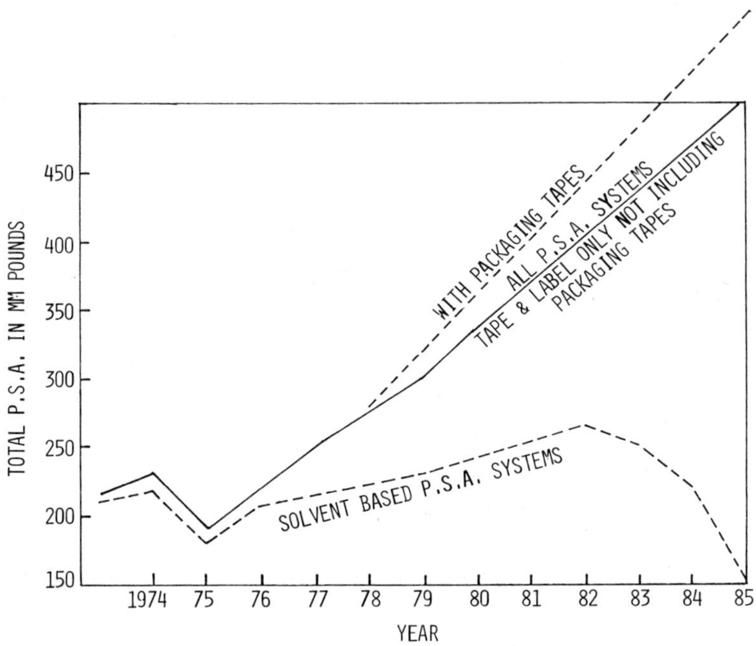

Fig. 1. Total PSA Growth 1973 to 1985

Elastomers have no inherent pressure sensitivity or tack and must be tackified with the appropriate compatible, hard resin to achieve the desired properties. The backbone polymer, however, supplies the strength or cohesiveness of the PSA.

Formulations containing styrene-butadiene rubber have made use of rosin or rosin derivatives as the tackifying resin, particularly rosin ester resins. A typical rosin ester is illustrated as follows:

ROSIN ESTER RESINS

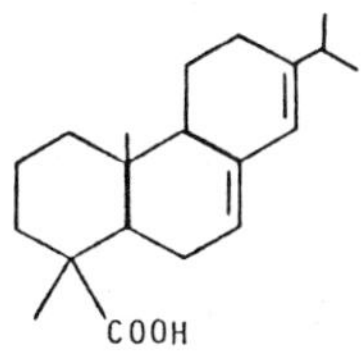

ABIETIC ACID

$$RCOOH + CH_3OH \longrightarrow RCOOCH_3 + H_2O$$

METHYL ESTER OF ROSIN

$$3RCOOH + \begin{matrix} H_2COH \\ | \\ HCOH \\ | \\ H_2COH \end{matrix} \longrightarrow \begin{matrix} H_2COR \\ | \\ HCOR \\ | \\ H_2COR \end{matrix} + 3H_2O$$

GLYCEROL ESTER OF ROSIN

R = THE HYDROCARBON RADICAL FROM ANY ROSIN ACID, HYDROGENATED ROSIN ACID, DEHYDROGENATED OR DISPROPORTIONATED ROSIN ACID.

Often a mixture of resins is used to supply the balance of properties necessary for the application. Formulation latitude has been expanded by the number of resins available to modify the elastomer. Today, even more polymers have become available to serve as the backbone polymer. Some elastomers whose usage has increased are: polyvinyl ethers, polyacrylates and styrene/butadiene or styrene/isoprene block copolymers. The resin population has changed as well and is keeping stride with the developments in elastomers. New modifications of resins and new resins have been produced to fit the compatibility profile established with the older SBR and NR based products.

The use of a class of resins with a particular elastomer has been derived somewhat empirically. The derivation was based on achieving the desired adhesive properties. A degree of compatibility is required between the resin and the elastomer in order to achieve those properties. The mechanism of tackification by resins is not entirely understood. Measurement of tack is a two-part process involving both bonding and debonding. Suffice it to say that the resin modifies the viscoelastic properties of the rubber component in such a way as to allow the adhesive composition to deform at low pressure so that immediate or almost immediate wetting and bonding is achieved when the tape or label is applied to a given substrate. The viscoelastic properties of the composition are also changed to the extent that it resists detachment from the substrate under stress or tension. The compatibility of the resin and elastomer allows these modifications of properties to occur.

An appreciable amount of resin is required to modify the elastomer's rheological properties sufficiently have "good" tack. If one plots resin content versus a specific adhesive property such as tack, the curve increases sharply to a maximum, then decreases sharply to very low levels. The drop in properties occurs when the resin becomes the dominant or continuous phase in a distinctly two-phase system or when the modulus of the composition is sufficiently high to prevent it from wetting the substrate in other systems. The level of resin at the maximum tack value depends upon the solubility and/or compatibility of the particular resin chosen with the particular elastomer used. In solvent-based adhesives this level has been on the order of 60% resin to 40% elastomer.

WATER-BASED ADHESIVES

What of water-based adhesives? We can apply some of the same principles and technology learned from solvent-applied adhesives to water-based systems. First, many of the polymers and elastomers available are the same (Table 2). Not only are these polymers available in latex form, but many are produced in latex form. The solid rubber used in solvent-based adhesives was derived from the latex, so one is actually eliminating a production step. Secondly, many of the resins the formulator has experience with and confidence in working with are available in dispersed form (Table 3). Some of these products have been available for a number of years and others are still in developmental stages.

Table 2. <u>Elastomers Available in Latex Form</u>

Natural Rubber
Styrene-Butadiene Rubber (SBR)
Modified SBR
Polychloroprene Rubber
Modified Polychloroprene
Butyl Rubber
Nitrile Rubber
Polybutadiene Rubber
Acrylics
Vinyl Acetate/Ethylene Copolymers

Table 3. Hercules Developmental Resin Emulsions

Resin	Resin Type	Resin S.P. (°C)
Hercolyn^{R} D	Hydrogenated methyl ester of rosin	Liquid
Cellolyn^{R} 21	Phthalate ester of technical hydroabietyl alcohol	63
Foral^{R} 85	Glycerol ester of highly hydrogenated rosin	82
Staybelite^{R} Ester 10	Glycerol ester of hydrogenated rosin	83
Pentalyn^{R} H	Pentaerythritol ester of hydrogenated rosin.	104
Foral^{R} 105	Pentaerythritol ester of highly hydrogenated rosin. methyl-styrene/styrene	104
Kristalex^{R} 3085	α-methyl-styrene/styrene copolymer resin	85
Piccotex^{R} 75	α-methyl-styrene/vinyl toluene copolymer resin	75
Piccolyte^{R} S-90	Polyterpene (beta)	90
Piccolyte^{R} A-85	Polyterpene (alpha)	85
Piccolyte^{R} C-85	Polyterpene (dipentene)	85
Piccolyte^{R} HM-110	Modified Polyterpene	110
Picco^{R} 6100	Aromatic C_9 Resin	100
Piccovar^{R} AP-25	Aromatic Plasticizer	Liquid
Hercotac^{R} 4085	Modified Aromatic	85
Piccopale^{R} 85	Aliphatic C_5 resin	85

The compositional chemistry of the resins and elastomers has not changed, so the same compatibility should be observed. However, there are additional factors which complicate the situation. Factors such as compatibility of the emulsifier systems must now be taken into consideration. The molecular weight of the polymer which is generally much higher than that found in solvent-based systems, plays a major role. The particulate configurations of the components must also be considered. However, even with all of these complicating factors, in general we can still say that the same types of resins used to tackify specific elastomers in solvent can also be used successfully to tackify those same polymers in water.

Let us now look at some specific resin/rubber combinations in order to illustrate the principles discussed thus far. The data presented were obtained on one-mil adhesives film, using two-mil Mylar film as the substrate.

1. Natural Rubber Latexes

First, we will look at natural rubber as the elastomer. As I mentioned previously, beta-pinene resin has been used extensively to tackify natural rubber in the past. In Figs. 2-4, the effects of a beta-pinene resin emulsion on the properties of a natural rubber latex are demonstrated. In these samples, the ratio of resin to rubber was varied and the properties of probe tack, quick stick, peel adhesion and shear adhesion were measured. The types of curves obtained illustrate the principles of increasing properties sharply to a maximum and then decreasing quickly as the resin becomes the continuous phase. The level of resin where optimum properties are obtained are in the same range as was indicated for solvent-based systems, i.e., approximately 60% resin to 40% elastomer. One rather unexpected result is observed in the shear adhesion values. We would not expect these high shear values in a solvent-based system. Shear adhesion depends largely on the cohesive strength of the composition which is contributed primarily by the elastomer. The use of the latex which has a much higher molecular weight than the milled down rubber samples used in solvents, gives the added advantage of improved shear adhesion. The extent to which this improvement is seen will still depend on the resin and elastomer. However, in general, we can expect to see greater cohesive strengths from water-based systems than those observed with solvent-applied systems.

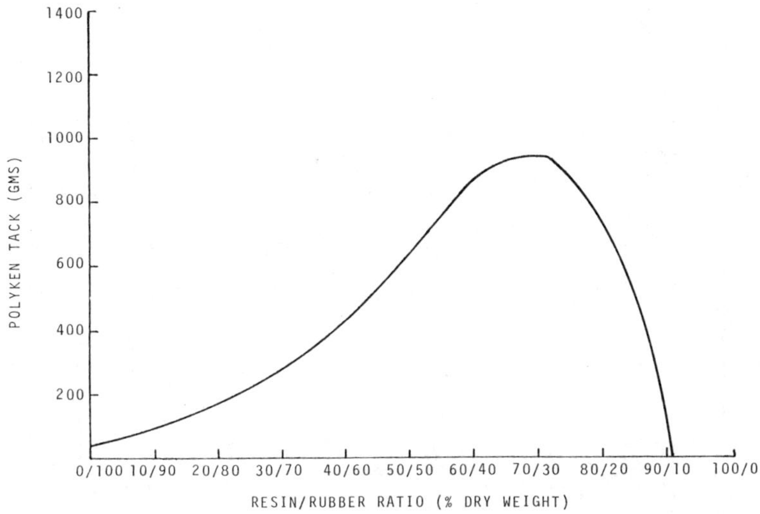

Fig. 2. Polyken Tack of Natural Rubber Latex Tackified with a Beta-Pinene Resin Emulsion

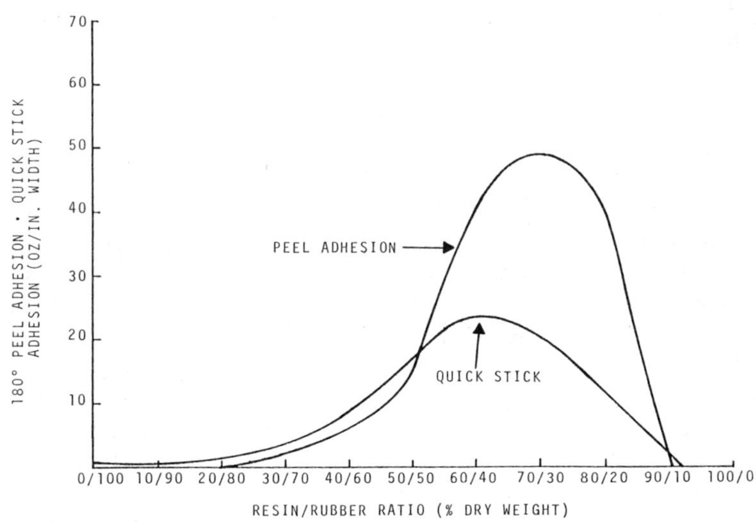

Fig. 3. Quick Stick and Peel Adhesion of Natural Rubber Latex Tackified with a Beta-Pinene Resin Emulsion

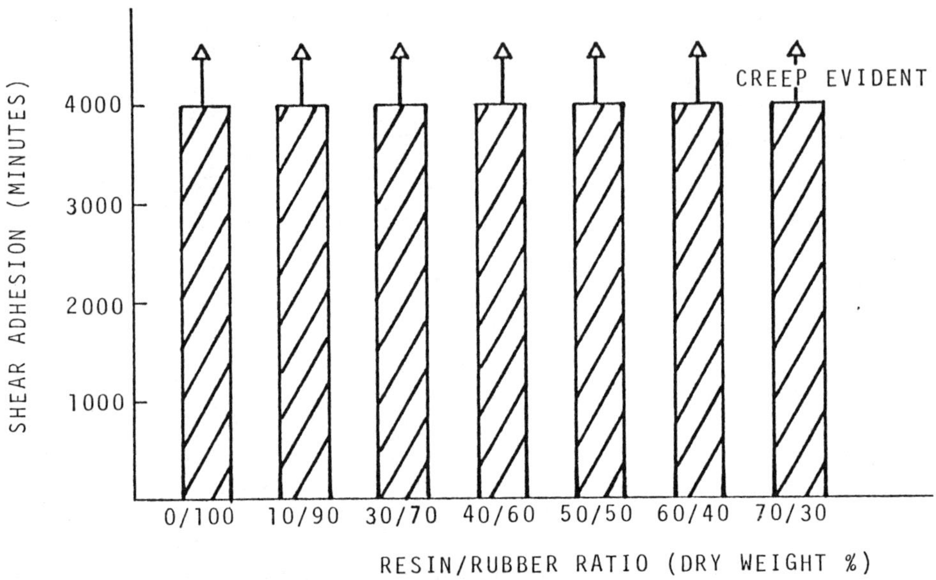

Fig. 4. Shear Adhesion of Natural Rubbrer Latex Tackified with a Beta-Pinene Resin Emulsion

The next series of Figs. 5-7 illustrate the use of a relatively new aromatic modified terpene resin, as a tackifier for natural rubber latex. Again, the properties of probe tack, quick stick, peel and shear adhesion were measured for various rubber/resin ratios. We observe that this resin tackifies the natural rubber latex quite well, yet the curves are substantially different from those generated with the beta-pinene resin emulsion. These differences can be attributed to differences in solubility parameter (caused by compositional differences) and molecular weight and distribution of the two resins.

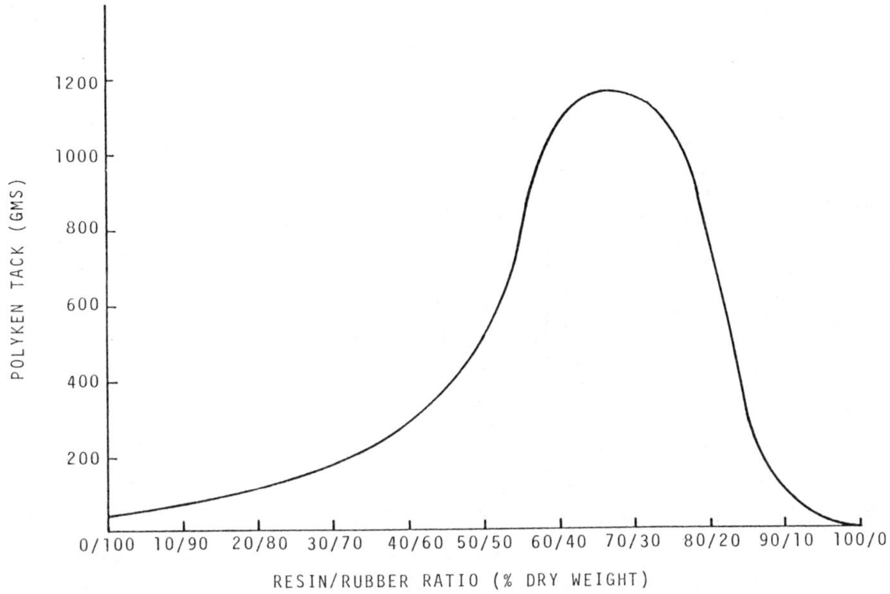

Fig. 5. Polyken Tack of Natural Rubber Latex Tackified with an Aromatic Modified Polyterpene Resin Emulsion

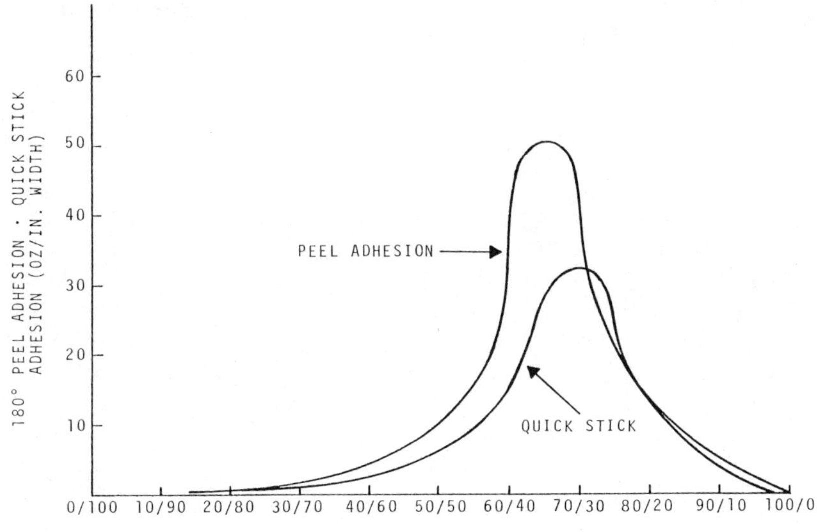

Fig. 6. Quick Stick and Peel Adhesion of Natural Rubber Latex Tackified with an Aromatic Modified Polyterpene Resin Emulsion

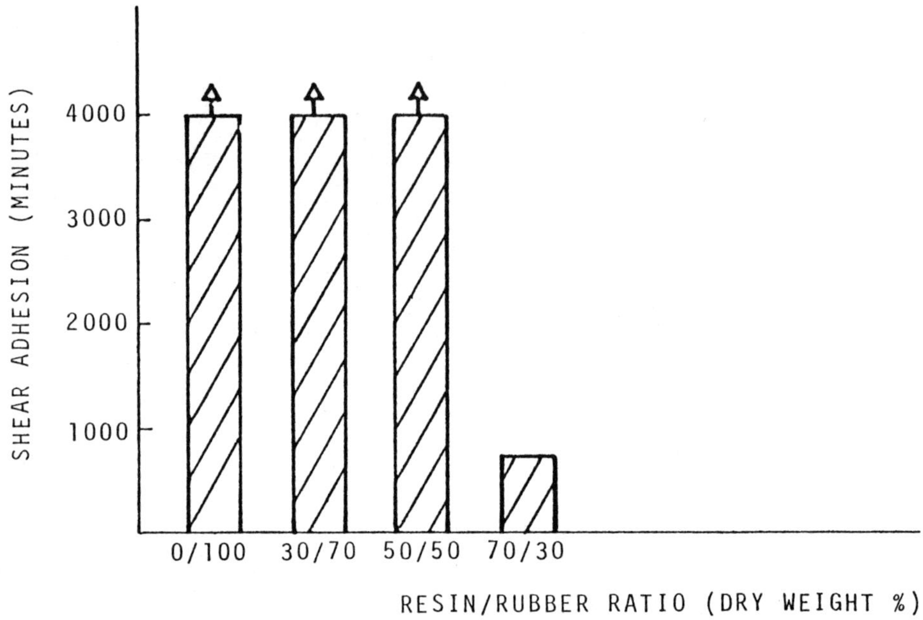

Fig. 7. Shear Adhesion of Natural Rubber Latex Tackified with an Aromatic Modified Polyterpene Resin

These two sets of curves illustrate the importance of examining a broad range of resin/rubber ratios when trying to achieve the proper balance of properties for a given application. For example, the aromatic modified terpene resin gives a higher tack value than the unmodified terpene resin at a ratio of 60% resin to 40% natural rubber. However, more formulation leeway is achieved using the terpene where the properties do not decrease as rapidly as you move away from the optimum concentration.

The shear adhesion values at very high levels (70% or greater) vary between the two resins also, with the modified terpene being lower. Cost/effectiveness will often determine which resin is used. Table 4 illustrates the differences observed between a solvent-applied system and a waterborne system using natural rubber as the elastomer and an alpha-pinene resin of very narrow molecular weight distribution. The numbers in the first three columns were obtained for the waterborne system while those in the last three columns illustrate the values for the same resin with the same natural rubber which was milled to allow its dissolution. Here again, we see the differences primarily in shear adhesion which was pointed out earlier.

While my emphasis in the previous examples has been on the use of polyterpene resins with natural latex, this is not to say that this is the only type of resin which can be used effectively as a tackifier for natural rubber. Table 5 illustrates the use of an emulsion of a highly hydrogenated glycerol ester of rosin with natural rubber latex. The table also demonstrates results obtained with a waterborne system versus the solvent-applied system. In this case, all values, including shear adhesion, fall in the expected ranges. Here we see that the rosin ester resin emulsion tackifies natural rubber so well that the high molecular weight ends can no longer reinforce the composition at a level of 60% resin and 40% rubber. If one desired to use a rosin ester as the tackifier for natural rubber, a lower level of resin would be suggested if the shear adhesion properties are critical.

The general classifications of resins which are recommended for use as tackifiers for natural rubber latex are polyterpene resins, modified polyterpene resins, modified aromatic hydrocarbons, and all rosin ester dispersions.

Table 4. Comparison of Solvent-Based and Water-Based PSA Compositions

Natural Rubber with β-Pinene Resin

	WATER BASED			SOLVENT BASED		
Resin/Rubber Ratio	40/60	50/50	60/40	40/60	50/50	60/40
Polyken Tack (grams)	1250	>1500	>1500	1000	>1500	>1500
90° Quick Stick (oz/in width)	10	30	83	8	25	49
180° Peel Adhesion (oz/in width)	17	40	93	19	40	90
Shear Adhesion	>3000	>3000	>3000	165	105	40

Table 5. Comparison of Solvent-Based and Water-Based PSA Compositions

Natural Rubber with the Glycerin Ester of Highly Hydrogenated Rosin

	WATER BASED	SOLVENT BASED
Polyken Tack (grams)	1031	1313
90° Quick Stick (oz/in width)	51	36
180° Peel Adhesion (oz/in width)	70	70
Shear Adhesion (minutes)	105	165

60/40 Resin Rubber Ratio

7101B

2. Styrene-Butadiene Latexes

Let us now look at the use of styrene-butadiene rubber as the elastomer. SBR is available in latex form covering a wide spectrum of properties. These latexes are generally prepared by one of two processes. The terminology hot or cold SBR stems from these processes. The hot SBR is prepared at a higher polymerization temperature and generally gives rise to a higher bound styrene content and a harder polymer. These SBR's are typically lower molecular weight polymers as compared to SBR's prepared by the cold polymerization process. The variety of polymers available just from these two processes is substantial and does not account for other types of modifications such as terpolymerization. Terpolymers with vinyl pyridine or with carboxylic acid bearing monomers broaden the range further. Though some of these polymers possess some inherent tack, this property is usually obtained by broadening the molecular weight distribution or lowering the average molecular weight at the expense of shear peoperties. The tack possessed by some of these polymers is also not sufficient to cover the wide variety of applications utilizing pressure-sensitive adhesives. The choice of resin is equally important in obtaining desired properties in the case of SBR latexes as it is in the case of natural rubber. Table 6 demonstrates the flexibility of formulation afforded the compounder using an SBR latex with the proper choice of resin. The values on this chart were generated by examining a range of resin/rubber ratios and plotting that data as was demonstrated in previous slides. The values on this chart were chosen to illustrate the variety of properties which can be obtained. While some of these compositions might be better suited for removable labels, others could be utilized as the basis for various types of tape applications.

Table 6. Resin Emulsion/SBR Rubber Latex Combinations

Combination Dry Parts	Polyken Tack (Grams)	90° Quick Stick (ozs./1")	180° Peel Adhesion (ozs./1")	178° Shear Adhesion 1/2" x 1" 1000 gms (min.)
60/40 Phthalate ester of technical hydro-abietylalcohol/SBR Latex-A (Hot Type)	1600	40	70	1200
40/60 Low molecular weight polystyrene/ SBR Latex-D (Hot Type)	1100	32	44	>3000
50/50 Polyterpene (d-limonene)/ SBR Latex-G (Cold Type)	1320	50	66	100
60/40 Phthalate ester of technical hydro-abietyl alcohol/ SBR Latex-H (Cold Type)	1320	20	92	100
50/50 Copolymer of methyl styrene and vinyl styrene/SBR Latex-J (vinylpyridine)	1600	64	92	>3000
50/50 Glycerin ester of highly hydro-genated rosin/SBR Latex-J (vinylpyridine)	1100	26	72	1620

The next series (Figs. 8-10) show the types of properties obtained with two different carboxylated SBR latexes tackified with the same ion-tolerant emulsion of the glycerol ester of highly hydrogenated rosin. Fig. 8 shows the property of probe tack. Both of the carboxylated latexes exhibit some inherent tack properties, but the probe tack is improved even further with increasing levels of the resin. The extent to which the tack is improved is dramatically different depending on which latex is used. Fig. 9 demonstrates the differences even further. Resin modification of Latex B, which is the lower molecular weight polymer,

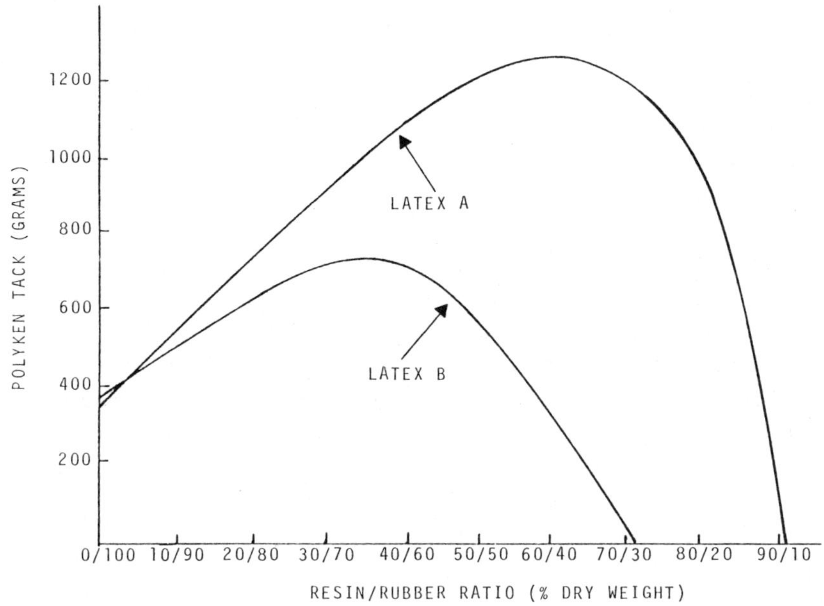

Fig. 8. Polyken Tack of Various Carboxylated SBR Latexes Tackified with an Emulsion of the Glycerol Ester of Highly Hydrogenated Rosin

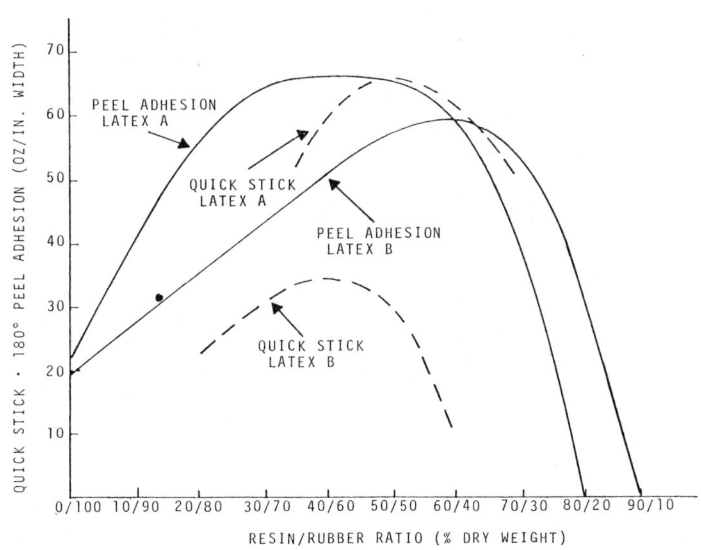

Fig 9. Quick Stick and Peel Adhesion of Various Carboxylated SBR Latexes Tackified with an Emulsion of the Glycerol Ester of Highly Hydrogenated Rosin

does not achieve the same level of properties as resin modification of Latex A. Fig. 10 shows shear adhesion. We could predict the results obtained with this property. The higher molecular weight latex tackified with the proper resin allows excellent shear adhesion values to be obtained at all levels. The samples containing the lower molecular weight polymer have lower values across the board. While these values are useful in some applications, one could not achieve very high cohesive strengths with this latex without cross-linking which is another complicating factor. These examples illustrate the importance of the choice of the proper polymer as well as resin and the importance of understanding the part each plays in achieving adhesive performance.

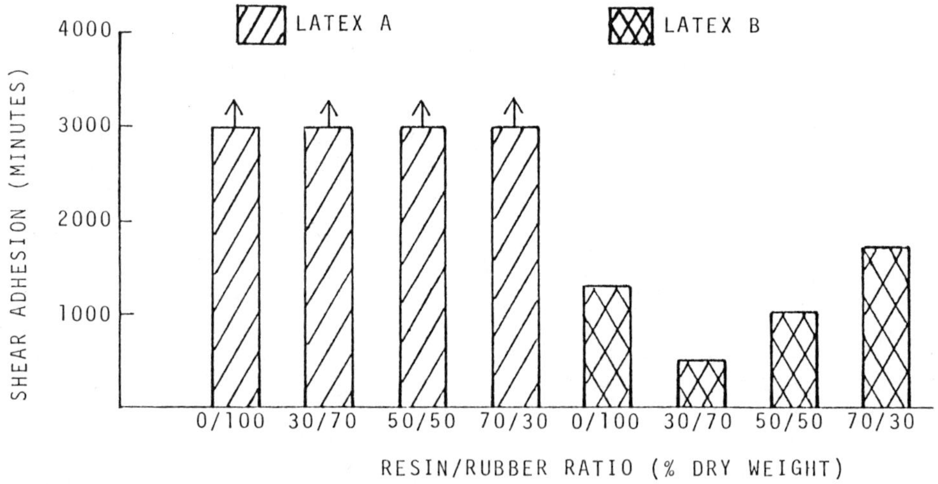

Fig. 10. Shear Adhesion of Various Carboxylated SBR Latexes Tackified with an Emulsion of the Glycerol Ester of Highly Hydrogenated Rosin

Figs. 11 and 12 show two other tackifier resins for carboxylated SBR's. The polymer tackified was Latex A of the previous figures. The two resin emulsions used in these figures are the glycerol ester of hydrogenated rosin and a C_9 hydrocarbon resin (both ion tolerant). Both of the resins can be used to give cost-effective adhesive formulations as the properties plotted in the accompanying figures indicate. Though rosin ester resins perform better as tackifiers for SBR, the modified aromatic resin is substantially compatible with this basic structure and also serves as a tackifier. Table 7 summarizes our recommendations for resins to be used with SBR latexes.

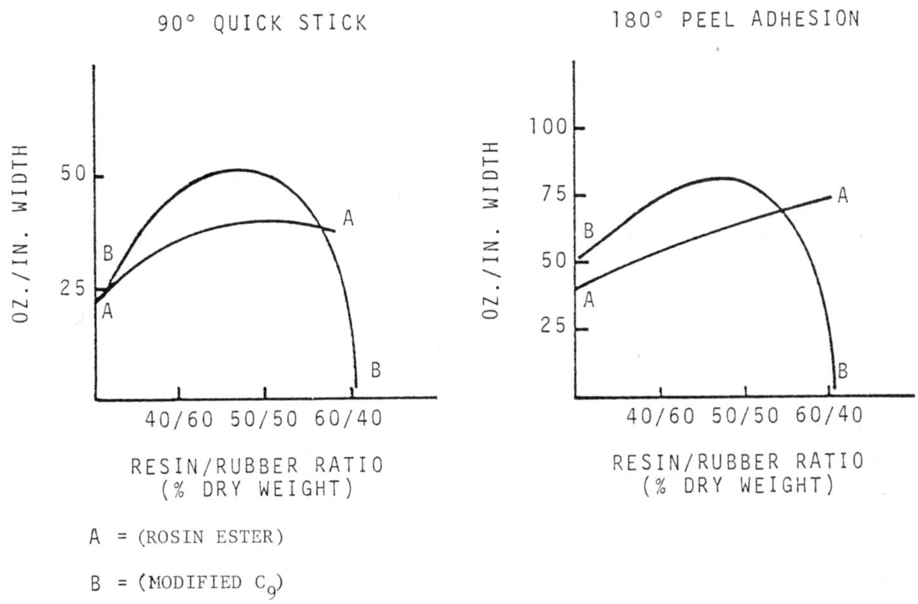

Fig. 11. Quick Stick and Peel Adhesion of a Carboxylated SBR Latex Tackified with an Emulsion of the Glycerol Ester of Hydrogenated Rosin or a Modified C_9 Aromatic Resin Emulsion

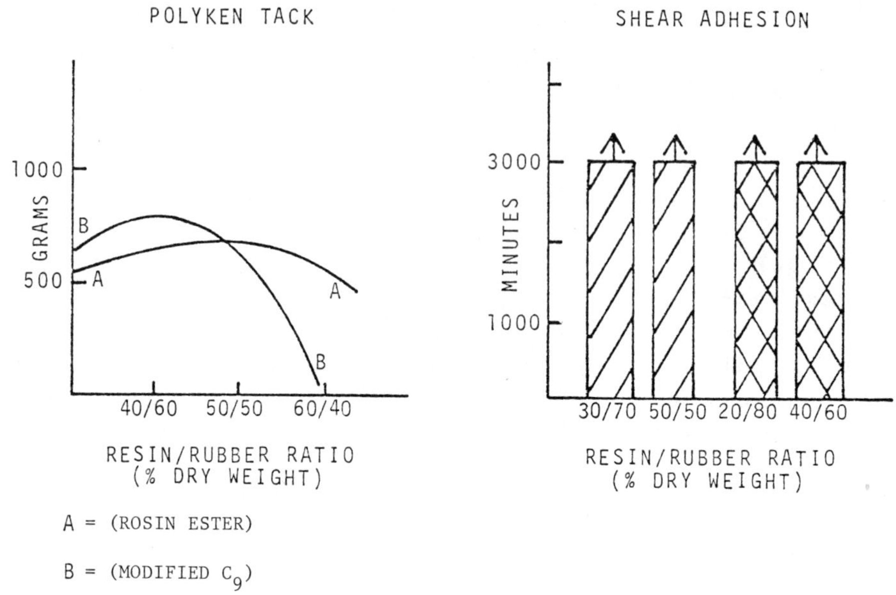

Fig. 12. Polyken Tack and Shear Adhesion of a Carboxylated SBR Latex Tackified with an Emulsion of the Glycerol Ester of Hydrogenated Rosin or a Modified C_9 Aromatic Resin Emulsion

Table 7. Resins Recommended for use with Styrene-Butadiene Latexes - Hot, Cold or Carboxylated *(2)

1. Rosin Esters
2. Alpha-Pinene
3. Beta-Pinene *(1)
4. Aromatic Plasticizers
5. Modified C_9 Aromatics
6. α-Methyl-Styrene/Vinyl Toluene Copolymers *(1)
7. α-Methyl-Styrene/Vinyl Toluene Copolymers *(1)

*(1) Where low R.T. tack is desired, tack will be developed at higher temperatures.

*(2) Most carboxylated lattices need ion-tolerant dispersions to prevent coagulation.

3. Acrylic Polymers

The next series of experimental results will be based on the use of acrylic polymers as the base polymer. A number of acrylic polymers are available for use in pressure-sensitive adhesives. These polymers may be used in solvent-based adhesives, hot melt or water-based adhesives. Their structures vary, though most are polymers of a higher alkyl acrylate such as 2-ethylhexyl acrylate and/or n-butyl acrylate along with small amounts of a polar monomer, such as acrylic acid, vinyl acetate, acrylamide, maleic anhydride or long-chain alkyl acrylamides.

These polymers have typically been used without tackifying resins or with quite low levels of tackifying resin. The rationale for this is that the desired properties can be built into an acrylic polymer by changes in monomer composition, ratios or final polymer molecular weight. However, literally hundreds of acrylic polymers would be required to fill all of the pressure-sensitive adhesive applications, making this a rather undesirable situation for the producers of acrylic polymers. A number of resins have been found to be compatible with acrylic polymers. The properties obtained with combinations of these resins with a number of acrylics exhibit desirable adhesive properties as well as lower cost. In all cases, an ion-tolerant resin emulsion was used to ensure compatibility.

Of particular interest are pure monomer resins. The purity of the monomers results in water-white resins. The use of these resins as tackifiers maintains the advantages of water-whiteness and clarity for which the acrylics are so well known.

Fig. 13 exhibits the type of probe tack which can be obtained using either an emulsion of a copolymer of α-methyl-styrene and vinyl toluene or an emulsion of a copolymer of α-methyl-styrene and styrene with two different acrylic polymers. The two polymers represent significant differences in inherent tack properties, yet both resins show improvements in tack with up to 50 phr resin.

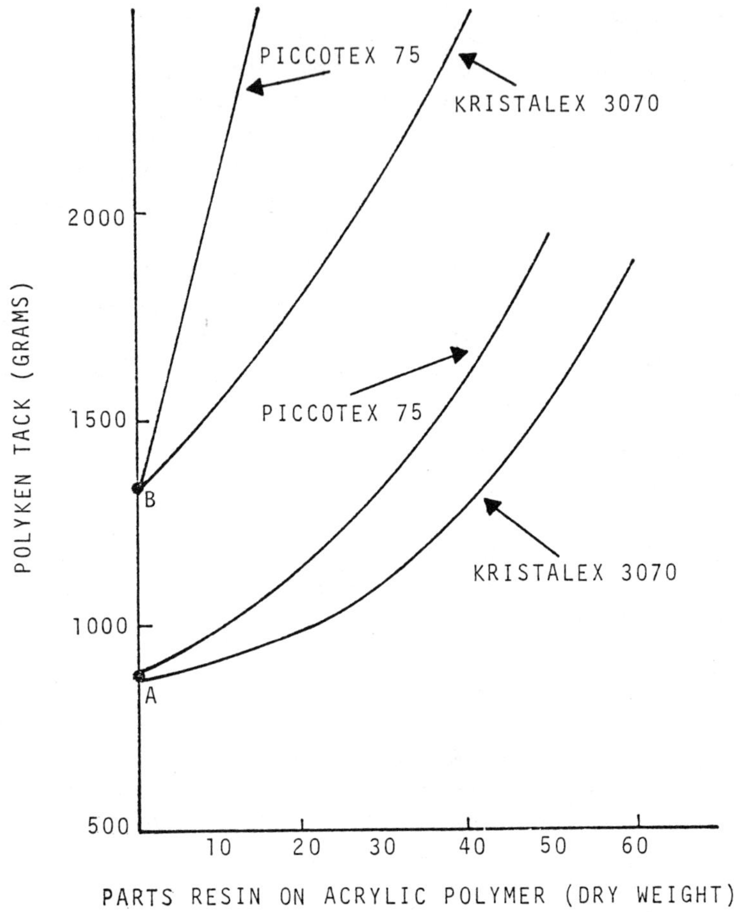

α-METHYL STYRENE/VINYL TOLUENE COPOLYMER (PICCOTEX® 75)

α-METHYL STYRENE/STYRENE COPOLYMER (KRISTALEX® 3085)

Fig. 13. Polyken Tack of Various Acrylic Latexes Tackified with Emulsions of Either an α-Methyl Styrene/Styrene Copolymer or an α-Methyl Styrene/Vinyl Toluene Copolymer

Fig. 14 represents what can be achieved using an emulsion of the copolymer resin of α-methyl-styrene and vinyl toluene with an acrylic having relatively high inherent tack and peel such as Acrylic Latex C. In this case, reinforcement is obtained in shear adhesion.

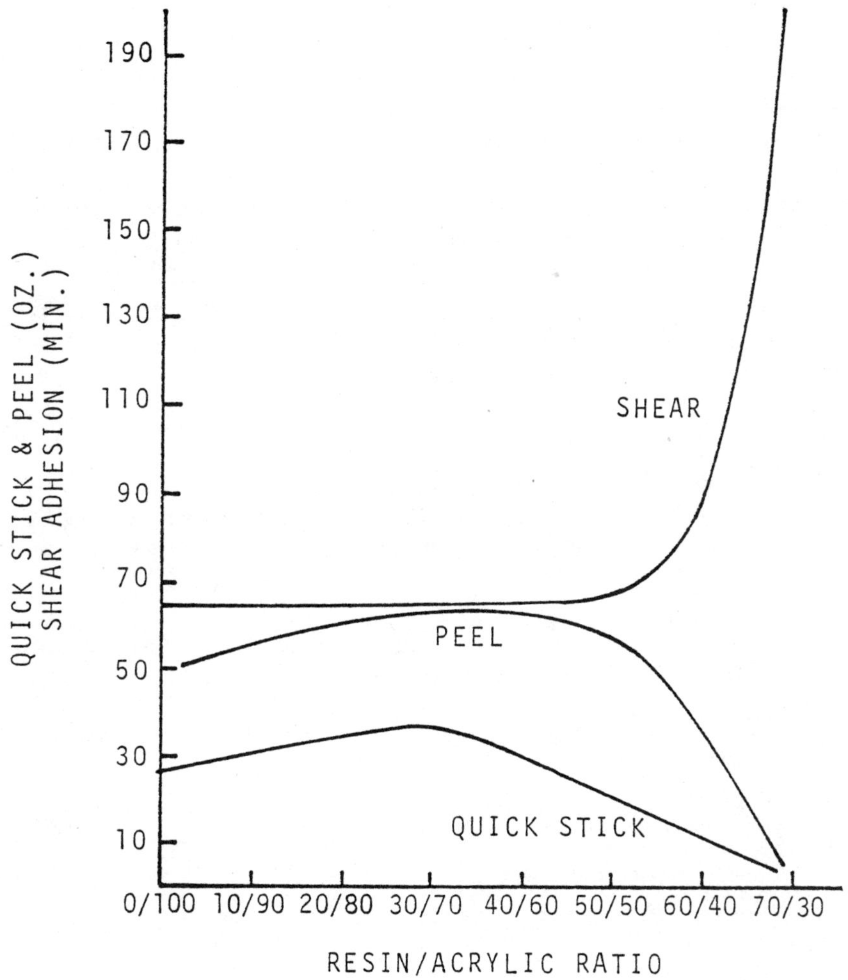

Fig. 14. Quick Stick, Peel and Shrear Adhesion of an Emulsion of an α-Methyl Styrene/Vinyl Toluene Copolymer Resin with Acrylic Latex C

The use of pure monomer resins as tackifiers for acrylics is not limited to any particular type of acrylic as Table 8 demonstrates. This chart gives a wide range of acrylics with different properties, all of which can be effectively tackified with the α-methyl-styrene/vinyl toluene resin. If water-white color is not a requirement, several other resins also improve the versatility of acrylic polymers. Rosin esters are again of particular interest. The glycerol ester of highly hydrogenated rosin and the phthalate ester of hydroabietyl alcohol in emulsion form contributes significantly to probe tack, peel and quick stick with a variety of polymers. Some significant values are given in Table 9. Our recommendations for resins to modify most acrylic polymers are rosin ester resins and pure monomer aromatic copolymer resins.

Table 8. Ion-Tolerant Emulsion of a α-Methyl-Styrene/ Vinyl Toluene Copolymer in Combination with Acrylic Latex

Combination Dry Parts	Polyken Tack (Grams)	90° Quick Stick (oz. in width)	180° Peel Adhesion (oz. in width)	Shear Adhesion (minutes)
40/60 Aromatic Copolymer Resin/ Acrylic Latex D	>1500 (380)	23 (H) (24)	80 (36)	>3000 (>3000)
50/50 Aromatic Copolymer Resin/ Acrylic Latex E	>1500 (1340)	67 (15)	85 (32)	>3000 (670)
50/50 Aromatic Copolymer Resin/ Acrylic Latex F	>1500 (1300)	90 (28)	137 (60)	600 (160)

H (Harsh)

Numbers in parentheses are the properties of the acrylic alone.

Table 9. Tackification of Acrylic Latexes with Emulsions of the Glycerol Ester of Highly Hydrogenated Rosin or the Phthalate Ester of Hydroabietyl Alcohol

Combination Dry Parts	Polyken Tack (Grams)	90° Quick Stick (oz. in width)	180° Peel Adhesion (oz. in width)	Shear Adhesion (minutes)
50/50 Foral[R] 85/ Acrylic Latex G	>1500 (1040)	43 (4)	83 (28)	>3000 (>3000)
50/50 Foral[R] 85/ Acrylic Latex E	>1500 (1340)	20 (15)	62 (32)	3000 (670)
50/50 Cellolyn[R] 21/ Acrylic Latex E	1500 (1340)	68 (15)	80 (32)	1100 (67)
50/50 Cellolyn[R] 21/ Acrylic Latex D	1500 (380)	68 (24)	80 (36)	>1100 (>3000)

Numbers in parentheses are the properties of the acrylic alone.

4. Polychloropene

Polychloroprene has been used as a polymer in contact adhesives for many years. Its use in pressure-sensitive adhesives is not as widespread, however, the availability and properties of this polymer make it another attractive alternative for water-based PSA's. Polychloroprene is available in both conventional uncarboxylated or carboxylated forms.

Table 10 exhibits the use of an ion-tolerant emulsion of the pentaerythritol ester of hydrogenated rosin with both a carboxylated polychloroprene latex and an uncarboxylated polychloroprene latex. Both systems exhibit pressure-sensitive properties which would be of interest in many PSA applications.

Table 10. Ion-Tolerant Emulsion of the Pentaerythritol Ester of Hydrogenated Rosin with Polychloroprene Latex

	Noncarboxylated Polychloroprene		Carboxylated Polychloroprene	
Resin/Rubber Ratio	40/60	50/50	40/60	50/50
90° Quick Stick (Oz/In Width)	55	83	12	27
180° Peel Adhesion (Oz/In Width)	125	>125	35.1	54.2
Shear Adhesion (Minutes)	1750	2400	>4000	>4000

Table 11 compares the glycerol ester of highly hydrogenated rosin with the pentaerythritol ester of hydrogenated rosin as a tackifier for carboxylated polychloroprene latex. In both cases an ion-tolerant emulsion was used. The lower softening point resin, the glycerol ester, gives higher values at higher resin levels. This is probably due to its ability to penetrate the latex particle more readily. The formulator may wish to take advantage of this phenomenon by using a higher level of resin to achieve the properties required for a given end-use.

Table 11. Comparison of Two Different Softening Point Rosin Ester Resin Emulsions as Tackifiers for Carboxylated Polychloroprene

	85°C Softening Point Resin			105°C Softening Point Ester		
Resin/Rubber Ratio	40/60	50/50	60/40	40/60	50/50	60/40
90° Quick Stick (Oz/In Width)	8.1	17.8	36	12	27	18.1 (H)
180° Peel Adhesion (Oz/In Width)	29.2	45	66.1	35.1	54.2	57
Shear Adhesion (Minuites)	>4000	>4000	>4000	>4000	>4000	>4000

H = Harsh

The types of resins generally recommended for use with polychloroprene are depicted in Table 12.

Table 12. Resins Recommended for use with Polychloroprene Latex Dispersions

1. All rosin ester dispersions (ion-tolerant dispersions preferred).

2. Hydrocarbon resins (ion-tolerant dispersions preferred).

   * α-Pinene
   B-Pinene — Polyterpene Resins
   d-limonene

   * Aromatic Plasticizers

   Pure Monomer Copolymer Resins

*Highest Tack

Many other polymers are available in latex form which we have not explored as thoroughly, such as polybutadiene.

5. Miscellaneous

The basic resin and dispersion technology has been developed to meet the tackification needs of the industry as it moves forward to water-based products, no matter what backbone polymer is chosen.

The following examples exhibit some general data obtained on substrates other than Mylar.

Table 13 gives an example which might be considered as a permanent label adhesive. The substrate in this example is Lusterkote label stock. The paper tear observed during the 180° peel test is an indication of the permanency one could expect from the formulation. The shear adhesion of this formulation is not extremely high, but it is as high as some commercial labels we have tested in the lab.

Table 13. Properties of Water-Based PSA Formulation on Lusterkote Label Stock

Formulation

Component	% Dry Weight
Carboxylated SBR	55.5
Aromatic Plasticizer Resin (Ion-tolerant)	16.7
Glycerol Ester of Hydrogenated Rosin (Ion-tolerant)	237.8
	100.0

Properties

Polyken Tack	-	516 (grams)
90° Quick Stick	-	32 (oz/in width)
180° Peel	-	Paper Tear
Shear Adhesion	-	450 (minutes)

The formulations in Table 14 were examined using treated, oriented polypropylene film as the tape substrate. These formulations might be useful as either box closing tapes or in specialty label applications.

Table 14. Properties of Water-Based PSA Formulations on Treated Polypropylene Film

Formulation, % Dry Weight	1	2	3	4
Natural Rubber Latex	40	40	-	-
X Polychloroprene	-	-	50	-
X SBR	-	-	-	40
Aromatic Modified Polyterpene Resin	60	-	-	-
β-Pinene Resin	-	60	-	-
Pentaerythritol Ester of Hydrogenated Rosin	-	-	50	-
Glycerol Ester of Highly Hydrogenated Rosin	-	-	-	20
Aromatic Plasticizer Resin	-	-	-	40
Properties				
90° Quick Stick (oz/in width)	12.9	19.4	83	16.1
180° Peel Adhesion (oz/in width)	19.5	30.3	Film Tear	47
Shear Adhesion (minutes)	12,000	3,000	2,400	2,000

These are but a few examples of the formulation possibilities.

CONCLUSIONS

As we have seen, the technology of water-based, pressure-sensitive adhesives is an outgrowth of the solvent-applied system. While the governing principles are the same, the technology based on waterborne compositions is actually an expansion which allows greater versatility and flexibility in achieving adhesive performance.

REFERENCES

1. D.W. Aubrey and M. Sherriff, "Peel Adhesion and Viscoelasticity of Rubber-Resin Blends", J. Polym. Sci., Chem. Ed., 18, pp. 2597-2608 (1980).

2. R. Bates, "Studies in the Nature of Adhesive Tack", J. App. Polym. Sci., 20, pp. 2941-2954 (1976).

3. Gardner L. Brown, Resins In Rubber, Pennsylvania Industrial Chemical Corp., Clairton, PA (1969).

4. G. Krause, F.B. Jones, O.L. Marrs, and K.W. Rollmann, "Morphology and Viscoelastic Behavior of Styrene-Diene Block Copolymers in Pressure Sensitive Adhesives", J. Adhesion, 8, pp. 235-258 (1977).

5. L.H. Lee, Adhesion Science and Technology, Part A, Plenum Press, New York, NY (1975). "The Chemistry of Tackifying Terpene Resins", pp. 395-412.

6. Thomas Sanderson, "Water-Based, Acrylic Pressure-Sensitive Adhesives", Adhesives Age, 21, No. 12, 31 (1978).

7. G.L. Schneberger, "Polymer Structure and Adhesive Behavior", Adhesives Age, 17, No. 4, 17 (1974).

8. M. Sherriff, R.W. Knibbs, and P.G. Langley, "Mechanism for the Action of Tackifying Resins in Pressure-Sensitive Adhesives", J. Appl. Polym. Sci., 17, pp. 3423-3438 (1973).

9. I. Skeist (ed.), "Pressure Sensitive Tapes & Labels", Handbook of Adhesives, 2nd ed., Van Nostrand Reinhold Company, New York (1977), Chap. 47, pp. 724-735.

10. F.H. Wetzel, "Rosins and Rosin Derivatives in Pressure-Sensitive Adhesives", Rubber Age, (November 1957).

"Aerobic" Acrylic Adhesives – New Technology in Acrylic Adhesives

Andrew G. Bachmann

American Chemical and Engineering Company
51 Greenwoods Road
Torrington, Connecticut 06790

ABSTRACT

Acrylic adhesives provide certain advantages in assembly operations such as fast, room temperature fixtures, the ability to bond a wide variety of surfaces, even oily metal, bond-line flexibility and good environmental and solvent resistance. Aerobic acrylic adhesives, due to their lack of noxious odors, low flammability, and lower toxicity parameters, overcome many of these disadvantages previously associated with acrylic adhesives. In addition, one-component ultraviolet-light curing grades with superior structural properties, and full cure in as little as one second, offer new opportunities in potting, bridge-bonding, wire-tacking, and glass and plastic assembly.

I. SURFACE ACTIVATED CURING ADHESIVES

A new type of nonvolatile reactive acrylic adhesive, bridging the gap between anaerobic acrylic and volatile reactive acrylic adhesives, has recently been developed. The term "aerobic" acrylic adhesives has been coined solely to identify and to set apart these adhesives from other acrylics; it does not necessarily describe their cure mechanism. Their term "aerobic" will be used to refer to a diminished sensitivity to air inhibition of thick layer curing properties and the ability to cure between two surfaces regardless of the presence or the absence of air.

Aerobic acrylics will not cure by themselves due to the absence of air. They require the use of pre-applied activators to initiate the cure mechanism. This property is distinct from anaerobic adhesives which are intrinsically single-component products. Even when accelerators or primers are used to increase their cure rates to meet the demands of rapid assembly techniques, anaerobic adhesives require the absence of air as a necessary condition of cure (1).

1. Types of Acrylic Adhesives

The types of adhesives commonly considered to be in the acrylic class are anaerobic, cyanoacrylate, and reactive acrylics. The differences in the types of acrylic adhesives that lead to their suitability for different end-uses lies in the initiation and the mechanism of polymerization. All but the cyanoacrylates polymerize by a free-radical mechanism (2).

The growing polymer chain can be terminated by a number of mechanisms including two radicals joining to form a molecular species, abstraction of hydrogen, or radical transfer. Oxygen is particularly effective in terminating free-radical polymerization and is responsible for the familiar air inhibition property (wet or tacky surface) noted by users of acrylic adhesives. The absence of oxygen or air is, therefore, required in order to obtain the most complete cure (polymerization) and hence, the best properties of acrylic adhesives.

Cyanoacrylates

Cyanoacrylates polymerize by anionic mechanism initiated by moisture or basic ions. The polymers formed tend to be more brittle than those formed from other acrylics, and the bond-lines are usually not resistant to degradation by moisture. Adhesive formulations are single-component, fast-curing products, however, and suited to many hard-to-bond surfaces such as rubber and many thermoplastics.

Anaerobics

Anaerobic adhesives take extreme advantage of the air inhibition of acrylic polymerization. They are catalyzed in such a manner that they depend upon the presence of oxygen to remain an unpolymerized liquid. Cure is affected only in the absence of oxygen (1).

Anaerobic adhesives cure as single-component products only on certain "active" metal surfaces when ions from the surface act as accelerators for the rate of adhesive cure. Single-component, anaerobic threadlocking adhesives are examples, deriving their speed of cure from both the anaerobic catalysts and the accelerating effect of transition metal ions derived from copper, iron, or steel surfaces.

For most other surfaces, however, accelerators of various kinds similar to the activators of two-component acrylics are often required to achieve cure rates of practical lengths. Structural anaerobic adhesives cured without the use of pre-applied accelerators or primers frequently cure too slowly for rapid assembly techniques (3).

Most anaerobic acrylic adhesives require surfaces that are clean, free from dirt or grease, and usually mechanically abraded (4).

Reactive Acrylics

Reactive acrylics differ significantly from anaerobics in that they depend upon the presence of a chemical initiator, not the absence of oxygen, for their cure. These adhesives exhibit a high solvent action allowing for the bonding of many unprepared surfaces, even oil-contaminated metals. Less surface preparation is required for bonding plastics with reactive acrylics than with most other adhesives. In part, this is due to the presence of monomers in the reactive acrylic formulations that may dissolve contaminants and may also attack the polymer substrate (3).

Compositions are suggested in the patent literature (4-8). The concept is essentially, that an elastomer is colloidally dispersed in a monomer or in a monomer/-oligomer/polymer solution. The resultant polymer is toughened by "elastomeric domains" resulting from the dispersed elastomer.

A reactive adhesive is used by applying the adhesive (polymer solution) to one of the surfaces to be bonded; and the activator, frequently in solution, is applied in a thin film to the opposing surface. In some types of reactive adhesives, the two films may be left open and unbonded for many hours without adverse effects on the resulting bond (9).

"2nd-Generation" Acrylic Adhesives

In the mid-seventies, spurred by developments from the DuPont Corporation, a new kind of two-component acrylic adhesive system was introduced to the joining industry. The DuPont technology has been widely licensed and is now referred to by numerous terms such as "second-generation acrylics", "reactive adhesives", "modified acrylics", "toughened acrylics", etc. These adhesives seemed to answer the need for improvements over some of the major disadvantages found in anaerobic structural adhesives, namely those of cost, the need for scrupulous surface preparation, and the ability to bond plastic surfaces.

Reactive acrylic adhesives do not require the long-term or sophisticated fixturing techniques usually required for epoxies and urethanes. They do not require the application of heat, nor is scrupulous surface preparation required in the majority of instances.

In the aerospace, aircraft, automotive, electrical, instrument, construction and marine industries, among others, modified acrylics offer distinct assembly advantages over high performance epoxies and urethanes, as well as anaerobic structural adhesives and cyanoacrylates.

In spite of this impressive list of characteristics, the use of reactive acrylic adhesives has been limited to selected high-performance adhesive markets. Their growth, otherwise, has been less rapid than expected (10). One of the principal reasons is due to the toxicity characteristics associated with methyl methacrylate and methacrylic acid monomers, major formulation constituents of "second-generation" acrylic adhesives, with regard to inhalation, ingestion and skin contact. Compositions containing large quantities of these monomers have been rejected on certain production lines such as automotive assembly. The low flash point and resultant flammability hazard have also been important contributors to the restricted use of modified acrylic adhesives (9 and 11).

Moreover, preferred activators for some reactive acrylics, dimethyl aniline and generically similar materials, are suspected carcinogens. Other reactive adhesives use Schiff-base activators, which are considered to be less toxic (9).

Recently, technique or "use" problems have been cited as inhibitions to the growth of reactive acrylic adhesives. The main problems are described as a high sensitivity to the concentration of activator and movement in the substrate prior to fixturing, both of which cause less than optimal

Table 1. Advantages of Modified Acrylic Adhesives

- Simple no-mix application
- Broad applicability to a wide variety of substrates (metal, plastic) with minimal surface preparation
- Tolerance for oil-contaminated surfaces
- Rapid bonding at room temperature
- Excellent bondline flexibility
- High peel strength, lap shear strength, and impact resistance
- Excellent solvent resistance
- Wide latitude in cure rate
- Ability to bond heat sensitive substrates
- Low shrinkage during polymerization
- Excellent reproducibility and reliability
- Good low and high temperature properties
- Wide formulating flexibility for specific end-use requirements

bond strength. Of course, disturbance of parts during fixturing will have the effect of reducing bond strength of any adhesive (10).

This article examines a new reactive acrylic adhesive that overcomes most of these disadvantages. This new family of adhesives, called aerobic acrylics, maintains, and in some cases improves upon, the performance advantages of second-generation acrylics. The products are low in odor, essentially nonflammable, and lower in toxicity. Their higher speed of cure and fixture rate of 10 to 60 seconds and their solvent-diluted activators have minimized the other problems associated with acrylic adhesives; namely, sensitivity to high concentrations of activator and movements in the substrates prior to fixture.

Aerobic Acrylic Adhesives

Aerobic acrylic adhesives are composed or proprietary catalysts, "elastomeric domain" fillers, and low vapor-pressure monomers. Formulations derived from this technology do not exhibit the same sensitivity toward air inhibition shown by other acrylic adhesives. The result is that aerobic acrylics are useable on more porous surfaces and in wider gaps than previously considered practical.

Some of the physical properties that are recognized as being affected by air inhibition mechanisms are compared in Table 2. Cure through a gap is considered a function of air inhibition, as well as diffusion of accelerators and the nature of monomers used in the adhesive formulation. Air beaten into several adhesive formulations and cured for only part of the recommended optimum time shows the effect of this type of inhibition. The aerobic acrylic appears to be only marginally affected.

An adhesive fillet, or the amount of excess adhesive squeezed outside of a lap shear joint on assembly, can also be used as a measure of air inhibition. The less the fillet cure, the greater the inhibition. A porous material, pine wood, was chosen as a substrate that would allow air to remain in intimate contact with the curing adhesive. The ability to lock the threads of an iron nut and bolt without use of an activator is a recognized test for determining whether or not a formulation is capable of anaerobic cure (1,4,5).

Table 2. Effects of Air Inhibition

Tensile shear Steel, psi (ASTM D-1002) 24 hour cure	Anaerobic Structural Acrylic	2nd Generation Acrylic	Aerobic Acrylic
2 mils	4,500	5,000	4,000
20 mils	2,500	3,200	3,500
60 mils	-0-	1,000	2,500
20 mils (air beaten into adhesive)	1,000	2,000	3,000
Length of fillet cure from the edge of the joint, steel, inches	.05	.15	.25
Cure on pine wood, 40 mil gap	500 psi	partial substrate failure	substrate failure
Time to fixture iron nut and bolt at 75°F. No accelerator used	2-6 hr.	4-24 hr.	no fixture after 30 days; test discontinued

2. Properties and Performance

Today's increasingly severe regulatory constraints on industries' use of chemicals make the properties of uncured adhesive systems of paramount concern to the end-user. The methyl methacrylate base of most of the second-generation acrylic adhesives introduces significant plant handling problems that are not present in aerobic acrylic adhesives. Table 3 also presents a summary of curing information of second-generation and aerobic acrylic adhesives, important factors to be considered in the selection of adhesives for assembly. While both second-generation and aerobic acrylic adhesives are classified as 100% solids non-solvent systems, the volatile base of the former can also cause loss of strength due to evaporation during "open time." A slowing of the fixture rate, but not loss of the ultimate strength, is noted after long "open times" with aerobic acrylics.

Table 3. Typical* Properties of Acrylic Adhesives

	2nd Generation	Aerobic
Toxicity of monomers	moderate to high	low
Odor	substantial	faint
Flash point	45°-60°F	over 200°F
Fixture time		
2 mil gap	5-12 min.	10-60 sec.
20 mil gap	10-20 min.	5 min.
50% strength	4-8 hr.	15-90 min.
100% strength	24 hr.	4-24 hr.
Effect of "open time"		
30 sec.	up to 25% strength loss	no effect
5 min.	up to 50% strength loss	no effect
24 hr.	no bond	10% slower fixture rate
Maximum gap tolerance, mills	30-40	60-80
Tensile shear		
Degreased, sand-blasted steel	to 5,000 psi	to 4,500 psi
Oily steel (as received)	to 4,000 psi	to 3,500 psi
Torsional impact		
Automotive steel, aluminum	to 20 in. lb.	to 60 in. lb. (limit of test)
Heat resistance	to 250°F	to 450°F
Solvent resistance	good	excellent
Creep resistance	fair	excellent

*Properties vary depending upon grade of adhesive chosen and ambient temperature.

Although acrylics are slightly less versatile with respect to bondable substrates than are modified acrylics, they are more versatile than anaerobic adhesives, which are usually limited to the assembly of cleaned or prepared, closely mating metal, glass, and ceramic parts (3).

Bondable substrates for aerobic and second-generation acrylics include clean as well as oily or "as received" metal polyurethanes, filled polycarbonates, filled nylon, phenolics, thermosets, wood, concrete, ceramics, glass, and other common materials of manufacture. Some thermoplastic materials are bonded better by using a second-generation acrylic adhesive. Examples are unfilled nylon, ABS, and rubber.

Table 3 also provides a comparison of cured properties of aerobic with second-generation acrylic adhesives. The data shows that, while tensile strengths are similar, the toughness and durability parameters are superior for the aerobics.

A torsional impact test, developed by the GMC Technical Center, Warren, Mich., was used to compare the relative toughness and impact resistance of structural adhesives. In this test, a 3.6-lb. weight falls in a 20-inch, 90°-arc to impact a one-half inch overlap bonded lap shear positioned at a right angle to the bond. The impact scale reads directly in inch-pounds of force needed to break the bond. Sixty inch-pounds is the limit of the test. Steel and aluminum lap shears severely deform near this limit. Table 4 illustrates the typical values obtained.

In cured and curing properties, aerobic acrylic adhesives have a clear advantage for use as a structural adhesive.

The cost per pound of adhesives is an important, though not a limiting, consideration on the choice of an acrylic adhesive. The cost of assembly usually far outweighs adhesive cost per pound. Generally, cyanoacrylates cost the most per pound, followed by anaerobics, aerobics, and then second-generation acrylics.

3. Bonding Methods

The bonding techniques for aerobic acrylic adhesives are similar to other acrylics. First, the activator is applied to one of the surfaces to be joined as a thin film (quasi monomolecular layer). Roller, brush, and spray methods work equally well as long as the activator layer is kept very thin. While the permissible ratio of activator to adhesive is quite wide, too much activator will lower bond strength. Activators are available in solvents to help

Table 4. Typical Torsional Impact*

Adhesive	Values (in-lb)
Polyamide epoxy	7 + 5
Cyanoacrylate	2 ± 2
Anaerobic	12 ± 5
"2nd-Generation acrylic"	20 ± 7
General purpose aerobic acrylic	40 ± 10 (metal deformation)
High impact aerobic acrylic	60 +

*On degreased, sandblasted steel, 1/2 inch overlap.

assure thin film applications. Second, the adhesive is applied to the mating surface, and the two are brought together and held until handling time is reached. This is the preferred technique for bonding motor magnets, metal cladding and wood veneers, stepping motors, golf clubs and other objects of similar geometries.

Application of activator to both substrates is recommended where one substrate is porous or where gaps more than 20 mils are to be filled. Activator and adhesive may be applied to the same surface, but this initiates the bonding action so that the parts must be joined immediately. For example, in automobile hem bonding, adhesive is applied in "chocolate drops" and activator is sprayed over the drops. The hem is formed immediately, squeezing activator and adhesive into intimate contact.

4. Aerobic Applications

Aerobic adhesives have a wider latitude in formulating for specific end-use requirements than do second-generation acrylic adhesives. Flexibility, opacity, cure speed, and surface compatibility may be controlled over a wide range to produce adhesives for metal, glass, many plastics, ceramics, and hard woods. The cured properties of specific aerobic formulations are scarcely affected by efficient thickening agents such as fumed silica. As a consequence, it is possible for an end-user to pick a set of cured properties and then have the adhesive's viscosity modified to specific requirements. The properties of three representative cured formulations are outlined in Table 5.

Table 5. Effect of Aerobic Formulation Variations

	A	B	C
Fixture rate on glass*	5 min.	25 sec.	7 sec.
Complete room temperature cure**	24-72 hr.	4-24 hr.	2-8 hr.
Bondline appearance	white, hazy	white, hazy	transparent
Gap filling	70 mils	50 mils	30 mils
Shore D hardness	50	75	10
Texture	tough, resiliant	rigid	very flexible
Typical end-uses	ferrite bonding, shim bonding, fiberglass assembly	sheet metal, galvanized steel, Kevlar, graphite, glass composites	lens bonding "plexiglass" window mounts
Tensile, cold rolled steel	2,000 psi	3,500 psi	1,800 psi

* Fixture rates are defined as the time required for two microscope slides, bonded in a 1/2-inch overlap, to resist movement from light finger pressure. Gaps estimated at 1-3 mils.

**Varies with ambient temperature and condition of the substrate.

Formulation "A" is frequently used when a substantial amount of time is required for alignment and adjustments during assembly operations. A typical end-use is the production of precision optical gauges. The parts are in the form of concentric circles of glass and stainless steel. They must be carefully rotated and adjusted to exact tolerances before the adhesive fixtures the parts.

A manufacturer of high speed printers also requires five minutes of assembly time before fixture occurs in a 10-mil shimmed bondline gap. In this application, a number of ferrite magnets are bonded to metal frames and assembled in the proper geometrical arrangement in one step. This adhesive process replaces a combination of mechanical clamps, spacers, and holding screws.

High levels of noxious odors and concerns over worker safety have resulted in acrylic adhesives being removed from consideration in automotive hem bonding plans (9). They were rejected despite a five-year development program that showed that replacing spot welding with a rapidly fixturing adhesive would substantially lower hood and lid assembly costs. However, an aerobic acrylic similar to the "B" formulation shown in Table 5 has been approved for trial production because of its low odor and nonflammable properties.

Aerobic acrylics also show enhanced cure speed, tensile strength, moisture resistance, and the ability to resist automotive bake cycles, which can be as high as 450°F. The operation also requires fixturing through a variable 10 to 20-mil gap in less than 30 seconds. Thixotropic adhesive is applied to the area to be bonded and activator sprayed directly on top of the adhesive. The metal is immediately folded to form a "hem" joint.

A photocopier manufacturer uses an aerobic adhesive to attach large glass lenses to metal frames. The application requires an adhesive that can absorb the forces generated from the large differences in expansion characteristics between glass and metal. Type "B" formulated adhesives were shown to provide adequate bonds with tensile strength exceeding the deformation strength of the metal frames. However, when subjected to a 300°F temperature range, the glass lenses frequently broke. Increasing the bondline thickness to 10 mils and using a flexible type "C" formulation produced shock absorbing bonds and the lenses now pass the thermal cycle requirements.

An aerobic acrylic adhesive has also been used to increase both product performance and reliability while reducing assembly costs in the manufacture of motorcycle alternators. A major manufacturer of high performance motorcycles had been holding ceramic magnets in place with a plastic retaining ring. The manufacturer had to increase the alternator's output from 17 to 22.5 Amps at 3600 rpm without increasing the size of the alternator. It was known that replacing the retaining ring with a bonding process would permit more magnet mass in the same space.

A number of attempts with anaerobic adhesives, epoxies, urethanes, and second-generation acrylic adhesives had proved unsuccessful. The bonded alternator had to withstand a -50° to $+360^{o}F$ operating temperature range and the rigors of constant vibration. The nature of the production process required rapid fixture times and could not tolerate noxious or toxic vapors. A type "B" formulation aerobic acrylic adhesive is now used because it meets necessary production and performance criteria.

Medallions and handles are mechanically attached to metal knife frames by combination of welding, riveting and staking. Several years ago, reactive acrylic adhesives were used to replace riveting in medallions with a resulting 65% savings on assembly costs. Controlling vapors in the labor-intensive assembly plant and 15-minute fixture times limited this use to a small part of the total assembly. Ease of handling and lack of noxious vapors have led to additional savings.

A fuse manufacturer replaced a 30-minute epoxy used in a heat-cure process to assemble stainless steel inserts into phenolic holders. By utilizing an aerobic adhesive bonding process, the assembly was complete in 30 seconds at room temperature. The resulting increase in productivity more than offset the higher adhesive cost. A cyanoacrylate bonding process was even faster, but did not resist aging in a moist atmospheric environment or pass a drop test for impact resistance.

Many kinds of adhesives are utilized in the bonding of loudspeaker magnets, DC motor assemblies, and other types of magnet bonding. Some manufacturers still prefer to use mechanical clamps. The factors that contribute to the choice of a joining process include the cost of the adhesive, labor rates, speed of adhesive fixture, acceptability of odor level, and energy costs. Aerobic acrylic adhesives have been used primarily for their rapid fixture (as little as 10 seconds) resulting in faster assembly rates.

A manufacturer of restaurant appliances fastens a stainless steel top to a combined counter/refrigerator/stove unit. The standard attachment technique included both welding and the use of self-tapping screws. An aerobic acrylic adhesive has replaced these techniques resulting in a lower cost assembly requiring less skilled labor. The same adhesive is used to tack refrigerator coils in place.

A similar application involved the use of an aerobic adhesive to assemble large insulated aluminum doors for storage rooms. The replacing of a difficult aluminum welding process led to more efficient techniques.

5. Conclusion

Acrylic structural adhesives have been specified in a number of applications due to their ability to increase productivity through rapid cure while providing equal or better performance than mechanical means of fastening. Second-generation reactive acrylics overcome the limitations imposed by scrupulous cleaning or preparation of surfaces prior to bonding and the tendency toward requiring small, thin bondlines. Their use has been limited, however, by concerns arising from the use of potentially toxic, flammable, noxious monomers.

Aerobic acrylic adhesives have most of the advantages of other acrylic adhesives while offering improvements in cure time, toughness, ease of use, and gap curing ability. The lack of noxious odors are a benefit to adhesive users.

II. ULTRAVIOLET LIGHT CURING - 'AEROBIC' ACRYLIC ADHESIVES

Aerobic acrylic adhesive technology has proven itself remarkably well suited to cures effected by ultraviolet light. Aerobic acrylics cured by UV light have the potential for vastly broadening the market for acrylic adhesives.

Acrylic adhesives have heretofore been limited by their inability to cure through gaps above 20 mils. Aerobic acrylics have extended this capacity to gaps as large as 60 mils, however, for larger gaps (most potting projects) slower curing epoxies have been the sole option for many manufacturers. Fast-curing UV-aerobic acrylic adhesives can now perform many potting operations and can cure to depths of 1/4 inch in areas accessible to UV light.

The concept of cross-linking polymers with UV light attained full commercial acceptance in the coatings and ink industries in the late 1970s. The extension of UV-curable technology to smooth, nonporous substrates, however, created problems in adhesion. Surfaces that had poor adhesion qualities included metals, which have smooth, nonporous surfaces and are frequently coated with lubricants during manufacture; plastics, which have low energy surfaces or are heavily plasticized; and glass, another smooth, nonporous surface (13). Early attempts at formulating structural adhesives reflected these limitations. Consequently, the use of UV technology with adhesives gained little acceptance.

Ultraviolet curing grades of "aerobic" acrylic adhesives, however, have been developed to provide structural bonds within seconds at room temperature on a wide variety of substrates. Parts may be precisely positioned and then structurally joined "on demand" at a time convenient to the assembly process.

Tough, impact resistant structural adhesives with tensile strengths up to 3,000 psi or more may join metal, plastic, glass, ceramics, and other substrates. Adhesive potting to 1/4 inch can be produced in a single exposure to high-intensity UV light. Transparent optical adhesives with good hydrolytic stability and resistance to mechanical and thermal shock are also derived from this technology.

1. Advantages of 'Aerobic' Acrylic

The advantages of "aerobic" acrylic adhesives are retained, and in a few cases, improved upon in their ultraviolet light curing analogues.

UV-aerobic acrylics require the presence of long-wave ultraviolet light for cure. Some grades will also cure by the pre-application of activator to the substrate that is to be bonded.

UV-curing aerobic acrylic adhesives have structural properties that appear to be significantly different from other UV-curing products. For example, poor adhesion to substrates exacerbated by high shrinkage on cure has been cited as a prime deficiency of UV-cured adhesives and coatings (13,14,18). This could have been a greater factor for limiting the acceptance of adhesives than of UV-cured coatings because of the higher performance generally expected of adhesives. Shrinkage on cure of UV coatings type products can be 10% to 25%. One UV-aerobic acrylic formulation has a shrinkage on cure of 1.5%, although 3 to 5% shrinkage is a more typical rate (Table 6).

The cured properties of UV-aerobic acrylics include high tensile strengths, heat resistance, impact strength, and other properties expected of aerobic acrylics. Formulations have been developed with properties ranging from hard, tough adhesives with a Shore-D hardness of 80 to soft, elastomeric materials with a Shore-A of 10, similar to silicone rubber.

Surface dryness and depth of cure are dependent on both the lamp used and the individual formulation. High intensities are required for the most rapid and deepest cures. A potting grade UV-aerobic will cure to 1/4 inch in 30 seconds under a 100,000 microwatt/cm^2 mercury vapor lamp. Table 7 shows curing times with various lights. Fig. 1 demonstrates a typical cure/intensity relationship of a UV-aerobic acrylic adhesive.

2. Curing Lights and Mechanisms

Some studies have indicated that UV lights represent a significant capital expense (3,19). While this is still the case for high speed UV-curable ink and coatings "dryers", many low and modest cost lights will cure UV-aerobic acrylic adhesives (Table 7). Significant advances have also been made in fiber optics which will transmit UV light, and small "penlight" lamps have become available, too. These light sources are ideal for hand assembly of small parts such as electronic components (16).

Ultraviolet light consisting of electromagnetic radiation in the 200-400 nanometer range has proven useful for producing free radicals from sensitive molecules such as ketones. The radicals can then initiate free-radical polymer reactions (15). It follows that cure with UV light depends on the specific wavelength absorption of the catalyst, the ease of radical formation, the intensity of the light, and the ability of the formulation to cure after being initiated.

Only initiators sensitive to long-wave ultraviolet light over 325 nanometers are used in UV aerobic acrylic

Table 6. Comparative Cured Properties of Selected Adhesives

	Optical Adhesive (acrylic)	UV-Curing Epoxy	UV-Curing Anaerobic	UV-Curing 'Aerobic' Acrylic Adhesive		
				Rigid	Elastomeric	Flexible
Commercial availability	mid 1970s	late 1970s	mid 1970s	1981	1982	1980
Apparent gap filling (inches)	0.15	0.020	0.050	0.15	0.25	0.30
Tensile on cold rolled steel using optional activator or primer	-	-	2,500 psi	3,800 psi	3,000 psi	1,000 psi
Shore D hardness	85	80	80	80	50	10
Thermal shock* time to failure	15 sec.	15 sec.	5 min.	2 hr.	4+ hr.	4+ hr.
Impact resistance on glass	poor	poor	good	excellent	excellent	excellent
Adhesion to oily cold rolled steel	poor	poor	poor	excellent	good	fair
Relative cost index+ (epoxy = 100)	2,000	100	1,000	500	400	400
Shrinkage on cure	10-25%	10-15%	8-15%	1-5%	1-5%	1-5%
Temperature resistance	N.A.	250°F	250°F	400°F	300°F	300°F

*1/4" X 1" square soft glass was bonded over a 1/2" area. The glass was not treated with adhesion promoters or cleaned. Each specimen was then cured for 5 minutes under a 275 watt mercury arc lamp and cooled to 0°C. After 30 minutes, specimens were dropped into boiling water and tested to failure with tongs. The test was terminated after 4 hours.

+Typical values taken from published technical data sheets.

Table 7. Some Sources of UV Light

Source	Wavelength	Intensity at 350-400 Nanometers (watts/cm^2)	Approximate Price Range	Seconds to Cure[1] (2 mil gap between glass slides)	Seconds to Cure[1] (1/8 inch deep wire tack)
Sunlight	full spectrum	very low	-	10-30	not recommended
Mineral light, "black light"	350-400 nanometers	100-7,000	$75-500	5-20	15-100 (higher intensities only)
275 watt consumer style sunlamp*	full spectrum + heat	5,000-7,000	$40-75	1-10	5-30
Industrial grade mercury arc lamp Type I	300-400 nanometers + visible light	5,000-10,000	$200-500	1-15	5-30
Industrial grade mercury arc lamp Type II#	full spectrum + + heat	1,500,000	$800-5,000	0.5-4	2-10
Pulsed xeron arc lamps; solar simulators#	full spectrum +	very high	$50,000 +	fast	1-5

*Shielding recommended.

#Shielding required.

[1]Speed of cure varies according to formulation.

adhesives. Long wavelength UV light is sometimes referred to as "black light" and is generally considered harmless. Short-wave radiation, on the other hand, can produce burns. (Sunburn from overexposure is a familiar example.) The high-intensity lamps shown in Table 7 all require shielding to protect workers from the short wavelengths that they emit in addition to long-wave UV light.

3. Bonding Categories

The ways in which UV-aerobic acrylic adhesives are used have been divided into categories (Table 8). One category is the bonding and lamination of clear materials such as glass and plastics.

Because UV aerobics exhibit high strength, toughness, and good structural properties, the second category has been designated "light welding": a process of joining parts by bridging the joint with cured adhesive, replacing long cure times or mechanical fasteners.

A third category is that of adhesive potting and specialty coatings on smooth surfaces like metal.

'Light Welding'. The ability to bridge bond many surfaces while retaining structural properties has led to a new concept for the joining of abutting surfaces. "Welding with light" is the process of laying a bead of liquid adhesive across the joint formed by abutting metal, glass, ceramic, thermoplastic, paper, and other surfaces and joining them by polymerizing (curing) the adhesive in a few seconds with exposure to long-wave UV light.

A laboratory example will serve to illustrate the process of "welding with light." Cold rolled steel joints were prepared using the same aerobic acrylic adhesive. A 1-in. bonded overlapped joint was measured at 3,500-lb. tensile pull. The "welded" joint showed 400-lb. tensile pull. Since the same adhesive was used, the difference in absolute tensile strengths can be ascribed to the differences in surface areas. Depending upon design strength requirements, therefore, "light welding" can rapidly join a large number of diverse parts. Properties can range from hard, thermoset structural "welds," which can exceed the strength of some substrates, to soft, pliable elastomeric adhesives, potting and tacking systems (Tables 6 and 7).

Table 8.

Bonding Categories Using UV 'Aerobic' Acrylic Adhesives

Bonding	'Welding'	Potting/Coating (Other)

• glass window assembly • lens bonding • polycarbonate shield assembly • plastic film laminations • costume jewelry • glass/plastic stemware • lamp assembly	• chip bonding • speaker software assembly • wire tacking • temporary bonding • ferrite magnet assembly • electrical leads to lamp housing • sealing motor parts • tamper proofing of mechanical fasteners • costume jewelry	• potting to ¼" deep • abrasion resistant coatings • solder mask • encapsulating

4. Industrial Applications

Parts to be joined can be positioned to exacting tolerances. The adhesive may be applied either before or after positioning and then fixtured or "welded" on demand by exposure to UV light. The low shrinkage of UV-aerobics helps assure non-movement of parts due to the curing cycle. Small parts can be joined in 1-10 seconds at room temperature. Thick bondlines require longer times or more powerful light sources.

Some ferrite parts require bridge bonding because adhesive between the surfaces interferes with electrical properties. The common technique of using a paste epoxy requiring long time-cycles or heat-cures can be replaced by a fast-curing "light welding" technique. The parts are held in position, a bead of thixotropic UV-aerobic adhesive is applied across the joint and cured with UV light. A 200-watt per inch industrial light required 2 seconds, while a low-power "black light" effected cures in 30 seconds.

A fabricated plastic holder for an electronic pick-up head is assembled and the concentric parts are adjusted to a fine tolerance. An adhesive is applied across gaps that may vary to 1/8-in. wide and cured. A filled UV-curing aerobic adhesive has been chosen due to its rapid cure "on demand" and low shrinkage on cure, keeping the parts within the required alignment of six seconds of arc.

A common method of wire-tacking today is to use cyanoacrylate adhesive and primer. In this process, a drop of cyanoacrylate is placed on top of the wire to be tacked, followed by a drop of primer. The resulting wire-tack is usually serviceable in 5 - 15 seconds although several hours can be required for full cure. The usage of the wire tack, although limited by a lack of moisture, temperature and impact resistance, has been finding increasing application in the electrical industry.

UV-curing aerobic acrylic adhesives, while less expensive on a per pound basis than cyanoacrylates, provide a structural wire-tack that resists environmental conditions. A useful property is that the aerobic acrylics fluoresce to optimum brightness only upon completion of cure, providing a visual means of end-point determination. An elastomer formulation is recommended for permanent wire-tacking in a high vibration environment. Removable wire-tacking formulations are also available (Fig. 1).

Loudspeaker software is frequently assembled and bonded with solvent-based and epoxy adhesives requiring long drying times or heat-assisted cures. UV-aerobic adhesives have been used to increase the speed of some assembly operations. Many different materials are used in loudspeakers. Several kinds of papers or fabric are bonded to each other, to coated foil, to aluminum, and to other metal surfaces. Some speakers use engineering plastics instead of paper for a sound generating surface. The bondlines must be rigid enough to transmit vibrations from the speaker coil to the sound generation surfaces, tough enough to withstand long-term vibration, and in some cases, able to withstand temperatures of 300°F or more.

In one assembly process, a plastic cone is placed over a metal coil. A UV-acrylic adhesive is applied to the mating surfaces and cured in 4 minutes using a low-power "black light." A higher volume producer uses a similar procedure; however, four 3-second passes under a high-intensity light is required due to the nature of the substrate and the design of the mating coated paper surfaces. Adhesive that was absorbed by the black paper surface could not be completely cured because of the blocking effects of the paper. To assure cure in this and other "shadow" situations, a pre-application of activator was required.

An electronics manufacturer attaches a plastic base to a 1-in. square ceramic circuit holder with a UV adhesive. Cure time is 10 seconds, which is about the same as the previously used cyanoacrylate. The lower cost, higher performance, and worker safety features justified the change in adhesive.

Beryllium-copper bellows have typically been joined by applying a two-part epoxy and accelerating the cure with heat. A single-component structural UV-aerobic acrylic performs the same operation in 30 seconds with the added advantage of having eliminated surface preparation.

Electronic chips are temporarily bonded with a 1-second UV-curing adhesive. The bonds hold the chip to a PC board prior to wave soldering. The adhesive must continue to bond during the soldering process, which could reach up to 900°F for 2-5 seconds. A special adhesive was developed that would stand up to these conditions, as well as bond even under the chips where light rays could not penetrate.

UV Transparent Surfaces. Bonding UV transparent surfaces is easily accomplished depending on the surfaces chosen. Some grades of transparent plastics contain UV absorbers that prevent light-induced curing, and therefore, require care in substrate selection.

Glass is efficiently bonded with UV-aerobic acrylics, and replaces epoxies, anaerobics, or heat-fusing techniques. A manufacturer of pressed glass stemware required an adhesive that would resist repeated exposures to detergent and water while providing a faster cure rate than an epoxy bonding process. A transparent UV acrylic was developed that cured a 2-mil gap completely in 2.5 seconds under a 250-watt mercury lamp. It did not discolor on aging and was moisture resistant (Fig. 2).

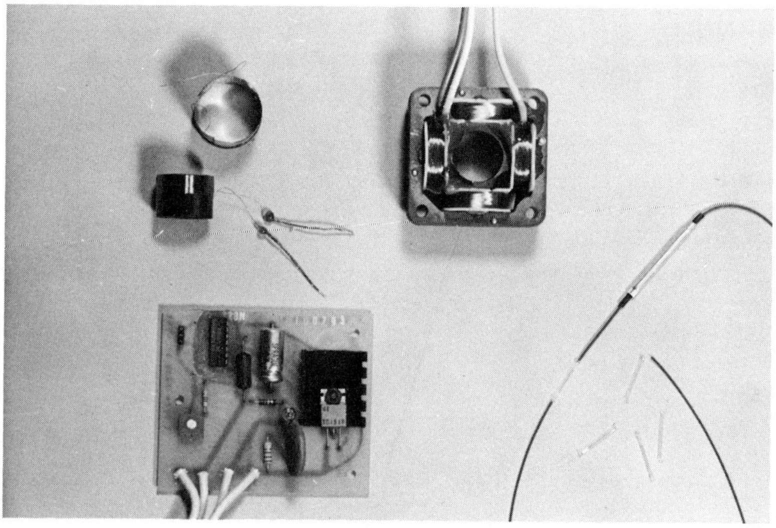

Fig. 1. Wires and Parts Tacked Down, Fiber Optics Bend using UV-curable Aerobic Acrylic.

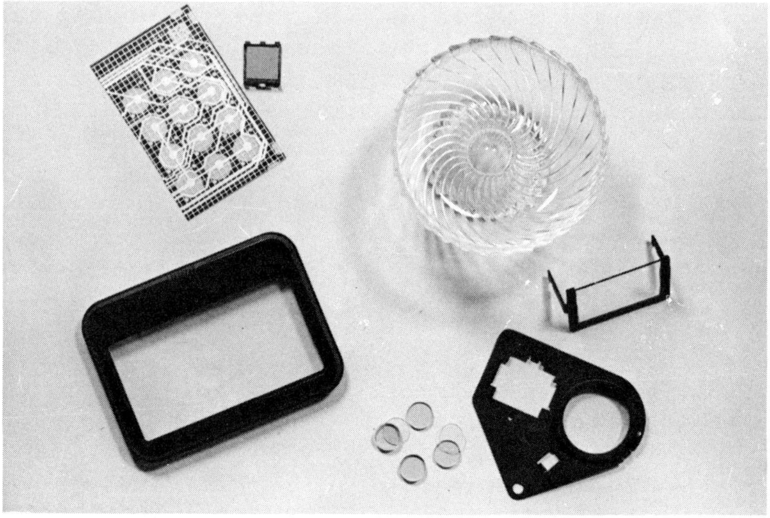

Fig. 2. Glass and Plastics Bended with UV-curable Aerobic Acrylic.

A copier manufacturer assembles a glass lens system to an aluminum base. The parts must be aligned to close tolerances, a process which can take variable amounts of time, and then bonded. A UV-aerobic acrylic accomplished this task in 15 seconds using a low power "black light."

A non-UV absorbing grade of polycarbonate window is bonded to a cast aluminum case by a manufacturer of off-road vehicles. The 30-second UV-curing system replaces a process in which holes were drilled and tapped into the aluminum and in which the window was sealed with a rubber o-ring before screwing on a metal retaining ring.

Colored polyacrylate windows are bonded over instrument gauges designed for an airplane instrument panel.

Automotive headlamp manufacturers usually heat-fuse glass to join lenses to reflectors and to seal assembly holes. Heat-curing epoxy adhesives provide some energy savings over the fusion process, but still require long heating times and substantial capital investment.

Manufacturers have begun using UV adhesives because of the very large differences in the cost of energy used for cure and the increase in productivity associated with faster cure cycles. Specially designed UV-aerobic acrylics cure rapidly, adhere to borosilicate glass, and withstand the severe environmental and vibration resistance required of automotive parts.

Potting and Sealing. Adhesive potting is achieved through the proper choice of adhesive, UV light, and substrate selection. Hard or elastomeric potting formulations can be used depending upon requirements. Depths of more than 1/4 inch are obtained by using a step-wise technique. In a structural adhesive potted plug 3/4 inch deep, three applications and curing cycles of adhesive (30 seconds each) were faster than a two-component epoxy with its long curing time.

Properties of the UV-curing aerobic acrylics--good adhesion and environmental resistance--suggest their use as sealing and coating agents. Speed of sealing and energy savings over the cost associated with two-part or heat-curing systems are two of the productivity-increasing advantages. Use of these systems may be limited; however, by the thickness of the desired coating and the ability to deliver UV light to all surfaces involved.

Design Considerations. Glass and clear plastics must not contain UV light absorbers that will prevent the light from reaching the bondline. Cure of aerobic acrylic adhesives only proceeds when UV radiation is directly contacting the bondline. If the light is turned off or dimmed, or if shadows come between the light and the adhesive, cure may not occur. UV light must directly strike the total surface of the bond (adhesive) in order for polymerization to occur.

Surfaces can be activated (3) to assure cure in applications where shadows are present or if the depth of the adhesive is such that UV light cannot penetrate completely.

5. Conclusion

The structural adhesive properties of UV curing aerobic acrylic adhesives have the potential of providing new techniques of assembly by bridge bonding or "welding" parts together. Ambient temperature cures allow the "welding" of substrates not suitable for joining by other assembly techniques. The speed of UV curing systems provides the opportunity for significant increases in productivity over slower curing methods of assembly. UV-curing grades of aerobic acrylic adhesives can be cured on demand in seconds. The combination of high tensile strength, good adhesion to many surfaces, toughness, and impact and moisture resistance significantly expands joint design capabilities.

REFERENCES

1. B. H. Burnam, G. S. Haviland, and R. G. Nystrom, "Anaerobic Adhesives" in "Handbook of Adhesive Bonding," Charles V. Cagle, ed., McGraw Hill, New York (1973).

2. J. D. Roberts and M. C. Caseria, "Basic Principles of Organic Chemistry," W.A. Benjamin, Inc., New York (1964).

3. L. C. Manhoney, "Structural Adhesives for Rapid Cure Applications," Adhesives Age, 22(12), 26-30 (1979).

4. U.S. Patent 3,616,040. "Process for Bonding," October 26, 1971 (to Loctite Corporation).

5. U.S. Patent 3,890,407. "Novel Adhesive Compositions," June 17, 1975 (to DuPont Corporation).

6. U.S. Patent 2,895,950. "Compositions Containing Hydroperoxide Polymerization Catalyst and Acrylate Acid Diester," July 21, 1955 (to Loctite Corporation).

7. U.S. Patent 3,724,504. "Fast Curing Polychloroprene Acrylic Adhesives," April 3, 1973 (to Lord Corporation).

8. U.S. Patent 3,832,274. "Fast Curing Adhesives," August 27, 1974 (to Lord Corporation).

9. J. W. Prane, "Reactive Adhesives," Adhesives Age, 23(8), 35-37 (1980).

10. R. E. Brown and R. E. McCrea, "Competition for Growth for Epoxy Adhesive Markets," Adhesives Age, 25(2), 21-25 (1982).

11. T. J. Wilkinson and D. P. Tyler, "Acrylics Improve for Bonding Automotive Alloys," Adhesives Age, 24(12), 34-38 (1981).

12. "Second-Generation Acrylic Adhesives," Adhesives Age, 19(9), 21-24 (1976).

13. D. C. Armbruster, "Advancements in Adhesion of Ultraviolet Curable Coatings," Technical Bulletin of Celanese Chemical Co., Dallas, Texas (January 1981).

14. A. Pryce, "The Technological Literature Relating to Polymerization Photoinitiators," Journal of Oil Color Chemistry, 59, 166-170 (1976).

15. Skiest Laboratories, "Survey on Adhesives" (1979).

16. C. Bluestein, "Reactive Cure System," Lyndhurst, NJ, Captan Associates (1980).

17. "Shrinkage Characteristics of Photomers," Technical Bulletin of Diamond Shamrock, Morristown, NJ.

18. E. G. Shur, "Ultraviolet Cured Coatings," Amer. Paint J., 2-7 (September 4, 1972).

19. R. W. Harris, "New Developments in Low Energy Reactive Adhesives," Adhesives Age, 21(2), 19-22 (1978).

Radiant Energy Curable Adhesives

Claire Bluestein

Captan Associates
218 Stuyvesant Avenue
Lyndhurst, NJ 07071

ABSTRACT

This overview of radiant energy curable technologies for adhesives includes discussions of the form of radiant energies, where they are used, their market size, chemical nature, and properties. The article centers on ultraviolet, infrared, and electron beam technologies.

INTRODUCTION

Modern adhesive technologies are growing in so many different directions - new applications, new materials, new techniques, new specialties - that a technologist seeking to design a new product or to improve a process may sometimes overlook one of the new technologies in reactive systems that yield improved bonding and aging properties - radiation curing.

Radiation curing of adhesives has nothing to do with materials that give off radiation. Generally, formulations currently available are safer to use and handle than the bulk of adhesive formulations on the market, in particular those containing significant volumes of volatile solvents. Radiation curing is a production technique for drying and curing of adhesives through the use of a form of radiant energy.

The conventional drying and curing techniques have meant the use of some form of heat, mostly convective heat. The newer drying and curing methods utilize radiant energy and cover a broad range of the available electromagnetic spectrum. The basic theory of the electromagnetic spectrum as well as the photon power available and the power from energetic particles (accelerated), such as electrons, ionized molecules, and free radicals, have been known for years.

The advances in engineering over the past 20-30 years have made it possible to develop practical plant equipment with capabilities for providing more efficient conversion of the power source; orientation and configuration for proper exposure of the substrate; focus or no focus for the light waves (photons) or particles; application techniques for laying down of inks, coatings, and adhesives; and removal of heat or vapors if formed. The radiation curable adhesives contain essentially no volatile materials and are designed to set instantly on exposure to the radiant energy source. Additional engineering developments have come about through the need for thinner adhesive levels because nonsolvent materials cure at the thickness level of application.

FORMS OF RADIANT ENERGY

Ultraviolet (UV), the most popular of the new cure techniques today, starts at 10 nm and goes up to 400 nm. The commonly found industrial lamps range from 200-400 nm or 0.2-0.4 microns. These lamps have power in the 4-6 electron volt range. The wavelengths of visible light range from 400-1000 nm (0.4-1 micron).

Equipment that produces particles for curing, rather than photons for curing (as in UV), is becoming more accepted industrially. Electron beams (EB) have been used industrially since the late 1940s. Beam power can range from 150,000 electron volts to 5,000,000 electron volts. With EB equipment possessing a range from 300,000 eV to 5,000,000 eV, the beam is focused or scans the substrate, and requires extensive shielding. A limited number of special processing facilities are available for toll manufacturing. The 150,000 to 300,000 eV equipment is the electron curtain or linear cathode type, and it is this type that is designed for installation in a pressure sensitive tape converting plant, for example.

Dry plasmas are a recent development and can be expected to become more widely accepted in the near future. Plasmas are ionized molecules and free radicals that are present in gases generated by radio frequency equipment. Current equipment produces low temperature plasmas by radio frequency ionization of either metallic or organic sources. These can be used for very efficient cleaning, smoothing of surfaces, and improvement of adhesion to substrates. Newer developments are focusing on learning how to use these free radical sources for grafting and other polymerizations.

WHERE USED

The apparent cost of effective radiation curable adhesives has limited their use. Because radiation curable adhesives contain no solvents, water, or low cost waxes their prices are higher than many other types of adhesives. However, material costs alone may give a false impression of cost effectiveness. To measure cost effectiveness in a manufacturing operation one must also consider that using a radiation curable adhesive can help to lower utility bills, reduce waste effluents and polluting gases and liquids, and also produce a more reliable product with a lower rejection rate and very likely lower labor costs.

Furthermore, the understanding of how and which materials to assemble to make an adhesive that will cure to suit a given application is currently limited to a relatively narrow circle of practitioners. There have been just enough radiation curable adhesives with inadequate performance marketed prematurely to make some manufacturers assume that such adhesives won't work properly. Availability of worthwhile products is limited, although their number is constantly growing.

Out of the large overall U.S. market that exists for all adhesives - $4 billion in 1981 - less than 15% is in specialized uses that would readily accept typical pricing, which is currently $6-$100 per lb.

Moreover, many prospective users have sizeable capital investment in application equipment. If these users can be shown that their present equipment is usable with simple retrofitting of a cure unit, they may readily accept a radiation curing adhesive. UV and IR cure units are more easily added to many production operations and may range in price from several hundred dollars to more than $100,000 depending on the operation, its needs, and complexities. It can be expected that an IR process will require more line length or a slower production speed than a UV process, and that at the same time, the substrate can withstand some heat. An EB installation can be expected to cost at least $250,000 and to be a unit constructed with shielded housing and an inerting chamber. If a new production line is planned, radiation curing should seriously be considered because frequently the overall line may be designed in a simpler, space-saving, and less costly way.

Examples of industry areas that are currently using or evaluating radiation curable adhesives are electronics, communications, transportation, dental, medical, consumer, graphic arts, and packaging. Special applications that are currently using radiation curable adhesives include wire tacking, potting, glass bonding such as lamp assembly, decorator objects, stemware and instruments assembly, speaker assembly, fiber optics assembly, electrical leads assembly, locking of mechanical fasteners, costume jewelry, dental fillings, orthodontics, laminations such as film-to-film and film-to-glass, nameplate bonding, flocking, membrane switches, conductive adhesives, structural assembly for automotive and aircraft, footwear, electronic tapes, and wallboard laminates. These are summarized in Table 1.

An application that is becoming of increasing interest to the electronics and communications industries is that of encapsulation or conformal coatings. These coatings are better classified as adhesives or elastomers because normal types of surface coatings do not meet the end-use-requirements, which include resistance to environmental stress, electricity and aging.

MARKET SIZE

From the listing of the wide range of specialized applications, many of which are in early stages of commercialization, it is quite a task to estimate the total market with precision. A 1979 estimate placed the use of UV curable adhesives at $100,000 and EB curables at $400,000. The 1981 usage of UV curables may have grown to more than $1 million; by the end of 1982, it should be about $2 million, and it should continue to grow more rapidly within the next few years. In the referenced report, the major market named was pressure sensitive tapes. The authors did not foresee the effectiveness of some of the formulated adhesives in the other applications already described. Moreover, the cure unit investment in many of these newer applications is minimal with respect to throughput of produced product with high add-on or performance value.

The EB curables market is growing nearer to the predicted pace and is currently near $3 million. This is up from the $400,000 in 1979. Under current economic conditions, large capital investment by companies for EB equipment will be cautious despite the efficiencies and effective products that can be gained from this method of manufacture. However, a single EB line operating at a full working schedule will consume a relatively large volume of adhesive. Many of these web systems are 60 inches or more and run at 1000 feet per minute. The market should be about $34 million in 1987 (Table 2).

Table 1. Applications Summary for Radiation Curable Adhesives

End-Use Markets	Applications
Electronics	Wire tacking Conductive inks Potting Electronic tapes Membrane switches Ceramic binders Conformal coating
Communications	Wire tacking Instrument assembly Fiber optics adhesives Speaker assembly
Consumer	Lamp assembly Jewelry Nameplates adhesives Appliance & wallboard laminates Flocked fabric Footwear
Transportation	Lamp assembly Electrical leads assembly Wire tacking Locking of mechanical fasteners
Graphic arts & packaging	Labels Transfer letters Packaging film laminates
Dental & medical	Orthodontic adhesives Dental fillings Instrument assembly

The IR and microwave, or dielectric, curing adhesives processes have not been well characterized. IR can either dry solvent or water, or it may cause cure; microwave or dielectric heat is frequently used to soften thermoplastics which bond. Materials in current use with this type of

Table 2. Summary of Radiation Curable Adhesives Markets (millions of dollars)

	1979	1981	1987 (est)
UV curables	0.1	1.0	6.0
EB curables	0.4	3.0	28.0
TOTAL	0.5	4.0	34.0

equipment, therefore, are conventional or slightly modified conventional adhesives. Future formulations with true IR reactivity should be more nearly similar to the UV and EB formulations.

CHEMICAL NATURE

Radiation curable adhesives can be based on raw materials that are not greatly different from those used in acrylic adhesives (first, second, and third generations), epoxy adhesives, urethane adhesives, and anaerobic adhesives. The formulation approach, however, is somewhat different. Very important is that no solvents are used to thin the resins; a monomer reactive diluent is used instead. The monomer must be matched with the resin to give the desired set of properties with respect to adhesion, substrate, flexibility or stiffness, cure behavior, and durability of the cured product. Most of the radiation curable adhesives have an acrylate (or methacrylate) basis; many of the acrylics are modified.

Monomers are molecules of relatively low molecular size and simple structure which are capable of reacting with themselves or other molecules to build a larger, more complex structure called a polymer. The process is called polymerization. These molecules are chiefly those characterized by having either an unsaturation or an oxirane (epoxy) contained within. Acrylates, methacrylates, and unsaturated polyesters all contain unsaturation. Oligomers are generally partially polymerized combinations of monomers ranging in molecular weight from about 600 to 8000. Oligomers contain a reactive unsaturation or epoxy group. Resins of generally nonreactive types may sometimes be included in adhesives formulation. Monomers and oligomers that have multiple reactive groups are cross-linkable in nature. These react more rapidly in radiation curing formulations and build up network strength, but they may cause embrittlement in the cured adhesive. The art of formulation thus requires a balancing of ingredients.

Epoxy systems can also be used for radiation curables, and although their use is more limited presently because of patent conflicts more of them will be seen in the future. Radiation curable resins, oligomers and monomers that are designed specifically for their adhesive properties will not become readily available to formulators until the markets grow much larger.

Even more important to the formulation is a photoinitiator for the UV curables or an accelerator for the EB curables. These do not necessarily have to be a separate ingredient, but they may be a molecular modification of a resin or monomer ingredient. A typical photoinitiator for a UV curable acrylic system is based on an aromatic keto compound, and for an epoxy system it is based on a diazonium or a sulfonium compound.

Table 3 contains a listing of commercial photoinitiator chemicals. For the EB systems, accelerators may belong to a fairly wide range of chemicals, but they are all quite polar in nature. For IR systems peroxides or azobisisobutyronitrile are utilized. Some of the UV curables contain a combination of UV and IR initiators to take advantage of the IR output that many UV lamps generate. Table 4 shows an example of a UV curable formulation.

PROPERTIES

Radiation curable adhesives are available in a wide range of viscosities. The thinnest of these are not as low in viscosity as solvent or water-based products, but they have a different rheology. They may be sprayed, or they may be run on gravure coaters. The thickest may have a paste-like viscosity.

A key feature of radiation curable systems is that they are one-part systems with shelf lives ranging from several weeks to six months or more. They should be stored in the dark and away from heat if possible. Some of these adhesives

Table 3. Commercial Photoinitiator Chemicals

BE	Benzoin ethers - isopropyl, ethyl, or isobutyl mixed
B	Benzil
BP	Benzophenone
CA	Chloroanthraquinone
CTX	Chlorothioxanthone
DBK	Dibenzylketone
DBS	Dibenzosuberone
DEAP	Diethoxyacetophenone
DMPA	Dimethoxyphenylacetophenone
EA	Ethylanthraquinone
ITX	Isopropylthioxanthone
MTX	Methylthioxanthone
PDO	1-Phenyl-1,2-propanedione-1-(o-ethoxy carbonyl)-oxime
Michler's Ketone	4,4-Bis(dimethylamino)benzophenone
Cationics	Triaryl sulfonium, diazonium, or phosphonium salts, usually fluoborates

Table 4. Example of UV Curable Adhesives Formulation

Conformal Coating	Parts by Weight
Urethane acrylate	33
Isobornyl acrylate	33
Trimethylolpropane triacrylate	25
Pigment and stabilizer	5
Photoinitiator system	4
	100

do not cure well in the presence of air, but others will. If a vendor of a UV curable product states that air exclusion or nitrogen blanketing is a necessary prerequisite, then this technology is of an older variety.

Safety, in terms of skin irritation and toxicity of most systems in 1982 should be good and greatly improved from the 1970s. Uncured adhesives may be rated by a Draize value, though this is not a true standard. However, a rule of thumb states that values of 3 or less are unlikely to cause most people problems. The odor of some of the materials may be unpleasant, but it will not be harmful and can be vented with simple installations. Also, the better products are not flammable.

Other properties of radiation curable adhesives are reported by the manufacturer according to the application. If unexposed to light and heat, they will remain open and nonclogging. The user who is unfamiliar with formulation techniques should be sure to request information from the supplier on suitable thinner and cleanup solvents. Extension with fillers and additives should also not be attempted without understanding of these systems.

REFERENCES

1. R.M. Kossoff and J.W. Prane, Radiation Curable Adhesives, R.M. Kossoff & Associates, Inc., New York (1980).

2. "Reactive Cure Systems: UV-IR-EB." CAPTAN Associates, Inc., Lyndhurst, NJ (1982-83).

3. A.H. Keough and J. Lacey, "Radiation Curing: A Discussion of Advantages, Features, and Applications," Association for Finishing Processes/Society of Manufacturing Engineers, Dearborn, MI (1980).

4. Radiation Processing, Literature, and Patents Survey, K & M Publications, Inc., Louisville, KY (1977).

5. S.P. Pappas, Editor, UV Curing, Science, and Technology, Technology Marketing Corp., Norwalk, CT (1978).

6. D. Bryce Smith, Advances in Photochemistry, American Chemical Society, Vol. 1-12, Washington, DC (1968-80).

7. T. L. Vigo and L. J. Nowacki, eds., Energy Conservation in Textile and Polymer Processing, American Chemical Society, Washington, DC (1979).

8. "Bibliography 117 - Radiation of Elastomers," "Bibliography 127 - UHF, HF, and IR Curing of Rubber and Plastics," Rubber Division Library & Information Service, University of Akron, Akron, OH (1976).

9. "Safe Handling and Use of Ultraviolet/Electron Beam Curable Coatings," National Paint and Coatings Association, Washington, DC (1980).

10. Handbook of Radiation Measurement and Protection, Section A; "General Scientific and Engineering Information," Section B; "Radiation Protection Principles and Practice", CRC Press, Inc., Boca Raton, FL (1982).

Isocyanatoethyl Methacrylate: A Latent Crosslinker for Coating Adhesive Resins

Paul E. Cranley

The Dow Chemical Company
Specialty Chemicals Department
Midland, Michigan 48640

ABSTRACT

Isocyanatoethyl methacrylate (IEM)© is a difunctional monomer with an aliphatic isocyanate functionality and a vinyl polymerizable double bond. Either end of the molecule can be reacted first, leaving the other functionality for latent reaction. The isocyanate group reacts with active hydrogen compounds at rates of reaction similar to other commercial aliphatic isocyanates. The methacrylate functionality allows copolymerization with other common vinyl monomers at reaction rates similar to methyl methacrylate. Applications for IEM fall into three categories: polyisocyanates made by polymerizing the methacrylate group, vinyl functionalized resins made by reacting the isocyanate group with polyfunctional molecules and polymerizable derivatives made by reacting the isocyanate group with monofunctional reagents. In this last application category, the blocked-IEM derivatives allow the benefits of IEM to be extended into water-based and one-component urethane systems. The majority of applications reported to date utilize IEM as a latent cross-linker or graft site to make resins suitable for coatings or adhesives.

INTRODUCTION

Isocyanatoethyl methacrylate (IEM) is a difunctional monomer with an aliphatic isocyanate functionality and a vinyl polymerizable double bond. Either end of the molecule can be reacted first, leaving the other functionality for latent reaction. Both functionalities are well behaved; they react as one would predict from similar functionalities in other molecules. Either end of IEM can be reacted at low temperatures, thus IEM finds utility in latent low-temperature cross-linking of resins for coating or adhesive applications. Using IEM, the resin chemist can conveniently tie together condensation polymers with addition polymers. This permits the bridging of two well studied technologies and allows the design of coating and adhesive resins with novel and useful properties.

PROPERTIES

Some of the physical properties (1) of neat isocyanatoethyl methacrylate are shown in Table 1. Commercial development material is a low viscosity, water-white liquid with 150 ppm IONOL® inhibitor added to stabilize the methacrylate group against polymerization. The monomer should always be stored in the presence of dry air, which keeps the inhibitor active. In addition, the monomer should be kept cool or refrigerated; under these conditions IEM has a shelf life of at least six months.

IEM REACTIONS: ACTIVE HYDROGEN COMPOUNDS

Isocyanatoethyl methacrylate (IEM) reacts with active hydrogen compounds, i.e., compounds with hydrogens replaceable by sodium. For example, IEM monomer reacts with alcohols at low temperatures to form the corresponding urethane:

$$\underset{\text{IEM}}{CH_2{=}CCH_3CO_2(CH_2)_2NCO} + \underset{\text{Alcohol}}{ROH} \longrightarrow \underset{\text{IEM Urethane}}{CH_2{=}CCH_3CO_2(CH_2)_2NH\overset{O}{\overset{\|}{C}}OR}$$

The relative reactivity of 1-butanol with IEM and other commercial isocyanates was studied by Frisch, Kresta and Wongkamolsech (2). As seen in Table 2, IEM reacts at a rate that is almost twice as fast as hexamethylene diisocyanate under these conditions. The presence of a low level (200-400 ppm) hydrolyzable chloride impurity in the IEM appears to accelerate this reaction.

Table 1. Characteristics of Isocyanatoethyl Methacrylate*

(Note: These characteristics are laboratory results on neat compound or material typical of the product, but they should not be confused with, or regarded as, specifications.)

Appearance	Clear, Colorless Liquid
Molecular Weight	155.15
Percent Isocyanate (Functionality), weight	27.1% (Theoretical)
Boiling Point	211°C
Freezing Point	-45°C
Density, 25°C	1.096 g/cc
Viscosity, 25°C	1.44 cps (water-like)
Flash Point (TCC)	97°C (206°F)
Solubility	Miscible with methylene chloride, chlorobenzene, toluene, acetone, ethyl acetate and ethoxyethyl acetate
T_g (homopolymer)	60°C (DSC)
Inhibitor	150 ppm IONOL® (2,6-dibutyl,4-methylphenol)
Heat Capacity, 20°C	0.43 (cal/g°C)
Thermal Conductivity, 20°C	3.81×10^{-4} cal/(cm sec°C)

*Developmental monomer available from The Dow Chemical Company (XAS-10743.00)

®IONOL is a trademark of the Shell Chemical Company

Table 2. Relative Reactivity of NCO with 1-Butanol*

	Isocyanate	k_{cat}
$H_{12}MDI$	4,4'-dicyclohexylmethane diisocyanate	6
IPDI	Isophorone diisocyanate	8
HDI	Hexamethylene diisocyanate	10
IEM	Isocyanatoethyl methacrylate**	19
TDI	Toluene diisocyanate 80/20	28
MDI	4,4'-diphenylmethane diisocyanate	150

CONDITIONS

Solvent:	2-ethoxyethyl acetate
Catalyst:	dibutyl tin dilaurate
Temperature:	25°C
Alcohol:	1-butanol
k_{cat} in units:	$kg^{1.5}/mole^{1.5}min.$

*Frisch, Kresta and Wongkamolseqh, University of Detroit, Unpublished Data

**345 ppm Hydrolyzable chloride impurity

Addition copolymers containing IEM retain much of this reactivity with primary alcohols. As seen in Table 3, the reaction rate constant of 1-butanol with polymerized IEM is about half that of monomeric IEM under the same conditions (3). Thus the reactivity of IEM-containing polymers is similar to that of other primary aliphatic isocyanates such as hexamethylene diisocyanate. This very good retention of NCO reactivity allows the use of IEM to incorporate NCO functionality into a variety of vinyl copolymers. In general, IEM-containing copolymers retain >90% of the theoretical isocyanate functionality.

The hydrolysis of IEM with water forms the disubstituted urea:

$$CH_2{=}C(CH_3)CO_2(CH_2)_2NCO + H_2O \longrightarrow CH_2{=}CCH_3CO_2(CH_2)_2NH\overset{\overset{\displaystyle O}{\|}}{C}NH(CH_2)_2CO_2C(CH_3){=}CH_2$$

This reaction is slow compared to the catalyzed reaction of IEM with primary alcohols but it can be accelerated with triethylene diamine. Moisture must be kept away from IEM monomer and IEM-containing polymers to avoid gel formation due to this urea cross-linker. The presence of water will slow the tin catalyzed reactions with alcohols. Presumably, this inhibition results from hydrolysis of the catalyst.

Table 3. Reactivity of IEM Based Polymeric Isocyanate*

Isocyanate	k_{cat}
IEM Monomer	19
IEM Homopolymer	10
Copolymer of IEM-MMA	9

CONDITIONS

Solvent:	2-ethoxyethyl acetate
Catalyst:	dibutul tin dilaurate
Temperature:	25°C
Alcohol:	1-butanol
k_{cat} in units:	$kg^{1.5}/mole^{1.5}min.$

*Frisch, Kresta, Wongkamolsech, University of Detroit, Unpublished Data

Blocking And Deblocking Of IEM

Certain active hydrogen compounds will react with IEM to form derivatives which will dissociate at elevated temperatures to regenerate the free isocyanate. The reaction of IEM with caprolactam exhibits this reversibility:

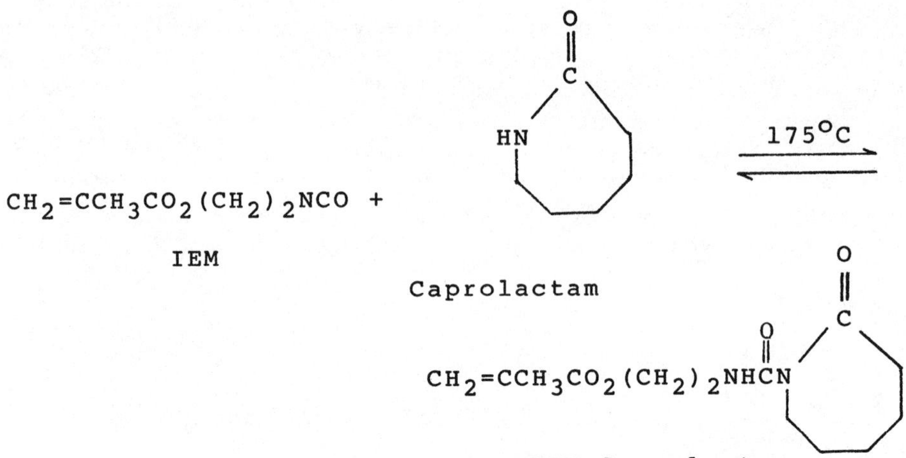

This deblocking reaction is accelerated by certain catalysts and by the presence of a polyol which can compete with the caprolactam for the NCO group via a transesterification type reaction. The blocking of the isocyanate group allows IEM-containing polymers to be formulated with polyols to give room temperature-stable formulations which deblock and cross-link at higher temperatures.

Regulski and Thomas (3,4) recently reported the synthesis of over 20 blocked-IEM derivatives. Fourteen of these derivatives were solution copolymerized with methyl methacrylate and ethyl acrylate to give terpolymers with predictable molecular weights and narrow $\overline{M}_w/\overline{M}_n$ distributions. The terpolymers were then formulated with a commercial acrylic polyol and dibutyl tin diacetate and baked for 30 minutes to give cross-linked enamels. The minimum baking temperatures required to give enamels that withstood at least 50 double-rubs of methyl ethyl ketone solvent was determined. Some of these minimum deblocking temperatures are shown in Table 4. Except for ethyl acetoacetate, all formulations gave baked enamels with excellent solvent resistance while maintaining adequate gloss, adhesion and pencil hardness. Thus blocked IEM derivatives can be used to design stable, one-component resin systems which deblock and cross-link at a variety of temperatures. As we shall discuss later, the blocking of IEM also allows its use in emulsion copolymerizations.

Table 4. Deblocking/Cross-linking Temperatures of IEM Derivatives

Group	Specific Derivative	Temperature
Phenol		
	Methyl salicylate	110°-120°C
	Methyl p-hydroxy benzoate	130°C
Imidazole	Imidazole	110°-130°
Oxime*		
	Methyl ethyl ketoxime	140°-150°C
	Acetone oxime	130°-140°C
N-Hydroxyimides		
	N-Hydroxyphthalimide	170°-175°C
	N-Hydroxysuccinimide	160°C
Alcohols		
	Methoxy propanol	175°-200°C
	Ethyl hexanol	>200°C
	Pentol	200°C
	Ethyl lactate	200°C
Lactams		
	Caprolactam	175°-200°C
	Pyrrolidinone	250°C
Active Methylene Compounds	Ethyl acetoacetate (Never Gave Sufficient Cure)	>175°C

*Caution: Heating <u>neat</u> IEM-blocked with oximes at or about 100°C may produce an exotherm which may initiate another exotherm at a higher temperature (175°C). The second exotherm is due to the degradation of the oxime yielding heat and gaseous products. <u>The Handbook of Reactive Chemicals Hazards</u>(5) lists this potential hazard as characteristic of oximes.

IEM REACTIONS: ADDITION POLYMERIZATION

The methacrylate group of IEM allows it to be free radical polymerized. Thomas (1) reports IEM can be solution polymerized in dry solvents using AIBN (azobis-isobutyronitrile) initiator to give a stable poly-isocyanate. The homopolymer of IEM retains 90-95% of its isocyanate functionality and has a glass transition temperature of about 60°C (DSC).

Likewise, IEM can be copolymerized with a variety of acrylic or styrenic comonomers. Frisch, Kresta, and Wongkamolsech (2) studied the copolymerization behavior of IEM with styrene, butyl acrylate and methyl methacrylate. As seen in Table 5, IEM has vinyl reactivity similar to methyl methacrylate when copolymerized with styrene or butyl acrylate.

Table 5. Vinyl Group Relative Reactivity Ratios

Temperature	M_1	M_2
110°C*	IEM $R_1 = 0.49$	Styrene $R_2 = 0.54$
99°C	Methyl methacrylate $R_1 = 0.48$	Styrene $R_2 = 0.54$
115°C*	IEM $R_1 = 1.62$	Butyl acrylate $R_2 = 0.13$
60°C	Methyl methacrylate $R_1 = 1.74$	Butyl acrylate $R_2 = 0.20$
115°C*	IEM $R_1 = 1.04$	Methyl methacrylate $R_2 = 0.74$
60°C	Methyl methacrylate $R_1 = 1.09$	Ethyl methacrylate $R_2 = 0.98$
60°C	Methyl methacrylate $R_1 = 0.79$	Butyl methacrylate $R_2 = 1.27$

*Frisch, Kresta, Wongkamolsech, University of Detroit, Unpublished Data

Many more copolymers of IEM have been reported; these will be reviewed in the application section. Thus IEM can be used to make a whole range of copolymeric isocyanate hardeners. The variety of comonomers available allows design of the glass transition temperature, solubility parameter and cross-link density of resins to meet the needs of the coating or adhesive application.

IEM REACTIONS: TRIMERIZATION

Burdett (6) found that IEM can be trimerized to form the tris-substituted isocyanurate derivative:

$3\ CH_2{=}CCH_3CO_2(CH_2)_2NCO$

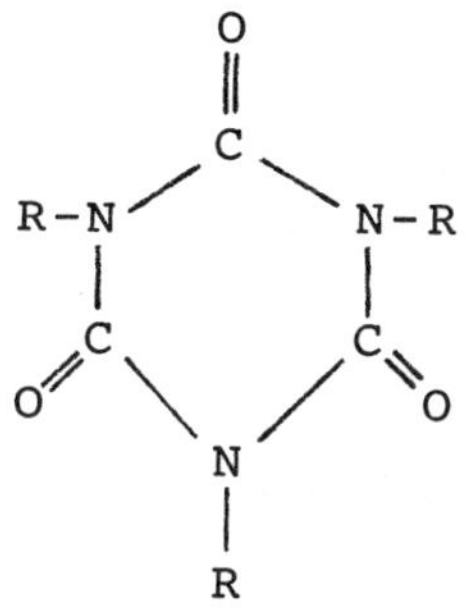

Where R = $CH_2{=}C(CH_3)CO_2(CH_2)_2$

The trimerization has been conducted in acetonitrile catalyzed by trialkyl- or triarylphosphines. This trimer has a crystalline melting point of 81.5 - 82.5^{o}C and can be used to provide cross-linking and high thermal stability to thermoset polymers.

IEM APPLICATIONS

The applications for IEM which have appeared in the literature can be grouped into three areas:

1. Applications where the methacrylate group is polymerized first, leaving the NCO group for a later reaction.

2. Applications where the NCO group is reacted with a polyfunctional material first, leaving the vinyl group for a later reaction.

3. Applications where the NCO group is reacted with a monofunctional reagent first to make a new monomer which can be later copolymerized.

Applications Where The Methacrylate Group Is Polymerized First

The synthesis of IEM was first reported by Bortnick in 1955 (7). In this same reference he taught the copolymerization of IEM with methyl methacrylate in the presence of .05% azobisisobutyronitrile (AIBN) to give cross-linked polymers. The cross-linking was probably due to residual moisture in the reaction. Bortnick reported the use of these polymers to render wool resistant to shrinkage.

Polymethyl methacrylate/polystyrene graft copolymers were made by Graham (8) by reaction of the terminal amino group of $NaNH_2$-initiated polystyrene with the pendant NCO groups of methyl methacrylate/IEM copolymers.

Photo-cross-linkable polymers described in a Bayer patent (9) were made by grafting cinnamic acid derivatives to IEM homopolymer and IEM/vinyl acetate or IEM/styrene/maleic anhydride copolymers.

Resins useful for thermoplastic forming are made from IEM copolymers cross-linked with polyphenols (10). This application takes advantage of the deblocking ability of IEM-phenol derivatives.

Tough, moisture-curable films based on polyisocyanates made from IEM are suggested for use in coatings, leather impregnants and seamless floor coverings (11,12).

Several recent patents issued to DuPont teach the copolymerization of IEM to controlled molecular weight oligomers useful in polyurethane enamels. Simms (13) achieves these low molecular weights ($\bar{M}_n$<4000) in one case by using 20-40:1 mole ratios of monomer to initiator. The initiators particularly suitable are those having a half-life at 120°C of ≥1 min, for example tert-butyl peracetate. In another route low molecular weight oligomers containing IEM were made by Brixius and Simms (14) by polymerizing with AIBN in the presence of thioalkyl chain-transfer agents. These thioalkyl-terminated IEM oligomers are useful for making moisture-curable coatings (15,16), and "high solids" two-component enamels.

A unique isocyano-functionalized polymer reported by Hertler was made by reacting 6-hydroxyhexyl isocyanide with IEM-containing acrylic or styrenic copolymers. These polymers can be formulated with trialkylborane cross-linkers (17) or Ni salt oligomerization catalysts (18) to give room-temperature cured enamels.

Diols such as 1,4-butanediol (19) have been reported as cross-linkers for IEM/methyl methacrylate/ethyl acrylate terpolymers to make glossy baked enamels with excellent impact strength.

Finally, epoxy resins containing a plurality of OH groups were cross-linked by Oriel et al. (20) with IEM-containing acrylic copolymers to give coatings which are both tough and weatherable.

In summary, by polymerizing the methacrylate group on IEM with a variety of comonomers, polyisocyanates are formed which are useful in making low temperature-cure resins. While most authors apply these resins to coatings, the same resins could be used in a variety of adhesive applications. The large variety of comonomers available allows one to design the key properties of glass transition temperature, solubility parameter and cross-link density into the resin to meet the requirements of the application.

Applications Where The NCO Group Is Reacted With A Polyfunctional Material First

The first reported use of IEM to graft to polymers was in 1972 by Rantz et al.(21). Hydroxypropyl cellulose was reacted with IEM to produce a polymer which was formulated into a UV-cured photo-release coating. Since then many workers have modified various substrates with IEM to achieve radiation curable resins.

Hydroxypropyl methacrylate/methyl acrylate/methyl methacrylate terpolymers and phthalic acid/trimethylol-propane copolymers were grafted with IEM by Rosenkranz et al. (22,23). These polymers, when formulated with 2-tert-butylanthraquinone, are irradiated with UV light to give insoluble coatings useful in the manufacture of printed circuits and relief printing plates.

Similar utility can be found for the polyisocyanates made from IEM that were discussed in the first application section. For example, hydroxyethyl methacrylate (24) or hydroxyethyl acrylate (25) can be grafted to an IEM-containing acrylic polyisocyanate to give vinyl functionalized resins for electron beam (EB) curing.

Fluorine containing polyols were derivatized with IEM by Newell and Wolf (26) to give novel radiation-hardenable urethane acrylates which are especially useful as adhesives for electro-optical components or as protective coatings.

H_{12}MDI-Poly(butylene isophthalate) diol prepolymers were partially capped with IEM by Darling (27) to form an intermediate which was further modified with 1,6-hexamethylene diamine chain extender. When diluted with other acrylic monomers and formulated with benzoin methyl ether a UV-cured film can be made with excellent tear strength and elongation.

Polyether polyols were reacted with IEM by Tefertiller et al. (28) to give radiation-curable resins.

Water-dispersible, UV-curable polymers were reported by Wendling (29). A polyester made from 1,3-diglycidyl-5, 5-dimethyl-hydantoin and itaconic acid was partially (50%) capped with IEM to give a water-dispersible polymer. This polymer was formulated with 2,2-dimethoxy-2-phenyl-acetophenone to give a photo-release coating that is developed with 10% aqueous ethanol to give an image having excellent resolution.

An acrylic-alkyd graft copolymer which is water-dispersible was prepared by Walus (30) by copolymerizing methyl methacrylate, butyl methacrylate and methacrylic acid in the presence of an epoxy-modified dehydrated castor oil alkyd which was pre-reacted with IEM. When formulated with NH_3 this gives an air-dry water-dispersible alkyd enamel.

Anaerobic adhesives can be made from tetraethylene glycol-IEM adducts. Hoffman (31) formulated this adduct with cumene hydroperoxide and dimethylaniline and applied it to cap screws and nuts. Upon full cure a torque of 3.5 ft-lbs was needed to remove the nut. Polyamines were also reacted with IEM by Frisch Jr. (32,33) without the use of catalysts to give similar anaerobic adhesive resins.

Peroxide-curable resins were prepared by Burton (34) by reacting varying amounts of IEM with the secondary hydroxyl groups on tris-(4-glycidylphenyl)methane trimethacrylate. This gives a hexafunctional epoxy cross-linker which can be formulated with styrene and peroxide to give a high heat-distortion temperature (>230°C at 264 psi) thermoset.

In summary, IEM can be used to graft vinyl functionality onto a variety of polyfunctional materials of both high and low molecular weight. This results in materials which can be later cured using peroxides, UV radiation, electron beam radiation and even air-dry mechanisms. Thus the use of IEM allows a variety of curing mechanisms which are especially useful for high solids or 100% solids formulation of coating and adhesive resins.

APPLICATIONS WHERE THE NCO GROUP IS REACTED WITH A MONOFUNCTIONAL REAGENT FIRST TO MAKE A NEW POLYMERIZABLE MONOMER

Isocyanatoethyl methacrylate can be reacted with monofunctional reagents to give polymerizable monomers. The blocked IEM derivatives which were discussed earlier are one class of compounds which fall into this application area. The first use of blocked-IEM was taught by Sueling and Kuntz in 1964. These workers emulsion copolymerized acrylontrile with IEM derivatives made from ethyl acetoacetate, phenol, sodium bisulfate and tertiary butanol. Cross-linking polyols such as hydroxymethylcellulose, polyvinyl alcohol and sorbitol were present during the emulsion polymerization (35). Alternatively (36), hydroxy-functionalized monomers such as methacrylic acid dihydroxyethylamide, hydroxypropyl methacrylate or 2-methacryloylamino-2-hydroxymethyl-propane-1:3-diol were copolymerized with the blocked-IEM and acrylonitrile. Fibers and films prepared from these emulsion polymers became insoluble in dimethylformamide after heating at 130-140°C.

Coating resins were made by solution polymerization of acetoxime-blocked IEM (37) with 2-hydroxypropyl methacrylate, ethyl acrylate, styrene and mercaptan chain transfer agents. When formulated with TiO_2 and baked at

150°C for 30 minutes hard glossy enamels were formed which were very resistant to boiling water, detergents, alkalis and acids.

Sodium bisulfite (38) was used to block IEM to give a water-soluble monomer which can be homopolymerized or copolymerized with 4-sodium sulphonatobutyl methacrylate. This gives polymers which are room temperature cross-linkers for $-NH_2$ and -OH containing photographic gelatins.

The tough and weatherable epoxy-acrylic graft copolymers made by Oriel et al. (20) that were cited earlier were also made from methyl ethyl ketoxime blocked IEM.

Recently, Saunders (39) conducted a detailed study of emulsion latices made from methyl ethyl ketoxime-blocked IEM (IEM-MEKO). Compositions investigated were styrene/-butyl acrylate/IEM-MEKO, styrene/butyl acrylate/IEM-MEKO latex formulated with active hydrogen compounds and styrene/butyl acrylate/IEM-MEKO copolymerized with vinyl acids or hydroxy monomers. The effect of urethane catalysts on the deblocking/curing reaction was also studied.

Saunders found that deblocking and cross-linking appear to proceed concurrently at temperatures of 150-170°C. Hydroxy-containing monomers such as hydroxyethyl acrylate incorporated into the IEM-MEKO polymer react with the NCO groups formed under deblocking conditions. Formulation of the latex with a polyether polyol results in an isocyanate hydroxyl reaction under these same deblocking conditions. Dibutyl tin dilaurate catalyst appears to promote these cross-linking reactions. Polymer compositions containing IEM-MEKO and acrylic acid or methacrylic acid show an increase in the amount of deblocked NCO reacted with increasing acid content. Catalysts did not significantly change the deblocking or curing reactions of these compositions.

Other derivatives which have been made vinyl polymerizable with IEM can be included into copolymers as an additive rather than as a latent cross-linker. For example, a variety of reagents which improve hydrophilicity as well as acid or base dyeability of acrylonitrile fibers were synthesized by Bahr et al. (40,41). Pharmacologically active derivatives of 1-adamantaneamine (42), polymerizable 2,1,3-benzostannathiazolines (43), polymeric chelating ligands derived from 1,3 bis-(2'-pyridylimino) isoindolines (44) and polymerizable polyglycol surfactants (45,46) are other examples of this application for IEM.

Thus the versatility of IEM suggests a whole range of applications in polyisocyanates, poly vinyl-functionalized resins and polymerizable derivatives. In all of these applications IEM allows latent cross-linking or grafting because of its two independent, low-temperature cure functionalities. The majority of applications reported to date utilize IEM as a latent cross-linker or graft site to make resins suitable for use in coatings or adhesives.

TOXICITY

Inhalation

Acute vapor inhalation tests show that short exposures could be harmful. The LC50 for rats is 25 ppm. Extreme care should be taken to avoid breathing vapors. IEM vapor has been shown to be an upper respiratory irritant at very low concentrations.

At our present level of experience, it is recommended that the average eight-hour working environment not exceed 0.025 ppm.

Eye Contact

IEM has an extremely severe effect upon the eye. Direct contact with this material may result in tissue destruction leading to permanent impairment of vision. If contact occurs, contaminated eyes must be flushed immediately with plenty of flowing tap water for at least 30 minutes. Medical attention must be immediately obtained for the patient. Special and particular precautions must be taken to prevent eye contact with this test material. Chemical worker's goggles are recommended whenever one works with IEM.

Skin Contact

This material has a severe effect on the skin. Brief skin contact with this material, especially if confined under clothing, might result in moderate redness, moderate swelling, and a moderate chemical burn. The LD50 for rabbits is between 1000 and 2000 mg/kg. This material could be absorbed in toxic amounts; especially upon gross confined exposure. Precautions must be taken to prevent any skin contact with this material. Protective clothing, impervious to the material, must be worn whenever the likelihood of skin contact exists. CAUTION; RUBBER SUITS, IF WET WITH IEM, WILL BE PENETRATED IN 10 MINUTES. If skin contact should occur, all contaminated clothing, including shoes, must be removed immediately and the affected skin area flushed thoroughly with water from a safety shower or other suitable device and cleaned with soap and plenty of water. Medical attention should be obtained. Contaminated clothing or shoes must be destroyed since complete decontamination is doubtful. Some isocyanate compounds are known to be skin sensitizers and may produce allergic asthma.

Ingestion

The single dose oral toxicity of the test material is low; the LD50 for rats is approximately 1000 mg/kg when dosed as a 10% solution in corn oil. The undiluted material would be irritating, and might burn the mucous membranes.

THESE CONCLUSIONS ARE BASED UPON RANGE FINDING TOXICOLOGICAL TESTS AND ARE LIMITED TO PRECAUTIONS FOR INDUSTRIAL HANDLING OF THE MATERIAL. DEVELOPMENT OF SPECIFIC USES WILL REQUIRE CONSIDERATION OF POTENTIAL FOR HUMAN EXPOSURE AND OF THE NEED FOR FURTHER TOXICOLOGICAL STUDIES.

FOR SAFE PRACTICE, USERS OF THIS MONOMER SHOULD CONSULT THE BROCHURE "SAFE HANDLING AND STORAGE OF ISOCYANATOETHYL METHACRYLATE," AND THE MATERIAL SAFETY DATA SHEET ON IEM AVAILABLE FROM THE DOW CHEMICAL COMPANY, SPECIALTY CHEMICALS DEPARTMENT, MIDLAND, MI 48640.

ACKNOWLEDGMENTS

The author would like to acknowledge Norm L. Madison and Mary R. Thomas and the many other technical and management personnel who have championed IEM over the last 8 years.

REFERENCES

1. M.R. Thomas, "Isocyanatoethyl Methacrylate: A Heterofunctional Monomer for Polyurethane and Vinyl Polymer Systems," paper given at the Las Vegas ACS Meeting, March 1982.

2. K.C. Frisch, J. Kresta, and K. Wongkamolsech, University of Detroit, unpublished data.

3. T. Regulski and M.R. Thomas, "Isocyanatoethyl Methacrylate II: The Blocked Isocyanate Derivatives, Preparation and Deblocking," paper given at the Seattle ACS Meeting, March 1983.

4. T. Regulski and M.R. Thomas, "Isocyanatoethyl Methacrylate III: Polymerization, Formulation and Evaluation of Blocked IEM Derivatives," paper given at the Seattle ACS Meeting, March 1983.

5. L. Bretherick, Handbook of Reactive Chemical Hazards, Butterworths, London-Boston (1979) p. 142.

6. K. Burdett, The Dow Chemical Company, unpublished data.

7. N.M. Bortnick (Rohm & Haas Co.) U.S. 2,718,516 (1955); Chem. Abstr., 50, 1371c.

8. R.K. Graham (Rohm & Haas Co.) U.S. 2,882,259; Chem. Abstr., 53, 1274i.

9. Farbenfabriken Bayer A.-G. Brit. 822,861 (1959); Chem. Abstr., 54, 9574b.

10. Farbenfabriken Bayer A.-G. Fr. 1,567,523 (1969), Brit. 1,187,194 (1970), Brit. 1,187,195 (1970); Chem. Abstr., 72, 113693n.

11. W.D. Emmons (Rohm & Haas Co.) Ger. Offen. 2,018,233 (1971), U.S. 3,743,626 (1973); Chem. Abstr., 76, 4548b.

12. B. Quiring, K. Wagner (Farbenfabriken Bayer A-G) Ger. Offen. 2,031,408 (1971), Brit. 1,354,783 (1974); Chem. Abstr., 76, 114926f.

13. J.A. Simms (duPont deNemours, E.I., and Co.) Ger. Offen. 3,007,792 (1980), U.S. 4,219,632 (1980); Chem. Abstr., 93, 241267t.

14. D.W. Brixius, J.A. Simms (duPont deNemours, E.I., and Co.) U.S. 4,222,909 (1980); Chem. Abstr., 94, 158473g.

15. duPont deNemours, E.I., and Co., Belg. 881,844 (1980); Chem. Abstr., 94, 67421x.

16. D.W. Brxius and J.A. Simms (duPont deNemours, E.I., and Co.) U.S. 4,351,755 (1982); Chem. Abstr., 97, 212801r.

17. W.R. Hertler (duPont deNemours, E.I., and Co.) U.S. 4,251,422 (1981); Chem. Abstr., 94, 176894z.

18. W.R. Hertler (duPont deNemours, E.I., and Co.) U.S. 4,251,421 (1981); Chem. Abstr., 94, 210482a.

19. Anon, Res. Disclosures, 195,272 (1980); Chem. Abstr., 94, 141287b.

20. S.L. Oriel, N.B. Tefertiller, J.W. Bozzelli, and T. Regulski (The Dow Chemical Company) U.S. 4,264,748 (1981); Chem. Abstr., 95, 44870z.

21. E. Ranz, W. Himmelmann, P. Berthaller, and J. Hocker (Agfa-Gevaert A.-G.) Ger. Offen. 2,112,141 (1972), U.S. 3,782,950 (1974); Chem. Abstr. 78, 153703s.

22. H.J. Rosendranz, H. Rudolph, E. Wolff, and H. VonRintelen (Farbenfabriken Bayer A.-G.) Ger. Offen. 2,121,252 (1972), U.S. 3,928,299 (1975); Chem. Abstr., 78, 112781x.

23. H.J. Rosenkranz, H. Rudolph, E. Wolff, and H. VonRintelen (Agfa-Gevaert A.-G.) Ger. Offen. 2,121,253 (1972), Brit. 1,384,343 (1975); Chem. Abstr., 78, 112782y.

24. G. Paal, U.D. Straehle, and G. Kielhorn, J. Electrochem. Soc., 120 (12), 1714-11. (1973); Chem. Abstr., 80, 71509k.

25. N.B. Tefertiller and R.F. Harris (The Dow Chemical Company) U.S. 4,343,919 (1982); Chem. Abstr., 97, 164700g.

26. R.G. Newell and F.S. Wolf (Minnesota Mining and Mfg. Co.) Ger. Offen. 3,013,827 (1980); Chem. Abstr., 94, 48340d.

27. T.R. Darling (duPont deNemours, E.I., and Co.) U.S. 4,097, 439 (1978); Chem. Abstr., 89, 181360y.

28. N.B. Tefertiller, K.A. Burdett, and R.F. Harris (The Dow Chemical Company) U.S. 4,233,425 (1980), U.S. 4,287,323 (1981); Chem. Abstr., 94, 48956.

29. L.A. Wendling (Minnesota Mining and Mfg. Co.) Eur. Pat. Appl. EP 40,002 (1981); Chem. Abstr., 96, 152850x.

30. A.N. Walus (duPont deNemours, E.I., and Co.) Ger. Offen. 3,038,215 (1981, U.S. 4,273,690 (1981); Chem. Abstr., 95, 44858b.

31. D.K. Hoffman (The Dow Chemical Company) U.S. 4,320,221 (1982);, Chem. Abstr., 96, 163867j.

32. K.C. Frisch, Jr. (The Dow Chemical Company) U.S. 4,374,969 (1983).

33. The Dow Chemical Company, Belg. BE 891,620 (1982); Chem. Abstr., 97, 111042n.

34. B.L. Burton (The Dow Chemical Company) U.S. 4,338,242 (1982); Chem. Abstr., 97, 128687z.

35. C. Sueling and E. Kuntz (Farbenfabriken Bayer A-G) Belg. 652,949 (1964), U.S. 3,453,223 (1969); Chem. Abstr., 64 14361c.

36. Farbenfabriken Bayer A.-G. Neth Appl. 6,409,962 (1965), U.S. 3,299,007 (1967); Chem. Abstr., 63, 5813g.

37. Farbenfabriken Bayer A.-G. Fr. 1,551,838 (1968), U.S. 3,542,739 (1970); Chem. Abstr., 74, 40315d.

38. H.W. Wood (Ilford Ltd.) Brit. 1,322,971 (1973); Chem. Abstr., 79, 141506.

39. F.L. Saunders, The Dow Chemical Company, published anonymously at Res. Discl. 209, 10 Sept. 10, 1981.

40. U. Bahr, H. Wieden, H.A. Rinkler, and G. Nischk, Makromol. Chem., 161, 1-47 (1972); Chem. Abstr., 78, 85675lw.

41. H. Rinkler, H. Krimm, G. Nischk, and H. Wieden (Bayer A.-G.) Ger. 1,518,903 (1973); Chem. Abstr., 79, 126114q.

42. H. Ringsdorf, H. Ritter and H. Rolly, Makromol. Chem. 177(3), 741-6 (1976); Chem. Abstr., 86, 50559k.

43. H. Belmares-Sarabia, U.S. 4,038,300 (1977); Chem. Abstr., 88, 7650n.

44. W.D. Siegl, Chem. Ind. (London) 8,291-2 (1981); Chem. Abstr., 95, 116041e.

45. Anon, Res. Discl. 208, 309-10; Chem. Abstr., 95, 152485m.

46. D.K. Hoffman, R.F. Harris, N.B. Tefertiller, and C. Randall (The Dow Chemical Company) Eur. Pat. Appl. 30,307 (1981); Chem. Abstr., 95, 204881j.

Moisture Permeation of Polymer Sealants and Interface Modifying Films

Bonnie L. Rathbun and Philipp wh. Schuessler

Process Engineering Department
IBM - Federal Systems Division
Owego, NY 13827

ABSTRACT

The original intent of this project was to identify and/or perfect vendor-supplied, state-of-the-art hydrophobic polymer sealants, such that the end product could meet military criteria for hermeticity in hybrid microelectronic devices. Solubility and diffusivity constants and permeabilities for several types of hydrophobic polymers were determined. Data from the laboratory and theory from literature both indicate that moisture impermeable polymers are not attainable, at least to meet military standards. The project objective was then redirected towards finding a means of passivating the surface of a microelectronic device against moisture. It was found that interface modification, i.e., passivation of the chemically reactive surface species, provides significant protection in high temperature and humidity environments. Plasma polymerization of hexamethyldisilazane (HMDS) afforded phenomenal protection to moisture sensitive thin film nichrome resistors even under the most severe test conditions.

INTRODUCTION

The efforts and results discussed herein were taken from the first two phases (1,2) of a multiphase project dedicated to identifying a means of making lower cost military microelectronic devices hermetic to moisture. Hermeticity, as defined by Military Standard 883, Method 1018 and its associated limits, would be achieved when less than 5000 ppm of moisture are found in a device tested at 100°C at any time in the service life of the device.

Microelectronic devices for military use are subjected to stringent mechanical and environmental tests during various stages of their fabrication. The study presented in this paper was mainly concerned with hybrid devices which contain thin metallic film circuitry and discrete components such as capacitors, transistors and diodes bonded to the circuitry. These hybrid devices must be sealed to protect the internal components from exposure to moisture and other environmental agents which could degrade the microelectronics. At present, military hybrid packages are most often hermetically sealed by glass seals, soldering or welding. These are costly processes. Also, hybrid packages thus sealed are not easily, nor inexpensively, reopened if repair to interior parts becomes necessary. Most hybrids for military use are custom-made in small quantities and usually quite costly compared with commercial counterparts. Thus, it becomes cost-conscious to repair a faulty military hybrid, if possible, rather than scrap it.

The original objective for the research presented here was to find a low cost, repairable seal for hybrid devices which would meet military requirements for hermeticity. Some polymer sealants were claimed to be "impermeable to water" or "hydrophobic". If this proved to be the case, and they were also easily removed for repair, then a low cost hybrid package might be constructed using polymer sealants.

HYDROPHOBIC POLYMERS

Experimental Procedures

Literature searches and studies of product literature from sealant manufacturers produced a list of six types of potential hydrophobic polymers to be tested. Polymers were selected for either having a highly nonpolar nature with a resultant low modulus or for being moderately polar with high cross-link density. Either of these characteristics was believed to reduce or eliminate the diffusion of moisture into the polymer.

Various test methods for moisture were screened. An optimum method for determining moisture absorption and diffusion through the polymer was selected. This method involved tracking the weight changes of a small polymer sample on a McBain balance when subjected to 100% relative humidity at constant temperature. The sample was first dried by vacuum in the balance. Any weight gain by the sample when exposed to 100% relative humidity would allow for the calculation of the material's moisture permeability rate.

Each material tested was processed into a thin film approximately 0.1 - 1.0 mm thick when cured. The sample was placed on a quartz spring in a McBain balance (Fig. 1) and evacuated to less than 1 torr until it appeared stabilized, i.e., no further loss of weight for a minimum of 48 hours. The McBain balance consisted of a one-meter long, jacketed Pyrex® glass cylinder containing the quartz spring. Weight changes of the sample were monitored via the vertical displacement of the sample using a cathetometer. There was a detection limit for the cathetometer of 0.05 mm which translated to weight changes of 0.005%.

Upon sample stabilization, moisture, at 100% relative humidity for the column temperature, was released into the balance chamber. Weight changes due to absorption of water were noted. After equilibrium was established (sample was presumed to be saturated), the sample was returned to vacuum to dry and weight changes due to loss of moisture were recorded.

Raw data obtained during periods of absorption and desorption was ultimately translated into the constants S (solubility), D (diffusivity), and P (permeability), via the equation of Springer and Shen (3).

Solubility was defined as the maximum amount of water absorbed by the material as shown by the asymptote in Fig. 2, a typical absorption curve for one of the polymers tested. The diffusivity constant was calculated from the above

$$D = \left[\frac{-2.3L^2}{\pi^2}\right]\left[\frac{\Delta \log \quad (M_{max}-M)}{\Delta t}\right]$$

where M_{max} = Maximum % moisture

M = % moisture at time, t

Δt = Initial small increment of time (hrs)

L = Thickness of sample (mm)

D = Diffusivity mm^2/hr

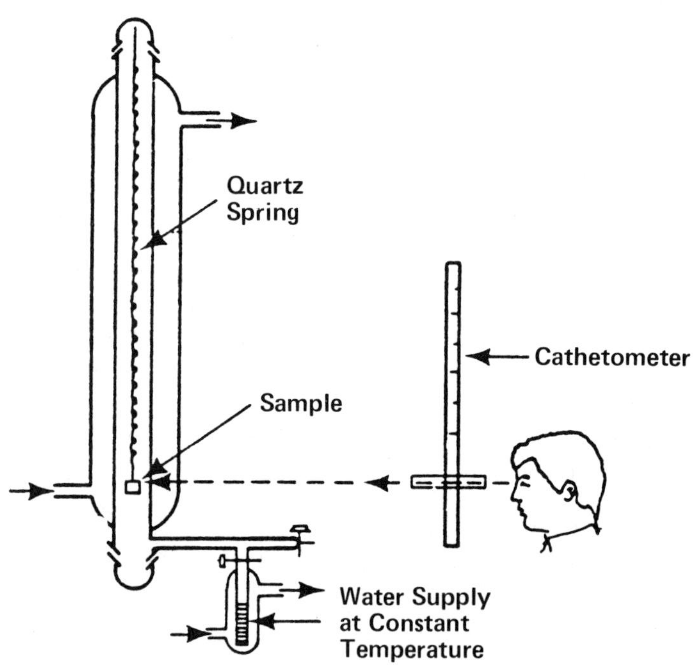

Fig. 1. Typical McBain Balance: Total spring excursion approximates 1 meter.

formula using data generated in the first few hours of the test. Permeability is the product of the solubility and diffusivity constants in the appropriate units.

Results and Discussion

Most of the hydrophobic polymers tested exhibited absorption curves similar to the fluoroacrylic in Fig. 2. The resultant constants and a list of the polymers tested are given in Table 1. It is interesting to note that all materials tested, except the silastyrene, had discernible moisture permeabilities and thus cannot be considered hydrophobic or impermeable by military standards.

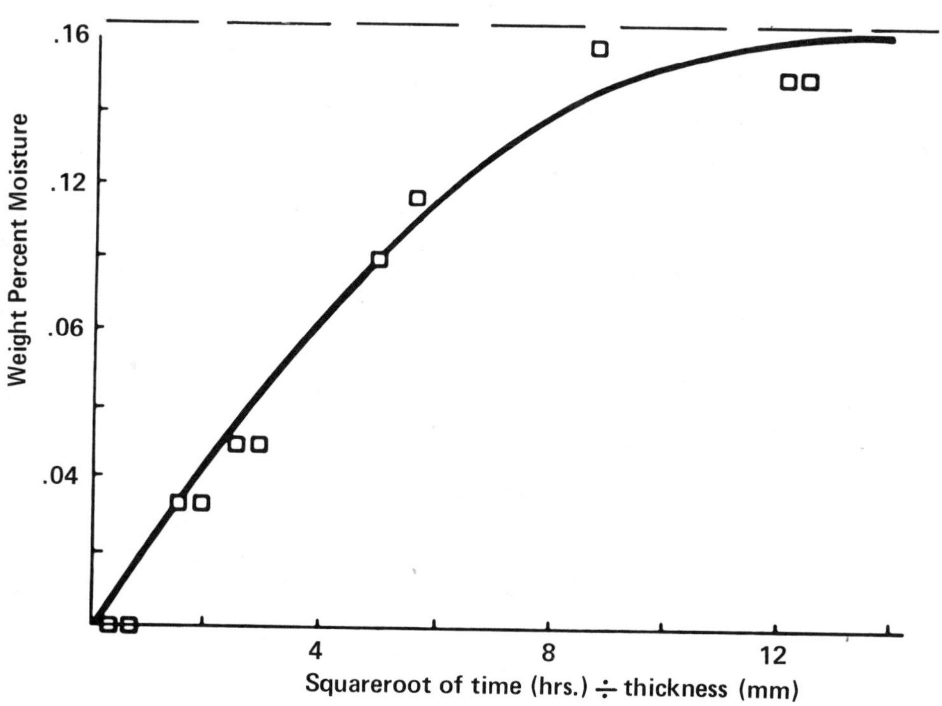

Fig. 2. Typical absorption curve with data points (□) for the fluoroacrylic polymer tested.

Table 1. Moisture Solubility, Diffusivity and Permeation Constants

Material	Diffusivity $\left(D\ \frac{cm^2}{sec}\right)$	Solubility $\left(S\ \frac{cm^3\ H_2O}{cm^3\ polymer\text{-}cm\ Hg}\right)$	Permeability (P=D x S)
Siloxyimides	$1.5\text{-}9.4\times10^{-6}$	0.069-0.100	$1.035\text{-}9.21\times10^{-7}$
Fluorosilicone	Not calculable due to anomalies		
Epoxy Novolac	3.8×10^{-8}	0.50	1.9×10^{-8}
Fluoroacrylic	3×10^{-7}	0.06	5.4×10^{-8}
Phenylated Silicone	3.8×10^{-5}	0.025	9.5×10^{-7}
Silastyrene	No absorption detected		

The range of values given for the siloxyimides covers three different materials tested of that type. Fluorosilicone demonstrated various anomalies which did not permit application of the given equation. The silastyrene showed no detectable moisture solubility. This material was selected as potentially hydrophobic due to its apparent low polarity. It contains methyl groups bonded to a silicon backbone. Although silastyrene has potential, its inventors (4) indicate that the material has UV and solvent degradation problems, i.e., environmental and process instabilities. Thus this polymer was not tested further.

The foregoing moisture solubility data shows that the term "hydrophobic polymer" is really not applicable for the materials tested, except possibly the silastyrene. A brief review of some theoretical presentations by L.H. Lee and others (5,6) indicate that the initial objective of finding a hermetic polymer sealant may never be realized. Fig. 3 from a presentation by Lee (5) reveals the reason for this assumption.

Essentially, the term "dispersion forces" relates to van der Waals forces which exist between all molecules regardless of their polarity. The wide energy spectrum from 10^{-1} to 10 Kcal/mole per bond for these forces indicates intermolecular attraction between nonpolar molecules, such as a hydrophobic polymer, and a polar molecule, such as water, cannot be minimized sufficiently to prevent moisture from absorbing and penetrating a polymer.

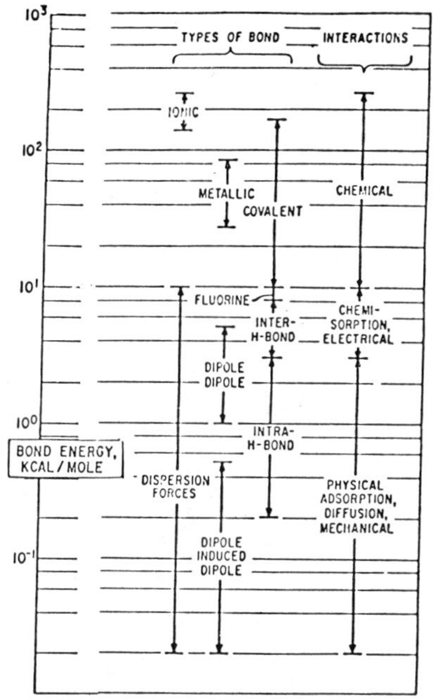

Fig. 3. Physical and Chemical Interactions at Coupling Agent Interfaces (Ref. 5).

The results of the McBain balance study support this concept. Although the polymers tested were chosen to reduce dipole-dipole interactions, moisture still permeated the polymers. Even those assumed to have a high cross-link density failed to block water molecules from diffusing into the polymer structure.

Thus, experimental and theoretical evidence supports the conclusion that hydrophobic polymer sealants for potential use in a military hybrid device do not exist at this time.

Based on this conclusion, a new or different packaging approach for microelectronic devices was sought. It was one that used relatively low temperatures and was amenable to repair processes. One probable solution which was studied was applying the moisture barrier to the surface of all the various materials/devices within the package. Surface passivation or interface modification with a reactive species which would not be subsequently desorbed by moisture was the focus for the balance of the research presented here.

INTERFACE MODIFICATION

Experimental Procedures

To test the effectiveness of the surface passivation, sample hybrid substrates without attached components were exposed to 100% relative humidity at 95°C for 1,000 hours while connected to a power supply. Both coated and uncoated substrates were monitored for change in resistance due to degradation, i.e., both electrolytic and galvanic, of circuit parts by moisture. Substrates (Fig. 4) consisted of six parallel circuits on alumina, each containing two to four nickel-chromium thin film resistors in series. This type of resistor is well known to be highly moisture sensitive and thus should be a good indicator of the passivation effects of the coatings. A 12-Volt DC bias, with sporatic disruptions, was applied to each circuit. The change in voltage of a known resistance in series with each circuit was monitored over 1,000 hours. The change in the circuit resistances was then calculated and plotted versus time for each circuit.

Materials selected as interface modifiers for surface passivation were chosen by their potential to react with surface oxides and hydroxyls where moisture would likely interact. The three which were ultimately tested are polyphenylsilsesquioxane (PSS), hexamethyldisilazane (HMDS), and N-β-(N-vinylbenzylamino) ethyl-γ-aminopropyltrimethoxysilane-monohydrogen chloride ("silane"). The last material proved to be corrosive to the test nichrome resistors due to the chloride part of the molecule. Therefore, most of the study centered around the first two materials.

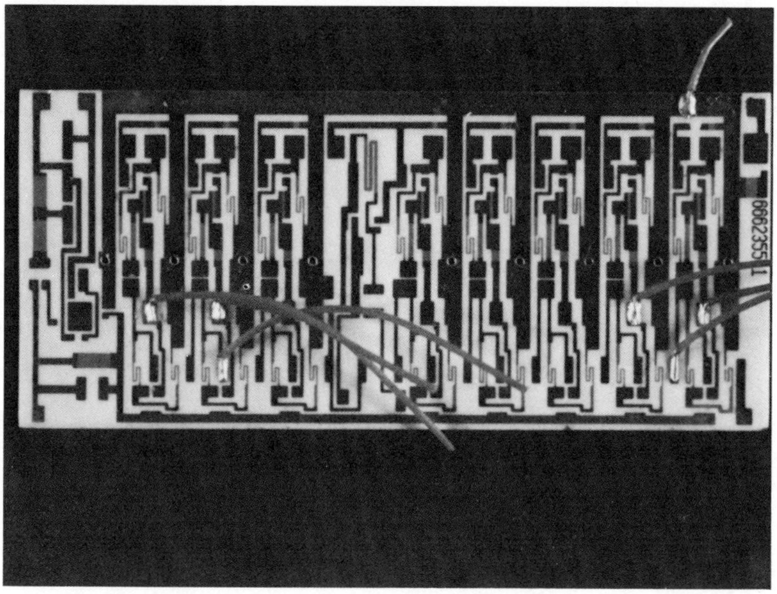

Fig. 4. Photograph of Sample Substrate Circuitry

Two methods of application of the materials were used: dip coating and plasma polymerization. Several test substrates were dipped into diluted solutions of the monomer followed by solvent evaporation at 100°C for several hours. Dilution provided a relatively thin coating compared with the as-received concentration of the material.

The second method of applying the interface modifier was plasma polymerization. This process was known to provide a uniform, pinhole-free conformal coating. In plasma polymerization, a "monomer", the starting organic material, was exposed to an induced, low power (10-30W) radio frequency (13.56 MHz). This served to cleave the material into energetic particles which polymerized on the surface of the substrate. The process was conducted at room temperature with a 1000 Å layer deposited in about one hour. With proper cleaning of equipment and refining of variables, higher rates of deposition could be obtained.

Fig. 5 is a schematic of the plasma polymerization system used to deposit the HMDS on the test substrates (7). A quartz crystal oscillator served as a thickness monitor.

The "monomer" used was 98% pure HMDS with no carrier gas employed. This process is sometimes used for depositing HMDS as an adhesion promoter in photolithographic techniques. It gives a pinhole-free coating of highly cross-linked polymer.

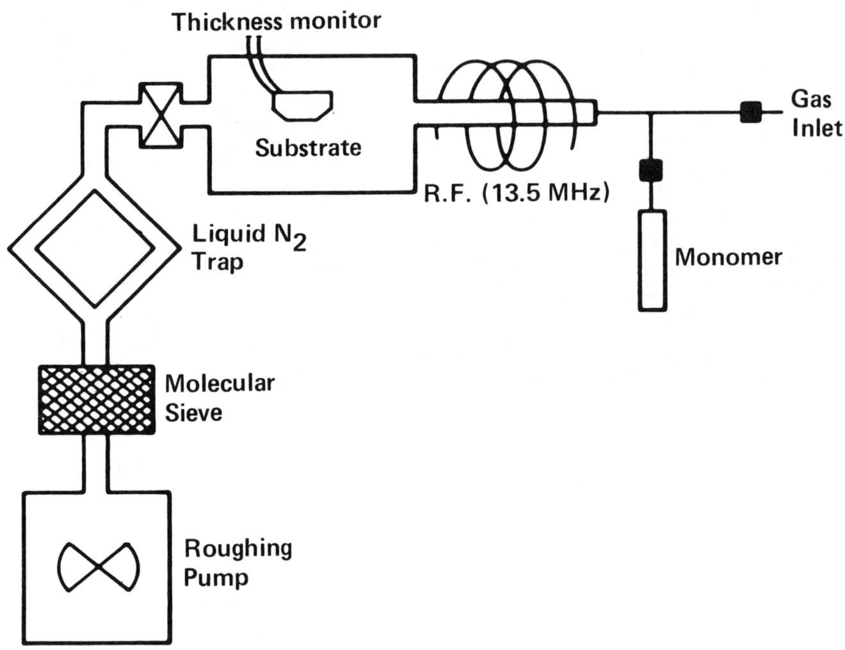

Fig. 5. Schematic of a Plasma Polymerization System

In addition to the temperature and humidity testing, several repairability tests were done using the more promising interface modifiers. The ideal coating was desired to be a good moisture-passivating agent, yet be easily removed should repair work be required. To remove the coatings, a soft eraser abrasion was used.

Results and Discussion

The results for the temperature and humidity tests are given in Table 2. Failures were indicated by a high resistance value due to an open circuit. Usually this occurred as a nichrome resistor disintegrated from a reaction with the moisture, which was further enhanced by the power-on condition. Slight drifting of the circuit

Table 2. Effects of Surface Passivation

Material	Coating Method	T&H Test
Control	No Coating	50% Failed 15.4% Drift
HMDS	Dip	33% Failed 7.5 Drift
HMDS	Plasma	No Failures 1.7% Drift
PSS	Dip	No Failures 4.2% Drift
"Silane"	Dip	50% Failed 11.3% Drift

resistance is tolerable, but large changes under power-on conditions inevitably lead to circuit failures. Drift indicated in Table 2 is the average for the remainder of the six circuits per substrate which did not fail during the test period.

Note that the unprotected circuits of the control had substantial drift of the three that did not fail during the test run. Failures were evident upon inspection of the part as occurring from disintegrated resistors. Dip coated PSS and HMDS had fewer failures and less drift. The best protection from moisture was offered by the plasma-polymerized HMDS. The indicated 1.7% drift is close to the acceptable limits of the design for power-on thermal aging effects.

Fig. 6 illustrates the average resistance change as a function of time under test conditions for the control, PSS and HMDS (plasma) coated substrates. This graph is based on the circuits which survived the 1000 hours at high temperature and humidity.

Based on these results, the plasma-polymerized HMDS has been shown to be effective in passivating moisture-sensitive thin film resistors. The method of application did appear to give a pinhole-free coating, compared to the dip method for the same material. Also HMDS has been widely used as a primer or coupling agent to aid in the adhesion of photoresists to various surfaces. Thus used and as employed in this study, HMDS chemically bonds to the surface sites (oxide and hydroxyls) where water reacts thus preventing interaction of the surface with moisture. In the application of resist, the coupling agent also provides reaction sites for the adhesion of the resist. For our study, no additional layer was put over the interface modifier.

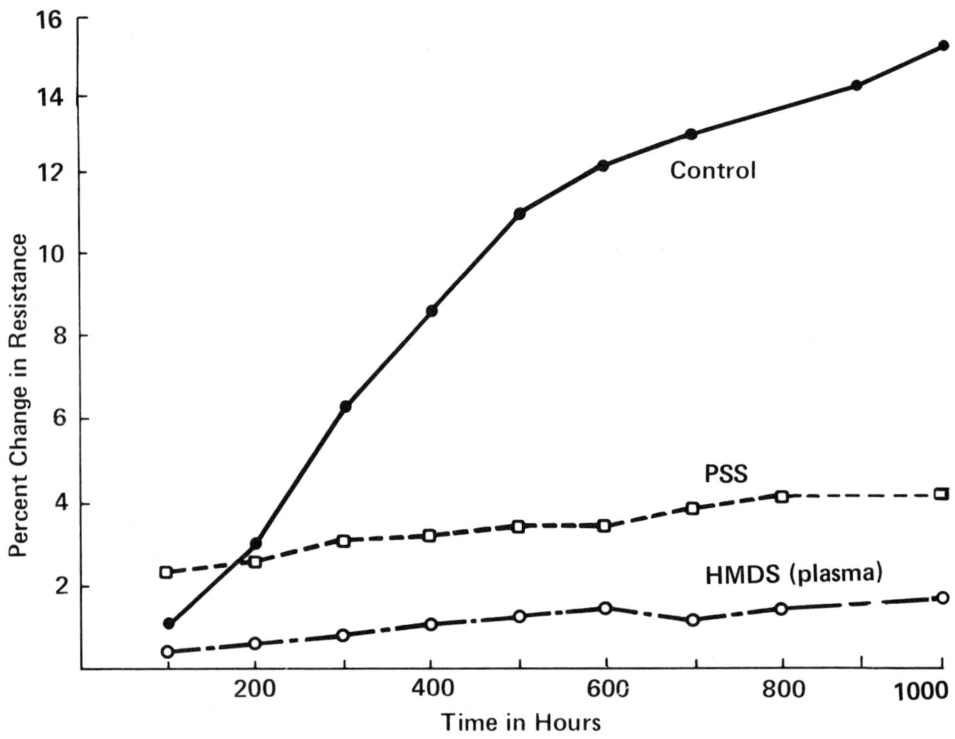

Fig. 6. Average change in resistance of coated and uncoated circuits with time exposed to 95°C, 100% relative humidity.

The repairability tests did not show as significant a difference as did the temperature and humidity tests. Plasma-deposited HMDS was easily removed by abrasion and did not leave a residue to interfere with repair work. However, the PSS left a gummy residue which did adversely affect electrical integrity of a repaired sample part.

The thin 1000 Å layer of plasma deposited HMDS was fragile to probing and scratches. However, once applied to the surface of the completed hybrid, it would be sealed by a cover within the hybrid package. This seal need not be hermetic, providing the interface modification film continues to prove effective in preventing moisture degradation of hybrid parts.

CONCLUSIONS

The first part of this project verified the theoretical conclusion that no polymer can be truly hydrophobic (allowing no water penetration) by nature of dispersion forces which attract nonpolar polymer molecules and polar water molecules. These forces cannot be eliminated. All of the "hydrophobic" polymer sealants tested exhibited diffusivity and solubility constants. Thus, sealing a hybrid with such polymers would eventually allow moisture penetration to the interior of the hybrid which subsequently could not meet MIL-STD-883, Method 1018 requirements of a hermetic hybrid.

As a result of these findings, the thrust of the project shifted toward identifying a way to prevent moisture from reacting with sensitive parts within a hybrid. By reacting a polymer with the surface sites where water might interact, the moisture inside the hybrid would be prevented from harming the parts. This passivation technique has been termed "interface modification".

Several such interface modifiers were found to have the potential of meeting the requirements of hermeticity and repairability. One, however, was outstanding in both categories. A plasma-polymerized hexamethyldisilazane (HMDS) coating provided effective protection for nichrome thin film resistors in high temperature and humidity, power-on conditions for 1,000 hours. HMDS was a very thin coating (approximately 1000 Å) which was easily removed for repair purposes.

Combining the knowledge gained from both parts of this study, it is possible that a low cost, hermetic hybrid could be assembled using polymers. If diffusion of moisture is not blocked by a polymeric seal, the moisture which does penetrate the interior of the hybrid may be prevented from damaging the internal parts by coating them with a passivating material.

Further tests of interface modifiers and their use on hybrids for passivation is the focus of the next phase of this project.

REFERENCES

1. P. wh. Schuessler, "Hydrophobic Resins", NASC Phase I Report, May 1981.

2. P. wh. Schuessler and B.L. Rathbun, "Hydrophobic Resins and Interface Modifiers", NASC Phase II Report, February 1982.

3. G.W. Springer and C.H. Shen, "Moisture Absorption and Desorption of Composite Materials", AFML-TR-76-102, June 1976.

4. Robert West, et al., Department of Chemistry, University of Wisconsin, Madison.

5. L.H. Lee, "The Adhesion of Solids Derived from Van der Waals Forces", Fundamentals of Adhesion - Theory, Practice and Application, SUNY New Paltz, October 7-9, 1981.

6. F.M. Fowkes, "Forces at the Interface", Polymer Conference Series, University of Utah, June 24-28, 1974.

7. N.K. Eib, K.L. Mittal, and A. Friedricks, "Wettability Studies of Plasma Polymerized Hexamethyldisilazane", J. Appl. Polym. Sci., 25, 2435-2438 (1980).

ESCA and AES Studies of the Interfacial Chemical Bonding Between Aluminum and Chromium (III) Fumarato-Coordination Compound

Yang Qi-Xiang*

Department of Chemistry
Kansas State University
Manhattan, KS 66506

and Zhou Qin-Li

Department of Chemistry
Nanking University
Nanking, People's Republic of China

ABSTRACT

The chromium(III) fumarato-coordination compound, commercially known as Volan 82, is a coupling agent developed by DuPont in the early 1970s. It has proven to be a very successful bonding agent between aluminum and polyethylene. Aluminum foil treated by aqueous chromium(III) fumarato-coordination compound solution can be coated with polyethylene to form water-resistant and durable composite film which is a useful packaging material and a composite shielding for wire cable. We wish to report the application of ESCA and AES in the study of the interfacial chemical bonding between aluminum and chromium(III) fumarato-coordination compound. The ESCA and SIMS analyses of the aluminum surface treated by aqueous chromium(III) fumarato-coordination compound solution show that chromium is retained on the surface. The spatial distribution of chromium on the aluminum surface was studied by Scanning Auger Microprobe. The Auger images of Al and Cr show that Cr Silver was partly deposited on the aluminum foil followed by treatment with chromium(III) fumarato-coordination compound. Auger line scan shows that Al, O and Cr have the same distribution on this line. The AES depth profile indicates that chromium(III) fumarato coordination compound is bonded not only with the outermost layers of aluminum oxide but also with the sub-surface layers of aluminum oxide on the aluminum surface. The ESCA technique was also used to study the Al 2p binding energy chemical shift. On the untreated aluminum foil surface, the aluminum atoms exist in two chemical forms: metallic Al with Al 2p binding energy of 72.4 eV, and Al_2O_3 with Al 2p binding energy of 74.8 eV. After the aluminum foil was treated by aqueous chromium(III) fumarato solution, the ESCA Al 2p peak was obviously broadened, the reacted Al_2O_3 binding energy shifted to 75.2 eV. Apparently, the chemical bonding between Al_2O_3 and chromium(III) fumarato coordination compound causes a chemical shift of Al_2O_3 Al 2p binding energy by 0.4 eV. The compositional change within the aluminum-polyethylene interface of the composite film was measured by AES depth profile technique. ESCA was used to measure the Al 2p binding energy at different levels within the interface. The results indicate that in the aluminum-polyethylene interface aluminum atoms exist in three chemical forms: metallic Al, unreacted Al_2O_3 and Al_2O_3 chemically bonded with chromium(III) fumarato coordination compound. The existence of these three forms of aluminum atoms confirms the chemical bonding mechanism between aluminum and the coupling agent.

*This paper is a part of Mr. YANG Qi-Xiang's Master thesis at the Department of Chemistry, Nanking University, Nanking, People's Republic of China.

INTRODUCTION

The aluminum-polyethylene composite film has been found to be a useful packaging material. Because polyethylene is chemically inert and has low polarity, it is difficult to bind polyethylene to an aluminum surface to form a durable composite film. This is particularly true for the aluminum-polyethylene interface of the composite film which may be easily attacked by water. In the early 1970s, DuPont Company developed a new coupling agent, aqueous chromium(III) fumarato – coordination compound solution known commercially as Volan-82 (1-5), which proved to be very successful in promoting the adhesion between aluminum and polyethylene. With the application of Volan-82, the aluminum-polyethylene composite film manufactured has not only obtained high bonding strength but it also has excellent water-resistant properties. In this research, the composite films were soaked respectively in dilute NaOH solution, dilute acetic acid solution, sea water and tap water for three years, and after three years' immersion the films were still in good physical condition (Fig. 1).

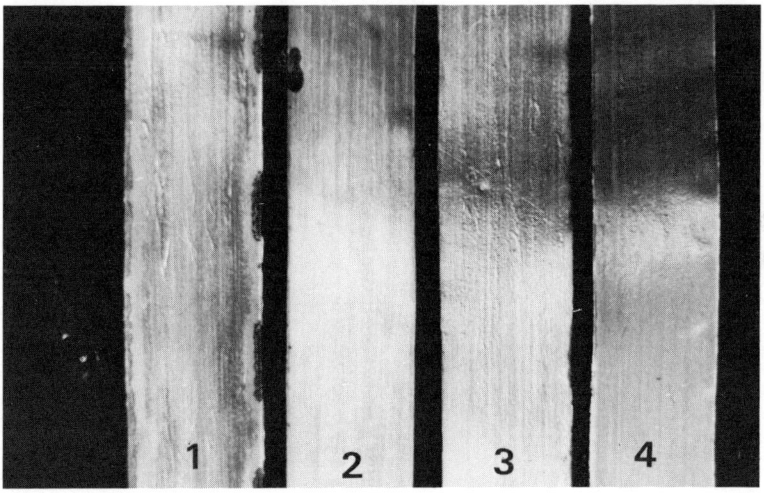

Fig. 1. Aluminum-Polyethylene Composite Films Soaked in Different Mediums for Three Years.

1. In 0.6% aqueous NaOH solution.
2. In 0.5% aqueous acetic acid solution.
3. In sea water.
4. In tap water.

Because of the superior water-resistant property, the aluminum-polyethylene composite film has also been used as shielding of wire cable to replace lead in the People's Republic of China.

Chromium(III) fumarato-coordination compound has a complex structure which is pH-dependent. The Cr^{3+} ions are linked by fumaric group or hydroxy bridge to form polymers: (2,3)

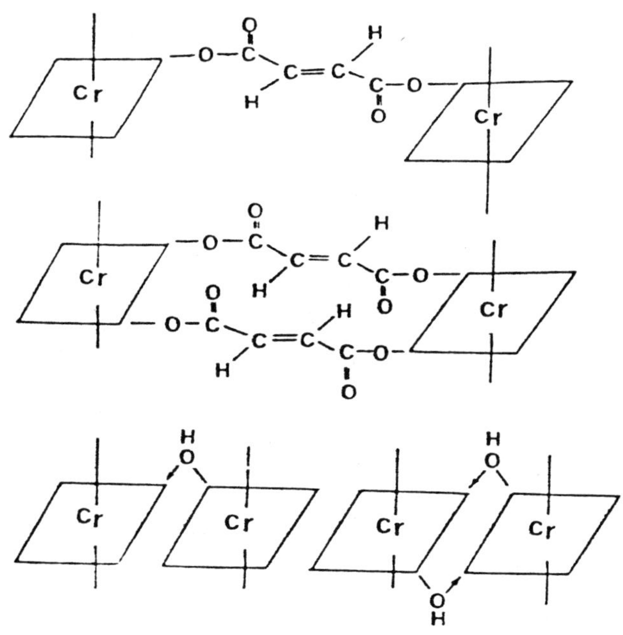

The water molecule coordinated with Cr^{3+} can be converted into hydroxy group by elimination of a proton. After chromium(III) fumarato-coordination compound was precipitated from its aqueous solution at pH of 2.7, the elemental analysis showed that the ratio of chromium to fumaric acid was 1:1.29.

The excellent water-resistant property of aluminum-polyethylene composite film was the result of chemical treatment of aluminum foil with the coupling agent. This treatment is now used in the commercial manufacturing of these composite films. This excellent adhesion suggests the formation of chemical bonding at the aluminum-polyethylene interface. It is generally believed that coupling agent can act as a "molecular bridge", which bonds the incompatible inorganic materials and organic polymers with a chain of primary bonds (6,7). J. A. Robertson and J. W. Trebilock suggested the formation of a new chromium-oxygen-aluminum bond between chromium(III) fumarato-coordination compound and aluminum oxide on the aluminum surface, but no evidence was given to support this assumption (8). We chose ESCA and AES to study the surface of aluminum foil treated by aqueous chromium(III) fumarato coordination compound solution and the aluminum-polyethylene interface. Scanning Auger Microprobe provided information about the distribution of chromium on the aluminum surface and the change of elemental composition of the aluminum-polyethylene interface, while ESCA was used to measure the chemical shift of Al 2p binding energy. Direct evidences have been obtained in this study to support the suggested chemical bonding mechanism between chromium(III) fumarato- coordination compound and aluminum. The combination of ESCA and AES is a powerful technique to elucidate the chemical bonding at the interface of the composite film.

EXPERIMENTAL

1. The Preparation of Chromium(III) Fumarato-Coordination Compound

Chromium(III) fumarato coordination compound (Volan-82) was prepared by the commercial method developed by DuPont Company (1). It can be described as follows: 20g $Cr(NO_3).9H_2O$ and 40g fumaric acid were dissolved in 940 ml distilled water, refluxed for one hour. When it was cooled to room temperature, 33 ml 1N $NaHCO_3$ solution was added to the reactants, another 33 ml and 34 ml 1N $NaHCO_3$ solutions were added following the same process of refluxing and cooling, the reactants were finally refluxed for 3 hours. Upon cooling, the unreacted furmaric acid was filtered, the aqueous chromium(III) fumarato-coordination compound solution (Volan-82) was obtained.

2. The Treatment of Aluminum Foil by Aqueous Chromium(III) Fumarato Solution

The industrially manufactured aluminum foil (typical thickness of 0.15 mm) was used. The aluminum foil was immersed in 8% NaOH solution at the temperature of 40°C for 30 seconds, rinsed first with tap water then with distilled water. The aqueous chromium(III) fumarato solution (Cr concentration = 2.43 mg/ml) was diluted five-fold, heated to 70°C. The clean aluminum foil was immersed in the solution for 5 minutes, rinsed with distilled water and ethyl alcohol, then dried at 100°C.

3. The Manufacturing of Aluminum-Polyethylene Composite Film

The aluminum-polyethylene composite film was manufactured by the industrial equipment at Nanking Communication Cable Company, People's Republic of China. The aluminum foil was passed through the tank containing 10% NaOH solution at 40°C, rinsed with water, dried by oven, then passed through a tank containing aqueous chromium(III) fumarato solution (Cr concentration = 0.24 mg/ml) at 70°C, dried again by oven, then passed through the extrusion molding where molten low density polyethylene (LDPE) was heated to 300-330°C, the molten polyethylene was combined with aluminum foil to form composite film at the pressure of the roller. The thickness of polyethylene on aluminum was 0.08 mm.

4. The ESCA, AES and SIMS Instrumentation

PHI 550 ESCA/SAM electron spectrometer was used to conduct all the ESCA and AES experiments. Mg Kα x-ray source (400 W) was used for ESCA. PHI 2500 Secondary Ion Mass Spectrometer was used for SIMS experiment. The ESCA experimental deviation of the measurmeent of binding energy is $\pm$ 0.1 eV.

RESULTS AND DISCUSSION

1. Elemental Analysis of the Aluminum Surface Treated by Aqueous Chromium(III) Fumarato Solution

Figures 2 and 3 illustrate the ESCA and SIMS spectra of

the surface of aluminum foil treated by aqueous chromium(III) fumarato solution. In Fig. 2, the double peaks between 580 eV and 590 eV are Cr 2p electron binding energy. These peaks show that after the aluminum foil was treated by the coupling agent, which remained on the surface. In the SIMS spectrum (Fig. 3), beside Cr peaks, peaks of the ligand fragments of the coupling agent were also found, such as $CH=CHCO^+$ (Mass 54), $C=CHCO^+$ (Mass 53), $COOH^+$ (Mass 45). The SIMS spectrum confirms that the chromium(III) fumarato-coordination compound was retained on the aluminum surface.

For comparison, aluminum foil was treated by chromium nitrate solution (Cr concentration = 0.48 mg/ml) in the same way as it was treated by the coupling agent. The ESCA spectrum of the aluminum surface treated by chromium nitrate solution is shown in Fig. 4. Most important is the fact that no Cr signal is observed in this spectrum; thus, only when the chromium(III) ion is coordinated by fumaric acid, it can become so active as to be successfully bonded to the aluminum surface.

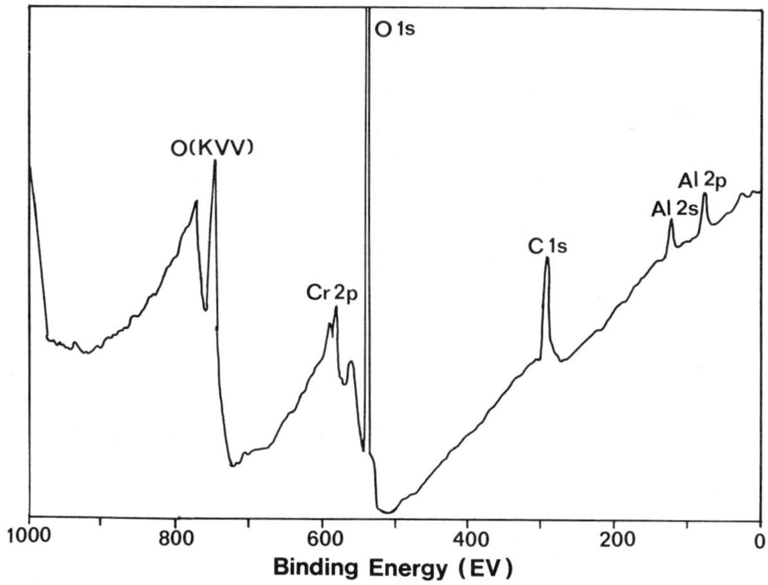

Fig 2. ESCA Spectrum of the Aluminum Foil Treated by Aqueous Chromium(III) Fumarato-Coordination Compound Solution

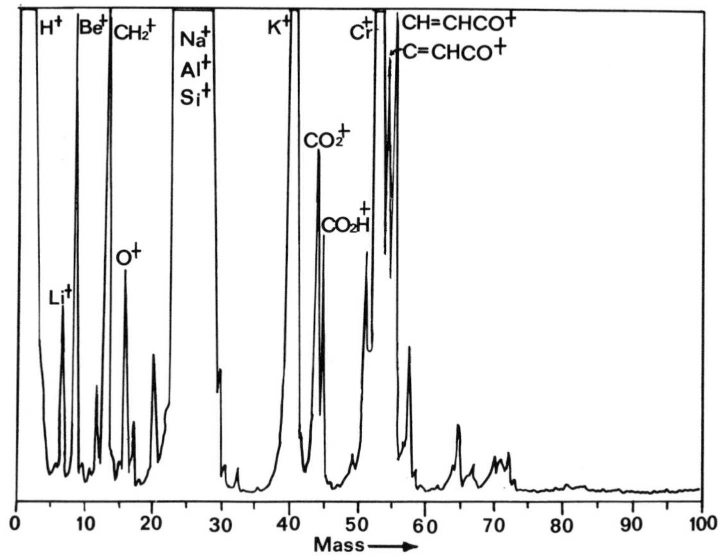

Fig. 3. SIMS Spectrum of the Aluminum Foil Treated by Aqueous Chromium(III) Fumarato-Coordination Compound Solution.

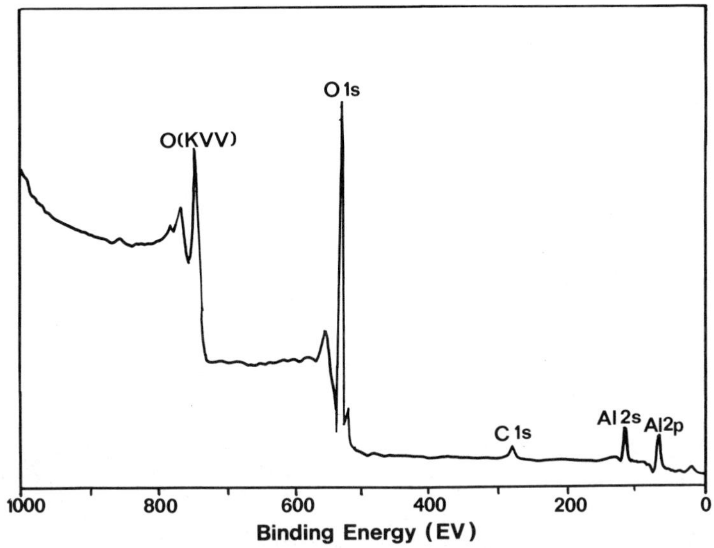

Fig. 4. ESCA Spectrum of the Aluminum Foil Treated by Aqueous Chromium Nitrate Solution.

2. Spatial Distribution of Chromium on the Aluminum Surface Treated by Aqueous Chromium(III) Fumarato Solution

Scanning Auger Microprobe was employed to study the spatial distribution of chromium(III) fumarato-coordination compound on the aluminum surface. Since we have previously known that once the aluminum foil was treated by Volan-82, the coupling agent was retained on the surface; and if chromium(III) fumarato-coordination compound is chemically bonded with Al_2O_3 on the aluminum surface, then the chromium should be distributed on the aluminum surface as homogeneously as aluminum itself. The Al and Cr Auger images in Fig. 5 show the distribution of Al and Cr on the surface of aluminum foil treated by aqueous chromium(III) fumarato solution. It is obvious from Fig. 5 that chromium was homogeneously distributed on the aluminum surface.

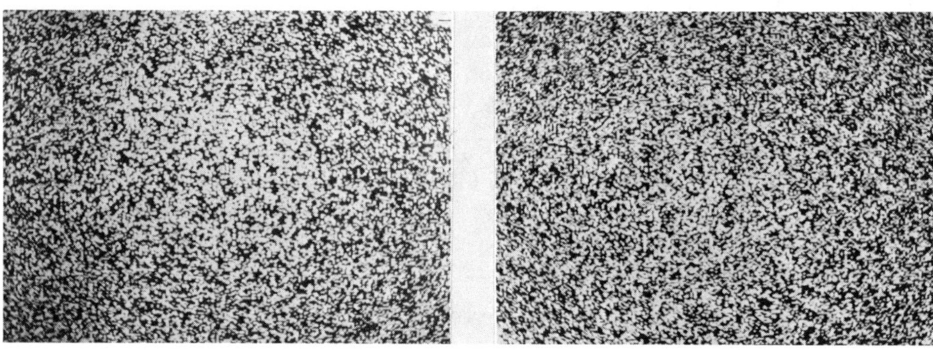

Al Cr

Fig. 5. Cr and Al Auger Images of the Aluminum Foil Surface Treated by Aqueous Chromium(III) Fumarato-Coordination Compound Solution.

The previous results lead us to believe that chromium(III) fumarato coordination compound was bonded with Al_2O_3 on the aluminum surface. In order to confirm this, we partly deposited aluminum foil with silver, then treated it with aqueous chromium(III) fumarato solution by the same method previously described. The AES spectrum of the surface, Fig. 6, demonstrates the presence of Al, O, Ag, Cr and C. ESCA studies indicate that silver is in the metallic form. Figure 7 is the Auger Line Scan diagram of Al, Cr and O, which shows that Al, Cr and O have the same distribution on the line, thus proving that the bonding only takes place between aluminum oxide and the chromium-containing coupling agent on the aluminum surface. This is in accordance with the chemical bonding mechanism that the OH groups of chromium(III) fumarato-coordination compound react with OH groups on the aluminum surface to form chromium-oxygen-aluminum bonds.

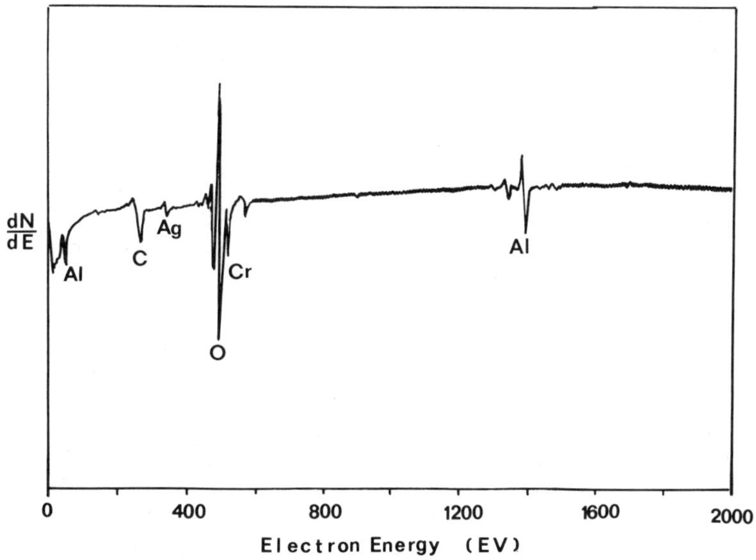

Fig. 6. AES Spectrum of the Aluminum Foil Partly Deposited by Ag and Treated by Aqueous Chromium(III) Fumarato-Coordination Compound Solution.

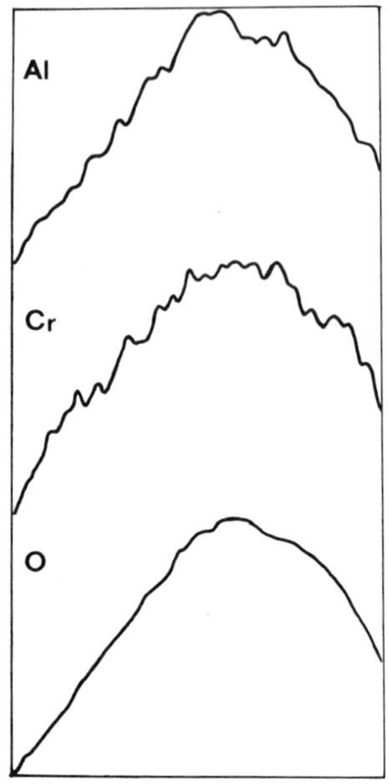

Auger Line Scan

Fig. 7. Al, Cr and O Auger Line Scan of the Aluminum Foil Partly Deposited by Ag and Treated by Aqueous Chromium(III) Fumarato-Coordination Compound Solution.

We also use Argon ion sputtering and Auger depth profile technique to study the distribution of chromium on the surface layers of aluminum foil. Figure 8 is the Auger depth profile of the aluminum surface treated by aqueous chromium(III) fumarato solution. The result from Fig. 8 indicates the variation of relative atomic concentration of Al, Cr, O and C as a function of sputtering time. It was also observed that the chromium-containing coupling agent was bonded not only with the outermost layers of Al_2O_3, but also with the sub-surface layers of Al_2O_3 of aluminum foil. After the surface underwent Ar^+ sputtering for 20 minutes, the relative atomic concentration of Al, O and C was 85.4%, 10.1%, and 2.8% respectively, the Cr concentration was 1.8%. These results confirm our belief that the chromium(III) fumarato-coordination compound can penetrate into the sub-surface layers and bond at these sub-surfaces. Figure 9 shows the variation of the relative atomic concentration ratio of Cr to C as a function of

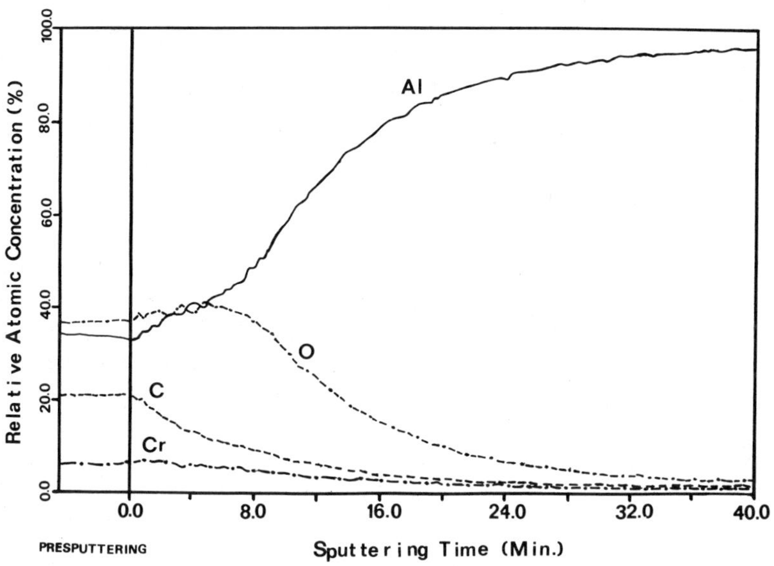

Fig. 8. AES Depth Profile of the Aluminum Foil Treated by Aqueous Chromium(III) Fumarato-Coordination Compound Solution.

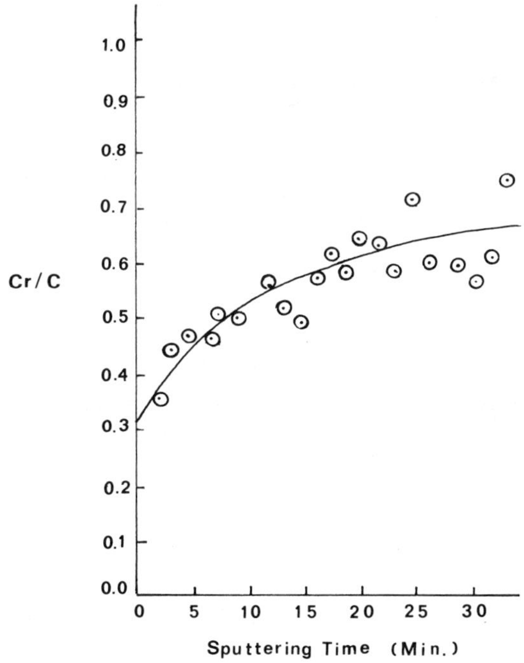

Fig. 9. Relative Atomic Concentration Ratio of Cr to C as a Function of Argon Ion Sputtering Time.

sputtering time. Figure 9 indicates that Cr/C ratio increases as sputtering time increases. The increase in Cr/C ratio with sputtering time can be explained as follows: Because it is the OH group coordinated with Cr^{3+} which reacts with Al_2O_3 on the aluminum surface, OH group and Cr^{3+} will penetrate deeper into the sub-surface layer to combine with the unreacted Al_2O_3 than the "non-active" fumaric group of the coordination compound, so Cr/C ratio increases as the depth increases. This is good yet indirect evidence to support the chemical bonding mechanism between chromium(III) fumarato-coordination compound and aluminum oxide. This phenomenon will be discussed in more detail in a future paper.

3. Chemical Shift of Al 2p Electron Binding Energy on the Aluminum Surface Treated by Aqueous Chromium(III) Fumarato Solution.

The Al 2p binding energy shift was studied by ESCA, and the chemical bonding between aluminum oxide and chromium(III) fumarato-coordination compound was directly confirmed from the experimental evidence. Figure 10 is the Al 2p ESCA spectrum of the untreated aluminum surface used as a comparison in this study. Before observing the ESCA spectrum, we use Ar^+ sputtering to etch the aluminum surface for 5 minutes; thus, the increased intensity of metallic Al 2p peak can be used as a standard in the measurement of binding energy. On the untreated aluminum surface, there are two forms of aluminum atoms: metallic Al with Al 2p binding energy of 72.4 eV and Al_2O_3 with Al 2p binding energy of 74.8 eV. Figure 11 is the Al 2p ESCA spectrum of the aluminum surface treated by aqueous chromium(III) fumarato solution. The sample surface was first sputtered by Ar^+ for 3 minutes before examination. When this surface was examined, the spectrum showed Al 2p binding energy of the metallic Al remained at 72.4 eV, but Al 2p peak of Al_2O_3 is obviously broadened. The Al 2p binding energy of unreacted Al_2O_3 is 74.6 eV. (The difference of approximately 0.2 eV in the measurement of Al 2p binding energy of unreacted Al_2O_3 is mainly due to the experimental deviation), the Al 2p peak at 75.2 eV represents the Al 2p binding energy of Al_2O_3 which is chemically bonded to the chromium(III) fumarato-coordination compound. So we conclude that the chemical

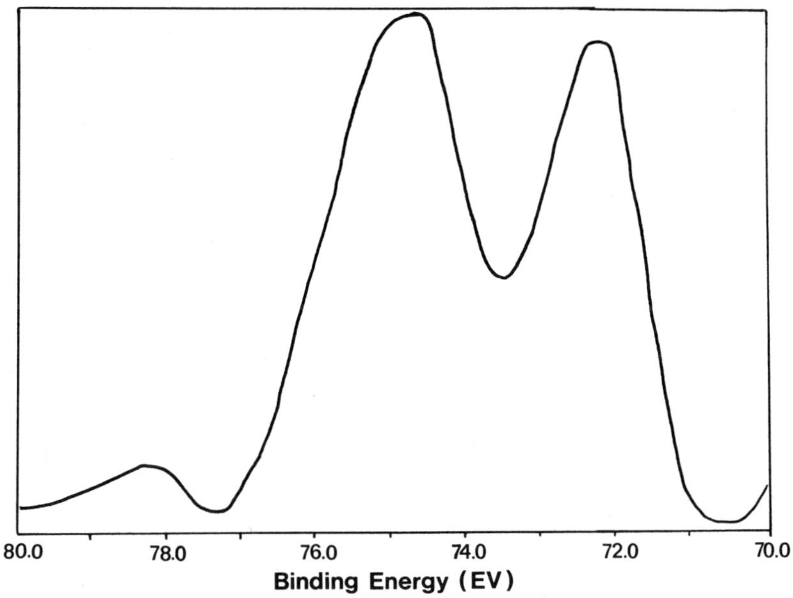

Fig. 10. Al 2P ESCA Spectrum of Untreated Aluminum Foil.

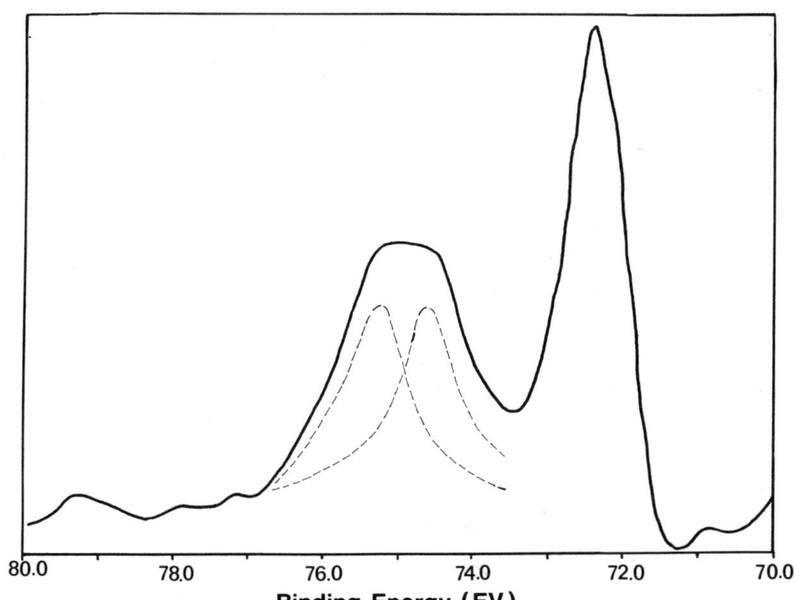

Fig. 11. Al 2p ESCA Spectrum of the Aluminum Foil Treated by Aqueous Chromium(III) Fumarato-Coordination Compound Solution.

shift of Al 2p binding energy of aluminum oxide on the aluminum surface is the result of formation of a new primary bond. The surface of aluminum foil is covered with aluminum oxide; the aluminum oxide formed at room temperature on the aluminum foil surface is proved to be amorphous Al_2O_3 (9) which contains a great number of hydroxy groups. The hydroxy group coordinated with Cr^{3+} can react with the hydroxy group on the aluminum surface by the elimination of a water molecule:

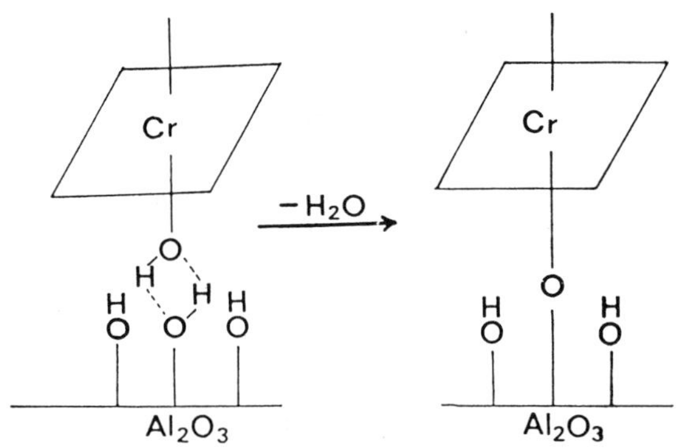

It can be explained that because the molecule of chromium(III) fumarato-coordination compound is much bigger than hydrogen atom, when the hydrogen of the hydroxy group on the aluminum surface is replaced by Chromium(III) fumarato-coordination compound, the sp^3 orbital of the oxygen bonded with aluminum (Al^{3+}) on the surface is converted into linear sp orbital. The more electronegative sp orbital decreases the effective nuclear charge of Al^{3+} of the aluminum surface, which increases the Al 2p binding energy by 0.4 eV. This is a direct evidence to confirm the chemical bonding between Al_2O_3 on the aluminum surface and chromium(III) fumarato-coordination compound.

4. Chemical Shift of Al 2p Binding Energy in the Aluminum-Polyethylene Interface.

The commercial aluminum-polyethylene composite film was kept in a soxhlet extractor and xylene was used in a 14-hour extraction to remove the polyethylene from the composite film. After removal of the polyethylene film the remaining film was rinsed by ethyl alcohol and dried. The dried film was used in the study of aluminum-polyethylene "free" interface by AES depth profile technique.

The advantage of AES depth profile is that the change of elemental composition can be detected readily, but little chemical information can be obtained. Therefore, the AES study was made along with ESCA measurements. The ESCA is capable of providing chemical shift information by measuring the Al 2p binding energy. By the combination of these two techniques, we can learn about chemical bonding at different levels in the aluminum-polyethylene "free" interface and confirm the chemical bonding mechanism between aluminum oxide and chromium(III) fumarato-coordination compound.

Figure 12 gives the results of the AES depth profile of aluminum-polyethylene interface. At five different levels, the Ar^+ sputtering was intervened, and Al 2p binding energy was measured by ESCA. Argon sputtering time, relative atomic concentration of carbon and aluminum at different levels are listed as follows:

POSITION NUMBER	Ar^+ SPUTTERING TIME (MIN.)	C(%)*	Al(%)*
1	0	85	12
2	2.6	37	37
3	5.7	18	70
4	8.9	10	86
5	20	3	95

*Relative atomic concentration

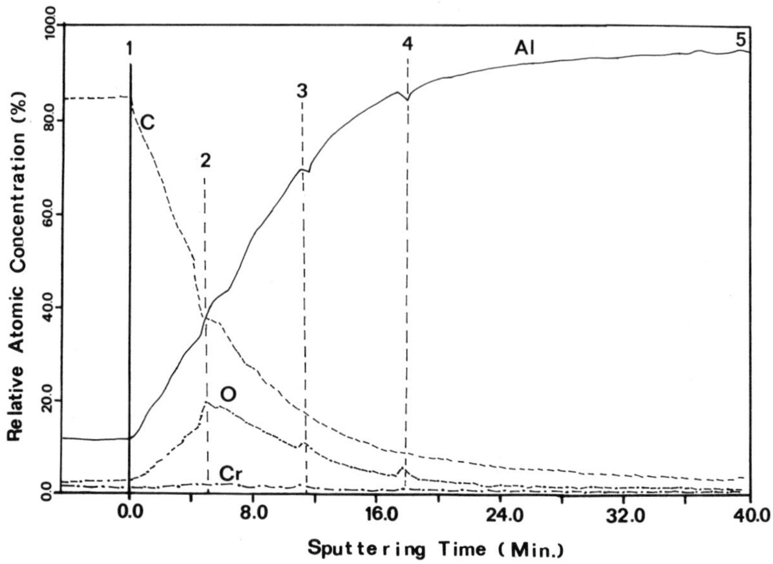

Fig. 12. AES Depth Profile of the Aluminum-Polyethylene Interface.

Figure 13 is the Al 2p ESCA spectra taken at 5 different levels at the aluminum-polyethylene interface. The first ESCA spectrum was taken before Ar^+ sputtering. The Al 2p peak of Al_2O_3 is broad. The peak at 74.6 eV represents the Al 2p binding energy of unreacted Al_2O_3 and the "shoulder" at 75.2 eV represents the Al 2p binding energy of Al_2O_3 chemically bonded with chromium(III) fumarato coordination compound. In the second ESCA spectrum, the Al 2p peak of Al_2O_3 is at 75.2 eV. In the third and fourth ESCA spectra, Al 2p peak of Al_2O_3 remains broad. In the fifth ESCA spectrum, Al 2p binding energy of Al_2O_3 is 74.8 eV; at this level, Al relative atomic concentration is 96% and most aluminum exists in metallic form. It is obvious that in the aluminum-polyethylene interface, aluminum atoms exist in three forms: metallic Al, unreacted Al_2O_3 and Al_2O_3 chemically bonded with chromium(III) fumarato-coordination compound.

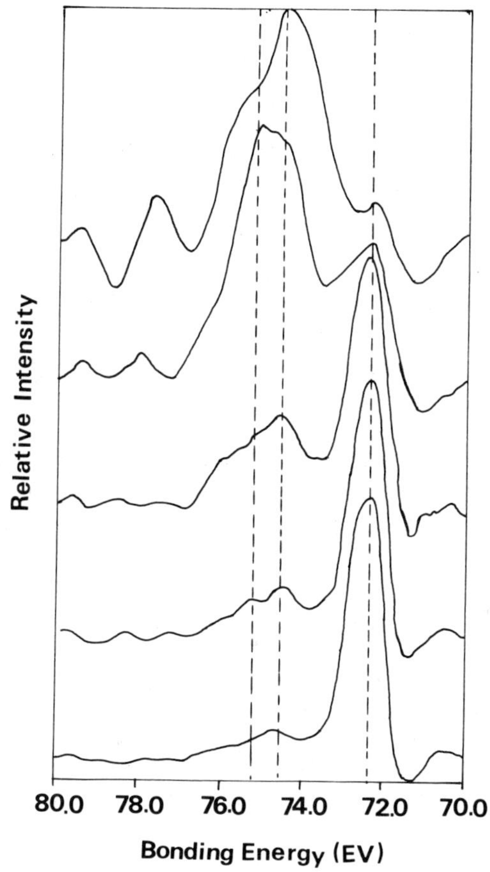

Fig. 13. Al 2p ESCA Spectra of 5 Different Levels of the Aluminum-Polyethylene Interface.

For comparison, we refer to the Al 2p ESCA of untreated aluminum foil (Fig. 10), which was observed after the surface was etched by Ar^+ for 5 minutes. It is evident that on the aluminum surface the aluminum atoms exist in two chemical forms: metallic Al and Al_2O_3, and the corresponding binding energies are 72.4 eV and 74.8 eV respectively. If we compare Fig. 13 with Fig. 10, the following conclusion is well justified.

CONCLUSION

Our studies suggest that chemical bonding does exist between aluminum and chromium(III) fumarato-coordination compound at the aluminum-polyethylene interface.

ACKNOWLEDGMENTS

Special acknowledgment must be made to Mrs. LU Jin and Mr. CHENG Wen-Feng of the Research Institute of Nanking Chemical Industries, People's Republic of China for their continued assistance with the ESCA and AES experiments. The authors also would like to thank the assistance of Mr. GU Ten-Cheng of Nanking Communication Cable Company, People's Republic of China. The authors are in debt to Dr. Wm. J. Fateley for his help in the preparation of the final manuscript.

REFERENCES

1. A. J. Deyrup, U.S. 3,787,326 (assigned to duPont Company), Jan. 22, 1974.

2. H. B. West, U.S. 3,725,448 (assigned to duPont Company), Apr. 3, 1973; U.S. 3,843,699 (assigned to duPont Company), Oct. 22, 1974.

3. C. C. Cumbo, U.S. 3,775,455 (assigned to duPont Company), Nov. 27, 1973; U.S. 3,950,506 (assigned to duPont Company), Apr. 13, 1976.

4. H. E. Bergna, U.S. 3,804,802 (assigned to duPont Company), Apr. 16, 1974.

5. F. I. Bunger and E. P. Moore, Jr., U.S. 3,705,183 (assigned to duPont Company), Dec. 5, 1972.

6. S. J. Monte, Additives for Plastics, Vol. 1, p. 69, Academic Press, New York, 1978.

7. E. P. Pleuddemann, Silane Coupling Agent, p. 17, Plenum Press, New York, 1982.

8. J. A. Robertson and J. W. Treibilock, TAPPI, 58 (4), 106-109, 1975.

9. F. Keller and J. D. Edwards, "The Behavior of Oxide Film on Aluminum", Proceedings of Intern. Conf. Sufr. React., Pittsburgh, Pennsylvania, 1975.

Adhesive Research in China

Lieng-Huang Lee

Webster Research Center
Xerox Corporation
Webster, New York 14580

ABSTRACT

In May and early June of 1983, I was invited by the National Academy of Sciences and the Chinese Academy of Sciences to lecture in the People's Republic of China. During that trip, I observed some impressive progress in adhesive research. This is a brief account of their research activities and their general interest in adhesion. The institutes mentioned in this report are: Guangzhou Institute of Chemistry, Academia Sinica, Shanghai Institute of Organic Chemistry, Academia Sinica, Institute of Chemistry, Academia Sinica, Heilongjiang Institute of Petroleum Research and Beijing Institute of Aeronautics and Astronautics (BIAA).

I. INTRODUCTION

In the past five years, the People's Republic of China (PRC) has been striving to make progress under the Four Modernizations movement (1). There has been a frenzied pace of construction in public buildings, public dwellings, and highways. Consequently, there has been a constant demand for materials such as adhesives and sealants for construction.

During May and the first few days in June, I traveled and lectured in the PRC under the Distinguished Scholar Exchange Program administered by the National Academy of Sciences and the Chinese Academy of Sciences. I visited eight research institutes and nine universities. During that trip, I was able to find some information about their research in adhesives. In this paper, I shall briefly describe some of their research activities and their general interest in adhesion.

II. MAJOR ADHESIVE RESEARCH ACTIVITIES AT INSTITUTES AND UNIVERSITIES

A. Guangzhou Institute of Chemistry, Academia Sinica

This Institute has 410 technical and research members of whom 80 are research scientists and 120 are assistants. Professor Ye Zuozhou, a chemist and an archivist, was one of the four deputy directors. He was the host for my recent visit. This Institute has nine divisions engaged in: cellulose chemistry, marine organic chemistry, turpene chemistry, polymeric adhesives, polymer composites and polymeric grouting compounds.

Their work on adhesives was not a major effort. Some anaerobic adhesives were developed by the Institute, and later produced by the Oriental Scientific Instruments Import & Export Corporation in Guangzhou. Two of the series are GY-340 anaerobic adhesive-sealant and GY-168 anaerobic flexible sealant. Both have a polyacrylate-base, with a shelf-life of one year. The curetime for GY-340 was 2 to 6 hours at room temperature; the curetime for GY-168 was 12 to 24 hours.

Their interesting work on polymer grouting compounds under the direction of Prof. Ye has grown from a laboratory curiosity to developed practical applications, e.g., protecting statuary from decay and preventing stress fractures in buildings and dams. One of their approaches (2) appeared to be filling sand and rocks with an acrylic acid salt solution containing a small amount of 1,3,5-triacryloyl-s-triazine.

B. Shanghai Institute of Organic Chemistry, Academia Sinica

This Institute has made its name from the synthesis of Bovine insulin and one of the nucleic acids. Its polymer research is only a small effort in comparison with such other activities as the synthesis of nucleic acids, polypeptides, and the characterization and purification of an abortion drug. Their better-known polymer research has been on the fluoropolymers. This Institute has 800 technical staff including 34 senior members and 280 middle-level research personnel. Currently it has accepted 100 graduate students.

They have made contributions on fluoropolymers especially on high-temperature and low-temperature adhesives and elastomers. One interesting copolymer is ethylene and tetrafluoroethylene (1:1), which is emulsion polymerized in the presence of a persulfate catalyst. This copolymer, unlike PTFE, is radiation-resistant. Another fluoropolymer project is the copolymerization between tetrafluoroethylene and hexafluoropropene. One high-temperature polymer, perfluoro poly-p-phenylene was prepared by this Institute through the following reaction (3):

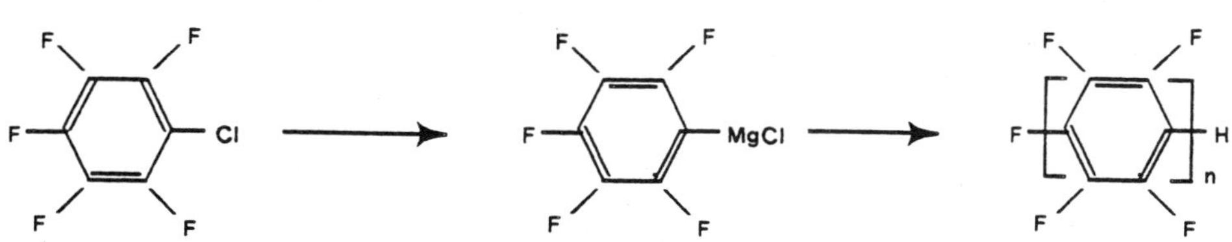

Elastomers which can withstand cryogenic temperatures, (e.g., $-320^{\circ}F$) are not readily available. Previously, this Institute reported (3) success in preparing perfluorinated polyurethanes through the reaction between perfluorophenyl diisocyanate and a perfluoropolyether:

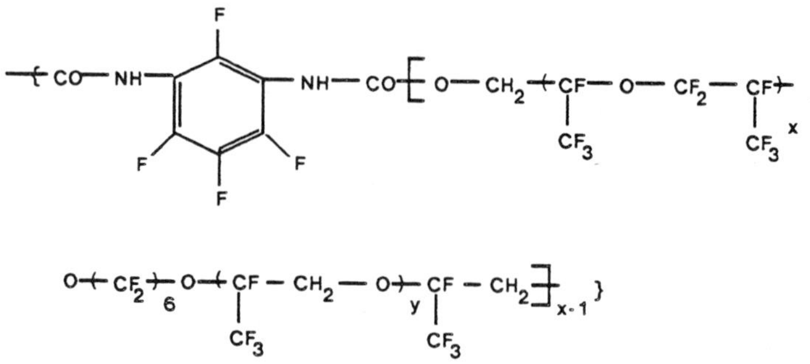

C. Institute of Chemistry, Academia Sinica, Beijing

This Institute has twelve departments and approximately 780 research and supportive personnel (approximately 450 research staff). Recently, they have taken approximately 30 college graduates as advanced degree candidates. Six of the departments of this Institute are polymer-oriented: Polymer Chemistry, Organosilicone, Polymer Physics, Specialized Polymers, Functional Polymers and Polymer Composites. Several projects in high-temperature polymers, adhesives, and composites have been carried out under the direction of Professor Qian Renyuan, a well-known polymer physicist in PRC. A brief description of these projects follow:

1. High-temperature Polymers. Huang and Ou (4) of this Institute synthesized five aromatic dinitriles:

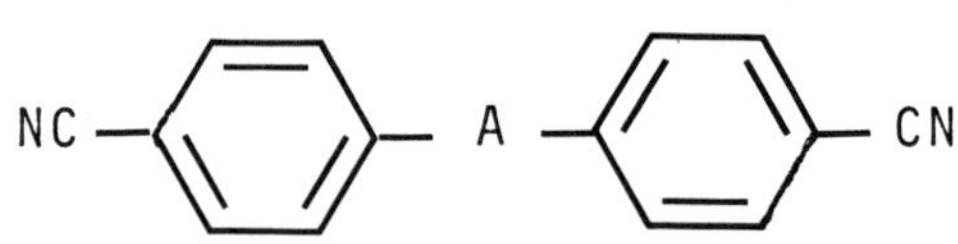

Besides the dinitrile without A, the others have A = -O-, $-CH_2-$, -CO-, and $-SO_2-$. These dinitriles can be cross-linked to form triazine units by heating with $ZnCl_2$. The resultant polymers possess high heat-resistance with thermal deformation temperatures over 450°C. These polymers resemble TSTR synthesized by Hsu (5) at NASA-Lewis Research Center.

Huang, et al. (6), also synthesized five thiazole-containing dinitriles:

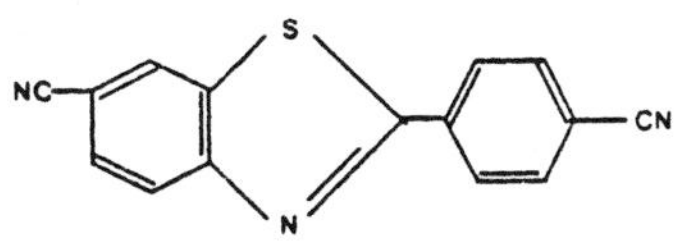

(I)

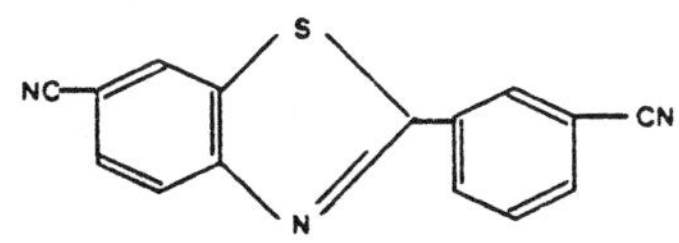

(II)

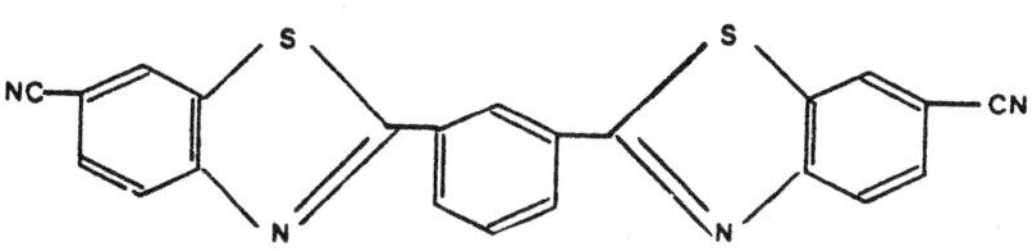

(III)

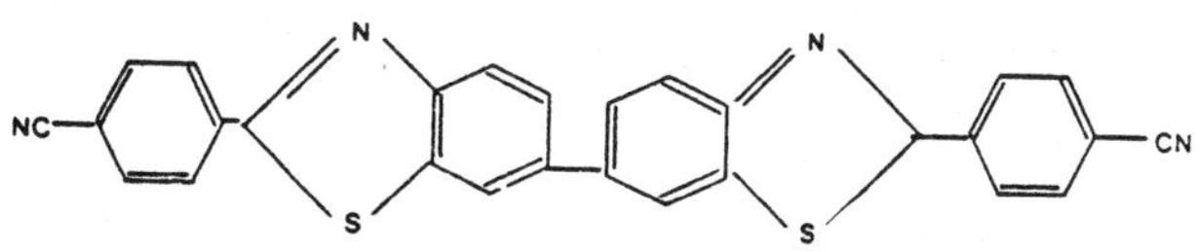

(IV)

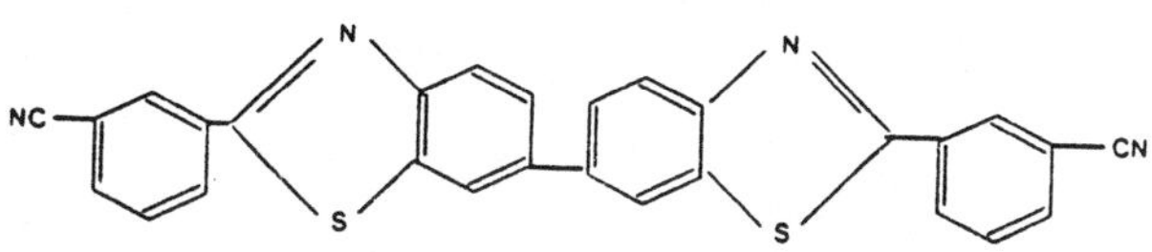

(V)

The polymers (7) formed from these dinitriles have high thermal and oxidative stabilities and are comparable to graphite in their weight loss pattern in air. However, none of these polymers have been evaluated as high-temperature adhesives.

Besides dinitriles, Huang, et al. (8), also synthesized carboximides. The carboximides are the starting materials for high temperature polymers (or adhesives). For example, bicylco(2.2.1) hept-5-ene-2,3-dicarboximide is an NA(nadic)-capped PMR(Polymerization of the Monomer Reactant) or PI, polyimide precursor.

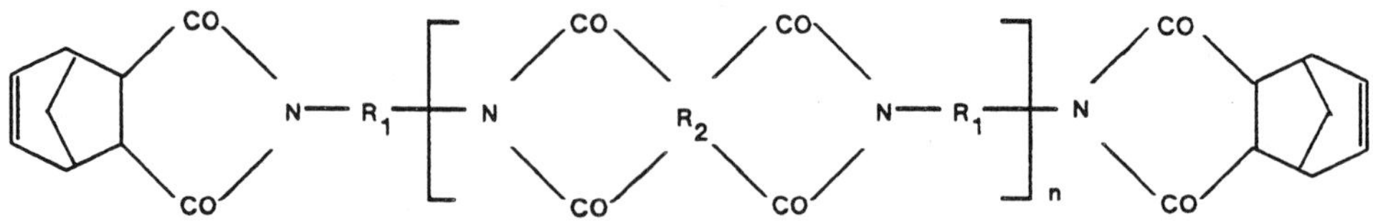

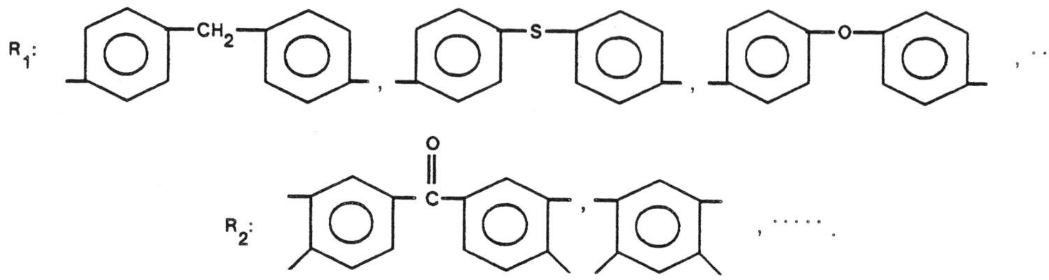

Lu, et al. (9), of the Institute reported that polyphenyl-as-triazines could be prepared from terephthalamidrazone. These polymers were claimed by them to be both heat- and water-resistant, but they have not evaluated these polymers as adhesives.

2. Adhesives. The adhesive research at the Institute is led by Dr. Yu Yunzhao, who accompanied me during my visit at Beijing. His work was in the epoxy resins toughened with CTBN, carboxy-terminated butadiene-nitrile rubber. In 1974, when I visited the Institute, this phase of work had already received a good start. In the meantime, Ge Zhengpei developed a method of preparing high-epoxy content resins (10). Besides epoxy, Ge carried out research on α-cyanoacrylates (11). In the same group, Nie Xuzong, et al. (12), studied the use of 2-ethyl-4-methyl-imidazole (EMI-2,4) as the curing agent for the rubber-toughened epoxy resin. This curing agent is especially suitable for room temperature hardening of the resin. One of their research papers is included in this volume.

A radiation-curable adhesive (CX Adhesive), jointly developed by this Institute and Northwest Optical

Instrument factory (13), is currently manufactured by Beijing Chun Shu Rubber Company. This adhesive is based on the glycerylester of perhydrophathalic acid and is used to adhere optical materials. Its refractive index , n_D^{23} is 1.505 ~ 1.515, and the light transmission is > 92%.

3. Polymer composites. The composite research at the Institute is led by Prof. Wu Renjie, Deputy director of the Institute. Chen, et al. (14), studied the effect of oxidation of carbon fiber on the wettability by the binder resin. With the aid of ESCA, they showed that the Q/C ratio on the fiber surface increased with the oxidation time. The interlaminar shear strength also increased correspondingly. Cai Weizhen and her colleagues showed me their exceptional setup for a carbon-fiber composite study. They built their own torsion pendulum for the study of dynamic mechanical properties of the composite and a contact angle goniometer for the study of the composite interface. It was apparent that surface treatment of carbon fiber was their major concern.

D. Heilongjiang Petroleum Reserch Institue, Harbin

My trip did not bring me to the Northeastern region of China. However, during my visit I was told that much adhesive research and development was carried out at this Institute in Harbin. Actually, the Journal of Adhesion and Chemistry (in Chinese) has been published by this Institute since September, 1982. Later, through correspondence with Mr. Wang Zhi-Lu (15), Deputy director of the Institute, I became acquainted with their research activities.

In this Institute, there are approximately 200 research personnel and one fourth of them are working on adhesives in two laboratories, one pilot-plant, one testing group and one adhesive-film manufacturing and testing facility. Their research and development broadly covered structural adhesives, strain gauge adhesives, wood adhesives, second-generation acrylics, anaerobic adhesives and composite binders. Their products varied from J-01 to J-53. There were at least thirty different kinds of adhesive and sealant, for example, nitrile-phenolics, modified epoxies, acrylics and epoxy-phenolics.

Their J-39 adhesives are room-temperature curable, second-generation acrylics (16). Methyl methacrylate is the monomer, and a nitrile rubber and Dow ABS resin are used as toughening agents. This family of adhesives have been used in automotive, machinery and aerospace structures. Their J-50 adhesives (17) are two-component, fast-cure acrylics. In addition, they also developed a special adhesive, J-15 structural adhesive, for hydroplanes.

J-29 is a structural foam adhesive-ribbon (tape) (18) specifically prepared for honeycomb structure. The epoxy resin is foamed by a diazo compound, and metallic powder filler is used to increase the thermal conductivity.,

A series of nitrile-phenolics, e.g., J-01, J-02, J-03, J-04, J-15, JX-9, JX-10, KH-506, were developed for aerospace structural adhesives (19). In addition, an epoxy-phenolic, J-44-1 (20) was supplied to adhere polyimide (Nomex) film to honeycomb structures. Thus, their structural adhesive line appears to be quite complete.

E. Beijing Institute of Aeronautics and Astronautics (BIAA)

The Institute has ten departments: 1) Flight Vehicle Design and Applied Mechanics, 2) Propulsion, 3) Materials Science and Engineering, 4) Electronic Engineering, 5) Automatic Control, 6) Computer Science and Engineering, 7) Manufacturing Engineering, *) Systems Engineering for the Aeronautical Industry, 9) Mathematics and Physics, and 10) Mechanical and Electrical Engineering: the last also serves as the Division of Basic Technological Courses. The departments offer 31 specialties. In addition the BIAA has an associated research institute. During my visit, Professor Shen Yuan, a renowned scholar in the field of aerodynamics, was the President of the Institute.

On June 3, 1983, I gave a seminar at the Institute on "Recent Research in Adhesives including Aerospace Adhesives." After the seminar, Prof. Chang Chihching led a tour through their Materials Department laboratories. He showed me his work on inorganic strain gauge adhesives based on aluminum di-H- phosphate which can be used at high temperatures. The testing laboratory at their Department was also equipped to test honeycomb structures. A double cantilever-beam (DCB) was used to determine the fracture energy of the adhesive.

Shortly after my return, Prof. Chang wrote about his plan to offer a course in adhesion at BIAA. "Adhesion" has been considered to be too practical a course for most universities, especially in China. His attempt could be an important beginning for future adhesion research in China.

III. GENERAL INTEREST IN ADHESION RESEARCH

Since my major purpose was to lecture, I had little time to gather information for a comprehensive survey. Thus, the above accounts on research activities are by no means complete or representative. The great numbers of research institutes, large and small, and the lack of published work would have made a comprehensive survey impossible even had I been able to devote full time to that purpose. Thus, it is difficult for me to assess their stage of achievements in adhesive research.

Among much of their successes, I would like to single out just one to illustrate their general concern on population control and the ingenious use of one medical adhesive to solve the problem. The Chinese medical personnel and scientists are experimenting with the use of α-cyanoacrylate for male contraception. At least one province, Shaanxi, has approved this particular adhesive as an injection for plugging up the sperm duct, or vas deferens. Up to May, 1983, in that province alone, at least 20,000 men had received the treatment. The initial clinical test results were very encouraging. There were no adverse reports on cancer, abnormalities or abrupt changes of health or behavior. So far, this treatment has been viewed as a safe means of contraception. However, no mention has been made regarding the reversibility of the process or its superiority to vasectomy. If the process were reversible, it would be more appealing to the West, and general worldwide acceptance could be assured.

Besides their ingenuity, their enthusiasm in adhesion research is contagious. There are over a dozen local sections of the Chinese Adhesion Society organized in large cities. Their National Adhesion Society will be formally established in 1984 with an expected membership of over two thousand. Besides local meetings, the Second National Adhesion Conference will be held in Harbin in August of

1984. It appears that their enthusiasm has set adhesion research in motion throughout China.

IV. CONCLUSIONS

As a developing country, China has come a long way in research and development of adhesives and sealants in meeting the growing needs of the Four Modernizations program. The enthusiasm of researchers and industrial workers in organizing adhesion societies and conferences and in publishing adhesion journals has prompted the exchange of ideas and technological developments. For their future, they may need some institutes or universities where the interdisciplinary adhesion science can be taught to their next generation of technologists.

REFERENCES

1. L. H. Lee, "China-Lecture Trip Report (May, 1983): Some Progress in Science, Technology and Education Since the Start of Four Modernizations in the People's Republic of China," a report submitted to the National Academy of Sciences and the Chinese Academy of Sciences, August 29, 1983.

2. Guangdong Chemistry Research Institute, presented to the High Polymer Research Conference held in Changchun, 1972.

3. Shanghai Institute of Organic Chemistry, presented to the High Polymer Research Conference, held in Changchun, 1972.

4. Zhi-tang Huang and Yuchun Ou, Gaofenzi Tongxin (in Chinese), 1, 1 (1978).

5. Li-Chen Hsu, and W. H. Philipp, a paper presented to the IUAAC Macro Conference, University of Massachusetts, Amherst, MA, July, 1982.

6. Zhi-tang Huang, Li-ben Wang, Jin-zhen Wang, and Shu-ying Zhang, Gaofenzi Tongxin (in Chinese), 4, 225 (1979).

7. Jing Liu, Shu-ying Zhang, Liben Wang, and Zhitang Huang, Gaofenzi Tongxin (in Chinese), 2, 116 (1981).

8. Chongli Zhang and Zhitang Huang, Gaofenzi Tongxin (in Chinese) 4, 198 (1980).

9. Feng-cai Lu, Lan-min Xing and Yu-lan Wang, Polymer Preprints, 22, No. 1, 62 (1981).

10. Zhengpei Ge, et al, Chemistry and Adhesion (in Chinese), 1, No. 1, 1 (1982).

11. Zhengpei Ge, Beijing Adhesion (in Chinese), 1, 1 (1983).

12. Yi-ming Li, Ke-ying Tan, Xiu-mai Zhang, and Xuzong Nie, Chemistry and Adhesion (in Chinese), 1, No. 1, 26 (1982).

13. Li-lin Shen and Ai-lin Duan, Beijing Adhesion (in Chinese), 1, 54 (1983).

14. Chang-ming Chen, Mu-jin Sun, and Renjie Wu, Chemistry and Adhesion (in Chinese), 1, No. 1, 1 (1982).

15. Zhi-lu Wang, Private communication, Jan. 1984.

16. Qi-ting Lu, Yi-kun Liu, Ming Yuan, and Gong-xin Shen, Chemistry and Adhesion (in Chinese), 1, No. 1, 15 (1982).

17. Huang Qian and Yu-lin Huang, Chemistry and Adhesion (in Chinese), 1, No. 1, 55 (1982).

18. An-tian Zhang, Chemistry and Adhesion (in Chinese), 1, No. 2 (1982).

19. Zheng-yun Lu, Chemistry and Adhesion (in Chinese), 2, No. 3, 218 (1983).

20. Liang Mao, Chemistry and Adhesion (in Chinese), 2, No. 2, 151 (1983).

21. Tian Lin, Chemistry and Adhesion (in Chinese), 2, No. 3, 217 (1983).

Discussion

On the Paper by A. G. Bachman

L. H. Lee (Xerox Corporation): What kind of initiators are being used in the system?

A. G. Bachman (American Chemical & Engineering Company, CT): Peroxyesters.

L. H. Lee: Please give sample formulations about the system. If it has been patented, give examples in the patent.

A. G. Bachman: Please see the enclosed patent* for formulation details to the extent that they can be disclosed. Four other patents are pending (*U.S. 4,348, 503).

L. H. Lee: Who are the suppliers of the aerobic adhesives? Please give trade names along with suppliers.

A. G. Bachman: American Chemical & Engineering Co. is the exclusive worldwide manufacturer of Aerobic Acrylic Adhesives. Elosol, Ltd. of Zurich, Switzerland, supplies our formulations to the European Economic Community. We are negotiating with a number of potential licensees in the Far East.

L. H. Lee: What is the percentage of the market share of the aerobic acrylic adhesives among other acrylics?

A. G. Bachman: We have no figures of the market share percentage because aerobics are so new. If certain customers' projections hold up, by 1987 we should have 50-75% market share in the U.S. Of course, this is a big "if."

CONTRIBUTORS

Lieng-Huang Lee received his Ph.D. degree (1955) in Chemistry from Case Institute of Technology, Cleveland, Ohio. Dr. Lee is a Senior Scientist at Xerox Corporation. He has edited seven books on adhesion, friction and wear, published over 50 papers and holds 26 U.S. patents. He is a Member of the Editorial Advisory Board, Journal of Adhesion. He participated in 1983 U.S.-China Distinguished Scholars Exchange.

D. Maugis is Maitre de Recherche at the Centre National de La Recherche Scientifique (CNRS). Graduated in Natural History and in Physics, he started working at Thomson-CSF on Surface Ionization and Microwaves, and received his Doctorate in Siences (1967) on Plasma Physics. At the CNRS since 1968, his area of interest is Tribology, mechanics of contacts and adherence of solids.

Michael C. H. Lee received his Ph.D. degree (1979) in mechanical engineering from University of California at Berkeley. He is a staff research scientist of GM Corporation. His research interests cover elastomers, polymer-filler interaction, polymer mixing and polymer thermodynamics.

Fasi Liang graduated from Sun Yatson University, China, in 1965 as a specialist in Polymer Science. He is a member of the staff of the Shanghai Rubber Research Institute in the People's Republic of China. He spent two years, 1981-1983, at the Institute of Polymer Science, the University of Akron, as a Visiting Scientist, conducting research on fundamental aspects of adhesion.

Patricia Dreyfuss received a Ph.D. in Polymer Science from the University of Akron in 1964. She has been an AAUW International Postdoctoral Fellow at the University of Liverpool and a Visiting Research Fellow at the University of Bristol, both in England. She is a Research Associate and member of the general faculty at the University of Akron. Her research on the effects of chemical bonding on adhesion started in 1974.

Yona Eckstein is currently employed by Owens-Corning Fiberglas Technical Center in Granville, Ohio. She received her Ph.D. degree from the Hebrew University in Jerusalem, Israel, in 1977 and held a post-doctoral position in the Institute of Polymer Sciences prior to joining Owens-Corning in 1981. She is the author or co-author of over 35 scientific papers in the field of glass-lasers spectroscopy, polymer characterization and polymer/glass adhesion.

Elisabeth J. Berger received her B.A. degree in Chemistry in 1975 from the University of Kansas. She received her Ph.D. degree from Pennsylvania State University in 1980. She is currently employed at Owens-Corning Fiberglas Corporation in Granville, Ohio.

Louis H. Sharpe received his Ph.D. degree in Physical Chemistry from Michigan State University in 1957. He is the Supervisor of Adhesive Engineering and Development at Bell Laboratories. Dr. Sharpe is the Editor-in-Chief of the *Journal of Adhesion*. His research interests have been adhesion, adhesives and mechanical properties of polymers.

Dave Briggs is a Science Associate with ICI Petrochemicals and Plastics Division, Middlesbrough, UK, responsible for the development and application of surface analytical techniques. For several years a major interest has been modification of polymer surfaces for adhesion enhancement. More recently he has developed SIMS for high resolution surface analysis of organic materials. He has edited two books on surface analytical techniques and authored over 60 publications. He is the founder and editor-in-chief of the international journal "*Surface and Interface Analysis*."

J. Thomas Dickinson received his Ph.D. in Physics from

the University of Michigan in 1968. He then joined the Washington State University Department of Physics as an Assistant Professor. Dr. Dickinson has performed research as a Visiting Scientist at Stanford University, the Oregon Graduate Center, and NASA-Ames Research Center. He is currently a Full Professor of Physics and Chemical Physics at W.S.U. Recent research has focused on fracto-emission from polymers, composites, adhesives, ceramics, and coatings.

R. L. Levy received a B.S. in Chemistry from Israel Institute of Technology in 1961 and a Ph.D. from University of Manitoba, Canada, in 1967. Currently working on various aspects of polymer durability and polymer characterization as a Scientist with the McDonnell Douglas Research Laboratories (MDRL).

D. P. Ames received a B.S. in Chemistry in 1944 and a Ph.D. in Physical Chemistry in 1949 from University of Wisconsin. At McDonnell Douglas, he is Staff Vice President-McDonnell Douglas Research Laboratories.

Bruce C. Copley graduated with a B.S. in Chemistry and a B.S. in Chemical Engineering from Massachusetts Institute of Technology in 1978. He joined the Industrial Tape Division of 3M in 1979. He has been working in adhesive research since that time.

Jay B. Class received his Ph.D. from the Pennsylvania State University in Organic Chemistry (1952). He joined Hercules Incorporated in 1952. Dr. Class' research has been primarily concerned with resins, elastomers and adhesives. He has been head of both the Rubber Laboratory and Adhesives Laboratory at Hercules. He is presently a Research Scientist in the Polymer Science Division at the Hercules Research Center.

Sung Gun Chu received his B.S. from Han Yang University, Seoul, Korea and his Ph.D. from the University of Texas (1978) in Polymer Physical Chemistry. He joined Hercules Incorporated in 1981 and is presently a Research Chemist in the Adhesives Laboratory at the Hercules Research Center.

Thomas C. Ward received his M.S. and Ph.D. in Physical Chemistry from Princeton University in 1966. He joined VPI

& SU as an Assistant Professor in 1968. He was promoted to Professor in 1981. Professor Ward has received several teaching honors, including the Wine Award and is a member of the Academy of Teaching Excellence at VPI & SU. His publications have been principally in the areas of polymer characterization, testing, adhesion and long-term property evaluation of polymeric materials.

K. C. Stueben is a Senior Research Scientist at Union Carbide's Technical Center in Bound Brook, New Jersey. He received his Ph.D. from the Polytechnic Institute of New York in 1960. His research experience has encompassed a number of areas, including micellar catalysis, synthetic paper, polymer flammability, exploratory adhesives, radiation curable pressure sensitive adhesives and both laboratory and plant scale preparation of condensation polymers, latexes and industrial chemicals. He is the author of numerous papers, chapters and patents in these fields. Dr. Stueben is currently Director of the Adhesion Society's course on Adhesion.

James Crivello is a Research Chemist at the General Electric Corporate Research and Development Center in Schenectady, New York. He received a Ph.D. in Organic Chemistry from the University of Notre Dame in 1966. Dr. Crivello's major interests lie in organic reaction chemistry, particularly in aromatic nitrations and oxidations. For the past eight years, he has been involved in various aspects of the study of photopolymerizations and in the design of new photoinitiators.

Vincent D. McGinniss received his Ph.D. in Chemistry from the University of Arizona. Prior to joining Battelle, Dr. McGinniss was associated with the Glidden Durkee Dwight P. Joyce Research Center. He holds over 40 U.S. patents and was twice a recipient of the Federation of Societies for Coatings Technology Roon Award. He has authored or co-authored over 50 publications and chapters for books on radiation technology.

Francis J. Campbell is a Physical Chemist at the Naval Research Laboratory. He heads the Materials Section of the Condensed Matter and Radiation Sciences Division. He graduated from the University of Toledo in 1948 and worked at Dow Chemical and Dow Corning before joining NRL in 1958.

Leslie R. Gatechair received a Ph.D. degree from the North Dakota State University Department of Polymers and Coatings (1980). Dr. Gatechair is currently engaged in applications development of photoinitiator and stabilizing additives for radiation cured coatings and imaging systems at the CIBA-GEIGY Corporation, Ardsley, New York. Since 1982, he has served on the Editorial Board of the Journal of Radiation Curing.

Paul M. Hergenrother received a B.S. degree in Chemistry from Geneva College and took graduate chemistry work at the University of Pittsburgh, Carnegie-Mellon University, and San Diego State University. In 1975, he joined the staff of Virginia Polytechnic Institute and State University as an Adjunct Research Professor working at NASA Langley Research Center. In 1980, he became associated with the NASA Langley Research Center as a Senior Polymer Scientist. His work has been primarily in the area of high performance polymers where he has authored more than 60 papers and holds several patents.

Anne K. St. Clair is a Senior Research Scientist in the Materials Division at NASA Langley Research Center, Hampton, Virginia. In 1969, she received a B.A. in Chemistry from Queens College and was a graduate of the Student Honors Program in Chemistry from Argonne National Laboratory. She received an M.S. in Inorganic Chemistry from Virginia Polytechnic Institute & State University (VPI & SU) in 1972. Her research interests at NASA have included the development of high-temperature polymers for aerospace applications.

Terry L. St. Clair is a Senior Research Scientist at NASA Langley Research Center where he has been since 1972. After working for E.I. duPont in Waynesboro, Virginia for two years he obtained a Ph.D. in Organic Chemistry from VPI & SU in 1972. His research at NASA has involved the development of high-temperature polymers for adhesive and matrix resin applications.

Stephen A. Ezzell is presently a graduate student in the Chemistry Department at Virginia Polytechnic Institute & State University, Blacksburg, Virginia. He graduated from St. Andrews Presbyterian College in 1981 with a B.S. in Chemistry. He was employed at NASA Langley Research Center as a Student Research Assistant during the summers of 1980 and 1981.

James P. Wightman is Professor of Chemistry at VPI & SU. He received his B.S. degree from Randolph-Macon College and his Ph.D. degree in Chemistry from Lehigh University. He was a Research Associate at Penn State prior to joining the faculty at Virginia Tech in 1962 and was a Visiting Professor (1975-76) at the University of Bristol (U.K.). He is co-author of 65 publications and has been a member of the Advisory Board of the *Journal of Colloid and Interfce Science*. His research interests include the adsorption of gases and liquids on solids and the characterization of solid surfaces with application to adhesion, desalination, lubrication, coal liquefaction and heavy metal abatement.

Surani Dias is a native of Sri Lanka and received her B.A. degree in Chemistry from Randolph-Macon Woman's College. She is a graduate student at VPI & SU working on the characterization of adherend surfaces.

Herbert J. Leary, Jr. is the manager of the Surface/Materials Annalysis Department in the IBM East Fishkill-Product Assurance Organization located in Poughkeepsie, New York. His Ph.D. in Physical Chemistry was earned from the University of Iowa. Dr. Leary has spent the past eight years developing methods for applying ESCA to the solution of semiconductor manufacturing and development problems, and to the study of the nature of polymer surfaces.

David S. Campbell is a graduate of the Franklin Institute, Boston, Massachusetts. His background includes 16 years in semiconductor activities at IBM in Essex Junction, Vermont. In addition to semiconductor processing responsibilities held in the areas of photolithography and diffusion, Mr. Campbell has applied SEM, Auger, Microprobe and ESCA to the solution of semiconductor problems.

T. Helminiak is a Scientist in the Polymers Branch of the Materials Laboratory at Wright-Paterson Air Force Base. Dr. Helminiak is involved in developing the methods by which improvements are made in advanced amorphous and ordered polymers. His background in solution properties and polymer physical chemistry has facilitated the development of high performance polymers for structural applications.

W. B. Jones is a Professor in Mechanical Engineering. Dr. Jones teaches courses in structural materials and is developing fracture mechanics characterization methods for polymers, adhesives and composites.

C. L. Hendricks obtained a B.S. in Chemistry from the University of Washington and an MBA with honors from City University, Seattle, Washington. He is currently lead engineer for adhesives and composites development and applications in the Boeing Aerospace Company Materials and Processes Technology Organization. He has conducted research programs in this technical area for the last 20 years.

Robert Buchanan received the B.S. degree from the University of Strathclyde in Scotland, and the Ph.D. from Cambridge University, England (where the original work was done that resulted in the development of a commercial SEM), where he worked in the Engineering Department. Dr. Buchanan has some 20 years of experience with both SEMs and TEMs. Most recently, he transferred to ISI in Santa Clara, Califiornia, where he is currently Vice President Research and Development.

Roscoe A. Pike received a Ph.D. in Organic Chemistry from Massachusetts Institute of Technology in 1953. He has been with the United Technologies Research Center as a Senior Materials Scientist in the High Temperature Materials Group from 1967 to 1974. He is presently with the Manufacturing Technology and Process Research Group.

Foster P. Lamm received his Ph.D. in Organometallic Chemistry from Wesleyan University in 1979. Since graduation he has been with the United Technologies Research Center as a Research Scientist in the Manufacturing Technology and Process Research Group.

Louis J. Baccei received his Ph.D. degree from the University of Maryland. He is currently Associate Director of Research at Loctite. His main research interests are the structure-property relationships of new materials and the study of new, ambient temperature, cure systems.

Bernard M. Malofsky received his Ph.D. from the University of Washington. After joining Loctite Corporation in 1972, he progressed through a number of research positions to his current job as Vice President, Research and Development for the Industrial Group.

Robert W. R. Humphreys received his B.Sc. and Ph.D. in Photochemistry and ESR Spectroscopy with Professor Donald R. Arnold at the University of Western Ontario. After 1 1/2 years postdoctoral work with Professor Cheves Walling at the University of Utah, he worked as a Research Chemist at Loctite Corporation in anaerobic adhesives chemistry. He recently joined Lever Research, Inc. as a Research Scientist. His current research interests include free radical and peroxide chemistry and metal catalyzed redox reactions.

Alphonsus V. Pocius obtained his Ph.D. in Physical Chemistry from the University of Illinois in Champaign-Urbana in 1974. He joined 3M in 1974. At 3M, his work has been in the areas of polymer characterization, surface chemistry, corrosion and electrochemistry and adhesive product development.

David A. Wangsness obtained his B.A. in Chemistry from Luther College in Iowa in 1961 and attended graduate school at the University of North Dakota in 1962. He joined 3M in 1962 as a Chemist and worked on the development of aerospace structural adhesives. Presently, he is Manager of the Aerospace Products Laboratory at 3M.

Carl J. Almer joined 3M in 1957. He obtained a Bachelor of Chemistry degree from the University of Minnesota Institute of Technology in 1964. His current work is in the area of structural adhesive films for the aerospace industry.

Alan G. McKown obtained a Bachelor of Chemical Engineering degree from Cleveland State University in 1963. He joined the Adhesives, Coatings and Sealers Division of 3M in 1964. At 3M he has been involved in epoxy and polyurethane structural adhesive development.

Ralph Drake has degrees from Case-Western Research University. He has been with B. F. Goodrich Chemical Group for 26 years and spent much of that time associated with new products, many of them adhesive raw materials. He is currently Marketing Manager for Reactive Liquid Polymers. He has over 40 patents and technical publications.

A. R. Siebert received a Ph.D. in Physical Chemistry from Case-Western Reserve University in 1957. He has been with B. F. Goodrich Company for 28 years where a broad background exists in polymerization and characterization of reactive liquid polymers and their utility to modify thermoset resins. He has a number of related patents and has presented and published numerous papers at national and international conferences.

Xuzong Nie graduated from Peking University, Beijing, in 1956. He has been Research Assistant and Research Associate at the Institute of Chemistry, Academia Sinica. Now he is Associate Research Professor at this Institute.

Yunchao Yu graduated from Fudan University in 1962. He has been Research Associate at the Institute of Chemistry, Academia Sinica. He was a Research Fellow of Av Humboldt Foundation, and received his Ph.D. degree from University Stuttghart (FRG). Now he is Vice Supervisor of the 8th Lab of the Institute of Chemistry.

Marlene Jones received her B.A. degree in Chemistry from Thiel College. She joined Hercules Incorporated in 1972 and is currently a Sales Development Supervisor in the Resins Division. Prior to this, Marlene held the position of Staff Chemist in the Adhesives Laboratory at the Hercules Research Center. Her experience in adhesives is primarily in the pressure-sensitive field.

Andrew G. Bachmann has served as President of the American Chemical & Engineering Company since founding it in December of 1979. He previously worked at Allied Product Corp. and Loctite Corp. Mr. Bachmann has authored or co-authored several patents in the field of structural adhesives and microencapsulation, and has extensive experience in the chemistry and medical uses of cyanoacrylates as surgical adhesives. He earned his B.S. and M.S. in Chemistry at the University of Richmond and his M.B.A. in Finance and Marketing from the University of Hartford.

Claire Bluestein is currently President of CAPTAN Associates, Inc., Lyndhurst, New Jersey, which she formed in 1976. Dr. Bluestein received her Ph.D. in Organic Chemistry at the University of Illinois and before that an A.B. with honors in Chemistry at the University of Pennsylvania. In addition, she had over 20 years of experience in industrial R&D before CAPTAN.

Paul Cranley is a Project Leader in the Specialty Chemicals Department, Dow Chemical U.S.A. He has been with Dow for nine years, the last five of which have been spent in the development of specialty monomers, intermediates and additives, particularly for coatings applications.

Bonnie L. Rathbun is an Associate Engineer with IBM, Federal Systems Division in Owego, New York. She has a Bachelor of Arts degree in Chemistry from Illinois Wesleyan University and Masters degrees in Science Education from Colgate University and in Chemical Engineering from Cornell University. Her main areas of experience have been in process engineering for hybrid microelectronics, hydrophobic materials evaluation, and plasma polymerization.

Philipp wh Schuessler is an Advisory Chemist presently working for IBM, Federal Systems Division. He is a graduate of the Rochester Institute of Technmology with Bachelors and Masters degrees in Analytical and Synthetic Inorganic Chemistry. Since joining IBM in 1963 he has worked primarily in the areas of materials and process engineering. He holds several patents and invention disclosures in these areas.

Qi-Xiang Yang received his B.S. degree in Chemistry in 1969 from Peking University, Beijing, People's Republic of China. From 1974 to 1977 he worked on the synthesis of polymer adhesives. Mr. Yang received his Masters degree in Polymer Chemistry in 1981 from Nanking University, Nanking. He is currently studying for his Ph.D. degree at Kansas State University.

Qin-Li Zhou received his B.S. degree in Chemistry in 1953 from Nanking University. He is currently an Associate Professor of Polymer Chemistry at Nanking University. His research interests have mainly been the syntheses and applications of coupling agents.

AUTHOR INDEX

T

SUBJECT INDEX

D